T0320931

INTEGRAL AND
DISCRETE TRANSFORMS
WITH APPLICATIONS AND
ERROR ANALYSIS

PURE AND APPLIED MATHEMATICS

A Program of Monographs, Textbooks, and Lecture Notes

MONOGRAPHS AND TEXTBOOKS IN
PURE AND APPLIED MATHEMATICS

91. *A. E. Fekete,* Real Linear Algebra (1985)
92. *S. B. Chae,* Holomorphy and Calculus in Normed Spaces (1985)
93. *A. J. Jerri,* Introduction to Integral Equations with Applications (1985)
94. *G. Karpilovsky,* Projective Representations of Finite Groups (1985)
95. *L. Narici and E. Beckenstein,* Topological Vector Spaces (1985)
96. *J. Weeks,* The Shape of Space: How to Visualize Surfaces and Three-Dimensional Manifolds (1985)
97. *P. R. Gribik and K. O. Kortanek,* Extremal Methods of Operations Research (1985)
98. *J.-A. Chao and W. A. Woyczynski, eds.,* Probability Theory and Harmonic Analysis (1986)
99. *G. D. Crown, M. H. Fenrick, and R. J. Valenza,* Abstract Algebra (1986)
100. *J. H. Carruth, J. A. Hildebrant, and R. J. Koch,* The Theory of Topological Semigroups, Volume 2 (1986)
101. *R. S. Doran and V. A. Belfi,* Characterizations of C*-Algebras: The Gelfand-Naimark Theorems (1986)
102. *M. W. Jeter,* Mathematical Programming: An Introduction to Optimization (1986)
103. *M. Altman,* A Unified Theory of Nonlinear Operator and Evolution Equations with Applications: A New Approach to Nonlinear Partial Differential Equations (1986)
104. *A. Verschoren,* Relative Invariants of Sheaves (1987)
105. *R. A. Usmani,* Applied Linear Algebra (1987)
106. *P. Blass and J. Lang,* Zariski Surfaces and Differential Equations in Characteristic p > 0 (1987)
107. *J. A. Reneke, R. E. Fennell, and R. B. Minton,* Structured Hereditary Systems (1987)
108. *H. Busemann and B. B. Phadke,* Spaces with Distinguished Geodesics (1987)
109. *R. Harte,* Invertibility and Singularity for Bounded Linear Operators (1988)
110. *G. S. Ladde, V. Lakshmikantham, and B. G. Zhang,* Oscillation Theory of Differential Equations with Deviating Arguments (1987)
111. *L. Dudkin, I. Rabinovich, and I. Vakhutinsky,* Iterative Aggregation Theory: Mathematical Methods of Coordinating Detailed and Aggregate Problems in Large Control Systems (1987)
112. *T. Okubo, Differential Geometry* (1987)
113. *D. L. Stancl and M. L. Stancl,* Real Analysis with Point-Set Topology (1987)
114. *T. C. Gard,* Introduction to Stochastic Differential Equations (1988)
115. *S. S. Abhyankar,* Enumerative Combinatorics of Young Tableaux (1988)
116. *H. Strade and R. Farnsteiner,* Modular Lie Algebras and Their Representations (1988)
117. *J. A. Huckaba,* Commutative Rings with Zero Divisors (1988)
118. *W. D. Wallis,* Combinatorial Designs (1988)
119. *W. Więsław,* Topological Fields (1988)
120. *G. Karpilovsky,* Field Theory: Classical Foundations and Multiplicative Groups (1988)
121. *S. Caenepeel and F. Van Oystaeyen,* Brauer Groups and the Cohomology of Graded Rings (1989)
122. *W. Kozlowski,* Modular Function Spaces (1988)
123. *E. Lowen-Colebunders,* Function Classes of Cauchy Continuous Maps (1989)
124. *M. Pavel,* Fundamentals of Pattern Recognition (1989)
125. *V. Lakshmikantham, S. Leela, and A. A. Martynyuk,* Stability Analysis of Nonlinear Systems (1989)
126. *R. Sivaramakrishnan,* The Classical Theory of Arithmetic Functions (1989)
127. *N. A. Watson,* Parabolic Equations on an Infinite Strip (1989)
128. *K. J. Hastings,* Introduction to the Mathematics of Operations Research (1989)
129. *B. Fine,* Algebraic Theory of the Bianchi Groups (1989)
130. *D. N. Dikranjan, I. R. Prodanov, and L. N. Stoyanov,* Topological Groups: Characters, Dualities, and Minimal Group Topologies (1989)
131. *J. C. Morgan II,* Point Set Theory (1990)
132. *P. Biler and A. Witkowski,* Problems in Mathematical Analysis (1990)
133. *H. J. Sussmann,* Nonlinear Controllability and Optimal Control (1990)
134. *J.-P. Florens, M. Mouchart, and J. M. Rolin,* Elements of Bayesian Statistics (1990)
135. *N. Shell,* Topological Fields and Near Valuations (1990)
136. *B. F. Doolin and C. F. Martin,* Introduction to Differential Geometry for Engineers (1990)

Additional Volumes in Preparation

INTEGRAL AND DISCRETE TRANSFORMS WITH APPLICATIONS AND ERROR ANALYSIS

Abdul J. Jerri

Clarkson University
Potsdam, New York

Marcel Dekker, Inc.   New York • Basel • Hong Kong

Library of Congress Cataloging-in-Publication Data

Jerri, Abdul J.
  Integral and discrete tranforms with applications and error
analysis / Abdul J. Jerri.
      p. cm. -- (Monographs and textbooks in pure and applied
  mathematics ; 162)
  Includes bibliographical references and index.
  ISBN 0-8247-8252-6
  1. Transformations (Mathematics) 2. Error analysis (Mathematics)
I. Title. II. Series
QA601.J45 1992
515'.723--dc20   92-13680
                                                       CIP

MARCEL DEKKER, INC.
270 Madison Avenue, New York, New York 10016

Current printing (last digit):
10 9 8 7 6 5 4 3 2 1

PRINTED IN THE UNITED STATES OF AMERICA

To my wife Suad,
and the children,
Saad, Iman, Nadia,
Fourat, and Huda

PREFACE

This book deals with the basic elements of the integral, finite, and discrete transforms, with emphasis on their use for solving boundary (and/or initial) value problems as well as facilitating the representation of signals and systems. The analysis of the discrete Fourier transforms is stressed here because they represent a form for which an efficient means of computation, the well-known fast Fourier transform (FFT) algorithm, exists. However, the discrete Fourier transform is only an approximation to the Fourier integral or Fourier series that we seek to compute. Thus, we stress the importance of detailed analysis of the errors incurred in this approximation as a prelude to taking advantage of the (efficient) FFT algorithm.

In comparison with other books on transform methods, this book has the added features of including the discrete Fourier transform along with a detailed analysis of its inherent errors, the applications to representing signals and systems, and an emphasis on how to choose the appropriate or "compatible" transform for solving a boundary value problem. The latter involves choosing, at the outset, the transform that will algebraize a given differential operator, and at the same time includes all the auxiliary conditions given for the boundary and/or initial value problem.

Certain sections are more introductory and are designed for an undergraduate course in boundary and initial value problems, where the (operational) transform method is stressed instead of the usual separation of variables. In addition, there is some emphasis on the transform (spectral) representation of systems and signals, which should be of particular interest to electrical engineers. The major part of the book can serve as a text for a senior or graduate course on integral and discrete transforms for engineers and scientists, and as a reference for the researcher with interest in the applications, error analysis, and basic theory of these transforms. The portions of this book suitable for an undergraduate or a graduate course are indicated in the guide to course adoption following this preface. This division in the levels of coverage is often reflected in an increase in difficulty of successive exercises in each section, where sometimes they are designated as optional for the undergraduate course.

The treatment here will be self-contained and is planned for engineering and science students with sophomore differential equations preparation. Nevertheless, a review of basic differential equations, along with their (series) solutions as special functions, as well as examples and exercises, is provided in Appendix A. Appendix B covers detailed modeling of the basic partial differential equations. A preparatory course in complex variables is not required, but a familiarity with the basic elements of complex numbers is helpful. The few brief sections that need complex variables are included for completeness and are clearly designated as optional.

The contents include the basic elements of familiar transforms such as the Laplace, Fourier, and Hankel (Bessel) transforms, with due emphasis on how to choose among such transforms, or their respective variations, for tackling a given problem. This approach is at variance with that used in most texts on boundary value problems, where the method of separation of variables is emphasized and complemented later by Fourier analysis. In this respect, students may wonder about the coherence of the separation of variables—Fourier series (or integrals) combination. By contrast, the transform approach starts with the appropriate tools and continues to the final solution. To the student as well as the instructor, the present approach should represent a more logical continuation of the operational method of the Laplace transform that is introduced in the sophomore differential equation course that generally precedes the junior boundary value problems course.

Thus, the first chapter of this text deals with how to choose a suitable transform for solving boundary and/or initial value problems. This is done first by way of many examples. It is followed by the theory behind finding these compatible transforms for general or new problems, which we recommend for a graduate course. The second chapter is dedicated to the study of integral transforms, in particular the Laplace, Fourier, multiple Fourier, and Hankel (Bessel) transforms. The emphasis here is on applications to boundary and/or initial value problems (on an infinite domain) and the representation of signals and systems. The important operational properties of these transforms are stated very precisely, illustrated, and then often proved rigorously. Most of the proofs here are meant for completeness and should be easily comprehended by graduate students or ambitious undergraduate students. The third chapter covers finite transforms and includes trigonometric Fourier and general orthogonal series expansions. The basic theorems here are also stated clearly and illustrated, and then almost all are proved. Several examples of solving boundary value problems on finite domains follow. There is also an application to signal analysis via the sampling series expansion, which relates a continuous signal, viewed as a "finite limit" Fourier transform in Chapter 2, to its samples viewed as the Fourier coefficients in Chapter 3.

The second part of the book deals with the practical approximation of computing the resulting Fourier series or integral representation of the final solution. This is done via a careful preparation on correct use of the efficient (fast) discrete algorithm of such Fourier transforms, namely the FFT. Understandably, emphasis is placed on a very detailed treatment of the various errors incurred in the use of such discrete algorithms for approximating the Fourier series or integrals.

In all the discussions, we include a good number of detailed examples with appropriate graphical illustrations, supported by related exercises (with guiding hints and answers to almost all of them). These exercises are divided according to the material covered in each section, with the division of the undergraduate- and graduate-level courses in mind, and are located at the end of each chapter. The answers are given at the end of each exercise.

The appendices in this book are devoted to supporting material such as basic differential equations (in particular, second-order differential equations and their series solutions as special functions) in Appendix A. Appendix B covers a detailed derivation or modeling of partial differential equations such as the wave equation and the diffusion equation, with examples and exercises. Appendix C contains a reasonable collection of tables of the various transforms.

A few basic references for each chapter of this book are listed at the end of that chapter, and a list of all texts, relevant references, and books on integral and discrete transforms and their applications is given in the bibliography at the end of the book.

I would like to acknowledge many helpful suggestions from students and colleagues during teaching and during the preparation of this book.

I owe deep appreciation to the anonymous reviewer who made very critical and detailed constructive suggestions for improving the first draft of the book.

I would like to thank my colleagues Professors M. L. Glasser and R. L. Herman for reading the prefinal draft of most of the manuscript. Dr. K. W. Tse deserves thanks for his help with the first draft.

Thanks are also due Professor M. Z. Nashed for encouraging the project all along and for his very valuable suggestions. The staff of Marcel Dekker, Inc., deserves thanks for their keen understanding and effective cooperation. Mr. J. Hruska, Jr., deserves thanks for making the drawings with patience and care.

This project put a considerable burden on my family, especially my wife Suad and my little daughter Huda, to whom I am very grateful.

Abdul J. Jerri

GUIDE TO COURSE
ADOPTION

In addition of being a reference, this book can be used as a text to accommodate two levels of courses in the applications and analysis of transform methods. At both levels, the errors incurred in practical computations of the transforms are illustrated and discussed. The difference is that the advanced course includes most of the proofs of the basic results, almost all the applications, and the detailed treatment of the discrete transforms.

Undergraduate Course in Transform Methods for Boundary and Initial Value Problems

Appendix A.1 and its exercises: Review of the simple basic differential equations.

Chapter 1, Sections 1.1–1.4: Illustrations of how to choose the proper transform for a boundary value problem.

Appendix B: Modeling of the wave equation, heat equation, and other equations.

In Chapters 2 and 3, precise statements and illustrations of the basic theorems, and not necessarily the proofs, are emphasized.

Chapter 2: Integral transforms.

Section 2.1: Laplace transforms—a fast review leading to solving partial differential equations with initial conditions.

Sections 2.2–2.5: Fourier transforms and their applications to solving boundary value problems on an infinite domain, and the representation of systems and signals (in Section 2.4). Section 2.2.1, especially the part on the proof of the Fourier integral formula, should be excluded.

Section 2.6: Double Fourier transforms for two-dimensional problems.

Appendix A.5: Bessel functions and their properties as needed in Section 2.7 on the Hankel transform and in Section 3.4 for the Bessel–Fourier series. Legendre polynomials as needed for the Legendre–Fourier series in Section 3.5.

Section 2.7: Hankel transform for solving boundary value problems with radial dependence (a few illustrations are sufficient).

Chapter 3: Finite transforms (Fourier series).

Sections 3.1–3.5: Illustration of Fourier series and its error analysis, then its application to boundary value problems on a finite domain.

Chapter 4: Discrete transforms.

Sections 4.1.1–4.1.4 and 4.1.7: Discrete Fourier transforms (DFTs) and the (illustrative) analysis of their errors in approximating Fourier series and integrals, their fast algorithm, the fast Fourier transform (FFT) in Section 4.1.3. Choices from Sections 4.1.6B–I recommended for electrical and computer engineering students.

Graduate (or Senior) Course in Analysis and Applications of Integral and Discrete Transforms

The necessary background material from the contents of the above undergraduate course *plus*:

A selection from Section 1.5 including the Lagrange identity, then examples of constructing compatible transforms in Section 1.6.

A good selection of the proofs of the most basic theorems of Fourier integrals in Section 2.2, and those for the Fourier series and general orthogonal expansions in Section 3.1.3. Almost all the applications in Sections 2.2–2.7 and 3.2–3.6.

Selection from Section 4.1.5 for the application of the z-transform and the DFT. Section 4.1.6 with at least one parallel exercise done for each of the illustrations of the topics of interest.

Section 4.2: Discrete orthogonal polynomial transforms; an overview if time permits.

CONTENTS

xi

1

COMPATIBLE TRANSFORMS

For most students of mathematics, science, and engineering, one's first exposure to integral transforms occurs with the Laplace transform, typically used to solve initial value problems associated, primarily, with ordinary differential equations having constant coefficients. The next encounter is when one needs to solve boundary value problems associated with partial differential equations, or when one needs the simpler frequency space representation of the Fourier transform for time signals. This is typical in the junior course on boundary value problems, in elementary quantum physics, or in courses in signal analysis.

For boundary value problems on finite domains and primarily homogeneous boundary conditions, the method of separation of variables is usually employed, while Fourier series expansion is used to treat the nonhomogeneous conditions. It is when the domain becomes infinite that students are made aware of the (infinite) Fourier integral transform as a necessary tool, and it is usually introduced as a limit of the Fourier series (of the finite domain).

Since this chapter is dedicated to the proper choice of integral transforms, we will concentrate our efforts on the clear illustration of this idea. In this fashion, we may not stop to justify all kinds of "mathematical subtleties" such as questions of convergence, but will rather follow the manipulative or "formal" way to advance our operations to the result desired. The omitted justifications of these operations, and very clear and precise statements of such necessary theorems (results), will be presented

1

in Chapters 2 and 3, on integral and finite transforms, respectively. Briefly, we will limit ourselves in those chapters to a certain (but nevertheless wide) class of functions that would satisfy the conditions of the stated theorems.

The proofs of such important results (or theorems) will be carried completely and will often be summarized as clear and concise theorems or corollaries. Others that may follow easily or that lie outside the range of this text's main goals will be left for the exercises provided.

To illustrate the importance of deciding on a suitable transform in advance, we will review in Example 1.1 of Section 1.1 the method of separation of variables (or product method) and its close relationship to Fourier analysis in general.

1.1 The Method of Separation of Variables and the Integral Transforms

With suitable geometry, the method of separation of variables assumes, at the outset, the form of the solution as a product of functions of a single variable. For example, the function $u(x,t)$ is written as a product of two functions in each of the variables x and t: $u(x,t) = X(x)T(t)$. This, as we shall see next, works well primarily for homogeneous equations, with each of the boundary (or auxiliary) conditions given at a constant value of one of the variables.

In the following example of the temperature distribution in a semi-infinite bar, we will illustrate this method and attempt to compare it with the transform method; it should be helpful even for readers familiar with this problem.

EXAMPLE 1.1: The Method of Separation of Variables

Consider the following initial and boundary value problem for $u(x,t)$, the temperature inside a semi-infinite thin bar [$x \in (0,\infty)$], where heat diffusion is governed by the one-dimensional *heat equation* $\partial u/\partial t = \partial^2 u/\partial x^2$,[†] the end $x = 0$ is kept at zero temperature, and the initial temperature is $u(x,0) = f(x)$. Of course, we shall assume that the temperature vanishes at infinity, i.e., $\lim_{x \to \infty} u(x,t) = 0$, $t > 0$, and that it is a bounded function in time and space, i.e., $|u(x,t)| < \infty$, $0 < x < \infty$, $t > 0$. The mathematical

---

[†]The heat equation is derived in detail in Appendix B.2; see also its exercises.

model for this problem is

$$\frac{\partial u}{\partial t} = \frac{\partial^2 u}{\partial x^2}, \qquad 0 < x < \infty, \ t > 0 \tag{E.1}$$

$$u(0,t) = 0, \qquad t > 0 \tag{E.2}$$

$$u(x,0) = f(x), \qquad 0 < x < \infty \tag{E.3}$$

The method of separation of variables assumes $u(x,t) = X(x)T(t)$ to be substituted into the homogeneous equation (E.1) and uses the homogeneous condition (E.2); then Fourier analysis is used to manage the nonhomogeneous condition (E.3). If we substitute $u(x,t) = X(x)T(t)$ into the partial differential equation (E.1), we obtain

$$T'(t)X(x) = T(t)X''(x) \tag{E.4}$$

an equation that involves only ordinary derivatives dT/dt and d^2X/dx^2. We will only consider $u(x,t) = X(x)T(t) \not\equiv 0$, as we are not interested in the trivial solution $u(x,t) \equiv 0$, so both sides of (E.4) can be divided by $X(x)T(t)$ to obtain

$$\frac{T'(t)}{T(t)} = \frac{X''(x)}{X(x)} \tag{E.5}$$

Since x and t are independent of each other, the two sides of (E.5) cannot be equal unless both sides are equal to a constant μ. Thus,

$$\frac{T'(t)}{T} = \frac{X''(x)}{X} = \mu \tag{E.6}$$

that is,

$$T'(t) - \mu T = 0 \tag{E.7}$$

and

$$X''(x) - \mu X = 0 \tag{E.8}$$

The solutions to these and other related differential equations are reviewed in Appendix A.1 and its exercises.

The partial differential equation is now reduced to two simple differential equations, (E.7) and (E.8), in t and x, respectively. By contrast, as we shall discuss in this book, the method of transforms will attempt to algebraize one or even both of the partial differential operations in (E.1), resulting, respectively, in an ordinary differential equation or an algebraic equation. The latter is easy to solve and, more importantly, there is no need to assume the special product form of the solution $u(x,t) = X(x)T(t)$ for (E.1) in advance. The transform method can also deal with problems that are not necessarily separable. To give a simple example, we have the final solution (E.15), (E.16) to this problem, which is not a product of two functions of a single variable. A much clearer example is the final solution (E.13) to the heat equation in Example 2.17:

$$u(x,t) = \frac{1}{\sqrt{4\pi t}} \int_{-\infty}^{\infty} f(\xi) \exp \frac{-(x - \xi)^2}{4t} \, d\xi$$

which is clearly not the product of two functions of the independent single variables x and t, respectively.

For completeness, and since we are going to consider this problem again in the next chapter, we shall follow the method of separation of variables through to its final solution.

The solutions to (E.7) and (E.8), respectively, are

$$T(t) = c_1 e^{\mu t} \tag{E.9}$$

and

$$X(x) = c_2 e^{x\sqrt{\mu}} + c_3 e^{-x\sqrt{\mu}} \tag{E.10}$$

so that the final solution to (E.1) is

$$u(x,t) = X(x)T(t) = c_1 e^{\mu t} \left[c_2 e^{x\sqrt{\mu}} + c_3 e^{-x\sqrt{\mu}} \right] \tag{E.11}$$

but since $u(x,t)$ must be finite as $t \to \infty$, we cannot allow μ to have a positive real value; to ensure this we choose μ to be negative, $\mu = -\lambda^2$, λ real. This will give a solution to (E.1) which is finite in t and vanishes as $t \to \infty$:

$$u_\lambda(x,t) = c_1 e^{-\lambda^2 t} \left[c_2 e^{i\lambda x} + c_3 e^{-i\lambda x} \right], \qquad i = \sqrt{-1} \dagger \tag{E.12}$$

There is, of course, no difficulty with $x \to \infty$ since $e^{i\lambda x}$ and $e^{-i\lambda x}$ can be written, via the following *Euler's identity*:

$$e^{i\lambda x} = \cos \lambda x + i \sin \lambda x$$

as combinations of $\sin \lambda x$ and $\cos \lambda x$, which are bounded for all x. Now, with this solution and its arbitrary constants c_2 and c_3, we should try to satisfy the homogeneous boundary condition (E.2):

$$u_\lambda(0,t) = c_1 e^{-\lambda^2 t} [c_2 + c_3] = 0 \tag{E.13}$$

---

†The very basic elements of complex numbers are reviewed briefly in Exercise 1.6 of this chapter.

Since we cannot make $c_1 = 0$, as it will render our general solution (E.12) a trivial one $u_\lambda(x,t) \equiv 0$, we must make $c_3 = -c_2$ to have

$$u_\lambda(x,t) = c_1 c_2 i e^{-\lambda^2 t} \left[e^{i\lambda x} - e^{-i\lambda x} \right] = 2c_1 c_2 i e^{-\lambda^2 t} \sin \lambda x \qquad \text{(E.14)}$$

We must note here how the boundary condition $u(0,t) = 0$ in (E.2) at $x = 0$ affected only the x-dependent factor of the solution in (E.12) to give the result in (E.14). We say that this boundary condition is associated with the boundary value at $x = 0$. Thus, it influences the x-dependent factor of the product solution as it did in (E.13), and we say that this boundary condition at $x = $ constant is compatible with $X(x)$, the x-dependent factor of the product solution $u(x,t) = X(x)T(t)$. In general, this remark applies to the use of all the transforms covered in this book. The same can be said about the initial condition given at $t = 0$, which will be associated later with the t factor of the solution. In Exercises 1.4 we will cover some sufficient conditions on the partial differential equation and its boundary conditions for the method of separation of variables to apply.

From this, we can say that in order to ensure that an auxiliary or boundary condition allows separation of the variables, it must be given at a boundary described by a constant value of the variable of that factor. In other words, we must pick the right coordinates which allow describing a given boundary of the domain by a constant value of one of the coordinates. A simple example is the study of the temperature inside a disc of radius R, where we must adopt the polar coordinates (r,θ) in order to describe the circumference (boundary) by a constant value of one of the variables, namely $r = R$ (valid for all values of θ in the domain of $u(r,\theta)$). Otherwise, in Cartesian coordinates (x,y) this boundary is described by $x^2 + y^2 = R^2$, which is not separable in either x or y. For an ultimate in such attempts, see Exercise 1.5, and for the detailed sufficient conditions for the separation of variables method to work see Exercise 1.4.

With the result in (E.14), which satisfies the homogeneous equations (E.1), (E.2), the role of the method of separation of variables has ended. However, to reach the final solution, we still have the nonhomogeneous condition $u(x,0) = f(x)$ in (E.3) left to be satisfied. What is usually done here is to appeal to Fourier analysis, where we consider a superposition of the solutions $u_\lambda(x,t)$ in (E.14) with respect to the parameter λ. For example, in the present case, with the sine waves $\sin \lambda x$ of (E.14), we use the Fourier sine integral representation where we integrate over λ, with a variable amplitude $(2/\pi)B(\lambda)$, instead of $2c_1 c_2 i$, to have

$$u(x,t) = \frac{2}{\pi} \int_0^\infty B(\lambda) e^{-\lambda^2 t} \sin \lambda x \, d\lambda \qquad \text{(E.15)}$$

The theory of the Fourier sine transform, which we will cover in the next chapter (Sec. 2.5), states that the nonhomogeneous condition (E.3) will be satisfied, that is, $\lim_{t \to 0} u(x,t) = f(x)$, provided that $B(\lambda)$ is chosen to be the following Fourier sine integral of the given function $f(x)$:

$$B(\lambda) = \int_0^\infty f(x) \sin \lambda x \, dx \tag{E.16}$$

The Fourier integrals are the subject of Sections 2.2 and 2.5, while the Fourier series is covered in Chapter 3.

In the following we will present a comparison of the integral transforms method with the above method of the products.

1.1.1 Integral Transforms

Instead of assuming the product form of the solution in advance, the integral transform method (to be described in this chapter) first decides what transform is compatible with the problem (E.1)–(E.3) in order to algebraize the differential equation (E.1) in Example 1.1. At the same time, it directly involves the auxiliary conditions (E.2), (E.3). One then solves the resulting algebraic equation and finally transforms it back to obtain the desired solution in the original space. In contrast to the method of separation of variables and resembling what is done via the Laplace transform, this method directly incorporates the nonhomogeneous condition $u(x,0) = f(x)$ in the transformed equation. In general, by means of one successful (suitable), or *compatible*, integral transform, a boundary or an initial value problem involving a partial differential equation in n variables is reduced to another one in $n - 1$ variables. This process, if carried out $n - 1$ times, results in an ordinary differential equation. This is easier to solve, since the theory of ordinary differential equations has been more extensively studied than that of partial differential equations. Another transformation, then, may reduce the differential equation to an algebraic equation.

Integral transforms are also used to solve compatible *integral equations*,[†] that is, equations where the unknown $u(x)$ appears under the integral sign

$$f(x) + u(x) = \int_a^{b(x)} K(x,t)u(t)\,dt \tag{1.1}$$

[†] For an elementary exposition of integral equations see the author's book (1985) or his UMAP self-contained paper (1986).

This will be illustrated in Section 2.1.4, where the Laplace transform is used for solving Volterra integral equations.

The question now is what constitutes a compatible transform. This, as we shall see next, depends on the differential operator in the given ordinary or partial differential equations and, of course, its (auxiliary) boundary conditions. In the case of integral equations in $u(x)$, like (1.1), such determination may be obtained easily from observing the limits of the integral and the kernel (or nucleus) $K(x,t)$.

1.2 Compatible Transforms

Consider the function $f(x)$, defined on the interval (a,b), and its integral transforms $F(\lambda)$

$$F(\lambda) = \int_a^b \rho(x)\overline{K(\lambda,x)}f(x)\,dx \equiv \kappa\,\{f\} \tag{1.2}$$

where $\overline{K(\lambda,x))}$ is called the *direct kernel*, or nucleus of the integral transform, and $\kappa\,\{f\}$ is an operation notation indicating the transformation of f via the integral transform associated with the kernel $\overline{K(\lambda,x)}$. $\overline{K(\lambda,x)}$ is the *complex* conjugate[†] of $K(\lambda,x)$, which is obtained by changing the imaginary number $i = \sqrt{-1}$ to $-i$ in $K(\lambda,x)$. Many references use $K(\lambda,x)$ for the kernel, and we may do so later in this book in order to be close to such notation. The weight function $\rho(x)$ in the integral (1.2) is introduced to facilitate working with some important variable coefficient differential equations such as the *Bessel differential equation* in (1.6) (see Appendix A.5).

For most of our examples of integral transforms in this section we will consider the usual cases, where $f(x)$ in (1.2) is defined on a semi-infinite $(a,b) = (0,\infty)$ or an infinite $(a,b) = (-\infty,\infty)$ domain and where the parameter λ is continuous. The classification of the different types of transforms in the next section will center primarily on whether the interval (a,b) is finite or infinite and also on whether λ and/or x is discrete or continuous.

The transformation in (1.2) is linear, that is, $\kappa\,\{c_1f_1 + c_2f_2\} = c_1\kappa\,\{f_1\} + c_2\kappa\,\{f_2\}$, where c_1 and c_2 are constants; we leave its proof as a simple exercise (see Exercise 2.5a). Since we are interested in eliminating differentiation with respect to one variable at a time in a partial differential

---

[†] See Exercise 1.6 for a simple review of complex numbers.

equation, and since most of the equations in mathematical physics are of second order, we shall consider here the most general second-order linear differential equation

$$L_2u \equiv A_0(x)\frac{d^2u}{dx^2} + A_1(x)\frac{du}{dx} + A_2(x)u = f(x) \tag{1.3}$$

where L_2 denotes the differential operator of the equation. Higher-order $(n > 2)$ differential equations will be presented for the interested reader in the last section of this chapter. The kernel $\overline{K(\lambda,x)}$ of the integral transform (1.2) is said to be *compatible* with the linear differential operator L_2 of (1.3) *if it transforms the differential equation (1.3) in $u(x)$ to an algebraic one in $U(\lambda)$, the integral transform (1.2) of $u(x)$.* Since physical models associated with differential equations usually have (auxiliary) boundary and/or initial conditions, it is not enough that the integral transform algebraizes the differential operator; it also must deal directly with and involves the conditions prescribed at the boundaries of the operator's domain.

The most frequently encountered integral transforms, which we shall discuss in the following chapter, are summarized in Table 1.1, along with their compatible differential operators L.

In Table 1.1, P refers to the *Cauchy principal value* of the integral, i.e.,

$$P\int_{-\infty}^{\infty} \frac{f(x)\,dx}{x - \lambda} = \lim_{A\to\infty}\int_{-A}^{A} \frac{f(x)\,dx}{x - \lambda} \tag{1.4}$$

and

$$\lim_{\epsilon\to 0^+}\left[\int_{-\infty}^{\lambda-\epsilon} + \int_{\lambda+\epsilon}^{\infty}\right]\frac{f(x)}{x - \lambda}\,dx \tag{1.5}$$

Also, $J_n(x)$ is the *Bessel function* of the first kind of order n, which is one of the two solutions of the Bessel differential equation,

$$x^2u'' + xu' + (x^2 - n^2)u = 0 \tag{1.6}$$

that is bounded at $x = 0$. It will be discussed and solved in detail in Appendix A.4. (See also Appendix A.5.1.) We may note that the Hankel transform in the last entry of Table 1.1 is with $\rho(x) = x$ and $K(\lambda,x) = J_n(\lambda x)$ in (1.2).

In the following examples, we will illustrate the concept of transform compatibility for the most familiar transforms, including the Laplace, Fourier, and Hankel (Bessel) transforms. The theory behind establishing these and other general transforms will be left for later sections, and the problems associated with nth $(n > 2)$ order differential equations are treated in the last section. This should help us in the following

TABLE 1.1 Integral Transforms and Their Compatible Differential Operators L

Transform	Compatible with L	Name
$F(s) = \mathcal{L}\{f\} = \int_0^\infty e^{-sx} f(x)\, dx$	$L \equiv \dfrac{d^n}{dx^n}$	Laplace transform (Sec. 2.1)
$F(\lambda) = \mathcal{F}\{f\} = \int_{-\infty}^\infty e^{-i\lambda x} f(x)\, dx$	$L \equiv \dfrac{d^n}{dx^n}$	Fourier exponential transform (Sec. 2.2)
$F_c(\lambda) = \mathcal{F}_c\{f\} = \int_0^\infty \cos \lambda x\, f(x)\, dx$	$L \equiv \dfrac{d^{2n}}{dx^{2n}}$	Fourier cosine transform (Sec. 2.5)
$F_s(\lambda) = \mathcal{F}_s(f) = \int_0^\infty \sin \lambda x\, f(x)\, dx$	$L \equiv \dfrac{d^{2n}}{dx^{2n}}$	Fourier sine transform (Sec. 2.5)
$F(\lambda) = \mathcal{M}\{f\} = \int_0^\infty x^{\lambda-1} f(x)\, dx$	$L \equiv x^n \dfrac{d^n}{dx^n}$	Mellin transform (Sec. 2.9.2)
$F(\lambda) = \mathcal{H}\{f\} = \dfrac{1}{\pi} P \int_{-\infty}^\infty \dfrac{f(x)\, dx}{x - \lambda}$	See relation to Fourier transform in Section 2.9.1	Hilbert transform (Sec. 2.9.1)
$F_n(\lambda) = \mathcal{H}_n\{f\} = \int_0^\infty x J_n(\lambda x) f(x)\, dx$	$L \equiv \dfrac{d^2}{dx^2} + \dfrac{1}{x}\dfrac{d}{dx} - \dfrac{n^2}{x^2}$	Hankel (Bessel) transform (Sec. 2.7)

chapters as we plan a strategy for selecting the most appropriate, or completely compatible, transform for solving initial and/or boundary value problems.

1.2.1 Examples of Compatible Transforms

EXAMPLE 1.2: Compatibility of the Laplace Transform with the Differential Operators $L \equiv d/dx$, d^n/dx^n, $0 < x < \infty$

The kernel $K(\lambda,x) = e^{-\lambda x}$ of the Laplace transform

$$\mathcal{L}\{f(x)\} \equiv F(\lambda) = \int_0^\infty e^{-\lambda x} f(x)\, dx \tag{1.7}$$

is compatible with the differential operator $L \equiv d/dx$, since if we perform one integration by parts with $u = e^{-\lambda x}$ and $dv = df$, for the Laplace transform of df/dx we obtain

$$\int_0^\infty e^{-\lambda x} \frac{df}{dx}\, dx = f(x)e^{-\lambda x}\, \Big|_0^\infty + \lambda \int_0^\infty e^{-\lambda x} f(x)\, dx \tag{E.1}$$

$$= \lim_{x \to \infty} f(x)e^{-\lambda x} - f(0) + \lambda F(\lambda) \tag{E.2}$$

We note the expression in (E.2) is algebraic in $F(\lambda)$, which makes the Laplace kernel compatible with the differential operator d/dx. However, the algebraic expression is not complete until we are given $f(0)$, the value of the original function at $x = 0$, and some information about $f(x)$ as x approaches infinity, ensuring that the limit in the first term of (E.2) exists. The availability of such information makes the compatibility complete.

The usual condition on $f(x)$ for the Laplace transform of df/dx is to give its initial condition $f(0)$ and assume that as $x \to \infty$ its growth is slower than $e^{\lambda x}$, that is, $\lim_{x \to \infty} e^{-\lambda x} f(x) = 0$. Hence, the final result of Laplace transforming df/dx, with these conditions, is

$$\int_0^\infty e^{-\lambda x} \frac{df}{dx}\, dx = \lambda F(\lambda) - f(0) \tag{1.8}$$

which is algebraic in $F(\lambda)$ and involves the initial condition $f(0)$ of the transformed function $f(x)$. The existence of the Laplace transform in (1.7) and the justification of the result in (1.8) will be stated more carefully and proved in Chapter 2 as Theorems 2.1 and 2.3, respectively.

It will be left as an exercise to show that the Laplace transform is compatible with the differential operator d^2/dx^2 when $u(0)$ and $u'(0)$ are given and where $\lim_{x \to \infty} u(x)e^{-\lambda x} = 0$ and $\lim_{x \to \infty} u'(x)e^{-\lambda x} = 0$, that is,

$$\mathcal{L}\left\{ \frac{d^2u}{dx^2} \right\} \equiv \int_0^\infty e^{-\lambda x} \frac{d^2u}{dx^2} \, dx = \lambda^2 U(\lambda) - \lambda u(0) - u'(0) \tag{1.9}$$

and, in general,

$$\mathcal{L}\left\{ \frac{d^nu}{dx^n} \right\} = \lambda^n U(\lambda) - \sum_{k=0}^{n-1} \lambda^k u^{(n-k-1)}(0) \tag{1.10}$$

where $u^{(n)}(0)$ indicates the nth derivative at $x = 0$. The precise statements and conditions of (1.9) and (1.10) will be given in Chapter 2 as corollaries to Theorem 2.3

EXAMPLE 1.3: Compatibility of the Exponential Fourier Transform with the Differential Operators d/dx, d^n/dx^n, $-\infty < x < \infty$

The exponential Fourier transform

$$\mathcal{F}\{u(x)\} = U(\lambda) = \int_{-\infty}^\infty e^{-i\lambda x} u(x) \, dx \tag{1.11}$$

is compatible with the differential operator d/dx provided that a condition is given for $u(x)$ as $x \to \mp\infty$, which is usually $\lim_{x \to \mp\infty} u(x) = 0$. The compatibility can be established from the definition of $\mathcal{F}\{du/dx\}$. After integrating by parts once we have

$$\mathcal{F}\left\{ \frac{du}{dx} \right\} = \int_{-\infty}^\infty e^{-i\lambda x} \frac{du}{dx} \, dx = e^{-i\lambda x} u(x) \Big|_{-\infty}^\infty + i\lambda \int_{-\infty}^\infty e^{-i\lambda x} u(x) \, dx$$

$$= i\lambda U(\lambda) \tag{1.12}$$

is algebraic in $U(\lambda)$, and it is clear that $\lim_{x \to \mp\infty} e^{-i\lambda x} u(x) = 0$ when $\lim_{x \to \mp\infty} u(x) = 0$ since $e^{-i\lambda x}$ is bounded as $x \to \mp\infty$ for real λ. The conditions for the existence of the Fourier transform in (1.11) for a special class of functions $u(x)$ will be presented completely in Chapter 2 as Theorem 2.12. The precise conditions for the validity of the result in (1.12) is given as Theorem 2.15.

It will be left as an exercise to show that the exponential Fourier transform is compatible with the nth-order differential operator d^n/dx^n, provided that the function and all its derivatives up to d^{n-1}/dx^{n-1} vanish as $x \to \mp\infty$, that is,

$$\mathcal{F}\left\{\frac{d^n u}{dx^n}\right\} = \int_{-\infty}^{\infty} e^{-i\lambda x}\frac{d^n u}{dx^n}\,dx = (i\lambda)^n U(\lambda) \tag{1.13}$$

The complete statement of the special case for du/dx of this result will be presented and proved in Chapter 2 as Theorem 2.15. This general case (1.13) will be given as Corollary 1 to Theorem 2.15.

EXAMPLE 1.4: Compatibility of Fourier Sine and Cosine Transforms with the Differential Operator d^{2n}/dx^{2n}

The Fourier sine and cosine transforms

$$\mathcal{F}_s\{u(x)\} \equiv U_s(\lambda) = \int_0^{\infty} u(x)\sin\lambda x\,dx \tag{1.14}$$

and

$$\mathcal{F}_c\{u(x)\} \equiv U_c(\lambda) = \int_0^{\infty} u(x)\cos\lambda x\,dx \tag{1.15}$$

are both compatible with the differential operator of *even* order d^{2n}/dx^{2n}; however, the required conditions at $x = 0$ are different. For example, in the case of the second-order differential operator d^2/dx^2, the Fourier sine transform after two integrations by parts gives

$$\begin{aligned}
\mathcal{F}_s\left\{\frac{d^2 u}{dx^2}\right\} &= \int_0^{\infty} \frac{d^2 u}{dx^2}\sin\lambda x\,dx \\
&= \left(\frac{du}{dx}\sin\lambda x\right)\Big|_0^{\infty} - \lambda\int_0^{\infty}\frac{du}{dx}\cos\lambda x\,dx \\
&= \left(\frac{du}{dx}\sin\lambda x\right)\Big|_0^{\infty} - \lambda\left[(u(x)\cos\lambda x)\Big|_0^{\infty}\right. \\
&\qquad\qquad \left. + \lambda\int_0^{\infty} u(x)\sin\lambda x\,dx\right]
\end{aligned} \tag{E.1}$$

As in the case of the exponential Fourier transform, we need $\lim_{x\to\infty} u(x) = 0$ and $\lim_{x\to\infty} u'(x) = 0$ (it will be the same for the Fourier cosine transform (1.15)), which reduces (E.1) to

$$\begin{aligned}
\mathcal{F}_s\left\{\frac{d^2 u}{dx^2}\right\} &= 0 - \lambda[-u(0) + \lambda U_s(\lambda)] \\
&= \lambda u(0) - \lambda^2 U_s(\lambda)
\end{aligned} \tag{1.16}$$

This is algebraic in $U_s(\lambda)$ but requires the value of the function $u(0)$ at $x = 0$. Hence, the Fourier sine transform is compatible for problems on $(0, \infty)$ with differential operators (with constant coefficients) of *even order* ($n = 2$ here) and where the *value of the function is given at* $x = 0$. The even order applies to the number of integrations by parts necessary to bring the kernel to $\sin \lambda x$ again to give us the transform $U_s(\lambda)$ as the last term in (E.1) or (1.16). To make a comparison, the Fourier sine and Laplace transforms cover the same domain $(0, \infty)$; however, the Laplace transform is compatible with any order differential operator d^n/dx^n and for $n = 2$ requires $u'(0)$ in addition to $u(0)$, as in (1.9). An example is the displacement of a semi-infinite vibrating string as a function of time, for which we know the initial displacement $u(x, 0)$ and the initial velocity $\partial u(x, 0)/\partial t$ at $t = 0$, compared to the temperature in a semi-infinite heated bar as a function of x, for which we have only the temperature $u(0, t)$ at $x = 0$. We would apply Laplace and sine transforms, respectively, to these two problems.

In the same way, we can show that the Fourier cosine transform is compatible with d^{2n}/dx^{2n}, and for $n = 1$ it requires the *derivative* of the function $u'(0)$ at $x = 0$, instead of $u(0)$ in the case of the sine transform,

$$\mathcal{F}_c\left\{\frac{d^2u}{dx^2}\right\} = \int_0^\infty \frac{d^2u}{dx^2} \cos \lambda x \, dx = -\lambda^2 U_c(\lambda) - u'(0) \tag{1.17}$$

With regard to the semi-infinite heated bar, the Fourier cosine transform is suitable when the end $x = 0$ is insulated, $\partial u(0, t)/\partial x = 0$, while the Fourier sine transform is suitable when the end $x = 0$ is kept at zero temperature, $u(0, t) = 0$, as we indicated in Example 1.1.

Following the same analysis, we can show that the Fourier sine and cosine transforms are compatible only with even-order differential operators d^{2n}/dx^{2n}, and that the sine transform requires all the even-order derivatives up to $2n - 2$ at $x = 0$ while the cosine transform requires all the odd-order derivatives up to $2n - 1$ at $x = 0$:

$$\mathcal{F}_s\frac{\{s^{2n}u\}}{dx^{2n}} = (-1)^n \lambda^{2n} U_s(\lambda) - \sum_{k=1}^{n}(-1)^k \lambda^{2k-1} u^{(2n-2k)}(0) \tag{1.18}$$

$$\mathcal{F}_c\frac{\{d^{2n}u\}}{dx^{2n}} = (-1)^n \lambda^{2n} U_c(\lambda) - \sum_{k=0}^{n-1}(-1)^k \lambda^{2k} u^{(2n-2k-1)}(0) \tag{1.19}$$

These results, (1.18) and (1.19), are left for exercises.

Note that the above three examples illustrate the compatibility of Laplace and Fourier transforms for differential operators with constant coefficients, along with their proper conditions on the function $u(x)$. In the next example we will illustrate how the Laplace transform, as a representative of the above transforms, associated with exponential and trigonometric kernels, fails to be compatible with the same differential operator having variable coefficients. This is so in the sense that it will not result in algebraizing such an operator, but may even increase the order of the differentiation of the resulting transform $U(\lambda)$. Unseen difficulties can arise, which we will attempt to explain. Of course, these difficulties should not discourage us completely from tackling some very important differential equations with certain variable coefficients via the Laplace or Fourier transforms (see Example 2.9 and Exercise 2.11c).

EXAMPLE 1.5:   Incompatibility of the Laplace Transform with Differential Operators That Have Variable Coefficients

We will show here that the Laplace transform of the variable coefficient differential expression $x^7 \, du/dx$ is

$$(-1)^7 \frac{d^7}{d\lambda^7} [\lambda U(\lambda) - u(0)]$$

It is a derivative of order seven of the transform $\lambda U(\lambda) - u(0)$. This is accomplished by first writing the Laplace transform $\lambda U(\lambda) - u(0)$ of du/dx from Example 1.2:

$$\lambda U(\lambda) - u(0) = \int_{-\infty}^{\infty} e^{-\lambda x} \frac{du}{dx} \, dx \tag{1.8}$$

Then, by taking the seventh-order derivative of both sides with respect to λ, and provided that the integral (1.8) allows the exchange of differentiation and integration,

$$\frac{d^7}{d\lambda^7} [\lambda U(\lambda) - u(0)] = \frac{d^7}{d\lambda^7} \int_0^{\infty} e^{-\lambda x} \frac{du}{dx} \, dx$$

$$= \int_0^{\infty} \frac{d^7}{d\lambda^7} (e^{-\lambda x}) \frac{du}{dx} \, dx$$

$$= \int_0^{\infty} (-x)^7 \frac{du}{dx} e^{-\lambda x} \, dx \tag{E.1}$$

Departing somewhat from our approach at the beginning of this chapter—that of emphasizing the operational or *formal* procedures without stopping to justify all our steps—we are obliged to stop here and give a *warning* concerning the very important *operation of exchanging differentiation with integration in (E.1)*. As shall be stated carefully in Chapter 2 as Theorem 2.6, it turns out that such an exchange is allowed provided that the resulting integral in (E.1) is uniformly convergent in λ. This turns out to be the case if we impose the condition on the function $u(x)$ that $(-x)^7 du/dx$ is absolutely integrable on $(0, \infty)$; that is, $\int_0^\infty | -x^7 du/dx | dx < \infty$. In Chapter 2 we will define our particular class of functions $u(x)$ and show that this condition is satisfied.

The result in (E.1) says that the Laplace transform of $-x^7 du/dx$ is the seventh-order derivative $(d^7/d\lambda^7)[\lambda U(\lambda) - u(0)]$. This example should make it clear that even such a simple variable coefficient increases the order of the differentiation for the transform $U(\lambda)$. In general, we can easily show that (see the complete derivation of equation (2.21) and Theorems 2.4 and 2.5 in Chapter 2)

$$\mathcal{L}\{x^n f(x)\} = (-1)^n \frac{d^n}{d\lambda^n} F(\lambda) \tag{1.20}$$

which suggests that we should avoid using Laplace or Fourier transforms for differential operators with variable coefficients, and in particular polynomial coefficients with a degree n higher than the order m of the derivative involved, that is, $x^n d^m u/dx^m$, $n > m$, since this would result in higher-order derivatives than m. This is the case unless we know, or discover, that the resulting higher-order differential equation in $U(\lambda)$ happens to be easier to solve than the original lower-order one in $u(x)$.

Looking at the result in (E.1) and (1.20), we may be curious about the case where the variable coefficient is not the integer power x^7 of x, but instead $x^{1/7}$. The surmised answer may be new to some of us, but we expect instead of a derivative of order 7 a derivative of *fractional* order $d^{1/7}/dx^{1/7}$. Indeed, fractional calculus for derivatives and integrations of order ν, where ν is not necessarily an integer, has been around for some time. We shall attempt to discuss this subject briefly in Exercise 1.8d, and we may have to revisit it again in Section 2.1 when we discuss the Laplace transform. (See Exercise 2.20h, g, f and Example 2.13.)

Before attempting (in Section 1.5) to answer the general question of how to find the compatible transform for the most general differential operator in (1.3), we shall present another example which will lead us in that direction. This will involve the Bessel differential equation whose compatible Hankel integral transform has a Bessel function kernel

$K(\lambda,x) = J_n(\lambda x)$ with a weight $\rho = x = x$ in (1.2). The Bessel equation and other important differential equations, with their solutions as special functions, are discussed in detail in Appendix A.5 and its exercises.

EXAMPLE 1.6: Compatibility of the Hankel (Bessel) Transform with the Part of the Bessel Differential Operator
$$d^2/dr^2 + (1/r)d/dr - n^2/r^2$$

We will show, using integration by parts and the Bessel differential equation (1.6) in $u(r)$,

$$\frac{d^2u}{dr^2} + \frac{1}{r}\frac{du}{dr} + \left(1 - \frac{n^2}{r^2}\right)u(r) = 0, \qquad 0 < r < \infty \tag{1.21}$$

that the J_n-Hankel transform

$$\mathcal{H}_n\{u(r)\} = U_n(\lambda) = \int_0^\infty rJ_n(\lambda r)u(r)\,dr \tag{1.22}$$

will algebraize the Bessel differential operator

$$L \equiv \frac{d^2}{dr^2} + \frac{1}{r}\frac{d}{dr} - \frac{n^2}{r^2}$$

provided again that the following conditions (of boundedness) for $u(r)$ and du/dr, at $r = 0$ and as $r \to \infty$, are met:

$$\lim_{r \to 0} ru'(r)J_n(\lambda r) = 0, \qquad \lim_{r \to \infty} ru'(r)J_n(\lambda r) = 0 \tag{E.1}$$

and

$$\lim_{r \to 0} ru(r)\frac{d}{dr}J_n(\lambda r) = 0, \qquad \lim_{r \to \infty} ru(r)\frac{d}{dr}J_n(\lambda r) = 0 \tag{E.2}$$

The final result we are after is

$$\mathcal{H}_n\left\{\frac{d^2u}{dr^2} + \frac{1}{r}\frac{du}{dr} - \frac{n^2}{r^2}u\right\}$$

$$= \int_0^\infty r\left[\frac{d^2u}{dr^2} + \frac{1}{r}\frac{du}{dr} - \frac{n^2}{r^2}u\right]J_n(\lambda r)\,dr = -\lambda^2 U_n(\lambda) \tag{E.3}$$

Since it is known that the Bessel function $J_n(\lambda r)$ is bounded at $r = 0$ and as $r \to \infty$ (see Problem A.25e in Appendix A.5) the boundary conditions (E.1), (E.2) are easily met if we assume that $ru(r)$ and $r\,du/dr$ vanish as $r \to 0$ and as $r \to \infty$, that is,

$$\lim_{r \to 0} ru(r) = 0, \qquad \lim_{r \to 0} ru'(r) = 0 \tag{E.4}$$

$$\lim_{r \to \infty} ru(r) = 0, \qquad \lim_{r \to \infty} ru'(r) = 0 \tag{E.5}$$

with the knowledge that $dJ_n(\lambda r)/dr$ is also bounded as $r \to 0$ and as $r \to \infty$, since it can be expressed in terms of a difference of two Bessel functions (see Exercise A.23c in Appendix A.5).

To facilitate the integration by parts, we note that the Bessel differential operator can be written in the more suitable form

$$\frac{d^2u}{dr^2} + \frac{1}{r}\frac{du}{dr} - \frac{n^2}{r^2}u = \frac{1}{r}\frac{d}{dr}\left(r\frac{du}{dr}\right) - \frac{n^2}{r^2}u$$

called the *self-adjoint form*, which will play a major role in determining compatible transforms for more general problems (Sections 1.5, 1.6.).

Now we use (1.23) in (E.3) and perform one integration by parts on the first term of the Bessel differential operator on the right side of (1.23) (with $dv = d(r\,du)$) to obtain

$$\mathcal{H}_n\left\{\frac{1}{r}\frac{d}{dr}\left(r\frac{du}{dr}\right) - \frac{n^2}{r^2}u(r)\right\}$$

$$= \int_0^\infty rJ_n(\lambda r)\left[\frac{1}{r}\frac{d}{dr}\left(r\frac{du}{dr}\right) - \frac{n^2}{r^2}u(r)\right]dr$$

$$= \int_0^\infty J_n(\lambda r)\left[\frac{d}{dr}\left(r\frac{du}{dr}\right) - \frac{n^2}{r}u(r)\right]dr$$

$$= \left[r\frac{du}{dr}J_n(\lambda r)\right]\Big|_0^\infty - \int_0^\infty r\frac{d}{dr}[J_n(\lambda r)]\frac{du}{dr}\,dr$$

$$- \int_0^\infty J_n(\lambda r)\frac{n^2}{r}u(r)\,dr \tag{E.6}$$

With $J_n(\lambda r)$ bounded as r approaches 0 or ∞, if we use the second parts of the conditions (E.4), (E.5) on $u'(r)$, the first term in (E.6) vanishes. Next we perform another integration by parts on the first integral of (E.6) (with $du = du$ and $v = r\,dJ_n(\lambda r)/dr$,

$$\mathcal{H}_n\left\{\frac{1}{r}\frac{d}{dr}\left(r\frac{du}{dr}\right) - \frac{n^2}{r^2}u(r)\right\}$$

$$= 0 - \left[u(r)r\frac{d}{dr}J_n(\lambda r)\right]\Big|_0^\infty + \int_0^\infty \frac{d}{dr}\left[r\frac{dJ_n(\lambda r)}{dr}\right]u(r)\,dr$$

$$- \int_0^\infty J_n(\lambda r)\frac{n^2}{r}u(r)\,dr$$

$$= -\left[u(r)r\frac{d}{dr}J_n(\lambda r)\right]\Big|_0^\infty$$

$$+ \int_0^\infty ru(r)\left[\frac{1}{r}\frac{d}{dr}\left(r\frac{d}{dr}J_n(\lambda r)\right) - \frac{n^2}{r^2}J_n(\lambda r)\right]dr \tag{E.7}$$

With $dJ_n(\lambda r)/dr$ bounded as $r \to 0$ or ∞, if we use the first parts of the conditions on $u(r)$ in (E.4), (E.5), the first term in (E.7) vanishes. Now in putting the two integrals together, realizing for the resulting integral that $J_n(\lambda r)$ is by definition a solution of the Bessel equation,

$$\frac{1}{r}\frac{d}{dr}\left(r\frac{dJ_n(\lambda r)}{dr}\right) + \left(\lambda^2 - \frac{n^2}{r^2}\right)J_n(\lambda r) = 0 \tag{1.23}$$

where the appearance of λ^2 in (1.23), instead of the 1 in (1.21), is the result of changing r to λr in (1.21) as verified in Exercise 1.9. We then have the desired result (E.3):

$$\mathcal{H}_n\left\{\frac{1}{r}\frac{d}{dr}\left(r\frac{du}{dr}\right) - \frac{n^2}{r^2}u\right\}$$

$$= 0 + \int_0^\infty \left[\frac{1}{r}\frac{d}{dr}\left(r\frac{dJ_n(\lambda r)}{dr}\right) - \frac{n^2}{r^2}J_n(\lambda r)\right]ru(r)dr$$

$$= \int_0^\infty -\lambda^2 J_n(\lambda r)ru(r)dr = -\lambda^2 U_n(\lambda) \tag{1.24}$$

which is algebraic in $U_n(\lambda)$. This result (1.24) will be discussed again in Section 2.7, where we will cover the Hankel transforms with detailed theory and applications.

Before we start classifying the different transforms in the next section, we note again that all the examples of compatible transforms we have presented up to now have *infinite* domains and *continuous* spectral variable λ. We present now an example of the (*finite*) sine transform with finite domain $(0, \pi)$ and *discrete* spectral parameter $\lambda_n = (n - \frac{1}{2})$.

EXAMPLE 1.7: Finite Fourier Sine Transform

Consider the following boundary value problem in $y(x)$ on the domain $(0, \pi)$:

$$\frac{d^2y}{dx^2} = g(x), \qquad 0 < x < \pi \tag{E.1}$$

$$y(0) = \alpha \tag{E.2}$$

$$y'(\pi) = \beta \tag{E.3}$$

We will show here that the finite sine transform

$$Y_n = f_s \{y\} = \int_0^\pi y(x)\sin(n - \tfrac{1}{2})x\,dx \tag{E.4}$$

is compatible with the problem (E.1)–(E.3); that is, it will algebraize the differential equation (E.1) and include the two boundary conditions (E.2), (E.3). The reason behind choosing the sine kernel $\sin(n - \tfrac{1}{2})x$ is the subject of Section 1.6 and Example 1.10 concerning the illustrations on how to construct the compatible finite transform for a boundary value problem.

If we multiply both sides of (E.1) by $\sin(n - \tfrac{1}{2})x$, integrate from $x = 0$ to $x = \pi$, and perform two integrations by parts—as we did in most of the previous examples—we have

$$\int_0^\pi \sin(n - \tfrac{1}{2})x\,\frac{d^2y}{dx^2}\,dx$$

$$= -(n - \tfrac{1}{2})^2 Y_n + (n - \tfrac{1}{2})y(0) - (-1)^n y'(\pi) \tag{E.5}$$

an algebraic equation in the finite sine transform Y_n. We note that such a transform includes the boundary conditions (E.2), (E.3) as required by (E.5); i.e., the transform has algebraized the differential equation as well as involved the given boundary conditions.

With reference to the next section, Classification of the Transforms, the above sine transform (E.4) is called *finite transform* (or the Fourier sine coefficients), which is the inverse of the *Fourier sine series* of $y(x)$,

$$y(x) = 2/\pi \sum_{n=1}^{\infty} Y_n \sin(n - \tfrac{1}{2})x \tag{E.6}$$

Chapter 3 will be devoted to such finite transforms or Fourier series, while Chapter 2 will cover the usual (infinite) transforms discussed in the preceding Examples 1.2–1.6.

In Section 1.5, we will summarize the results of these six Examples 1.2–1.7 in an attempt to motivate some general theory that will encompass these and other problems for determining the kernel of the transform that is compatible with given boundary or initial value problems. The emphasis will be on second-order differential equations.

1.2.2 Nonlinear Terms

Along with the difficulty, or incompatibility, of the Laplace and Fourier transforms in handling variable coefficients, we find that all integral transforms are used almost exclusively for linear problems. This is evident from the absence of nonlinear problems in all basic texts and references on the subject of integral transforms or the general operational calculus. In our discussion, nonlinear problems may appear, but we will still refer to the compatible transform of the linear part of a given nonlinear ordinary or partial differential equation. For example, the Korteweg–deVries (KdV) partial differential equation in $u(x,t)$,

$$\frac{\partial u}{\partial t} + 6u\frac{\partial u}{\partial x} + \frac{\partial^3 u}{\partial x^3} = 0, \qquad -\infty < x < \infty, \; 0 < t < \infty \tag{1.25}$$

is nonlinear because of the term $6u\,\partial u/\partial x$; however, we may say that the exponential Fourier transform (1.11) is compatible with the linear term $\partial^3 u/\partial x^3$ since according to (1.13) we have

$$\mathcal{F}\left\{\frac{\partial^3 u}{\partial x^3}\right\} = (i\lambda)^3 U(\lambda,t), \qquad U(\lambda,t) = \mathcal{F}\{u(x,t)\} \tag{1.26}$$

Also, if we are given the initial condition

$$u(x,0) = f(x), \qquad -\infty < x < \infty \tag{1.27}$$

then, according to (1.8), the Laplace transform is compatible with the linear term $\partial u/\partial t$ in (1.25),

$$\mathcal{L}\left\{\frac{\partial u}{\partial t}\right\} = \lambda U(x,\lambda) - u(x,0) = \lambda U(x,\lambda) - f(x),$$

$$U(x,\lambda) = \mathcal{L}\{u(x,t)\} \tag{1.28}$$

An attempt at using integral transforms, along with an iterative method, for solving nonlinear problems will be presented very briefly in Chapter 3 in the optional section 3.7. (See also Exercises 3.67–3.72, and our review paper (1991) on the subject.)

1.3 Classification of the Transforms

In this section we will classify many transforms with emphasis on the differences between them in order to point out the errors involved in their practical computation. For these comparisons we will take the Fourier transform

$$\mathcal{F}\{u\} \equiv U(\lambda) = \int_{-\infty}^{\infty} e^{-i\lambda x} u(x)\, dx \tag{1.11}$$

and its variations.

1.3.1 Integral Transforms

In the last section we introduced

$$U(\lambda) = \int_a^b \rho(x)\overline{K(\lambda,x)}u(x)\, dx \tag{1.2}$$

as the most general (linear) integral transform of $u(x)$ and pointed out that, in general, the domain (a,b) is infinite and the parameter λ is continuous. In Table 1.1 and in most of the examples of the last section (Examples 1.2–1.6), the variable λ was continuous while the domain (a,b) was semi-infinite $(0,\infty)$ in the cases of the Laplace (1.7), Fourier sine (1.14), Fourier cosine (1.15), and Hankel (1.22) transforms, while (a,b) was infinite $(-\infty,\infty)$ in the case of the (exponential) Fourier transform (1.11).

1.3.2 Band-Limited Functions (or Transforms)

A very important class of functions $u(x)$ are those which actually vanish identically beyond a finite domain $I = (c,d)$, which are called *band-limited* (sometimes written "bandlimited") functions, and we shall designate their transforms by $U_I(\lambda)$,

$$U_I(\lambda) = \int_I \rho(x)\overline{K(\lambda,x)}u(x)\, dx \tag{1.29}$$

We note that the variable λ is still continuous. The term "band-limited" was coined in the electrical engineering literature for $u(x)$, in (1.11), as a function of frequency x, where x is limited to a finite band $I = (-a,a)$ and vanishes identically outside this interval. In other words, there are no frequencies beyond this band

$$U_I \equiv U_a(\lambda) = \int_{-a}^{a} e^{-i\lambda x} u(x)\, dx \tag{1.30}$$

where λ here stands for time t. A very simple example of a band-limited function is the Fourier transform of the *gate function* $p_a(x)$,

$$p_a(x) = \begin{cases} 1, & -a < x < a \\ 0, & |x| > a \end{cases} \tag{1.31}$$

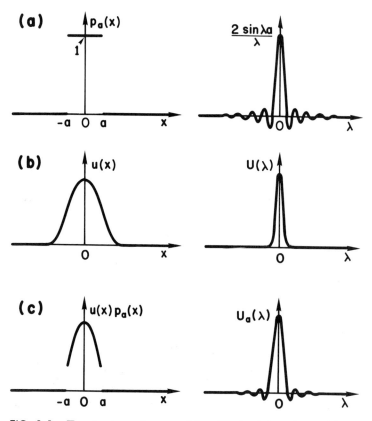

FIG. 1.1   Fourier transform and bandlimited functions. (a) The gate function and its Fourier transform. (b) $u(x)$ and its Fourier transform $U(\lambda)$. (c) The truncated $u(x)$ and its bandlimited transform $U_a(\lambda)$.

as illustrated in Fig. 1.1a. The Fourier transform of $p_a(x)$, according to (1.11), is

$$U_a(\lambda) = \int_{-\infty}^{\infty} e^{-i\lambda x} p_a(x)\,dx = \int_{-a}^{a} e^{-i\lambda x}\,dx$$

$$= \frac{e^{-i\lambda x}}{-i\lambda}\bigg|_{x=-a}^{a} = \frac{e^{-i\lambda a} - e^{i\lambda a}}{-i\lambda} = \frac{2\sin\lambda a}{\lambda} \tag{1.32}$$

which is also illustrated in Fig. 1.1a.

Band-limited functions will be the subject of detailed discussions in Section 2.4.2 of Chapter 2, along with their important use for systems and signal analysis (in Section 2.4.1).

Comparing the Fourier transform $U(\lambda)$ of (1.11) and its band-limited case $U_a(\lambda)$ of (1.30), we see that (1.30) is the result of multiplying $u(x)$ in the integral of (1.11) by the gate function $p_a(x)$, so that

$$
U_a(\lambda) = \int_{-\infty}^{\infty} e^{-i\lambda x} p_a(x) u(x)\, dx
$$

$$
= \int_{-a}^{a} e^{-i\lambda x} u(x)\, dx \tag{1.33}
$$

This may also be described as *truncating* $u(x)$ to $(-a, a)$. Figure 1.1b illustrates a $u(x)$ and its Fourier transform $U(\lambda)$, while Fig. 1.1c illustrates $p_a(x)u(x)$, the truncation of $u(x)$, and its band-limited Fourier transform. It is clear that $U_a(\lambda)$ is different from $U(\lambda)$; we note, however, that the difference between $U_a(\lambda)$ and $U(\lambda)$ should decrease as we increase a. So in the example of Fig. 1.1, if we are forced to approximate $U(\lambda)$ by a band-limited function, we should at least make the width of the gate function as large as possible, $2a$. Figure 1.2 shows the actual effect of doubling the width or relaxing the truncation—note that the wiggles in $U_{2a}(\lambda)$ have practically disappeared; however, $U_{2a}(\lambda)$ is still clearly wider than $U(\lambda)$. These important (approximation) topics are discussed in detail in Chapter 4, where they are essential to the use of its discrete Fourier transforms in approximating Fourier series and integrals.

1.3.3 Finite Transforms—the Fourier Coefficients

In Fig. 1.1b note that $u(x)$ is defined along the whole real line and is definitely not periodic. However, another very important operation, namely the restriction of λ (for $U_a(\lambda)$) to the *discrete* values $\lambda_n = n\pi/a$, corresponding to its Fourier (exponential) series coefficients $U_a(n\pi/a)$,

$$
U_a\left(\frac{n\pi}{a}\right) = \int_{-a}^{a} \exp\left(-\frac{in\pi}{a} x\right) u(x)\, dx \tag{1.34}
$$

results in the Fourier series representation for $u(x)$ on $(-a, a)$,

$$
u(x) = \frac{1}{2a} \sum_{n=-\infty}^{\infty} U_a\left(\frac{n\pi}{a}\right) \exp\left(i\frac{n\pi}{a} x\right), \qquad -a < x < a \tag{1.35}
$$

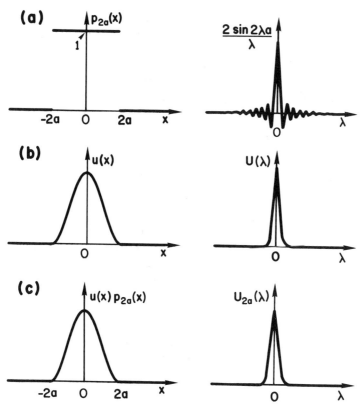

FIG. 1.2 Wide enough truncation window for tolerable truncation error (windowing effect).

which is clearly *periodic* with period $2a$ since

$$u(x + 2a) = \frac{1}{2a} \sum_{n=-\infty}^{\infty} U_a \left(\frac{n\pi}{a} \right) \exp \left(i \frac{n\pi}{a} (x + 2a) \right)$$

$$= \frac{1}{2a} \sum_{n=-\infty}^{\infty} U_a \left(\frac{n\pi}{a} \right) \exp \left(i \frac{n\pi}{a} x \right) e^{2in\pi}$$

$$= \frac{1}{2a} \sum_{n=-\infty}^{\infty} U_a \left(\frac{n\pi}{a} \right) \exp \left(i \frac{n\pi}{a} x \right) = u(x), \qquad -a < x < a,$$

$$u(x) = u(x + 2a)$$

(when we use $e^{2\pi in} = \cos 2\pi n + i \sin 2\pi n = 1$). The samples $U_a(n\pi/a)$ of the band-limited function $U_a(\lambda)$ in (1.33) with samples spacing π/a is called the *finite (exponential) Fourier transform* of $u(x)$. Figure 1.3 illustrates how the sampling $U_a(n\pi/a)$ of $U_a(\lambda)$ in Fig. 1.1 leads to the representation of $u(x)$ which is periodic with period $2a$ and where $U_a(k\pi/a) \equiv c_k$ is termed the *Fourier series coefficient* of the *Fourier series* (1.35) of $u(x)$ on $(-a, a)$.

In Section 2.4.2 we will discuss band-limited functions and the *sampling theorem* which uses samples $U_a(n\pi/a)$ to construct the continuous band-limited function $U_a(\lambda)$.

In general, for the usually finite, but sometimes infinite, domain I, the integral transform $U(\lambda)$ with *discrete* λ_m can be constructed as follows:

$$U(\lambda_m) = \int_I \rho(x)\overline{K(\lambda_m, x)}u(x)\,dx \qquad (1.36)$$

$U(\lambda_m)$ is called the *Fourier coefficient* of the Fourier series expansion of $u(x)$ on the interval I in terms of the *orthonormal set* $\{K(\lambda_m, x)\}$; that is,

$$u(x) = \sum_m U(\lambda_m)K(\lambda_m, x) \qquad (1.37)$$

where $\overline{K}$ is the *complex conjugate*[†] of K; that is, $\overline{K}$ is obtained from K by changing i to $-i$. By saying $\{K(\lambda_m, x)\}$ is an orthonormal set we mean

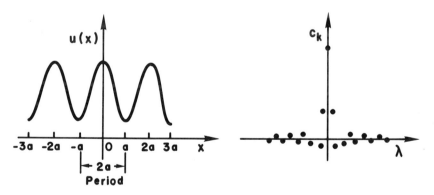

FIG. 1.3 The periodic function $u(x)$ with period $2a$ and its (finite transform) Fourier coefficients c_k.

[†]A brief review of the basic elements of complex numbers is the subject of Exercise 1.6.

that

$$\int_I \rho(x)K(\lambda_m,x)\overline{K(\lambda_n,x)}\,dx = \begin{cases} 0, & \lambda_m \neq \lambda_n \\ 1, & \lambda_m = \lambda_n \end{cases} \tag{1.38}$$

Such orthonormal functions and their series expansion (1.37) are the subjects of Sections 1.5.2 and 3.1.3. We note that, in contrast to the exponential (trigonometric) series, the general series expansion is not necessarily periodic since $K(\lambda_m,x)$ may not be periodic; an example is the Bessel function (see Fig. A.1 of Exercise A.22e of Appendix A). Also, this general (orthonormal or orthogonal) series expansion is not limited to expanding functions $u(x)$ defined only on a finite domain, since there exist examples of Fourier series expansion of $u(x)$ on $(0,\infty)$ or $(-\infty,\infty)$ in terms of orthonormal functions such as Laguerre and Hermite polynomials on such intervals (see Sections 3.5.2, 3.5.3). In these two examples, the Fourier coefficients in (1.36) would be represented by infinite integrals. Thus, the name *finite transform*, or the *Fourier coefficients*, should be identified with λ_m, the discrete values of λ for $U(\lambda_m)$ and not necessarily with a finite domain for $u(x)$. For the more general definition of orthogonality, see (1.66), and (1.70).

1.3.4 The Truncation and Discretization (Sampling) Errors

With reference to the actual numerical computation of the Fourier series—assuming that we have $U_a(n\pi/a)$ in an exact form from integrating (1.34)—we must recognize that we can add only a finite number of terms (for the infinite Fourier series in (1.35));

$$S_N(x) = \frac{1}{2a}\sum_{n=-N}^{N} u_a\left(\frac{n\pi}{a}\right)\exp\left(i\frac{n\pi}{a}x\right) \tag{1.39}$$

$S_N(x)$ stands for the "truncated" Fourier series approximation to $u(x)$ of (1.35), with a truncation error

$$\epsilon_N = |u(x) - S_N(x)| \tag{1.40}$$

In case we do not have the coefficients $U_a(n\pi/a)$ in a closed form, we also have to approximate its integral (1.34) as an infinite sum, which is to be truncated again to a finite sum; that is, the computation of the Fourier coefficient integral (1.34) will involve two errors: that of its *discretization* to a sum and that of the *truncation* of this (infinite) to a finite sum. This contrasts with only a truncation error for the Fourier series. In relation to such errors, we may inquire into the computation for the infinite Fourier integral (1.11) of $U(\lambda)$. In this case, we first approximate (or truncate) the infinite integral by a (finite) band-limited integral like $U_a(\lambda)$. Then this integral has to be discretized to an infinite sum which is then truncated to a finite sum. So we may say that we had to truncate the domain of $u(x)$ to $(-a, a)$, discretize the resulting band-limited integral to a series, and then truncate the infinite series. These (discretization and truncation errors) along with other related errors will be discussed and illustrated in detail in Sections 4.1.1 and 4.1.6 of Chapter 4.

1.3.5 The Discrete Transforms

In practical applications, the desired function $u(x)$ may be defined only on a discrete finite set of points x_k, $k = 0, 1, 2, \ldots, N - 1$. In this case, the differential equation will be replaced by a *difference equation,*[†] and the transform may still be defined as a sum instead of an integral:

$$\tilde{U}(\lambda_n) = \sum_{k=0}^{N-1} \overline{K(\lambda_n, x_k)} \bar{u}(x_k), \qquad n = 0, 1, 2, \ldots, N - 1 \tag{1.41}$$

This is called a "discrete transform," which we designate by $\tilde{U}$ to distinguish it from $U(\lambda_n)$, the finite transform in (1.36).

In the case of the Fourier exponential transform (1.11), the discrete version is

$$\tilde{U}\left(\frac{n\pi}{a}\right) = \sum_{k=0}^{N-1} \bar{u}\left(\frac{ka}{\pi N}\right) \exp\left(-\frac{ink}{N}\right), \qquad n = 0, 1, \ldots, N - 1 \tag{1.42}$$

or, as usually seen in electrical engineering texts,

$$\tilde{U}\left(\frac{n}{NT}\right) = \sum_{k=0}^{N-1} \bar{u}(kT) \exp\left(-\frac{i2\pi nk}{N}\right),$$

$$n = 0, 1, \ldots, N - 1, \quad T = \frac{a}{\pi N} \tag{1.43}$$

Section 4.1 is devoted to the detailed study of these discrete Fourier transforms (DFTs), their error analysis, and their fast algorithm, the fast Fourier transform (FFT).

We note that the computation of this discrete Fourier transform involves N additions as k goes from $k = 0$ to $k = N - 1$, and $(N - 1)^2$

[†] See Section 4.1.5 for solving difference equations associated with boundary values via the discrete Fourier transforms and those associated with initial values via the z-transform (the discrete version of the Laplace transform).

multiplications as k and n go through 0, 1, 2, ..., $N - 1$. The reason we used $(N - 1)^2$ instead of N^2 is that for $n = k = 0$ the kernel has the value unity and we don't need to perform the multiplication.

We recognize that the kernel $\exp(-i2\pi nk/N)$ is periodic with period N; hence it has the same value as nk runs through integer multiples of N. Capitalizing on this important (*cyclic*) property of the discrete Fourier transform kernel, a very efficient algorithm was established in 1965 for evaluating this discrete transform that made it one of the important tools for a variety of applied problems which involve Fourier analysis. This celebrated algorithm reduces the required $(N - 1)^2$ multiplications of (1.43) to $\frac{1}{2} N \log_2 N$ and is called the *fast Fourier transform (FFT)*. We can easily see the advantage here when N is very large; such is the case when we compare N^2 to $(N/2)\log_2 N$, as illustrated in Table 1.2. What is important here is that if we are to use the FFT algorithm for the above different variations of the Fourier representations, we must first put the given transform in the discrete form (1.43) as an approximation, knowing that in our attempt to improve the approximation, N can be increased with very reasonable extra cost.

If we review the different Fourier transforms, we find that the one closest to the above discrete transforms is the Fourier series (1.35). Since it is already discretized, it will suffer only from the truncation error (1.40), which generally can be improved with increasing N. However, we should also see that what we obtain from (1.43) are only the samples of the continuous periodic function. The fast Fourier transform algorithm is discussed in detail in Section 4.1.3 of Chapter 4.

In the case of the Fourier coefficient integral (1.34), it is first discretized to an infinite sum and then that sum is truncated to the discrete transform

TABLE 1.2   Number of Multiplications for the FFT

Order of computations N	FFT number of multiplications $(N/2)\log_2 N$	Direct method number of multiplications $(N - 1)^2$
8	12	49
16	32	225
32	80	961
64	192	3,969
128	448	16,129
256	1,024	65,025

(1.43). Hence, it will suffer from a *discretization* (*sampling*) error as well as a *truncation* error.

In the case of the infinite integral $U(\lambda)$ of (1.11), we must approximate it by a band-limited function $U_a(\lambda)$, which is to be discretized to a sum which, in turn, is to be truncated. The discrete transforms, and the above errors incurred in their use for approximating Fourier series and integrals, are the subject of very detailed discussions in Chapter 4.

1.4 Comments on the Inverse Transforms—Tables of the Transforms

---

Having illustrated in Section 1.2 the idea of compatible transforms for cases where the transformed function $U(\lambda)$ is easy to find from the resulting algebraic equation, the problem now is how to find the original solution $u(x)$ from $U(\lambda)$:

$$U(\lambda) = \int \rho(x)\overline{K(\lambda,x)}u(x)\,dx \equiv \kappa\{u\} \tag{1.44}$$

in other words how to find $u(x) = \kappa^{-1}\{U(\lambda)\}$, the inverse of the κ-transform.

Such a task, of finding the inverse Laplace or Fourier transform as solving integral equation in $u(x)$, generally involves complex integration. This is especially true in the case of the inverse Laplace transform, which we will allude to briefly in Section 2.8.

In practice, and for most of our examples in this book, for the inverse Laplace transform we will rely on the availability of extensive tables for the Laplace transform pairs like those in Erdelyi et al. (1954, Vol. 1), Ditkin and Prudnikov (1965), and Roberts and Kaufman (1956). For convenience, we include in Appendix C a reasonable collection of tables of integral transforms with Section C.1 covering the Laplace transforms. These will be combined with the working knowledge of what we are going to develop in how operations are transformed—for example, how d^2u/dx^2 is Laplace transformed to $\lambda^2 U(\lambda) - \lambda u(0) - u'(0)$ in (1.9). In the case of the Fourier transform, after exhausting such available tables, we may resort to computing numerically the inverse Fourier transform with the help of the FFT algorithm which we introduced at the end of the last section, and which we shall develop in Chapter 4 and in particular Section 4.1.3.

Since we have the inverse as an unknown $u(x)$ in (1.44) under the integral sign, to find it means solving (1.44) as (a special case of) an integral equation in $u(x)$. This, as we shall see, is a related topic which we shall allude to here very briefly, and we refer the interested reader to the author's (1985) introductory book on the subject of integral equations for undergraduates.

1.4.1 Integral Equations—Basic Definitions

An equation with the unknown $u(x)$ under the integral sign,

$$u(x) = f(x) + \int_a^{b(x)} K(x,t)u(t)\,dt \tag{1.44a}$$

is called an integral equation in $u(x)$, where $K(x,t)$ is its *nucleus* or *kernel* and $f(x)$ represents the *nonhomogeneous* term. With the limit $b(x)$ depending on the variable x, it is called the *Volterra integral equation*, and when $b(x) = b$, a constant, it is termed the *Fredholm integral equation*. The equation is said to be of the *first kind* when $u(x)$ does not appear outside the integral, as in

$$g(x) = \int_a^{b(x)} K(x,t)u(t)\,dt \tag{1.44b}$$

which is the case that concerns us the most here, where we see our integral transforms (1.7), (1.11), and (1.22) as special cases of this equation. The integral equation is called *singular* if the kernel $K(x,t)$ is not defined at some points of its domain or the limits of integration involve infinity. The latter is clearly the case with the Laplace (1.7), Fourier (1.11), and Hankel (1.22) transforms. The first type of singularity will be illustrated with the important *Abel integral equation* (2.61) with kernel $K(x,t) = 1/\sqrt{x-t}$, which is singular at $x = t$.

A very important special case of an integral equation that may be amenable to the Laplace or Fourier transform method of solution is one in which the kernel $K(x,t)$ depends on the difference $x - t$ only, that is, $K(x,t) = K(x - t)$, and we term this a *difference kernel*. We may remark here that such compatibility is again a consequence of the special feature of the exponential kernel of the Laplace and Fourier transforms.

1.5 The Compatible Transform and the Adjoint Problem

For the purpose of motivating the outline of a theory to determine the compatible transform for a differential equation along with its given auxiliary conditions, let us first summarize the basic results of the concepts as illustrated in Examples 1.2 to 1.7 in Section 1.2.

We started with a differential equation in $u(x)$ associated with a simple differential operator L_n like the second-order L_2 in (1.3). Then we applied the special cases of the integral transform (1.2) and used repeated integration by parts to reduce the differential operation L_2u on $u(x)$ to an algebraic expression (principally $\lambda U(\lambda)$) on the transformed function $U(\lambda) = \kappa\{u\}$.

In the process we had to make sure that we accounted for the auxiliary conditions on $u(x)$ (which included either the qualitative boundedness condition or a specified value of the function and/or its derivative at a boundary point). For example, in the case of the exponential Fourier transform (Example 1.3) with the basic operator $L_1 = d/dx$ of order one, we required that the function vanishes as $x \to \mp\infty$, while for the Laplace transform (Example 1.2), with the same first-order operator, we required a specific value of the function $u(0)$ at $x = 0$ and the vanishing of $u(x)e^{-\lambda x}$ as $x \to \infty$. When we Laplace transformed d^2/dx^2 in (1.9), we needed $u(0)$ and $u'(0)$, and we assumed the vanishing of $u(x)e^{-\lambda x}$ and $u'(x)e^{-\lambda x}$ as $x \to \infty$. In the case of the Fourier sine transform (Example 1.4), where the basic operator $L_2 = d^2/dx^2$ is of order 2, we required the vanishing of $u(x)$ and $u'(x)$ as $x \to \infty$ and a specific value $u(0)$ at $x = 0$. The case of the cosine transform required, in a similar manner, the vanishing of $u(x)$ and $u'(x)$ as $x \to \infty$, but it specified $u'(0)$ at $x = 0$. The Hankel transform (Example 1.6) treated an operator of order 2, with variable coefficients, where essentially some type of boundedness was required on $u(r)$ and $u'(r)$ at $r = 0$ and as $r \to \infty$. Example 1.7 of the finite sine transform treated a second-order operator $L_2 = d^2/dx^2$ on the finite domain $(0, \pi)$, where, in general, a linear combination of u and u' is given at both boundary points: $\alpha_1 u(0) + \alpha_2 u'(0)$ at $x = 0$ and $\beta_1 u(\pi) + \beta_2 u'(\pi)$ at $x = \pi$, and where in this very specific (simple) example we chose to let $\alpha_2 = 0$ an $\beta_1 = 0$ to have $u(0) = \alpha$ and $u'(\pi) = \beta$, respectively.

We will clarify at this point that we speak of the basic differential operator $L_1 = d/dx$ of order 1 for the Laplace and exponential Fourier transforms, the basic $L_2 = d^2/dx^2$ operator of order 2 for the sine and cosine transforms, and another for the Hankel transforms. This means that other higher-order operators like d^n/dx^n can be treated by applying the Laplace or exponential Fourier transforms n times in succession, provided that at each basic operation we supply the necessary auxiliary conditions on the transformed function. For example, in the case of Laplace transforming d^2y/dx^2 we can write it as $dy'/dx = dv/dx$, that is, as the first-order operator to be treated for v with the Laplace transform so as to have

$$\mathcal{L}\left\{\frac{d^2y}{dx^2}\right\} = \mathcal{L}\left\{\frac{dv}{dx}\right\} = \lambda V(\lambda) - v(0) = \lambda V(\lambda) - y'(0),$$

$$V(\lambda) = \mathcal{L}\{v\} \tag{1.45}$$

after employing (1.8). But we also have

$$V(\lambda) = \mathcal{L}\{v\} = \mathcal{L}\{y'\} = \lambda Y(\lambda) - y(0) \tag{1.46}$$

which can be combined with (1.45) to give

$$\mathcal{L}\left\{\frac{d^2y}{dx^2}\right\} = \lambda[\lambda Y(\lambda) - y(0)] - y'(0) = \lambda^2 Y(\lambda) - \lambda y(0) + y'(0) \tag{1.47}$$
$$\tag{1.9}$$

the result we derived in (1.9). The same thing can be done to obtain $\mathcal{L}\{d^n y/dx^n\}$ of (1.10) by repeating the Laplace transformation n times on the basic operator d/dx. Similar steps can be followed to derive the Fourier transform $\mathcal{F}\{d^n y/dx^n\}$ of (1.13) from the basic result $\mathcal{F}\{du/dx\}$ of (1.12).[†]

1.5.1 The Adjoint Differential Operator

The above discussion is only a summary, and as such it may serve as a prelude to the main problem of determining the compatible kernel $K(\lambda,x)$[‡] of the transform for a given general differential equation and its auxiliary conditions. In particular, as we face a nonhomogeneous differential equation of order n, i.e.,

$$L_n u = f \tag{1.48}$$

we will ask the following questions:

1. What are the general boundary conditions that go with the differential equation (1.48) which will still allow us to find the compatible transform that results in an algebraic equation in the transformed function $U(\lambda)$?
2. How do we find the desired compatible kernel $K(\lambda,x)$?

[†] The rest of this chapter, aside from the regular Sturm–Liouville problem (1.63)–(1.71) and its Example 1.9, is considered optional for a one-semester course in boundary value problems.
[‡] In this general treatment, we use $K(\lambda,x)$ instead of $\overline{K(\lambda,x)}$ for the kernel to be close to the notation of other sources.

Our analysis and illustrations will center on the special important case of L_2, which we will cover right after presenting two basic identities. It will shortly become very clear that the following *Lagrange* and *Green identities* prove to be essential tools to our development.

A. The Lagrange and Green Identities

After examining Example 1.2–1.7 of Section 1.2, it would be natural to expect that we would start algebraizing the differential operator L_n by multiplying (1.48) by a function v, to stand for the kernel $K(\lambda,x)$, then repeat the integration by parts to reduce the order of the differentiation of $L_n u$. But, as we expect from the process of integrating by parts, the differentiation operation would actually move, in an understandable new form, onto the new variable $v = K(\lambda,x)$. Such a new operation on v is designated by $L_n^* v$, where L_n^* is called the *adjoint* of L_n, which we will define in (1.113). Fortunately, all such important processes of integration by parts have been done for L_n and can easily be obtained from the following result, the *Lagrange identity*:

$$\bar{v}L_n u - u\overline{L_n^* v} = \frac{d}{dx} H_n\left(u, u', \ldots, u^{(n-1)};\ v, v', \ldots, v^{(n-1)}\right) \tag{1.49}$$

where the right-hand side is an *exact differential* of the function H_n in u and v and their derivatives up to order $n-1$. Upon integration of (1.49) we have the result of the n repeated integrations by parts started as *Green's identity*:

$$\int_a^b \bar{v}L_n u\, dx = \int_a^b u\overline{L_n^* v}\, dx + H_n\left(u, u', \ldots, u^{(n-1)};\ v, v', \ldots, v^{(n-1)}\right)\Big|_a^b \tag{1.50}$$

where the differential operation L_n on u has moved as a new differential operation $\overline{L_n^* v}$ onto the other variable v. The explicit form for L_n, L_n^* and the form of H_n in (1.49), (1.50) will be presented, respectively, in (1.112), (1.113), and (1.118). The special important case for $n = 2$ is given in (1.51), (1.52), and (1.79). The new operator L_n^* is called the *adjoint* of L_n. In the particular case when the operator is equal to its adjoint, $L_n = L_n^*$, we term it a *self-adjoint* operator. The theory for such self-adjoint operators is simpler and fortunately, as we shall show in Example 1.8 (in the coming section on L_2), the most general second-order differential operator L_2 of (1.3),

$$L_2 u \equiv A_0(x)\frac{d^2 u}{dx^2} + A_1(x)\frac{du}{dx} + A_2(x)u \tag{1.51}$$

can be made self-adjoint by multiplying it by

$$p(x) = \frac{r(x)}{A_0(x)} = \frac{1}{A_0(x)} \exp\left(\int \frac{A_1(x)}{A_0(x)} \, dx\right)$$

At the level of this text we will concentrate in this section and the following Section 1.6 on second-order differential equations, since most of the differential equations we often meet are second order. These are, of course, special (fortunate) cases, since their differential operator L_2 can be made self-adjoint. We also feel it instructive to give at least a brief introduction to the general problem, associated with nth-order differential equations, which necessitates familiarity with the adjoint operator L_n^* in (1.113). This will, of course, serve our treatment of the second-order differential equations, where we will need at least the explicit definition of the adjoint operator L_2^* and the explicit form of H_2 in (1.50) from (1.113) and (1.118), respectively, for $n = 2$. This general problem is presented at the end of the chapter in Section 1.7, where the Lagrange identity (1.49) is illustrated for L_3 and L_4. We leave this topic for the interest of the reader.

In summary, it is the integral on the right-hand side of (1.50) that determines the condition (differential equation) on the kernel $v = K(\lambda,x)$ in order to give an algebraic expression in $U(\lambda)$. This will happen if we choose v as a solution of the (adjoint) problem $L_n^* v = \lambda v$, as can be seen for the case $n = 2$ in (1.58) and the discussion that follows it. Also, it is the particular form of H_n in (1.50) (with its explicit expression in (1.118)), and for $n = 2$ in (1.120), that determines what auxiliary conditions on the unknown u and its derivatives can be accommodated, or involved directly in the process, as well as what conditions can be imposed on the (unknown) desired kernel $v = K(\lambda,x)$ and its derivatives.

B. The Second-Order Differential Operator

In (1.51) we have the usual second-order operator L_2 in the form

$$L_2 u = A_0(x)\frac{d^2u}{dx^2} + A_1(x)\frac{du}{dx} + A_2(x)u \tag{1.51}$$

Its adjoint L_2^* is

$$L_2^* u = \frac{d^2}{dx^2}[A_0(x)u(x)] - \frac{d}{dx}[A_1(x)u(x)] + A_2(x)u \tag{1.52}$$

which can be seen as a special case of L_n^* in (1.113).

We will show in Example 1.8 that the second-order operator L_2 of (1.51) can be made self-adjoint by multiplying it by

$$p(x) = \frac{r(x)}{A_0(x)} = \frac{1}{A_0(x)} \exp\left(\int \frac{A_1(x)}{A_0(x)} dx\right)$$

giving

$$L_s u \equiv p(x)L_2 u = \frac{d}{dx}\left[r(x)\frac{du}{dx}\right] - q(x)u \qquad (1.53)$$

where $q = -r(x)A_2(x)/A_0(x)$.

EXAMPLE 1.8: The Second-Order Operator L_2 Can Be Made Self-Adjoint

Consider L_2, the differential operator of order two in (1.51), and divide it by $A_0(x) \neq 0$, that is,

$$\frac{1}{A_0(x)}L_2 u = u''(x) + \frac{A_1(x)}{A_0(x)}u'(x) + \frac{A_2(x)}{A_0(x)}u \qquad (E.1)$$

then multiply both sides by the integrating factor $r(x)$

$$r(x) = \exp\left[\int \frac{A_1(x)}{A_0(x)} dx\right], \qquad \frac{dr}{dx} = \frac{A_1(x)}{A_0(x)}r(x) \qquad (E.2)$$

to have

$$\frac{r(x)}{A_0(x)}L_2 u = r(x)u'' + r(x)\frac{A_1(x)}{A_0(x)}u'(x) + r(x)\frac{A_2(x)}{A_0(x)}u(x) \qquad (E.3)$$

Remembering that $r'(x) = [A_1(x)/A_0(x)]r(x)$, the right-hand side can be put in the special form $r(x)u'' + r'(x)u' - q(x)u$, which as we will show is self-adjoint, where $q(x) = -r(x)A_2(x)/A_0(x)$,

$$\frac{r(x)}{A_0(x)}L_2 u = \frac{d}{dx}\left[r(x)\frac{du}{dx}\right] - q(x)u(x) \equiv L_s$$

$$= r(x)\frac{d^2u}{dx^2} + r'(x)\frac{du}{dx} - q(x)u(x)$$

$$= r(x)u'' + \frac{A_1(x)}{A_0(x)}r(x)u' - \left(-r(x)\frac{A_2(x)}{A_0(x)}\right)u$$

$$= ru'' + r(x)\frac{A_1(x)}{A_0(x)}u' + r(x)\frac{A_2(x)}{A_0(x)}u \qquad (E.4)$$

which is (E.3).

Next we show that the form

$$L_s u = \frac{d}{dx}\left[r(x)\frac{du}{dx}\right] - q(x)u(x) \tag{E.5}$$

is self-adjoint. We first write it in the form (1.51), then write its adjoint L_s^* to show that $L_s = L_s^*$

$$L_s = r(x)u'' + r'u' - qu$$

and its adjoint, according to (1.52), is

$$\begin{aligned}
L_s^* &= \frac{d^2}{dx^2}[r(x)u] - \frac{d}{dx}[r'(x)u] - qu \\
&= ru'' + 2r'u' + r''u - r''u - r'u' - qu \\
&= ru'' + r'u' - qu
\end{aligned}$$

Hence, $L_s^* = L_s$ is a *self-adjoint operator*.

1.5.2 The Two Eigenvalue Problems

In this section we will present the two eigenvalue problems that are associated with the second-order operator, and which are the general source, for the orthogonal functions that we shall use in the Fourier series expansions. This includes the familiar *Sturm–Liouville problem*.

A. Second-Order Differential Equations

Here we give two general boundary conditions associated with the second-order operator L_2, which consist of linear combinations of u and u' at the two boundary points $x = a$ and $x = b$:

$$\Gamma_j u = \sum_{k=1}^{2} \left[M_{jk} u^{(k-1)}(a) + N_{jk} u^{(k-1)}(b) \right], \qquad j = 1, 2 \tag{1.54}$$

$$\Gamma_1 u \equiv M_{11}u(a) + N_{11}u(b) + M_{12}u'(a) + N_{12}u'(b) \tag{1.54a}$$

$$\Gamma_2 u \equiv M_{21}u(a) + M_{21}u(b) + M_{22}u'(a) + N_{22}u'(b) \tag{1.54b}$$

where M_{jk} and N_{jk} are constants. Let $\Gamma = \{\Gamma_1, \Gamma_2\}$ represent all the boundary conditions on u signified by $j = 1, 2$.

The generalization of these boundary conditions to n general boundary conditions associated with the nth-order L_n should be straightforward. It consists of linear combinations of all the derivatives of u up to $u^{(n-1)}$ at the two boundary points $x = a$ and $x = b$ (see (1.126) in Section 1.7, and for more detailed discussion consult Coddington and Levinson (1955)). The eigenvalue problem associated with L_2 is the following differential equation subject to the above two (homogeneous) boundary conditions:

$$L_2 u = \lambda u \qquad (1.55)$$

$$\Gamma u = 0 \qquad (1.56)$$

where λ is called the *characteristic value* or *eigenvalue* and the corresponding solutions $\{u\}$ are the *characteristic functions* or *eigenfunctions*. To ensure that such eigenfunctions are orthogonal so that we may use them for the general Fourier series expansion (1.37) or finite transforms, it is not enough that L_2 be self-adjoint; the whole eigenvalue problem (1.55), (1.56) must be self-adjoint! The *eigenvalue problem* (1.55), (1.56) is called *self-adjoint* if

$$\int_a^b v L_2 u - \int_a^b u L_2 v = 0 \qquad (1.57)$$

for all u, v that are continuously differentiable real-valued functions which satisfy the boundary conditions $\Gamma v = 0$.

To show what this condition means in terms of the boundary H_2 form of Green's identity (1.50), we consider

$$\int_a^b [v L_2 u - u L_2^* v]\, dt = H_2[u(b), u'(b); v(b), v'(b)]$$

$$- H_2[u(a), u'(a); v(a), v'(a)] \qquad (1.50a)$$

We can see clearly that the eigenvalue problem (1.55), (1.56) is self-adjoint if in the above we have (for real-valued functions)

(i) $L_2 = L_2^*$
(ii) The boundary conditions (1.56) on both u and v; that is, $\Gamma u = \Gamma v = 0$ imply the vanishing of H_2 at the boundary points on the right side of (1.50a).

Such conditions, (i) and (ii), guarantee orthogonal solutions to the eigenvalue problem (1.55), (1.56) (where they are needed in Chapter 3). We shall prove this important result shortly in Example 1.9 for the case of the *Sturm–Liouville problem* associated with a second-order operator L_2. We also note that this problem is closely related to our purpose of finding the compatible kernel v as a solution of the same type (but the adjoint) differential equation; i.e., for $v = K(\lambda,x)$ (not necessarily real-valued function),

$$\overline{L_2^* v} = \overline{\lambda} \overline{v} \quad \text{or} \quad \overline{L_2^* K} = \overline{\lambda K} \tag{1.58}$$

This makes the second integral of (1.50a) algebraic in the transform

$$\int_a^b u \overline{L_2^* v} = \int u \overline{\lambda v} = \overline{\lambda} \int \overline{K(\lambda,x)} u(x) \, dx = \overline{\lambda} U(\lambda) \tag{1.59}$$

We purposely left v and K in (1.58) as possibly complex-valued functions to remember the very important case of the Fourier transform kernel $K(\lambda,x) = e^{i\lambda x}$, which is a complex-valued function.

For real λ and $K(\lambda,x)$ we have

$$U(\lambda) = \int K(\lambda,x) u(x) \, dx \tag{1.60}$$

There is a difference between the familiar eigenvalue problem (1.55), (1.56) and our present (adjoint) problem for finding the compatible kernel like that in (1.58a) and (1.62). This difference is that for the transform kernel $K(\lambda,x)$, we require only boundary conditions on the unknown function u in (1.54a), (1.54b) to accommodate what is required for H_2 in the above Green's identity (1.50a); if this requirement is not met, then the unknown kernel $v = K$ must take the "hurdle" to make sure that H_2 is defined for all its variables. This does not mean that H_2 must necessarily vanish, which is in contrast to (1.57) in the case of the eigenvalue problem (1.55), (1.56). As we shall see in the following illustrations, for the second-order operator, such conditions on $v = K$ will amount to exactly the same boundary condition (1.56) on the original function u, that is, for $v = K(\lambda,x)$, so that

$$\Gamma v = 0 \quad \text{or} \quad \Gamma K(\lambda,x) = 0 \tag{1.61}$$

So with the two conditions (1.58) and (1.61) on the (real) kernel $K(\lambda,x)$, we say that the compatible kernel, in general, must satisfy the following *adjoint eigenvalue problem*:

$$L_2^* K = \lambda K \tag{1.58a}$$

$$\Gamma K = 0 \tag{1.62}$$

where $\Gamma \equiv \{\Gamma_1, \Gamma_2\}$ as defined in (1.54a), (1.54b).

Before we start our illustrations for finding compatible transforms we will present a brief account of the eigenvalue problem (1.55), (1.56) for the important special case associated with L_s of (1.53), the second-order differential operator (in its self-adjoint form).

The Regular Sturm–Liouville Problem

The self-adjoint eigenvalue problem (1.55), (1.56) associated with the second-order (self-adjoint) operator L_s is one of the most important problems in applied mathematics and is called the *Sturm–Liouville* problem. It is usually written in the form

$$L_s u = \frac{d}{dx}\left[r(x)\frac{du}{dx}\right] - q(x)u(x) = \lambda\rho(x)u(x), \qquad a < x < b \qquad (1.63)$$

$$\alpha_1 u(a) + \alpha_2 u'(a) = 0 \qquad (1.64)$$

$$\alpha_3 u(b) + \alpha_4 u'(b) = 0 \qquad (1.65)$$

where α_1, α_2, α_3, and α_4 are constants. We note here that we shall consider this Sturm–Liouville problem with good or "regular" coefficients, which puts it within our means to prove the orthogonality of its solutions $\{u_n(x)\}$ on $[a,b]$. For regular Sturm–Liouville problems, we assume that $r(x)$, $r'(x)$, $q(x)$, and $\rho(x)$ are continuous on the closed interval $a \leq x \leq b$, and that $r(x) > 0$, $\rho(x) > 0$ on $[a,b]$. However, in the applications we will meet functions like the Bessel functions and the Legendre polynomials, which are solutions of a nonregular or "singular" Sturm–Liouville problem, where the coefficients violate the above regularity conditions. In these two cases it turns out that the coefficients are regular in the interior of the interval, but are singular at one end point $x = 0$ in the case of the Bessel functions and at the two end points $x = \mp 1$ for the Legendre polynomials. Fortunately, for these two cases and other important special functions, the way the coefficients are singular is not the worst type of singularity, which is called "regular singularity." Here a known method, the *Frobenius method*, can be applied to the differential equation to obtain its solutions in an infinite series form. The "regular" and "regular singular" Sturm–Liouville problems and their respective methods of solution, namely an infinite power series and the Frobenius infinite series solutions, are discussed in Appendixes A.3 and A.4, respectively.

The boundary conditions (1.64), (1.65) are sufficiently general to include many of the physical conditions that are encountered. For example, with $\alpha_2 = 0$ in (1.64) we have the value of the dependent variable specified at the boundary and for $\alpha_1 = 0$ the derivative (the flux of a conserved property in physical applications) is specified as zero at the boundary $x = a$. With α_1, $\alpha_2 \neq 0$, we have an equation of the same form as *Newton's law of cooling*, which describes convective heat transfer from a solid surface, that is,

$$h(T_s - T_o) + k\left.\frac{\partial T}{\partial x}\right|_s = 0$$

where h is called the heat transfer coefficient, $(T_s - T_o)$ is the temperature difference between the surface and the surrounding fluid, and k is the thermal conductivity of the solid.

It is obvious that for any value of the parameter λ there is a so-called *trivial solution* $u(x) \equiv 0$ to the Sturm–Liouville problem. As the trivial solution is devoid of physically significant information, we are interested in nontrivial solutions $u(x) \not\equiv 0$. As we shall see, there may be only λ_n, $n = 1, 2, 3, \ldots$, discrete values of λ for which the differential equation and the boundary conditions can be satisfied nontrivially. We refer to the nontrivial solutions as *characteristic functions* or *eigenfunctions*, and we call the discrete values $\{\lambda_n\}$ of λ the *characteristic values* or *eigenvalues* of the equation. The solutions corresponding to the discrete values $\{\lambda_n\}$ are referred to as $\{u_n\}$. We show next that they are orthogonal.

Orthogonality of the Eigenfunctions for the Regular Sturm–Liouville Problem

The solutions $\{u_n(x)\}$ to this famous problem (1.63)–(1.65) constitute the main source of orthogonal functions, $u_n(x)$, that we referred to in (1.38) and which we shall need again for the kernels $K(\lambda_n, x) = u_n(x)$ of the finite transforms of Chapter 3.

An extremely important result of the Sturm–Liouville problem (1.63)–(1.65) and its self-adjoint operator is that under some conditions of continuity and differentiability of the coefficients of L_s in (1.63), that we shall state again shortly, the solutions $\{u_n(x)\}$ are orthogonal on the interval (a, b). This means that for any two solutions $u_n(x)$ and $u_m(x)$ of (1.63)–(1.65), corresponding to two distinct values λ_n of (1.63),

$$\int_a^b \rho(x) u_n(x) \overline{u_m(x)} \, dx = 0, \qquad n \neq m \tag{1.66}$$

Here $\rho(x)$ is the weight function appearing in (1.63) and the general integral transform (1.2) or (1.36) and a, b are the limits of the interval on which the problem is defined. Note here that sometimes we may use $\sqrt{\rho(x)} u_n(x)$ as the orthogonal set with weight function $\sqrt{\rho(x)}$.

Consider the (regular) Sturm–Liouville problem (1.63)–(1.65) and assume that the functions $r(x)$, $r'(x)$, $q(x)$, and $\rho(x)$ are continuous on the interval $[a, b]$, $r(x) > 0$, $\rho(x) > 0$ on $[a, b]$, and $\alpha_1^2 + \alpha_2^2 > 0$, $\alpha_3^2 + \alpha_4^2 > 0$. These conditions term such a Sturm–Liouville problem as the *regular* one and are given to ensure the existence of continuous solutions (eigenfunctions $u_n(x)$) to the problem. When these conditions, especially those of the continuity of the coefficients, are not met, the problem is termed *irregular* and solutions may be found. A special example in this direction is the Bessel equation (1.6). For our purpose we will state and prove the very important property of such solutions, i.e., their orthogonality property, in the following.

EXAMPLE 1.9: Orthogonality of the Eigenfunctions of the Regular Sturm–Liouville Problem

Even though the definition of orthogonality covers complex-valued functions, for simplicity in this example we *will limit* our proof to real-valued functions, where in (1.66) we have $\overline{u_m(x)} = u_m(x)$. The general case of complex-valued functions follows in a similar way.

Let $u_n(x)$ and $u_m(x)$ be solutions of the regular Sturm–Liouville equations (1.63)–(1.65) corresponding to distinct eigenvalues λ_n and λ_m, $\lambda_n \neq \lambda_m$, and let the derivatives $u_n'(x)$ and $u_m'(x)$ be continuous; then $u_n(x)$ and $u_m(x)$ are orthogonal on the interval (a,b) with respect to the weight function $\rho(x)$. That is, we shall prove that if $\lambda_n \neq \lambda_m$

$$\int_a^b \rho(x) u_n(x) u_m(x)\, dx = 0^\dagger$$

Proof: To show this, we substitute the solutions $u_n(x)$ and $u_m(x)$ in (1.63), manipulate the equations, and integrate. Thus we write

$$[r(x) u_n'(x)]' - [q(x) + \rho(x)\lambda_n] u_n(x) = 0 \qquad (\text{E.1})$$

and

$$[r(x) u_m'(x)]' - [q(x) + \rho(x)\lambda_m] u_m(x) = 0 \qquad (\text{E.2})$$

Multiplying the first equation by $u_m(x)$, the second by $u_n(x)$, and subtracting the second from the first gives

$$[r u_n']' u_m - [r u_m']' u_n - (\lambda_n - \lambda_m)\rho u_n u_m = 0$$

The first two terms can be combined into an exact differential. We note that

$$[r u_n' u_m - r u_m' u_n]' = (r u_n')' u_m + r u_n' u_m' - (r u_m')' u_n - r u_m' u_n'$$

---

†This is if $u_n(x)$ is a real-valued function. For complex functions the integral is

$$\int_a^b \rho(x) u_n(x) \overline{u_m(x)}\, dx = 0, \qquad \lambda_n \neq \lambda_m$$

where $\overline{u_m}$ is the complex conjugate of u_m, and the proof follows in a similar manner. An important special case to remember is $u_n(x) = \exp\left(\frac{in\pi x}{a}\right), i = \sqrt{-1}$.

$$= (ru'_n)'u_m - (ru'_m)'u_n$$

Hence

$$-(\lambda_n - \lambda_m)\rho u_n u_m = [r(u'_m u_n - u'_n u_m)]' \tag{E.3}$$

Now we integrate both sides of (E.3) from $x = a$ to $x = b$ to obtain

$$(\lambda_m - \lambda_n) \int_a^b \rho(x)u_n(x)u_m(x)\,dx = [r(x)\{u'_m(x)u_n(x) - u'_n(x)u_m(x)\}]_a^b$$

$$= r(b)[u'_m(b)u_n(b) - u'_n(b)u_m(b)]$$

$$- r(a)[u'_m(a)u_n(a) - u'_n(a)u_m(a)] \tag{E.4}$$

We now appeal to the boundary conditions to show that the right-hand side vanishes. The boundary conditions on u_n are (assuming that $\alpha_2 \neq 0$, $\alpha_4 \neq 0$)

$$\alpha_1 u_n(a) + \alpha_2 u'_n(a) = 0, \qquad u'_n(a) = -\frac{\alpha_1}{\alpha_2}u_n(a)$$

$$\alpha_3 u_n(b) + \alpha_4 u'_n(b) = 0, \qquad u'_n(b) = -\frac{\alpha_3}{\alpha_4}u_n(b)$$

and on u_m are

$$\alpha_1 u_m(a) + \alpha_2 u'_m(a) = 0, \qquad u'_m(a) = -\frac{\alpha_1}{\alpha_2}u_m(a)$$

$$\alpha_3 u_m(b) + \alpha_4 u'_m(b) = 0, \qquad u'_m(b) = -\frac{\alpha_3}{\alpha_4}u_m(b)$$

So if we substitute these values of $u'_n(a)$, $u'_n(b)$, $u'_m(a)$, and $u'_m(b)$ in the right side of (E.4) we obtain

$$(\lambda_m - \lambda_n) \int_a^b \rho(x)u_n(x)u_m(x)\,dx$$

$$= r(b)\left[-\frac{\alpha_3}{\alpha_4}u_m(b)u_n(b) + \frac{\alpha_3}{\alpha_4}u_n(b)u_m(b)\right]$$

$$- r(a)\left[-\frac{\alpha_1}{\alpha_2}u_m(a)u_n(a) + \frac{\alpha_1}{\alpha_2}u_n(a)u_m(a)\right]$$

$$= r(b)[0] - r(a)[0] = 0$$

$$(\lambda_m - \lambda_n) \int_a^b \rho(x)u_n(x)u_m(x)\,dx = 0 \tag{E.5}$$

Hence, if $\lambda_n \neq \lambda_m$ then

$$\int_a^b \rho(x)u_n(x)u_m(x)\,dx = 0 \tag{E.6}$$

Special Cases

Note that for some systems the orthogonality of the eigenfunctions can be proved without recourse to the boundary conditions, and in other cases boundary conditions are required at only one boundary to establish orthogonality.

The General Case: $r(a), r(b) \neq 0$. In this case, the above proof applies and boundary conditions are required at $x = a$ and $x = b$.

Case 1: $r(a) = 0$ and $r(b) \neq 0$. In this event, it is not necessary to specify the boundary conditions at $x = a$ to prove orthogonality, for the second term on the right-hand side of (E.4) vanishes without applying the boundary conditions.

Case 2: $r(a) \neq 0$ and $r(b) = 0$. By similar reasoning to that of case 1, it is clear that it is not necessary to specify boundary conditions at $x = b$ since the first term on the right-hand side of (E.4) vanishes.

Case 3: $r(a) = r(b) = 0$. No boundary conditions are required to establish the orthogonality in this case, for the right hand side of (E.4) vanishes without considering the boundary conditions.

Next, we will illustrate how the orthogonality property of the solutions $\{u_n(x)\}$ can be employed to expand given functions on (a, b) in infinite series of these functions. Hence, the name *orthogonal* or *Fourier series expansion*.

Sturm–Liouville Problem and the Orthogonal (Fourier) Series Expansion

Here, we concentrate on the role of the orthogonality of the solutions to the Sturm–Liouville problem in determining the coefficients c_n of the *orthogonal*, or *Fourier series*, expansion

$$f(x) = \sum_{n=1}^{\infty} c_n u_n(x), \qquad a < x < b \tag{1.67}$$

of a function $f(x)$ defined on the interval (a, b). In Section 3.1.1 of Chapter 3, we discuss and illustrate this vital subject for a variety of functions $f(x)$ along with various types of convergence for this series. It is followed in Section 3.1.3 by proving most of the basic theorems on the convergence of such series expansions. One of the advantages of the orthogonality property lies in determining the coefficients c_n of the orthogonal series expansion (1.67) as

$$c_n = \frac{\int_a^b \rho(x) f(x) \overline{u_n(x)} \, dx}{\int_a^b \rho(x) |u_n(x)|^2 \, dx} \tag{1.68}$$

This is obtained when we "formally" multiply both sides of (1.67) by $\rho(x)\overline{u_m(x)}$, integrate from $x = a$ to $x = b$, and use the orthogonality property (1.66),

$$\int_a^b \rho(x)\overline{u_m(x)}f(x)\,dx = \int_a^b \rho(x)\overline{u_m(x)} \sum_{n=1}^{\infty} c_n u_n(x)\,dx$$

$$= \sum_{n=1}^{\infty} c_n \int_a^b \rho(x)\overline{u_m(x)}u_n(x)\,dx$$

after allowing the exchange of the integration with the infinite summation on the right side. The latter is an operation which we shall show is justified for a particular (but wide) class of functions $f(x)$ in Chapter 3, Theorems 3.5 and 3.7. The essence of this theorem is that the function $f(x)$ be square integrable on (a,b) with respect to the weight function $\rho(x)$, that is, $\int_a^b \rho(x)|f(x)|^2\,dx < \infty$. Because of the orthogonality property of $\{u_n(x)\}$, the integral inside the series above will vanish unless $n = m$; in this case we have only the mth term of the series

$$\int_a^b \rho(x)\overline{u_m(x)}f(x)\,dx = c_m \int_a^b \rho(x)|u_m(x)|^2\,dx$$

which is the result for c_m in (1.68).

We may remark here that in Chapter 3 we will use the integral on the left to define the *finite integral transform* $F(\lambda_n)$ of $f(x)$, $a < x < b$, as in (3.6). The orthogonal series expansion (1.67) for $f(x)$, $a < x < b$, will stand for the *inverse finite transform* of $F(\lambda_n)$ as in (3.7) (see Examples 1.10–1.12). These general finite transforms are also called *Sturm–Liouville transforms*.

We mention here, for future reference, that the integral in the denominator of (1.68) is called the *norm square* of the function $u_n(x)$ and is denoted by

$$\|u_n\|^2 = \int_a^b \rho(x)|u_n(x)|^2\,dx \tag{1.69}$$

When the orthogonal functions $u_n(x)$ are divided by their norm $\|u_n\|$,

$$\phi_n(x) = \frac{u_n(x)}{\|u_n\|} \tag{1.70}$$

the resulting functions $\phi_n(x)$ are called *orthonormal* functions because their norm is unity. For example,

$$\int_a^b \rho(x)|\phi_n(x)|^2\,dx = \int_a^b \rho(x)\frac{|u_n(x)|^2}{\|u_n\|^2}\,dx$$

$$= \frac{1}{\|u_n\|^2} \int_a^b \rho(x)|u_n(x)|^2 \, dx = \frac{\|u_n\|^2}{\|u_n\|^2} = 1$$

where (1.69) has been employed. These were used in (1.38).

Also, the solutions $u_n(x)$ are called the *characteristic functions* or *eigenfunctions* of the differential operator. For example, the operator L_s in (1.53),

$$L_s u_n(x) = \frac{d}{dx} \left[r(x) \frac{du_n}{dx} \right] - q(x)u_n(x) = \lambda_n \rho(x)u_n(x) \tag{1.71}$$

has eigenfunctions $u_n(x)$ corresponding to the values λ_n which are called the *characteristic values* or *eigenvalues*. Hence, we will often refer to the orthogonal expansion (1.67) as the *eigenfunction expansion*. We stress that it is the weight function $\rho(x)$ of (1.63) that must be used for the general (finite) transform (1.36) with kernel $\overline{K(\lambda_n,x)} = \overline{u_n(x)}$, which is the subject of Chapter 3.

In the following section we will concentrate on the method of determining the compatible transforms for problems associated with second-order differential operators. We stress that for the first part we will leave L_2 in its usual form (1.51), which is not self-adjoint, in order to illustrate the role of the adjoint operator L_2^* in determining the compatible kernel. The rest of the discussion and the applications will employ the self-adjoint form L_s of (1.53). This can be covered in a more compact and elegant way as a special case of the Green's identity (1.50) for $n = 2$.

B. The Adjoint Problem for Determining the Compatible Transform—Second-Order Differential Equations

We first note that the function

$$p(x) = \frac{r(x)}{A_0(x)} = \frac{1}{A_0(x)} \exp \left(\int \frac{A_1(x)}{A_0(x)} \, dx \right)$$

that we use to make $p(x)L_2 = L_s$ self-adjoint is not necessarily the same as the weight function $\rho(x)$ of the integral transform (1.2). Even though many texts assume this for convenience, it is clear that L_2 could have been given in its self-adjoint form L_s in (1.53), that is, where $p(x) = 1$. But, $\rho(x)$ must still be the weight in the final differential equation like (1.71) that we have to solve to find the compatible kernel $K(\lambda_n,x)$. We will leave $p(x)$ and $\rho(x)$ different for the remainder of this section. However, where we use the L_s form we will also adopt $p(x) = \rho(x)$, as we shall do in Section 1.6.

We will show here that, in general, for the transform (1.2) to be compatible with the second-order differential operator L_2 (which is not in its self-adjoint form) and its associated boundary conditions,

$$A_0(x)\frac{d^2y}{dx^2} + A_1(x)\frac{dy}{dx} + A_2(x)y \equiv L_2 y = f(x), \qquad a < x < b \qquad (1.72)$$

$$\alpha_1 y(a) + \alpha_2 y'(a) \equiv B_a y \qquad (1.73)$$

$$\beta_1 y(b) + \beta_2 y'(b) \equiv B_b y \qquad (1.74)$$

the *compatible kernel* $K(\lambda,x)$ must satisfy the following *adjoint boundary value problem* (for real K):

$$(\rho L_2)^* K = \lambda \rho K \qquad (1.75)$$

$$\alpha_1 \frac{\rho(a)}{p(a)} K(\lambda,a) + \alpha_2 \left(\frac{\rho}{p}K\right)'(a) = 0 \qquad (1.76)$$

$$\beta_1 \frac{\rho(b)}{p(b)} K(\lambda,b) + \beta_2 \left(\frac{\rho}{p}K\right)'(b) = 0 \qquad (1.77)$$

Even though this is a special case of the adjoint problem (1.58a), (1.62), we shall repeat most of the discussion that led to such a general problem. This is done to further reinforce understanding as well as to introduce the weight function $\rho(x)$ of the transform (1.2). In addition, it will elucidate the difference between $p(x)$, the factor that made L_2 self-adjoint ($L_s = p(x)L_2$), and $\rho(x)$.

In applying the integral transform (1.2) on the differential operator L_2 of (1.51),

$$\int_a^b \rho(x)K(\lambda,x)(L_2 y)(x)\,dx = \int_a^b K(\lambda,x)(\rho L_2 y)(x)\,dx \qquad (1.78)$$

our aim is to free the y from the differentiation operation of L_2. This can be done, in principle, by two repeated integrations by parts, or in a more efficient way through the use of the Lagrange identity for $n = 2$ in (1.121). However, and more importantly, this integration operation should involve only what is given to us as boundary conditions on the unknown $y(x)$, namely the conditions $B_a y$ and $B_b y$ of (1.73), (1.74). As we have seen for the general problem of L_n, and as we shall see next, the algebraization of the differential operator will be satisfied with condition (1.75) on K, while the involvement of only the given conditions $B_a y$ and $B_b y$ will be satisfied by conditions (1.76) and (1.77) on K. Through two integrations by parts or by using the Lagrange identity for $n = 2$ in (1.121),

$$\bar{v}L_2 u - u\overline{L_2^* v} = \frac{d}{dx}[A_0(x)\bar{v}u' - (A_0(x)\bar{v})'u + A_1\bar{v}u] \qquad (1.79)$$

we have that

$$\int_a^b K(\lambda,x)(\rho L_2 y)(x)\,dx = \int_a^b y(x)[(\rho L_2)^* K](x)\,dx$$

$$+ [A_0\rho Ky' - (A_0\rho K)'y + A_1\rho Ky]\Big|_a^b \quad (1.80)$$

From this it is clear that the operator L_2 is algebraized if we choose K to be the solution of the adjoint problem (1.75). Hence, (1.80) becomes

$$\int_a^b K(\lambda,x)(\rho L_2 y)(x)\,dx$$

$$= \lambda \int_a^b y(x)\rho(x)K(\lambda,x)\,dx + [A_0\rho Ky' - (A_0\rho K)'y + A_1\rho Ky]\Big|_a^b$$

$$= \lambda Y(\lambda) + [A_0\rho Ky' - (A_0\rho K)'y + A_1\rho Ky]\Big|_a^b \quad (1.81)$$

which is algebraic in the transform function $Y(\lambda)$. What remains is to make sure that the boundary values in the brackets should involve $B_a y$ and $B_b y$ at the boundary points a and b and not only the individual y or y'. This is not yet clear in the present form of (1.81); however, if we write the bracket terms in the more suitable form

$$[A_0\rho Ky' - (A_0\rho K)'y + A_1\rho Ky]_a^b = r\left[\frac{A_0\rho}{r}Ky' - \left(\frac{A_0\rho K}{r}\right)'y\right]_a^b$$

$$= r\left[\frac{\rho}{p}Ky' - \left(\frac{\rho K}{p}\right)'y\right]_a^b \quad (1.82)$$

we have an easier way seeing the conditions (1.76), (1.77) on K.

Even this form (1.82) still involves y and y' individually at the boundary points a and b instead of $B_a y$ and $B_b y$. So we will substitute for $y'(b)$ and $y'(a)$ from the boundary conditions (1.73), (1.74) (assuming that $\beta_2 \neq 0$ and $\alpha_2 \neq 0$),

$$y'(b) = \frac{1}{\beta_2}[B_b y - \beta_1 y(b)]$$

$$y'(a) = \frac{1}{\alpha_2}[B_a y - \alpha_1 y(a)]$$

in (1.82) to have

$$r\left[\frac{\rho}{p}Ky' - \left(\frac{\rho K}{p}\right)'y\right]_a^b$$

$$= r(b) \left[\frac{\rho(b)}{p(b)} K(\lambda,b) \cdot \frac{1}{\beta_2} \{B_b y - \beta_1 y(b)\} - \left(\frac{\rho}{p} K\right)' (b) y(b) \right]$$

$$- r(a) \left[\frac{\rho(a)}{p(a)} K(\lambda,a) \cdot \frac{1}{\alpha_2} \{B_a y - \alpha_1 y(a)\} \right.$$

$$\left. - \left(\frac{\rho}{p} K\right)' (a) y(a) \right] \qquad (1.83)$$

$$= \frac{r(b)}{\beta_2} \left[\frac{\rho(b)}{p(b)} K(\lambda,b) B_b y - \left\{ \beta_1 \frac{\rho(b)}{p(b)} K(\lambda,b) \right. \right.$$

$$\left. + \beta_2 \left(\frac{\rho}{p} K\right)' (b) \right\} y(b) \right] - \frac{r(a)}{\alpha_2} \left[\frac{\rho(a)}{p(a)} K(\lambda,a) B_a y \right.$$

$$\left. - \left\{ \alpha_1 \frac{\rho(a)}{p(a)} K(\lambda,a) + \alpha_2 \left(\frac{\rho}{p} K\right)' (a) \right\} y(a) \right] \qquad (1.84)$$

Here again we can use only $B_b y$ and $B_a y$ but not the individual $y(b)$ and $y(a)$; the only way to eliminate this difficulty is to assign both coefficients of $y(a)$ and $y(b)$ in the parentheses of (1.84) to zero. That is,

$$\beta_1 \frac{\rho(b)}{p(b)} K(\lambda,b) + \beta_2 \left(\frac{\rho}{p} K'\right)' (b) = 0 \qquad (1.77), (1.85)$$

$$\alpha_1 \frac{\rho(a)}{p(a)} K(\lambda,a) + \alpha_2 \left(\frac{\rho K}{p}\right)' (a) = 0 \qquad (1.76), (1.86)$$

which are the desired conditions (1.77), (1.76) imposed on the kernel $K(\lambda,x)$ along with (1.75) to make it compatible with the boundary value problem (1.72)–(1.74).

We note here that the form (1.82) is actually a consequence of the fact, as shown in Example 1.8, that any second-order differential operator can be made self-adjoint. This note is to warn against expecting this when we deal with differential equations of order $n > 2$, when, in general, they are not self-adjoint and we must work with the original form (1.81).

With these conditions (1.85), (1.86) on $K(\lambda,x)$, we can solve for the kernel $K(\lambda,x)$ of the final integral transform (1.78) that is compatible with the differential equation (1.72). Incorporating the boundary conditions (1.73), (1.74), the integral transformation (1.78), (1.81) of the boundary value problem (1.72)–(1.74) gives

$$\int_a^b \rho(x) K(\lambda,x) (L_2 y)(x) \, dx = \lambda Y(\lambda) + \frac{r(b)}{\beta_2} \frac{\rho(b)}{p(b)} K(\lambda,b) B_b y$$

$$- \frac{r(a)}{\alpha_2} \frac{\rho(a)}{p(a)} K(\lambda,a)B_a y, \qquad \beta_2 \neq 0, \ \alpha_2 \neq 0 \quad (1.87)$$

For convenience, if $p(x)$ is taken as $\rho(x)$, we have a simpler result:

$$\int_a^b \rho(x)K(\lambda,x)(L_2 y)(x)\,dx$$

$$= \lambda Y(\lambda) + \frac{r(b)}{\beta_2} K(\lambda,b)B_b y - \frac{r(a)}{\alpha_2} K(\lambda,a)B_a y \qquad (1.88)$$

1.6 Constructing the Compatible Transforms for Self-Adjoint Problems—Second-Order Differential Equations

---

As we mentioned at the end of part A of Section 1.5.2, we will treat those boundary value problems (1.72)–(1.74) which are associated with the second-order differential operators that can always be put into self-adjoint form. Hence, we will be dealing primarily with regular Sturm–Liouville problems or with the singular case which arises when the interval (a,b) is unbounded or the coefficients of the operator have a singularity at one end point. The Laplace and Fourier transforms (1.7), (1.11) are cases where the domain is unbounded, while Hankel transforms (1.22) have a kernel $J_n(\lambda r)$, which is the bounded solution at $r = 0$, of the Bessel differential equation (1.21) with singularity at $r = 0$. Generally, for the regular case, we will be working with the Sturm–Liouville eigenvalue problem (1.63)–(1.65) and its associated orthonormal expansion, which we classified in Section 1.3, as the finite integral transform. The singular case (of the infinite domain) would correspond, then, to the integral transforms. Discrete transforms correspond to the discrete problem associated with the discretized Sturm–Liouville problem (difference equation—see Sections 1.3.5 and 4.1.5).

Next we will discuss the compatibility question in more detail with specific examples and by adhering to the self-adjoint form L_s of L_2. The complete treatment is found in Churchill (1972) and also in Sneddon (1972). The following treatment should be considered as a special case of the above treatment where we take L_2 as L_s the self-adjoint form and $p(x) = \rho(x)$. It is done here for emphasis and also because it is the usual treatment found in other texts; thus, it should be optional.

Consider the second-order differential operator L_2,

$$L_2 y = A_0 y''(x) + A_1 y'(x) + A_2(x)y(x), \qquad a < x < b \qquad (1.72)$$

with its self-adjoint form L_s, letting

$$p(x) = \rho(x) \quad \text{for } p(x)L_2 = L_s$$

$$L_2 y = \frac{1}{\rho(x)} \left[\frac{d}{dx} \left(r\frac{dy}{dx} \right) - qy \right] = \frac{1}{\rho(x)} L_s y \tag{1.72a}$$

Now we consider the general boundary value problem associated with this differential operator for which we seek to find a compatible transform:

$$\frac{1}{\rho(x)} L_s y = f(x) \tag{1.89}$$

$$L_s y = \frac{d}{dx} \left(r\frac{dy}{dx} \right) - qy = \lambda \rho y \tag{1.90}$$

$$B_a y = \alpha_1 y(a) + \alpha_2 y'(a) \tag{1.91}$$

$$B_b y = \beta_1 y(b) + \beta_2 y'(b) \tag{1.92}$$

With this *self-adjoint* form in (1.90), the kernel of the compatible transform $K(\lambda,x)$ is to be determined as the solution of the adjoint problem of $(1/\rho(x))L_s K(\lambda,x) = \lambda K(\lambda,x)$ and the same boundary conditions (1.91), (1.92), as we have shown in Section 1.5. Let us remember that our purpose in finding the compatible transform $\mathcal{K}$ for the differential operator L_2,

$$\mathcal{K}\{L_2 y\} = \int_a^b \rho(x)K(\lambda,x)L_2 y(x)\,dx = \int_a^b \rho(x)K(\lambda,x)\frac{L_s}{\rho(x)} y\,dx$$

$$= \int_a^b K(\lambda,x)L_s y(x)\,dx \tag{1.93}$$

is to free the solution y from the differential operator L_s to obtain an algebraic term in $Y(\lambda)$, the transform of $y(x)$,

$$Y(\lambda) = \int_a^b \rho(x)K(\lambda,x)y(x)\,dx \tag{1.94}$$

For this we appeal to the Lagrange identity (1.79) in the same way we did for (1.80), except here in (1.93) we have a self-adjoint operator L_s. Thus, instead of the operator L_2 in (1.80) we have L_s and also that the second (algebraic) term in the right of (1.80) becomes the one term in the right of (1.82) with $\rho = p$

$$\int_a^b K(\lambda,x)[L_s y(x)]\,dx = \int_a^b y(x)[L_s K(\lambda,x)]\,dx + r[Ky' - K_x y]\Big|_a^b \tag{1.95}$$

where $K_x \equiv \partial K(\lambda,x)/\partial x$.

The integral on the right can be made linear in $Y(\lambda)$ if we let $K(\lambda,x)$ be the solution of the Sturm–Liouville differential equation (1.90), i.e.,

$$L_s K = \lambda \rho K \tag{1.96}$$

to give

$$\int_a^b K(\lambda,x)[L_s y(x)]dx = \lambda Y(\lambda) + r[Ky' - K_x y] \Big|_a^b$$

$$= \lambda Y(\lambda) + r(b)[K(\lambda,b)y'(b) - K_x(\lambda,b)y(b)]$$

$$- r(a)[K(\lambda,a)y'(a) - K_x(\lambda,a)y(a)] \tag{1.97}$$

where the right-hand side is now algebraic in the transform $Y(\lambda)$. What remains is to have the boundary values in the brackets in (1.97) involve only the given boundary conditions $B_a y$ and $B_b y$ of (1.91), (1.92), and not y and y' individually:

$$B_a y = \alpha_1 y(a) + \alpha_2 y'(a) \tag{1.91}$$

$$B_b y = \beta_1 y(b) + \beta_2 y'(b) \tag{1.92}$$

To do this we solve for $y'(a)$ and $y'(b)$ from (1.91), (1.92), assuming that $\alpha_2 \neq 0$, $\beta_2 \neq 0$, and then substitute in (1.97) to obtain

$$\int_a^b K(\lambda,x)[L_s y(x)]dx$$

$$= \lambda Y(\lambda) + r(b)\left[K(\lambda,b)\frac{B_b y - \beta_1 y(b)}{\beta_2} - K_x(\lambda,b)y(b)\right]$$

$$- r(a)\left[K(\lambda,a)\frac{B_a y - \alpha_1 y(a)}{\alpha_2} - K_x(\lambda,a)y(a)\right]$$

$$= \lambda Y(\lambda) + \frac{r(b)}{\beta_2}[K(\lambda,b)B_b y - \{\beta_1 K(\lambda,b) + \beta_2 K_x(\lambda,b)\}y(b)]$$

$$- \frac{r(a)}{\alpha_2}[K(\lambda,a)B_b y - \{\alpha_1 K(\lambda,a) + \alpha_2 K_x(\lambda,a)\}y(a)] \tag{1.98}$$

Here we can still use the involvement of the individual values of $y(b)$ and $y(a)$ although they are not given explicitly. The only way to resolve this difficulty is to assign zero to their coefficients in the parentheses, i.e.,

$$\beta_1 K(\lambda,b) + \beta_2 K_x(\lambda,b) = 0 \tag{1.99}$$

$$\alpha_1 K(\lambda,a) + \alpha_2 K_x(\lambda,a) = 0 \tag{1.100}$$

These are then the boundary conditions the kernel $K(\lambda,x)$ must satisfy, along with the differential equation (1.96), in order to be compatible with the differential equation

$$L_s y = f(x) \tag{1.101}$$

and its boundary conditions (1.91), (1.92).

With these conditions (1.99), (1.100) on the kernel K, the final result in (1.98) becomes

$$\int_a^b K(\lambda,x)[L_s y(x)]\,dx = \lambda Y(\lambda) + \frac{r(b)}{\beta_2} K(\lambda,b)B_b y - \frac{r(a)}{\alpha_2} K(\lambda,a)B_a y$$
$$\tag{1.88), (1.102}$$

To summarize, for a boundary value problem associated with second-order operator L_2 we should first put the operator in a self-adjoint form, that is, $L_s = p(x)L_2$. Then the compatible kernel must satisfy the following eigenvalue problem associated with L_s and the same type of boundary conditions $B_a y$ and $B_b y$ given on the unknown function $y(x)$:

$$L_s K = \lambda \rho K \tag{1.96}$$

$$B_b K = 0 \tag{1.99}$$

$$B_a K = 0 \tag{1.100}$$

As a final remark, we indicate that if we succeed in finding a compatible transform for a problem associated with the differential operator L_2 and its boundary conditions $B_a y$ and $B_b y$, we can still use it in succession to algebraize operators that can be expressed as powers of L_2, like L_2^2. In this special case, for example, we can use the transformation twice, and as we expect we would need the boundary conditions $B_a y$ and $B_b y$ on the unknown as well as on the output of operator L_2, that is, $B_a(L_2 y)$ and $B_b(L_2 y)$.

1.6.1 Examples of the Sturm–Liouville and Other Transforms—Boundary Value Problems

In the following examples we illustrate the method of finding compatible Fourier-type, or Sturm–Liouville, transforms to boundary value problems. This includes the finite sine, the Hankel, and the exponential Fourier transforms.

EXAMPLE 1.10: Finite Fourier Sine Transforms

Consider the following boundary value problem and compare it with (1.90)–(1.92).

$$L_2 y = \frac{d^2 y}{dx^2} = F(x), \qquad 0 < x < \pi, \ \rho(x) = 1 \tag{E.1}$$

$$B_0 y = y(0), \qquad \alpha_2 = 0 \tag{E.2}$$

$$B_\pi y = y'(\pi), \qquad \beta_1 = 0 \tag{E.3}$$

From the above treatment for the compatible kernel in (1.96), (1.99), (1.100) the kernel $K(\lambda, x)$ must satisfy the following boundary value problem:

$$\frac{d^2 K}{dx^2} = \lambda K \tag{E.4}$$

$$K(\lambda, 0) = 0 \tag{E.5}$$

$$K_x(\lambda, \pi) = 0 \tag{E.6}$$

The orthogonal solutions to this new Sturm–Liouville eigenvalue problem are

$$K(\lambda_n, x) = K(-(n - \tfrac{1}{2})^2, x) = \sin(n - \tfrac{1}{2})x, \qquad n = 1, 2, \ldots \tag{E.7}$$

So our *finite sine* transform is

$$Y_n = \mathcal{F}_n\{y\} = \int_0^\pi y(x) \sin(n - \tfrac{1}{2})x \, dx \tag{E.8}$$

and according to (1.97), the resulting transformation of $d^2 y / dx^2$ in (E.1) is

$$\int_0^\pi \sin(n - \tfrac{1}{2})x \, \frac{d^2 y}{dx^2} \, dx$$

$$= -(n - \tfrac{1}{2})^2 Y_n + (n - \tfrac{1}{2})y(0) - (-1)^n y'(\pi) \tag{E.9}$$

This is the desired algebraic result in Y_n, which we gave a direct illustration of in Example 1.7.

The case of the finite cosine transform, where $f'(0)$ is given at the first end $x = 0$ and the more general boundary condition that involves a linear combination of the function and its derivative $h f(a) + f'(a)$, given at the other end $x = a$, is covered in Section 3.2 of Chapter 3 (see (3.107)–(3.111)) and is illustrated in Example 3.11.

EXAMPLE 1.11: Finite Hankel Transform

Consider the differential operator L,

$$Ly = \frac{d^2y}{dx^2} + \frac{1}{x}\frac{dy}{dx} \equiv \frac{1}{x}\frac{d}{dx}\left[x\frac{dy}{dx}\right] = \frac{1}{x}L_sy,$$

$$0 < x < b, \ \rho(x) = x, \ p(x) = x \tag{E.1}$$

which is the differential part of the Bessel differential equation

$$\frac{d^2y}{dx^2} + \frac{1}{x}\frac{dy}{dx} - y = 0 \tag{E.2}$$

We note that this differential operator is singular at $x = 0$. The given boundary condition at $x = b$ is

$$B_b y = y'(b) \tag{E.3}$$

and $y(x)$ is twice continuously differentiable on $[0, b]$. From our analysis, the kernel $K(\lambda, x)$ must satisfy $L_s K = \lambda \rho(x) K$ as in (1.90),

$$L_s K = \frac{d}{dx}\left(x\frac{dK}{dx}\right) = \lambda x K \tag{E.4}$$

and

$$B_0 K \text{ bounded} \tag{E.5}$$

$$B_b K = K_x(\lambda, b) = 0 \tag{E.6}$$

(E.3) can be employed in (1.102) to give

$$\int_0^b K[L_s y(x)]dx = \int_0^b K(\lambda, x)\frac{d}{dx}\left(x\frac{dy}{dx}\right)dx$$

$$= \lambda Y(\lambda) + bK(\lambda, b)y'(b) \tag{E.7}$$

where $r(x) = x, \ r(0) = 0$.

The bounded solution of (E.4) is the Bessel function of the first kind of order zero, $J_0(x\sqrt{-\lambda})$, and to satisfy the boundary condition (E.6) we must have

$$K_x(\lambda, b) = -\sqrt{-\lambda}J_1(b\sqrt{-\lambda}) = 0 \quad \text{with } -\lambda = \nu^2, \ \nu \text{ real} \tag{E.8}$$

where

$$b\sqrt{-\lambda} = j_{1,n}, \quad \lambda_n = -\frac{j_{1,n}^2}{b^2}, \quad J_1(j_{1,n}) = 0, \quad n = 1, 2, \ldots \tag{E.9}$$

It is customary that we use $\{j_{m,n}\}_{n=1}^{\infty}$ for the zeros of $J_m(x)$. Here we used the fact that

$$\frac{dJ_0(x)}{dx} = -J_1(x) \tag{E.10}$$

Hence, the kernel for the desired finite J_0-Hankel transform is

$$K(\lambda_n,x) = J_0\left(\frac{j_{1,n}x}{b}\right) \tag{E.11}$$

and the transform of Ly in (E.1) becomes algebraic in $Y(j_{1,n})$,

$$h_n\left\{\frac{1}{x}\frac{d}{dx}\left(x\frac{dy}{dx}\right)\right\} = -\frac{j_{1,n}^2}{b^2}Y(j_{1,n}) + bJ_0(j_{1,n})y'(b) \tag{E.12}$$

The case of using the finite Hankel transform with a more general boundary condition that involves a linear combination of the function and its derivative $hy(a) + y'(a)$ is covered in Section 3.4.1 of Chapter 3 and is applied in Example 3.14.

EXAMPLE 1.12: Exponential Fourier Transform

Let us consider the differential operator $L = d^2y/dx^2$ for the following singular problem:

$$Ly = \frac{d^2y}{dx^2}, \qquad \rho(x) = 1, \quad -\infty < x < \infty \tag{E.1}$$

$$\lim_{x \to \mp\infty} y(x) = 0, \qquad \lim_{x \to \mp\infty} y'(x) = 0 \tag{E.2}$$

From (1.97) we get

$$\int_{-\infty}^{\infty} K(\lambda,x)\frac{d^2y}{dx^2}\,dx = \lambda Y(\lambda) + [Ky' - yK_x]_{-\infty}^{\infty} \tag{E.3}$$

provided that $K(\lambda,x)$ satisfies

$$\frac{d^2K}{dx^2} = \mu K \tag{E.4}$$

and $|K(\mu,x)|$ is bounded for all μ and x.
    The bounded solution to (E.4) has $\mu = -\lambda^2$,

$$K(\lambda,x) = C_1 \cos \lambda x + C_2 \sin \lambda x \tag{E.5}$$

$$= C_3 e^{i\lambda x} + C_4 e^{-i\lambda x} \tag{E.6}$$

Since either $e^{i\lambda x}$ or $e^{-i\lambda x}$ is a combination of $\sin \lambda x$ and $\cos \lambda x$, we may choose the kernel $K(\lambda, x)$ as $e^{i\lambda x}$ or $e^{-i\lambda x}$. We should be able to show that a sine-cosine combination Fourier transform is equivalent to either one of the Fourier exponential $e^{i\lambda x}$ or $e^{-i\lambda x}$ transforms, which is done at the beginning of Section 2.2 of Chapter 2. (See (2.64), (2.65) and (2.64a)–(2.65b).) Section 2.2 is devoted to this Fourier exponential transform, or its equivalent the trigonometric Fourier sine-cosine transforms, along with precise theorems for the existence of the transform and its inverse and the various operational properties. It is followed by many applications to boundary and initial value problems and to the analysis of signals and systems in Sections 2.3 and 2.4, respectively.

It will be left for Exercise 1.20 to show the kernels $K(\lambda, x) = \sin \lambda x$ and $\cos \lambda x$ are the compatible kernels to the following singular cases of boundary value problems on the semi-infinite interval $(0, \infty)$. Respectively,

1. $L_s y = \dfrac{d^2 y}{dx^2}, \qquad 0 < x < \infty$ (E.7)

     $B_0 y = y(0)$ (E.8)

     $\lim_{x \to \infty} y(x) = 0, \qquad \lim_{x \to \infty} y'(x) = 0$ (E.9)

2. $L_s y = \dfrac{d^2 y}{dx^2}, \qquad 0 < x < \infty$ (E.10)

     $B_0 y = y'(0)$ (E.11)

     $\lim_{x \to \infty} y(x) = 0, \qquad \lim_{x \to \infty} y'(x) = 0$ (E.12)

1.6.2 A Remark Concerning Initial Value Problems

We must stress here that what we have been dealing with is a boundary value problem where, for example, y and y' are not necessarily independent of each other at the boundary points as seen in (1.91), (1.92). Even for the semi-infinite interval, boundary conditions like (1.91) are imposed at $x = 0$ and qualitative conditions regarding boundedness of the solution are imposed at $x = b = \infty$. We mention this boundary value problem on the semi-infinite domain to differentiate it from another distinct class of problems, namely the *initial value problems*, where (for second-order differential equations) $y(0)$ and $y'(0)$ are given *independently* at the one initial point $x = 0$.

$$y(0) = C_1 \tag{1.103}$$

$$y'(0) = C_2 \tag{1.104}$$

A simple example of initial conditions for an initial value problem is that of the semi-infinite vibrating string, where the initial displacement $u(x, 0)$ and velocity $u_t(x, 0)$ are given independently at time zero. On the other hand, for the semi-infinite heated bar the temperature $u(0, t)$ and its gradient $u_x(0, t)$ at $x = 0$ must be related to the outside temperature T_0 through the law of heat convection,

$$c_1 u(0, t) + c_2 u_x(0, t) = T_0 \qquad (1.105)$$

which makes it a boundary condition at $x = 0$. Again, compatible transforms for such boundary values are called Sturm–Liouville or Fourier-type transforms, in contrast to the Laplace transform and its associates (Mellin), which, as we have illustrated in the first section, are for initial value problems. This means that for initial value problems we should modify our result (1.102) and return to the first line of (1.97). If our intention is to algebraize the differential operator of the initial value problem, we can proceed with integration by parts in an attempt to do so and then involve the appropriate initial conditions. For second-order operators, we can still use the Lagrange identity on the semi-infinite interval $(0, \infty)$ to have

$$\int_0^{\infty} K(\lambda, x) L_2 y \, dx = \lambda Y(\lambda) + r(x)[K(\lambda, x)y'(x) - y(x)K_x(\lambda, x)] \Big|_0^{\infty}$$

$$= \lambda Y(\lambda) - r(0)[K(\lambda, 0)y'(0) - y(0)K_x(\lambda, 0)] \qquad (1.106)$$

after imposing boundedness of K and K_x at $x = 0$ and ∞ and assuming that y and y' vanish as x approaches infinity.

For higher-order differential operators, we can still consult the next Section 1.7 for the general Lagrange identity (1.49) with its H_n in (1.118) for the appropriate initial conditions. An example is to verify the result in (1.10).

Next we will present a simple example that illustrates the use of the Laplace transform for initial value problems.

EXAMPLE 1.13: Initial Value Problem and the Laplace Transform

Consider the operator $L = d^2/dx^2$, which is obviously self-adjoint with $\rho(x) = 1$, and its associated initial value problem

$$Ly = \frac{d^2 y}{dx^2}, \qquad 0 < x < \infty \qquad (\text{E.1})$$

$$y(0) = c_1 \qquad (\text{E.2})$$

$$y'(0) = c_2 \qquad (\text{E.3})$$

From (1.96), we must have the kernel satisfy the adjoint problem

$$\frac{d^2 K}{dx^2} = \mu K \tag{E.4}$$

whose bounded solution is $\phi(\lambda, x) = e^{-\lambda x}$, $\mu = \lambda^2$. Compared to the boundary value problem result (1.102), the initial value form (1.106) directly involves the initial conditions (E.2), (E.3) on y along with the values of the kernel K. In this case, we have a kernel of the Laplace transform, $e^{-\lambda x}$,

$$\int_0^\infty e^{-\lambda x} \frac{d^2 y}{dx^2} dx = \int_0^\infty \frac{d^2}{dx^2}(e^{-\lambda x}) y(x) dx + [e^{-\lambda x} y'(x) + \lambda e^{-\lambda x} y(x)]_0^\infty$$

$$= \lambda^2 Y(\lambda) + 0 - [y'(0) + \lambda y(0)]$$

$$= \lambda^2 Y(\lambda) - \lambda y(0) - y'(0) \tag{E.5}$$

after we assume that $\lim_{x\to\infty} y(x)$ and $\lim_{x\to\infty} y'(x)$ are bounded. This is also the result we obtained in (1.9).

1.7 The *n*th-Order Differential Operator[†]

As we mentioned earlier, the material in this section is considered optional, since we have already placed our emphasis on the important special case of second-order differential equations in the last two sections, 1.5 and 1.6. The very brief treatment of the present general case of L_n will run very much in parallel to that of L_2, except, of course, that L_n, $n > 2$, in general is not self-adjoint.

To stress the importance of the adjoint operator L_n^*, we point out another of its many advantages, this one in the direction of reducing the order of the differential equation

$$L_n u = f \tag{1.107}$$

by one to a new differential equation of order $n - 1$. This can be accomplished by making w a solution of the homogeneous equation

$$L_n^* w = 0 \tag{1.108}$$

[†] Optional section, with Section 1.5 almost a prerequisite.

which, when substituted in the Green identity (1.50), will give us

$$\int \overline{w} L_n u \, dx = \int \overline{w} f(x) \, dx \equiv g(x)$$

$$= H_n(u, u', \ldots, u^{(n-1)}; w, w', \ldots, w^{(n-1)}) \tag{1.109}$$

after using (1.107). This is a differential equation of order $n - 1$ in the unknown u, as is apparent from the form H_n, assuming that we have found w from (1.108). This makes w, the solution of the (*homogeneous*) *adjoint problem* (1.108), stand as an *integrating factor* to $L_n u = f$, where through multiplying by $\overline{w}$ we reduced its order by one to (1.109). This is a very useful result, if repeated n times, for integrating the problem $L_n u = f$ to its end, i.e., to having the solution $u(x)$. It is not, however, exactly in our direction of seeking a final algebraic expression to $L_n u = f$ in the transform $U(\lambda)$ and not in $u(x)$. Indeed, as we have seen in the case of L_2 in (1.58), the equation associated with the adjoint operator L_n^* that we have to solve for finding the compatible kernel is a different one, namely

$$\overline{L_n^* v} = \overline{\lambda v} \tag{1.110}$$

For the present general treatment of L_n we will choose the weight function $\rho(x)$ of the integral transforms (1.2) to be 1 for the sake of simplifying the algebra. $\rho(x)$ was included in the important special case of L_2 in the last two sections. The necessity of solving this equation (1.110) for the kernel $v = K(\lambda,x)$ of (1.2) (with $\rho(x) = 1$ for now) is clearly seen from the Green identity (1.50) with $v = K(\lambda,x)$; the integral on the right-hand side becomes algebraic in $U(\lambda)$; i.e.,

$$\int_a^b \overline{K(\lambda,x)} L_n u \, dx$$

$$= \int_a^b u \overline{L_n^* K(\lambda,x)} \, dx + H_n(u, u', \ldots, u^{(n-1)}, K, K_x, \ldots, K_x^{(n-1)}) \Big|_a^b$$

$$= \int_a^b u \overline{\lambda} \, \overline{K(\lambda,x)} \, dx + H_n(u, u', \ldots, u^{(n-1)}, K, K_x, \ldots, K_x^{(n-1)}) \Big|_a^b$$

$$= \overline{\lambda} U(\lambda) + H_n(u, u', \ldots, u^{(n-1)}, K, K_x, \ldots, K_x^{(n-1)}) \Big|_a^b \tag{1.111}$$

where $K_x^{(i)} \equiv \partial^i K(\lambda,x)/\partial x^i$.

Next we present the general form of L_n, the differential operator of order n, its adjoint L_n^*, and the explicit form of their Lagrange identity

(1.49). In order to acquire more familiarity with the adjoint L_2^*, we will illustrate finding its compatible transform before reducing L_2 to its self-adjoint form L_s (which we have illustrated as the very important convenient special case of the last two sections). Again, we must stress that we will do this for L_2^* to emphasize the importance of the adjoint operator, since, in general, operators of order $n > 2$ are not necessarily self-adjoint, and we must deal with the adjoint equation (1.110) for finding the compatible kernel $v = K(\lambda, x)$.

Definition: The Adjoint Operator L_n^*.   Consider the nth-order differential operator L_n,

$$L_n u = A_0(x) \frac{d^n u}{dx^n} + A_1(x) \frac{d^{n-1}}{dx^{n-1}} + \cdots + A_n(x)u \tag{1.112}$$

Its *adjoint* L_n^* is defined as

$$L_n^* u = (-1)^n \frac{d^n}{dx^n}(\overline{A}_0 u) + (-1)^{n-1} \frac{d^{n-1}}{dx^{n-1}}(\overline{A}_1 u) + \cdots + \overline{A}_n u$$

$$= \sum_{m=1}^{n} (-1)^m \frac{d^m}{dx^m}(\overline{A}_{n-m}(x)u(x)) + \overline{A}_n(x)u \tag{1.113}$$

Here the coefficients $A_j(x)$ are taken to be complex-valued functions, with continuous derivatives up to $n-j$ on the closed interval $a \le x \le b$ and $A_0(x) \ne 0$ on $[a,b]$. $\overline{A}(x)$ is the complex conjugate of $A(x)$; however, we shall be working mainly with real-valued functions, where $\overline{A}_j(x) = A_j(x)$. From (1.112) with $n = 2$, the usual second-order operator L_2 would take the form

$$L_2 u = A_0(x) \frac{d^2 u}{dx^2} + A_1(x) \frac{du}{dx} + A_2(x)u \tag{1.114}$$

with its adjoint L_2^* from (1.113) as

$$L_2^* u = \frac{d^2}{dx^2}[A_0(x)u(x)] - \frac{d}{dx}[A_1(x)u(x)] + A_2(x)u \tag{1.115}$$

In the case of L_3 we have from (1.112), with $n = 3$,

$$L_3 u = A_0 \frac{d^3 u}{dx^3} + A_1 \frac{d^2 u}{dx^2} + A_2 \frac{du}{dx} + A_3 u \tag{1.116}$$

and its adjoint L_3^* from (1.113)

$$L_3^* u = -\frac{d^3}{dx^3}(A_0(x)u(x)) + \frac{d^2}{dx^2}(A_1(x)u(x))$$

$$- \frac{d}{dx}(A_2(x)u(x)) + A_3(x)u \tag{1.117}$$

We have already shown in Example 1.8 that the second-order operator L_2 of (1.114) can be made self-adjoint. However, L_3 in its above general form cannot be made self-adjoint, as we will illustrate in (1.123).

Besides showing that $L_n^* = L_n$ (as in the special case of L_2 in Example 1.8) for a self-adjoint operator, another very useful way of showing that a differential operator L_n may be self-adjoint is via the Lagrange identity (1.49). This essentially means we can still get the exact differential term dH_n/dx on the right side of (1.49) even if we take $L_n^* = L_n$ on the left side. We will illustrate this after presenting the explicit form of the Lagrange identity (for L_n and its adjoint L_n^* as given in (1.49)). The Lagrange identity (1.49) states that

$$\bar{v}L_n u - u\overline{L_n^* v} = \frac{d}{dx} H_n[u, u', \ldots, u^{(n-1)}; v, v', \ldots, v^{(n-1)}] \tag{1.49}$$

where the explicit form of H_n is

$$H_n[u, u', u^{(n-1)}; v, v', v^{(n-1)}] = \sum_{m=1}^{n} \sum_{\substack{j+k=m-1 \\ j \geq 0, k > 0}} (-1)^j \frac{d^j}{dx^j} (A_{n-m}(x)\bar{v}) \frac{d^k u}{dx^k} \tag{1.118}$$

The special case of H_2 for L_2 is

$$H_2[u, u'; v, v'] = \sum_{m=1}^{2} \sum_{\substack{j+k=m-1 \\ j \geq 0, k \geq 0}} (-1)^j \frac{d^j}{dx^j} (A_{2-m}(x)\bar{v}) \frac{d^k u}{dx^k} \tag{1.119}$$

$$= A_1(x)\bar{v}u + A_0(x)\bar{v}u' - u\frac{d}{dx}(A_0\bar{v}) \tag{1.120}$$

Thus, the Lagrange identity for L_2 becomes

$$\bar{v}L_2 u - u\overline{L_2^* v} = \frac{d}{dx}[A_0(x)\bar{v}u' - (A_0(x)\bar{v})'u + A_1\bar{v}u] \tag{1.121}$$

We use here the form (1.118) for $n = 3$ to illustrate the Lagrange identity for L_3:

$$\bar{v}L_3 u - u\overline{L_3^* v}$$

$$= \frac{d}{dx}[uA_2\bar{v} + u'A_1\bar{v} - u(A_1\bar{v})' + u''A_0\bar{v} - u'(A_0\bar{v})' + u(A_0\bar{v})''] \tag{1.122}$$

which should give an indication that it is not self-adjoint. In other words, we cannot establish an exact differential on the right-hand side of (1.122) without using the adjoint operator $L_3^* u$ on the left-hand side. To illustrate this point further we show that even the simple case of $L_3 u = u''' + u$ with constant coefficients is not self-adjoint, since if we write $vL_3 u - uL_3 v$

while adding and subtracting terms to make exact differentials, we still have the following:

$$vL_3u - uL_3v = vu''' - uv'''$$

$$= vu''' + v'u'' - v'u'' - v''u' + v''u' + v'''u - v'''u - v'''u$$

$$= (vu'')' - (v'u')' + (v''u')' - 2v'''u$$

$$= d[vu'' - v'u' + v''u'] - 2v'''u \tag{1.123}$$

This is not an exact differential because of the last term. On the other hand, $L_4u = d^4u/dx^4 + u$ is self-adjoint since we can satisfy the Lagrange identity without having to involve its adjoint L_4^*; that is,

$$vL_4u - uL_4v = vu^{(4)} - uv^{(4)}$$

$$= vu^{(4)} + v'u''' - v'u''' - v''u'' + v''u'' + v'''u'$$

$$\quad - v'''u' - v^{(4)}u$$

$$= (vu''')' - (v'u'')' + (v''u')' - (v'''u)'$$

$$= \frac{d}{dx}(vu''' - v'u'' + v''u' - v'''u) \tag{1.124}$$

after adding and subtracting the appropriate terms to group differential terms to represent the right-hand side as an exact total differential.

The Lagrange identity for L_4 from (1.49) and (1.118) is

$$\bar{v}L_4u - u\overline{L_4v} = \frac{d}{dx}[uA_3\bar{v} - u(A_2\bar{v})' + u'A_3\bar{v} - u'(A_1\bar{v})'$$

$$\quad + u''(A_1\bar{v}) + u(A_1\bar{v})'' - u''(A_0\bar{v})' + u'(A_0\bar{v})''$$

$$\quad - u(A_0\bar{v})''' + u'''(A_0\bar{v})] \tag{1.125}$$

For further reference we give the general n boundary conditions associated with the nth-order operator L_n, which consist of a linear combination of all the derivatives of u up to $u^{(n-1)}$ at the two boundary points $x = a$ and $x = b$:

$$\Gamma_j u = \sum_{k=1}^{n}[M_{jk}u^{(k-1)}(a) + N_{jk}u^{(k-1)}(b)], \qquad j = 1, 2, \ldots, n \tag{1.126}$$

$$\equiv \Gamma u$$

where M_{jk} and N_{jk} are constants. Γ_j stands for the jth boundary condition on u, and let $\Gamma = \{\Gamma_j\}_{j=1,2,\ldots,n}$ represent all the boundary conditions on u signified by $j = 1, 2, 3, \ldots, n$, which is a clear generalization of the two boundary conditions associated with L_2 in (1.54a) and (1.54b).

The eigenvalue problem associated with L_n is the following differential equation subject to the above n (homogeneous) boundary conditions:

$$L_n u = \lambda u \tag{1.127}$$

$$\Gamma u = 0 \tag{1.128}$$

where λ is called the *characteristic value* or *eigenvalue* and the resulting corresponding solutions $\{u\}$ are the *characteristic functions* or *eigenfunctions*. The rest of the discussion follows in parallel to the discussion for L_2 in (1.51) to (1.62), and for more details the reader may consult Coddington and Levinson (1955).

Relevant References to Chapter 1

The basic references, among others, used in the illustrations and construction of the compatible transforms are Churchill (1972), Sneddon (1972), Tranter (1951), Miles (1971), and Weinberger (1965). Coddington and Levinson (1955) was also used, and the interested reader can rely on it heavily for the adjoint problem of Section 1.5 and especially for the nth-order operator of Section 1.7.

Exercises

Section 1.1 The Method of Separation of Variables and Integral Transforms

1.1. Use the method of separation of variables to show that

$$u(x,t) = [A \cos \lambda x + B \sin \lambda x][D \cos \lambda ct + E \sin \lambda ct] \tag{E.1}$$

where A, B, C, D, and λ are arbitrary constants, is a solution of the *wave equation* (see Appendix B.1 for its derivation)

$$\frac{\partial^2 u}{\partial x^2} = \frac{1}{c^2} \frac{\partial^2 u}{\partial t^2} \tag{E.2}$$

Hint: For the solutions of the resulting differential equations in this problem, the next Exercises 1.2 and 1.3, and other problems, see Exercises A.1, A.3, A.4, and A.2 in Appendix A.

1.2. Consider the boundary value problem for the potential distribution $u(x,y)$ inside a square plate of length π [as described by the Laplace equation (E.1)], where all the sides are kept at zero potential except for the lower edge ($y = 0$), which is kept at a given potential $f(x)$; that is, $u(x,0) = f(x)$, $0 < x < \pi$,

$$\nabla^2 u = \frac{\partial^2 u}{\partial x^2} + \frac{\partial^2 u}{\partial y^2} = 0, \qquad 0 < x < \pi,\ 0 < y < \pi \qquad \text{(E.1)}$$

$$u(0,y) = 0, \qquad 0 < y < \pi \qquad\qquad\qquad \text{(E.2)}$$

$$u(\pi,y) = 0, \qquad 0 < y < \pi \qquad\qquad\qquad \text{(E.3)}$$

$$u(x,\pi) = 0, \qquad 0 < x < \pi \qquad\qquad\qquad \text{(E.4)}$$

$$u(x,0) = f(x), \qquad 0 < x < \pi \qquad\qquad\qquad \text{(E.5)}$$

[For the derivation of the Laplace equation (E.1), see (B.2.8) in Appendix B.2.]

(a)   Use the method of separation of variables to find a solution that satisfies the homogeneous *Laplace equation* (E.1) and all the homogeneous boundary conditions (E.2)–(E.4). See hint to Exercise 1.1.

(b)   Use the result in (a) to find the solution to the problem (E.1)–(E.5) when $f(x) = \sin 2x$ in (E.5).

(c)   Consider the fact that the following *Fourier series* converges to $g(x)$ on the interval $(0,a)$:

$$g(x) = \sum_{n=1}^{\infty} b_n \sin \frac{n\pi x}{a} \qquad\qquad\qquad \text{(E.6)}$$

if b_n is evaluated as the *Fourier coefficients*

$$b_n = \frac{2}{a} \int_0^a g(x) \sin \frac{n\pi x}{a}\, dx \qquad\qquad\qquad \text{(E.7)}$$

Use this to find the solution to the problem (E.1)–(E.5) when $f(x) = x$, $0 < x < \pi$.

Hint: From part (a) you can show that

$$u(x,y) = \sum_{n=1}^{\infty} c_n \sin nx \sinh n(\pi - y) \qquad\qquad \text{(E.8)}$$

satisfies (E.1)–(E.4). Then substitute for (E.5) in (E.8) and evaluate c_n according to (E.6), (E.7).

ANS.   (a) $u(x,y) = \sin nx \sinh n(\pi - y)$

(b) $u(x,y) = \dfrac{\sin 2x \sinh 2(\pi - y)}{\sinh 2\pi}$

(c) $u(x,y) = 2\displaystyle\sum_{n=1}^{\infty} \dfrac{(-1)^{n+1}}{n} \dfrac{\sin nx \sinh n(\pi - y)}{\sinh n\pi}$

1.3. (a) Use the method of separation of variables to find a solution $u(r,\theta)$ to the Laplace equation in polar coordinates

$$\frac{\partial^2 u}{\partial r^2} + \frac{1}{r}\frac{\partial u}{\partial r} + \frac{1}{r^2}\frac{\partial^2 u}{\partial \theta^2} = 0 \tag{E.1}$$

Hint: Let $u(r,\theta) = R(r)\Theta(\theta)$ and follow the method of Example 1.1 [(E.4)–(E.11)]. Note that r^γ and $r^{-\gamma}$ are solutions of

$$r^2 R'' + rR' - \gamma^2 R = 0 \tag{E.2}$$

(the Euler–Cauchy equation—for a review see Exercise A.2f in Appendix A).

(b) Find a solution that also satisfies the *periodic boundary condition*

$$u(r,\theta) = u(r,\theta + 2\pi) \tag{E.3}$$

(c) Consider the Laplace equation (E.1) inside a unit disc $0 \leq r < 1$ and then find a *bounded* solution that satisfies (E.1) and (E.3).

(d) Do part (c) for a bounded solution outside the unit disc $(r > 1)$.

ANS. (a) $u(r,\theta) = [c_1 r^\gamma + c_2 r^{-\gamma}][c_3 \cos \gamma\theta + c_4 \sin \gamma\theta]$; γ, c_1, c_2, c_3, and c_4 are arbitrary constants.

(b) $u_n(r,\theta) = [c_1 r^n + c_2 r^{-n}][c_3 \cos n\theta + c_4 \sin n\theta]$, $n =$ integer.

(c) $u_n(r,\theta) = c_1 r^n [c_3 \cos n\theta + c_4 \sin n\theta]$, $0 \leq r < 1$, $n = 0, 1, 2, \ldots$

(d) $u_n(r,\theta) = c_2 r^{-n}[c_3 \cos n\theta + c_4 \sin n\theta]$, $r > 1$, $n = 0, 1, 2, \ldots$

1.4. In Example 1.1 and Exercises 1.1–1.3, we observed that the method of separation of variables was applicable to (1) *homogeneous* partial differential equations and their auxiliary conditions that (2) are evaluated at a *constant value* of the separated variable involved. In general, these two conditions (1), (2) are necessary but not sufficient. We state here a reasonable sufficient condition for the problem to be separable, i.e., for the method of the products to apply.

Consider the partial differential equation in $u(x,y,z,t)$ with a differential operator L

$$L[u(x,y,z,t)] = 0 \tag{E.1}$$

and its auxiliary conditions, as described by the operators L_i, $i = 1$, $2, \ldots, n$,

$$L_i[u(x,y,z,t)], \qquad i = 1, 2, \ldots, n \tag{E.2}$$

where each operator L_i is *evaluated at a constant value of the variable involved* and n depends on the given problem. An example of L is $Lu \equiv [\partial^2/\partial x^2 + \partial^2/\partial y^2]u(x,y) = 0$ in the Laplace equation (E.1) of Exercise 1.2. For L_i we may have $L_3[u(x,y)] = u(x,y)|_{y=\pi} = u(x,\pi) = 0$ in (E.4) of Exercise 1.2. For the above initial and boundary value problem (E.1), (E.2) to be separable, it is sufficient that

1. There is a function $g(x,y,z,t)$ such that if we let $u(x,y,z,t) = X(x)Y(y)Z(z)T(t)$, in (E.1) we have

$$\frac{L[X(x)Y(y)Z(z)T(t)]}{g(x,y,z,t)X(x)Y(y)Z(z)T(t)} = f_1(x) + f_2(y) + f_3(z) + f_4(t) \tag{E.3}$$

   that is, the partial differential operator is separable. Note that in most cases of interest $g(x,y,z,t) = $ constant.

The second important condition is that

2. Each boundary value operator L_i evaluated at the *constant value* of one variable *should not have partial derivatives with respect to the other variables, and its coefficients should be independent of the other variables.* For example, if L_2 is the boundary value operator evaluated at $x = a$, then it should not involve partial derivatives in y, z, and t, and its coefficients should be independent of y, z, and t.

(a) For Example 1.1 and the above Exercises 1.1–1.3, verify that in each of these problems the *homogeneous* partial differential equation and its corresponding auxiliary conditions, *which must be evaluated at a constant value of the (separated!) variable involved*, do satisfy the above sufficient conditions 1 and 2 for the method of the products (separation of variables) to apply.

(b) Determine whether the following boundary and initial value problem is separable:

$$Lu \equiv \frac{\partial^2 u}{\partial x^2} - x^3 t \frac{\partial^2 u}{\partial t^2} = 0, \qquad a < x < b, \ t > 0 \tag{E.3}$$

$$L_1 u = u(a,t) = 0, \qquad t > 0 \tag{E.4}$$

$$L_2 u = u(b,t) + t \frac{\partial u(b,t)}{\partial x} = 0, \qquad t > 0 \tag{E.5}$$

$$L_3 u = u(x,0) = f(x), \qquad a < x < b \tag{E.6}$$

$$L_4 u = \frac{\partial u}{\partial t}(x,0) = g(x), \qquad a < x < b \tag{E.7}$$

Hint: For (E.3) use $g(x,t) = x^3$, then watch for the above condition 2 on operator L_2 of (E.5).

ANS. With $g(x,t) = x^3$, the partial differential equation (E.3) is separable, and all the auxiliary conditions of L_1 to L_4 in (E.4)–(E.7) are evaluated at a constant value of one of the variables. Also, L_1, L_3, and L_4 satisfy condition 2 above; however, L_2, which is taken at the constant value of the variable $x = b$, has the coefficient of $\partial u(b,t)/\partial x$ involving the other (not fixed for L_2) variable t. Note that $L_3 u = f(x)$ and $L_4 u = g(x)$ are separable in the sense that they are associated with the ordinary differential equation in $X(x)$, but Fourier analysis is needed to satisfy them (see how we had to do this for the nonhomogeneous condition (E.3) in Example 1.1, where we moved from (E.14) to (E.15), (E.16) of the Fourier sine transform analysis).

1.5. This exercise illustrates how far we have to go sometimes in order to have each boundary condition represented at a *constant value* of one of the coordinates used for the problem. This is to satisfy the sufficient conditions in the separation of variables method, as given in Exercise 1.4, or for the values needed at the constant limits of integration when the integral transforms are used. The simplest case is our resorting to polar coordinates (r, θ) to accommodate, for example, a given potential at the rim of a unit disc, $u(1, \theta) = f(\theta)$, $-\pi < \theta < \pi$.

Consider the problem of Exercise 2.27 in Section 2.7 for the potential distribution due to an electrified unit disc in the xy-plane with center at the origin. The disc is kept at a constant potential

u_0; thus, of course, the potential in space would be symmetric with respect to the xy-plane. For such a circular disc we choose the compatible cylindrical coordinates $u(r,z)$ with no θ dependence, as it is clear from the above condition. Also, it is very reasonable to assume that $u(r,z)$ vanishes as z tends to $\mp\infty$. The preceding boundary conditions now can be stated as

$$u(r,0) = u_0, \qquad 0 \leq r < 1 \tag{E.1}$$

$$\frac{\partial u}{\partial z}(r,0) = 0, \qquad 1 < r < \infty \tag{E.2}$$

Note that even though these conditions are given at a constant value of $z = 0$, we still have neither the function nor its derivative given *for all values* of the other coordinate r. Indeed, we have u given on one part of the domain in r, namely $0 \leq r < 1$, and $\partial u/\partial r$ on the remaining other part: $1 < r < \infty$. As we shall see in Example 2.27, such a situation, although accommodated by the Hankel transform, will lead to a *dual set of integral equations* to be solved, which is not such an easy situation! In order to have the representation of the boundary condition (E.1) be given at a constant value of one coordinate that *is valid for all values of the second coordinate*, and the boundary condition (E.2) be given for a constant value of the second coordinate (that is valid for all values of the first coordinate), a change from the cylindrical coordinates (r,z) to the following *oblate spherical* coordinates (μ,ζ) was necessary:

$$z = \mu\zeta, \qquad -1 < \mu < 1, \ -\infty < \zeta < \infty \tag{E.3}$$

$$r = (1 - \mu^2)(1 + \zeta^2)^{1/2}, \qquad -1 < \mu < 1, \ -\infty < \zeta < \infty \tag{E.4}$$

(a) Verify that the surface $\zeta = 0$, which is called a *spheroid* (for all values of μ), is the circular disc $0 \leq r < 1$ in our cylindrical coordinates. Also, that the surface $\mu = 0$, which is a *hyperboloid of revolution* (for all values ζ), is the remainder of the xy-plane outside the unit disc.

(b) With the information in part (a), show that the boundary conditions (E.1) and (E.2) are now associated with the respective *constant values of the coordinates* $\zeta = 0$ and $\eta = 0$ (valid for all values of the other coordinate), i.e.,

$$v(\mu,0) = u_0, \qquad -1 < \mu < 1 \tag{E.5}$$

and

$$v_\mu(0,\zeta) = 0, \qquad -\infty < \zeta < \infty \tag{E.6}$$

(c) The Laplace equation in $u(r,z)$,

$$\frac{\partial^2 u}{\partial r^2} + \frac{1}{r}\frac{\partial u}{\partial r} + \frac{\partial^2 u}{\partial z^2} = 0, \qquad 0 < r < \infty, \; -\infty < z < \infty$$

(E.7)

becomes for $v(\mu, \zeta)$

$$\frac{\partial}{\partial \mu}\left\{(1-\mu^2)\frac{\partial v}{\partial \mu}\right\} + \frac{\partial}{\partial \zeta}\left\{(1+\zeta^2)\frac{\partial v}{\partial \zeta}\right\} = 0,$$

$$-1 < \mu < 1, \; -\infty < \zeta < \infty \qquad \text{(E.8)}$$

Verify that $v(\mu, \zeta) = (2u_0/\pi)\cot^{-1}\zeta$ is a solution to (E.8) with its boundary conditions in (E.5), (E.6).

(d) Suggest a transform that is compatible with the differential operation of (E.8) with respect to the variable μ.

Hint: For the differential operator in the first term of (E.8), see the Legendre equation in (3.148) and in Appendix A.5.2; then consult the use of the finite Legendre transform (3.150) in Section 3.5 of Chapter 3, as in Example 3.15 and Exercises 3.47a of the same section.

1.6. *Review of Complex Numbers.* A complex number z is defined as an ordered pair (x,y), $z = (x,y)$, and is represented by a point (x,y) in the xy-plane, or what we will call the *(complex) z-plane*, as shown in Figure 1.4. The pure imaginary number $i = \sqrt{-1}$ is represented by the point $(0, 1)$ on the y-axis, while the real number 1 is represented

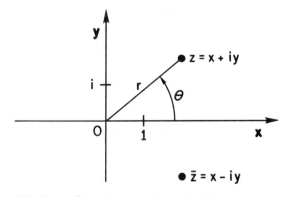

FIG. 1.4 Complex numbers, for Exercise 1.6.

by $(1,0) \equiv 1$ on the x-axis. The complex conjugate $\bar{z}$ of z is defined as $\bar{z} = (x, -y)$ as shown in Fig. 1.4.

(a) Let $z_1 = (x_1, y_1)$, $z_2 = (x_2, y_2)$, and c be a real number. We define

$$cz = (cx, cy) \tag{E.1}$$

$$z_1 + z_2 = (x_1 + x_2, y_1 + y_2) \tag{E.2}$$

$$z_1 z_2 = (x_1 x_2 - y_1 y_2, x_1 y_2 + x_2 y_1) \tag{E.3}$$

Show that

(i) $i \equiv (0, 1)$ satisfies $i^2 = -1$.

   Hint: Use (E.3).

(ii) $(x, y) = x + iy$

   Hint: $(x, y) = (x, 0) + (0, y) = x(1, 0) + y(0, 1)$.

(iii) $|z|^2 = z\bar{z} = x^2 + y^2 = r^2$

   Hint: Use (E.3) or multiply $(x + iy)(x - iy)$ using $i^2 = -1$ of part (i).

(iv) $\overline{z_1 z_2} = \bar{z}_1 \bar{z}_2$

(v) $|z_1 z_2| = |z_1||z_2|$

(b) Write the polar representation of the complex number $z = re^{i\theta}$.

   Hint: $z = x + iy = r\cos\theta + ir\sin\theta = r(\cos\theta + i\sin\theta)$; then use the *Euler identity*: $e^{i\theta} \equiv \cos\theta + i\sin\theta$. We must note that the proof of this identity involves the heart of the theory of complex numbers, which is to use a Taylor series expansion for the *complex-valued function* $e^{i\theta}$, but we take it here as a shortcut in the name of a fast necessary review.

(c) Show that z can also be written as $z = re^{i(\theta + 2\pi k)}$ where k is any integer.

   Hint: Use the Euler identity.

(d) Show that $(\cos\theta + i\sin\theta)^n = \cos n\theta + i\sin n\theta$.

   Hint: Use the Euler identity for $e^{i\theta}$ and then for $e^{in\theta}$.

(e) Find the real $\text{Re}f(z)$ and imaginary $\text{Im}f(z)$ parts of the following complex-valued functions, $\omega = f(z) = \text{Re}f + i\,\text{Im}f$ or $f = f_R + if_I$. When $j = \sqrt{-1}$ is used, we write $f = f_r + jf_i$.

(i) $\omega = z^2$.

   Hint: $(x + iy)^2 = x^2 - y^2 + i2xy$.

(ii) $\omega = e^z$.

   Hint: $e^z = e^{x+iy} = e^x(\cos y + i \sin y)$.

(iii) $\omega = \frac{1}{z}$, $z \neq (0,0)$.

   Hint: multiply numerator and denominator by the complex conjugate $\bar{z}$

$$\omega = \frac{\bar{z}}{|z|^2} = \frac{x - iy}{x^2 + y^2}, \qquad z \neq (0,0) \equiv 0$$

(f)   If a function of two variables $u(x,y)$ satisfies the Laplace equation,

$$\frac{\partial^2 u}{\partial x^2} + \frac{\partial^2 u}{\partial y^2} = 0, \qquad (x,y) \in D$$

we term it *harmonic* in D. Show that the real and imaginary parts of the functions $f(z) = z^2$ and $f(z) = e^z$ in part (e) are harmonic. What can you say about the case of $f(z) = 1/z$?

Hint: Avoid the origin, which is called the *singularity* of $f(z) = 1/z$.

(g)   Use the fact that $z = re^{i(\theta + 2\pi k)}$, $k = 0, \mp 1, \mp 2, \ldots$, to generate the expected m roots ω of the algebraic equation, $\omega^m = z$, or $\omega = z^{1/m}$ for a given z.

Hint: Let the desired roots ω be represented by the complex number $\omega = \rho e^{i\phi}$, and the known z by $z = re^{i\theta}$. Note that for $[\rho e^{i\phi}]^m = re^{i(\theta + 2\pi k)}$, $k = 0, \mp 1, \mp 2, \ldots$, the different k values don't make any difference, but for $\rho e^{i\phi} = r^{1/m} \exp[i(\frac{\theta + 2\pi k}{m})]$, $k = 0, \mp 1, \mp 2, \ldots$, they do for $k = 0, 1, 2, 3, \ldots, m - 1$ only, which will give the m values for ω.

(h)   (i)   Find the two square roots of -1; that is, solve $\omega^2 = -1$.

   Hint: $z = -1 = e^{i\pi}$, $\omega = \rho e^{i\theta} = z^{1/2} = e^{i(\pi + 2\pi k)/2}$, $k = 0, 1$ as given in part (g) with $m = 2$.

   (ii)   Find the three roots of 1, that is, $\sqrt[3]{1}$.

   Hint: $\omega = \rho e^{i\phi} = (1)^{1/3} = e^{i(2\pi + k)/3}$, $k = 0, 1, 2$.

(i)   (i)   Show that the Fourier transform $F(\lambda) = \int_{-\infty}^{\infty} e^{-i\lambda z} f(x) dx$ of a real even function $f(x)$ is a real function.

Hint: Write $e^{-i\lambda x} = \cos\lambda x - i\sin\lambda x$, and note that the integral of the odd part of the function $e^{-i\lambda x}f(x)$ vanishes.

(ii) Generate the rest of the related results in Table 4.1.

ANS. (e) (i) $\operatorname{Re}z^2 = x^2 - y^2$, $\operatorname{Im}z^2 = 2xy$

(ii) $\operatorname{Re}e^z = e^x \cos y$, $\operatorname{Im}e^z = e^x \sin y$

(iii) $\operatorname{Re}\dfrac{1}{z} = \dfrac{x}{x^2 + y^2}$, $\operatorname{Im}\dfrac{1}{z} = \dfrac{-y}{x^2 + y^2}$

(h) (i) $\omega_1 = i$, $\omega_2 = -i$

(ii) $\omega_1 = 1$, $\omega_2 = \left(-\dfrac{1}{2} + i\dfrac{\sqrt{3}}{2}\right)$, $\omega_3 = \left(-\dfrac{1}{2} - i\dfrac{\sqrt{3}}{2}\right)$

Section 1.2 Compatible Transforms

1.7. (a) Prove the result (1.9)

$$\mathcal{L}\left\{\frac{d^2u}{dx^2}\right\} = \lambda^2 U(\lambda) - \lambda u(0) - u'(0), \qquad U(\lambda) = \mathcal{L}\{u\}$$

with the conditions given on $u(x)$, $u'(x)$: $\lim_{x\to\infty} e^{-\lambda x}u(x) = 0$ and $\lim_{x\to\infty} e^{-\lambda x}u'(x) = 0$.

(b) Use integration by parts and the fact that $u(x)$ and du/dx vanish as $x \to \mp\infty$ to prove the special case of (1.13):

$$\mathcal{F}\left\{\frac{d^2u}{dx^2}\right\} = -\lambda^2 U(\lambda), \qquad U(\lambda) = \mathcal{F}\{u\}$$

(c) Use integration by parts and the fact that $u(x)$ and du/dx vanish as $x \to \infty$, to prove (1.17)

$$\mathcal{F}_c\left\{\frac{d^2u}{dx^2}\right\} = -\lambda^2 U_c(\lambda) - u'(0), \qquad U_c(\lambda) = \mathcal{F}_c\{u\}$$

1.8. Consider the following formula for the nth derivative of x^p, p integer ≥ 0.

$$\frac{d^n}{dx^n}x^p = \frac{p!}{(p-n)!}x^{p-n}, \qquad p = 0, 1, 2, 3, \ldots \tag{E.1}$$

The $p!$ definition can be extended, to p being any real number except for negative integers, via the definition of the *gamma function* (see (2.16)–(2.18) of Section 2.1.1 in Chapter 2 and Exercise 2.6 with Fig. 2.15) as

$$\Gamma(\nu + 1) = \int_0^\infty e^{-t} t^\nu \, dt, \qquad \nu \neq \text{negative integer} \tag{E.2}$$

with its basic properties being

(i) $\quad \nu\Gamma(\nu) = \Gamma(\nu + 1) \equiv \nu!, \ \nu > 0$ (E.3)

(ii) $\quad \Gamma(\nu)\Gamma(1 - \nu) = \pi/\sin \pi\nu, \ 0 < \nu < 1$ (E.4)

(a) Use property (E.3) for the gamma function to show that $\Gamma(p + 1) = p!, \ p = 0, 1, 2, 3, \ldots$.

(b) Use (E.2) to show that $0! = 1! = 1$.

(c) Use (E.4) to show that $\Gamma(\frac{1}{2}) = \sqrt{\pi}$.

(d) One of the many definitions of the *fractional* derivative $d^\alpha f / dx^\alpha$ of order α, $\alpha \neq$ negative integer, is given here for the case of $f(x) = x^\nu$, $\nu > 0$,

$$\frac{d^\alpha x^\nu}{dx^\alpha} = \frac{\Gamma(\nu + 1)}{\Gamma(\nu - \alpha + 1)} x^{\nu - \alpha}, \qquad \nu > 0, \ \alpha \neq -1, -2, \ldots$$
$$\tag{E.5}$$

Use (E.5) to show that the derivative of order half $d^{1/2}/dx^{1/2}$ of $f(x) = c = $ constant *is not zero*.

ANS. (d) $\quad \dfrac{d^{1/2} c}{dx^{1/2}} = \dfrac{c}{\sqrt{\pi x}}$

1.9. For the Bessel differential equation (1.21), with solution $J_n(r)$,

$$\frac{d^2 u}{dr^2} + \frac{1}{r}\frac{du}{dr} + \left(1 - \frac{n^2}{r^2}\right) u(r) = 0 \tag{E.1}$$

Show that $J_n(\lambda x)$ is a solution of

$$\frac{d^2 u}{dx^2} + \frac{1}{x}\frac{du}{dx} + \left(\lambda^2 - \frac{n^2}{x^2}\right) u(x) = 0 \tag{E.2}$$

1.10. (a) Consider the second-order differential equation

$$\frac{d^2 y}{dx^2} + B_1(x)\frac{dy}{dx} + B_2(x)y(x) = f(x) \tag{E.1}$$

Multiply by the integrating factor $r(x) = \exp[\int B_1(x)dx]$ to reduce it to the *self-adjoint* form

$$\frac{d}{dx}\left[r(x)\frac{dy}{dx}\right] - q(x)y(x) = F(x) \tag{E.2}$$

where $q(x) = -r(x)B_2(x)$ and $F(x) = r(x)f(x)$. See Example 1.8.

(b) Use results (E.1) and (E.2) to establish the self-adjoint form of the differential operator in (1.23) of the Bessel differential equation.

1.11. Consider the Fourier transform (1.11) of $u(x)$,

$$U(\lambda) = \int_{-\infty}^{\infty} e^{-i\lambda x}u(x)dx = \mathcal{F}\{u\} \tag{E.1}$$

and its inverse

$$u(x) = \frac{1}{2\pi}\int_{-\infty}^{\infty} e^{i\lambda x}U(\lambda)d\lambda = \mathcal{F}^{-1}\{U\} \tag{E.2}$$

(a) Find the Fourier transform of the gate function,

$$p_a(x) = \begin{cases} 1, & |x| < a \\ 0, & |x| > a \end{cases} \tag{E.3}$$

(See (1.31), (1.32).)

(b) The Dirac delta "function" $\delta(x)$ is defined through the following identity (see Section 2.4.1B in Chapter 2)

$$\int_{-\infty}^{\infty} f(x)\delta(x - x_0)dx = f(x_0) \tag{E.4}$$

This essentially says that $\delta(x - x_0)$ acts like an infinite pulse at $x = x_0$ and is zero elsewhere. Use the result in part (a) to find $\mathcal{F}\{1\}$.

(c) From (b) establish another possible definition for $\delta(x)$.

Hint: See the Fourier transform of $p_a(x)$ as $a \to \infty$ in part (a) and the result in part (b). Also see Section 2.4.1B of Chapter 2 on this and other possible definitions of the delta (impulse) function $\delta(x)$.

ANS. (a) $\dfrac{2\sin\lambda a}{\lambda}$

(b) $\lim\limits_{a\to\infty} \dfrac{2\sin a\lambda}{\lambda} = \mathcal{F}\{1\}$. This transform does not exist in our usual sense of a function but as $\delta(\lambda)$, a generalized (Dirac delta) function (see Sec. 2.4.1B).

(c) $\lim\limits_{a\to\infty} \dfrac{\sin ax}{\pi x} \equiv \delta(x)$.

Section 1.3 Classification of the Transforms

1.12. (a) Prove that the set $\{e^{(in\pi/a)x}\}_{n=-\infty}^{\infty}$ is orthogonal on the interval $(-a,a)$; that is,

$$\int_{-a}^{a} e^{in\pi x/a} e^{-im\pi x/a}\, dx = 0, \qquad n \neq m \tag{E.1}$$

(b) Prove that the set $\{\sin(n\pi x/a)\}_{n=1}^{\infty}$ is orthogonal on the interval $(0,a)$. See (1.66).

1.13. Consider the Fourier sine series of $f(x)$ on $(0,a)$

$$f(x) = \sum_{n=1}^{\infty} b_n \sin \frac{n\pi x}{x} \tag{E.1}$$

$$b_n = \frac{2}{a} \int_{0}^{a} f(x)\sin \frac{n\pi x}{a}\, dx \tag{E.2}$$

(a) Use the orthogonality property in Exercise 1.12b to derive expression (E.2) for the Fourier sine coefficients b_n.

Hint: Multiply both sides of (E.1) above by $\sin(k\pi x/a)$; then integrate from $x = 0$ to $x = a$, allowing the integration operation to enter inside the infinite sum on the right side of (E.1).

(b) Establish the Fourier sine series (E.1) for $f(x) = x$, $0 < x < \pi$. See Exercise 1.2c.

ANS. (b) $x = 2\sum\limits_{n=1}^{\infty} \dfrac{(-1)^{n+1}}{n} \sin nx$.

1.14. Consider the *discrete Fourier transform* $\tilde{U}(n/NT)$ and its inverse $\tilde{u}(kT)$,

$$\tilde{U}\left(\frac{n}{NT}\right) = \sum_{k=0}^{N-1} \tilde{u}(kT)e^{-2i\pi nk/N},$$

$$n = 0, 1, 2, \ldots, N-1, \quad i = \sqrt{-1} \tag{E.1}$$

$$\tilde{u}(kT) = \frac{1}{N}\sum_{n=0}^{N-1}\tilde{U}\left(\frac{n}{NT}\right)e^{2i\pi nk/N},$$

$$k = 0, 1, 2, \ldots, N-1 \tag{E.2}$$

(a) Prove the following *discrete orthogonality* of $\{e^{-2i\pi nk/N}\}_{n=1}^{N-1}$:

$$\sum_{k=0}^{N-1} e^{2i\pi lk/N}e^{-2i\pi nk/N} = \begin{cases} 0, & n \neq l \\ N, & n = l \end{cases} \tag{E.3}$$

Hint: Note that this series in (E.3) is a geometric series,

$$\sum_{k=0}^{N-1}\left[\exp\left(\frac{2i\pi(n-l)}{N}\right)\right]^k = \sum_{k=0}^{N-1}W^k = \frac{W^N - 1}{W - 1},$$

$$W \equiv \exp\left(\frac{2i\pi(n-l)}{N}\right) \neq 1$$

(b) Verify that $\tilde{u}(kT)$ of (E.2) is the inverse of the discrete transform $\tilde{U}(n/NT)$ of (E.1).

Hint: Substitute $\tilde{u}$ from (E.2) in (E.1), interchange the sums, and use the discrete orthogonality (E.3).

1.15. Consider the function $f(x) = 1$, $-a < x < a$,

(a) Find its exponential Fourier series

$$f(x) = \sum_{n=-\infty}^{\infty} c_n e^{-in\pi x/a}, \qquad -a < x < a \tag{E.1}$$

$$c_n = \frac{1}{2a}\int_{-a}^{a} f(x)e^{in\pi x/a}\,dx \tag{E.2}$$

(b) Use $p_a(kT)$ for $\tilde{u}(kT)$ in (E.1) of Exercise 1.14 to find its discrete Fourier transform $\tilde{U}(n/NT)$ (instead of c_n, the finite exponential Fourier transform). Use $a = 4$, 8, 16.

(c) Compare $\check{U}(n/NT)$, the *discrete* transform of part (b), to c_n of (a), the *finite transform* (E.2), and to the *finite limit* transform (*band-limited*) of $p_a(x)$ in Exercise 1.11a.

Section 1.4 Comments on the Inverse Transforms—Tables of the Transforms

1.16. (a) Verify that $u(x) = \cos x - \sin x$ is a solution to the (*Volterra*) *integral equation* (which is characterized by a variable limit of integration, here x)

$$\sin x = \int_0^x e^{x-t} u(t) \, dt$$

(b) Verify that $u(x) = 1$ is a solution to the (*Fredholm*) *integral equation* (which is characterized by constant limits of integration; here 0 and 1)

$$u(x) = \sin x + \int_0^1 (1 - x \cos xt) u(t) \, dt$$

(c) Solve the following Fredholm integral equation in $u(t)$:

$$\int_{-\infty}^\infty e^{i\lambda t} u(t) \, dt = \begin{cases} 1, & |\lambda| < a \\ 0, & |\lambda| > a \end{cases}$$

Hint: See Exercise 1.11a.

ANS. (c) $u(t) = \dfrac{\sin at}{\pi t}$

Section 1.5 The Compatible Transforms and the Adjoint Problem

1.17. (a) Establish that the following is a self-adjoint boundary value problem, as a special case of the Sturm–Liouville problem (1.63)–(1.65):

$$\frac{d^2 y}{dx^2} + \lambda^2 y = 0, \qquad 0 < x < a \tag{E.1}$$

$$y(0) = 0 \tag{E.2}$$

$$y(a) = 0 \tag{E.3}$$

(b) Find the set of solutions for the problem (E.1)–(E.3). Justify that the set is orthogonal on $(0, a)$.

ANS. (b) $\{y_n(x)\} = \left\{ \sin \dfrac{n\pi x}{a} \right\}$

1.18. (a) Let c_n and d_n, respectively, be the Fourier coefficients of $f(x)$ and $g(x)$ defined on (a,b) as in (1.67)–(1.68). Prove the following *Parseval's equality*:

$$\int_a^b \rho(x) f(x) \overline{g(x)}\, dx = \sum_{n=-\infty}^{\infty} c_n \overline{d_n} \|u_n\|^2$$

Hint: Write the Fourier series (1.67) for $f(x)$, multiply both sides by $\rho(x)\overline{g(x)}$, integrate from a to b, then use the definition of d_n as in (1.68).

(b) Use Parseval's equality in part (b) to evaluate $\sum_{n=1}^{\infty}(1/n^2)$.

Hint: See Exercise 1.13 and the result of b_n in part 1.13b.

(c) Evaluate $\sum_{n=1}^{\infty}(1/n^4)$.

Hint: See part (b) and a Fourier sine series for other functions like $f(x) = x^2$, $0 < x < \pi$.

ANS. (b) $\displaystyle\sum_{n=1}^{\infty} \dfrac{1}{n^2} = \dfrac{\pi^2}{6}$

(c) $\displaystyle\sum_{n_1}^{\infty} \dfrac{1}{n^4} = \dfrac{\pi^4}{90}$

1.19.* (a) Use the Lagrange identity (1.49) to show that L_3 in $L_3 u = u''' + u$ is not self-adjoint.

Hint: See (1.123).

Section 1.6 Constructing the Compatible Transforms for Self-Adjoint Problems—Second-Order Differential Equations

1.20.* Establish the result for the compatible kernels (supplying the necessary conditions on the kernel) of the two problems (E.7)–(E.9) and (E.10)–(E.12) at the end of Example 1.12.

---

*The rest of the exercises in the chapter are considered optional for a one-semester course in boundary value problems.

1.21.* Illustrate the adjoint problem (1.75)–(1.77) for determining the compatible kernel for the problem

$$\frac{d^2y}{dx^2} = f(x), \qquad 0 < x < a$$

$$y'(0) = 0$$

$$hy(a) + y'(a) = 0$$

ANS. $K(\lambda_n, x) = \cos \lambda_n x$, where λ_n are the zeros of

$$\tan \lambda_n a = \frac{h}{\lambda_n}, \qquad n = 1, 2, \ldots$$

Also see (3.107)–(3.111) and Example 3.11 of Section 3.2.

1.22.* Illustrate the adjoint problem (1.75)–(1.77) for determining the compatible kernel for the problem

$$\frac{1}{x} \frac{d}{dx} \left(x \frac{dy}{dx} \right) - \frac{n^2}{x^2} y = g(x), \qquad 0 < x < a$$

with boundary conditions

(a) $y(0)$ bounded, $y(a) = 0$

(b) $y(0)$ bounded, $hy(a) + y'(a) = 0$

ANS. (a) See (3.126)–(3.129) in Chapter 3.

(b) See (3.134)–(3.139) in Chapter 3.

Section 1.7 The nth-Order Differential Operator

1.23.* (a) Establish that the general L_3 of (1.116) is not self-adjoint.

Hint: See (1.123).

(b) Establish that the following even-order differential operator L_{2n} is self-adjoint:

$$L_{2n}u = a_0 u^{(2n)} + a_1 u^{(2n-1)} + \cdots + a_{2n-1}u' + a_{2n}(x)u$$

where all $a_0, a_1, \ldots, a_{2n-1}$ are constants and $a_{2n}(x)$ is the only variable coefficient.

2

INTEGRAL TRANSFORMS

In this chapter, we will develop the Laplace, Fourier (all the forms), and Hankel transforms as tools to solve boundary and initial value problems as well as for signal and systems analysis. Section 2.1 covers the Laplace transform and its applications to *initial value problems* associated with ordinary differential equations, integral equations, and partial differential equations. In Section 2.2, we cover the Fourier transform with applications in solving *boundary value problems* as well as in simplifying the analysis of time *signals and systems* in frequency (Fourier) space. All the basic theorems concerning the important properties of the Fourier (exponential or trigonometric) transforms are stated precisely and then illustrated. The most important of these results, such as the *Fourier integral theorem* and the *convolution theorem*, are proved in detail. Although these proofs are important and very instructive, we suggest their inclusion mainly for a graduate-level course on transform methods. Section 2.3 is dedicated to very illustrative applications of the Fourier exponential transform for boundary and initial value problems, including those associated with the wave equation, the heat equation, the Schrödinger equation, and the Laplace equation. Section 2.4 deals with the use and advantages of the Fourier transform in the representation of signals and linear systems analysis and especially in relation to band-limited signals. This is followed in Section 2.5 by introducing and applying the Fourier sine and cosine transforms, which are suitable for odd and even functions, respectively. Sections 2.6 and 2.7 deal, respectively, with the two-dimensional Fourier transform and the Hankel (Bessel) transform, along with their relations and practical applications. Again, as we mentioned above, the proofs of the basic theorems in these sections are suggested for a graduate course. Section 2.8[†] is dedicated to the relation between the Laplace transform and the Fourier transform $F(z)$ with complex variable argument $z = x + iy$, which is needed for a better hold on the derivation of the Laplace transform inversion formula. Sections 2.8.1 and 2.8.2 are not suggested as a requirement for an undergraduate course on transform methods and may even be considered optional in a graduate course for students without some preparation in functions of complex variables. Other transforms, like Hilbert's and Mellin's, are briefly introduced in the last section. In addition, we have a brief discussion of the *z-transform*, which is the discrete analog of the Laplace transform and which we will have occasion in Section 4.1.5B to apply to a number of relevant discrete initial value problems. The chapter is concluded by commenting on the most relevant references (mostly texts) for supporting the subject of integral transforms and their applications.

2.1   Laplace Transforms

As we have seen in Section 1.1, the Laplace transform

$$F(s) = \int_0^\infty e^{-sx} f(x) \, dx \tag{2.1}$$

of the function $f(x)$, $0 < x < \infty$ is compatible with differential operators of all orders with constant coefficients. Also, the limits of the integral, as it appears from transforming df/dx [see (1.8)],

$$\int_0^\infty e^{-sx} \frac{df}{dx} \, dx = sF(s) - f(0) \tag{2.2}$$

make it suitable for initial value problems where $f(0)$ is given.

From the definition of the Laplace transform (2.1), the transformed function $f(x)$ needs to be defined only for $x > 0$. In many applications of the Laplace transform, this is the case. For example, the current $i_c(t)$ in a circuit as a function of time is not there before the circuit is switched on at $t = 0$, that is,

$$i_c(t) = \begin{cases} 0, & t < 0 \\ f(t), & t \geq 0 \end{cases} \tag{2.3a}$$

[†]Optional section.

Such functions are called *causal*; they have values different from zero only after the cause, which is similar to the switching of the circuit at $t = 0$, where the circuit may then have a current $i(t) = f(t)$, $t \geq 0$. In the sequel, when working with the Laplace transform, we shall drop the subscript c in $i_c(t)$ or $f_c(t)$. If the switching is delayed by $\tau > 0$, then there is no current for $t - \tau < 0$ and we have a *delayed causal signal* $i_d(t)$

$$i_d(t) = \begin{cases} 0, & t < \tau \\ f(t), & t > \tau \end{cases} \tag{2.3b}$$

Sometimes we may need to speak of $i_s(t)$, the *causal signal* $i(t)$ of (2.3a) *shifted* by a time $t_1 > 0$, where

$$i_s(t) = \begin{cases} 0, & t < t_1 \\ f(t - t_1), & t > t_1 \end{cases} \tag{2.3c}$$

We illustrate a causal sine wave in Fig. 2.1a, a causal sine wave delayed by $\tau = \pi/3$ in Fig. 2.1b, and a causal sine wave shifted by $t_1 = \pi/4$ in Fig. 2.1c. To facilitate the representation of the above variations of causal functions, we define the *unit step function* $u_a(t)$ by

$$u_a(t) = \begin{cases} 0, & t < a \\ 1, & t > a \end{cases} \tag{2.4}$$

as illustrated in Fig. 2.1d, which is itself a causal function shifted by $t_1 = a$ with a constant value of 1 for $t > a$.

The Laplace transform (2.1) of the unit step function $u_a(t)$ in (2.4) is easily computed as

$$F(s) = \mathcal{L}\{u_a(t)\} = \int_0^\infty e^{-st} u_a(t) \, dt = \int_a^\infty e^{-st} \, dt$$

$$= \int_0^\infty e^{-s(a+\tau)} \, d\tau = e^{-sa} \int_0^\infty e^{-s\tau} \, d\tau = e^{-sa} \left. \frac{e^{s\tau}}{-s} \right|_{\tau=0}^\infty$$

$$= \frac{e^{-sa}}{s}, \qquad s > 0$$

after letting $t = a + \tau$ in the second integral.

Now, with the help of the unit step function $u_a(t)$ in (2.4) we can write the above expressions in (2.3a), (2.3b), and (2.3c) for the causal $i(t)$, the delayed $i_d(t)$, and the shifted $i_s(t)$ currents, respectively, as

$$i_c(t) = f(t)u_0(t) \tag{2.3a}$$

$$i_d(t) = f(t)u_\tau(t) \tag{2.3b}$$

$$i_s(t) = f(t - t_1)u_{t_1}(t) \tag{2.3c}$$

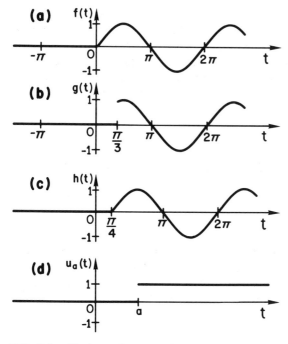

FIG. 2.1   Variety of causal functions.

(a)   A causal sine function

$$f(t) = \begin{cases} 0, & t > 0 \\ \sin t, & t > 0 \end{cases}$$

(b)   A *delayed* causal sine function (by $t_1 = \pi/3$):

$$g(t) = \begin{cases} 0, & t \leq \pi/3 \\ \sin t, & t > \pi/3 \end{cases}$$

(c)   A *shifted* causal sine function (by $t_1 = \frac{1}{4}$):

$$h(t) = \begin{cases} 0, & t < \pi/4 \\ \sin(t - \pi/4), & t > \pi/4 \end{cases}$$

(d)   The unit step function (causal):

$$u_a(t) = \begin{cases} 0, & t < a \\ 1, & t > a \end{cases}$$

(See also Figs. 4.23 and 4.24.)

A very useful function related to the unit step function (2.4) is the *signum function* sgn(t),

$$\text{sgn}(t) = \frac{t}{|t|} = \begin{cases} -1, & t < 0 \\ 1, & t > 0 \end{cases} \tag{2.5}$$

which is illustrated in Fig. 2.2a. As can be shown easily (we will need this as well in Chapter 4), both the unit step function $u_a(t)$ and the gate function $p_a(t)$ can be expressed in terms of the signum function as

$$u_o(t) = \tfrac{1}{2} + \tfrac{1}{2}\,\text{sgn}\,t,$$

$$u_a(t) = \tfrac{1}{2} + \tfrac{1}{2}\,\text{sgn}(t - a)$$

$$p_a(t) = \tfrac{1}{2}[\text{sgn}(t + a) - \text{sgn}(t - a)]$$

which are illustrated in Fig. 2.2b–d.

Clearly, these three basic functions are continuous on $(-\infty, \infty)$ except for a (finite) jump discontinuity at $t = 0$ and $t = a$ for $u_0(t)$ and $u_a(t)$, respectively, and two jump discontinuities at $t = \mp a$ for $p_a(t)$.

We note that the domain of the Laplace transform $F(s)$ in (2.1) requires that $f(x)$ assures the convergence of the integral. We will shortly present Theorem 2.1 for the precise conditions on a very reasonable and applicable class of functions $f(x)$ to guarantee the existence of their Laplace transform $F(s)$ in (2.1). Such a class of functions is described as (i) *sectionally continuous* on each bounded interval $0 < x < R$ and (ii) of *exponential order* as $x \to \infty$; that is, $|f(x)|$ does not grow faster than $Me^{\alpha x}$, where M and α are constants. Condition (ii) is often written in the asymptotic notation as $f(x) = O(e^{\alpha x})$, which means $\lim_{x \to \infty}(f(x)/e^{\alpha x}) = M$.

Definition $f(x)$ is called *sectionally (or piecewise) continuous* on an interval $a < x < b$ if this interval can be subdivided by a finite number of points $a \le x_0 \le x_1 \le x_2 \le \cdots \le x_n = b$ into n subintervals $x_{i-1} \le x \le x_i$, $i = 1, 2, 3, \ldots, n$),

(i) the function $f(x)$ is continuous on $x_{i-1} \le x \le x_i$, and
(ii) $f(x)$ approaches a finite limit as x approaches the limits of the subinterval, x_{i-1} and x_i, from the interior. $\qquad \square$

Figure 2.3 illustrates a function $f(x)$ which is sectionally continuous on the interval (a, b); that is, it is continuous on each of the open subintervals (a, x_1), (x_1, x_2) and (x_2, b). Note, for example, that the left- and right-hand

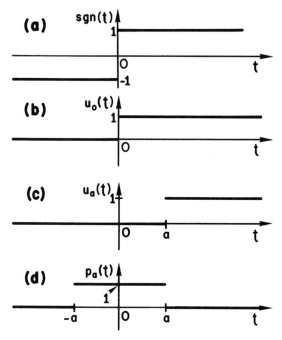

FIG. 2.2   Basic useful functions as combinations of the signum function.

(a)   The signum function

$$\text{sgn}(t) = \frac{t}{|t|} = \begin{cases} -1, & t < 0 \\ 1, & t > 0 \end{cases}$$

(b)   The unit step function $u_0(t) = \frac{1}{2} + \frac{1}{2}\,\text{sgn}\,t$.
(c)   The unit step function (shifted by $t_1 = a$) $u_a(t) = \frac{1}{2} + \frac{1}{2}\,\text{sgn}(t - a)$
(d)   The gate function

$$p_a(t) = \frac{1}{2}[\text{sgn}(t + a) - \text{sgn}(t - a)] = \begin{cases} 1, & |t| < a \\ 0, & |t| > a \end{cases}$$

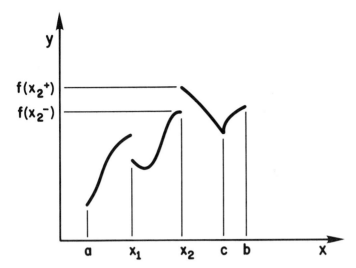

FIG. 2.3   A sectionally continuous function $f(x)$ on (a,b) with two jump discontinuities at x_1 and x_2.

limits $f(x_2-)$ and $f(x_2+)$, as x approaches x_2, are not equal, and we say that $f(x)$ has a *jump discontinuity* at x_2 of magnitude $J = f(x_2+)-f(x_2-)$.

In Figs. 2.1b,d and 2.2a,d, we note that the delayed causal sine function, the unit step function, the signum function, and the gate function are all examples of simple sectionally continuous functions.

For the theory of Fourier series and integrals, we shall need a more restricted class of functions than the above piecewise continuous functions, namely the piecewise smooth functions.

Definition   A piecewise continuous function in an interval $a < x < b$ is termed piecewise (or sectionally) smooth if, in addition to being piecewise continuous,

(1)   its first derivative df/dx is continuous on each of the subintervals $x_{i-1} < x < x_i$, $i = 1, 2, 3, \ldots, n$, and

(2) df/dx approaches a finite limit as x approaches the limits of the subinterval x_{i-1} and x_i from the interior; i.e., the right- and left-side limits $f'(x_i+)$, $f'(x_i-)$ exist at each limit of the subintervals x_i, $i = 1, 2, 3, \ldots, n$.   □

For example, the function in Fig. 2.3 is piecewise continuous on (a,b) but it is not piecewise smooth because of condition (2) above in the subinterval (c,b), where its derivative df/dx does not approach a limit $f'(c+)$ as x approaches the end point c from the right. For completeness, we may mention that the function is sectionally smooth on (a,c), and it is smooth on (a,x_1).

Clearly, all the functions in Figs. 2.1 and 2.2 are piecewise smooth on $(-\infty, \infty)$. Also, the truncated function $p_a(x)u(x)$ in Fig. 1.1c is piecewise smooth on $(-\infty, \infty)$. On the other hand, the familiar (not causal) sine function $u(t) = \sin t$ is smooth on $(-\infty, \infty)$; also $u(t) = \sin t$ in Fig. 2.1a is smooth on the semi-infinite interval $(0, \infty)$.

Definition $f(x)$ is termed of *exponential growth*, or of *exponential order*, as $x \to \infty$ if there exists a constant α such that $e^{-\alpha x}|f(x)|$ is bounded for all x greater than a finite number A. In other words, $|f(x)| \leq Me^{\alpha x}$ for $x > A$, with M, α constants, and we say $f(x)$ is $O(e^{\alpha x})$. For example, the function $f(x) = e^{3x} \sin x$ is of exponential order $O(e^{3x})$ since with $M = 1$, $|f(x)| = |e^{3x} \sin x| \leq e^{3x}$ for all x. However, functions like $f(x) = e^{3x^2}$ are not of exponential order. Polynomial functions like $f(x) = x^n$ are clearly of exponential order with $\alpha > 0$. □

We will now state and prove a very basic theorem on the existence of the Laplace transform $F(s)$ in (2.1) for the class of functions $f(x)$ which are sectionally continuous and of exponential order.

THEOREM 2.1: The Existence of Laplace Transforms

If $f(x)$ is

(i) sectionally continuous on the interval $0 \leq x \leq A$, and
(ii) of exponential order $e^{\alpha x}$, that is, $|f(x)| \leq Me^{\alpha x}$ for $x > A$,

then the Laplace transform $F(s)$ of $f(x)$ in (2.1) exists for $s > \alpha$. □

Proof:

$$F(s) = \int_0^\infty e^{-sx}f(x)\,dx = \int_0^A e^{-sx}f(x)\,dx + \int_A^\infty e^{-sx}f(x)\,dx \qquad (E.1)$$

The first integral on the finite interval $(0,A)$ clearly converges since $e^{-sx}f(x)$ is bounded for the sectionally continuous $f(x)$. For the convergence of the second integral, we will use the result of comparison of improper integrals, along with the exponential growth of $f(x)$, to show that it converges provided that $s > \alpha$.

$$\left| \int_A^\infty e^{-sx} f(x)\,dx \right| \leq \int_A^\infty |e^{-sx} f(x)|\,dx \leq \int_A^\infty e^{-sx} |f(x)|\,dx$$

$$\leq M \int_A^\infty e^{-(s-\alpha)x}\,dx < \infty \quad \text{for } s > \alpha \qquad (E.2)$$

The last improper integral clearly converges for $s > \alpha$, which concludes our proof. ∎

In this fashion, we have shown not only that the Laplace transform exists by proving that $\int_0^\infty e^{-sx} f(x)\,dx < \infty$ for $s > \alpha$ but also that $e^{-sx} f(x)$ is absolutely integrable, that is, $\int_0^\infty |e^{-sx} f(x)|\,dx < \infty$, for $s > \alpha$.

We must also note that the above two conditions (i) and (ii), that $f(x)$ be sectionally continuous on $(0,A)$ and of exponential growth as $x \to \infty$, are sufficient but not necessary. An example of $f(x)$ not sectionally continuous is $f(x) = 1/x^{1/2}$, which is infinite as $x \to 0$, but we can show (see the definition of the gamma function for $\nu = -\frac{1}{2}$ in (2.15) and Exercise 2.6b of this chapter) that $\mathcal{L}\{1/x^{1/2}\} = \sqrt{\pi/s}$, $s > 0$.

To be precise, the definition (2.1), or the results of the Laplace transform, should always be reported with the essential condition $s > \alpha$, which is what we notice when we consult the Laplace transform tables. Also, sometimes we may extend the definition of the Laplace transform $F(s)$ to complex numbers $\sigma = s + i\lambda$, where the condition for the existence of $F(\sigma)$ will read as $s = \text{real } \sigma > \alpha$. This extension will be needed when we try in Section 2.8 to derive the inverse Laplace transform formula (2.6) below with the aid of the Fourier transform of complex argument.

We must mention here that some references may use another form for the Laplace transform such as $\bar{F}(p) = p \int_0^\infty e^{-px} f(x)\,dx$, which is adopted to make the Laplace transform of $f(x) = 1$, $\bar{F}(p) = 1$, instead of $F(s) = 1/s$ as in (2.1). We shall adhere to (2.1) in this book.

In the following few sections, we will illustrate applications of the Laplace transform for solving initial value problems associated with ordinary differential equations, partial differential equations, and integral equations. In this section, we will rely entirely on tabulated Laplace transform pairs to transform back to the original function $f(x)$ of (2.1). At the same time we will develop a few of the most important pairs.

The basic integral transform tables include Roberts and Kaufman (1966) and Oberhettinger and Baddi (1974) for Laplace transforms, and Erdelyi et al. (1954), Vols. 1, 2, Ditkin and Prudnikov (1965), and Abramowitz and Stegun (1964) for the Laplace, Fourier, Hankel, and other transforms.

In Section 2.8[†] we will relate the Laplace transform to the Fourier transform and then derive *its inverse transform*,

$$f(x) = \mathcal{L}^{-1}\{F\} = \frac{1}{2\pi i} \lim_{L \to \infty} \int_{\gamma-iL}^{\gamma+iL} e^{zx} F(z)\,dz \qquad (2.6)$$

as a complex integral, where γ is greater than the real part of any singularity of $F(z)$. This development, which we consider optional at this stage, requires familiarity with complex integration, which will enable us to find the inverse Laplace transform $f(x)$ directly through (2.6).

A very important property of the Laplace transform is its *linearity property*, since the Laplace integration in (2.1) is a linear operation. This means that we can show easily for k, c_1, and c_2 constants that

(i) $\quad \mathcal{L}\{kf(x)\} = kF(s)$ \hfill (2.7)

or, more generally,

(ii) $\quad \mathcal{L}\{c_1 f_1(x) + c_2 f_2(x)\} = c_1 F_1(s) + c_2 F_2(s)$ \hfill (2.8)

(Clearly, (i) is a special case of (ii).) This is done by the simple application of the definition of the Laplace transform (2.1). The above result may be stated as the following theorem:

THEOREM 2.2: The Linear Property of the Laplace Transform

Let $f_1(x)$, $f_2(x)$ be of the class of functions in Theorem 2.1, i.e., sectionally continuous and of exponential order. Let $F_1(s)$, $F_2(s)$ be their respective Laplace transforms and k, c_1, c_2 be constants; then

(i) $\quad \mathcal{L}\{kf(x)\} = kF(s)$ \hfill (2.7)

(ii) $\quad \mathcal{L}\{c_1 f_1(x) + c_2 f_2(x)\} = c_1 F_1(s) + c_2 F_2(s)$ \hfill $\square$ (2.8)

We have the proof as Exercise 2.5a.

We may note that the second property above is no more than the distributive property of the (linear) integration operation over the addition of two functions $c_1 f_1(x)$ and $c_2 f_2(x)$. Since we know well that there is no such distributive property of integration over the product of two functions $f_1(x)f_2(x)$, it is no wonder that we do not rush into presenting a result in this line. Indeed, there is no such direct result. However, the Laplace transform is distributive over a new kind of product of $f_1(x)$ and $f_2(x)$

[†]Section 2.8 is optional for both undergraduate and graduate courses for students without preparation in functions of complex variables.

which is called the *convolution product* $(f_1 * f_2)(x)$ of $f_1(x)$ and $f_2(x)$ and is defined as follows.

Definition Let $f_1(x)$ and $f_2(x)$ be causal (vanish identically for $x < 0$) and defined on $(0, \infty)$; then their *Laplace convolution product* is defined as

$$(f_1 * f_2)(x) \equiv \int_0^x f_1(x - \xi) f_2(\xi) d\xi \tag{2.9}$$

It is easy to show that this convolution product is commutative, that is,

$$f_1 * f_2 = f_2 * f_1,$$

$$(f_1 * f_2)(x) = \int_0^x f_1(x - \xi) f_2(\xi) d\xi$$

$$= \int_0^x f_1(\eta) f_2(x - \eta) d\eta \equiv (f_2 * f_1)(x)$$

after letting $x - \xi = \eta$, and where we used the fact that $f_2(x)$ is a causal function where $f_2(x - \eta) \equiv 0$ for $\eta > x$.

For such a convolution product, we will have a result of the very important convolution theorem of Laplace transforms in Theorem 2.11, which essentially says that

$$\mathcal{L}\{(f_1 * f_2)(x)\} = F_1(s) F_2(s) \tag{2.10}$$

where $F_1(s)$ and $F_2(s)$ are, respectively, the Laplace transforms of $f_1(x)$ and $f_2(x)$.

2.1.1 Transform Pairs and Operations

First we will derive a few Laplace transforms to prepare the Laplace transform pairs that we will need for our illustrations.

$$\mathcal{L}\{e^{ax}\} = \int_0^\infty e^{-sx} e^{ax} \, dx = \frac{e^{-(s-a)x}}{-(s-a)} \bigg|_0^\infty = \frac{0-1}{-(s-a)} = \frac{1}{s-a}, \quad s > a$$

where $s > a$, since if $s \leq a$ the above integral will obviously diverge, i.e.,

$$\mathcal{L}\{e^{ax}\} = \frac{1}{s-a}, \quad s > a \tag{2.11}$$

In particular,

$$\mathcal{L}\{1\} = \frac{1}{s}, \quad s > 0 \tag{2.12}$$

To find the Laplace transform of $\sin ax$, take

$$\mathcal{L}\{\sin ax\} = \int_0^\infty e^{-sx} \sin ax\, dx = \int_0^\infty e^{-sx} \frac{[e^{iax} - e^{-iax}]}{2i}\, dx$$

$$= \frac{1}{2i}\left[\frac{1}{s-ia} - \frac{1}{s+ia}\right] = \frac{a}{s^2 + a^2}, \qquad s > 0$$

after using (2.8), (2.11) and noting that $s > 0$, since for $s = 0$ the integral does not converge,

$$\mathcal{L}\{\sin ax\} = \frac{a}{s^2 + a^2}, \qquad s > 0 \tag{2.13}$$

The case for $\cos ax$ is similar and gives

$$\mathcal{L}\{\cos ax\} = \frac{s}{s^2 + a^2}, \qquad s > 0 \tag{2.14}$$

since

$$\mathcal{L}\{\cos ax\} = \int_0^\infty e^{-sx} \cos ax\, dx = \frac{1}{2} \int_0^\infty e^{-sx}\left[e^{iax} + e^{-iax}\right] dx$$

$$= \frac{1}{2}\left[\frac{1}{s-ia} + \frac{1}{s+ia}\right] = \frac{s}{s^2 + a^2}, \qquad s > 0 \tag{2.14}$$

The Laplace transform for a combination of the trigonometric functions can be obtained with the help of the trigonometric identities. For example,

$$\mathcal{L}\{\cos^2 x\} = \mathcal{L}\left\{\frac{1 + \cos 2x}{2}\right\} = \frac{1}{2}[\mathcal{L}\{1\} + \mathcal{L}\{\cos 2x\}]$$

$$= \frac{1}{2}\left[\frac{1}{s} + \frac{s}{s^2 + 4}\right] = \frac{s^2 + 2}{s(s^2 + 4)}, \qquad s > 0$$

after using (2.12) and (2.14).

As seen here, the Laplace transforms of x^ν can easily be related to the *gamma function* after letting $sx = y$,

$$\mathcal{L}\{x^\nu\} = \int_0^\infty e^{-sx} x^\nu\, dx = \int_0^\infty e^{-y} \frac{y^\nu}{s^{\nu+1}}\, dy$$

$$= \frac{\Gamma(\nu + 1)}{s^{\nu+1}}, \qquad s > 0,\ \operatorname{Re}\nu > -1 \tag{2.15}$$

where $\Gamma(\nu + 1)$ is the gamma function, defined as

$$\Gamma(\nu + 1) = \int_0^\infty e^{-x} x^\nu\, dx, \qquad \nu \neq -1, -2\ldots \tag{2.16}$$

From (2.16) we can show, using integration by parts with $u = x^\nu$ and $dv = e^{-x}\,dx$, that

$$\Gamma(\nu + 1) = \nu\Gamma(\nu) \tag{2.17}$$

which makes it stand as the generalization of the *factorial function* $n!$; that is, when $\nu = n$, a nonnegative integer, $\Gamma(n + 1) \equiv n!$. We also find that $\Gamma(\frac{1}{2}) = \sqrt{\pi}$ from the relation

$$\Gamma(\nu)\Gamma(1 - \nu) = \frac{\pi}{\sin \pi\nu} \tag{2.18}$$

(which is proved in Exercise 2.58h). See Fig. 2.15, the graph of $\Gamma(\nu)$ in Exercise 2.6.

Next we may give a few very important operational pairs of Laplace transforms. In Example 1.2 of Chapter 1 we showed that

$$\mathcal{L}\left\{\frac{df}{dx}\right\} = sF(s) - f(0) \tag{2.2}$$

Now we state this result more precisely in the following theorem:

THEOREM 2.3: The Laplace Transform of Derivatives

Let $f(x)$ be a real function which is

(i) continuous for $x \geq 0$ and of exponential order $e^{\alpha x}$, and let
(ii) df/dx be sectionally (piecewise) continuous in every finite closed interval $0 \leq x \leq A$. Then $\mathcal{L}\{df/dx\}$ exists for $s > \alpha$ and (2.2) results,

$$\mathcal{L}\left\{\frac{df}{dx}\right\} = sF(s) - f(0), \qquad s > \alpha \qquad \Box \tag{2.2}$$

Proof: We will show that under conditions (i) and (ii) above the improper integral

$$\mathcal{L}\left\{\frac{df}{dx}\right\} = \lim_{A \to \infty} \int_0^A e^{-sx} f'(x)\,dx \tag{E.1}$$

exists, and that the limit is $sF(s) - f(0)$. Just as in the proof of Theorem 2.1, in any closed interval $0 \leq x \leq A$ the sectionally continuous df/dx has at most a finite number n of discontinuities, which we may assume to be at $x_1, x_2, \ldots, x_n$. Then we can write the integral in (E.1) as

$$\int_0^A e^{-sx} f'(x)\,dx = \int_0^{x_1} e^{-sx} f'(x)\,dx + \int_{x_1}^{x_2} e^{-sx} f'(x)\,dx + \cdots$$

$$+ \int_{x_n}^A e^{-sx} f'(x)\,ds \tag{E.2}$$

The integrand in each of the above n finite integrals is continuous; hence we can perform one integration by parts to have

$$\int_0^A e^{-sx} f'(x)\,dx = e^{-sx} f(x) \Big|_0^{x_1-} + s \int_0^{x_1} e^{-sx} f(x)\,dx$$

$$+ e^{-sx} f(x) \Big|_{x_1+}^{x_2-} + s \int_{x_1}^{x_2} e^{-sx} f(x)\,dx$$

$$+ \cdots + e^{-sx} f(x) \Big|_{x_n+}^{A-} + s \int_{x_n}^A e^{-sx} f(x)\,dx \qquad \text{(E.3)}$$

Since $f(x)$ is continuous, so is $e^{-sx} f(x)$, hence the terms outside the integrals in (E.3) will cancel out except for the term $-f(0)$ at $x = 0$ and $e^{-sA} f(A)$ at $x = A$. The integrals, on the other hand, will add to $s \int_0^A e^{-sx} f(x)\,dx$, and (E.3) becomes

$$\int_0^A e^{-sx} f'(x)\,dx = s \int_0^A e^{-sx} f(x)\,dx + e^{-sA} f(A) - f(0)$$

Since $f(x)$ is of exponential order $e^{\alpha x}$, the integral on the right will converge as $A \to \infty$ for $s > \alpha$ as we have shown in Theorem 2.1. Hence, we have

$$\lim_{A \to \infty} \int_0^A e^{-sx} f'(x)\,dx = \int_0^\infty e^{-sx} f'(x)\,dx$$

$$= s \int_0^\infty e^{-sx} f(x)\,dx - f(0) = sF(s) - f(0) \qquad \text{(2.2)}$$

since $\lim_{A \to \infty} e^{-sA} f(A) = 0$ for $s > \alpha$. ∎

If we apply the operation (2.2) again, we can derive the Laplace transform of $d^2 f/dx^2$, as in (1.9), as being

$$\mathcal{L}\left\{ \frac{d^2 f}{dx^2} \right\} = \mathcal{L}\left\{ \frac{d}{dx}\left(\frac{df}{dx} \right) \right\}$$

$$= s\mathcal{L}\left\{ \frac{df}{dx} \right\} - \frac{df}{dx}(0) = s[sF(s) - f(0)] - f'(0)$$

$$\mathcal{L}\left\{ \frac{d^2 f}{dx^2} \right\} = s^2 F(s) - sf(0) - f'(0) \qquad \text{(2.19)}$$

In these operations we must remember that we have applied the Laplace transform to $d^2 f/dx^2$ as well as df/dx. Thus, according to Theorem 2.3,

d^2f/dx^2 is assumed sectionally continuous on $(0,\infty)$ and of exponential order, to receive the existence of its Laplace transform in (2.19).

A more precise statement of this result and a concise proof are given here as a corollary to Theorem 2.3.

Corollary 1: If $f(x)$ is such that df/dx is continuous on $(0,\infty)$ and

(i) $\lim\limits_{x\to\infty} f(x)e^{-sx} = 0$ $\hspace{3cm}$ (E.1)

(ii) $\lim\limits_{x\to\infty} f'(x)e^{-sx} = 0$ $\hspace{3cm}$ (E.2)

then (2.19) follows:

$$\mathcal{L}\left\{\frac{d^2f}{dx^2}\right\} = s^2F(s) - sf(0) - f'(0) \hspace{2cm} \square \ (2.19)$$

Proof: The proof starts with letting $g(x) = f'(x)$, continuous on $(0,\infty)$, where from Theorem 2.3 and condition (E.2) we have (with $G(s) = \mathcal{L}\{g(x)\}$)

$$\mathcal{L}\{f''(x)\} = \mathcal{L}\{g'(x)\} = sG(s) - g(0) = sG(s) - f'(0) \hspace{1cm} (E.3)$$

Repeating the use of Theorem 2.3 on $g(x)$ with condition (E.1),

$$G(s) = \mathcal{L}\{g(x)\} = \mathcal{L}\{f'(x)\} = sF(s) - f(0) \hspace{1.5cm} (E.4)$$

If we use $G(s)$ from (E.4) in (E.3) we have (2.19):

$$\mathcal{L}\{f''(x)\} = s[sF(s) - f(0)] - f'(0) = s^2F(s) - sF(0) - f'(0) \hspace{0.5cm} \blacksquare \ (2.19)$$

From now on we shall denote $d^j f(x)/dx^j$, the jth-order derivative of $f(x)$, by $f^{(j)}(x)$.

The Laplace transform for $d^n f/dx^n$ can be developed in the same way as above to give

$$\mathcal{L}\left\{\frac{d^n f}{dx^n}\right\} = s^n F(s) - \sum_{k=0}^{n-1} s^k f^{(n-k-1)}(0) \hspace{2cm} (2.20)$$

under an appropriate set of assumptions on f and its first $n-1$ derivatives in the fashion of the above Corollary 1.

An accurate statement of the result (2.20) is presented next as the following Corollary 2 to Theorem 2.3. The proof, which parallels that of Corollary 1 for $n = 2$, will be left for the interested reader as an exercise.

Corollary 2: Let $f(x)$ be such that $d^{n-1}f(x)/dx^{n-1}$ is continuous on $(0,\infty)$;

(i) all its derivatives up to $n - 1$ are given at $x = 0$: $d^j f(0)/dx^j \equiv f^{(j)}(0)$ is given for all $j = 0, 1, 2, 3, \ldots, n - 1$; and

(ii) $\lim_{x \to \infty} e^{-sx} f^{(j)}(x) = 0$ for $j = 0, 1, 2, 3, \ldots, n - 1$.

Then (2.20) results. □

This means then that the Laplace transform algebraizes the differential operator $d^n f/dx^n$ of any order n, as we have shown for (1.10) of Example 1.2 in Chapter 1, provided that (i) all the derivatives of order up to $n - 1$ are continuous, and given at $x = 0$, and that all these derivatives should be of exponential growth $e^{\alpha x}$ as $x \to \infty$, that is, $\lim_{x \to \infty} e^{-sx} d^j f/dx^j = 0$ for $j = 0, 1, 2, 3, \ldots, n - 1, s > \alpha$.

EXAMPLE 2.1: Laplace Transforms of Differential Equations

(a) Find the Laplace transform of the differential equation

$$\frac{d^2 y}{dx^2} - 2 \frac{dy}{dx} - 8y = 0 \tag{E.1}$$

We let $Y(s)$ be the Laplace transform of $y(x)$ and use (2.2), the linearity property (2.8) of the Laplace transform, and (2.19) to obtain

$$s^2 Y(s) - sy(0) - y'(0) - 2[sY(s) - y(0)] - 8Y(s) = 0$$

$$= [s^2 - 2s - 8]Y(s) - (s - 2)y(0) - y'(0) = 0 \tag{E.2}$$

In Example 2.1, the Laplace transformation was applied to a differential equation with all its terms having constant coefficients. In this case, Laplace operation pairs (2.2) and (2.19) were used to algebraize the differential equation, and we termed the Laplace transform compatible with such a differential equation. On the other hand, when the differential equation has variable coefficient terms, even like the simple $x^n f(x)$, the Laplace transform would be more complicated. In this case, as we showed in Example 1.5, Chapter 1, equation (1.20),

$$\mathcal{L}\{x^n f(x)\} = (-1)^n \frac{d^n}{ds^n} F(s) \tag{2.21}, (1.20)$$

Before we give the detailed and justified proof of this result, we shall stop to give an illustration in the following example.

EXAMPLE 2.2:

(a) To find the Laplace transform of $f(x) = x^2 \sin ax$, we have from (2.13), $\mathcal{L}\{\sin ax\} = a/(s^2 + a^2)$; hence, according to the above result (2.21), we have

$$\mathcal{L}\{x^2 \sin ax\} = (-1)^2 \frac{d^2}{ds^2}\left(\frac{a}{s^2 + a^2}\right)$$

$$= \frac{d}{ds}\left[\frac{-2as}{(s^2 + a^2)^2}\right] = \frac{6as^2 - 2a^3}{(s^2 + a^2)^3}$$

(b) The Laplace transform of the following variable-coefficient differential equation in $y(x)$:

$$\frac{d^2y}{dx^2} - 2\frac{dy}{dx} - 8x^3y = 0, \qquad 0 < x < \infty \tag{E.1}$$

is a third-order differential equation in $Y(s) = \mathcal{L}\{y(x)\}$,

$$s^2Y(s) - sy(0) - y'(0) - 2[sY(s) - y(0)] - 8(-1)^3\frac{d^3Y}{ds^3} = 0$$

$$8\frac{d^3Y}{ds^3} + s(s-2)Y(s) - (s-2)y(0) - y'(0) = 0 \tag{E.2}$$

Here (E.1) is the same as the differential equation of Example 2.1, except for the third term having the variable coefficient $-8x^3$, where we applied operation (2.21) on it.

As in Example 1.5 of Chapter 1, to conclude the above result (2.21) we differentiate the Laplace integral in (2.1)

$$\frac{d^n}{ds^n}F(s) = \frac{d^n}{ds^n}\int_0^\infty e^{-sx}f(x)\,dx$$

$$= \int_0^\infty f(x)\frac{d^n}{ds^n}e^{-sx}\,dx = \int_0^\infty (-x)^n f(x)e^{-sx}\,dx$$

giving the desired result

$$\mathcal{L}\{x^n f(x)\} = (-1)^n\frac{d^n}{ds^n}F(s) \tag{2.21}$$

after allowing the interchange of differentiation with integration. As we mentioned then, such an interchange of the two operations is allowed (according to the following Theorem 2.5) provided that the resulting integral (after differentiating inside it) is uniformly convergent. This turns out to be the case (see Theorem 2.5 and Weinberger (1965, p. 324)) if we require $(-x)^n e^{-sx}f(x)$ to be absolutely integrable on $(0, \infty)$, that is, $\int_0^x |-x^n e^{-sx}f(x)|\,dx < \infty$. Indeed, this condition results in the transform $F(s)$ being continuously differentiable, which allows the computation of

$d^n F(s)/ds^n$. This is based on a theorem for the Laplace transform $F(\sigma)$ with complex argument $\sigma = s + i\lambda$:

THEOREM 2.4: Laplace Transform of Variable (Polynomial) Coefficient Terms

If $e^{-s_1 x} g(x)$ is absolutely integrable, its Laplace transform $(G(\sigma), \sigma = s + i\lambda))$ is analytic (indefinitely differentiable) for real $\sigma = s > s_1$, and $\lim_{\sigma \to \infty} G(\sigma) = 0$. □

Hence, this theorem can be applied for $g(x) = x^n f(x)$, and so its transform $(-1)^n \, d^n F(s)/ds^n$ is indefinitely differentiable; i.e., all the differentiation with respect to s, up to order n and even more, is allowed on the Laplace integral representing $F(s)$ above, thus allowing the differentiation to enter inside the integral to result in (2.21).

The above justification for the result (2.21) can now be stated as the following theorem for our class of sectionally continuous and of exponential order functions.

THEOREM 2.5: Laplace Transform of Polynomial Coefficient Terms (Sectionally Continuous and of Exponential Order)

Let $f(x)$ be in the class of functions as specified in Theorem 2.1; that is, it is sectionally continuous on the interval $0 \leq x \leq A$ and of exponential order $e^{\alpha x}$: $|f(x)| \leq M e^{\alpha x}$ for $x > A$. Then the result (2.21) is valid. □

Proof: In the proof of Theorem 2.1, we showed for the above class of functions that $\int_0^\infty e^{-sx} f(x) \, dx < \infty$, $s > \alpha$. Even more we showed that $e^{-sx} f(x)$ is absolutely integrable, that is, $\int_0^\infty |e^{-sx} f(x)| \, dx < \infty$. For the result of (2.21) we speak of the Laplace transform of $g(x) = x^n f(x)$, which is, of course, of the above class of functions since $f(x)$ is. Hence, according to Theorem 2.4, $(-1)^n \, d^n F(s)/ds^n$ is indefinitely differentiable. This means that the differentiation on $F(s)$ in the integral leading to (2.21) is valid, and (2.21) results. ∎

In Theorem 2.6 we state without proof more general conditions for the interchange of differentiation, with respect to a parameter inside the improper integral, and the integral. This encompasses two Theorems 2.4 and 2.5, and it will be very useful in the development of the Fourier integrals, in particular the rigorous proof of the Fourier integral Theorem 2.14 in Section 2.2.

THEOREM 2.6: General Theorem for Allowing Differentiation, with Respect to a Parameter, Inside an Improper Integral

Consider the improper integral with parameter y, $c \leq y \leq d$,

$$J(y) = \int_a^\infty g(x)h(x,y)\,dx \qquad (2.22)$$

If $g(x)$ is absolutely integrable on (a, ∞), and if the function $h(x,y)$ and its partial derivative $\partial h(x,y)/\partial y$ are continuous and bounded for $a \leq x < \infty$ and $c \leq y \leq d$, then the integral in (2.22) is a differentiable function of the parameter y for $c \leq y \leq d$ and differentiation with respect to y can be interchanged with the integration to have

$$\frac{dJ(y)}{dy} = \int_a^\infty g(x)\frac{\partial h}{\partial y}(x,y)\,dx, \qquad c \leq y \leq d \qquad \Box \ (2.23)$$

In justifying the interchange of operations, on which we shall insist in the next chapters, we group here most of the theorems in this direction, whose knowledge, we hope, will minimize the "handwaving" usually used in the *formal* derivations of results for transform analysis and other important mathematical analysis. Theorem 2.6 states the conditions for allowing the interchange of differentiation and (infinite) integration. Theorem 2.7 states the conditions for interchanging two integration operations,

$$\int_0^c q(x)\,dx = \int_0^c \left[\int_0^\infty f(x,\xi)\,d\xi\right] dx = \int_0^\infty \left[\int_0^c f(x,\xi)\,dx\right] d\xi \qquad (2.24)$$

Theorem 2.8, the *Weierstrass test* for uniform convergence of the above integral for $q(x)$, will complement Theorem 2.7 as a complete test for allowing the interchange of the integrals in (2.24). An important example is what we are going to need next to prove a very important theorem (the convolution Theorem 2.11) for the Laplace transform. There, the double integrations that result from multiplying two Laplace integrals must be interchanged to arrive at the desired result in (2.35). *Lebesgue convergence*, Theorem 2.10, gives the conditions for interchanging the limit process of a sequence $a_n(x)$ with the integration with respect to x operation over that sequence,

$$\lim_{n\to\infty} \int_{-\infty}^\infty a_n(x)\,dx = \int_{-\infty}^\infty \lim_{n\to\infty} a_n(x)\,dx = \int_{-\infty}^\infty a(x)\,dx \qquad \begin{matrix}(2.25)\\(2.34)\end{matrix}$$

for which we shall state the conditions preceding Eq. (2.34).

Theorem 3.5 of Chapter 3 states the conditions for allowing the interchange of the infinite summation with the integration (with generally finite limits a, b)

$$\int_a^b \sum_{k=0}^{\infty} c_k(x) \, dx = \sum_{k=0}^{\infty} \int_a^b c_k(x) \, dx \tag{2.26}$$

This theorem is of importance in the analysis of general infinite series in Section 3.1.2 and, in particular, the orthogonal (or Fourier) series expansion with $c_k(x) = c_k u_k(x)$ in Section 3.1.3. Here $\{c_k\}$ are the Fourier coefficients and $\{u_k(x)\}$ is the orthogonal set of functions on (a, b) as given in (1.67), (1.68) or (3.53), (3.54).

Of course, Theorem 3.5 may be considered as a special case (except for the usually finite limits of integration) of Theorem 2.10 when we look at it as

$$\lim_{n \to \infty} \int_a^b \sum_{k=0}^{n} c_k(x) \, dx = \lim_{n \to \infty} \sum_{k=0}^{n} \int_a^b c_k(x) \, dx = \sum_{k=0}^{\infty} \int_a^b c_n(x) \, dx$$

$$\tag{2.27}$$

with the sequence $a_n(x)$ being the above nth partial sum of the infinite series

$$a_n(x) = \sum_{k=0}^{n} c_k(x)$$

In this section we will present the *convolution theorem for the Laplace transform*, which gives the expression for the inverse Laplace transform of a product $F_1(s)F_2(s)$ of two Laplace transforms as the convolution product

$$(f_1 * f_2)(x) \equiv \int_0^x f_1(x - \xi) f_2(\xi) \, d\xi \tag{2.9}$$

of $f_1(x)$ and $f_2(x)$ as we defined it in (2.9). For the proof of this theorem we need to interchange a finite integral with an improper (infinite limit) integral in a double integral that results from multiplying the two Laplace integrals of $F_1(s)$ and $F_2(s)$. We shall state without proof two theorems or tests (Theorems 2.7, 2.8) that will spell out the conditions needed to allow the interchange of the two integrations in the resulting double integral (for the proofs see Churchill (1972)).

THEOREM 2.7: Conditions for Interchanging Double Integrals

If the integral

$$\int_0^{\infty} f(x, \xi) \, d\xi = q(x), \qquad 0 \le x \le c \tag{2.28}$$

is uniformly convergent with respect to x, that is, convergent for all $x \in [0,c]$, and its integrand is continuous by subregions over the rectangle $0 \le x \le c$, $0 \le \xi \le A$ for each positive A, then the integral represents a sectionally continuous function $q(x)$; also

$$\int_0^c q(x)\,dx = \int_0^c \int_0^\infty f(x,\xi)\,d\xi\,dx = \int_0^\infty \int_0^c f(x,\xi)\,dx\,d\xi \qquad \square \quad (2.29)$$

This theorem says that the uniform convergence in x of the integrand of the double integral in (2.29) is sufficient to allow the interchange of the finite integration over x with the other infinite (improper) integral over ξ.

What remains is a test for the uniform convergence of $q(x)$ in (2.28). This is given as the following Weierstrass test (see Churchill (1972), p. 41):

THEOREM 2.8: Weierstrass Test for Uniform Convergence of the Integral in (2.28)

Let $f(x,\xi)$ in (2.28) be continuous by subregions over the region $0 \le x \le c$, $0 \le \xi \le A$ for each positive constant A. If a sectionally continuous function $M(\xi)$, independent of x, exists such that

$$f(x,\xi) \le M(\xi) \qquad (0 \le x \le c,\ \xi > 0)$$

and such that $\int_0^\infty M(\xi)\,d\xi$ exists, then the integral in (2.28) is uniformly convergent with respect to x, where $0 \le x \le c$. $\qquad \square$

To illustrate this Weierstrass test we will consider an example of the application of Theorem 2.8, which has a result needed for the proof of the convolution Theorem 2.11.

EXAMPLE 2.3: Illustration of the Weierstrass Test (Theorem 2.8)

Let $f_1(x)$ and $f_2(x)$ be in our class of functions that are sectionally continuous and of exponential order $e^{\alpha x}$, that is,

$$|f_1(x)| \le N_1 e^{\alpha x}, \qquad |f_2(x)| \le N_2 e^{\alpha x}, \qquad x > x_0$$

We will show that the function

$$f(\eta,t) = f_1(\eta)e^{-st}f_2(t-\eta) \tag{E.1}$$

which we shall face in the proof of the convolution theorem (Theorem 2.11), satisfies the Weierstrass test

$$|f(\eta,t)| = |f_1(\eta)e^{-st}f_2(t-\eta)| = e^{-st}|f_1(\eta)||f_2(t-\eta)|$$

$$\le e^{-st}N_1 e^{\alpha\eta}N_2 e^{\alpha(t-\eta)} = N_1 N_2 e^{-(s-\alpha)t}, \qquad s > \alpha$$

$$|f(\eta,t)| \le N e^{-(s-\alpha)t} = M(t), \qquad s > \alpha,\ N_1 N_2 = N \tag{E.2}$$

Hence, $f(\eta, t)$ is bounded by the function $M(t) = Ne^{-(s-\alpha)t}$, which is independent of the first variable η. The second part of the Weierstrass test is to show that $M(t) = Ne^{-(s-\alpha)t}$ is integrable on $(0, \infty)$, which is very clear, since

$$\int_0^\infty M(t)\,dt = N\int_0^\infty e^{-(s-\alpha)t}\,dt = \frac{N}{s-\alpha}, \qquad s > \alpha \qquad \square \text{ (E.3)}$$

EXAMPLE 2.4: Reversing the Order of Integration for Improper Double (Iterated) Integral

In this example, we present results that will be needed in the derivation of the inverse Fourier transform (Theorem 2.14) in Section 2.2.

Let $f(x)$ be absolutely integrable on $(-\infty, \infty)$, that is,

$$\int_{-\infty}^{\infty} |f(x)|\,dx < \infty \tag{E.1}$$

We will show, using Theorems 2.7 and 2.8, that the order of integration in the double integral

$$\frac{1}{\pi}\int_0^A \int_{-\infty}^{\infty} f(t)\cos\lambda(t-x)\,dt\,d\lambda \tag{E.2}$$

can be reversed to

$$\frac{1}{\pi}\int_{-\infty}^{\infty}\int_0^A f(t)\cos\lambda(t-x)\,d\lambda\,dt = \frac{1}{\pi}\int_{-\infty}^{\infty} f(t)\int_0^A \cos\lambda(t-x)\,d\lambda\,dt \tag{E.3}$$

since the inner (improper) integral in (E.2) is clearly uniformly convergent. This is so as

$$\frac{1}{\pi}\left|\int_{-\infty}^{\infty} f(t)\cos\lambda(t-x)\,dt\right| \leq \frac{1}{\pi}\int_{-\infty}^{\infty} |f(t)||\cos\lambda(t-x)|\,dt$$

$$\leq \frac{1}{\pi}\int_{-\infty}^{\infty} |f(t)|\,dt < \infty \tag{E.4}$$

after using $|\cos\lambda(t-x)| \leq 1$ and (E.1). This says that the inner integral in (E.2) converges independently of the parameters x and λ. The interchange of the double integral as in (E.3) will enable us to evaluate the new inner (finite) integral in (E.3) as

$$\int_0^A \cos\lambda(t-x)\,d\lambda = \left.\frac{\sin\lambda(t-x)}{t-x}\right|_{\lambda=0}^{A} = \frac{\sin A(t-x)}{t-x} \tag{E.5}$$

for the double integral in (E.2) to be replaced by the single (improper) integral

$$\int_{-\infty}^{\infty} f(t) \frac{\sin A(t-x)}{\pi(t-x)} dt \qquad \text{(E.6)}$$

We note that the improper integral in Theorem 2.7 is on $(0, \infty)$, while the one in (E.2) is on $(-\infty, \infty)$. We leave it as a simple exercise to write the latter integral as the sum of two integrals on $(-\infty, 0)$ and $(0, \infty)$ for Theorems 2.7 and 2.8 to be applied to both.

For the analysis of the Fourier transform of functions defined on $(-\infty, \infty)$, we will need a theorem similar to Theorems 2.7 and 2.8 to allow the interchange of double integrals of functions $f(x, y)$ for $x, y \in (-\infty, \infty)$. This is *Fubini's theorem*, which we shall state here without proof.

THEOREM 2.9: Fubini's Theorem for the Interchange of Two
Infinite Integrals

If the double integral

$$\int_{-\infty}^{\infty} \int_{-\infty}^{\infty} f(x, y) dx \, dy \qquad (2.30)$$

is absolutely convergent, that is,

$$\int_{-\infty}^{\infty} \int_{-\infty}^{\infty} |f(x, y)| dx \, dy < \infty \qquad (2.31)$$

then the integral $\int_{-\infty}^{\infty} f(x, y) dy$ exists for almost all x (i.e., allowing its nonexistence at a set of points) and is an integrable function of x. Moreover, the double integral (2.30) is equal to its iterated integrals

$$\int_{-\infty}^{\infty} \int_{-\infty}^{\infty} f(x, y) dx \, dy = \int_{-\infty}^{\infty} \left[\int_{-\infty}^{\infty} f(x, y) dy \right] dx \qquad (2.32)$$

$$= \int_{-\infty}^{\infty} \left[\int_{-\infty}^{\infty} f(x, y) dx \right] dy \qquad \square \ (2.33)$$

Before developing the convolution theorem for the Laplace transform, we pause here to state without proof Theorem 2.10, which allows the interchange of the limit process of a sequence with the infinite integration of that sequence $a_n(x)$.

Theorem 2.10: Lebesgue Convergence Theorem for
Interchanging Limits and Integration

Let the sequence $f_n(x)$ be integrable on $(-\infty, \infty)$. If $|f_n(x)| \le F(x)$ almost everywhere on $(-\infty < x < \infty,\ n = 1, 2, 3, \ldots)$ for some integrable function $F(x)$, and if $\lim_{n \to \infty} f_n(x) = f(x)$, then $f(x)$ is integrable and

$$\lim_{n \to \infty} \int_{-\infty}^{\infty} f_n(x)\,dx = \int_{-\infty}^{\infty} \lim_{n \to \infty} f_n(x)\,dx = \int_{-\infty}^{\infty} f(x)\,dx \qquad \Box \ (2.34)$$

By having a condition or a result valid *almost everywhere* on an interval we mean that such condition may be violated only on a set of points of that interval, and we emphasize that it is not on a subinterval of this interval. Such a set which is clearly identically zero on the interval (a,b) except for such a set of points is called the *null set* $n(x)$, which satisfies $\int_a^b |n(x)|^2\,dx = 0$.

An example that illustrates the use of this theorem is found in the main part of the proof of Theorem 2.12 on the existence of the Fourier transform of absolutely integrable functions.

2.1.2 The Convolution Theorem for Laplace Transforms

Now we shall introduce the convolution theorem for Laplace transforms, along with its detailed (rigorous) proof, which depends on Theorems 2.7 and 2.8.

Theorem 2.11: The Convolution Theorem for Laplace
Transforms

Let $F_1(s)$ and $F_2(s)$ be, respectively, the Laplace transforms of $f_1(x)$ and $f_2(x)$, which are in the class of functions we considered in Theorem 2.1—by being sectionally continuous on each interval $0 \le x \le A$ and of exponential order $e^{\alpha x}$ as $x \to \infty$ (i.e., $|f(x)| \le Me^{\alpha x}$ for $x > A$). Then the Laplace transform of the convolution product $(f_1 * f_2)(x)$, as defined in (2.9), exists as $F_1(s)F_2(s)$ for $s > \alpha$, that is,

$$\mathcal{L}\{(f_1 * f_2)(x)\} = F_1(s)F_2(s), \qquad s > \alpha \qquad \Box \ (2.35)$$

Here we shall present a "formal" proof followed by a second "rigorous" (i.e., completely justifiable) proof. This is to illustrate the difference between the "formal" proof that we used in Chapter 1—which is somewhat insensitive to justifying some basic operations, in this case the interchange of two integrals—and the mathematically complete proof. In both proofs we will assume that the functions $f_1(x)$ and $f_2(x)$ are *causal*; that is, they vanish identically for $x < 0$.

1. "Formal" Proof:

$$\mathcal{L}\{(f_1 * f_2)(x)\} = \int_0^\infty e^{-sx} \int_0^x f_1(x - \xi) f_2(\xi) d\xi \, dx$$

$$= \int_0^\infty e^{-sx} \int_0^\infty f_1(x - \xi) f_2(\xi) d\xi \, dx \tag{E.1}$$

since $f_1(x - \xi)$ as a causal function vanishes identically for $\xi > x$, and hence the inner integral has no contribution for $x \leq \xi < \infty$. The above integral is considered as a double iterated integration, whence if we allow interchange of the integrations we obtain

$$\mathcal{L}\{(f_1 * f_2)(x)\} = \int_0^\infty \int_0^\infty e^{-sx} f_1(x - \xi) f_2(\xi) d\xi \, dx$$

$$= \int_0^\infty f_2(\xi) d\xi \int_0^\infty e^{-sx} f_1(x - \xi) \, dx$$

$$= \int_0^\infty f_2(\xi) d\xi \int_0^\infty e^{-s(t + \xi)} f_1(t) \, dt$$

$$= \int_0^\infty e^{-s\xi} f_2(\xi) d\xi \int_0^\infty e^{-st} f_1(t) \, dt$$

$$= F_1(s) F_2(s) \tag{E.2}$$

after letting $t = x - \xi$, $dt = dx$, and noting that $f_1(t) \equiv 0$ for $t < 0$, since both $f_1(t)$ and $f_2(t)$ are assumed causal functions. ∎

2. Rigorous (Complete) Proof: Here we write the product of the Laplace integrals for $F_1(s)F_2(s)$ and spend some time to justify the uniform convergence of the inner improper integral, which is necessary for allowing the interchange of this integral with the finite limit one

$$F_1(s)F_2(s) = \int_0^\infty e^{-s\eta} f_1(\eta) d\eta \int_0^\infty e^{-s\xi} f_2(\xi) d\xi$$

$$= \int_0^\infty f_1(\eta) \int_0^\infty e^{-s(\eta + \xi)} f_2(\xi) d\xi \, d\eta$$

$$= \int_0^\infty f_1(\eta) \left[\int_0^\infty e^{-st} f_2(t - \eta) dt \right] d\eta \tag{E.3}$$

after letting $\eta + \xi = t$. We now look at the inside integral, with respect to t, as the improper integral and we write the second integral as the limit as $A \to \infty$

$$F_1(s)F_2(s) = \lim_{A \to \infty} \int_0^A \int_0^\infty f_1(\eta)e^{-st}f_2(t - \eta)\,dt\,d\eta \qquad \text{(E.4)}$$

To exchange the order of integration, according to Theorems 2.7 and 2.8, the inner improper integral over t should be uniformly convergent. This means, according to the first part of the Weierstrass test (Theorem 2.8), that $f(\eta,t) = f_1(\eta)e^{-st}f_2(t - \eta)$ of (2.28) must be bounded by a function $M(t)$ independent of η. This, as we have done in detail in Example 2.3, can be shown for our class of functions of exponential order $(e^{\alpha\eta})$, that is, $|f_1(\eta)| \le N_1 e^{\alpha\eta}$, $f_2(\eta) \le N_2 e^{\alpha\eta}$, $\eta > \eta_0$:

$$|f_1(\eta)e^{-st}f_2(t - \eta)| \le N_1 N_2 e^{\alpha\eta}e^{-st}e^{\alpha(t-\eta)} = N_1 N_2 e^{-(s-\alpha)t}$$

$$= M(t) \qquad \text{(E.5)}$$

The second part of the Weierstrass test is to show that $M(t) = N_1 N_2 e^{-(s-\alpha)t}$ should be integrable on $(0, \infty)$, which is the case since the result of this integral is easily shown, for $s > \alpha$, to be $N_1 N_2/(s - \alpha)$.

Having shown that the inner integral in (E.4) is uniformly convergent, we now can interchange the order of integration to write

$$F_1(s)F_2(s) = \lim_{A \to \infty} \int_0^\infty e^{-st} \int_0^A f_1(\eta)f_2(t - \eta)\,d\eta\,dt$$

$$= \lim_{A \to \infty} \left[\int_0^A e^{-st} \int_0^A f_1(\eta)f_2(t - \eta)\,d\eta\,dt \right.$$

$$\left. + \int_A^\infty e^{-st} \int_0^A f_1(\eta)f_2(t - \eta)\,d\eta\,dt \right] \qquad \text{(E.6)}$$

Using the fact that $f_1(t)$ and $f_2(t)$ are of exponential order, we can show that the second integral is bounded by

$$\frac{N_1 N_2 A}{s - \alpha} e^{-(s-\alpha)A}, \qquad s > \alpha$$

which will approach zero as $A \to \infty$. The first double integral in (E.6) is done on the square $0 \le \eta \le A$, $0 \le t \le A$ as in Fig. 2.4, but as in the first proof the causal function $f_2(t - \eta)$ vanishes identically for $t < \eta$. Thus, the double integral is carried out over the triangle $0 \le t \le A$, $t > \eta$, as shown in the shaded area of Fig. 2.4,

$$\int_0^A e^{-st} \int_0^A f_1(\eta)f_2(t - \eta)\,d\eta\,dt = \int_0^A e^{-st} \int_0^t f_1(\eta)f_2(t - \eta)\,d\eta\,dt$$

$$\text{(E.7)}$$

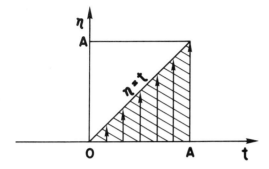

FIG. 2.4 The domain for the double integral in (E.7).

If we take the limit as in (E.6), we obtain our desired result

$$F_1(s)F_2(s) = \int_0^\infty e^{-st} \int_0^t f_1(\eta)f_2(t-\eta)\,d\eta\,dt$$

$$= \mathcal{L}\left\{\int_0^t f_1(\eta)f_2(t-\eta)\right\} = \mathcal{L}\left\{(f_1 * f_2)(t)\right\} \qquad (2.35)$$
$$\qquad (2.10)$$

Because there is no simple symmetry between the Laplace transform (2.1) and its inverse (2.6), we do not expect such a simple formula as the convolution product (2.9) in the s-space for the transform of the product $f_1(x)f_2(x)$. ■

EXAMPLE 2.5: Laplace Transform Inverse—Use of the
                 Convolution Theorem

Find

$$\mathcal{L}^{-1}\left\{\frac{1}{s(s^2+1)}\right\}$$

We have here a product of two Laplace transforms, $F_1(s) = 1/s$ and $F_2(s) = 1/(s^2 + 1)$, where according to (2.12) and (2.13) their corresponding inverse Laplace transforms are $f_1(x) = 1$ and $f_2(x) = \sin x$. Hence, according to (2.35) the result is the convolution product of 1 and $\sin x$, that is,

$$\mathcal{L}^{-1}\left\{\frac{1}{s(s^2+1)}\right\} = \int_0^x 1\sin(x-\xi)\,d\xi = \cos(x-\xi)\Big|_0^x = 1 - \cos x$$

$$\mathcal{L}^{-1}\left\{\frac{1}{s(s^2+1)}\right\} = 1 - \cos x \qquad (E.1)$$

This result can also be obtained in a simpler way by using partial fractions, where we can write

$$\frac{1}{s(s^2 + 1)} = \frac{1}{s} - \frac{s}{s^2 + 1} \tag{E.2}$$

as the difference between two known Laplace transforms; hence, according to (2.12) and (2.14), we have

$$\mathcal{L}^{-1}\left\{\frac{1}{s(s^2 + 1)}\right\} = \mathcal{L}^{-1}\left\{\frac{1}{s} - \frac{s}{s^2 + 1}\right\} = \mathcal{L}^{-1}\left\{\frac{1}{s}\right\} - \mathcal{L}^{-1}\left\{\frac{s}{s^2 + 1}\right\}$$

$$= 1 - \cos x \tag{E.3}$$

The most complete inverse Laplace transform tables of such rational functions are those of Roberts and Kaufman (1966) and Nixon (1965).

A. Other Properties and Laplace Transform Pairs

Now we present the *shifting property* of the Laplace transform

$$\mathcal{L}\{e^{-ax}f(x)\} = \int_0^\infty e^{-(s+a)x}f(x)\,dx = F(s + a), \qquad s + a > \gamma \tag{2.36}$$

where γ is such that $f(x)$ grows at most as $e^{\gamma x}$ as $x \to \infty$ (or, $f(x) = O(e^{\gamma x})$). Hence multiplying the function $f(x)$ by e^{-ax} means shifting the argument of $F(s)$ by $-a$. For example,

$$\mathcal{L}\{e^{-ax}\cos bx\} \frac{(s + a)}{(s + a)^2 + b^2}$$

The simple proof of (2.36) is left for an exercise.

There is a complementary result for a causal function $f(x - a)$ which vanishes identically for $-\infty < x < a$.

$$\mathcal{L}\left\{g(x) = \begin{cases} 0, & -\infty < x < a \\ f(x - a), & x > a \end{cases}\right\} = e^{-as}F(s) \tag{2.37}$$

Figure 2.1c illustrates the causal sine function,

$$f(x) = \begin{cases} \sin x, & x > 0 \\ 0, & x < 0 \end{cases}$$

shifted by $x = \pi/4$ to make

$$g(x) = \begin{cases} 0, & -\infty < x < \pi/4 \\ \sin\left(x - \dfrac{\pi}{4}\right), & x > \pi/4 \end{cases} \equiv u_{\pi/4}(x)\sin\left(x - \frac{\pi}{4}\right)$$

This result (2.37) can be shown easily by direct integration with the aid of the unit step function $u_a(x)$ in (2.4),

$$\mathcal{L}\{g(x)\} = \mathcal{L}\{u_a(x)f(x-a)\} = \int_0^\infty e^{-sx}u_a(x)f(x-a)dx$$

$$= \int_a^\infty e^{-sx}f(x-a)dx = \int_0^\infty e^{-s(\xi+a)}f(\xi)d\xi$$

$$= e^{-sa}\int_0^\infty e^{-s\xi}f(\xi)d\xi = e^{-as}F(s)$$

after letting $x = \xi + a$ in the third integral.

As an example, the Laplace transform of the above shifted causal sine function is $e^{-(\pi/4)s}/(1+s^2)$

$$\mathcal{L}\left\{\begin{matrix} 0, & -\infty < x < \dfrac{\pi}{4} \\ \sin\left(x - \dfrac{\pi}{4}\right), & \dfrac{\pi}{4} < x < \infty \end{matrix}\right\} = \frac{e^{-(\pi/4)s}}{1+s^2}, \qquad s > 0$$

after using (2.37) with $F(s) = \mathcal{L}\{f(x)\} = \mathcal{L}\{\sin x\} = 1/(1+s^2)$, $s > 0$.

Another simple example would be the Laplace transform of $u_a(t - t_0)$, the unit step function itself $u_a(t)$ shifted (delayed) by time t_0,

$$\mathcal{L}\{g(t)\} = \mathcal{L}\left\{\begin{matrix} 0, & -\infty < t < t_0 \\ u_a(t - t_0), & t_0 < t < \infty \end{matrix}\right\}$$

$$= e^{-t_0s}F(s) = e^{-t_0s}\frac{e^{-as}}{s} = \frac{e^{-(t_0+a)s}}{s}, \qquad s > 0$$

after using (2.37) with $f(t) = u_a(t)$, and, as we showed following its definition in (2.4), its Laplace transform $F(s) = e^{-as}/s$, $s > 0$.

A summary of the above Laplace transform pairs and a few other important ones is presented in Table 2.1, and further in Appendix C.1. In Table 2.1, $J_0(x)$ is the *Bessel function* of the first kind of order zero, and it is the bounded solution at $x = 0$ of the Bessel differential equation with $n = 0$,

$$x^2\frac{d^2y}{dx^2} + x\frac{dy}{dx} + (x^2 - n^2)y = 0 \tag{2.38}$$

Also, erf(x) is the *error function*, defined as

$$\text{erf}(x) = \frac{2}{\sqrt{\pi}}\int_0^x e^{-\xi^2}d\xi \tag{2.39}$$

TABLE 2.1 Laplace Transform Pairs

$f(x)$	$F(s) = \int_0^\infty e^{-sx} f(x) \, dx$
Pairs	
x^ν, $\mathrm{Re}\,\nu > -1$	$\dfrac{\Gamma(\nu + 1)}{s^{\nu+1}}$, $s > 0$
e^{ax}	$\dfrac{1}{s-a}$, $s > a$
$\sin ax$	$\dfrac{a}{s^2 + a^2}$
$\cos ax$	$\dfrac{s}{s^2 + a^2}$
$\mathrm{erfc}\left(\dfrac{a}{2\sqrt{x}}\right)$	$\dfrac{1}{s} e^{-a\sqrt{s}}$, $a > 0$
$\mathrm{erf}(\sqrt{x})$	$\dfrac{1}{s\sqrt{s+1}}$, $s > -1$
$J_0(ax)$	$\dfrac{1}{\sqrt{s^2 + a^2}}$
Operations	
$f_1(x) + f_2(x)$	$F_1(s) + F_2(s)$
$f(ax)$	$\dfrac{1}{a} F\left(\dfrac{s}{a}\right)$
$\dfrac{d^n f}{dx^n}$	$s^n F(s) - \displaystyle\sum_{k=0}^{n-1} s^k f^{(n-k-1)}(0)$
$\dfrac{\partial}{\partial y} f(x,y)$	$\dfrac{\partial}{\partial y} F(s,y)$
$x^n f(x)$	$(-1)^n \dfrac{d^n}{ds^n} F(s)$
$\displaystyle\int_0^x f(\xi)\,d\xi$	$\dfrac{1}{s} F(s)$
$e^{-ax} f(x)$	$F(s + a)$
$u_a(x) f(x - a)$	$e^{-as} F(s)$
$\displaystyle\int_0^x f_1(x - \xi) f_2(\xi)\,d\xi \equiv f_1 * f_2$	$F_1(s) F_2(s)$
$f(t)$, $f(t + T) = f(t)$, $t \geq 0$	$(1 - e^{-sT})^{-1} \displaystyle\int_0^T e^{-s\tau} f(\tau)\,d\tau$
$F(x)$	See the inversion formula (2.6)

where $\int_0^\infty e^{-\xi^2} d\xi = \sqrt{\pi}/2$, and erfc$(x)$ is the *complementary error function*,

$$\text{erfc}(x) = 1 - \text{erf}(x) = \frac{2}{\sqrt{\pi}} \int_x^\infty e^{-\xi^2} d\xi \qquad (2.40)$$

Also, the *gamma function* $\Gamma(\nu + 1)$ was defined in (2.16) as

$$\Gamma(\nu + 1) = \int_0^\infty e^{-x} x^\nu dx \qquad (2.16)$$

These and other special functions are discussed in Appendix A.5 and its exercises.

The Laplace transform for most of the entries in Table 2.1 can be done easily (and some of them with a bit of guidance), which we shall leave for exercises.

A very important Laplace transform operational pair is that of the integral of a function $f(x)$,

$$\mathcal{L}\left\{\int_0^x f(\xi) d\xi\right\} = \frac{F(s)}{s}, \qquad s > 0 \qquad (2.41)$$

This pair complements our very important Laplace transform pair of the derivative df/dx as given in Theorem 2.3,

$$\mathcal{L}\left\{\frac{df}{dx}\right\} = sF(s) - f(0) \qquad (2.2)$$

The result in (2.41) can be proved easily by letting $g(x) = \int_0^x f(\xi) d\xi$, with its Laplace transform $G(s)$, and clearly $g(0) = 0$. From the fundamental theorem of calculus we have $dg/dx = f(x)$, and if we use (2.2) above for $\mathcal{L}\{dg/dx\}$ we have

$$F(s) = \mathcal{L}\{f(x)\} = \mathcal{L}\left\{\frac{dg}{dx}\right\} = sG(s) - g(0) = sG(s)$$

$$G(s) = \frac{F(s)}{s} \qquad (2.41)$$

2.1.3 Solution of Initial Value Problems Associated with Ordinary and Partial Differential Equations

A. Ordinary Differential Equations

To illustrate the use of the Laplace transform in solving initial value problems associated with differential equations with constant coefficients, we consider the following example.

EXAMPLE 2.6: Initial Value Problem Associated with Homogeneous Differential Equations

$$\frac{d^2y}{dx^2} - 2\frac{dy}{dx} - 8y = 0 \qquad \text{(E.1)}$$

$$y(0) = 5 \qquad \text{(E.2)}$$

$$y'(0) = 10 \qquad \text{(E.3)}$$

We Laplace transform (E.1) with

$$Y(s) = \int_0^\infty e^{-sx} y(x)\,dx \qquad \text{(E.4)}$$

and use the pairs from (2.2) and (2.19), as we did in Example 2.1, to obtain

$$s^2 Y(s) - sy(0) - y'(0) - 2[sY(s) - y(0)] - 8Y(s) = 0$$

$$= s^2 Y(s) - 5s - 10 - 2sY(s) + 10 - 8Y(s) = 0$$

$$[s^2 - 2s - 8]Y(s) = 5s$$

$$Y(s) = \frac{5s}{s^2 - 2s - 8} \qquad \text{(E.5)}$$

So the solution to the problem (E.1)–(E.3) is the inverse Laplace transform of $Y(s)$ in (E.5). At this stage, we may consult the extensive Laplace transform tables available or merely simplify (E.5) by using partial fractions, whence

$$Y(s) = \frac{5s}{s^2 - 2s - 8} = \frac{5s}{(s - 4)(s + 2)} = \frac{10}{3}\frac{1}{s - 4} + \frac{5}{3}\frac{1}{s + 2}$$

Hence, the solution is

$$y(x) = \mathcal{L}^{-1}\{Y(s)\} = \frac{10}{3}\mathcal{L}^{-1}\left\{\frac{1}{s - 4}\right\} + \frac{5}{3}\mathcal{L}^{-1}\left\{\frac{1}{s + 2}\right\}$$

$$= \frac{10}{3}e^{4x} - \frac{5}{3}e^{-2x} \qquad \text{(E.6)}$$

after using the simple Laplace transform pair (2.11)

$$\mathcal{L}\{e^{ax}\} = \frac{1}{s - a} \qquad \text{(2.11)}$$

EXAMPLE 2.7: Initial Value Problem Associated with Nonhomogeneous (with Driving Force) Differential Equations

(a) Here we consider the displacement $x(t)$ of an oscillator with spring constant k and mass m under the influence of an external force $f(t)$ in the positive direction of $x(t)$, as shown in Fig. 2.5.

$$m\frac{d^2x}{dt^2} + kx = f(t), \qquad t > 0 \tag{E.1}$$

Let the initial displacement and velocity be x_0 and v_0, respectively:

$$x(0) = x_0 \tag{E.2}$$

$$v(0) = v_0 \tag{E.3}$$

We may divide (E.1) by m and let $\omega = \sqrt{k/m}$, the frequency of the system,

$$\frac{d^2y}{dx^2} + \omega^2 x = \frac{1}{m}f(t) = g(t), \qquad 0 < t < \infty \tag{E.4}$$

where now $g(t)$ is the *external force per unit mass* of the system. Usually it is assumed that the mass of the spring m_s is negligible compared to the mass m of the oscillating body. If we take the Laplace transform of equation (E.4), using the basic operation pair (2.19) for transforming d^2x/dt^2, and invoking the initial conditions (E.2), (E.3) that are needed for this transform in (2.19), we end up with an algebraic equation in $X(s) = \mathcal{L}\{x(t)\}$,

$$s^2 X(s) - sx_0 - v_0 + \omega^2 X(s) = G(s) \tag{E.5}$$

where $G(s) = \mathcal{L}\{g(t)\}$.

The solution to this simple equation is

$$X(s) = x_0 \frac{s}{s^2 + \omega^2} + v_0 \frac{1}{s^2 + \omega^2} + \frac{1}{s^2 + \omega^2} G(s) \tag{E.6}$$

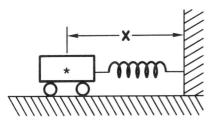

FIG. 2.5   Illustration for Example 2.7.

In trying to find $x(t)$, as the inverse Laplace transform of this $X(s)$, we note, remembering the Laplace transform pairs in (2.13) and (2.14), that the first two terms are easy to invert in terms of $\cos \omega t$ and $\sin \omega t$, respectively. However, the third term, as a product of a simple $1/(s^2 + \omega^2)$ and a general $G(s)$, will need the use of the result of the convolution Theorem 2.11 in (2.35), to obtain

$$\mathcal{L}^{-1}\left\{ \frac{1}{s^2 + \omega^2} G(s) \right\} = \left(\frac{1}{\omega} \sin \omega t \right) * g(t)$$

after using (2.13) for the inverse Laplace transform of $1/(s^2 + \omega^2)$. If we use these results, the inverse Laplace transform of (E.6) becomes

$$x(t) = x_0 \mathcal{L}^{-1}\left\{ \frac{s}{s^2 + \omega^2} \right\} + \frac{v_0}{\omega} \mathcal{L}^{-1}\left\{ \frac{\omega}{s^2 + \omega^2} \right\}$$

$$+ \frac{1}{\omega} \mathcal{L}^{-1}\left\{ \frac{\omega}{s^2 + \omega^2} \cdot G(s) \right\}$$

$$= x_0 \cos \omega t + \frac{v_0}{\omega} \sin \omega t + \frac{1}{\omega} \int_0^t \sin \omega(t - \tau) g(\tau) d\tau$$

$$= x_0 \cos \omega t + \frac{v_0}{\omega} \sin \omega t + \frac{1}{m\omega} \int_0^t \sin \omega(t - \tau) f(\tau) d\tau \qquad (E.7)$$

(b) If we take the very special case of a constant force $f(t) = f_0$, the convolution integral in (E.7) can be easily evaluated to give

$$f_0 \int_0^t \sin \omega(t - \tau) d\tau = \frac{f_0}{\omega} \cos \omega(t - \tau) \Big|_{\tau=0}^t = \frac{f_0}{\omega}(1 - \cos \omega t) \qquad (E.8)$$

$$x(t) = x_0 \cos \omega t + \frac{v_0}{\omega} \sin \omega t + \frac{f_0}{m\omega^2}(1 - \cos \omega t),$$

$$= x_0 \cos \omega t + \frac{v_0}{\omega} \sin \omega t + \frac{f_0}{k}(1 - \cos \omega t), \qquad (E.9)$$

(c) In the case that this constant force is turned off at time t_1—that is, it is applied only for $0 < t < t_1$—we can express this force $f_1(t)$ in terms of the shifted unit step (Heaviside's) function u_{t_1} as defined in (2.4) with its Laplace transform

$$\mathcal{L}\{u_{t_1}(t)\} = \frac{e^{-st_1}}{s}$$

$$f_1(t) = 1 - u_{t_1}(t) = \begin{cases} 1, & 0 < t < t_1 \\ 0, & t > t_1 \end{cases} \qquad (E.10)$$

Hence $G(s)$ for (E.7) is the Laplace transform of $(1/m)f_1(t)$, where

$$\mathcal{L}\{1 - u_{t_1}(t)\} = \frac{1}{s} - \frac{e^{-st_1}}{s} \tag{E.11}$$

and we have

$$G(s) = \frac{1}{m}\mathcal{L}\{1 - u_{t_1}(t)\} = \frac{1}{m}\left[\frac{1}{s} - \frac{e^{-st_1}}{s}\right] \tag{E.12}$$

To find the displacement $x(t)$ for this special case, we return to the first part of (E.7), where the first two terms stay the same while the third term now becomes

$$\frac{1}{m\omega}\mathcal{L}^{-1}\left\{\frac{\omega}{s^2 + \omega^2}\left[\frac{1}{s} - \frac{e^{-st_1}}{s}\right]\right\}$$

$$= \frac{1}{m\omega}\mathcal{L}^{-1}\left\{\frac{1}{s}\cdot\frac{\omega}{s^2 + \omega^2} - (e^{-st_1})\cdot\frac{1}{s}\frac{\omega}{s^2 + \omega^2}\right\}$$

$$= \frac{1}{m\omega}\left[\int_0^t 1\sin\omega\tau\,d\tau + u_{t_1}(t)\int_0^{t-t_1} 1\sin\omega\tau\,d\tau\right]$$

$$= \frac{1}{m\omega^2}(1 - \cos\omega t) - \frac{1}{m\omega^2}[1 - \cos\omega(t - t_1)]u_{t_1}(t) \tag{E.13}$$

after computing the above convolution integrals and then invoking the shifting pair (2.37), associated with the multiplication by e^{-st_1}, in the second term. If we use the result (E.13) for the third term in (E.7), we obtain the final solution $x(t)$, the displacement of the oscillator under the influence of a constant unit force that lasted for only t_1 seconds from the initial time $t_0 = 0$.

$$x(t) = x_0\cos\omega t + \frac{v_0}{\omega}\sin\omega t + \frac{1}{k}\{(1 - \cos\omega t)u_0(t)$$

$$- [1 - \cos\omega(t - t_1)]u_{t_1}(t)\} \tag{E.14}$$

where in the last part we remember that $m\omega^2 = mk/m = k$.

EXAMPLE 2.8: System of Linear (Coupled) Differential Equations

In this example we apply the Laplace transform to solve the following system of coupled (simultaneous) first-order differential equations in $x(t)$ and $y(t)$:

$$\frac{dx}{dt} = 1 - 2y, \qquad t > 0 \tag{E.1}$$

$$\frac{dy}{dt} = \frac{1}{2}x - \frac{1}{2}t, \qquad t > 0 \tag{E.2}$$

$$x(0) = 0 \tag{E.3}$$

$$y(0) = 1 \tag{E.4}$$

Let $X(s)$ and $Y(s)$ be the Laplace transforms of $x(t)$ and $y(t)$, respectively. If we apply the Laplace transform on both (E.1) and (E.2), using its operational property (2.2) on the first-order derivatives and invoking its needed initial conditions in (E.3), (E.4), we have

$$sX(s) - x(0) = \frac{1}{s} - 2Y(s)$$

$$sX(s) = \frac{1}{s} - 2Y(s)$$

$$sX(s) + 2Y(s) = \frac{1}{s} \tag{E.5}$$

and

$$sY(s) - y(0) = \frac{1}{2}X(s) - \frac{1}{2s^2}$$

$$sY(s) - \frac{1}{2}X(s) = 1 - \frac{1}{2s^2} \tag{E.6}$$

If we solve for $X(s)$ and $Y(s)$ in the two simultaneous algebraic equations (E.5), (E.6) by multiplying (E.5) by $\frac{1}{2}$ and (E.6) by s, we have

$$\frac{1}{2}sX(s) + Y(s) = \frac{1}{2s} \tag{E.7}$$

$$s^2Y(s) - \frac{1}{2}sX(s) = s - \frac{1}{2s} \tag{E.8}$$

We now add (E.7) and (E.8) to eliminate $X(s)$,

$$(s^2 + 1)Y(s) = s, \qquad Y(s) = \frac{s}{s^2 + 1} \tag{E.9}$$

The inverse Laplace transform of this $Y(s)$, according to the Laplace pair (2.14), is

$$y(t) = \cos t \tag{E.10}$$

To find $x(t)$, we either solve for $X(s)$ in the same way and then find its inverse Laplace transform, or more simply substitute $y(t) = \cos t$ in the original differential equation (E.2) to obtain $x(t)$ directly,

$$x(t) = 2\frac{dy}{dt} + t = -2\sin t + t = t - 2\sin t$$

$$x(t) = t - 2\sin t \tag{E.11}$$

It should be easy to verify that the pair $x(t) = \cos t$ and $y(t) = t - 2\sin t$ constitute the solution to the coupled system of differential equations and their initial conditions in (E.1)–(E.4).

EXAMPLE 2.9:   Some Special Variable-Coefficient Differential
Equations

In Example 2.2 we used the following differential equation with a variable coefficient like x^3 in its third term:

$$\frac{d^2y}{dx^2} - 2\frac{dy}{dx} - 8x^3y = 0, \qquad 0 < x < \infty \tag{E.1}$$

to warn against, in general, using the Laplace transform for solving differential equations with variable coefficients. This is the case here, since according to the Laplace transform pair (2.21), this differential equation in $y(x)$ is transformed to a (higher) third-order differential equation (with variable coefficients in $Y(s)$)

$$8\frac{d^3Y}{ds^3} + s(s-2)Y(s) - (s-2)y(0) - y'(0) = 0 \tag{E.2}$$

instead of what we usually expect, an algebraic equation in $Y(s)$ after transforming constant-coefficient differential equations in $y(x)$. Of course, if the variable coefficient was x instead of x^3, the resulting equation in $Y(s)$ would be a first-order one and might not be as bad as the third-order equation with variable coefficients in (E.2). However, there are sometimes instances in which two such factors—the order and the variable coefficients in the differential equation—may, under the Laplace transformation, balance to some advantage. The following example illustrates such a point. We consider the second-order differential equation in $y(x)$ with no variable coefficients other than just x,

$$x\frac{d^2y}{dx^2} + \frac{dy}{dx} + xy = 0, \qquad 0 < x < \infty \tag{E.1}$$

with the needed initial conditions (for Laplace transform application) on $y(0)$ and $y'(0)$, for example,

$$y(0) = 1 \tag{E.2}$$

$$y'(0) = 0 \tag{E.3}$$

We note, after multiplying (E.1) by x, that the equation is that of the J_0 Bessel function in (2.38) with $n = 0$. So, if in Laplace transforming (E.1) we have struck a balance of a simpler (more familiar) differential equation in $Y(s)$ than that of the Bessel differential equation in $y(x) = J_0(x)$ of (E.1), then we have succeeded in deriving $Y(s)$ as the Laplace transform of the Bessel function $J_0(x)$.

If we Laplace transform (E.1), using (2.19), (2.2), and (2.21) for the Laplace transform of d^2y/dx^2, dy/dx, and covering the variable coefficient x, respectively, we have

$$-\frac{d}{ds}[s^2 Y(s) - sy(0) - y'(0)] + sY(s) - y(0) - \frac{d}{ds} Y(s) = 0 \qquad (E.4)$$

and if we use the initial conditions (E.2), (E.3) we have

$$-\frac{d}{ds}[s^2 Y(s) - s - 0] + sY(s) - 1 - \frac{dY}{ds} = 0$$

$$-s^2 \frac{dY}{ds} - 2sY(s) + 1 + sY(s) - 1 - \frac{dY}{ds} = 0$$

$$(s^2 + 1)\frac{dY}{ds} + sY(s) = 0 \qquad (E.5)$$

This is definitely a simpler (*separable*) *first-order homogeneous* differential equation in $Y(s)$ than the Bessel differential equation (E.1) in $y(x)$,

$$\int \frac{dY}{Y} = -\int \frac{s}{s^2 + 1}\, ds$$

$$\ln Y = -\tfrac{1}{2} \ln(s^2 + 1) + C$$

$$Y(s) = \frac{C}{\sqrt{s^2 + 1}} \qquad (E.6)$$

To find C we need a condition on $Y(s)$. But if $Y(s) = \mathcal{L}\{J_0(t)\}$, then we may look at $Y(s)$ in (E.6) as the Laplace transform of $J_0(t)$ within an arbitrary amplitude C,

$$y(t) = CJ_0(t)$$

To find C, we have in (E.2) that $y(0) = 1$, and we have from Appendix A.5, or simply from the series that defines $J_0(t)$ in (A.4.15), that $J_0(0) = 1$. Hence $y(0) = 1 = CJ_0(1) = 1$, $C = 1$. Thus we have derived the Laplace transform of the Bessel function $J_0(t)$ by use of the Laplace transform,

$$\mathcal{L}\{J_0(t)\} = \frac{1}{\sqrt{s^2 + 1}}$$

a Laplace pair, which is not as easy to get, as in the case of the trigonometric functions in (2.13) and (2.14), as we can see for other methods in

Exercise 2.11b. See Exercise 2.11c for a different way to find the above arbitrary constant C.

B. Partial Differential Equations

Here we will present the *diffusion equation in a semi-infinite domain* which describes the temperature distribution $u(x,t)$ in a semi-infinite bar ($0 < x < \infty$) where the end $x = 0$ is kept at constant temperature u_0 and the initial temperature of the bar is zero.

$$\frac{\partial u}{\partial t} = \frac{\partial^2 u}{\partial x^2}, \qquad 0 < x < \infty, \; 0 < t < \infty \tag{2.42}$$

$$u(0,t) = u_0, \qquad\qquad 0 < t < \infty \tag{2.43}$$

$$u(x,0) = 0, \qquad\qquad 0 < x < \infty \tag{2.44}$$

The diffusion or heat equation (2.42) is derived in detail in Appendix B.2 (see Fig. B.3).

In this problem, we will apply the Laplace transform to the derivative $\partial u/\partial t$, since we are given the initial condition $u(x,0) = 0$ necessary for the transform pair (2.2) of the first derivative in t but not the condition $u'(0,t)$ necessary for the transform pair (2.19) of the second derivative $\partial^2 u/\partial x^2$ in x.

Let

$$U(x,\mu) = \int_0^\infty u(x,t)e^{-\mu t}\,dt \tag{2.45}$$

and Laplace transform (2.42). By consulting the Laplace transform pair (2.2), we obtain

$$\mu U(x,\mu) - u(x,0) = \frac{\partial^2 U(x,\mu)}{\partial x^2} \tag{2.46}$$

But $u(x,0) = 0$; hence,

$$\frac{\partial^2 U}{\partial x^2} - \mu U(x,\mu) = 0$$

With μ as a parameter the partial derivative $\partial^2 U(x,\mu)/\partial x^2$ is sometimes written as a total derivative $d^2 U(x,\mu)/dx^2$ since the emphasis at this stage is on the remaining differential operation with respect to x. In this book we will use both notations.

We now have a second-order differential equation in $U(x,\mu)$ whose bounded solution as $x \to \infty$ is

$$U(x,\mu) = A(\mu)e^{-x\sqrt{\mu}} \tag{2.47}$$

To find $A(\mu)$ we use the boundary condition $U(0,\mu)$, which is the Laplace transform of $u(0,t)$, to obtain

$$U(0,\mu) = \int_0^\infty u(0,t)e^{-\mu t}\,dt = \int_0^\infty u_0 e^{-\mu t}\,dt = \frac{u_0}{\mu} = A(\mu)$$

With this result, the transform in (2.47) becomes

$$U(x,\mu) = \frac{u_0}{\mu}e^{-x\sqrt{\mu}} \tag{2.48}$$

To arrive at the solution $u(x,t)$ as the inverse Laplace transform, we need the Laplace transform pair (see Table 2.1)

$$\mathcal{L}\left\{\operatorname{erfc}\left(\frac{a}{2\sqrt{t}}\right)\right\} = \frac{e^{-a\sqrt{\mu}}}{\mu} \tag{2.49}$$

Hence, the solution becomes

$$u(x,t) = u_0 \operatorname{erfc}\frac{x}{2\sqrt{t}} \tag{2.50}$$

EXAMPLE 2.10: The Diffusion Equation in a Finite Rod

Consider the initial value problem for the temperature distribution in a finite rod of length π, where both ends are kept at zero temperature and the initial temperature is given by $u(x,0) = f(x)$, $0 < x < \pi$,

$$\frac{\partial u}{\partial t} = \frac{\partial^2 u}{\partial x^2}, \qquad 0 < x < \pi,\ t > 0 \tag{E.1}$$

$$u(0,t) = 0, \qquad t > 0 \tag{E.2}$$

$$u(\pi,t) = 0, \qquad t > 0 \tag{E.3}$$

$$u(x,0) = f(x), \qquad 0 < x < \pi \tag{E.4}$$

Clearly, the Laplace transform is compatible with $\partial u/\partial t$ of (E.1) and its initial condition (E.4). Let $U(x,s) = \mathcal{L}\{u(x,t)\}$; the Laplace transform of the heat equation in (E.1) is

$$sU(x,s) - u(x,0) = \frac{d^2 U(x,s)}{dx^2},^\dagger \qquad 0 < x < \pi \tag{E.5}$$

$$\frac{d^2 U(x,s)}{dx^2} - sU(x,s) = -f(x), \qquad 0 < x < \pi \tag{E.6}$$

---

†The notation for the derivative of the transform varies somewhat among the major references on transform methods. This ranges, for example, from $d^2 U(x,s)/dx^2$, to emphasize only the remain-

To solve this nonhomogeneous differential equation in $U(x,s)$, we usually resort to the method of variation of parameters as discussed in Appendix A.2, which also leads to the method of the Green function. However in the spirit of solving differential equations by the transform (or operational) methods of this book, we note that the finite Fourier sine transform of Section 3.2 is compatible with d^2U/dx^2 of equation (E.6) and its given (vanishing) boundary conditions as the Laplace transforms of (E.2) and (E.3). Of course, we could also apply the finite sine transform to the heat equation (E.1) in $u(x,t)$ to end up with a simpler first-order differential equation for its variation in time. This attempt is the subject of Example 3.10 in Section 3.2. The complete transform method of applying the Laplace transform with its result in (E.6), followed by a finite sine transform (or in the other order), is the subject of Exercise 3.20b in Section 3.2. Exercise 3.21b of Section 3.2 deals with another example which illustrates the possible preference for doing just one of these transforms over the other.

2.1.4 Applications to Volterra Integral Equations[†] with Difference Kernels

What we mean here by a *difference kernel* is that the kernel $K(x,\xi)$ depends on the difference $x-\xi$ only; that is, $K(x,\xi) = K(x-\xi)$. A Volterra integral equation of the second kind with difference kernel $K(x-\xi)$,

$$u(x) = f(x) + \lambda \int_0^x K(x-\xi)u(\xi)d\xi \qquad (2.51)$$

contains an integral in the form of a Laplace convolution product (2.9),

$$\int_0^x K(x-\xi)u(\xi)d\xi = K * u \qquad (2.52)$$

Hence, a Volterra equation with difference kernel (2.51) yields a solution by the Laplace method; in other words, the Laplace transform is compatible with such Volterra integral equations with a difference kernel. So, if we Laplace transform (2.51) letting $U(s)$, $F(s)$, and $\kappa(s)$ be the Laplace

---

ing differentiation with respect to x (considering s as a "constant" parameter), to $\partial^2 U(x,s)/\partial x^2$. We shall take the liberty of using either notation.

[†] A detailed "Introduction to Integral Equations and Their Applications" for undergraduates is found in Jerri (1985). See also the author's brief exposition of integral equations in Jerri (1986), which is a self-contained introduction to integral equations for sophomores.

transforms of $u(x)$, $f(x)$, and $K(x)$, respectively, and realize from the convolution theorem (2.35) that $\mathcal{L}\{K * u\} = \kappa U$, we obtain

$$U(s) = F(s) + \lambda\kappa(s)U(s) \tag{2.53}$$

$$U(s) = \frac{F(s)}{1 - \lambda\kappa(s)}, \qquad \lambda\kappa \neq 1 \tag{2.54}$$

The solution $u(x)$ of (2.51) is then the inverse Laplace transform of (2.54) or

$$u(x) = \mathcal{L}^{-1}\left\{\frac{F(s)}{1 - \lambda\kappa(s)}\right\}, \qquad \lambda\kappa(s) \neq 1 \tag{2.55}$$

This can be evaluated with the aid of Laplace transform pairs (Table 2.1), as we will illustrate in the following example.

EXAMPLE 2.11: Volterra Integral Equations of the Second Kind

(a) Use the Laplace transform to solve the Volterra equation

$$u(x) = f(x) + \lambda \int_0^x e^{x-t} u(t)\, dt \tag{E.1}$$

The kernel $K(x, t) = e^{x-t}$ is a difference kernel; hence, if we Laplace transform (E.1) using $\kappa(s) = \mathcal{L}\{K(x)\} = \mathcal{L}\{e^x\} = 1/(s - 1)$ in (2.54), we obtain

$$U(s) = \frac{F(s)}{1 - \lambda/(s - 1)} = F(s) \cdot \frac{s - 1}{s - 1 - \lambda}$$

$$= F(s)\left(\frac{s - 1 - \lambda + \lambda}{s - 1 - \lambda}\right) = F(s) + \frac{\lambda F(s)}{s - 1 - \lambda} \tag{E.2}$$

So the solution $u(x)$ of (E.1) is the inverse Laplace transform of (E.2); that is,

$$u(x) = \mathcal{L}^{-1}\left\{F(s) + \lambda\frac{F(s)}{s - (\lambda + 1)}\right\}$$

$$= \mathcal{L}^{-1}\{F(s)\} + \lambda\mathcal{L}^{-1}\left\{\frac{1}{s - (\lambda + 1)} \cdot F(s)\right\}$$

$$= f(x) + \lambda\mathcal{L}^{-1}\left\{\frac{1}{s - (\lambda + 1)} F(s)\right\} \tag{E.3}$$

Hence, if we use the convolution theorem (2.35) on the last term in (E.3) with $\mathcal{L}\{e^{(\lambda+1)x}\} = 1/(s - (\lambda + 1))$ from (2.11), we obtain

$$u(x) = f(x) + \lambda e^{(\lambda+1)x} * f(x) = f(x) + \lambda \int_0^x e^{(\lambda+1)(x-t)} f(t)\, dt$$

(b) Use the Laplace transform to solve the following problem:

$$u(x) = x - \int_0^x (x - t)u(t)\,dt \qquad (E.1)$$

The kernel $K(x,t) = x - t$ is a difference kernel; hence, we use the Laplace transform to obtain

$$U(s) = \frac{1}{s^2} - \frac{1}{s^2}U(s) \qquad (E.2)$$

after using the convolution theorem (2.35) and $\mathcal{L}\{x\} = 1/s^2$ from (2.15) with $\nu = 1$. From (E.2) we have

$$U(s) = \frac{1/s^2}{1 + 1/s^2} = \frac{1}{s^2 + 1} \qquad (E.3)$$

Hence, if we use (2.13) we have

$$u(x) = \mathcal{L}^{-1}\left\{\frac{1}{s^2 + 1}\right\} = \sin x \qquad (E.4)$$

which can be easily verified as the solution of (E.1).

We may remark here that the other main class of integral equations, namely the Fredholm integral equations (see Jerri, 1985), may sometimes appear with difference kernel, but as a convolution product of the Fourier transform in Section 2.2, a subject that we shall leave to the interest of the reader. (See Exercise 2.53.)

A. Integrodifferential Equations

When the function $u(x)$ in an equation occurs in a derivative as well as an integral, the equation is called an *integrodifferential equation*. For example,

$$\frac{d^2u}{dx^2} = f(x) + \int_0^x K(x,t)u(t)\,dt \qquad (2.56)$$

We note here that when $K(x,t)$ is a difference kernel the right-hand side of (2.56) is amenable to Laplace transformation, while the derivative on the left side, according to (2.19), incorporates the initial conditions $u(0)$ and $u'(0)$ in its Laplace transformation.

EXAMPLE 2.12: Initial Value Problem

(a) Solve the following initial value problem associated with the integrodifferential equation

$$\frac{d^2u}{dx^2} = e^{2x} - \int_0^x e^{2(x-t)} \frac{du}{dt} dt \tag{E.1}$$

$$u(0) = 0 \tag{E.2}$$

$$u'(0) = 0 \tag{E.3}$$

We apply the Laplace transform to (E.1) using (2.19),

$$\mathcal{L}\left\{\frac{d^2u}{dx^2}\right\} = s^2U(s) - su(0) - u'(0) \tag{E.4}$$

on the left side, the convolution theorem (2.33) for the integral, and (2.2),

$$\mathcal{L}\left\{\frac{du}{dx}\right\} = sU(s) - u(0) \tag{E.5}$$

for the derivative inside the integral to obtain

$$s^2U(s) - su(0) - u'(0) = \frac{1}{s-2} - \frac{1}{s-2}[sU(s) - u(0)] \tag{E.6}$$

With the use of the initial conditions (E.2) and (E.3), this becomes

$$s^2U(s) = \frac{1}{s-2} - \frac{sU(s)}{s-2}$$

$$U(s) = \frac{1}{s(s-1)^2} = \frac{1}{(s-1)^2} - \frac{1}{(s-1)} + \frac{1}{s}$$

Hence, the solution to (E.1) is

$$u(x) = \mathcal{L}^{-1}\left\{\frac{1}{(s-1)^2} - \frac{1}{s-1} + \frac{1}{s}\right\}$$

$$= xe^x - e^x + 1$$

after using the Laplace transform pair (2.11) with $a = 1$, 0 and (2.21) with $f(x) = x$, $n = 1$.

B. Volterra Integral Equation of the First Kind with Difference Kernel—Laplace Transform

A Volterra integral equation of the *first kind* is of the form

$$f(x) = \lambda \int_0^x K(x,t)u(t) dt \tag{2.57}$$

where the unknown $u(x)$ appears under the integral sign only. Its special case with difference kernel $K(x,t) = K(x-t)$,

$$f(x) = \lambda \int_0^x K(x-t)u(t)dt \tag{2.58}$$

has the integral in the form of the convolution product of the Laplace transform. So we can apply the Laplace transform to (2.58), as we did for (2.51), to obtain

$$F(s) = \lambda\kappa(s)U(s) \tag{2.59}$$

where $F(s)$, $\kappa(s)$, and $U(s)$ are the Laplace transforms of $f(x)$, $K(x)$, and $u(x)$, respectively; that is,

$$U(s) = \frac{1}{\lambda}\frac{F(s)}{\kappa(s)} \tag{2.59a}$$

The solution to (2.58) is obtained by taking the inverse Laplace transform of both sides of (2.59a) to obtain

$$u(x) = \frac{1}{\lambda}\mathcal{L}^{-1}\left\{\frac{F(s)}{\kappa(s)}\right\} \tag{2.60}$$

To give a simple illustration, consider the integral equation with difference kernel

$$\sin x = \lambda \int_0^x e^{x-t}u(t)dt \tag{E.1}$$

This Volterra equation of the first kind has the difference kernel $K(x,t) = e^{x-t}$. Thus, we Laplace transform it using

$$\kappa(s) = \mathcal{L}\{K(x)\} = \mathcal{L}\{e^x\} = \frac{1}{s-1}, \qquad \mathcal{L}\{\sin x\} = \frac{1}{s^2+1}$$

on (2.59) so as to have

$$\frac{1}{s^2+1} = \lambda\frac{1}{s-1}U(s)$$

$$U(s) = \frac{1}{\lambda}\frac{1/(s^2+1)}{1/(s-1)} = \frac{1}{\lambda}\frac{s-1}{s^2+1} = \frac{1}{\lambda}\left(\frac{s}{s^2+1} - \frac{1}{s^2+1}\right)$$

To obtain the solution $u(x)$ of the integral equation, we take the inverse Laplace transform of the $U(s)$ in the above equation and employ the Laplace transform pairs for $\cos x$ and $\sin x$ in (2.14) and (2.13), respectively, to have

$$u(x) = \frac{1}{\lambda}\mathcal{L}^{-1}\left\{\frac{s}{s^2+1} - \frac{1}{s^2+1}\right\}$$

$$= \frac{1}{\lambda}(\cos x - \sin x)$$

C. Abel's Generalized Integral Equation

Abel's integral equation

$$-\sqrt{2g}f(x) = \int_0^x \frac{\phi(t)}{\sqrt{x-t}}\, dt \tag{2.61}$$

is a *singular* Volterra equation of the first kind with difference kernel $K(x,t) = 1/\sqrt{x-t}$ (which is not defined at $x = t$; that is, singular at $t = x$ as we alluded to it in Section 1.4.1); hence, we can use the Laplace transform to solve it. We may mention that Abel's integral equation is in the unknown $\phi(t)$, the shape of the path a particle under gravity should follow to reach the ground in a prescribed time $f(x)$. See Jerri (1985, p. 56).

The generalized Abel integral equation

$$f(x) = \int_0^x \frac{u(t)}{(x-t)^\alpha}\, dt, \qquad 0 < \alpha < 1 \tag{2.62}$$

is also of the convolution type and can be solved by using the Laplace transform. In the next example we will use the Laplace transform to solve (2.61) and leave for an exercise (see Exercise 2.25) obtaining the solution of (2.62), which is

$$u(x) = \frac{\sin \alpha \pi}{\pi} \frac{d}{dx} \int_0^x (x-t)^{\alpha-1} f(t)dt, \qquad 0 < \alpha < 1 \tag{2.63}$$

EXAMPLE 2.13: Abel's Integral Equation

Use the Laplace transform to solve Abel's equation (2.61).

We let $F(s)$ and $\Phi(s)$ be the Laplace transforms of $f(x)$ and $\phi(x)$, respectively, and use $\mathcal{L}\{K(x)\} = \mathcal{L}\{1/\sqrt{x}\} = \sqrt{\pi/s}$ from (2.15), noting that $\Gamma(\tfrac{1}{2}) = \sqrt{\pi}$ from (2.18), to obtain

$$\rightarrow \sqrt{2g}F(s) = \sqrt{\frac{\pi}{s}}\, \Phi(s) \tag{E.1}$$

$$\Phi(s) = -\sqrt{\frac{2g}{\pi}}\, \sqrt{s}F(s) \tag{E.2}$$

Hence the solution to Abel's equation (2.61) is

$$\phi(x) = -\sqrt{\frac{2g}{\pi}}\, \mathcal{L}^{-1}\{\sqrt{s}F(s)\} \tag{E.3}$$

In trying to use the convolution theorem for $\mathcal{L}^{-1}\{\sqrt{s}F(s)\}$ in (E.3), we have difficulty in finding $\mathcal{L}^{-1}\{\sqrt{s}\}$ because the condition $\nu > -1$ on the Laplace transform pair (2.15) prevents us from letting $\nu = -\frac{3}{2}$ to have a pair for $s^{1/2}$. On the other hand, the inverse Laplace transform of $s^{-1/2}$ is $1/\sqrt{\pi x}$; hence, we may multiply and divide the right-hand side of (E.2) by $\sqrt{s}$ to obtain

$$\Phi(s) = -\sqrt{\frac{2g}{\pi}}\, s \cdot \frac{1}{\sqrt{s}}F(s) \tag{E.2}$$

$$= -\sqrt{\frac{2g}{\pi}} \cdot sH(s)$$

$$H(s) \equiv \frac{F(s)}{\sqrt{s}}$$

and we can then easily use the convolution theorem (2.35) to write

$$\mathcal{L}^{-1}\{H(s)\} = \mathcal{L}^{-1}\left\{\frac{1}{\sqrt{s}}F(s)\right\} = \frac{1}{\sqrt{\pi}}\int_0^x \frac{f(t)}{\sqrt{x-t}}\, dt = h(x) \tag{E.4}$$

Now to find the inverse Laplace transform of $\Phi(s)$ in (E.2) we have, according to (2.2),

$$\mathcal{L}^{-1}\{sH(s) - h(0)\} = \frac{dh}{dx} \tag{E.5}$$

$$= \mathcal{L}^{-1}\{sH(s)\} - \mathcal{L}^{-1}\{h(0)\}$$

$$= \mathcal{L}^{-1}\{sH(s)\} = \frac{dh}{dx}$$

since from (E.4) we have $h(0) = 0$.

Hence we use $h(x)$ from (E.4) to obtain $\phi(x)$, the inverse Laplace transform of $\Phi(s) = -\sqrt{2g/\pi} \cdot sH(s)$ in (E.2),

$$\Phi(x) = -\sqrt{\frac{2g}{\pi}}\frac{d}{dx}\left\{\frac{1}{\sqrt{\pi}}\int_0^x \frac{f(t)}{\sqrt{x-t}}\, dt\right\} \tag{E.6}$$

which is the solution to (2.61), as a special case of (2.63) with $\alpha = \frac{1}{2}$.

2.1.5 The z-Transform

In this book our plan is to cover the finite Fourier, Hankel, and other transforms in Chapter 3 to solve general boundary value problems on a finite domain. Then in Chapter 4 we will cover the discrete version of these transforms along with associated efficient algorithms like the fast

Fourier transform (FFT). The z-transform is the finite or discrete version of the Laplace transform, which, as expected, is valuable for solving (discretized) *initial value problems* whose continuous analogs can be treated by the Laplace integral transform. We will introduce the z-transform in Section 2.9.3 and present some of its pairs (see Example 2.32); then we will apply it in Section 4.1.5B of Chapter 4. We postpone applications to the latter section in order to gain experience with discrete transforms and their discrete (or difference) operational calculus in Sections 4.1.1 to 4.1.4. In summary, we will be using the z-transform for solving difference equations that are approximations to the differential equations of the initial value problems treated by the Laplace integral transform.

The application of the z-transform in Section 4.1.5B is illustrated by its use for solving the difference equations of initial value problems in Example 4.5. Another application is to a problem of combinatorics—for example, how to shelve boxes of different sizes on a shelf of a given length in Example 4.6.

2.2 Fourier Exponential Transforms

The Fourier exponential transform of the function $f(x)$ defined on $(-\infty, \infty)$ is

$$F(\lambda) = \mathcal{F}\{f\} = \int_{-\infty}^{\infty} e^{-i\lambda x} f(x)\, dx \qquad (2.64)$$

As we have shown in Section 1.2, the Laplace transform kernel and the exponential kernel $K(\lambda, x) = e^{-i\lambda x}$ are *compatible* with differential operators of all orders with *constant coefficients*. The limits of integration in (2.64) makes it suitable for problems defined on the whole real line $(-\infty, \infty)$. In contrast to the Laplace transform, the Fourier exponential transform has a simple and symmetric formula for the inverse

$$f(x) = \mathcal{F}^{-1}\{F\} = \frac{1}{2\pi} \int_{-\infty}^{\infty} e^{i\lambda x} F(\lambda)\, d\lambda \qquad (2.65)$$

which we shall state more precisely in the next section as Theorem 2.14.

As mentioned in Section 1.4.1, finding the Fourier transform inverse in (2.65) amounts to no more than solving for $f(x)$ in the *singular (Fredholm) integral equation* (2.64) in $f(x)$, assuming $F(\lambda)$ is known.

A. The Different Notation in Texts, References, and Tables of Fourier Transforms

We must stress here that there is variation among the different references and texts as to what is called the Fourier transform or its inverse. This is due to the symmetry between (2.64) and (2.65); hence, different authors use the notation that is most commonly used in a particular field. For example, a typical variation from (2.64), (2.65) is that found in electrical engineering texts:

$$f(t) = \frac{1}{2\pi} \int_{-\infty}^{\infty} F(\omega)e^{-j\omega t}\, d\omega, \qquad j = \sqrt{-1} \tag{2.66}$$

for the time function signal $f(t)$ as the Fourier transform of $F(\omega)$;

$$F(\omega) = \int_{-\infty}^{\infty} f(t)e^{j\omega t}\, dt \tag{2.67}$$

Most often the electrical engineers deal with $\omega = 2\pi f$, where f is the frequency in cycles per second (cps, hertz) where the above two forms are written as

$$g(t) = \int_{-\infty}^{\infty} G(f)e^{-j2\pi ft}\, df \tag{2.66a}$$

$$G(f) = \int_{-\infty}^{\infty} g(t)e^{j2\pi ft}\, dt \tag{2.67a}$$

In Chapter 4 we will present the *discrete Fourier transform* (DFT) and its fast algorithm, the *fast Fourier transform* (FFT), where we shall use the above notation. This is for the obvious reason that the literature and the algorithm of the FFT were developed and are currently presented in this notation.

It should then be no surprise for us to see the following pairs of Fourier integrals:

$$f(t) = \int_{-\infty}^{\infty} F(\omega)e^{i\omega t}\, d\omega \tag{2.68}$$

$$F(\omega) = \frac{1}{2\pi} \int_{-\infty}^{\infty} f(t)e^{-i\omega t}\, dt \tag{2.69}$$

or

$$f(x) = \frac{1}{\sqrt{2\pi}} \int_{-\infty}^{\infty} F(p)e^{-ipx}\, dp \tag{2.70}$$

$$F(p) = \frac{1}{\sqrt{2\pi}} \int_{-\infty}^{\infty} f(x)e^{-ipx}\, dx \tag{2.71}$$

although the latter pairs, (2.70) and (2.71), is most often seen in physics texts.

What is most important here is for us to be consistent in the use of the pair we choose, and we should clearly indicate any variation from such notation. In this text, we will adhere mainly to (2.64), (2.65) for most of our treatment, but we are likely to turn to (2.68), (2.69) for signal analysis or to (2.70), (2.71) for some applications in physics. Among the basic references on the subject, Churchill (1972), a basic reference on transforms, uses our notation in (2.64), (2.65); Sneddon (1972), in his treatise with applications, uses the symmetric notation (2.70), (2.71); and Papoulis (1977), for signal analysis, uses (2.66), (2.67), which is close to ours. Davies (1985) uses (2.68), (2.69), while Wolf (1978), with applications in physics, uses the symmetric notation (2.70), (2.71). Weinberger (1965) and Powers (1987), as two examples of texts for applications to partial differential equations and boundary value problems, use the form (2.68), (2.69). In particular, we should mention that the well-known reference on the transforms, *with exhaustive tables*, by Erdelyi and co-authors (1954, Vol. 1) uses our notation, while that of Ditkin and Prudnikov (1965) uses our notation for its tables and the symmetric notation for the discussion in the text.

B.   The Trigonometric Fourier Sine-Cosine Transforms

We must stress here that all the above Fourier integral representations for $f(x)$ on $(-\infty, \infty)$ are called the *(complex) exponential Fourier integral representations*. This is natural because of the complex exponential function used for the kernel of these Fourier integrals. When the subject of Fourier integral representation is to be motivated, many texts speak of the *Fourier (trigonometric) sine and cosine integral representation*. In our notation (2.65), for example, $f(x)$ on $(-\infty, \infty)$ is represented as

$$f(x) = \int_{0}^{\infty} [A(\lambda)\cos \lambda x + B(\lambda)\sin \lambda x]\, d\lambda \tag{2.64a}$$

with $A(\lambda)$ and $B(\lambda)$ called, respectively, the *Fourier cosine* and *sine transforms* of $f(x)$:

$$A(\lambda) = \frac{1}{\pi} \int_{-\infty}^{\infty} f(x)\cos \lambda x\, dx \tag{2.65a}$$

$$B(\lambda) = \frac{1}{\pi} \int_{-\infty}^{\infty} f(x)\sin \lambda x\, dx \tag{2.65b}$$

From these two integrals, we can easily show for the two special cases of *odd* and *even* functions $f_o(x)$, $f_e(x)$ on $(-\infty, \infty)$, that $A(\lambda) \equiv 0$ and $B(\lambda) \equiv 0$ for the odd and even functions, respectively. This is so, since for the odd function $f_o(x)$ the integrand $f_o(x) \cos \lambda x$ in (2.65a) for $A(\lambda)$ is an odd function on the symmetric interval $(-\infty, \infty)$, and thus the integral vanishes,

$$A(\lambda) \equiv 0$$

while the integrand $f_o(x) \sin \lambda x$ in (2.65b) for $B(\lambda)$ is an even function on the symmetric interval $(-\infty, \infty)$, which gives

$$B(\lambda) = \frac{2}{\pi} \int_0^\infty f_o(x) \sin \lambda x \, dx$$

Thus the Fourier (trigonometric) integral representation for odd functions $f_o(x)$ on $(-\infty, \infty)$ is only a Fourier sine integral,

$$f_o(x) = \int_0^\infty B(\lambda) \sin \lambda x \, d\lambda$$

$$B(\lambda) = \frac{2}{\pi} \int_0^\infty f(x) \sin \lambda x \, dx$$

Note how we dropped the subscript o in $f_o(x)$ in this integral, as it is no longer needed for it on $(0, \infty)$.

For even functions $f_e(x)$ on $(-\infty, \infty)$, the same thing can be done to show that $B(\lambda)$ vanishes and we have only a Fourier cosine integral representation,

$$F_e(x) = \int_0^\infty A(\lambda) \cos \lambda x \, d\lambda$$

$$A(\lambda) = \frac{2}{\pi} \int_0^\infty f(x) \cos \lambda x \, dx$$

In Section 2.5 we will have detailed discussions of these Fourier sine and cosine transforms, followed by illustrating their use for solving boundary value problems on the half-line $(0, \infty)$.

We will show here, with the aid of the *Euler identities*

$$\cos \lambda x = \frac{e^{i\lambda x} + e^{-i\lambda x}}{2} \qquad (e^{i\lambda x} = \cos \lambda x + i \sin \lambda x)$$

$$\sin \lambda x = \frac{e^{i\lambda x} - e^{-i\lambda x}}{2i} \qquad (e^{-i\lambda x} = \cos \lambda x - i \sin \lambda x)$$

that this *trigonometric* Fourier integral representation (2.64a), (2.65a), (2.65b) is equivalent to the *exponential* Fourier representation (2.65),

(2.64):

$$f(x) = \frac{1}{\pi} \int_0^\infty \left[\left(\frac{e^{i\lambda x} + e^{-i\lambda x}}{2} \right) A(\lambda) + \left(\frac{e^{i\lambda x} - e^{-i\lambda x}}{2i} \right) B(\lambda) \right] d\lambda$$

$$= \frac{1}{2\pi} \int_0^\infty \left[e^{i\lambda x} (A(\lambda) - iB(\lambda)) + e^{-i\lambda x} (A(\lambda) + iB(\lambda)) \right] d\lambda$$

$$= \frac{1}{2\pi} \int_0^\infty e^{i\lambda x} [A(\lambda) - iB(\lambda)] d\lambda - \int_0^{-\infty} e^{i\lambda x} [A(-\lambda) + iB(-\lambda)] d\lambda$$

after making a simple change of variable $\lambda \to -\lambda$ in the second integral. If we define $F(\lambda)$ as

$$F(\lambda) = \begin{cases} \pi[A(\lambda) - iB(\lambda)] & \lambda > 0 \\ \pi[A(-\lambda) + iB(-\lambda)] & \lambda < 0 \end{cases}$$

we have

$$f(x) = \frac{1}{2\pi} \int_0^\infty e^{i\lambda x} F(\lambda) d\lambda + \frac{1}{2\pi} \int_{-\infty}^0 e^{i\lambda x} F(\lambda) d\lambda$$

$$f(x) = \frac{1}{2\pi} \int_{-\infty}^\infty e^{i\lambda x} F(\lambda) d\lambda \qquad (2.65)$$

If we use $A(\lambda)$, $B(\lambda)$ as in (2.65a), (2.65b) for the above $F(\lambda)$,

$$F(\lambda) = \pi[A(\lambda) - iB(\lambda)]$$

$$= \frac{\pi}{\pi} \int_{-\infty}^\infty f(x)[\cos \lambda x - i \sin \lambda x] dx \qquad \lambda > 0$$

$$F(\lambda) = \frac{\pi}{\pi} \int_{-\infty}^\infty f(x)[\cos \lambda x - i \sin \lambda x] dx \qquad \lambda < 0$$

so

$$F(\lambda) = \int_{-\infty}^\infty f(x) e^{-i\lambda x} dx \qquad -\infty < \lambda < \infty \qquad (2.64)$$

2.2.1† Existence of the Fourier Transform and Its Inverse—the Fourier Integral Formula

Before we start establishing the most important results and properties of the Fourier transform $F(\lambda)$, of the function $f(x)$ defined on the whole real

---

†The proofs in this section are not required for the undergraduate course in boundary value problems.

line $(-\infty, \infty)$, in (2.64), we stop to observe a major difference between this Fourier transform and the Laplace transform $F(s)$, of (the usually causal functions) $f(x)$ defined on $(0, \infty)$ in (2.1),

$$F(s) = \int_0^\infty e^{-sx} f(x)\,dx \qquad (2.1)$$

The difference is in the direction of more restriction on the class of functions $f(x)$ for their Fourier integral (2.64) to converge, to say they are Fourier transformable. There will be more restrictions on $f(x)$ if it is to be Fourier transformed to $F(\lambda)$ via (2.64), and for this $F(\lambda)$ to be also Fourier invertible via (2.65) back to the original function $f(x)$. This will be the subject of the *Fourier integral formula* to be stated and proved as Theorem 2.14.

We note that both Fourier and Laplace transforms have an exponential kernel which is bounded on the domain of $f(x)$, that is $|e^{-i\lambda x}| = 1$, $e^{-sx} \leq 1$, $s > 0$. However, the kernel e^{-sx}, $s > 0$, is much more helpful for the convergence of the (improper) integral of the Laplace transform than the kernel $e^{i\lambda x}$ is for that of the Fourier transform. For example, when $f(x) = 1$, its Laplace transform (2.1) exists with a value $F(s) = 1/s$, $s > 0$. On the other hand, the Fourier transform (2.64) for $f(x) = 1$, $-\infty < x < \infty$, does not exist in the usual sense, meaning that the improper integral (2.64) will not converge to a usual function we know. If we must Fourier transform this function, we have to resort to a more general sense of *generalized functions*, or what we call *distributions* (see Section 2.4.1B). The Fourier transform of 1 then can be shown to be a special case of a generalized function, which is a constant multiple of the *Dirac delta function* $F(\lambda) = 2\pi\delta(\lambda)$ that we attempted to define in detail in Exercise 1.11b,c for Section 1.2 in Chapter 1, and which we will discuss again and apply to the analysis of systems and signals in Section 2.4.1. Such difficulty can easily be illustrated in terms of our usual functions (not distributions) if we try, for example, to do the improper integral (2.64) at $\lambda = 0$ for $f(x) = 1$. In this case, we have $\int_{-\infty}^\infty 1\,dx = \infty - \infty$ as an indeterminate form. Another simple example is that of $f(x) = x$, $0 < x < \infty$, with $F(s) = 1/s^2$, $s > 0$, while the Fourier transform of $f(x) = x$, $-\infty < x < \infty$, in (2.64) will result in similar difficulties at $x = \mp\infty$ if we are to perform the usual integration by parts,

$$\int_{-\infty}^\infty x e^{-i\lambda x}\,dx = \frac{x e^{-i\lambda x}}{-i\lambda}\bigg|_{x=-\infty}^\infty - \int_{-\infty}^\infty \frac{e^{-i\lambda x}}{-i\lambda}\,dx$$

$$= -\frac{x}{\lambda} e^{-i\lambda x}\bigg|_{x=-\infty}^\infty + \frac{1}{\lambda^2} e^{-i\lambda x}\bigg|_{x=-\infty}^\infty$$

Again, the difficulty here is that the Fourier kernel $e^{-i\lambda x}$ is not helpful when it comes to $x = \mp\infty$, but it is only bounded, where the x factor in the first term above will raise difficulties at $x = \mp\infty$ similar to the ones we saw in the first example above of $f(x) = 1$, $-\infty < x < \infty$. This means that if we are to speak of the Fourier transform for such functions as $f(x) = 1$, x we have to accept interpreting our results in the more general sense of *distributions* or *generalized functions*. Even though we will revisit this important subject (see Sec. 2.4.1B), we shall proceed, at the level of this book, by staying with results of the Fourier integral (2.64) as ordinary functions. In this case, we restrict our class of functions $f(x)$, $-\infty < x < \infty$ (severely compared to those being Laplace transformed) in such a way that their Fourier integral in (2.64) converges for $F(\lambda)$ to exist. After presenting some essential preliminaries, we shall state and prove in Theorem 2.12 the sufficient condition of $f(x)$ being absolutely integrable on $(-\infty, \infty)$ for its Fourier transform $F(\lambda)$ in (2.64) to exist. By *absolutely integrable $f(x)$* on $(-\infty, \infty)$ we mean that

$$\int_{-\infty}^{\infty} |f(x)| \, dx < \infty \tag{2.72}$$

In Theorem 2.14, as the important *Fourier integral formula*, we shall impose even more restrictive conditions on $f(x)$ for it to have a Fourier transform $F(\lambda)$ in (2.64), at the same time ensuring that this $F(\lambda)$ has an inverse Fourier transform $f(x)$ as in (2.65). There are variations of those more restrictive conditions, in the theory of Fourier transform, but we shall choose here the most suitable one for our development in this book. This means that $f(x)$ would be in the class of functions that are *absolutely integrable* on $(-\infty, \infty)$ as well as *sectionally (or piecewise) smooth* on every finite interval of $(-\infty, \infty)$ for the Fourier integral Theorem 2.14 to be valid. Another version for the Fourier integral formula—i.e., to be able to Fourier transform $f(x)$ to $F(\lambda)$ via (2.64) and then back to $f(x)$ via (2.65)—is to have $f(x)$ square integrable on $(-\infty, \infty)$. By $f(x)$ *square integrable* on $(-\infty, \infty)$ we mean

$$\int_{-\infty}^{\infty} |f(x)|^2 \, dx < \infty \tag{2.73}$$

For completeness, we may define the class of functions $f(x)$ which are p-integrable on the general interval (a, b) and with a weight function $\rho(x) > 0$,

$$\int_{-a}^{b} \rho(x) |f(x)|^p \, dx < \infty \tag{2.74}$$

Here we mean the integral (2.74) and its special cases in (2.72) and (2.73) to converge in the usual sense, where the integral, as introduced in elementary calculus, is a limit of a sum of rectangular areas, which is called the integral in the Riemann sense, or the *Riemann integral*. In the literature, however, we often see that the class of functions defined by (2.74) is referred to as $L_p(a,b)$, $L^p(a,b)$, $L_p(a,b)_p$, or just L_p. This L refers to an integration in a more general sense than the Riemann one we are familiar with from elementary calculus. This is called integration *in the sense of Lebesgue*, for which the L in $L_p{}^\dagger$ stands. To be brief, the essence of this generalization is that what makes a contribution to the Lebesgue integral of the function $f(x)$ is its value on a "support," that is, an interval, and not values of the function at a finite set of discrete points. This will lead to two functions $f(x)$ and $g(x)$ that can differ; that is, $f(x_i) - g(x_i) \neq 0$ (and can be very large), $i = 1, 2, 3, \ldots, n$ on this finite set of discrete points $x_1, x_2, \ldots, x_n$ in the interval (a,b), but still $f(x)$ and $g(x)$ have the same integral on (a,b). This can be said in another way—that $f(x) - g(x) = n(x)$ is a null function on (a,b), but they still have the same integral on (a,b). A *null function* $n(x)$ on (a,b), as explained above, is identically zero on (a,b) except at a finite (or countable) set of points such that

$$\int_a^b |n(x)|^2 \, dx = 0 \tag{2.75}$$

Functions $f(x)$ and $g(x)$ that differ only by a null set (or function) on (a,b) are called $f(x) = g(x)$ *almost everywhere* on (a,b), or in brief we use a.e. for *almost everywhere*, whence $f(x) = g(x)$.

In the Lebesgue sense of integration, we say that $f(x)$ and $g(x)$ differ only on a set of "support" of "measure" zero, while in our language we say $f(x) - g(x) = n(x)$, where $n(x)$ has a "support" of zero width.

With the above note of caution regarding the integration in the sense of Lebesgue, we may for simplicity adopt that notation and term the absolutely integrable functions of (2.72) as $L_1(-\infty, \infty)$ and the square-integrable functions of (2.73) as $L_2(-\infty, \infty)$ class.

Clearly, the condition of $f(x)$ being absolutely integrable or square integrable on $(-\infty, \infty)$ will prevent our above two examples of $f(x) = 1$ and $f(x) = x$ from having a Fourier transform (in the sense of our usual functions and not the sense of generalized functions or distributions). It is trivial to show that the two functions are not square integrable on

---

$\dagger$ The special (important) case L_2 of this L_p should not be confused with L_2, the second-order differential operator, used extensively in Chapter 1.

$(-\infty, \infty)$,

$$\int_{-\infty}^{\infty} 1\, dx = \infty$$

$$\int_{-\infty}^{\infty} x^2\, dx = \left.\frac{x^3}{3}\right|_{x=-\infty}^{\infty} = \infty$$

With such a severe limitation on the Fourier-transformed functions $f(x)$, compared to only *piecewise continuous* and *exponential order* functions on $(0, \infty)$ for the Laplace transform, there is, however, still a great advantage to the Fourier transform of such functions. This is that the inverse of the resulting Fourier transform $F(\lambda)$ exists with a formula (2.65) that is very symmetric with that of the Fourier transform (2.64). This contrasts with the nonsymmetric formula for Laplace transform inversion (2.3), along with the usually more complicated complex contour integration necessary for its computation.

Indeed, Theorem 2.14 as the *Fourier integral theorem* or formula, presented in (2.86), will summarize all the above for the existence of the Fourier transform as well as its inverse for functions $f(x)$ which are *absolutely integrable* and *piecewise smooth* on $(-\infty, \infty)$.

We should mention that the theory of Fourier transforms will provide more elegant results for the class of functions $f(x)$ that are both absolutely integrable and square integrable on $(-\infty, \infty)$, that is, $f(x) \in L_1 \cap L_2$. Then the first question that will come to mind is to show that on $(-\infty, \infty)$ there are square-integrable functions which are not absolutely integrable and vice versa. A simple example of a function which is square integrable but not absolutely integrable on $(-\infty, \infty)$ is $f(x) = (1+x^2)^{-1/2}$, since

$$\int_{-\infty}^{\infty} |f(x)|^2\, dx = \lim_{A \to \infty} \int_{-A}^{A} \frac{1}{1+x^2} = \lim_{A \to \infty} [\tan^{-1} A - \tan^{-1} -A]$$

$$= \frac{\pi}{2} + \frac{\pi}{2} = \pi$$

while

$$\int_{-\infty}^{\infty} |f(x)|\, dx = \lim_{A \to \infty} \int_{-A}^{A} \frac{1}{\sqrt{1+x^2}}\, dx$$

$$= \lim_{A \to \infty} [\sinh^{-1} A - \sinh^{-1} -A] = \infty + \infty = \infty$$

We shall leave it as an exercise (see Exercise 2.35, with some hints) to show that the function $f(x) = |x|^{-1/2}(1+x^2)^{-1}$ is absolutely integrable but is not square integrable on $(-\infty, \infty)$.

Even though we shall work mainly with absolutely integrable (on $(-\infty, \infty)$) and piecewise smooth functions on every finite subinterval of

$(-\infty, \infty)$, the theory of *square-integrable* functions is still very important for Fourier transforms. It will be felt even more when we discuss the Fourier series, on the finite interval (a,b), in Section 3.1 of Chapter 3, and especially the "special" convergence of this series. The square-integrable functions and the Fourier series, with its convergence, will be of importance if we are to derive the Fourier integral formula in Theorem 2.14 by considering its integral as a limit of a Fourier series, which is what many texts do.

A very important tool that goes with square-integrable functions is Schwarz's inequality, which we will state here and leave its proof as an exercise (see Exercise 2.36) with detailed hints.

A. Schwarz's Inequality

For the two functions $f(x)$ and $g(x)$ defined and square integrable on the finite interval (a,b), Schwarz's inequality states that

$$\left| \int_a^b f(x)g(x)\, dx \right|^2 \le \int_a^b |f(x)|^2\, dx \int_a^b |g(x)|^2\, dx \tag{2.76}$$

The proof is left for Exercise 2.36a, with detailed leading steps. This can be extended to infinite integrals to have

$$\left| \int_{-\infty}^{\infty} f(x)g(x)\, dx \right|^2 \le \int_{-\infty}^{\infty} |f(x)|^2\, dx \cdot \int_{-\infty}^{\infty} |g(x)|^2\, dx \tag{2.77}$$

which is proved in Exercise 2.36c with the necessary details.

A simple generalization of this Schwarz inequality is to have the integrals with a real nonnegative weight function $\rho(x) > 0$ (see Exercise 2.36d)

$$\left| \int_{-\infty}^{\infty} \rho(x)f(x)g(x)\, dx \right|^2 \le \int_{-\infty}^{\infty} \rho(x)|f(x)|^2\, dx \int_{-\infty}^{\infty} \rho(x)|g(x)|^2\, dx \tag{2.78}$$

A needed tool that results easily from (2.77) (see Exercise 2.36e) is the *triangle inequality*,

$$\left[\int_{-\infty}^{\infty} |f(x) + g(x)|^2\, dx \right]^{1/2} \le \left[\int_{-\infty}^{\infty} |f(x)|^2\, dx \right]^{1/2} + \left[\int_{-\infty}^{\infty} |g(x)|^2\, dx \right]^{1/2} \tag{2.79}$$

With such tools as (2.77) we can easily show that *if both $f(x)$ and $g(x)$ are square integrable on $(-\infty, \infty)$, then their product $f(x)g(x)$ is absolutely* *integrable*, since

$$\left[\int_{-\infty}^{\infty} |f(x)g(x)| \, dx\right]^2 = \left[\int_{-\infty}^{\infty} |f(x)||g(x)| \, dx\right]^2$$

$$\leq \int_{-\infty}^{\infty} |f(x)|^2 \, dx \int_{-\infty}^{\infty} |g(x)|^2 \, dx$$

after the use of $|f(x)g(x)| = |f(x)||g(x)|$ and Schwartz's inequality on the middle integral. Thus,

$$\int_{-\infty}^{\infty} |f(x)g(x)| \, dx \leq \left[\int_{-\infty}^{\infty} |f(x)|^2 \, dx\right]^{1/2} \left[\int_{-\infty}^{\infty} |g(x)|^2 \, dx\right]^{1/2} < \infty$$

Also, for the special case of the *finite interval* (a,b) we can show, using (2.76), *that a square-integrable function $f(x)$ on (a,b) is absolutely integrable too*,

$$\int_a^b |f(x)| \, dx = \int_a^b 1 \cdot |f(x)| \, dx \leq \left[\int_a^b 1 \, dx\right]^{1/2} \left[\int_a^b |f(x)|^2 \, dx\right]^{1/2}$$

$$\leq (b-a)^{1/2} \left[\int_a^b |f(x)|^2 \, dx\right] < \infty$$

As we mentioned above, the Fourier integral Theorem 2.14 will ensure at the same time the existence of the Fourier transform $F(\lambda)$ in (2.64), for the class of functions $f(x)$ that are both absolutely integrable on $(-\infty, \infty)$ and sectionally smooth on every finite subinterval of $(-\infty, \infty)$, as well as the existence of $f(x)$ itself as the inverse Fourier transform of $F(\lambda)$, given in (2.65). That is, the Fourier integral formula (2.85) in Theorem 2.14 will validate both (2.64) and (2.65) for absolutely integrable and sectionally (piecewise) smooth $f(x)$ and $(-\infty, \infty)$.

Since, at least in our formal proofs or exercises, we may derive the Fourier transform or its inverse as a limit of a Fourier series, we should define the usual types of convergence to such a limit, which we will cover in more detail in Section 3.1 on Fourier series and orthogonal expansion. In particular, we may speak of the limit of the *nth partial sum*

$$S_n(x) = \sum_{k=1}^{n} c_k(x) \tag{2.80}$$

to $f(x)$ as $n \to \infty$, or the limit of the (finite limit) Fourier integral

$$F_A(\lambda) = \int_{-A}^{A} e^{-i\lambda x} f(x) \, dx \tag{2.81}$$

to $F(\lambda)$ as $A \rightarrow \infty$.

Definitions Pointwise convergence: the sequence $S_n(x)$ is said to *converge to $f(x)$ pointwise* if

$$\lim_{n\to\infty} S_n(x) = f(x) \quad \text{at each } x \tag{2.82}$$

If such convergence is valid independent of x in (a,b), the convergence of $S_n(x)$ to $f(x)$ is termed *uniform convergence*. Specifically, given $\epsilon > 0$, there is $N(\epsilon)$ independent of x such that $|f(x) - S_N(x)| < \epsilon$ for $n > N(\epsilon)$.

For the class of square-integrable functions on (a,b), there is another suitable and more practical type of convergence, which is called *convergence of S_n in the mean*. $S_n(x)$ is said to *converge to $f(x)$ in the mean on $[a,b]$* if

$$\lim_{n\to\infty} \int_a^b |f(x) - S_n(x)|^2 \, dx = 0 \tag{2.83}$$

In contrast to pointwise convergence, where $\lim_{n\to\infty} S_n(x) = f(x)$ for each x in $[a,b]$, we note here that for convergence in the mean $S_n(x)$ will get close to $f(x)$ as $n \rightarrow \infty$ but still may be much different from $f(x)$ on a set of separate points. This is clear because the integral in (2.83) is not affected by such a difference when its width is only zero.

For completeness, we present the *Cauchy criterion for the convergence of a sequence*. The sequence $S_n(x)$ satisfies the *Cauchy criterion* if

$$\lim_{\substack{n\to\infty \\ m\to\infty}} \int_a^b |S_n(x) - S_m(x)|^2 \, dx = 0 \tag{2.84}$$

This is the Cauchy criterion we adopt here for our purposes when working with square-integrable functions. The usual definition is

$$\lim_{\substack{n\to\infty \\ m\to\infty}} |S_n - S_m| = 0, \qquad n, \, m > N. \qquad \square$$

The triangle equality is very useful here to show that if *a sequence $S_n(x)$ converges in the mean to $f(x)$, then it satisfies the Cauchy criterion*. The proof depends on using $[S_n(x) - f(x)] + [f(x) - S_m(x)]$ in (2.84) instead of $S_n(x) - S_m(x)$, then involving the triangle inequality (2.79) for the sum of the two functions $S_n(x) - f(x)$ and $f(x) - S_m(x)$,

$$\int_a^b |S_n(x) - S_m(x)|^2 \, dx = \int_a^b |[S_n(x) - f(x)] + [f(x) - S_m(x)]|^2 \, dx$$

$$\leq \int_a^b |S_n(x) - f(x)|^2 \, dx + \int_a^b |f(x) - S_m(x)|^2 \, dx$$

and (2.84) results as $n,m \to \infty$, since it is assumed that the sequence $S_n(x)$ (or $S_m(x)$) converges in the mean to $f(x)$, which makes the last two integrals vanish.

It is time now to present a theorem for the existence of the Fourier transform $F(\lambda)$ in (2.64). This will be followed by the Fourier integral Theorem 2.14 for the existence of both $F(\lambda)$ and its inverse $f(x)$.

THEOREM 2.12: Existence of the Fourier Transform of Absolutely Integrable Functions

If $f(x)$ in (2.64) is absolutely integrable, its Fourier transform $F(\lambda)$ exists and, moreover, it is continuous. ◻

Proof:

$$|F(\lambda)| = \left| \int_{-\infty}^{\infty} e^{-i\lambda x} f(x)\, dx \right| \le \int_{-\infty}^{\infty} |e^{-i\lambda x}| |f(x)|\, dx$$

$$= \int_{-\infty}^{\infty} |f(x)|\, dx < \infty$$

after using the fact that $f(x)$ is absolutely integrable on $(-\infty, \infty)$. Hence, $F(\lambda)$ is bounded. To show that $F(\lambda)$ is continuous, we will show that $\lim_{h\to 0}|F(\lambda + h) - F(\lambda)| = 0$:

$$|F(\lambda + h) - F(\lambda)| = \left| \int_{-\infty}^{\infty} e^{-i(\lambda+h)x} f(x)\, dx - \int_{-\infty}^{\infty} e^{-i\lambda x} f(x)\, dx \right|$$

$$= \left| \int_{-\infty}^{\infty} e^{-i\lambda x}(e^{-ihx} - 1)f(x)\, dx \right|$$

$$\le \int_{-\infty}^{\infty} |e^{-ihx} - 1||f(x)|\, dx \qquad \text{(E.1)}$$

What we need now is to take the limit as $h \to 0$ and allow it to enter inside the last integral. As we know from Lebesgue convergence Theorem 2.10, this requires that the integrand $|e^{-ihx} - 1||f(x)|$ be bounded by a function $G(x)$ independent of the parameter of the limit process (here h, a positive real number), and $G(x)$ is integrable. Even though the main statement of Theorem 2.10 concerns $\lim_{n\to\infty} f_n(x)$, it is known to be valid for $\lim_{A\to\infty} f_A(x)$ for A a positive real number, instead of just the integer n. Indeed, in (E.1) we have Theorem 2.10 satisfied, since $|e^{-i\lambda h} - 1| < 2$, $|e^{-i\lambda h} - 1||f(x)| < G(x) = 2|f(x)|$, and $G(x) = 2|f(x)|$ is integrable since $|f(x)|$ is. Allowing the limit as $h \to 0$ inside the last integral will make $\lim_{h\to 0} |e^{-i\lambda h} - 1| = 0$ and the integral will vanish. Hence, $\lim_{h\to 0} F(\lambda + h) - F(\lambda) = 0$, and $F(\lambda)$ is continuous. ∎

We must note here that the Fourier transform $F(\lambda)$, of the absolutely integrable function $f(x)$, is not necessarily absolutely integrable. To give an example, we consider the function

$$f(x) = \begin{cases} e^{-x}, & x \geq 0 \\ 0, & x < 0 \end{cases}$$

which is absolutely integrable on $(-\infty, \infty)$

$$\int_{-\infty}^{\infty} |f(x)| \, dx = \int_{0}^{\infty} e^{-x} \, dx = 1$$

with Fourier transform

$$F(\lambda) = \int_{0}^{\infty} e^{-i\lambda x} e^{-x} \, dx = -\frac{1}{1+i\lambda} e^{-i\lambda x} e^{-x} \Big|_{0}^{\infty} = \frac{1}{1+i\lambda}$$

after knowing that $e^{-i\lambda x} = \cos \lambda x - i \sin \lambda x$ is bounded at $x = \infty$. $F(\lambda)$ here is a complex-valued function whose absolute value $|F(\lambda)| = \sqrt{F(\lambda)\overline{F(\lambda)}}$, where $\overline{F(\lambda)}$ is the complex conjugate of $F(\lambda)$, which is obtained from $F(\lambda)$ by replacing i by $-i$,

$$|F(\lambda)| = \sqrt{F(\lambda)\overline{F(\lambda)}} = \sqrt{\frac{1}{(1+i\lambda)(1-i\lambda)}} = \frac{1}{\sqrt{(1+\lambda^2)}}$$

This $F(\lambda)$ is not absolutely integrable, since

$$\int_{-\infty}^{\infty} |F(\lambda)| \, d\lambda = \int_{-\infty}^{\infty} \frac{1}{\sqrt{1+\lambda^2}} \, d\lambda = \sinh^{-1} \lambda \Big|_{\lambda=-\infty}^{\infty} = \infty - (-\infty) = \infty$$

So, if we look at the symmetric form of the inverse Fourier transform (2.65), we should not expect existence of this integral for such an input $F(\lambda)$, which is the output of (2.64) as the Fourier transform of an absolutely integrable function $f(x)$. The reason is that $F(\lambda)$ here is, in general, not necessarily absolutely integrable to guarantee the integral in (2.65) to exist and define $f(x)$. But, in practice, we would like to Fourier transform back and forth from $f(x)$ to $F(\lambda)$ in (2.64) and then in a very symmetric way from $F(\lambda)$ to $f(x)$ via (2.65). Indeed, when we speak of signals, $f(x)$ is the representation in the time space while $F(\lambda)$ is its representation in the frequency, or Fourier, space. In quantum mechanics $f(x)$ is in the coordinate space, while $F(\lambda)$ is the representation in the momentum (λ) space.

To be able to utilize the Fourier transform (2.64) and its inverse (2.65) as convergent integrals, we must restrict our class of transformed functions $f(x)$ to more than just absolutely integrable. One of the simplest versions of such restrictions, which satisfies our needs here and which we shall adopt in this book, is that $f(x)$ must be *sectionally smooth* in addition to being *absolutely integrable* on $(-\infty, \infty)$. This is the statement of the Fourier integral Theorem 2.14 (or formula) that we shall state next and prove soon. For such an (elaborate!) proof we will need an important result, the *Riemann–Lebesgue lemma* [see (2.91)], which we designate as Theorem 2.13 and whose proof is given as Exercise 2.37a.

The *Fourier integral Theorem 2.14* that we are seeking to prove, and have prepared for so much, is the following:

THEOREM 2.14:   The Fourier Integral Theorem (the Fourier Transform Inversion Formula)

If $f(x)$ is a piecewise smooth function on every finite interval of the real line $(-\infty, \infty)$ and is absolutely integrable on $(-\infty, \infty)$, then

$$
\begin{aligned}
f(x) &= \frac{1}{2\pi} \int_{-\infty}^{\infty} d\lambda e^{i\lambda x} \int_{-\infty}^{\infty} f(\xi) e^{-i\lambda\xi} d\xi \\
&= \frac{1}{2\pi} \int_{-\infty}^{\infty} dx \int_{-\infty}^{\infty} f(\xi) e^{i\lambda(x-\xi)} d\xi
\end{aligned}
\tag{2.85}
$$

at x where $f(x)$ is continuous, and to $\frac{1}{2}[f(x+)+f(x-)]$ at points x where there is a jump discontinuity.   □

Before we start the proof, it is instructive first to develop a few simple identities and results, along with that developed in Example 2.4, that are necessary for our proof. This will help us to concentrate on the essence of the proof and note the side derivations of such results.

1.   First we note that the double integral in (2.85) is equivalent to the integral

$$
\lim_{A\to\infty} \frac{1}{\pi} \int_{0}^{A} d\lambda \int_{-\infty}^{\infty} f(\xi) \cos \lambda(x - \xi) d\xi = \frac{1}{2}[f(x+) + f(x-)]
\tag{2.86}
$$

which is what we usually see in most statements of the Fourier integral formula. Such an equivalence is shown easily when we use the Euler identity $e^{i\lambda(x-\xi)} = \cos \lambda(x - \xi) + i \sin \lambda(x - \xi)$ in the inner integral of (2.85) to obtain

$$
\begin{aligned}
\int_{-\infty}^{\infty} e^{i\lambda(x-\xi)} f(\xi) d\xi &= \int_{-\infty}^{\infty} f(\xi) \cos \lambda(x - \xi) d\xi \\
&\quad + i \int_{-\infty}^{\infty} f(\xi) \sin \lambda(x - \xi) d\xi
\end{aligned}
$$

The first integral is an even function of λ and the second is an odd function of λ. So, on integrating with respect to λ over the symmetric interval $(-\infty, \infty)$ in the outer integral of (2.85), the odd function in λ would give a zero contribution while the even one would result in doubling the integration on the same interval $(0, \infty)$ of λ:

$$\frac{1}{2\pi} \int_{-\infty}^{\infty} d\lambda \int_{-\infty}^{\infty} f(\xi) e^{i\lambda(x-\xi)} d\xi$$

$$= \frac{1}{2\pi} \int_{-\infty}^{\infty} d\lambda \left[\int_{-\infty}^{\infty} f(\xi) \cos \lambda(x - \xi) d\xi \right.$$

$$\left. + i \int_{-\infty}^{\infty} f(\xi) \sin \lambda(x - \xi) d\xi \right]$$

$$= \frac{1}{\pi} \int_{0}^{\infty} d\lambda \int_{-\infty}^{\infty} f(\xi) \cos \lambda(x - \xi) d\xi \qquad (2.87)$$

2. We will also need the result

$$\frac{1}{\pi} \int_{0}^{\infty} \frac{\sin A\xi}{\xi} d\xi = \frac{1}{2}, \qquad A > 0 \qquad (2.88)$$

This is usually obtained from knowing that $2(\sin A\lambda)/\lambda$ is the Fourier transform of the gate function $p_A(x)$, but it cannot be used here since it uses the symmetry of Fourier transforming (2.64) and Fourier inverting (2.65), which is the essence of Theorem 2.14 that we are in the process of proving. Instead, we will compute the above result (2.88) as a limiting case for $\alpha \to 0+$ of the following result:

$$J(y) = \int_{0}^{\infty} e^{-\alpha\xi} \frac{\sin y\xi}{\xi} d\xi = \arctan \frac{y}{\alpha}, \qquad \alpha \geq 0, \ -B \leq y \leq B \qquad (2.89)$$

$$\int_{0}^{\infty} \frac{\sin y\xi}{\xi} d\xi = \lim_{\alpha \to 0+} \int_{0}^{\infty} e^{-\alpha\xi} \frac{\sin y\xi}{\xi} d\xi$$

$$= \lim_{\alpha \to 0+} \arctan \frac{y}{\alpha} = \begin{cases} \pi/2, & y > 0 \\ 0, & y = 0 \\ -\pi/2, & y < 0 \end{cases} \qquad (2.90)$$

and in particular for the case $y = A > 0$ we have

$$\frac{1}{\pi} \int_{0}^{\infty} \frac{\sin A\xi}{\xi} d\xi = \frac{1}{2} \qquad (2.88)$$

Result (2.89) can be found in the tables of Laplace transforms where $F(\alpha) = \mathcal{L}\{f(x)\} = \mathcal{L}\{(\sin yx)/x\} = \arctan(y/\alpha)$, $\alpha > 0$ and y is a parameter. However, its full derivation here, as the following Example 2.14, is instructive since it will put us face to face with the conditions of allowing the differentiation, with respect to a parameter, inside an improper integral. This makes a good example for Theorem 2.6, which will be consulted in a somewhat different mode than that which we used for differentiating inside the Laplace integral to obtain the result (2.21) in the last section.

EXAMPLE 2.14: Laplace Integral Evaluation in (2.89)

For the result (2.89) we may set to evaluate

$$J(y) = \int_0^\infty e^{-\alpha x} \frac{\sin yx}{x} \, dx, \qquad \alpha > 0, \; -B \le y \le B \tag{E.1}$$

which can be seen as the Laplace transform $F(\alpha) = \mathcal{L}\{f(x)\} = \mathcal{L}\{(\sin yx)/x\}$ with y as a parameter. We note that we did not encounter it in Section 2.1, but it will be familiar if something can be done to remove the x in the denominator of $(\sin yx)/x$. This is possible if we can differentiate both sides of (E.1) with respect to the parameter y and if the convergence of the resulting integral allows the differentiation operation to be interchanged with the infinite integral according to Theorem 2.6,

$$\frac{dJ(y)}{dy} = \frac{d}{dy} \int_0^\infty e^{-\alpha x} \frac{\sin yx}{x} \, dx$$

$$= \int_0^\infty e^{-\alpha x} \frac{d}{dy} \left(\frac{\sin yx}{x} \right) dx = \int_0^\infty e^{-\alpha x} \cos yx \, dx$$

$$\frac{dJ(y)}{dy} = \frac{\alpha}{\alpha^2 + y^2}, \qquad \alpha > 0 \tag{E.2}$$

after recognizing the last integral as the Laplace transform of $\cos yx$, considering the parameter y as constant inside this last integral.

We shall leave the justification of the interchange of the two operations to the end of the example and continue the computations to arrive at (2.89). Integrating (E.2) gives

$$J(y) = \int_0^y \frac{\alpha}{\alpha^2 + y^2} \, dy = \arctan \frac{y}{\alpha} + C, \qquad \alpha > 0 \tag{E.3}$$

where $C = 0$, since $J(0) = 0$ from the direct substitution of $y = 0$ inside the integral of (E.1),

$$J(y) = \arctan \frac{y}{\alpha}, \qquad \alpha > 0, \; -B \le y \le B \tag{E.4}$$

as the desired result in (2.89).

What remains now is the justification for the differentiation with respect to the parameter y inside the integral of (E.1). Here we appeal to Theorem 2.6, which requires that

(i) $h(x,y) = (\sin xy)/x$ and its partial derivative with respect to y, $\partial h(x,y)/\partial y = \cos xy$ are continuous and bounded on $0 \leq x < \infty$ and $-B \leq y \leq B$, which is the case here, and

(ii) $g(x) = e^{-\alpha x}$, $\alpha > 0$, be absolutely integrable on $0 \leq x < \infty$, which is also clearly the case.

---

3.   Last we will comment on the following Riemann–Lebesgue lemma. In preparing for the proof of Theorem 2.14, we need a very important result (the Riemann–Lebesgue lemma), which is also instrumental in the proofs for the convergence of Fourier series in Chapter 3. We will state one particular version here and leave its simple proof for Exercise 2.37a, with ample leading hints.

THEOREM 2.13:   A Version of the Riemann–Lebesgue Lemma

If $f(x)$ is a piecewise smooth function on the closed interval $[a,b]$, then

$$\lim_{\lambda \to \infty} \int_a^b f(x) \sin \lambda x \, dx = 0 \qquad \qquad \square \ (2.91)$$

The intuitive explanation of this result is that for a piecewise smooth function $f(x)$, $\sin \lambda x$ with a very high frequency λ (large) will oscillate very fast, causing the areas of the strips $\Delta x f(x) \sin \lambda x$ to oscillate very fast between positive and negative values. This "formally" results in the areas under the highly oscillating function $f(x) \sin \lambda x$ on $[a,b]$ becoming zero.

Another statement of the Riemann–Lebesgue lemma for Fourier integrals is given in Exercise 2.37b. Also, with a different mode of proof than that used in Exercise 2.37a, where integration by parts is used, the above lemma is valid for the less restrictive condition of $f(x)$ piecewise continuous instead of piecewise smooth (see Exercise 2.37c). The reason here lies in our (simple) mode of proof in Exercise 2.37a, where an integration by parts is performed in the process for $f(x)$ on the closed subintervals of $[a,b]$, where $f(x)$ and $f'(x)$ must be continuous for such an integration operation. This is the case above when we assume $f(x)$ to be sectionally smooth on (a,b). The familiar version of the Riemann–Lebesgue lemma requires only that $f(x)$ is absolutely integrable (see Exercise 2.37b).

Now we should be ready to prove the Fourier integral Theorem 2.14 in its equivalent form of (2.86).

THEOREM 2.14: The Fourier Integral Theorem

If $f(x)$ is piecewise smooth on every finite interval of the real line $(-\infty, \infty)$ and is absolutely integrable on $(-\infty, \infty)$, then

$$\tfrac{1}{2}[f(x+) + f(x-)]$$

$$= \lim_{A \to \infty} \frac{1}{\pi} \int_0^A d\lambda \int_{-\infty}^{\infty} f(\xi) \cos \lambda(x - \xi) d\xi \qquad \square \quad (2.86)$$

Proof: To prove (2.86) we start by using the result we developed in Example 2.14, for the above double integral to reduce it to the following single integral:

$$\int_0^A d\lambda \int_{-\infty}^{\infty} f(\xi) \cos \lambda(x - \xi) d\xi = \int_{-\infty}^{\infty} f(\xi) \int_0^A \cos \lambda(x - \xi) d\lambda d\xi$$

$$= \int_{-\infty}^{\infty} f(\xi) \frac{\sin A(\xi - x)}{\xi - x} d\xi \qquad (E.1)$$

This was derived in Example 2.4 by allowing interchange of the order of the double integral, as it is allowed according to Theorems 2.7 and 2.8.

Now we want to show that

$$\lim_{A \to \infty} \int_{-\infty}^{\infty} f(\xi) \frac{\sin A(\xi - x)}{\pi(\xi - x)} d\xi = \tfrac{1}{2}[f(x+) + f(x-)] \qquad (E.2)$$

which after the simple change of variable $t = \xi - x$ becomes

$$\lim_{A \to \infty} \int_{-\infty}^{\infty} f(t + x) \frac{\sin At}{\pi t} dt = \tfrac{1}{2}[f(x+) + f(x-)] \qquad (E.3)$$

We will show first that

$$\lim_{A \to \infty} \int_0^{\infty} f(t + x) \frac{\sin At}{\pi t} dt = \tfrac{1}{2} f(x+) \qquad (E.4)$$

and the other needed part for (E.3),

$$\lim_{A \to \infty} \int_{-\infty}^0 f(t + x) \frac{\sin At}{\pi t} dt = \tfrac{1}{2} f(x-) \qquad (E.5)$$

will follow in the same way.

To show (E.4) we take advantage of the identity we developed in (2.88),

$$\int_0^{\infty} \frac{\sin A\xi}{\pi \xi} = \frac{1}{2}, \qquad A > 0 \qquad (2.88)$$

to write

$$\frac{f(x+)}{2} = f(x+) \int_0^\infty \frac{\sin At}{\pi t} \, dt = \frac{1}{\pi} \int_0^\infty f(x+) \frac{\sin At}{\pi t} \, dt \tag{E.6}$$

where $f(x+)$ was entered inside the integral as constant function of t. If we use the integrals in (E.4) and (E.6) we can write

$$\frac{1}{\pi} \int_0^\infty f(t+x) \frac{\sin At}{t} \, dt - \tfrac{1}{2} f(x+)$$

$$= \frac{1}{\pi} \int_0^\infty [f(t+x) - f(x+)] \frac{\sin At}{t} \, dt$$

$$= \frac{1}{\pi} \int_0^\infty \frac{f(t+x) - f(x+)}{t} \sin At \, dt \tag{E.7}$$

What remains in showing (E.4) is to show that the integral in (E.7) tends to zero as $A \to \infty$. To do this we shall divide the interval of integration $0 \le t \le \infty$ into three integrals, $0 \le t \le \delta$, $\delta \le t \le T$, and $T \le t < \infty$. Denoting each of these integrals on its interval $[a,b]$ by $I_{a,b}$, this would say that

$$I_{0,\infty} = I_{0,\delta} + I_{\delta,T} + I_{T,\infty} \tag{E.8}$$

$$|I_{0,\infty}| \le |I_{0,\delta}| + |I_{\delta,T}| + |I_{T,\infty}|$$

We will show that for sufficiently small $\delta > 0$ and all A, $|I_{0,\delta}| < \epsilon/3$ for any $\epsilon > 0$ in (E.13), and for sufficiently large T and all $A \ge 1$, $|I_{T,\infty}| < \epsilon/3$ in (E.15) for any $\epsilon > 0$. With the help of the Riemann–Lebesgue lemma (Theorem 2.13), we will also show that $|I_{\delta,T}| < \epsilon/3$ in (E.10) for all sufficiently large $A \ge 1$. If this is the case, then we have

$$|I_{0,\infty}| < |I_{0,\delta}| + |I_{\delta,T}| + |I_{T,\infty}| < \epsilon \tag{E.9}$$

for sufficiently small δ, sufficiently large T, and sufficiently large $A \ge 1$. So, if we let $A \to \infty$ in (E.7), we obtain (E.4). What remains, of course, is to validate the estimates of the three integrals mentioned above.

In the middle integral $I_{\delta,T}$, we note that for any fixed x, $(f(t+x) - f(x+))/t$ is a piecewise continuous function on $(\delta, T]$, where according to the Riemann–Lebesgue lemma (Theorem 2.13) the integral $I_{\delta,T}$ will vanish as $A \to \infty$, which means we will have

$$|I_{\delta,T}| < \frac{\epsilon}{3} \tag{E.10}$$

for large enough $A \ge 1$. For the integral $I_{0,\delta}$ we take δ very small so that $(f(t+x) - f(x+))/t$ for $t \in (0,\delta)$ is very close to its derivative $f'(x)$, which means that we can easily bound it as

$$\left| \frac{f(t+x) - f(x+)}{t} \right| < 1 + |f'(x+)| \qquad (\text{E.11})$$

With this, we can have

$$
\begin{aligned}
|I_{0,\delta}| &= \frac{1}{\pi} \left| \int_0^\delta \frac{f(x+t) - f(x+)}{t} \sin At \, dt \right| \\
&\leq \frac{1}{\pi} \int_0^\delta \left| \frac{f(x+t) - f(x+)}{t} \right| |\sin At| \, dt \\
&\leq \frac{1}{\pi} \int_0^\delta \left| \frac{f(x+t) - f(x+)}{t} \right| dt < \frac{\delta}{\pi} (1 + |f'(x)|) \qquad (\text{E.12})
\end{aligned}
$$

after using $|\sin At| \leq 1$. Hence, we can choose δ small enough that $\delta < (\pi/3(1 + |f'(x)|))\epsilon$, which makes

$$|I_{0,\delta}| < \frac{\epsilon}{3} \qquad (\text{E.13})$$

As to the third integral $I_{T,\infty}$, we write

$$
\begin{aligned}
|I_{T,\infty}| &= \left| \frac{1}{\pi} \int_T^\infty f(x+t) \frac{\sin At}{t} \, dt - \frac{1}{\pi} f(x+) \int_T^\infty \frac{\sin At}{t} \, dt \right| \\
&\leq \frac{1}{\pi} \left| \int_T^\infty f(x+t) \frac{\sin At}{t} \, dt \right| + \frac{1}{\pi} \left| f(x+) \int_T^\infty \frac{\sin At}{t} \, dt \right| \\
&\leq \frac{1}{\pi} \int_T^\infty |f(x+t)| \left| \frac{\sin At}{t} \right| dt + \frac{1}{\pi} |f(x+)| \left| \int_{AT}^\infty \frac{\sin \xi}{\xi} \, d\xi \right| \\
&\leq \frac{M}{\pi T} + \frac{1}{\pi} |f(x+)| \left| \int_{AT}^\infty \frac{\sin \xi}{\xi} \, d\xi \right| \qquad (\text{E.14})
\end{aligned}
$$

after letting $\xi = At$ in the second integral, using $|(\sin At)/t| = |(\sin At)|/|t| < 1/T$ for $t \in [T, \infty]$, and $M = \int_T^\infty |f(x + \xi)| \, d\xi$, since f is absolutely integrable on $(-\infty, \infty)$. The second integral in (E.14) is convergent, since we have $\int_0^\infty ((\sin \xi)/\xi) \, d\xi = \pi/2$ from (2.88). This means that, provided $A \geq 1$, for sufficiently large T the contribution of this integral can be made very small. Then along with $|f(x)|$ bounded on $[T, \infty]$, the second term in (E.14) can be bounded by $\epsilon/6$. The first term, for finite M and sufficiently large T, can also be bounded by $\epsilon/6$, which gives

$$|I_{T,\infty}| < \frac{\epsilon}{3} \qquad (\text{E.15})$$

If we observe (E.10), (E.13), and (E.15), we have (E.4), which, along with the similar result (E.5), concludes our detailed proof of the Fourier integral theorem in (2.86). ∎

B. Simple (but Not Rigorous) Proofs of the Fourier Integral Formula (2.86)

The reader may feel that the above proof of Theorem 2.14 for the Fourier integral formula (2.86) is long. However, it is the most complete, considering the mathematical background assumed or what we have prepared for it in this book. There are, however, other methods that, if presented in a fast or simple way, would lack either the background or the rigor in justifying a number of assumed limiting processes.

(i) The first such method depends on familiarity with generalized functions (or distributions) and in particular the Dirac delta function $\delta(x)$, which we have attempted to define (casually, not rigorously) in Exercise 1.11b,c of Chapter 1. Earlier in the discussion of Theorem 2.14, we were tempted to use this (not so well defined distribution here) $\delta(x)$ to "casually" prove (2.86) in the following fashion.

We assumed that, in taking the limit as $A \to \infty$ of the equivalent integral (2.86) of integral (2.85), we can interchange the limit process with the infinite integral. If we have to justify this, we must think of a generalization of the Lebesgue convergence Theorem 2.9, where the integer $n \to \infty$ is replaced by $A \to \infty$. Even more, we must be prepared to take the resulting limit inside the integral in terms of the more general sense of distributions, and in particular as a Dirac delta function $\delta(x - \xi)$ from one of its defining limit processes as $\delta(x-\xi) = \lim_{A\to\infty}(\sin A(x - \xi)/\pi(x - \xi))$. We note that this generalized function $\delta(x)$ is ∞, or an infinite pulse, at $x = 0$ if we look at it in the sense of our usual functions. If we assume the background necessary for justifying the above limit interchange and receiving a limit in this general sense of distributions, then the Fourier integral formula (2.86) results in a two-line step of taking the limit as $A \to \infty$ of the integral in (2.86)

$$\lim_{A\to\infty} \frac{1}{\pi} \int_0^A d\lambda \int_{-\infty}^{\infty} f(\xi) \cos \lambda(x - \xi) d\xi$$

$$= \int_{-\infty}^{\infty} f(\xi) \left[\lim_{A\to\infty} \frac{1}{\pi} \int_0^A \cos \lambda(x - \xi) d\lambda \right] d\xi$$

$$= \int_{-\infty}^{\infty} f(\xi) \lim_{A\to\infty} \frac{\sin A(\xi - x)}{\pi(\xi - x)} d\xi = \int_{-\infty}^{\infty} f(\xi)\delta(\xi - x)d\xi = f(x)$$

after using the above limit to define $\delta(\xi - x)$, then defining $\delta(\xi - x)$ as an "impulse at $\xi = x$" to give the result $f(x)$ for the last integral.

(ii) The second method, and the one found in most books, does not need such sophisticated concepts as the generalized functions above, but it does slide over some very important justifications of passing to the limits. Such justifications, if done very properly, may even make such a proof longer than the detailed one we gave above for (2.86) of Theorem 2.14. We shall present this method here for completeness and also to show the pitfalls of doing the proof so fast. The reader familiar with this may skip it.

This method uses the following very familiar Fourier trigonometric series, which we shall present in detail in Chapter 3, for a piecewise smooth function $f(x)$ on a finite interval $(-l,l)$,

$$\tfrac{1}{2}[f(x+)+f(x-)] = \frac{a_0}{2} + \sum_{n=1}^{\infty} a_n \cos \frac{n\pi x}{l} + b_n \sin \frac{n\pi x}{l} \tag{2.92}$$

and converges to $f(x)$, where $f(x)$ is continuous at the interior point x. The Fourier coefficients a_0, a_n, and b_n in (2.92) are defined as

$$a_0 = \frac{1}{l} \int_{-l}^{l} f(t)\,dt \tag{2.93}$$

$$a_n = \frac{1}{l} \int_{-l}^{l} f(t) \cos \frac{n\pi t}{l}\, dt \tag{2.94}$$

$$b_n = \frac{1}{l} \int_{-l}^{l} f(t) \sin \frac{n\pi t}{l}\, dt \tag{2.95}$$

Since we are after $f(x)$ on $(-l,l)$, we will assume that there it is to be defined, piecewise smooth, and absolutely integrable. Also, we look at the above Fourier trigonometric series as a representation of such $f(x)$ on every finite interval $(-l,l)$.

Proof: The proof starts by substituting the integrals in (2.93)–(2.95) for the Fourier coefficients in (2.92),

$$f(x) = \frac{1}{2l} \int_{-l}^{l} f(t)\,dt + \frac{1}{l} \sum_{n=1}^{\infty} \left[\cos \frac{n\pi x}{l} \int_{-l}^{l} f(t) \cos \frac{n\pi t}{l}\, dt \right.$$

$$\left. + \sin \frac{n\pi x}{l} \int_{-l}^{l} f(t) \sin \frac{n\pi t}{l}\, dt \right]$$

$$= \frac{1}{2l} \int_{-l}^{l} f(t)\,dt + \frac{1}{l} \sum_{n=1}^{\infty} \int_{-l}^{l} f(t) \cos \frac{n\pi}{l}(t-x)\,dt \tag{2.96}$$

after realizing that $\cos(n\pi x/l)$ and $\sin(n\pi x/l)$ entered the integrals (with respect to t) as constant functions of t.

Now starts either a very cautious or a fast (nonrigorous) process of taking the limit of both sides of (2.96) as $l \to \infty$. Clearly our function $f(x)$, defined on $(-l, l)$, will become defined on $(-\infty, \infty)$. If we take the limit of the right side of (2.96) as $l \to \infty$, we see that the first integral vanishes, since the integral

$$\int_{-\infty}^{\infty} f(t) \, dt = L < \infty$$

by virtue of $f(t)$ is now absolutely integrable on $(-\infty, \infty)$. Thus,

$$\lim_{l \to \infty} \frac{1}{2l} \int_{-l}^{l} f(t) \, dt = \lim_{l \to \infty} \frac{1}{2l} L = 0$$

Our result, after taking the limit of (2.96) as $l \to \infty$, becomes

$$f(x) = \lim_{l \to \infty} \frac{1}{l} \sum_{n=0}^{\infty} \int_{-l}^{l} f(t) \cos \frac{n\pi}{l} (t - x) \, dt \tag{2.97}$$

At this stage lies the key point of this proof where we either stop to justify the limiting process very carefully or go over it in a "cursory" (nonrigorous) way. This is to show that the above limit in (2.97) does indeed result in the desired double integral (2.86) of the Fourier integral formula. Our task here, as $l \to \infty$, involves the intricate combination of the finite limits integral going to an infinite limit integral, and the infinite sum becoming an infinite integral with respect to a continuous argument λ. For our purpose here of showing "caution" for a careful analysis of such a limiting process, we say that such justification may take even longer preparation than that which we did for the proof of Theorem 2.14. We leave it for the interested reader to consult the references at the end of the chapter, along with some on real analysis.

To show how it is usually done, we proceed to let $\lambda_n = n\pi/l$ with $\Delta \lambda_n = (n+1)\pi/l - n\pi/l = \pi/l$; the latter is prepared by having the $1/l$ factor inside the sum of (2.97). Now it is done "intuitively" that $\Delta \lambda_n = \pi/l \to d\lambda$ and $\lambda_n = n \Delta \lambda_n \to \lambda$ as $l \to \infty$. With this in mind, the limit of (2.97) is considered a double integral

$$f(x) = \lim_{\substack{l \to \infty \\ \Delta \lambda_n \to 0 \\ (\Delta \lambda_n \to d\lambda)}} \frac{1}{\pi} \sum_{n=1}^{\infty} \Delta \lambda_n \int_{-l}^{l} f(t) \cos \frac{n\pi}{l} (t - x) \, dt \tag{2.98}$$

$$= \frac{1}{\pi} \int_{0}^{\infty} d\lambda \int_{-\infty}^{\infty} f(t) \cos \lambda (t - x) \, dt \tag{2.86}$$

as in (2.86), but only with intuition, and not a really hard mathematical justification.   ∎

Next, we will turn to the two very special cases of Fourier transforms, namely those for $f(x)$ being *odd* or *even functions*, which will result in the *Fourier sine* and *cosine transforms*, respectively. The existence of such transforms should follow easily as simple corollaries to Theorem 2.14.

C. The Fourier Sine and Cosine Transforms (Fourier Transforms on the Half-Line)

In most of the applications on an infinite domain, the function $f(x)$ may be defined only on the semi-infinite interval $(0, \infty)$. In this case, we may extend its definition to $(-\infty, \infty)$ and then rely on the Fourier integral theorem 2.14 for its existence on $(-\infty, \infty)$. Such extension, of course, must be done in a sensible way, most importantly not to contradict our assumptions of Theorem 2.14. Two such reasonable extensions are those of the *odd* and *even extensions to* $(-\infty, 0)$. It would even be better if there were a reason for each of these choices. Such a reason is often present in many applications when, for example, a boundary condition is given at the end $x = 0$ for the function $f(x)$ to vanish, that is, $f(0) = 0$. In this case, we extend the function $x = 0$ as an odd function $f_o(x)$, where the Fourier integral formula will give

$$\tfrac{1}{2}[f_o(0+) + f_o(0-)] = \tfrac{1}{2}[f(0+) - f(0+)] = 0$$

even if the extended $f_o(x)$ has a jump discontinuity at $x = 0$. In case $f_o(x)$ is continuous at $x = 0$, we get $f_o(0) = 0$. With such odd extension, the Fourier integral (2.64) of $f_o(x)$ will reduce to a *Fourier sine integral*,

$$\mathcal{F}\{f_o\} = F(\lambda) = \int_{-\infty}^{\infty} e^{-i\lambda x} f_o(x)\,dx = \int_{-\infty}^{\infty} [\cos \lambda x - i \sin \lambda x] f_o(x)\,dx$$

$$= \int_{-\infty}^{\infty} f_o(x) \cos \lambda x\,dx - i \int_{-\infty}^{\infty} f_o(x) \sin \lambda x\,dx$$

$$= -2i \int_{0}^{\infty} f(x) \sin \lambda x = -2i\mathcal{F}_s\{f\}$$

after using the fact that the integrand of the first integral is an odd function, where the integral vanishes, and that the integrand in the second integral is an even function, which gave double the value of the integral on $(0, \infty)$. We define the *Fourier sine integral* for $f(x)$ on $(0, \infty)$ as

$$\mathcal{F}_s\{f\} = F_s(\lambda) = \int_{0}^{\infty} f(x) \sin \lambda x\,dx \tag{2.99}$$

and justify the existence of its inverse

$$f(x) = \mathcal{F}_s^{-1}\{F_s\} = \frac{2}{\pi} \int_0^\infty F_s(\lambda) \sin \lambda x \, d\lambda \qquad (2.100)$$

as an easily established corollary to Theorem 2.14.

Corollary 1 The Inverse Fourier Sine Transform: If $f(x)$ is an absolutely integrable function on $(0,\infty)$, and is piecewise smooth on every finite interval of $(0,\infty)$, then

$$\tfrac{1}{2}[f(x+) + f(x-)] = \frac{2}{\pi} \int_0^\infty \sin \lambda x \, d\lambda \int_0^\infty f(\xi) \sin \lambda \xi \, d\xi \qquad \square \quad (2.101)$$

Proof: The proof follows easily from (2.86) of Theorem 2.14, when used with $f(x) = f_o(x)$, the odd extension of our function, and the ideas of odd and even integrands similar to what we used in getting the above result $\mathcal{F}\{f_o\} = -2i\mathcal{F}_s\{f\}$.

$$\tfrac{1}{2}[f_o(x+) + f_o(x-)] = \frac{1}{\pi} \int_0^\infty d\lambda \int_{-\infty}^\infty f_o(\xi)[\cos \lambda \xi \cos \lambda x$$

$$+ \sin \lambda \xi \sin \lambda x] d\xi$$

$$= \frac{1}{\pi} \int_0^\infty d\lambda \int_{-\infty}^\infty f_o(\xi) \cos \lambda \xi \cos \lambda x \, d\xi$$

$$+ \frac{1}{\pi} \int_0^\infty d\lambda \int_{-\infty}^\infty f_o(\xi) \sin \lambda \xi \sin \lambda x \, d\xi$$

$$= 0 + \frac{2}{\pi} \int_0^\infty d\lambda \int_0^\infty f(\xi) \sin \lambda \xi \sin \lambda x \, d\xi$$

$$\tfrac{1}{2}[f(x+) + f(x-)] = \frac{2}{\pi} \int_0^\infty \sin \lambda x \, d\lambda \int_0^\infty f(\xi) \sin \lambda \xi \, d\xi \qquad (2.101)$$

where the o for "odd" is being dropped, as it is understood now for the sine transform. ∎

Besides the above-mentioned application of $f(0) = 0$ for the sine transform of $f(x)$ on $(0,\infty)$, it very often happens that a function $f(x)$ on $(-\infty,\infty)$ is itself, naturally, an odd function. In this case, we have a Fourier sine transform to represent it on $(-\infty,\infty)$ and in particular the domain of interest $(0,\infty)$.

Another reason for extending the definition of $f(x)$ to $(-\infty,0)$ occurs when we have a boundary condition at $x = 0$ where the derivative df/dx

vanishes. For this case of $f'(0) = 0$, we extend the function about $x = 0$ as an even function $f_e(x)$, where the Fourier integral formula will give $f_e(0)$ at $x = 0$. This is the case since here $f_e(x)$ is continuous at $x = 0$ when its derivative $f_e'(0)$ is given to exist as $f_e'(0) = 0$. This is in contrast to the odd extension for the sine transform, where we get $\frac{1}{2}[f_o(0+) + f_o(0-)] = 0$ at $x = 0$. We get $f_o(0)$ only if the odd extension $f_o(x)$ happened to be continuous at $x = 0$. For the even extension, the Fourier integral (2.64) of $f_e(x)$ becomes a *Fourier cosine integral*,

$$\mathcal{F}\{f_e\} = F(\lambda) = \int_{-\infty}^{\infty} e^{-i\lambda x} f_e(x)\,dx = \int_{-\infty}^{\infty} [\cos \lambda x - i \sin \lambda x] f_e(x)\,dx$$

$$= \int_{-\infty}^{\infty} f_e(x) \cos \lambda x\,dx - i \int_{-\infty}^{\infty} f_e(x) \sin \lambda x\,dx$$

$$= 2 \int_0^{\infty} f(x) \cos \lambda x\,dx = 2\mathcal{F}_c\{f\}$$

after using the properties of the even and odd integrands on $(-\infty, \infty)$ in the above first and second integrals, respectively. We define the Fourier cosine integral on $f(x)$ on $(0, \infty)$ as

$$\mathcal{F}_c\{f\} = F_c(\lambda) = \int_0^{\infty} f(x) \cos \lambda x\,dx \tag{2.102}$$

and we justify the existence of its inverse

$$f(x) = \mathcal{F}_c^{-1}\{F\} = \frac{2}{\pi} \int_0^{\infty} F(\lambda) \cos \lambda x\,d\lambda \tag{2.103}$$

as the following second corollary to Theorem 2.14.

Corollary 2 The Inverse Fourier Cosine Transform: If $f(x)$ is absolutely integrable on $(0, \infty)$ and is piecewise smooth on every finite interval of $(0, \infty)$, then

$$\frac{1}{2}[f(x+) + f(x-)] = \frac{2}{\pi} \int_0^{\infty} \cos \lambda x\,d\lambda \int_0^{\infty} f(\xi) \cos \lambda \xi\,d\xi \qquad \square \tag{2.104}$$

Proof: The proof follows easily from (2.86) of Theorem 2.14 when used with $f(x) = f_e(x)$, the even extension of $f(x)$, and follows steps similar to those that lead to (2.101) of the Fourier sine transform. ∎

EXAMPLE 2.15: Evaluation of Fourier Type Integrals

In this example we will show that

$$\int_0^\infty \frac{\sin x}{x} \cos \alpha x \, dx = \begin{cases} \dfrac{\pi}{2}, & 0 \le \alpha < 1 \\ \dfrac{\pi}{4}, & \alpha = 1 \\ 0, & \alpha > 1 \end{cases} \tag{E.1}$$

we recognize this as a Fourier-type integral from its trigonometric $\cos \alpha x$ kernel. We also note that the Fourier-transformed function $(\sin x)/x$ is an even function, and it is in the form of $(2 \sin \lambda a)/\lambda$, the exponential Fourier transform (2.64) of the (piecewise continuous) gate function

$$p_a(t) = \begin{cases} 1, & |t| < a \\ 0, & |t| > a \end{cases}$$

Clearly, $p_a(t)$ is absolutely integrable and piecewise smooth on $(-\infty, \infty)$, hence it satisfies the conditions of Theorem 2.14 for the existence of its Fourier transform and also the Fourier inverse of the latter as the following:

$$\int_0^\infty \frac{\sin x}{x} \cos \alpha x \, dx = \frac{1}{2} \int_{-\infty}^\infty \frac{\sin x}{x} e^{i\alpha x} \, dx = \frac{\pi}{2} \cdot \frac{1}{2\pi} \int_{-\infty}^\infty 2 \frac{\sin x}{x} e^{i\alpha x} \, dx$$

$$= \frac{\pi}{2} \cdot \frac{1}{2} [p_1(\alpha+) + p_1(\alpha-)] \tag{E.2}$$

This integral converges to $\pi/2$ where $p_1(\alpha)$ is continuous since $\frac{1}{2}[p_1(\alpha+)+p_1(\alpha-)] = p_1(\alpha) = 1$ for $0 \le \alpha < 1$. The integral converges to $(\pi/4)[p_1(\alpha+)+p_1(\alpha-)] = (\pi/4)[0+1] = \pi/4$ at the point of jump discontinuity $\alpha = 1$ of $p_1(\alpha)$, and it converges to $(\pi/4)[0+0] = 0$ for all $\alpha > 1$, where $p_1(\alpha)$ is continuous since it vanishes identically there. Hence we have the desired result in (E.1).

The particular special case of (E.1) for $\alpha = 0$ gives

$$\int_0^\infty \frac{\sin x}{x} \, dx = \frac{\pi}{2} \tag{E.3}$$

which is the well-known *Dirichlet integral*. We note that this result (E.3) cannot be reached with elementary integration methods. Indeed, we have already struggled with it, as we needed it in (2.88) for the proof of the Fourier integral theorem 2.14. Of course, then we could not have applied the method of this example, as it relies on the result of Theorem 2.14. Instead we obtained (E.3) as a special case of a Laplace transform pair in (2.89), again with the help of another (an improper) integral (Laplace) transform.

Looking at the Fourier integral formulas in (2.101) and (2.104) for the Fourier sine and cosine transforms, respectively, we can easily see that such transforms and their inverses can be put in a more symmetric form. This is done simply by distributing the $2/\pi$ factor as $\sqrt{2/\pi}$ in each of the transform and its inverse:

$$F_s(\lambda) = \sqrt{\frac{2}{\pi}} \int_0^\infty f(x) \sin \lambda x \, dx \tag{2.105}$$

$$f(x) = \sqrt{\frac{2}{\pi}} \int_0^\infty F_s(\lambda) \sin \lambda x \, d\lambda \tag{2.106}$$

$$F_c(\lambda) = \sqrt{\frac{2}{\pi}} \int_0^\infty f(x) \cos \lambda x \, dx \tag{2.107}$$

$$f(x) = \sqrt{\frac{2}{\pi}} \int_0^\infty F_c(\lambda) \cos \lambda x \, dx \tag{2.108}$$

D.   The Different Notations in References and Tables of Fourier Transforms

The above symmetric forms of the Fourier sine and cosine and even the exponential Fourier transforms are used in many references, in particular those oriented toward physics applications. These forms are to be compared to the ones we adopt here, in (2.99), (2.100), (2.102), and (2.103). As we remarked earlier, and repeat again here for more emphasis, in this respect among the well-known basic references on integral transforms, Churchill (1972) has our notation; Sneddon (1972) uses the above symmetric notation; Papoulis (1965) in signal analysis uses the notation in (2.66), (2.67) with, understandably $j = \sqrt{-1}$ instead of $i = \sqrt{-1}$; Davies (1978) uses notation similar to ours in (2.68), (2.69); and Wolf (1985) uses the above symmetric notation. Walker (1988) uses (2.66a), (2.67a).

With such varied notations among the different references, it becomes essential that we check the notation of any book we use in cross referencing, particularly when we consult the tables of Fourier transforms. In that regard, the book on integral transforms by Ditkin and Prudnikov (1965), with its expanded valuable tables, uses our notation in the tables and the symmetric form within the text. The very well-known main reference of two volumes on integral transforms and their extensive tables by Erdelyi and co-authors (Vol. 1, 1954) uses our notation.

In Section 2.5 we will have more on the analysis of these Fourier sine and cosine transforms, their compatibility with even-order derivatives, and their applications to physical problems defined on the half-real line $0 < x < \infty$.

2.2.2 Basic Properties and the Convolution Theorem

In this section we will establish the most basic properties of the exponential Fourier transform. These properties are summarized and presented as Fourier transform pairs in Table 2.2. The properties of the Fourier sine or cosine transforms will often follow as special cases.

As we shall see, the rigorous proofs of these properties will need clear justifications which will force us to limit the class of functions we work with for these conditions to be satisfied.

We will concentrate on proving the most essential results needing more care in their proofs, like the Fourier transform of derivatives and the convolution theorem. These proofs will bring to light the importance of the basic conditions needed for justifying the existence of the Fourier transform and its inverse, or a combination of them, that were the core of the Fourier integral formula (Theorem 2.14). The rest of the results will be stated clearly, with part of their proofs left as exercises with, often, liberal hints.

In the section following this one, we will illustrate the necessity of such properties for solving boundary value problems associated with partial differential equations on the whole line $(-\infty, \infty)$.

A. The Linear Property

As we have seen for the Laplace transform, the Fourier transform, when it exists, is an integral which is a linear operation; that is, it is a distributive operation over the addition of two integrable functions.

For $f_1(x)$ and $f_2(x)$ absolutely integrable on the real line $(-\infty, \infty)$, and c_1, c_2 constants we have the linearity property of Fourier transforms:

$$\mathcal{F}\{c_1 f_1(x) + c_2 f_2(x)\} = c_1 F_1(\lambda) + c_2 F_2(\lambda) \tag{2.109}$$

where $F_1(x)$ and $F_2(\lambda)$ are the exponential Fourier transforms of $f_1(x)$ and $f_2(x)$, respectively. The proof is straightforward when we use $f(x) = c_1 f_1(x) + c_2 f_2(x)$ in (2.64), knowing that $f(x)$ is absolutely integrable, since $f_1(x)$ and $f_2(x)$ are:

$$\mathcal{F}\{c_1 f_1(x) + c_2 f_2(x)\} = F(\lambda) = \int_{-\infty}^{\infty} e^{-i\lambda x}[c_1 f_1(x) + c_2 f_2(x)]\, dx$$

TABLE 2.2 Fourier (Exponential) Transform Pairs

$f(x)$	$F(\lambda)$
$f(x) = \dfrac{1}{2\pi} \displaystyle\int_{-\infty}^{\infty} e^{i\lambda x} F(\lambda) \, d\lambda$	$F(\lambda) = \displaystyle\int_{-\infty}^{\infty} e^{-i\lambda x} f(x) \, dx$ (see (2.65))

Pairs

$$f(x) = \begin{cases} A, & |x| \le a \\ 0, & |x| > a \end{cases} \qquad \frac{2A \sin \lambda a}{\lambda}$$

$$e^{i\alpha x^2} \qquad \sqrt{\frac{\pi}{\alpha}} e^{i\pi/4} e^{-i\lambda^2/4\alpha}$$

$$\frac{J_n(t)}{t^n} \qquad \begin{cases} \dfrac{2(1-\lambda^2)^{n-1/2}}{1 \cdot 3 \cdot 5 \cdots (2n-1)}, & |\lambda| < 1 \\ 0, & |\lambda| > 1 \end{cases}$$

Operations

$f(ax)$	$\dfrac{1}{|a|} F\left(\dfrac{\lambda}{a}\right)$
$e^{iax} f(x)$	$F(\lambda - a)$
$f(x - x_0)$	$e^{-ix_0\lambda} F(\lambda)$
$\dfrac{d^n f}{dx^n}$	$(i\lambda)^n F(\lambda)$
$\dfrac{\partial^n f(x,y)}{\partial x^n}$	$(i\lambda)^n F(\lambda, y)$
$\displaystyle\int_{\mp\infty}^{x} f(\xi) \, d\xi$	$\dfrac{F(\lambda)}{i\lambda}$
$\displaystyle\int_{-\infty}^{\infty} f_1(\xi) f_2(x - \xi) \, d\xi$	$F_1(\lambda) F_2(\lambda)$
$F(x)$	$2\pi f(-\lambda)$

$$= c_1 \int_{-\infty}^{\infty} e^{-i\lambda x} f_1(x) \, dx + c_2 \int_{-\infty}^{\infty} e^{-i\lambda x} f_2(x) \, dx$$

$$= c_1 F_1(\lambda) + c_2 F_2(\lambda)$$

$$\mathcal{F}\{c_1 f_1(x) + c_2 f_2(x)\} = c_1 F\{f_1(x)\} + c_2 F\{f_2(x)\} \qquad (2.109)$$

From now on, even when it is not mentioned explicitly, we will assume any function $f(x)$ to be transformed, as in the operations above, to be absolutely integrable for the existence of its transform. In addition, most often it is assumed to be piecewise smooth to guarantee the existence of the inverse of its transform. Also, it should be made clear in our justifications for the proofs that any new function $g(x)$ resulting from certain operations on $f(x)$ such as $g(x) = df/dx$ satisfies the above conditions on $(-\infty, \infty)$.

B. Scaling

$$\mathcal{F}\{cf(x)\} = \frac{1}{|c|} F\left(\frac{\lambda}{c}\right), \qquad |c| > 0 \tag{2.110}$$

We prove this result here for $c > 0$ and leave the case of $c < 0$ as an exercise.

$$\mathcal{F}\{f(x)\} = \int_{-\infty}^{\infty} f(cx)e^{-i\lambda x}\, dx$$

$$= \frac{1}{c} \int_{-\infty}^{\infty} f(u)e^{-i(\lambda/c)u}\, du = \frac{1}{c} F\left(\frac{\lambda}{c}\right), \qquad c > 0$$

after making the change of variable $u = cx$.

C. Shifting in the Physical x-Space

$$\mathcal{F}\{f(x - a)\} = e^{-i\lambda a} F(\lambda) \tag{2.111}$$

which is left for an exercise (see Exercise 2.41a).

D. Shifting in the Fourier λ-Space

$$\mathcal{F}\{e^{icx}f(x)\} = F(\lambda - c) \tag{2.112}$$

is also left for a simple exercise (see Exercise 2.41b). We are reminded here that the Fourier transform of $e^{icx}f(x)$ does exist, since

1. $e^{icx}f(x) = \frac{1}{2}[\cos cx + i \sin cx]f(x)$ is piecewise smooth, for it is a product of a smooth function $\frac{1}{2}(\cos cx + i \sin cx)$ and the piecewise smooth $f(x)$.
2. $e^{icx}f(x)$ is absolutely integrable, since

$$\int_{-\infty}^{\infty} |e^{icx}f(x)|\, dx = \int_{-\infty}^{\infty} |e^{icx}||f(x)|\, dx = \int_{-\infty}^{\infty} |f(x)|\, dx < \infty$$

for $f(x)$ absolutely integrable on $(-\infty, \infty)$.

E. Complex-Valued Functions $f(x)$[†]

Most of the functions we dealt with up to now are assumed to be real-valued functions $f(x)$ on $(-\infty, \infty)$. When $f(x)$ is complex-valued, we show the following result, which we shall need in establishing the Parseval equality in (2.136):

$$\mathcal{F}\{\overline{f(-x)}\} = \overline{F(\lambda)} \tag{2.113}$$

where $\overline{f(x)}$ stands for the complex conjugate of $f(x)$,

$$\mathcal{F}\{\overline{f(-x)}\} = \int_{-\infty}^{\infty} \overline{f(-x)}e^{-i\lambda x}\,dx = -\int_{\infty}^{-\infty} \overline{f(u)}e^{i\lambda u}\,du$$

$$= \int_{-\infty}^{\infty} \overline{f(u)}\,\overline{e^{-i\lambda u}} = \int_{-\infty}^{\infty} \overline{f(u)e^{-i\lambda u}}\,du = \overline{\int_{-\infty}^{\infty} f(u)e^{-i\lambda u}\,du}$$

$$= \overline{F(\lambda)}$$

$$\mathcal{F}\{\overline{f(-x)}\} = \overline{F(\lambda)} \tag{2.113}$$

In the following Theorem 2.15, we will prove one of the most useful properties of the Fourier transforms for solving differential equations (with constant coefficients!). This is the transforming of the derivative df/dx in the x-space to the algebraic $i\lambda F(\lambda)$ in the Fourier λ-space. The Fourier transform of higher derivatives $d^n f/dx^n$ will follow easily as a corollary to Theorem 2.15. These results were derived formally in Chapter 1 as (1.12) and (1.13), but we shall state them here more carefully along with sound proofs.

THEOREM 2.15: The Fourier Transform of df/dx

Let $f(x)$ be continuous and absolutely integrable on $(-\infty, \infty)$, and let df/dx be piecewise smooth and absolutely integrable on $(-\infty, \infty)$; then

$$\mathcal{F}\{df/dx\} = i\lambda F(\lambda) \qquad \qquad \square \tag{2.114}$$

The proof will follow the same (formal) steps as used in getting this result (1.12) in Chapter 1. The difference here is that we have to justify all these steps:

1. The minute we speak of the Fourier transform $G(\lambda)$ of $df/dx = g(x)$, we have to guarantee the existence of $G(\lambda)$, which is the case here since the theorem assumes such sufficient conditions of

---

[†]For a review of simple elements of complex numbers, see Exercise 1.6 in Chapter 1.

df/dx (piecewise smooth and absolutely integrable) on $(-\infty, \infty)$ as dictated by the Fourier integral Theorem 2.14. This theorem also guarantees the existence of the Fourier inverse df/dx of $i\lambda F(\lambda)$.

2. When we have our result (2.114) in terms of $F(\lambda)$, we also need assurance of its existence according to Theorem 2.12, which is the case since we assume here that the transformed function $f(x)$ is absolutely integrable. The Fourier integral Theorem 2.14 is valid here since $f(x)$ is absolutely integrable, and it is also piecewise smooth, since its derivative df/dx is piecewise smooth.

3. Before we start our steps, we will need

$$\lim_{|x| \to \infty} f(x) = 0 \tag{2.115}$$

to get rid of the boundary terms resulting from the integration by parts. In some books, for assurance, this condition may be put in the above theorem explicitly. However, this condition clearly follows as a necessary condition to our assumption that $f(x)$ is absolutely integrable on $(-\infty, \infty)$.

So we will write the formal steps which, combined with the above justifications (1)–(3), make it a rigorous proof. With one integration by parts on the following integral we have:

$$\mathcal{F}\left\{ \frac{df}{dx} \right\} = \int_{-\infty}^{\infty} e^{-i\lambda x} \frac{df}{dx}\, dx = e^{-i\lambda x} f(x) \Big|_{x=-\infty}^{\infty} - \int_{-\infty}^{\infty} -i\lambda e^{-i\lambda x} f(x)\, dx$$

$$= 0 + i\lambda \int_{-\infty}^{\infty} e^{-i\lambda x} f(x)\, dx = i\lambda F(\lambda) \tag{2.116}$$

after noting that $e^{-i\lambda x}$ is bounded at $x = \mp\infty$, and from (2.115) $f(x)$ vanishes at both limits $x = \mp\infty$.

The special very important case for second-order derivatives,

$$\mathcal{F}\left\{ \frac{d^2 f}{dx^2} \right\} = -\lambda^2 F(\lambda) \tag{2.117}$$

and the most general case,

$$\mathcal{F}\left\{ \frac{d^n f}{dx^n} \right\} = (i\lambda)^n F(\lambda) \tag{2.118}$$

can be proved by repeated integration by parts as was done for (1.12) and (1.13) in Chapter 1. We will state the result (2.118) as the following corollary. The conditions will make clear the mathematical justifications for the formal steps followed in a way very similar to the ones above used for the proof of the theorem. These conditions are to secure the existence of all resulting integrals from the repeated integrations by parts in addition to making the boundary terms, resulting from such integrations, vanish at $x = \mp\infty$.

Corollary 1 The Fourier transform of $d^n f/dx^n$: Let $f(x)$, $f'(x)$, $f''(x)$, ..., $f^{(n-1)}(x)$ be continuous and absolutely integrable on $(-\infty, \infty)$, and let the nth derivative $f^{(n)}(x)$ be piecewise continuous and absolutely integrable on $(-\infty, \infty)$. Then

$$\mathcal{F}\left\{\frac{d^n f}{dx^n}\right\} = (i\lambda)^n F(\lambda) \qquad \square \quad (2.118)$$

We note how the conditions on $d^n f/dx^n$ secure the existence of its Fourier transform, and the conditions on the preceding $n - 1$ derivatives $f^{(n-1)}(x)$, ..., $f''(x)$, $f'(x)$, $f(x)$ secure the existence of the succeeding Fourier integrals resulting from the successive n integrations by parts until we reach the Fourier integral of $f(x)$. It should be clear that the following necessary condition, for the vanishing of the boundary terms (of the integration by parts)

$$\lim_{|x| \to \infty} f^{(k)}(x) = 0, \qquad k = 0, 1, 2, ..., n - 1 \qquad (2.119)$$

is implied by the assumptions that the derivatives $f^{(k)}(x)$, $k = 1, 2, ...,$ $n - 1$, are absolutely integrable on $(-\infty, \infty)$.

F. The Fourier Transform of df/dx with Jump Discontinuities for $f(x)$

Corollary 2 $f(x)$ with jump discontinuities: Instead of $f(x)$ in Theorem 2.15 being continuous, assume that it is sectionally continuous with a jump discontinuity at $x = a$ of magnitude $J = |f(a+) - f(a-)|$. Then the result (2.114) (for the continuous $f(x)$) is generalized to

$$\mathcal{F}\left\{\frac{df}{dx}\right\} = i\lambda F(\lambda) - e^{-i\lambda a}[f(a+) - f(a-)] \qquad (2.120)$$

In case $f(x)$ has n jump discontinuities at $a_1, a_2, ..., a_n$, (2.120) becomes

$$\mathcal{F}\left\{\frac{df}{dx}\right\} = i\lambda F(\lambda) - \sum_{k=1}^{n} e^{-i\lambda a_k}[f(a_k+) - f(a_k-)] \qquad \square \quad (2.121)$$

We will prove (2.120) here, and we leave (2.121) as an exercise. Because of the jump discontinuity at $x = a$, we will write the Fourier integral of df/dx on two intervals, $(-\infty, a-)$ and $(a+, \infty)$,

$$\mathcal{F}\left\{\frac{df}{dx}\right\} = \int_{-\infty}^{a-} \frac{df}{dx} e^{-i\lambda x} \, dx + \int_{a+}^{\infty} \frac{df}{dx} e^{-i\lambda x} \, dx \qquad (2.122)$$

Then we perform an integration by parts on each of the above integrals, similar to what we did for (2.116), to have

$$= f(x)e^{-i\lambda x}\Big|_{-\infty}^{a-} + i\lambda \int_{-\infty}^{a-} f(x)e^{-i\lambda x} \, dx$$

$$+ f(x)e^{-i\lambda x}\Big|_{a+}^{\infty} + i\lambda \int_{a+}^{\infty} f(x)e^{-i\lambda x} \, dx$$

$$= -[f(a+) - f(a-)]e^{-i\lambda a} + i\lambda \int_{-\infty}^{\infty} f(x)e^{-i\lambda x} \, dx$$

$$= -[f(a+) - f(a-)]e^{-i\lambda a} + i\lambda F(\lambda) \qquad (2.120)$$

after noting that $e^{-i\lambda x}$ is continuous at $x = a$, namely $e^{-i\lambda a+} = e^{-i\lambda a-} = e^{-i\lambda a}$, and bringing the two integrals together to give the integral for $F(\lambda)$.

Now we state the corresponding result to (2.114) in the Fourier space,

$$\mathcal{F}\{-ixf(x)\} = \frac{dF}{d\lambda}(\lambda) \qquad (2.123)$$

or

$$\mathcal{F}^{-1}\left\{\frac{dF(\lambda)}{d\lambda}\right\} = -ixf(x) \qquad (2.124)$$

Just as was done formally in Chapter 1, this result is obtained by differentiating the Fourier integral

$$F(\lambda) = \int_{-\infty}^{\infty} e^{-i\lambda x} f(x) \, dx$$

$$\frac{dF}{d\lambda} = \frac{d}{d\lambda} \int_{-\infty}^{\infty} e^{-i\lambda x} f(x) \, dx = \int_{-\infty}^{\infty} \frac{d}{d\lambda} e^{-i\lambda x} f(x) \, dx$$

$$= \int_{-\infty}^{\infty} -ixf(x)e^{-i\lambda x} \, dx = \mathcal{F}\{-ixf(x)\} \qquad (2.123)$$

This step of interchanging differentiation, with respect to λ, with integration is justified if the middle integral above resulting from such interchange is uniformly convergent to allow such an operation. The conditions in the following Theorem 2.16, as a precise statement of our desired result (2.123), will justify this interchange. Hence, we consider the theorem proved after the steps in (2.123) are justified.

THEOREM 2.16:   Fourier Transform of Variable (Polynomial) Coefficient Terms

Let $f(x)$ be continuously differentiable (i.e., df/dx is continuous on $(-\infty, \infty)$) and $xf(x)$ be absolutely integrable on $(-\infty, \infty)$; then (2.123) results:

$$\mathcal{F}\{-ixf(x)\} = \frac{dF(\lambda)}{d\lambda} \tag{2.123}$$

We note that for $f(x)$ continuously differentiable, the integrand $(-ix) \times (f(x)e^{-i\lambda x})$ in the integral of (2.123) is obviously continuously differentiable on $(-\infty, \infty)$; also, it is absolutely integrable since $xf(x)$ is. So the integral is uniformly convergent and the interchange of differentiation and integration is permitted (also see Theorem 2.6). □

The result in (2.123) can be generalized to higher derivatives $d^n F/d\lambda^n$, as stated in the following corollary, whose proof we leave as an exercise.

Corollary 1: Let $f(x)$ be continuously differentiable and $x^n f(x)$ be absolutely integrable on $(-\infty, \infty)$; then

$$\mathcal{F}\{(-ix)^n f(x)\} = \frac{d^n F(\lambda)}{d\lambda^n} \qquad \square \tag{2.125}$$

Also, note the extension of this result to

$$(-i)^{n+k} \mathcal{F}\left\{x^n \frac{d^k f}{dx^k}\right\} = \frac{d^n}{d\lambda^n}(\lambda^k F(\lambda)) \tag{2.126}$$

in Exercise 2.42b.

G. The Convolution Theorem

We had shown for the Laplace transform, regarding its convolution Theorem 2.11, that the inverse Laplace transform of the product $F_1(s)F_2(s)$ of two Laplace transforms $F_1(s)$, $F_2(s)$ is the Laplace convolution product $(f_1 * f_2)(x)$ of their respective inverses $f_1(x)$ and $f_2(x)$ as given in (2.35)

$$\mathcal{L}\{(f_1 * f_2)(x)\} = F_1(s)F_2(s) \tag{2.35}$$

In this respect the Fourier transform (2.64) is also an integral with exponential kernel $e^{-i\lambda x}$, instead of e^{-sx} for the Laplace transform. Thus, it is not surprising to expect a similar result as a convolution theorem for the Fourier transform.

The Fourier Convolution Theorem 2.18
Consider $f_1(x)$ and $f_2(x)$ and their respective Fourier transforms $F_1(\lambda)$ and $F_2(\lambda)$. The Fourier convolution product of $f_1(x)$ and $f_2(x)$ is defined as

$$(f_1 * f_2)(x) \equiv \int_{-\infty}^{\infty} f_1(x-t)f_2(t)\,dt \tag{2.127}$$

and the Fourier convolution theorem states that

$$\mathcal{F}\{(f_1 * f_2(x))\} = \int_{-\infty}^{\infty} e^{-i\lambda x}\,dx \int_{-\infty}^{\infty} f_1(x-t)f_2(t)\,dt$$

$$= F_1(\lambda)F_2(\lambda) \tag{2.128}$$

Following the operations for deriving the convolution theorem of the Laplace transforms, we will first derive this result formally with a few steps, then follow it by a careful statement and a rigorous proof that will make very clear the justification of these steps for the class of functions considered. We do this along with a necessary study of some of the properties of the convolution product in (2.128) because of the importance of this theorem in many applications. This is in addition to generating other important tools of Fourier analysis such as the *Parseval equality* in (2.136). The convolution theorem application in solving ordinary or partial differential equations is clear. As in the case of the Laplace transform, we often end with a product of the Fourier transforms in the Fourier λ-space, which is to be transformed back to the physical x-space. This is often done via the convolution theorem. An illustration of such application is given in Example 2.18. However, sometimes the product may be reduced to a sum of simple transforms, as done with partial fractions, for example, for each of which we may find the inverse transform in available tables of transforms and their inverses.

To prove (2.128) formally, we write

$$\mathcal{F}(f_1 * f_2)(x) = \int_{-\infty}^{\infty} e^{-i\lambda x} \left[\int_{-\infty}^{\infty} f_1(x-t)f_2(t)\,dt \right] dx$$

$$= \int_{-\infty}^{\infty} \int_{-\infty}^{\infty} f_1(x-t)f_2(t)e^{-i\lambda x}\,dx\,dt \tag{2.129}$$

after noting that we have allowed the exchange of the order of the iterated integrals. This, as we have seen before, is a crucial point which must be justified. Such justification is usually attained at the expense of limiting our functions to a class that will allow such exchange. As we shall see in the following rigorous proof, it turns out that limiting our functions $f_1(x)$, $f_2(x)$ to be bounded, absolutely integrable on $(-\infty, \infty)$, and piecewise continuous is sufficient to allow such interchange of order of integration in (2.129).

With a simple change of variable $u = x - t$, $v = t$ in (2.129), we have

$$\mathcal{F}\{(f_1 * f_2)(x)\} = \int_{-\infty}^{\infty} \int_{-\infty}^{\infty} f_1(u)f_2(v)e^{-i\lambda(u+v)}\,du\,dv$$

$$= \int_{-\infty}^{\infty} f_1(u)e^{-i\lambda u}\,du \int_{-\infty}^{\infty} f_2(v)e^{-i\lambda v}\,dv \qquad (2.130)$$

$$\mathcal{F}\{(f_1 * f_2)(x)\} = F_1(\lambda)F_2(\lambda) \qquad (2.128)$$

Before we present the convolution theorem with its proof, we need to develop some necessary properties of the Fourier convolution product $(f_1 * f_2)(x) = h(x)$ in (2.127).

If $f_1(x)$ and $f_2(x)$ are continuously differentiable and absolutely integrable on $(-\infty, \infty)$, then their convolution product $h(x) = (f_1 * f_2)(x)$, as defined by (2.127), is also continuously differentiable and absolutely integrable. To show that $h(x)$ is absolutely integrable we have:

$$\int_{-\infty}^{\infty} |h(x)|\,dx = \int_{-\infty}^{\infty} \left| \int_{-\infty}^{\infty} f_1(x-t)f_2(t)\,dt \right| dx$$

$$\leq \int_{-\infty}^{\infty} dx \left[\int_{-\infty}^{\infty} |f_1(x-t)f_2(t)|\,dt \right]$$

$$= \int_{-\infty}^{\infty} |f_2(t)|\,dt \int_{-\infty}^{\infty} |f_1(x-t)|\,dx$$

$$= \int_{-\infty}^{\infty} |f_2(t)|\,dt \int_{-\infty}^{\infty} |f_1(u)|\,du < \infty \qquad (2.131)$$

since $f_1(x)$ and $f_2(x)$ are absolutely integrable. However, in the above formal manipulations for (2.131), we have also allowed the exchange of the order of integration in the second double integral. This is the same as what was faced above in the formal proof of the Laplace convolution Theorem 2.11. As we know from Theorem 2.7, this is fine if the inner integral is uniformly convergent. This is the case of the above class of functions as shown in the following Theorem 2.17, which we shall state without proof due to the limited space we have here (see Churchill, 1972), p. 393).

THEOREM 2.17: The Boundedness of the Convolution Product of Absolutely Integrable Functions

Let $f_1(x)$ and $f_2(x)$ be bounded, absolutely integrable on $(-\infty, \infty)$, and sectionally continuous on each bounded interval; then the convolution integral

$$\int_{-\infty}^{\infty} f_1(x-t)f_2(t)\,dt \equiv (f_1 * f_2)(x) = h(x) \qquad (2.127)$$

is absolutely and uniformly convergent to $h(x)$ which is bounded and continuous for all x. $\qquad\square$

The functions $f_1(x)$, $f_2(x)$ we used in the above result (2.131) are continuously differentiable, which makes them bounded and even continuous to satisfy that part of Theorem 2.17.

Next we leave it as an exercise to prove that the convolution product is commutative, distributive over addition, and associative:

Lemma For $f_1(x)$, $f_2(x)$, and $f_3(x)$ in the class of functions of Theorem 2.17,

(1) $(f_1 * f_2)(x) = (f_2 * f_1)(x)$ $\qquad\qquad\qquad\qquad$ (2.132)

(2) $f_1(x) * [f_2(x) + f_3(x)] = (f_1 * f_2)(x) + (f_1 * f_3)(x)$ $\qquad$ (2.133)

(3) $f_1 * (f_2 * f_3) = (f_1 * f_2) * f_3$ $\qquad\qquad\qquad$ $\square$ (2.134)

With the aid of Theorem 2.17 and the experience we have with the result (2.129) that allows the exchange of the iterated integrals, we feel ready to state and prove the following *Fourier convolution Theorem 2.18.*

THEOREM 2.18: The Fourier Convolution Theorem

If $f_1(x)$ and $f_2(x)$ are in the class of functions used in Theorem 2.17—namely, they are bounded, absolutely integrable on $(-\infty, \infty)$, and piecewise continuous over each bounded interval—then the Fourier transform of the convolution product exists, and its Fourier transform is the product $F_1(\lambda)F_2(\lambda)$, as in (2.128):

$$\mathcal{F}(f_1 * f_2)(x) = F_1(\lambda)F_2(\lambda) \qquad (2.128)$$

where $F_1(\lambda)$ and $F_2(\lambda)$ are the Fourier transforms of $f_1(x)$ and $f_2(x)$, respectively. Obviously, $F_1(\lambda)$, $F_2(\lambda)$ exist and $(f_1 * f_2)(x)$ is absolutely integrable. $\qquad\square$

Proof: The last statement of the existence of $F_1(\lambda)$ and $F_2(\lambda)$ is very clear from Theorem (2.12) as $f_1(x)$, $f_2(x)$ are absolutely integrable. The fact that $h(x) = (f_1 * f_2)(x)$ is absolutely integrable was shown in (2.131). The main result can now be established by following the steps of the formal proof in (2.129), (2.130), except here we are sure that we can exchange the order of the integrals in the double integral in (2.130). This is the case, since according to Theorem 2.17 the inner integral in the first double integral of (2.129) is uniformly convergent for the class of functions considered here. This is sufficient to allow the exchange of the iterated integrals according to Theorem 2.7. ∎

H. The Parseval Equality

The convolution theorem (2.128), which can be written as

$$\frac{1}{2\pi} \int_{-\infty}^{\infty} e^{i\lambda x} F_1(\lambda) F_2(\lambda) d\lambda = \int_{-\infty}^{\infty} f_1(x-t) f_2(t) dt \qquad (2.128)$$

can be used easily to derive the very important Parseval's equality for Fourier analysis. In (2.128), we let $x = 0$,

$$\frac{1}{2\pi} \int_{-\infty}^{\infty} F_1(\lambda) F_2(\lambda) d\lambda = \int_{-\infty}^{\infty} f_1(-t) f_2(t) dt \qquad (2.135)$$

Then we consider the special case with $f_1(-t) = \overline{f_2(t)}$, which can be written as $f_2(t) = \overline{f_2(t)} = \overline{f_1(-t)}$. In (2.113) we developed the result $\mathcal{F}\{\overline{f_1(-t)}\} = F_1(\lambda)$, and for our case this becomes $\mathcal{F}\{\overline{f_1(-t)}\} = \mathcal{F}\{f_2(t)\} = F_1(\lambda)$. If we use this $f_2(t) = \overline{f_1(-t)}$ in (2.135), we obtain the *Parseval equality*,

$$\frac{1}{2\pi} \int_{-\infty}^{\infty} F_1(\lambda) \overline{F_1(\lambda)} d\lambda = \int_{-\infty}^{\infty} f_1(-t) \overline{f_1(-t)} dt$$

$$= \int_{-\infty}^{\infty} f_1(u) \overline{f_1(u)} du$$

$$\frac{1}{2\pi} \int_{-\infty}^{\infty} |F_1(\lambda)|^2 d\lambda = \int_{-\infty}^{\infty} |f_1(t)|^2 dt \qquad (2.136)$$

after a simple change of variable $u = -t$ in the right integral and the use of $f\overline{f} = |f|^2$.

Among the applications of the Parseval equality as a tool in Fourier analysis (as we shall see in this chapter and the coming chapters) is its use for evaluating integrals that cannot be obtained by the methods of elementary calculus. The following Example 2.16 will illustrate this application for the evaluation of the integral

$$\int_0^{\infty} \frac{\sin^2 x}{x^2} dx = \frac{\pi}{2}$$

EXAMPLE 2.16:   Evaluation of Improper Integrals

In evaluating the integral $\int_0^\infty ((\sin^2 x)/x^2)\,dx$ we note that the integrand is an even function and we can write it now in a form suitable for the Parseval equality (2.136),

$$\int_0^\infty \frac{\sin^2 x}{x^2}\,dx = \frac{1}{2}\int_{-\infty}^\infty \frac{\sin^2 x}{x^2}\,dx \tag{E.1}$$

Looking at Parseval's equality, we find the opportunity of trade-off between the integral of the square of the function $f_1(t)$ and that of its transform $F_1(\lambda)$. So we can search for the possibility of one being easier to do than the other. In this case, as we showed in (1.32), $F_1(\lambda) = (2\sin \lambda a)/\lambda$ as the Fourier transform of the gate function

$$f_1(x) = p_a(x) = \begin{cases} 1, & |x| < a \\ 0, & |x| > a \end{cases} \tag{E.2}$$

whose square can easily be integrated. We again evaluate

$$F_1(\lambda) = \int_{-\infty}^\infty e^{-i\lambda x} f_1(x)\,dx = \int_{-a}^a e^{-i\lambda x}\,dx$$

$$= -\frac{1}{i\lambda} e^{-i\lambda x}\,\Big|_{x=-a}^a = -\frac{1}{i\lambda}\left[e^{-i\lambda a} - e^{i\lambda a} \right]$$

$$= \frac{2\sin \lambda a}{\lambda} \tag{E.3}$$

So, if we use $F_1(\lambda)$ of (E.3) and $f_1(x)$ of (E.2) in the Parseval equality (2.136), we have

$$\frac{1}{2\pi}\int_{-\infty}^\infty \frac{4\sin^2 \lambda a}{\lambda^2}\,d\lambda = \int_{-\infty}^\infty p_a^2(x)\,dx = \int_{-a}^a 1\,dx = 2a$$

$$\int_{-\infty}^\infty \frac{\sin^2 \lambda a}{\lambda^2}\,d\lambda = \pi a \tag{E.4}$$

and with $a = 1$ in (E.4) to be used in (E.1), we have

$$\int_0^\infty \frac{\sin^2 \lambda}{\lambda^2}\,d\lambda = \frac{\pi}{2} \tag{E.5}$$

We may mention here that we used the gate function in Example 2.16 to good advantage. Moreover, the gate function $p_a(x)$ in (E.2) is used very often in signal analysis (Section 2.4) to represent the system function $p_a(\omega)$ for an ideal low-pass filter where the spectrum $p_a(\omega)F(\omega)$, of its (bandlimited) output signal $f_a(t)$, vanishes identically outside the band $\omega \in (-a, a)$. The Fourier transform $(2 \sin at)/t$ of the gate function represents the impulse response of such an ideal low-pass filter to input signals at time zero. This will play an important role in interpolating the samples of bandlimited signals, as we shall see in Section 2.4.2; the basic elements of systems analysis are also covered in Section 2.4.1. More on the use of Fourier transforms in signal analysis may be found in Papoulis (1977) and other similar texts on the subject, such as Ziemer and co-authors (1983).

Next we illustrate the use of the convolution Theorem 2.18 in (2.128) for evaluating the inverse Fourier transform as a solution to the heat equation in an infinite domain, which we will need in Example 2.18 of Section 2.3.

EXAMPLE 2.17: Inverse Fourier Transform—Application of the Convolution Theorem

Find the inverse Fourier transform of

$$U(\lambda, t) = F(\lambda)e^{-\lambda^2 t} = \mathcal{F}\{u(x, t)\}, \qquad u(x, 0) = f(x), \quad -\infty < x < \infty \tag{E.1}$$

We note that this function is a product of $F(\lambda) = \mathcal{F}\{f\}$ and $e^{-\lambda^2 t} = \mathcal{F}\left\{e^{-x^2/4}t/\sqrt{4\pi t}\right\}$ (see Table 2.2); which is left as Exercise 2.46. So, according to the result (2.128) of the convolution Theorem 2.18, its inverse Fourier transform $u(x, t)$ is the convolution product of $f(x)$ and $e^{-x^2/4t}/\sqrt{4\pi t}$,

$$u(x, t) = \mathcal{F}^{-1}\{U(\lambda, t)\} = \mathcal{F}^{-1}\left\{F(\lambda)e^{-\lambda^2 t}\right\} = [f(x)] * \left[\frac{e^{-x^2/4t}}{\sqrt{4\pi t}}\right]$$

$$= \int_{-\infty}^{\infty} f(\xi) \frac{e^{-(x-\xi)^2/4t}}{\sqrt{4\pi t}} d\xi$$

$$u(x, t) = \frac{1}{\sqrt{4\pi t}} \int_{-\infty}^{\infty} f(\xi)e^{-(x-\xi)^2/4t} d\xi \tag{E.2}$$

which is, in fact, an interesting integral representation of $u(x, t)$. Here, we note that in (E.2) we must show that $\lim_{t \to 0} u(x, t) = f(x)$ to satisfy $u(x, 0) = f(x)$ in (E.1) and leave it as an exercise, as we shall need it for the applications of the Fourier transform for solving the heat equation in an infinite domain in the next section (see Example 2.18).

2.3   Boundary and Initial Value Problems—Solutions by Fourier Transforms

In this section we will illustrate the use of the Fourier transform in the following three examples, 2.18—2.21 for solving diffusion, vibration problems, the Schrödinger equation, and the potential equation in an infinite domain (the whole real line). This includes the temperature or neutron distribution in an infinite bar ($-\infty < x < \infty$), the displacement of a very long string, the (linearized) Schrödinger equation of quantum physics, and the potential distribution in the upper half-plane. A generalization of the Schrödinger equation with a simple constant-coefficient parameter will also be covered. With one value of this constant coefficient we will have the Schrödinger wave equation (Example 2.20) with its *oscillatory* characteristic, while another special value of this constant results in the diffusion (heat) equation with its *dissipation* nature in contrast to the first case of the oscillatory wave equation. These applications for boundary and initial value problems will be followed by the application of Fourier analysis to signals and systems representations, after we introduce the subject with basic definitions and important concepts of systems and their input-output signals.

2.3.1   The Heat Equation on an Infinite Domain

EXAMPLE 2.18:   Temperature Distribution in an Infinite Bar

In Section 2.1 we used the Laplace transform to solve for the temperature distribution in a semi-infinite bar ($0 < x < \infty$) in (2.42). Here we present the same problem for an infinite bar ($-\infty < x < \infty$):

$$\frac{\partial^2 u}{\partial x^2} = \frac{\partial u}{\partial t}, \qquad -\infty < x < \infty,\ 0 < t < \infty \qquad \text{(E.1)}$$

$$u(x, 0) = f(x), \qquad -\infty < x < \infty \qquad \text{(E.2)}$$

The heat equation (E.1) in one dimension for a bar and in two dimensions for a plate, along with other important partial differential equations such as the wave equation, are derived (or modeled) in detail in Appendix B.

It is clear that the Fourier transform is suitable for this problem in the coordinate variable x while the Laplace transform is the one for the time variable t.

Let $U(\lambda, t)$ be the Fourier transform of $u(x, t)$,

$$U(\lambda, t) = \int_{-\infty}^{\infty} e^{-i\lambda x} u(x, t)\, dx \qquad \text{(E.3)}$$

and Fourier transform both sides of (E.1) so that

$$\int_{-\infty}^{\infty} e^{-i\lambda x} \frac{\partial^2 u(x,t)}{\partial x^2} \, dx = \int_{-\infty}^{\infty} e^{-i\lambda x} \frac{\partial u(x,t)}{\partial t} \, dx \tag{E.4}$$

If we use (2.117) for the left side of (E.2), assuming that u and $\partial u / \partial x$ vanish as $x \to \mp\infty$, and interchange the integration with the differentiation in the right-hand side, we obtain

$$-\lambda^2 U(\lambda, t) = \frac{\partial}{\partial t} \int_{-\infty}^{\infty} e^{-i\lambda x} u(x,t) \, dx$$

$$-\lambda^2 U(\lambda, t) = \frac{\partial U}{\partial t}(\lambda, t) \tag{E.5}$$

Clearly the justification for the above interchange of operations is warranted, and Theorem 2.6 should be consulted. This, of course, will put restrictions on $u(x,t)$ and its derivative $u_t(x,t)$. The vanishing of $U(x,t)$ and $U_t(x,t)$ as $x \to \mp\infty$ is implied by their absolute integrability on $(-\infty, \infty)$. This is a simple first-order differential equation which is separable, and the bounded solution is

$$U(\lambda, t) = C(\lambda) e^{-\lambda^2 t} \tag{E.6}$$

Often the resulting differential equation is not as simple as (E.5). Hence, we may attempt Laplace transform on the derivative in t to reduce (E.5) to an algebraic equation which is easier to solve.

After solving (E.5), we are left to find the arbitrary constant $C(\lambda)$ and, then, the original solution $u(x,t)$, which is the inverse Fourier transform of $U(\lambda, t)$ in (E.6). To evaluate $C(\lambda)$ from (E.6) we need an initial condition for $U(\lambda, t)$ such as

$$U(\lambda, 0) = C(\lambda) \tag{E.7}$$

This condition can be obtained as the Fourier transform of the given initial condition, $u(x, 0) = f(x)$,

$$U(\lambda, 0) = C(\lambda) = \int_{-\infty}^{\infty} u(x, 0) e^{-i\lambda x} \, dx$$

$$C(\lambda) = \int_{-\infty}^{\infty} f(x) e^{-i\lambda x} \, dx = F(\lambda) \tag{E.8}$$

Hence, the solution (E.6) becomes

$$U(\lambda, t) = F(\lambda) e^{-\lambda^2 t} \tag{E.9}$$

Now we use the inverse Fourier transform (2.65) on $U(\lambda,t)$ to arrive at the solution $u(x,t)$:

$$u(x,t) = \frac{1}{2\pi} \int_{-\infty}^{\infty} e^{i\lambda x} F(\lambda) e^{-\lambda^2 t} d\lambda \qquad (E.10)$$

$$u(x,t) = \frac{1}{2\pi} \int_{-\infty}^{\infty} \left[\int_{-\infty}^{\infty} f(\xi) e^{-i\lambda\xi} d\xi \right] e^{-\lambda^2 t} e^{i\lambda x} d\lambda \qquad (E.11)$$

after substituting for $F(\lambda)$ in terms of $f(x)$ as in (2.64). This solution gives the temperature $u(x,t)$ in terms of the initial temperature $f(x)$ as a double integral—a formal solution, perhaps, and tedious to evaluate. As we have done in Example 2.17, we will again illustrate the convolution theorem, one of the many tools developed for the Fourier transforms, to emphasize its advantage in reducing the double integral to a single integral involving $f(x)$. If we compare (E.9) with the result (2.128) of the convolution Theorem 2.18, we find that $F_1(\lambda) = F(\lambda)$ with its inverse Fourier transform $f_1(x) = f(x)$ and $F_2(\lambda) = e^{-\lambda^2 t}$ with its inverse Fourier transform

$$f_2(x) = \frac{e^{-x^2/4t}}{\sqrt{4\pi t}} \qquad (E.12)$$

as given in Table 2.2 (or done in Exercise 2.46) and as used in Example 2.17. Hence, applying the convolution theorem as in (2.128) to (E.9) gives the solution to (E.1) and (E.2), the initial and boundary value problem of the heat equation, as

$$u(x,t) = \frac{1}{2\pi} \int_{-\infty}^{\infty} e^{i\lambda x} F(\lambda) e^{-\lambda^2 t} d\lambda$$

$$= \int_{-\infty}^{\infty} f(\xi) \frac{e^{-(x-\xi)^2/4t}}{\sqrt{4\pi t}} d\xi$$

$$u(x,t) = \frac{1}{\sqrt{4\pi t}} \int_{-\infty}^{\infty} f(\xi) e^{-(x-\xi)^2/4t} d\xi \qquad (E.13)$$

which involves $f(x)$ in a single integral instead of the double integral in (E.11). As mentioned at the end of Example 2.17, what remains now is to show that this solution, in addition to satisfying the partial differential equation (E.1), must satisfy the initial condition $u(x,0) = f(x)$; that is, we ought to show that $u(x,t)$ of (E.13) must have $\lim_{t\to 0} u(x,t) = f(x)$. This result, we warn, should be approached with extreme care; we need *Dirac delta function* behavior inside the integral as $t \to 0$, which is given in (2.161) and verified in Exercise 2.57d.

2.3.2 The Wave Equation

EXAMPLE 2.19: Wave Equation in One Dimension

We consider the following boundary and initial value problem for $u(x,t)$, the displacement of an infinite string:

$$\frac{\partial^2 u}{\partial x^2} = \frac{1}{c^2}\frac{\partial^2 u}{\partial t^2}, \qquad -\infty < x < \infty, \; 0 < t < \infty, \; c = \text{constant} \qquad (E.1)$$

$$u(x,0) = f(x), \qquad -\infty < x < \infty \qquad (E.2)$$

$$\frac{\partial u}{\partial t}(x,0) = g(x), \qquad -\infty < x < \infty \qquad (E.3)$$

with the initial displacement and velocity $f(x)$ and $g(x)$, respectively. The wave equation in one dimension (E.1) for a vibrating string and in two dimensions for a vibrating membrane is derived in Appendix B.1. (see also Exercise B.5).

This problem, in its x spatial variable dependence, is suitable for the Fourier transform. As we will see later, it also appears suitable for Laplace transform and both cosine and sine transforms in the case of the time variable t. However, it should be clear from the above initial conditions that it is compatible with Laplace transform only, as we discussed in Chapter 1.

If we let

$$U(\lambda,t) = \int_{-\infty}^{\infty} e^{-i\lambda x} u(x,t)\,dx \qquad (E.4)$$

and use the Fourier transform on the wave equation (E.1), following the same steps as we did for the heat equation (E.1) of Example 2.18, we obtain

$$-\lambda^2 U(\lambda,t) = \frac{1}{c^2}\frac{\partial^2 U}{\partial t^2}(\lambda,t) \qquad (E.5)$$

which is a simple second-order differential equation in t with the general solution

$$U(\lambda,t) = A(\lambda)\cos \lambda ct + B(\lambda)\sin \lambda ct \qquad (E.6)$$

To determine $A(\lambda)$ and $B(\lambda)$, we transform the initial conditions of the displacement and the velocity to give

$$U(\lambda,0) = \int_{-\infty}^{\infty} e^{-i\lambda x} u(x,0)\,dx = \int_{-\infty}^{\infty} e^{-i\lambda x} f(x)\,dx = F(\lambda) \qquad (E.7)$$

and

$$\frac{\partial U(\lambda, 0)}{\partial t} = \int_{-\infty}^{\infty} e^{-i\lambda x} \frac{\partial u(x, 0)}{\partial t} \, dx = \int_{-\infty}^{\infty} e^{-i\lambda x} g(x) \, dx = G(\lambda) \quad \text{(E.8)}$$

We then apply the initial conditions $U(\lambda, 0) = F(\lambda)$ and $U_t(\lambda, 0) = G(\lambda)$ to (E.6) to obtain

$$U(\lambda, 0) = A(\lambda) + 0 = F(\lambda), \qquad A(\lambda) = F(\lambda)$$

$$\frac{\partial U}{\partial t}(\lambda, 0) = 0 + \lambda c B(\lambda) = G(\lambda), \qquad B(\lambda) = \frac{G(\lambda)}{\lambda c}$$

Hence, with $A(\lambda) = F(\lambda)$ and $B(\lambda) = G(\lambda)/\lambda c$ in (E.6), we have

$$U(\lambda, t) = F(\lambda) \cos \lambda ct + \frac{G(\lambda)}{\lambda c} \sin \lambda ct$$

$$= F(\lambda) \left[\frac{e^{i\lambda ct} + e^{-i\lambda ct}}{2} \right] + \frac{G(\lambda)}{\lambda c} \left[\frac{e^{i\lambda ct} - e^{-i\lambda ct}}{2i} \right] \quad \text{(E.9)}$$

We must now find the inverse Fourier transform of (E.9) to obtain $u(x, t)$, the solution of the vibrating string (E.1)–(E.3). We note that (E.9) involves products like $F(\lambda) e^{i\lambda ct}$, which is the Fourier transform of $f(x + ct)$, that is,

$$\mathcal{F}\{f(x + ct)\} = e^{i\lambda ct} F(\lambda) \quad \text{(E.10)}$$

according to the Fourier pair of Table 2.2, or as a symmetric pair of (2.111),

$$\mathcal{F}\{f(x - x_0)\} = e^{-i\lambda x_0} F(\lambda) \quad \text{(E.11)}$$

The other more involved product is $e^{-i\lambda ct} G(\lambda)/\lambda$, the Fourier transform of $i \int_{\mp\infty}^{x-ct} g(\xi) d\xi$, since according to the Fourier pair of Table 2.2, and which can be developed easily (see Exercise 2.41d), we have

$$\mathcal{F}\left\{ i \int_{\mp\omega}^{x} g(\xi) d\xi \right\} = \frac{G(\lambda)}{\lambda} \quad \text{(E.12)}$$

Thus, with the results of (E.12) and (E.11) we have

$$\mathcal{F}\left\{ i \int_{\mp\infty}^{x-ct} g(\xi) d\xi \right\} = e^{-i\lambda ct} \frac{G(\lambda)}{\lambda} \quad \text{(E.13)}$$

If we now take the inverse Fourier transform of (E.9) and apply (E.10) and (E.13) to its appropriate terms we obtain

$$u(x, t) = \frac{1}{2} [f(x + ct) + f(x - ct)] + \frac{i}{2ic} \int_{\mp\infty}^{x+ct} g(\xi) d\xi - \int_{\mp\infty}^{x-ct} g(\xi) d\xi$$

$$= \frac{1}{2}[f(x+ct) + f(x-ct)] + \frac{1}{2c} \int_{x-ct}^{x+ct} f(\xi)\,d\xi \qquad (E.14)$$

which is the *D'Alembert* solution to the wave equation problem of (E.1)–(E.3). For a different method of arriving at this D'Alembert solution, see Exercise 2.49d and Exercise 2.32.

2.3.3 The Schrödinger Equation

EXAMPLE 2.20: The Schrödinger Wave Equation—A Generalization for Dispersive and Dissipative Processes

The previous two examples of application (Examples 2.18 and 2.19) represented the use of the Fourier integral transform for solving the dissipative (or diffusive) heat equation and the oscillatory (or harmonic) wave equation in the infinite domain $(-\infty, \infty)$. In this example, we present the famous equation of wave, or quantum, mechanics, namely the (linear) one-dimensional *Schrödinger equation* for the wave function $\phi(x, t)$,

$$i\frac{\partial \phi}{\partial t} = \frac{\partial^2 \phi}{\partial x^2}, \qquad -\infty < x < \infty,\ t > 0,\ i = \sqrt{-1} \qquad (E.1)$$

We note that it is a special case (with $\gamma = -i$) of the partial differential equation

$$\frac{\partial \phi}{\partial t} = \gamma \frac{\partial^2 \phi}{\partial x^2}, \qquad -\infty < x < \infty,\ t > 0 \qquad (E.2)$$

with the parameter γ that is allowed to have both complex and real values.

On the other hand, we observe that the heat equation in one dimension in Example 2.17,

$$\frac{\partial u}{\partial t} = \frac{\partial^2 u}{\partial x^2} \qquad (E.3)$$

is also a special case of equation (E.2) for the real and positive value of $\gamma = 1 > 0$ as the diffusivity of the medium.

Before we start applying the Fourier transform to solve the Schrödinger equation (E.1) along with its initial condition,

$$\frac{\partial \phi}{\partial t} = \gamma \frac{\partial^2 \phi}{\partial x^2}, \qquad -\infty < x < \infty,\ t > 0 \qquad (E.1)$$

$$\phi(x, 0) = f(x), \qquad -\infty < x < \infty \qquad (E.4)$$

we will elaborate a bit to show the difference the constant factor γ in (E.2) makes for the nature of the particular processes described by (E.2). The nature is purely dissipative when γ is real and positive, as the diffusivity of the heat equation; and it is a *dispersive* wave when $\gamma = -i$, a pure imaginary. By dispersive wave we mean that its velocity ω depends on the wavelength λ or, in other words, on the wave number $k = 2\pi/\lambda$. To show this, learning from our solution to the wave equation in Example 2.19, we can assume the harmonic function

$$\phi_k(x,t) = A e^{i(kx - \omega(k))t}, \qquad A = A(k), \text{ a constant function of } x \text{ and } t$$
(E.5)

as a solution to equation (E.2); clearly (E.2) is satisfied,

$$\frac{\partial \phi_k}{\partial t} = -i\omega(k)A e^{i(kx - \omega(k))t} = (ik)^2 \gamma A e^{i((kx - \omega)(k))t} = \gamma \frac{\partial^2 \phi_k}{\partial x^2}$$

provided that

$$\omega(k) = -ik^2\gamma \tag{E.6}$$

Equation (E.6) represents the dependence of the velocity $\omega(k)$ on the wave number k and is called in general the *dispersion relation*.

In the case of the Schrödinger equation we have $\gamma = -i$, and this relation becomes the (quadratic) dispersion relation of the waves represented by $\phi_k(x,t)$,

$$\omega(k) = k^2 \tag{E.7}$$

$$\phi_k(x,t) = A e^{ikx} e^{-ik^2 t} \tag{E.8}$$

with oscillatory (dispersive) behavior in time t and space x. On the other hand, for the heat equation $\gamma = \delta^2$,

$$\omega(k) = -ik^2\delta^2, \qquad \delta^2 > 0 \tag{E.9}$$

$$\phi_k(x,t) = A e^{ikx} e^{-k^2 \delta^2 t} \tag{E.10}$$

with a decaying (dissipative) behavior in time and oscillatory behavior in x.

We may also mention that the wave function $\phi(x,t)$, for the Schrödinger equation, has a different representation and meaning from the definite $u(x,t)$ as the temperature in the heat equation. In wave mechanics, the wave function $\phi(x,t)$ is used to give a measure of the probability $|\phi(x,t)|^2 \Delta x$ of finding the quantum or particle in the interval of space Δx. Thus, since the probability of finding the quantum over the whole real line is one, we assume that our solution, the solution $\phi(x,t)$ to the Schrödinger equation, must satisfy the following condition:

$$\int_{-\infty}^{\infty} |\phi(x,t)|^2 \, dx = 1 < \infty \tag{E.11}$$

that is, $\phi(x,t)$ must be square integrable in the x (physical) space $(-\infty, \infty)$.

We may appeal here to Parseval's equality (2.136), which will assure us of the same probability of one in the Fourier or the wave number k-space,

$$\int_{-\infty}^{\infty} |\phi(x,t)|^2 \, dx = 1 = \frac{1}{2\pi} \int_{-\infty}^{\infty} |\Phi(k,t)|^2 \, dk \tag{E.12}$$

where $\Phi(k,t)$ is the Fourier transform of $\phi(x,t)$ as in (2.70), (2.71).

Now we turn to solving the Schrödinger equation with its initial condition (E.1), (E.4) by use of the Fourier transform. Since this is an example which was originally presented and is still mainly found in physics texts, we shall, for variation in notation and experience, use their symmetric notation as in (2.70), (2.71),

$$\Phi(k,t) = \frac{1}{\sqrt{2\pi}} \int_{-\infty}^{\infty} \phi(x,t) e^{-ikx} \, dx \tag{E.13}$$

$$\phi(x,t) = \frac{1}{\sqrt{2\pi}} \int_{-\infty}^{\infty} \Phi(k,t) e^{ikx} \, dk \tag{E.14}$$

It is clear that if we are to stay with the waves $\{\phi_k(x,t)\}$ in (E.8) as solutions,

$$\phi_k(x,t) = A(k) e^{i(kx - k^2 t)} \tag{E.8}$$

none of these individual waves, in general, will satisfy the initial condition (E.4). $\phi(x,0) = f(x)$ is such a general (practical) function that is seldom purely harmonic as in the solution (E.8) which satisfies

$$\phi_k(x,0) = A(k) e^{ikx} \neq f(x) \tag{E.15}$$

So the idea here, as for the wave equation and the heat equation, is to appeal to Fourier analysis to sum over all such waves to have the sum $\phi(x,t)$, as a Fourier integral in (E.14), which converges to $\phi(x,0) = f(x)$. This means that we consider

$$\phi(x,t) = \frac{1}{\sqrt{2\pi}} \int_{-\infty}^{\infty} A(k) e^{i(kx - k^2 t)} \, dk \tag{E.16}$$

and insist on

$$\phi(x,0) = \frac{1}{\sqrt{2\pi}} \int_{-\infty}^{\infty} A(k) e^{ikx} \, dk = f(x) \tag{E.17}$$

This is the case, as we have reasoned in (E.11) that $f(x) = \phi(x,0)$ is square integrable, which guarantees convergence (in the mean) of the above Fourier integral of (E.17) to $f(x)$. Of course, $\lim_{x \to \infty} f(x) = 0$

is implied from this by the Riemann–Lebesgue lemma. The amplitude of the wave $A(k)$ in the k-space can be found as the inverse Fourier transform of $f(x)$,

$$A(k) = \frac{1}{\sqrt{2\pi}} \int_{-\infty}^{\infty} f(x)e^{-ikx}\, dx \tag{E.18}$$

as a special case of the Fourier transform $\Phi(k,t)$ of $\phi(x,t)$,

$$\Phi(k,t) = \mathcal{F}\{\phi(x,t)\} = \frac{1}{\sqrt{2\pi}} \int_{-\infty}^{\infty} \phi(x,t)e^{-ikx}\, dx \tag{E.19}$$

Now we apply the Fourier transform to the general equation (E.2) (to cover both the Schrödinger wave equation for $\gamma = -i$ and the diffusion equation for $\gamma = \delta^2 > 0$),

$$\frac{\partial}{\partial t}\, \Phi(k,t) = -k^2\gamma\Phi(k,t) \tag{E.18}$$

$$\Phi(k,t) = A(k)e^{-k^2\gamma t} \tag{E.19}$$

The Fourier transform inverse to $\Phi(k,t)$ is

$$\phi(x,t) = \frac{1}{\sqrt{2\pi}} \int_{-\infty}^{\infty} A(k)e^{-k^2\gamma t}e^{ikx}\, dx \tag{E.20}$$

To satisfy the initial condition $\phi(x,0) = f(x)$ of (E.4) we apply it here

$$\phi(x,0) = \frac{1}{\sqrt{2\pi}} \int_{-\infty}^{\infty} A(k)e^{ikx}\, dk = f(x) \tag{E.21}$$

which is satisfied if $A(k)$ is the Fourier transform of $f(x)$,

$$A(k) = \frac{1}{\sqrt{2\pi}} \int_{-\infty}^{\infty} f(x)e^{-ikx}\, dx \tag{E.22}$$

So, with this $A(k)$ in (E.20), we have the final solution to the initial value problem (E.2), (E.4) as

$$\phi(x,t) = \frac{1}{\sqrt{2\pi}} \int_{-\infty}^{\infty} dk \left[\frac{1}{\sqrt{2\pi}} \int_{-\infty}^{\infty} f(\xi)e^{-ik\xi}\, d\xi \right] e^{-k^2\gamma t}e^{ikx}\, dk$$

$$= \frac{1}{2\pi} \int_{-\infty}^{\infty} \left[\int_{-\infty}^{\infty} f(\xi)e^{ik(x-\xi)}\, d\xi \right] e^{-k^2\gamma t}\, dk \tag{E.23}$$

As in Examples 2.18 and 2.19 this integral may be reduced to a single interval after allowing reversal of the order of integration, which, of course, must be justified. If we allow such interchange of the two integrals in (E.23) we have

$$\phi(x,t) = \frac{1}{2\pi} \int_{-\infty}^{\infty} \left[\int_{-\infty}^{\infty} e^{ik(x-\xi)} e^{-k^2\gamma t} \, dk \right] f(\xi) d\xi \tag{E.24}$$

The inner integral is over a symmetric interval $(-\infty, \infty)$. If we use the Euler identity

$$e^{ik(x-\xi)} = \cos k(x - \xi) + i \sin k(x - \xi) \tag{E.25}$$

in this (inner) integral, the (odd) sine function's contribution to it is zero, whence the inner integral in (E.24) becomes a cosine one:

$$\int_{-\infty}^{\infty} \cos k(x - \xi) e^{-k^2\gamma t} \, dk \tag{E.26}$$

It will be verified as in Example 2.18 that this integral is uniformly convergent, which according to Theorem 2.9 is sufficient to allow the interchange of the order of integration in the above double integral of (E.24). We can also show as in Exercise 2.46 that the integral in (E.26) can be computed exactly, that is, in terms of elementary functions,

$$\int_{-\infty}^{\infty} \cos k(x - \xi) e^{-k^2\gamma t} \, dk = \sqrt{\frac{\pi}{\gamma t}} e^{-(x-\xi)^2/4\gamma t} \tag{E.27}$$

This is done in the same way as we did for the final solution of the heat equation in Example 2.18 with the help of Exercise 2.46.

With this result (E.27), the double integral (E.24) becomes a single integral,

$$\phi(x,t) = \frac{1}{\sqrt{4\pi\gamma t}} \int_{-\infty}^{\infty} f(\xi) e^{-(x-\xi)^2/4\gamma t} \, d\xi \tag{E.28}$$

as the final solution to the general Schrödinger equation (E.2) with its initial condition (E.4).

As was the case for the final solution (integral) of the heat equation ((E.13) in Example 2.18)), this integral, for the general data $f(x)$, cannot be simplified further to have it in terms of elementary functions. For this reason careful numerical integration may be done, or (specialized) approximation methods may be tried for the parameter γ. More on the interpretation of the solution (E.28) with illustrations for dissipative (for $\gamma = \delta^2 > 0$) and dispersive (for $\gamma = -i$) phenomena may be found in Dodd and co-authors (1982).

2.3.4 The Laplace Equation

The Laplace equation in the potential $u(x,y,z)$

$$\nabla^2 u \equiv \frac{\partial^2 u}{\partial x^2} + \frac{\partial^2 u}{\partial y^2} + \frac{\partial^2 u}{\partial z^2} = 0$$

controls the potential distribution $u(x,y,z)$ in a space (domain) free of charge. It also applied in the same way to the gravitational potential free of mass distribution. For the derivation of the Laplace equation, see (B.2.8)–(B.2.9) in Appendix B.2 and Exercise B.5.

As an example of the application of the exponential Fourier transform, we present a boundary value problem for the potential distribution $u(x,y)$ in the upper half-plane (Fig. 2.6), where we assume that the potential on the edge (the real line $y = 0$) is given as $u(x,0) = f(x)$, $-\infty < x < \infty$. We assume that $f(x)$ is absolutely integrable on $(-\infty, \infty)$. We may also make the reasonable assumption that the potential vanishes as we approach infinity. All these conditions, of course, are to satisfy the hypotheses of the basic theorems of the Fourier transform which assure its existence, along with its inverse, and the convolution theorem, to mention a few.

Boundary value problems with such a boundary condition, where the value of the function u is given on the boundary are called *Dirichlet problems*. Those where the derivative $\partial u/\partial y$ (or the gradient in general) of the function u (in the case $\partial u(x,0)/\partial y$) is given on the boundary are termed *Neumann problems*.

EXAMPLE 2.21: Potential Distribution in the Upper Half-Plane

We are to solve the following boundary value problem, associated with the Laplace equation in the upper half-plane:

$$\frac{\partial^2 u}{\partial x^2} + \frac{\partial^2 u}{\partial y^2} = 0, \qquad -\infty < x < \infty, \ 0 < y < \infty \tag{E.1}$$

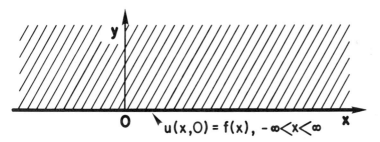

FIG. 2.6 Potential distribution in the upper half-plane.

$$u(x,0) = f(x), \qquad -\infty < x < \infty \tag{E.2}$$

with the conditions that $f(x)$ is absolutely integrable on $(-\infty, \infty)$, that is, $\int_{-\infty}^{\infty} |f(x)| dx < \infty$, and that $u(x,y)$ vanishes at infinity.

From the domain $(-\infty, \infty)$ in x and the vanishing of $u(x,y)$ as $x \to \mp\infty$, we note that the problem is compatible with the exponential Fourier transform, which is what we are going to do here. However, we also note that for the domain $(0, \infty)$ in y, the Fourier sine transform will do since the derivative in y is of even order, and the value of the function $u(x,0) = f(x)$ at $y = 0$ is given as required by the Fourier sine transform pair in (1.16),

$$\mathcal{F}_s \left\{ \frac{d^2 f}{dx^2} \right\} = \lambda f(0) - \lambda^2 F_s(\lambda) \tag{1.16}$$

But since the Fourier sine transform will be covered in detail in Section 2.5, we will leave this alternative until then for Example 2.23a. The Laplace transform, on the other hand, will not work for $\partial^2 u / \partial y^2$ since it requires having $u(x,0)$ as well as $\partial u(x,0)/\partial y$ at $y = 0$, where the latter is not given here.

We let

$$U(\lambda, y) = \int_{-\infty}^{\infty} e^{-i\lambda x} u(x,y) dx \tag{E.3}$$

and Fourier transform the Laplace equation (E.1) to have

$$-\lambda^2 U(\lambda, y) + \frac{\partial^2 U(\lambda, y)}{\partial y^2} = 0 \tag{E.4}$$

whose bounded solution is

$$U(\lambda, y) = A(\lambda) e^{-|\lambda| y}, \qquad 0 < y < \infty \tag{E.5}$$

We find the arbitrary amplitude $A(\lambda)$ by Fourier transforming the boundary condition (E.2),

$$\mathcal{F}\{u(x,0)\} = \mathcal{F}\{f(x)\} = F(\lambda) = U(\lambda, 0) = A(\lambda)$$

$$U(\lambda, y) = F(\lambda) e^{-|\lambda| y}, \qquad 0 < y < \infty \tag{E.6}$$

To find the original solution $u(x,y)$, we must find the inverse Fourier transform of $U(\lambda, y)$, which as we see is a product of two functions that will need the convolution Theorem 2.18. We know the inverse transform of $F(\lambda)$ is $f(x)$, but we must find the inverse Fourier transform of the second factor $e^{-|\lambda| y}$. At this stage we usually appeal to the Fourier transform tables in Appendix C.2 or the usual extensive references listed at the end of Section 2.2.1. However, the case here is a rather simple one which is amenable to a few steps of computations. The inverse Fourier transform of $e^{-|\lambda|y}$ is

$$g(x,y) = \frac{1}{2\pi} \int_{-\infty}^{\infty} e^{-i\lambda x} e^{-|\lambda|y} \, d\lambda$$

$$= \frac{1}{2\pi} \left[\int_{-\infty}^{0} e^{\lambda(y-ix)} \, d\lambda + \int_{0}^{\infty} e^{-\lambda(y+ix)} \, d\lambda \right]$$

$$= \frac{1}{2\pi} \left[\frac{1}{y-ix} + \frac{1}{y+ix} \right] = \frac{1}{\pi} \frac{y}{x^2+y^2} \qquad \text{(E.7)}$$

Thus, if we use the Fourier convolution Theorem 2.18 on the product of the transforms in (E.6) we have the final solution $u(x,y)$ to the above boundary value problem (E.1), (E.2).

$$u(x,y) = \mathcal{F}^{-1}\{U(\lambda,y)\} = \mathcal{F}^{-1}\{e^{-|\lambda|y}F(\lambda)\}$$

$$u(x,y) = (g*f)(x,y) = \frac{y}{\pi} \int_{-\infty}^{\infty} \frac{1}{(x-\xi)^2 + y^2} f(\xi) \, d\xi \qquad \text{(E.8)}$$

2.4 Signals and Linear Systems—Representation in the Fourier (Spectrum) Space

Consider the Fourier integral representation (2.68) of the signal $f(t)$

$$f(t) = \int_{-\infty}^{\infty} F(\omega) e^{i\omega t} \, d\omega \qquad \begin{array}{c} \text{(2.137)} \\ \text{(2.68)} \end{array}$$

where $F(\omega)$ is its representation in the frequency space ω. The frequency ω is often written as $\omega = 2\pi\nu$, where ν is the frequency in cycles per second (cps, hertz). It is known that signals are transmitted on a finite band of frequency, for example, $\omega \in (-a,a)$ or $\omega \in (-2\pi W, 2\pi W)$, where W is the maximum frequency in hertz. This means that $F(\omega)$ vanishes identically outside this band. Such signals are called *bandlimited* to the bandlimit a (or $2\pi W$) and we shall denote them by $f_a(t)$:

$$f_a(t) = \int_{-a}^{a} F(\omega) e^{i\omega t} \, d\omega = 2\pi \int_{-2\pi W}^{2\pi W} F(2\pi\nu) e^{i2\pi\nu t} \, d\nu \qquad \text{(2.138)}$$

This can be written as

$$f_a(t) = \int_{-\infty}^{\infty} p_a(\omega) F(\omega) e^{i\omega t} \, d\omega \qquad \text{(2.139)}$$

where

$$p_a(\omega) = \begin{cases} 1, & |\omega| < a \\ 0, & |\omega| > a \end{cases} \tag{2.140}$$

is called the *gate function*, which is the *system function* representation for the *ideal low-pass filter* and will be discussed next along with general systems. To continue our discussion of the important applications of Fourier analysis in the theory of systems and communications theory, we must stop here to present a brief but very clear introduction of some of the basic topics and definitions, such as the above system function, used in the analysis of systems and most often *linear systems*.

2.4.1 Linear Systems

We consider $f(t)$ to be an input signal to a system, designated by L, as a black box for now, and look at the resulting output of the system as another (related) signal $g(t)$. Symbolically, we consider the system to be an operator L acting on the input $f(t)$ to give the output $g(t)$,

$$g(t) = L[f(t)] \tag{2.141}$$

All of what we will consider here is for *linear systems* L. This means that given two input signals $f_1(t)$, $f_2(t)$ with corresponding outputs $g_1(t)$, $g_2(t)$, then a linear combination $c_1 f_1(t) + c_2 f_2(t)$ (with c_1, c_2 constants) of these two inputs will result, for the linear system, in the same type of linear combination $c_1 g_1(t) + c_2 g_2(t)$ of their respective outputs $g_1(t)$, $g_2(t)$, namely

$$L[c_1 f_1(t) + c_2 f_2(t)] = c_1 L[f_1(t)] + c_2 L[f_2(t)]$$
$$= c_1 g_1(t) + c_2 g_2(t) \tag{2.142}$$

We notice that this linear property of the system goes well with the Fourier, Laplace, and other integral transforms introduced in Chapter 1, as all these (and others to come) are linear (integral or discrete) transformations.

A. Some Examples of Linear Systems

We may also notice that the operation of each of these linear integral transforms, characterized by their particular kernels and their limits of integrations, represents a known linear system L. This linear operator L takes a known function $f(x)$ as the input to give us the transformed function $F(\lambda) = L[f(x)]$ as the output, or the result of this integral transformation operation. When we are interested in the inverse of the transform $f(x) = L^{-1}\{F(\lambda)\}$ we usually have a formula again, for this inverse linear operator L^{-1}. This contrasts with usual systems analysis, where we are interested in finding out about the characteristics of the "unknown" system L. We do that by having a known input signal $f(t)$ and measure the system's output $g(t)$ to study the characteristics of the system L in our formula $g(t) = L[f(t)]$.

Oil Exploration and Medical Testing

A very important example is provided by the exploration for oil (or water) underground, where the pockets of oil exhibit a particular characteristic absent in the otherwise oil-less solid ground. In this case, an ultrasonic wave signal $f(t)$ is sent deep into the ground, where oil is suspected on the basis of other data such as geological information. Then the reflected signal $g(t)$ as an output is studied to find information about L in $g(t) = L[f(t)]$. This information can then be correlated with the information obtained from experiments of the same type on already discovered oil reserves or wells.

Another familiar example is the injection of a certain concentration of a dye in a vein, with a sensitivity to the operation of the kidney, for example. We then receive a later sample of the blood from an artery with another concentration g of the dye as our output from the system to learn about L in $g = L[f]$, which characterizes the healthy or not-so-healthy kidney.

Differential and Integral Equations—Green's Function

Another very familiar example of known linear systems relates to what we covered in Chapter 1 for linear nonhomogeneous (second-order) differential equations. For example,

$$L[u(t)] = f(t)$$

$$= \left[a_0(t) \frac{d^2}{dt^2} + a_1(t) \frac{d}{dt} + a_2(t) \right] u(t) = f(t), \qquad a < t < b \quad (2.143)$$

Here the system L is known, since it is characterized by the differential operator with known coefficients $a_0(t)$, $a_1(t)$, and $a_2(t)$ as

$$L \equiv a_2(t) \frac{d^2}{dt^2} + a_1(t) \frac{d}{dt} + a_2(t) \tag{2.144}$$

Also, the output $f(t)$ is usually known; for example, in mechanics, it is the driving external force used for the nonhomogeneous term in such a differential equation. Consequently, we are able to find the unknown input $u(t)$ as the sought for solution to the differential equation. We would wish to write this "symbolically" in terms of the inverse operator L^{-1} as the "*inverse problem*" for (2.143),

$$u(t) = L^{-1}[f(t)]$$

where L^{-1} represents what we do to solve a particular nonhomogeneous differential equation. In courses on elementary differential equations, this involves the superposition of solutions, which is allowed because of the assumption that the differential operator L is linear, and the many integrations and auxiliary conditions used to arrive at the unique solution $u(t)$. In sum, we may say that L^{-1} here is the result of integration, and we may term it as an *integral operator* to contrast with and complement its inverse, the differential operator L in (2.143). In an advanced course on applied mathematics, after elementary differential equations, we learn that such an inverse, or integral operator, is characterized by an integration, as expected, over the input $f(t)$, but it is weighted by a function called the *Green's function* $G(t,\tau)$,

$$u(t) = \int_a^b G(t,\tau)f(\tau)d\tau \tag{2.145}$$

Here $G(t,\tau)$ can usually be constructed from two linearly independent solutions $\nu_1(t)$ and $\nu_2(t)$ of the associated homogeneous problem $L[u(t)] = 0$, and the two auxiliary conditions, given at $t = a, b$ for the nonhomogeneous problem.

If in a linear system as characterized by the differential equation (2.143) it is the external force or effect $f(t)$ that is unknown, such as weather conditions or wind speed, we may attempt to measure $u(t)$ for our system and appeal to (2.145) to relate this known $u(t)$ to the desired unknown $f(t)$,

$$u(t) = L^{-1}[f(t)] \equiv \int_a^b G(t,\tau)f(\tau)d\tau \tag{2.146}$$

This is done by solving for its unknown input $f(t)$, given the output $u(t)$ and the known mathematical characterization of this inverse system via the knowledge of its Green's function $G(t,\tau)$. Equation (2.146) with the unknown $f(t)$ inside the integral is called an *integral equation* in $f(t)$. For an elementary treatment of integral equations with varied applications in engineering, physics, the life sciences, and business, and with a complete chapter giving a clear treatment of Green's function and its construction, we refer the reader to the author's book (1985) on introductory integral equations.

Circuit Theory, Systems, and Signal Analysis

In circuit theory, the differential equation (2.143) with $t > 0$ is a representation for our system, where $f(t)$ is a known external electromotive force and the input $u(t)$ is the current that we can measure. What remains is to know the explicit description, or the mathematical characterization, of the linear system L. This must lead us to think about an (inverse) integral representation like (2.145),

$$u(t) = \int_0^t G(t,\tau) f(\tau) d\tau \tag{2.147}$$

which gives an answer to our dilemma of having to find the Green's function as characterizing the inverse operation L^{-1} of the sought for system L. This, as we shall see shortly, happened to be the case, where the Green's function $G(t,\tau)$ stands as the *impulse response* $h(t,\tau)$ of the system, which we will derive soon. Also, we will derive it for the special important case of *time-invariant* systems—systems whose responses $h(t,\tau)$ do not depend on t and τ as two independent variables but only on their difference $t - \tau$ as $h(t-\tau)$. In the language of integral equations, such an $h(t-\tau) = G(t-\tau)$ is called a *difference kernel*. Still, $h(t-\tau)$ is associated with the inverse system L^{-1}. Here, thanks to the Fourier transforms, as we shall see, the direct characterization of the system L is called the *system function* $H(\omega)$ and is related only to the Fourier transform of the impulse response $h(t-\tau)$ or $h(t)$ of the time-invariant system.

Time-Invarying and Spatially Invarying Systems

Most of systems analysis, and especially that which uses Fourier and Laplace transforms, is limited to the description of such *time-invarying systems* when we are considering $f(t)$ as a function of time. The same thing is assumed for $f(x)$ as a function of the spatial variable x, and the system is called a *spatially invarying system*. By time-invarying we meant that if the input $f(t)$ is shifted in time by t_0, to $f(t-t_0)$, then its output $g(t)$ will also depend only on the difference (shift) $\Delta t = t - t_0$ as $g(t-t_0)$. We emphasize that it does not depend on two independent times t and t_0 but only on their difference $t - t_0$. When the output of a system depends on t and t_0 as two independent variables, we call it a *time-varying systems* and designate the output dependence as $g(t,t_0)$ or $g(t\theta t_0)$, where θ may refer to a "general type of translation" or "transformation." Then at time-invarying system L is defined through the equation

$$g(t-t_0) = L[f(t-t_0)] \tag{2.148}$$

As we shall see, the compatibility of the Fourier or Laplace transforms with time-invarying (or spatially invarying) systems analysis stems from their particular exponential kernels. When, in our analysis, two different kernels like $e^{i\omega t}$ and $e^{-i\omega t_0}$ are multiplied, we have $e^{i\omega(t-t_0)}$, which is dependent on the shift $\Delta t = t - t_0$ only. In contrast, the Hankel (Bessel) transform with its nonexponential (Bessel function) kernel in (1.22)

$$U_n(\lambda) = \int_0^\infty r J_n(\lambda r) u(r) \, dr \tag{1.22}$$

is not useful for spatially invarying systems. However, the Hankel transform is needed for the analysis of relevant systems in optics, where the circular symmetry dictates the usefulness of polar coordinates, as evidenced by Papoulis's book on systems in optics (1968). We shall discuss this topic in Section 2.7 on the Hankel transforms. As we shall show in Section 2.7, it turns out that for functions of two variables $f(x,y)$ which are circularly symmetric—i.e., the function depends on (x^2+y^2) as $f(x^2+y^2)$—such a J_0-Hankel transform is equivalent to a two-dimensional Fourier transform. Thus, in order to avoid the above generalized translation, the Fourier transform is used in two dimensions for its compatibility with the spatially invarying property of the system in each of the two Cartesian coordinates x and y.

If we are now to present the *system function* for the linear system L in this limited space, we must deal with its reaction to incoming input *samples of the signal* $\{f(\tau_n)\}$. The sample $f(\tau_n)$ defined at the point of time τ_n is taken as a pulse of height $f(\tau_n)$ but with no width or "support," and hence in our usual sense of integration it has a zero area under it. The question now is how to relate these discrete signals to what the *analog system* accepts as inputs, such as waves, or alternatively signals, represented by continuous functions $f(t)$ defined on its usual support $(0, \infty)$ in time or a finite interval of time $(0, T)$.

B. The Dirac Delta Function (Impulses)—Distributions (Generalized Functions)

The way in which we can present the sample $f(\tau_n)$ with an association to some area under it again involves the *Dirac delta function* $\delta(\tau - \tau_n)$ as a special case of the generalization of our usual concept of functions, i.e., the *generalized functions* or *distributions*. As we introduced it in Exercise 1.11b and c of Chapter 1, the *Dirac delta function* $\delta(\tau)$ is a pulse, and it may be seen in our usual sense of functions as one with *infinite height*. But the following definition, through this generalized sense of distributions, gives a finite area of 1 under the Dirac delta function:

$$\int_{-\infty}^{\infty} \delta(\tau) \, d\tau = 1 \tag{2.149}$$

This is a special case of its definition, which for infinitely continuously differentiable functions $\phi(t)$ is

$$\int_{-\infty}^{\infty} \phi(\tau)\delta(\tau - t)d\tau = \phi(t) \qquad (2.150)$$

or more usually is

$$\int_{-\infty}^{\infty} \phi(t)\delta(t - t_0)dt = \phi(t_0) \qquad (2.150a)$$

where in (2.149) we have $\phi(t) = 1$, $-\infty < t < \infty$.

For the Dirac delta function $\delta(t - t_0)$, as a special case of a distribution or a generalized function, we can have some feeling about this distribution and its difference from our usual functions. We notice in (2.150a), for the definition of $\delta(t - t_0)$, that $\delta(t - t_0)$ does not multiply the continuously differentiable function $\phi(t)$ inside the integral in our usual sense and then sum over this product to arrive at the result of the integration of this product on $(-\infty, \infty)$. It is a part of the integral which just *assigns* a value $\phi(t_0)$ at t_0 to $\phi(t)$ out of all its values for $t \in (-\infty, \infty)$. The generalized function or a distribution $\gamma(t)$ is a *process, then, of assigning a value* (or a number $N(\gamma, \phi)$)—called the functional—to the infinitely continuously differentiable function $\phi(t)$ beside it in the integral,

$$\int_{-\infty}^{\infty} \phi(t)\gamma(t)dt = N(\gamma, \phi) \qquad (2.151)$$

This number $N(\gamma, \phi)$ (or functional) depends on the distribution $\gamma(t)$ and the function $\phi(t)$. In the special case of the Dirac delta function $\gamma(t) = \delta(t)$, its assignment or functional to $\phi(t)$ is just its value $\phi(0)$ at $t_0 = 0$. As they are called "generalized functions," the distributions come down to our ordinary functions if we take $\phi(t) = f(t)$ as the usual functions we work with, and the "assignment process" (functional) reduces to our expected number for the value of the (convergent) integral of the product of the two ordinary functions $\phi(t)$ and $f(t)$,

$$\int_{-\infty}^{\infty} \phi(t)f(t)dt = N(f, \phi) \qquad (2.152)$$

We are not planning to go into more of the theory of distributions in this book, and it may be sufficient to be familiar with the very important case of the Dirac delta function. To give some of the basic properties of the Dirac delta function, we take these as special examples to get a feeling for such generalized functions. For example, $\delta(t)$ has the analog of the derivative of our ordinary function $d\delta(t)/dt$, but as a functional it acts to assign $-\phi'(0)$ to the function $\phi(t)$ with it in the integral (2.150), that defines $\delta(t)$, from (2.151) with $\gamma(t) = d\delta(t)/dt$,

$$\int_{-\infty}^{\infty} \phi(t) \frac{d\delta(t)}{dt} \, dt = -\phi'(0) \tag{2.153}$$

This is proved by differentiating the integral (2.150), which defines $\delta(\tau - t)$, with respect to t,

$$\frac{d}{dt} \int_{-\infty}^{\infty} \phi(\tau)\delta(\tau - t) d\tau = \frac{d}{dt} [\phi(t)]$$

$$= \int_{-\infty}^{\infty} \phi(\tau) \frac{d}{dt} \delta(\tau - t) d\tau = -\int_{-\infty}^{\infty} \phi(\tau)\delta'(\tau - t) dt = \frac{d\phi}{dt}$$

$$\int_{-\infty}^{\infty} \phi(\tau)\delta'(\tau - t) d\tau = -\phi'(t) \tag{2.153a}$$

where the derivative $\delta'(\tau - t)$ had acted as a functional, now assigning a derivative $-\phi'(t)$ of $\phi(t)$ at $\tau = t$.

The distributions multiply our functions in an unusual way; for example, if $f(t)$ is a continuous function, multiplying it by $\delta(t - t_0)$ results in an assignment of a value of another distribution $f(t_0)\delta(t - t_0)$,

$$f(t)\delta(t - t_0) = f(t_0)\delta(t - t_0) \tag{2.154}$$

This can be proved by showing that both $f(t)\delta(t - t_0)$ and $f(t_0)\delta(t - t_0)$ act as distributions to assign the same value to $\phi(t)$ in the integral (2.150) or (2.151),

$$\int_{-\infty}^{\infty} \phi(t)f(t)\delta(t - t_0) dt = \int_{-\infty}^{\infty} [\phi(t)f(t)]\delta(t - t_0) dt$$

$$= \phi(t_0)f(t_0)$$

and

$$\int_{-\infty}^{\infty} \phi(t)f(t_0)\delta(t - t_0) dt = f(t_0) \int_{-\infty}^{\infty} \phi(t)\delta(t - t_0) dt$$

$$= f(t_0)\phi(t_0) \tag{2.155}$$

This particular property of the Dirac delta function $f(t)\delta(t - t_n) = f(t_n)\delta(t - t_n)$ will be counted on to a great extent in the analysis of the discretization $\{f(t_n)\}$ of the signal $f(t)$ in Chapter 4 on the discrete transforms. We will also use it in the following discussion regarding a simple derivation (or modeling) of the impulse response of linear systems to the continuous signal $f(t)$. In practice, the systems receive discrete samples $\{f(t_n)\}$ as an input, and we will show next how the (infinite)

FIG. 2.7 The impulse train $\sum_{n=-\infty}^{\infty} \delta(t - t_n)$.

impulses of the Dirac delta functions facilitate the representation of such discrete input samples. First we present the *impulse train*

$$\sum_{n=-\infty}^{\infty} \delta(t - t_n) \qquad (2.156)$$

shown in Fig. 2.7 as infinite impulses located at $\{t_n\} = \{nT\}$. The arrows in Fig. 2.7 indicate the infinite length of the impulse.

With the help of this impulse train and the property $f(t)\delta(t - t_n) = f(t_n)\delta(t - t_n)$ in (2.154), we can input the samples $\{f(t_n)\}$ to the system as the following *pulse train* of (finite) samples:

$$\sum_{n=-\infty}^{\infty} f(t_n)\delta(t - t_n) \qquad (2.157)$$

This is shown in Fig. 2.8 for $t_n = nT$, where the arrows of the infinite impulses are used to indicate only the exact locations of the samples $\{t_n\}$. We assign the arrow the size of the sample $f(t_n)$ in Fig. 2.8.

With the aid of the impulse train (2.156), we can reproduce all the periodic replicas $\overline{f}(t)$ or extensions (Fig. 2.7, Fig. 2.9; see also Fig. 4.1a),

$$\overline{f}(t) \equiv \sum_{n=-\infty}^{\infty} f(t - nT) \qquad (2.158)$$

of the function $f(t)$ defined on the interval $(0, T)$ and with period T. This is done simply by taking the Fourier convolution product of $f(t)$ and

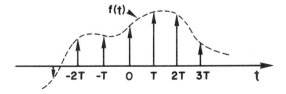

FIG. 2.8 The pulse train $\sum_{n=-\infty}^{\infty} f(nT)\delta(t - nT)$.

the impulse train of (2.156) with $t_n = nT$,

$$\bar{f}(t) = \int_{-\infty}^{\infty} f(\tau) \left[\sum_{n=-\infty}^{\infty} \delta(t - nT - \tau) \right] d\tau$$

$$= \sum_{n=-\infty}^{\infty} \int_{-\infty}^{\infty} f(\tau)\delta(t - nT - \tau) d\tau$$

$$= \sum_{n=-\infty}^{\infty} f(t - nT) \tag{2.158}$$

after using the definition of the Dirac delta function (2.150) for the last integral inside the infinite series.

In case the function $f(t)$ is defined on (say a symmetric interval) $(-T/2, T/2)$ and periodic of period T, then $\bar{f}(t)$ represents the periodic extension of $f(t)$ as replicas on all the periods of T along the whole real line. We note that the sum in (2.158) represents the superposition of all the translations of $f(t)$ by nT. Now suppose that $f(t)$ is defined on $(-T/2, T/2)$ and vanishes identically outside this interval; then the sum in (2.158) will represent $\bar{f}(t)$ as the periodic extension of the picture, or exact replicas, of $f(t)$ on $(-T/2, T/2)$ to the whole real line as shown in Fig. 2.9 (see also Fig. 4.1a). However, suppose that in practical applications we had to guess at such support of the function, but mistakenly estimated the actual support $(-T_1/2, T_1/2)$ by a smaller support $(-T/2, T/2)$ where $T < T_1$ as in Fig. 2.10 (see also Fig. 4.2a). In this case the function vanishes identically outside $(-T_1/2, T_1/2)$; thus it has, in general, a nonzero value on the left-out part of the support $(T/2, T_1/2)$

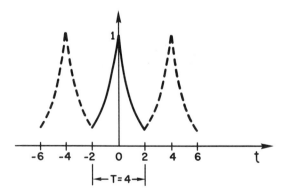

FIG. 2.9   The periodic extension $\bar{f}(t)$ of $f(t) = e^{-|t|}$, $-2 \le t \le 2$, with the correct desired period of 4.

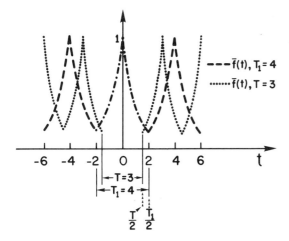

FIG. 2.10   The periodic extension $\bar{f}(t)$ of $f(t) = e^{-|t|}$ with the desired period $T_1 = 4$, and a *wrong* period of $T = 3$.

and $(-T_1/2, -T/2)$. In this case the sum in (2.158) gives the superposition of the translations of the actual $f(t)$ on $(-T/2, T/2)$ by NT, where there will be an overlap on both sides of each NT (see Figs. 2.10 and 4.2a). For example, around $t = T/2$ we are forgetting about the piece of the function on $(T/2 - T_1/2, T/2)$ to the left of $T/2$ and on $(T/2, T_1/2)$ to the right of $T/2$. So as the sum (2.158) adds up all these overlapped translations, we will get a wrong picture (see also Figs. 4.3 and 4.6). It is, of course, not the periodic extension $\bar{f}(t)$ of $f(t)$ with the actual period T_1. This is one of the main errors in signal analysis resulting from our shortcutting the support of the function $(-T_1/2, T_1/2)$ to $(-T/2, T/2)$, $T < T_1$. Such an error is termed an *aliasing error*, and its uncovering is due to using the pulse train (2.158) with due credit to the impulse train (2.157) of the delta functions, from which (2.158) was derived.

A very stern way of defining the impulse delta functions, though it may now sound familiar when we speak in the sense of distributions, appears when we have to accept it as the answer to the Fourier transform of $F(\omega) = 1$, $\omega \in (-\infty, \infty)$. This constant function $F(\omega) = 1$ is, of course, not integrable on $(-\infty, \infty)$, and thus it does not submit to our conditions of integrability on $(-\infty, \infty)$ for its inverse Fourier transform $f(t)$ in (2.137) to exist as a function in our usual sense and away from the general sense of distributions. But as we did in Exercise 1.11b and c in Chapter 1, we are able to give this $\delta(t)$ in terms of a limit of known function, where this limit along with integration will in a "formal" way make itself felt as behaving like the $\delta(t)$. What we did is to consider instead of $F(\omega) = 1$ on $(-\infty, \infty)$ the gate function

$$p_a(\omega) = \begin{cases} 1, & |\omega| < a \\ 0, & |\omega| > a \end{cases}$$

of (2.140) and Fig. 1.1a, whose Fourier transform is $f(t) = (2\sin at)/t$ as shown in Fig. 1.1a. Now we can look at the following limit:

$$\lim_{a \to \infty} \frac{1}{2\pi} \int_{-\infty}^{\infty} \frac{2\sin at}{t} \phi(t) \, dt$$

for continuously differentiable function $\phi(t)$, where we get the feeling that $(\sin at)/\pi t$ is highly oscillatory and decaying; its main contribution for large a is in the neighborhood of the origin. This feeling can be supported by looking at Fig. 1.2a, where we doubled a to $2a$ and where $(\sin at)/\pi t$ is narrowing more around the origin, pointing to its ultimate pulselike behavior as a becomes very large. For fixed large a, if we look at $(\sin at)/t$ as $t \to 0$, we can see, using L'Hôpital's rule, that $\lim_{t \to 0}((\sin at)/t) = a$. So it behaves like an impulse near $t = 0$ with its very large amplitude a, and we may write formally that

(i) $\quad \delta(t) = \lim_{a \to \infty} \dfrac{\sin at}{\pi t}$ $\qquad\qquad$ (2.159)

Of course, one more clear condition on such a definition of $\delta(t)$ is that:

$$\lim_{a \to \infty} \frac{1}{2\pi} \int_{-\infty}^{\infty} \frac{2\sin at}{t} e^{i\lambda t} \, dt = 1 = \mathcal{F}\{\delta(t)\}, \qquad -\infty < \omega < \infty,$$

$\qquad\qquad\qquad\qquad\qquad\qquad\qquad\qquad\qquad\qquad\qquad\qquad$ (2.160)

which makes $\delta(t)$ as the inverse Fourier transform of $F(\omega) = 1$, $-\infty < \omega < \infty$. This happened to be only one among such limit processes and integrations that give a feeling for more definitions of $\delta(t)$. Here we mention a few and leave their interpretation for the interest of the reader.

(ii) $\quad \delta(t) = \lim_{c \to 0} \dfrac{1}{c\sqrt{\pi}} e^{-t^2/c^2}$ $\qquad\qquad$ (2.161)

Of course, again this definition should satisfy

$$\lim_{c \to 0} \int_{-\infty}^{\infty} \frac{1}{c\sqrt{\pi}} e^{-t^2/c^2} \phi(t) \, dt = \phi(0) \qquad\qquad (2.162)$$

and the special case of it,

$$\lim_{c \to 0} \int_{-\infty}^{\infty} \frac{1}{c\sqrt{\pi}} e^{-t^2/c^2} \, dt = 1 \qquad\qquad (2.163)$$

Another way of defining $\delta(t)$, like our puzzlement with the Fourier transform of $F(\omega) = 1$, $\omega \in (-\infty, \infty)$, is when we may want to formally differentiate a function *at* its jump discontinuity. For example, in the case of the unit step function $U(t)$,

$$U(t) = \begin{cases} 1, & t > 0 \\ 0, & t < 0 \end{cases} \tag{2.164}$$

as illustrated in Fig. 2.2b as $u_0(t)$, how can we speak of the derivative at $t = 0$ as a usual limit?

$$\frac{dU}{dt} = \lim_{\Delta t \to 0} \frac{U(\Delta t+) - U(\Delta t-)}{2\,\Delta t}$$

where $U(\Delta t+) - U(\Delta t-) = 1$ even when $\Delta t \to 0$.

The formal way around this is to appeal to the "smoothing" effect of integration to show that

$$\int_{-\infty}^{\infty} \phi(t)\frac{dU}{dt}\,dt = \phi(0) \tag{2.165}$$

which gives to dU/dt the main property of $\delta(t)$ as presented in (2.150a) with $t_0 = 0$. We do this by integrating by parts in (2.165),

$$\int_{-\infty}^{\infty} \phi(t)\frac{dU}{dt}\,dt = \phi(t)U(t)\Big|_{t=-\infty}^{\infty} - \int_{-\infty}^{\infty} U(t)\phi'(t)\,dt$$

$$= [\phi(\infty) - 0] - \int_{0}^{\infty} \phi'(t)\,dt$$

$$= \phi(\infty) - [\phi(\infty) - \phi(0)] = \phi(0)$$

This result is reached after daring to write dU/dt, which is not defined in the usual sense, as we have tried to explain above. Thus such a definition is formal, but it is also correct based on the theory of distributions.

C. Impulse Response and the System Function

Having introduced the sample $f(\tau_n)$ (or samples $\{f(\tau_n)\}$) into the system as an input $f(\tau_n)\delta(t - \tau_n)$ (or inputs $\{f(\tau_n)\delta(t - \tau_n)\}$, we would now like to see how we can introduce the continuous signal $f(\tau)$ as an input to the system. We shall use τ here for the time of the signal $f(\tau)$, as we reserve t for the time of the system's response $h(t - \tau_n)$ to the sample $f(\tau_n)$ of the input signal, as we will show in (2.168)–(2.170). We shall take a very simplistic approach to this, which parallels what we do in elementary calculus for defining the integration of $f(\tau)$. We subdivide the domain of $f(\tau)$ into very small time subintervals of width $\Delta\tau$ with τ_n as the center of such a subinterval, as shown in Fig. 2.11.

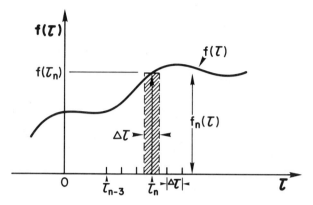

FIG. 2.11   A representative segment $f_n(\tau)$ of the signal $f(\tau)$.

Let us consider $f_n(\tau)$, the indicated representative *curvy segment*† *of the function* $f(\tau)$, defined on the nth subinterval with center at τ_n and very small width $\Delta\tau$. Then, as in Fig. 2.11, our continuous function $f(\tau)$ can be seen as the superposition of all these segments $(f_n(\tau))$ of the function $f(\tau)$),

$$f(\tau) = \sum_{n=-\infty}^{\infty} f_n(\tau) \tag{2.166}$$

We stress that in (2.166) we are adding the pieces, or segments, of the curve $f(\tau)$ and not the elements of the sequence $\{f(\tau_n)\}$ whose infinite sum, if it converges, gives a number

$$A = \sum_{i=-\infty}^{\infty} f(\tau_i).$$

This A for $f(t) \geq 0$ is the total area of the sum of the rectangular strips in Fig. 2.11 with heights

$$f(\tau_n), \qquad n = 0, \mp1, \mp2, \ldots$$

and constant (large!) width of $\Delta\tau = 1$. This area A is only a very *coarse* approximation to the area under $f(\tau) \geq 0$, or integral of $f(\tau)$ over the real line $-\infty < t < \infty$.

If we consider each segment $f_n(\tau)$ as an input to the linear system, then the output is $g_n(\tau) = L[f_n(\tau)]$. By the linearity of the system, the

---

†A small segement of the curve $f(\tau)$, not to be confused with the sequence $f_n(x)$, often used in Chapter 3.

output $f(\tau) = \sum_{n=-\infty}^{\infty} f_n(\tau)$ is $g(\tau) = \sum_{n=-\infty}^{\infty} g_n(\tau)$, since

$$g(\tau) = L[f(\tau)] = L\left[\sum_{n=-\infty}^{\infty} f_n(\tau)\right]$$

$$= \sum_{n=-\infty}^{\infty} L[f_n(\tau)] = \sum_{n=-\infty}^{\infty} g_n(\tau)$$

$$g(\tau) = \sum_{n=-\infty}^{\infty} g_n(\tau) \tag{2.167}$$

provided, of course, that the operation of L inside the series $\sum_{n=-\infty}^{\infty} f_n(\tau)$ above is justified; that is, $\sum_{n=-\infty}^{\infty} L[f_n(\tau)]$ remains convergent.

The above operations were symbolic, and now we would like to see the output $g(\tau)$ as a function of something explicitly related to what characterizes the symbolic action of the system $\underset{\sim}{L}$ to the input signal $f(\tau)$. To do this we may think of the average value $\tilde{f}_n$ of the $f_n(\tau)$ as being very close to the sampled value $f(\tau_n)$ at the middle of its nth subinterval (with very small width $\Delta\tau$). In this way the very narrow rectangular strip of height $f(\tau_n)$ has an area very close to the area under (the curvy) segment $f_n(\tau)$; that is, $f(\tau_n)\Delta\tau \simeq \tilde{f}_n(\tau)\Delta\tau$. The system may see this input $f(\tau_n)\Delta\tau$ at time τ and respond to it by affecting its amplitude by a factor that depends on the particular system considered and, of course, on the instants of (the input signals) time τ_n and its own response time t. If the system is time invarying, this dependence will be only on $t - \tau_n$, and we write the system's response of modulating the input pulse $f(\tau_n)$ as $h(t - \tau_n)$, which is termed the *impulse response* of the system L. Thus the output of the system for the input $f(\tau_n)\Delta\tau$ becomes

$$g_n(t) \simeq h(t - \tau_n)f(\tau_n)\Delta\tau \simeq h(t - \tau_n)\tilde{f}_n(\tau)\Delta\tau \tag{2.168}$$

The total response, as an output, of this linear system to all the pulses $\{f(\tau_n)\Delta\tau\}$ is then

$$g(t) = \sum_{n=-\infty}^{\infty} g_n(t) \simeq \sum_{n=-\infty}^{\infty} h(t - \tau_n)f(\tau_n)\Delta\tau \tag{2.169}$$

If we use the usual (Riemann) definition of integration, found in elementary calculus, the above infinite sum will define the following integral as $\Delta\tau \to 0$:

$$g(t) = \int_{-\infty}^{\infty} h(t - \tau)f(\tau)d\tau \tag{2.170}$$

where, of course, $f(\tau_n)\Delta\tau$ and $\tilde{f}_n(\tau)\Delta\tau$ became $f(\tau)d\tau$ as $\Delta\tau \to d\tau$. We note that the above integral is the Fourier convolution product $(h*f)(t)$, defined in (2.127), of the input signal $f(t)$ and the impulse response of the system $h(t)$. Thus an important application of Fourier integral analysis is to the analysis of linear systems, where the output $g(t)$ of the system is the convolution product of the system's impulse response $h(t)$ and the input signal $f(t)$, that is, $g(t) = (h * f)(t)$.

The above derivation or modeling of $g(t) = (h * f)(t)$ is very elementary and formal. Now we should wonder about the word "impulse" in the term "impulse response." Our derivation was with respect to the system's response to *finite* pulses $f(\tau_n)$, and it does not seem to reflect how this expression behaves as a response to an *infinite* impulse like the Dirac delta function (infinite) impulse $\delta(\tau - \tau_n)$. Now that we have this result (2.170), even if it is only through the formal way above, it is nevertheless exact. Furthermore, we can now easily show that $h(t - \tau_n)$ is the response of the system to the impulse $\delta(\tau - \tau_n)$ located at τ_n. We do this by taking in (2.170) the input function $f(\tau)$ as the impulse $\delta(\tau - \tau_n)$, which is, of course, a generalized function or distribution, to have

$$g(t) = \int_{-\infty}^{\infty} h(t - \tau)\delta(\tau - \tau_n)\,d\tau = h(t - \tau_n)$$

$$g(t) = h(t - \tau_n) \tag{2.171}$$

as the response (output) of the system L to the (input) impulse $\delta(\tau - \tau_n)$. If the impulse is located at $\tau_0 = 0$, then $g(t) = h(t)$. This explains how the term impulse response of $h(t)$ was coined in systems analysis, for the system's response to the special case of an impulse located at $\tau_0 = 0$.

In (2.170), the input $f(t)$ and the output $g(t)$ as functions of time are related to the impulse response $h(t)$ of the system through this complicated formula involving the convolution product $h * f$ of $h(t)$ and $f(t)$. The power of Fourier analysis for applications in systems stems from simplifying or "algebraizing" such a complicated relation when we analyze it in the "frequency" or Fourier transform space ω. This can easily be seen when we look at $F(\omega)$, $G(\omega)$, and $H(\omega)$ as the respective Fourier transforms of the input $f(t)$, the output $g(t)$, and the impulse response of the system $h(t)$, in order to see that the above convolution product relation (2.170) in the time space becomes a simple "algebraic" product in the frequency space as

$$G(\omega) = H(\omega)F(\omega) \tag{2.172}$$

Then we can find $H(\omega)$ from knowledge of the Fourier transforms of the input and output $F(\omega)$ and $G(\omega)$. Through the inverse Fourier transformation of $H(\omega)$, we find the system's desired impulse response $h(t) = \mathcal{F}^{-1}\{H(\omega)\}$.

With the algebraic relation (2.172) for the three components related to the system, it should be no wonder that the systems analyst works in frequency space, hence the usual statement of transmitting a signal $f(t)$ on a frequency band. This means that, even though we are speaking of or hearing the signal $f(t)$ as a function of time, our description of it above is related to its simpler representation $F(\omega)$ in the frequency or Fourier space ω, where the frequency ω is limited to the band (a,b); that is, $F(\omega)$ vanishes identically beyond $\omega \in (a,b)$.

A direct way of defining the *system function* $H(\omega)$ of the system L is to show that its amplitude is the amplitude of the output $H(\omega)e^{i\omega t}$ corresponding to the input $e^{i\omega t}$, or that

$$L[e^{i\omega t}] = H(\omega)e^{i\omega t} \tag{2.173}$$

To show this we take $f(t) = e^{i\omega_0 t}$ as input in (2.170) to receive its output as $g(t) = H(\omega_0)e^{i\omega_0 t}$

$$g(t) = \int_{-\infty}^{\infty} h(t - \tau)e^{i\omega_0 \tau} d\tau$$

$$= \int_{-\infty}^{\infty} h(t)e^{i\omega_0(t-\tau)} d\tau$$

$$= e^{i\omega_0 t} \int_{-\infty}^{\infty} h(\tau)e^{-i\omega_0 \tau} d\tau$$

$$g(t) = e^{i\omega_0 t} H(\omega_0) \tag{2.173a}$$

after using the commutativity of the above convolution product. Since this result is valid for arbitrary ω_0, (2.173) is proved.

The system function $H(\omega)$ coming from the definition of the Fourier transform will in general involve the complex number i (usually written $j = \sqrt{-1}$ in electrical engineering books and literature). Thus we write it as a complex-valued function with $H_R(\omega)$ and $H_I(\omega)$ as its real and imaginary components (we also use H_r and H_i when $j = \sqrt{-1}$ is used)

$$H(\omega) = H_R(\omega) + iH_I(\omega)$$

$$= |H(\omega)|e^{i\phi(\omega)} = \sqrt{H(\omega)\overline{H(\omega)}}e^{i\phi(\omega)}$$

$$= \sqrt{H_R^2(\omega) + H_I^2(\omega)}e^{i\phi(\omega)} \equiv A(\omega)e^{i\phi(\omega)} \tag{2.174a}$$

$$= A(\omega)[\cos \phi(\omega) + i \sin \phi(\omega)] \tag{2.174b}$$

where $|H(\omega)| = A(\omega)$ is the magnitude of the system function and $\phi(\omega)$ is its phase. Simple elements of complex numbers are discussed in Exercise 1.6.

Since $H(\omega)$, as the Fourier transform of the real-valued $h(t)$, is in general a complex function, we cannot exactly graph it for our computations of Fourier transforms via the fast Fourier transform in Chapter 4. The most usual method for such graphical illustration, which we will adopt in Chapter 4, is to plot its (real-valued) amplitude square $|H(\omega)|^2 = A^2(\omega)$. Sometimes the real and imaginary parts $H_R(\omega)$ and $H_I(\omega)$ are graphed separately, or $A(\omega)$ is graphed along with the phase dependence $\phi(\omega)$.

One of our examples in Chapter 4 ((4.29)–(4.31); see Figs. 3.8a and 4.11a) is the numerical computation of the Fourier transform (system function) $H(\omega)$ of the following causal function $h(t)$ (impulse response):

$$h(t) = e^{-at} U(t) = \begin{cases} e^{-at}, & t > 0 \\ 0, & t < 0 \end{cases} \tag{2.175}$$

as shown in Fig. 2.12, where $U(t)$ is the unit step function. Here we are able to evaluate $H(\omega)$ exactly; it is the complex-valued function $H(\omega) = 1/(a + i\omega)$,

$$H(\omega) = \int_{-\infty}^{\infty} h(t) e^{-i\omega t}\, dt = \int_0^{\infty} e^{-at} e^{-i\omega t}\, dt = \int_0^{\infty} e^{-(a+i\omega)t}\, dt$$

$$= -\frac{1}{a+i\omega} e^{-at} e^{-i\omega t}\Big|_{t=0}^{\infty} = -\frac{1}{a+i\omega}[0-1] = \frac{1}{a+i\omega}$$

since e^{-at} with $a > 0$ vanishes at $t = \infty$.

This $H(\omega) = 1/(a + i\omega)$ can be written in the two forms (2.74a), (2.74b) as

$$H(\omega) = \frac{1}{a+i\omega} = \frac{a-i\omega}{(a+i\omega)(a-i\omega)}$$

$$= \frac{1}{\sqrt{a^2+\omega^2}}\left[\frac{a}{\sqrt{a^2+\omega^2}} - i\,\frac{\omega}{\sqrt{a^2+\omega^2}}\right] \tag{2.176}$$

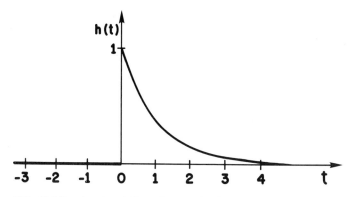

FIG. 2.12   A causal function as in (2.175).

where the amplitude $A(\omega) = 1/\sqrt{a^2 + \omega^2}$ and the phase angle $\phi(\omega)$ is $\phi(\omega) = -\tan^{-1}(\omega/a)$. Also, $H_R(\omega) = a/(a^2 + \omega^2)$ and $H_I(\omega) = -\omega/(a^2 + \omega^2)$.

EXAMPLE 2.22: System Functions for Differential Equations

In our examples of linear systems, we gave the differential operator L of the nonhomogeneous differential equation $L[u(t)] = f(t)$ in (2.143) as a linear system with the nonhomogeneous term $f(t)$ as the input and the desired solution $u(t)$ as the output. Here we illustrate it with the simple example

$$L[u(t)] = \frac{du}{dt} + au(t) = \cos bt \tag{E.1}$$

where $f(t) = \cos bt$ is the input and $u(t)$ is the output.

We have just shown in (2.173a) that the system function $H(\omega)$ is the coefficient in $H(\omega_0)e^{i\omega_0 t}$, the output of the system L to the input $e^{i\omega_0 t}$. So for $f(t) = \cos bt = (e^{ibt} + e^{-ibt})/2$, we consider it as the sum of two inputs $\frac{1}{2}e^{ibt}$ and $\frac{1}{2}e^{-ibt}$. For our linear system (differential equation), the output will be the superposition of their corresponding outputs $\frac{1}{2}H(b)e^{ibt}$ and $\frac{1}{2}H(-b)e^{-ibt}$ according to (2.173a) with $\omega_0 = b$ and $-b$, respectively. If we assume that the system function $H(\omega)$ is for a real system and is a real-valued function of the frequency ω, and if we assume that the response $h(t)$ is also real, we can show next that $H(\omega)$ will be an even function of ω.

All such properties will be needed in the actual numerical computations in Chapter 4, which are proved there and listed in Table 4.1.

$$H(\omega) = \int_{-\infty}^{\infty} e^{i\omega t} h(t)\, dt$$

$$= \int_{-\infty}^{\infty} h(t)\cos\omega t\, dt + i\int_{-\infty}^{\infty} h(t)\sin\omega t\, dt$$

$$= H_R(\omega) + iH_I(\omega) \tag{E.2}$$

Clearly, for $h(t)$ real, $H_R(\omega)$ is an even function in ω because of the $\cos\omega t$ in its integral, and $H_I(\omega)$ is an odd function because of the $\sin\omega t$ in its integral. Hence the real $H(\omega) = H_R(\omega)$ is an even function.

So with the result for the real response function $h(t)$ and its real even system function $H(\omega)$, we can conclude that the sum of the two outputs from our system is

$$\frac{1}{2}H(b)e^{ibt} + \frac{1}{2}H(-b)e^{-ibt} = H(b)\cos bt \tag{E.3}$$

where we used $H(-b) = H(b)$ for the even $H(\omega)$. But from (2.172) the Fourier transform of the system is $G(\omega) = H(\omega)F(\omega)$, so if we use the Fourier transform on the differential equation (E.1), letting $G(\omega) = \mathcal{F}\{u(t)\}$, we have

$$i\omega G(\omega) + aG(\omega) = F(\omega) \tag{E.4}$$

and for $\omega = \omega_0$ as constant for $\cos bt$,

$$i\omega_0 G(\omega_0) + aG(\omega_0) = F(\omega_0) = \mathcal{F}\{\cos bt\}$$

$$G(\omega_0) = \frac{1}{i\omega_0 + a} F(\omega_0) = H(\omega_0)F(\omega_0) \tag{E.5}$$

Thus with $\omega_0 = b$ in $\cos bt$, we have

$$H(b) = \frac{1}{i\omega_0 + a} = \frac{1}{a + ib} = \frac{a - ib}{a^2 + b^2}$$

$$= \frac{1}{\sqrt{a^2 + b^2}} e^{-i\tan^{-1} b/a}$$

$$H(\omega) = \frac{1}{\sqrt{a^2 + \omega^2}} e^{-i\tan^{-1} \omega/a} \tag{E.6}$$

Systems in Cascade

Suppose we have a complex system made out of two systems L_1 and L_2 connected in cascade (series). Let us assume that these two systems have respective impulse responses $h_1(t)$ and $h_2(t)$ and system functions $H_{(\omega)}$ and $H_2(\omega)$. We are after the impulse response $h(t)$ and the system function $H(\omega)$ for this combination of the two systems L_1 and L_2 in cascade. We will use the Fourier convolution Theorem 2.18 in (2.128), with the help of the impulse $\delta(t)$ (of course!) to show that $h(t) = (h_1 * h_2)(t)$, a Fourier convolution product, while the system function is a simple algebraic product $H(\omega) = H_1(\omega)H_2(\omega)$. We start by inputting an impulse $\delta(t)$ to the first system L_1 to have its output (its impulse response by definition) as $h_1(t)$. This output $h_1(t)$ will in turn be input to the second system L_2 to have an output, according to the convolution product representation (2.170) of this system, as

$$g(t) = \int_{-\infty}^{\infty} h_2(t - \tau)h_1(\tau)\,dt = \int_{-\infty}^{\infty} h(t - \tau)\delta(\tau)\,dt$$

$$= h(t) \tag{2.177}$$

In the last integral for obtaining this result we used the fact that the impulse response of the whole system is the output of it to an impulse $\delta(t)$. Thus $g(t) = (h_2 * h_1)(t) = (h_1 * h_2)(t)$. The system function $H(\omega)$

is the Fourier transform of this $h(t)$, and according to the convolution theorem it is

$$H(\omega) = \mathcal{F}\{(h_1 * h_2)(t)\} = H_1(\omega)H_2(\omega) \qquad (2.178)$$

Causal Functions and Causal Systems

A function $f(t)$ is called a *causal function*, as we mentioned in Chapter 1, if it is identically zero for $t < 0$. The signal $f(t)$ for real time t must be causal because $f(t) \equiv 0$, that is, identically zero before we switch the circuit on at $t = 0$; thus

$$f(t) = \begin{cases} f(t), & t \geq 0 \\ 0, & t < 0 \end{cases} \qquad (2.179)$$

We should remember that for the Laplace transform of $f(t)$, $t > 0$, we assumed $f(t)$ to be causal in the derivation of the Laplace convolution Theorem 2.11 in (2.35).

A system is also called a *causal system* if its response to a causal input $f(t)$ is a causal output $g(t)$. It is clear that the (real) physical system that receives signals $f(t)$ as a function of the real time is causal, since as it receives $f(t)$ after $t = 0$, its output $g(t)$ is generated only after this time $t = 0$ and nothing comes out for $t < 0$. Some very well-known physical systems, however, are not causal. For example, an optical system can have as input an object described by $f(x)$ only on the half-line $(x \geq 0)$; thus the input $f(x)$ is a causal input. However, the expected image, as the output $g(x)$ of the system, may not vanish identically for $x \leq 0$ to make it causal.

D. Advantages of Fourier (Spectrum) Space Representation in Other Fields

The benefits of the representation in the Fourier space are not limited to our present familiar example of communication systems but extend to almost all applied as well as theoretical fields. One very well-known example is that of spectroscopy, where the measurement is done for wavelength dependence, which is a Fourier space. In the same line is the work of the quantum physicist; even though the wave function $\psi(x)$ for the quantum is considered in the laboratory x-coordinate space, the analysis is done for its Fourier transform $\Psi(p)$ in the p-momentum space (or for $\Psi(k)$, where k is the wave number). This application was illustrated in detail in Example 2.20 with the Schrödinger equation.

We mention again that physicists use the Fourier representation in p-momentum space very intensively, and as a result a number of important tools for Fourier analysis have been established by physicists. There is perhaps no better witness to this than the Dirac delta function $\delta(t)$ in our Fourier analysis; Dirac, a physicist, invented it, and for us it resolves the dilemma of the Fourier transform of $F(\omega) = 1$, $\omega \in (-\infty, \infty)$. This important concept motivated the search by mathematicians, or physicists, to put it on sound mathematical grounds and led to the concepts of distributions or generalized functions. Unfortunately, there is no space in this book to go further into this subject, and we will limit ourselves to the use of such distributions as the impulse (Dirac delta) function $\delta(t)$ of Section 2.4.1C.

We emphasize that in their intensive use of Fourier analysis, physicists used the symmetric form (2.70), (2.71),

$$f(x) = \frac{1}{\sqrt{2\pi}} \int_{-\infty}^{\infty} F(p)e^{ipx}\,dx \tag{2.70}$$

$$F(p) = \frac{1}{\sqrt{2\pi}} \int_{-\infty}^{\infty} f(x)e^{-px}\,dx \tag{2.71}$$

or

$$f(x) = \frac{1}{\sqrt{2\pi}} \int_{-\infty}^{\infty} F(k)e^{ikx}\,dk \tag{2.70a}$$

$$F(k) = \frac{1}{\sqrt{2\pi}} \int_{-\infty}^{\infty} f(x)e^{-ikx}\,dx \tag{2.71a}$$

where p and k stand for momentum and wave number, respectively. This is done in most physics books, and the detailed mathematics references on transforms by Sneddon (1972) and Ditkin and Prudnikov (1965) (within the text but not the tables) give some examples of the use of this symmetric form (2.70), (2.71). We have chosen the form (2.64), (2.65) as a compromise because of our emphasis on using it for solving boundary value problems, as we did in Section 2.3 and in the communication systems analysis of this section. Besides, the form we have adopted here is close to the form (2.66), (2.67) most used for discrete Fourier transforms and their fast algorithm, the fast Fourier transform (FFT), which we shall discuss and apply in Chapter 4. Another very familiar application of Fourier integrals representation is in statistics.

2.4.2 Bandlimited Functions—the Sampling Expansion

In the last section we introduced the important class of bandlimited functions $f_a(t)$ in (2.138),

$$f_a(t) = \int_{-a}^{a} e^{i\omega t}F(\omega)\,d\omega = 2\pi \int_{-2\pi W}^{2\pi W} F(2\pi\nu)e^{i2\pi\nu t}\,d\nu \tag{2.138}$$

where W is the highest frequency (in cycles per second, or hertz) of the signal. Here we will present a more detailed discussion of these bandlimited functions, especially with regard to their useful property of deserving a sampling series expansion (or interpolation) and their use in communications systems in general. Moreover, in the section following we will show their relation as a building block to the *B-splines* (*hill functions*) of numerical analysis.

A. Finite Energy (Power!) Signals—Parseval's Equality

If we consider $f(t)$ as the voltage across one unit of resistance, then $\int_{-\infty}^{\infty} |f(t)|^2 dt$ should represent the total energy, which for all practical purposes is finite, and we call the signal a *finite energy signal* (or finite power signal). If we consult the Parseval equality (2.136), we can easily see that

$$\frac{1}{2\pi} \int_{-\infty}^{\infty} |F(\omega)|^2 d\omega = \int_{-\infty}^{\infty} |f(t)|^2 dt \tag{2.180}$$

Hence for a finite energy signal, the right-hand side of (2.180) is finite and we conclude that $F(\omega)$ is square integrable on $(-\infty, \infty)$; that is, $\int_{-\infty}^{\infty} |F(\omega)|^2 d\omega < \infty$. This condition, as we shall see in Chapter 3 in (3.121) is important for facilitating the analysis (proof) of the sampling series representation of bandlimited signals and for the convergence of the Fourier series expansion of $F(\omega)$ on $(-a, a)$ in general. (See (3.66), (3.124).)

It is easy to see from (2.180) and (2.138) that a bandlimited signal has finite energy (or power) when $F(\omega)$ is square integrable on $(-a, a)$; that is,

$$\int_{-\infty}^{\infty} |f_a(t)|^2 dt = \frac{1}{2\pi} \int_{-a}^{a} |F(\omega)|^2 d\omega < \infty \tag{2.181}$$

The following discussion will center around relating (the not necessarily bandlimited function) $f(t)$

$$f(t) = \int_{-\infty}^{\infty} e^{i\omega t} F(\omega) d\omega \tag{2.68}$$

in (2.68) to its bandlimited case $f_a(t)$ in (2.138)

$$f_a(t) = \int_{-a}^{a} e^{i\omega t} F(\omega) d\omega \tag{2.138}$$

Then we will, in the absence of the Fourier series of Chapter 3, establish by a "plausible" simple method the sampling series expansion for $f_a(t)$.

B. Filtering a Signal and the Sampling Expansion

Consider the bandlimited signal

$$f_a(t) = \int_{-a}^{a} F(\omega)e^{i\omega t}\, d\omega = \int_{-\infty}^{\infty} p_a(\omega)F(\omega)e^{i\omega t}\, d\omega \qquad (2.182)$$

Now we use the convolution theorem (2.128) with $F_1(\omega) = p_a(\omega)$ and $F_2(\omega) = F(\omega)$ with due care for the difference between our usual notation (2.64), (2.65) and the present notation (2.68) for $f(t)$, where

$$F(\omega) = \frac{1}{2\pi} \int_{-\infty}^{\infty} e^{-i\omega t} f(t)\, dt \qquad (2.69)$$

From (2.137) we have $f_2(t) = f(t)$ and from (1.32) we have

$$f_1(t) = \frac{2}{2\pi}\frac{\sin at}{t} = \frac{\sin at}{\pi t}$$

Hence, we can write (2.182) as

$$f_a(t) = \int_{-\infty}^{\infty} f_1(t - \tau)f_2(\tau)\, d\tau = \int_{-\infty}^{\infty} \frac{\sin a(t - \tau)}{\pi(t - \tau)} f(\tau)\, d\tau \qquad (2.183)$$

From (2.182) we can see that the output $g(t) = f_a(t)$ has a Fourier transform

$$G(\omega) = p_a(\omega)F(\omega)$$

with $F(\omega)$ the Fourier transform of the input signal $f(t)$. If we compare the above result with (2.170),

$$G(\omega) = H(\omega)F(\omega) \qquad (2.170)$$

where $H(\omega)$ is the *system function*, we conclude that $p_a(\omega)$ is the system function, which is known to describe the *ideal low-pass filter*. Thus from (2.182) we can say that the bandlimitedness of $f_a(t)$ is the result of passing its original version, the not necessarily bandlimited signal $f(t)$, through an ideal low-pass filter with system function as the gate function $p_a(\omega)$. The (convolution product) integral representation (2.183) of the bandlimited function $f_a(t)$ is an equivalent way of saying that $f_a(t)$ is the result of passing the original signal $f(t)$ through a filter, whose *impulse response* is $h(t) = (\sin at)/\pi t$, which corresponds to the same ideal low-pass filter system function $p_a(\omega)$. So we can look at the representation of $f_a(t)$ in (2.183) as telling the whole story as far as the physical interpretation of "bandlimitedness" is concerned.

However, this picture is given for a continuous input $f(t)$ and not the discrete set of samples that the system (most often) actually receives. The answer to this practical problem lies within the sampling theorem, which deals directly with the samples of the bandlimited function $\{f_a(n\pi/a)\}$ to give, in a way parallel to (2.183),

$$f_a(t) = \sum_{n=-\infty}^{\infty} f_a\left(\frac{n\pi}{a}\right) \frac{\sin a(t - n\pi/a)}{a(t - n\pi/a)} \tag{2.184}$$

This can be interpreted when we look for a discrete parallel to (2.183), where, of course, we have to be ready for a finite $\Delta\tau = \pi/a$ to replace $d\tau$ in the convolution integral of $f_a(t)$. Indeed, we had this $\Delta\tau = \pi/a$ originally in the discrete convolution sum (2.169) when we used it to arrive at the convolution integral (2.170). Here we shall appeal again to the pulse train of (2.157), and assign each pulse a weight of $\Delta\tau = \pi/a$ as explained above, to give the input samples

$$\sum_{n=-\infty}^{\infty} \frac{\pi}{a} f_a\left(\frac{n\pi}{a}\right) \delta\left(t - \frac{n\pi}{a}\right) \tag{2.185}$$

as the "discrete" input to the ideal low-pass filter as shown in Fig. 2.13. The output, according to the convolution theorem, now gives

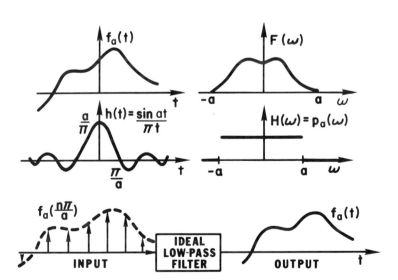

FIG. 2.13 Physical interpretation of the sampling expansion of bandlimited functions $f_a(t)$. (From Jerri (1977a), courtesy of the Proceedings of the IEEE.)

us the sampling expansion (2.184) for the bandlimited signal $f_a(t)$,

$$
f_a(t) = \int_{-\infty}^{\infty} \frac{\sin a(t - \tau)}{\pi(t - \tau)} \left[\sum_{n=-\infty}^{\infty} \frac{\pi}{a} f_a \left(\frac{n\pi}{a} \right) \delta \left(\tau - \frac{n\pi}{a} \right) \right] d\tau
$$

$$
= \sum_{n=-\infty}^{\infty} f_a \left(\frac{n\pi}{a} \right) \int_{-\infty}^{\infty} \frac{\sin a(t - \tau)}{a(t - \tau)} \delta \left(\tau - \frac{n\pi}{a} \right) d\tau
$$

$$
= \sum_{-\infty}^{\infty} f_a \left(\frac{n\pi}{a} \right) \frac{\sin a \left(t - \frac{n\pi}{a} \right)}{a \left(t - \frac{n\pi}{a} \right)}
$$

$$
f_a(t) = \sum_{n=-\infty}^{\infty} f_a \left(\frac{n\pi}{a} \right) \frac{\sin(at - n\pi)}{(at - n\pi)} \tag{2.184}
$$

In comparing the sampling expansion (2.184) and the convolution integral (2.183) representations of the (same) bandlimited signal $f_a(t)$, we recognize the clear parallel in their (convolution) forms. However, the sampling expansion (2.184) is an *infinite sum* over the *discrete time instants* $\{n\pi/a\}$ of the samples $\{f_a(n\pi/a)\}$ of the bandlimited signal $f_a(t)$, while the convolution product (2.183) is an integral over the *continuous time* τ of the (not necessarily bandlimited) input signal $f(\tau)$ to the ideal low-pass filter (with impulse response $\sin at/\pi t$). Thus the sampling expansion (2.184) stands to bridge the analysis of signals (in the continuum) as integrals and their analysis as an infinite sum of their more applicable discrete samples.

For the sampling expansion as a series, it will also bring us to the other very important branch of Fourier analysis, namely the Fourier series, which is the main subject of Chapter 3. We will have the Fourier series and its coefficients (which define the finite Fourier transforms) as an infinite sum to represent (periodic) functions on a finite interval $(-a, a)$. This contrasts with the present chapter's Fourier integral representation of functions $f(t)$ on the infinite domain $(-\infty, \infty)$.

With these notions of the sampling series expansion (2.184) related to the continuous convolution integral (2.183), and with the Fourier series and integrals in mind, we hope to have built the intuition for a possible Fourier series expansion for a function inside the integral of (2.182) to lead us to the sampling series (2.184). Indeed, this happens to be the case, but for completeness we shall postpone this development of the sampling series expansion to Section 3.3.1 in (3.121), as we shall need the Fourier series and its convergence in Section 3.1 in order to have a rigorous proof of this important result (2.184).

In that respect we consider the derivation of (2.184), as presented above with the help of the pulse train, as "plausible," since we did not even worry about the exchange of integration and summation that led to (2.184). We followed such a development in order to present at this time the sampling expansion and its important (discretizing) role for our present discussion of signals and systems analysis. The purpose here is also to bring up the practical necessity of discretizing the transmitted signals as $\{f_a(n\pi/a)\}$ instead of the (impossible to transmit) continuous signal $f_a(t)$. Moreover, we will have a chance to realize the consequences of errors that we have to watch for during the practical execution of this vital discretization process. The sampling expansion (2.184) expresses the continuous signal $f_a(t)$ in terms of the (infinite set of) samples $\{f_a(n\pi/a)\}$, the only practical form we can transmit or receive in the finite time we are allocated in our practical or realistic world. Our organization in this book is to cover the analysis of the various errors that occur in actual applications. As we shall discuss in full detail in Chapter 4, the error resulting from the necessary process of "discretization" may top the list of all the errors.

Another reason for wanting to expose the reader at this time to this discretization, through the sampling expansion, is that we plan to present in Chapter 4 fast algorithms like the fast Fourier transform (FFT) for the efficient numerical (practical!) computation of the (infinite) Fourier transform, its inverse, and the (infinite) Fourier series and its coefficients (as finite-limit Fourier integrals).

Such emphasis on the practical aspects of the error analysis as well as fast algorithms for the practical computation of the Fourier analysis tools, we hope, will contribute in a somewhat different way to a more complete picture of the transform methods on this level.

We emphasize again the role of the sampling expansion as a very basic tool in bridging the ideal continuous $f_a(t)$ and the practical discrete $f_a(n\pi/a)$. This will become very evident from the theory and the many detailed illustrations in Chapter 4.

Again, we should stress here that the importance of the sampling expansion lies not only in interpolating signals but also in specifying the required spacing $\Delta t = t_{n+1} - t_n = (n + 1)\pi/a - n\pi/a = \pi/a$ in terms of the bandlimit a. Such a spacing requirement plays an important role in developing the discrete Fourier transforms, whose efficient algorithm is the very useful FFT, which will be the main topics of Chapter 4 besides the error analysis of their applications.

Two obvious errors may be involved in the practical application of the sampling expansion (2.184) of signals, namely the *truncation* (ϵ_N) and *aliasing* (ϵ_A) *errors*. The first is due to the inclusion of only a finite number N of samples instead of the infinite number required by the series (2.184). The second is due to the uncertainty of knowing exactly the bandlimit a of the received signal $f_a(t)$. Research to find better estimates for such errors is ongoing. For most of the early error bounds for the truncation error

$$\epsilon_N(t) = \left| f_a(t) - \sum_{n=-N}^{N} f_a\left(\frac{n\pi}{a}\right) \frac{\sin a\,(t - n\pi/a)}{a\,(t - n\pi/a)} \right| \qquad (2.185)$$

complex integration is involved. Other methods may be employed, but they will become more tractable after we have the Fourier series analysis in Chapter 3. In the case of the aliasing error of the sampling expansion, we are supposed to use the samples $f_a(n\pi/a)$ of a bandlimited function $f_a(t)$, with sample spacing π/a, to generate $f_a(t)$ from (2.184). But if we are not sure about the bandlimit a, we don't know for sure whether $F(\omega)$ vanishes identically beyond $(-a, a)$. Of course, this $f(t)$ could very well be bandlimited to $b \sim a$, in which case we must have a good feeling about it, and the misuse of the sampling expansion with $a \sim b$ will not be so severe! In general, if we take the samples of $f(t)$ with sample spacing π/a (that of $f_a(t)$!)—that is, we use $f(n\pi/a)$ for the sampling expansion (2.184)—in the hope of generating the continuous function $f(t)$ of these samples, we are committing the *aliasing error*:

$$\epsilon_A = \left| f(t) - \sum_{n=-\infty}^{\infty} f\left(\frac{n\pi}{a}\right) \frac{\sin(at - n\pi)}{at - n\pi} \right| \qquad (2.186)$$

This error must be very obvious, since the sampling series (2.184) as a tool is meant to generate the bandlimited function $f_a(t)$ from its samples $f_a(n\pi/a)$, and with its own "characteristic sample space π/a." That is, (2.184) is never meant to generate any (not necessarily bandlimited) function $f(t)$ from its samples $\{f(n\pi/a)\}$, and with "the violent use of a" the bandlimit of $f_a(t)$. It is interesting to note that one of the earliest upper bound estimates for this aliasing error ϵ_A is

$$\epsilon_A \leq \frac{1}{\pi} \int_{|\omega| > a}^{\infty} |F(\omega)| \, d\omega \qquad (2.187)$$

This supports the intuition that if $f(t)$ is bandlimited to b a bit larger than a, where b and a are both large, the aliasing error is not so serious, since it is proportional to the small area under $|F(\omega)|$ on the very small interval (a, b). This is especially true where guided by the Riemann-Lebesgue lemma (Theorem 2.13) that $F(\omega)$, as the Fourier transform of sectionally smooth $f(t)$ on $(-\infty, \infty)$, decays to zero as $\omega \to \infty$.

We have already noted that the sampling expansion (2.184) for $f_a(t)$ requires an exact sample spacing $\Delta t = \pi/a$. This means that the rate of the sampling

$$\frac{1}{\pi/a} = \frac{a}{\pi} = \frac{2\pi W}{\pi} = 2\nu_{\max}$$

is twice the maximum frequency $\nu_{\max}$ (in cycles per second, hertz) of the band $(-a,a)$ where the spectrum $F(\omega)$ is defined. The rate $2\nu_{\max}$ is the condition in communications theory known as the *Nyquist rate*. The condition is usually stated in the field of transmitting (discrete) signals as, "the sample should be transmitted at a rate no less than twice the highest frequency of the band, i.e., no less than the Nyquist rate $2\nu_{\max}$."

C. Time-Limited Signals

We can also speak of time-limited signals

$$F_T(\omega) = \frac{1}{2\pi} \int_{-T}^{T} f(t)e^{-i\omega t}\, dt \tag{2.188}$$

where $f(t)$ is given in (2.137). According to the uncertainty principle (as may be seen from (2.181) and (2.138)), a signal cannot be bandlimited and time limited at the same time. So, in case we are forced into such a situation, we must admit an error, and we usually optimize to find the shape of the signal that minimizes such an error. An obvious situation may occur when we have a time-limited signal as in (2.188) but still want to use the sampling expansion (2.184) even though we cannot assume that $f(t)$ is bandlimited to a; that is, there is an uncertainty in knowing the required bandlimit a for (2.184). In practice, we assume a bandlimit a and admit an aliasing error, which requires a derivation of its upper bound as in (2.187), and report the results within the accuracy of such a bound.

D. Generalized Sampling Expansion for Other Transforms

In Chapter 3 we will also cover more general Fourier series or orthogonal expansions besides the usual trigonometric sine-cosine or exponential Fourier series, which are related to our present Fourier exponential transform. For example, we will cover the Fourier Bessel series as related to the Hankel transform. In this respect, a sampling expansion for the finite-limit (or bandlimited!) J_n-Hankel transform,

$$f_a(t) = \int_0^a xF(x)J_n(xt)\, dx \tag{2.189}$$

will be derived in Section 3.6 as a special case of a generalized sampling theorem. Such sampling series have possible applications in optics, where the Hankel transform is used and where "discrete" analysis via such a sampling expansion is warranted.

2.4.3 Bandlimited Functions and B-Splines (Hill Functions)

The gate function $p_a(\omega)$ in (2.140) that we used for the bandlimited signal $f_a(t)$ in (2.182) and (2.139) is known in numerical analysis as the *hill function (B-spline)* of *order one*. The Fourier transforms of higher-order hill functions, which we shall present next, have been used to speed up the convergence of the sampling series (2.184) and result in a better (tighter) upper bound for the truncation error ϵ_N of (2.185).

As a very good exercise and an important application to higher-order Fourier convolution multiplication in (2.127), we define the hill function (B-spline) of order $R + 1$, $\phi_{R+1}(a(R + 1), \omega)$, as the Rth fold (Fourier transform) convolution product; i.e.,

$$\phi_{R+1}(a(R + 1), \omega) = (\phi_1(a, -) * \overset{R}{\cdots} * \phi_1(a, -))(\omega)$$

$$= (\phi_1(a, -) * \phi_R(aR, -)) = \int_{-\infty}^{\infty} \phi_1(a, x)\phi_R(aR, \omega - x)\, dx \qquad (2.190)$$

of the gate function (the hill function of order one) $p_a(\omega)$:

$$\phi_1(a, \omega) = p_a(\omega) = \begin{cases} 1, & |\omega| < a \\ 0, & |\omega| > a \end{cases} \qquad (2.191)$$

The notation $a(R+1)$ for $\phi_{R+1}(a(R+1), \omega)$ indicates that the hill function of order $R + 1$ vanishes outside the larger interval $(-a(R + 1), a(R + 1))$ instead of $(-a, a)$ as for the gate function $\phi_1(a, \omega)$. This can be proved by using the convolution theorem, and we leave its illustration for $R = 1$ as an exercise (see Exercise 2.45a).

As another consequence of this convolution product (2.190), we recognize that $\phi_{R+1}(a(R + 1), \omega)$ is the Fourier transform of $\eta_{R+1}(t) = [(2\sin at)/t]^{R+1}$, since the Fourier transform of $\phi_1(a, \omega)$ is $\eta_1(t) = (2\sin at)/t$. Unless otherwise indicated, $\phi_{R+1}(\omega)$ will be written for $\phi_{R+1}(a(R + 1), \omega)$.

The hill functions have their usual application in numerical analysis, but they have also been used indirectly for the sampling expansion to speed its convergence for a lower truncation error bound. For example, the following sampling series was derived for the bandlimited signal $f_{ra}(t)$

$$f_{ra}(t) = \sum_{n=-\infty}^{\infty} f_{ra}\left(\frac{n\pi}{a}\right) \frac{\sin(at - n\pi)}{(at - n\pi)} \left[\frac{\sin(aq/m)(t - n\pi/a)}{(aq/m)(t - n\pi/a)}\right]^m$$

$$(2.192)$$

where $q = 1-r$, $0 < r < 1$. This series, when compared with (2.184), should clearly indicate the involvement of the Fourier transform of $\phi_m(aq,\omega)$, as the factor inside the series, with its behavior like $1/n^m$, which improves the convergence of the series and makes it "self-truncating" for larger m. The proof of (2.192) is an excellent exercise in the effective use of the convolution product of the Fourier transform (see Exercise 2.62e and also Exercises 2.45, 2.44, 3.30d, 3.34a). Let us again note that in (2.190) the Rth-fold self-convolution $\phi_{R+1}(a(R + 1),\omega)$ of $\phi_1(a,\omega) = p_a(\omega)$ has a domain with width that is $R + 1$ times that of $\phi_1(a,\omega)$; that is, it can be shown that the half-width of the domain of the Fourier convolution product of two functions, say $\phi_1(a,\omega)$ and $\phi_1(a,\omega)$, adds to $2a$ for $\phi_2(2a,\omega) = \phi_1(a,\omega) * \phi_1(a,\omega)$, and, as indicated above, we shall leave its proof for an exercise (see Exercise 2.45).

As mentioned earlier, we will give in Section 3.6 a very rigorous short and simple proof for the sampling expansion (or interpolation) (2.184) of the bandlimited function $f_a(t)$ in (2.182). The proof of the self-truncating sampling series (2.192) will follow in a very similar way, and represents a very good exercise for Section 3.6 in the application of Fourier series and also for the command of the Fourier convolution operation (2.127) (see Exercises 2.62 and 3.62). If we look at $f_a(t)$ in (2.182), it is the Fourier transform of $p_a(\omega)F(\omega) = \phi_1(\omega)F(\omega)$. The essence of the sampling series in (2.192) is that we will use $\phi_1(ra,\omega) * \phi_m(qa,\omega)F(\omega)$ instead of $\phi_1(\omega)F(\omega)$. We realize here that $\phi_1(ra,\omega) * \phi_m(qa,\omega)$ has a support (domain) of half-width that adds up to $ra + qa = (r + q)a = a$. Of course, $\phi_1(ra,\omega) * \phi_m(qa,\omega)$, even though like $p_a(\omega) = \phi_1(a,\omega)$ it vanishes identically outside the support $(-a,a)$, is only an approximation to $p_a(\omega) = \phi_1(a,\omega)$. That is, we don't expect $\phi_1 * \phi_m$, after all the convolution integrations, to look like the gate function with its very obvious discontinuities at $\mp a$. The difference is that while $\phi_1(a,\omega)$ contributes to the sampling function $\sin(at - n\pi)/(at - n\pi)$, which behaves like $1/n$ for n large, $\phi_1(ra,\omega) * \phi_m(qa,\omega)$ contributes to an added factor to the sampling function in (2.192) which behaves like $1/n^m$. The latter factor makes (2.192) a self-truncating series for large m.

2.5 Fourier Sine and Cosine Transforms

In Section 2.1 we discussed the Laplace transform, which is suitable for problems on the semi-infinite interval $(0,\infty)$ and is compatible with derivatives of order n provided that all the derivatives up to order $n-1$ are given at $x = 0$, as presented in (2.20) as the first corollary to Theorem 2.3. In Section 2.2 we presented the Fourier exponential transform (2.64), which is suitable for the infinite domain $(-\infty, \infty)$ and is compatible with all derivatives $d^n f / dx^n$ provided that all the derivatives up to order $n - 1$ vanish as $x \rightarrow \mp\infty$. This result in (2.118) is the first corollary to Theorem 2.15, which assumes that all the derivatives $d^i f / dx^i$, $i = 1, 2, \ldots,$ $n - 1$ are continuous and absolutely integrable on $(-\infty, \infty)$ and the transformed function $f(x)$ is continuous and absolutely integrable on $(-\infty, \infty)$. The necessary vanishing of all the derivatives up to order $n - 1$ is met by the condition of their absolute integrability on $(-\infty, \infty)$, as dictated by Corollary 1 of Theorem 2.15, which implies their vanishing as $x \rightarrow \mp\infty$.

We have already shown in our analysis leading to (2.99) and (2.102) that the Fourier sine and cosine transform (2.99), (2.102) are special cases of the Fourier exponential transform (2.64). This is in the sense that when $f(x)$ defined on $(-\infty, \infty)$ is an *odd function* $f_o(x)$ or an *even function* $f_e(x)$, the Fourier exponential transform (2.64) of $f(x)$ reduces, respectively, to the Fourier sine transform $F_s(\lambda)$ of (2.99) and the Fourier cosine transform $F_c(\lambda)$ of (2.102),

$$\mathcal{F}_s \{f\} = F_s(\lambda) = \int_0^\infty f(x) \sin \lambda x \, dx \tag{2.193}$$
$$\tag{2.99}$$

$$\mathcal{F}_c \{f\} = F_c(\lambda) = \int_0^\infty f(x) \cos \lambda x \, dx \tag{2.194}$$
$$\tag{2.102}$$

As special cases of the Fourier exponential transform, we expect to have the justification for many of their properties based on the theorems we proved for the Fourier exponential transform. Indeed, we have already stated and proved the existence of the Fourier sine transform and its inverse (2.101) as a simple Corollary 1 to the Fourier integral Theorem 2.14 of (2.86) or its equivalent (2.85). The case for the Fourier cosine transform (2.104) followed as Corollary 2 to the same Theorem 2.14. For emphasis we shall repeat the precise statements of these corollaries:

Corollary 1 to Theorem 2.14: The Existences of the Inverse Fourier Sine Transform: Let $f(x)$ be defined for $0 \leq x < \infty$ and extended as an *odd* function $f_o(x)$ on $(-\infty, \infty)$. Let $f(x)$ now satisfy the conditions of the Fourier integral Theorem 2.14 of being sectionally (piecewise) smooth and absolutely integrable on $(-\infty, \infty)$. Then at points λ where $F_s(\lambda)$,

$$F_s(\lambda) = \int_0^\infty f(x) \sin \lambda x \, dx \equiv \mathcal{F}_s \{f_0\} \tag{2.193}$$

is continuous we have its *inverse* as

$$\frac{1}{2} [f_o(x+) + f_o(x-)] = \frac{2}{\pi} \int_0^\infty F_s(\lambda) \sin \lambda x \, d\lambda \equiv \mathcal{F}_s^{-1} \{F_s\} \tag{2.195}$$

Of course, this integral converges to $f_o(x)$ at points x where $f(x)$ is continuous. □

Corollary 2 to Theorem 2.14: The Existence of the Inverse Fourier Cosine Transform: Let $f(x)$ be defined on $0 \le x < \infty$ and extended as an *even* function $f_e(x)$ on $(-\infty, \infty)$. Let $f(x)$ now satisfy the conditions of the Fourier integral Theorem 2.14 of being sectionally (piecewise) smooth and absolutely integrable on $(-\infty, \infty)$. Then at points λ where $F_c(\lambda)$,

$$F_c(\lambda) = \int_0^\infty f(x) \cos \lambda x \, dx \equiv \mathcal{F}_c \{f_e\} \tag{2.194}$$

is continuous we have its *inverse* as

$$\frac{1}{2} [f_e(x+) + f_e(x-)] = \frac{2}{\sqrt{\pi}} \int_0^\infty F_c(\lambda) \cos \lambda x \, d\lambda \equiv \mathcal{F}_c^{-1} \{F_c\} \tag{2.196}$$

and it gives $f_e(x)$ at points x where $f_e(x)$ is continuous. □

The formal way of deriving, for example, the inverse Fourier sine transform formula (2.195) is to depend on the Fourier integral formula (2.86) of Theorem 2.14 for the special case of $f(x) = f_o(x)$, an odd function, which is continuous on $(0, \infty)$

$$f_o(x) = \frac{1}{\pi} \int_0^\infty d\lambda \int_{-\infty}^\infty f_o(\xi) [\cos \lambda x \cos \lambda \xi + \sin \lambda x \sin \lambda \xi] \, d\xi$$

$$= 0 + \frac{2}{\pi} \int_0^\infty d\lambda \int_0^\infty f(\xi) \sin \lambda x \sin \lambda \xi \, d\xi$$

$$= \frac{2}{\pi} \int_0^\infty d\lambda \sin \lambda x \int_0^\infty f(\xi) \sin \lambda \xi \, d\xi$$

$$f(x) = \frac{2}{\pi} \int_0^\infty F_s(\lambda) \sin \lambda x \, d\lambda \equiv \mathcal{F}_s^{-1} \{F_s\} \tag{2.197}$$

after using the fact that $\int_{-\infty}^\infty f_o(\xi) \cos \lambda x \cos \lambda \xi \, d\xi = 0$, where the integrand is an odd function in ξ. The inversion formula for the Fourier cosine transform,

$$f(x) = \mathcal{F}_c^{-1} \{F_c\} = \frac{2}{\pi} \int_0^\infty F_c(\lambda) \cos \lambda x \, d\lambda \tag{2.198}$$

is also derived from (2.86) in the same simple way, except here we consider $f(x) = f_e(x)$ to be an even function which is continuous on $(0, \infty)$.

We mention again that the notations for the Fourier transforms differ from one reference to another. The usual symmetric notations for the Fourier sine and cosine transforms and their inverses, compared to ours in (2.193), (2.197) for the sine transform and (2.194), (2.198) for the cosine transform, are

$$F_s(\lambda) = \sqrt{\frac{2}{\pi}} \int_0^\infty f(x) \sin \lambda x \, dx \tag{2.193a}$$

$$f(x) = \sqrt{\frac{2}{\pi}} \int_0^\infty F_s(\lambda) \sin \lambda x \, d\lambda \tag{2.197a}$$

$$F_c(\lambda) = \sqrt{\frac{2}{\pi}} \int_0^\infty f(x) \cos \lambda x \, dx \tag{2.194a}$$

$$f(x) = \sqrt{\frac{2}{\pi}} \int_0^\infty F_c(\lambda) \cos \lambda x \, d\lambda \tag{2.198a}$$

2.5.1 Compatibility of the Fourier Sine and Cosine Transforms with *Even*-Order Derivatives

Using the method of repeated integration by parts, we showed in (1.16) and (1.17) of Chapter 1 that these sine and cosine transforms are compatible with derivatives of even order only, $d^{2n}f/dx^{2n}$. We have also shown that the Fourier sine transform requires the even-order derivatives up to order $2n - 2$ at $x = 0$ in (1.18), while the cosine transform requires the odd derivatives up to order $2n - 1$ at $x = 0$ in (1.19); that is,

$$\mathcal{F}_s\left\{\frac{d^{2n}f}{dx^{2n}}\right\} = (-1)^n \lambda^{2n} F_s(\lambda) - \sum_{k=1}^{n}(-1)^k \lambda^{2k-1}f^{(2n-2k)}(0) \tag{2.199}$$

$$\mathcal{F}_c\left\{\frac{d^{2n}f}{dx^{2n}}\right\} = (-1)^n \lambda^{2n} F_c(\lambda) - \sum_{k=0}^{n-1}(-1)^k \lambda^{2k}f^{(2n-2k-1)}(0) \tag{2.200}$$

As we did in Example 1.4, we will show again here that

$$\mathcal{F}_s\left\{\frac{d^2f}{dx^2}\right\} = -\lambda^2 F_s(\lambda) + \lambda f(0): \tag{2.201}$$

$$\mathcal{F}_s\left\{\frac{d^2f}{dx^2}\right\} = \int_0^\infty \frac{d^2f}{dx^2} \sin \lambda x \, dx$$

$$= \frac{df}{dx} \sin \lambda x \bigg|_0^\infty - \lambda \int_0^\infty \cos \lambda x \, \frac{df}{dx} \, dx$$

As in the case of the Fourier exponential transform, we assume that df/dx and $f(x)$ vanish as $x \to \infty$ for the right-hand side of the above equation to become

$$\mathcal{F}_s\left\{\frac{d^2f}{dx^2}\right\} = 0 - \lambda\left[f(x)\cos\lambda x\,\Big|_0^\infty - \int_0^\infty -\lambda f(x)\sin\lambda x\,dx\right]$$

$$= \lambda f(0) - \lambda^2\int_0^\infty f(x)\sin\lambda x\,dx$$

$$= -\lambda^2 F_s(\lambda) + \lambda f(0) \qquad (2.201)$$

As we mentioned above, we will, as much as it is permissible, depend on the basic theorems of the Fourier exponential transforms to cover these sine and cosine transforms as special cases. In this regard we will work with the class of functions suitable for these theorems. In our present case, as we consult Corollary 1 of Theorem 2.15 in (2.118), it is assumed that all the derivatives up to order $n-1$ (here f, $f'(x)$) are continuous and absolutely integrable on $(0,\infty)$, The vanishing of $f(x)$ and $f'(x)$ at $x = \infty$, as needed above by the step leading to (2.201), is met since it is implied by the absolute integrability of $f(x)$ and $f'(x)$ on $(0,\infty)$. We may say, in general, that the validity of (2.201) stems from the first corollary of Theorem 2.15. In the same way, after two integrations by parts, we can show that

$$\mathcal{F}_c\left\{\frac{d^2f}{dx^2}\right\} = -\lambda^2 F_c(\lambda) - f'(0) \qquad (2.202)$$

We note that (2.201) shows that the Fourier sine transform is suitable for initial value problems, with second-order differential equations in $f(x)$, when the value $f(0)$ of the function is given at $x = 0$, while (2.202) shows that the cosine transform is suitable when the derivative $f'(0)$ is given at $x = 0$.

Clearly, the remark following (2.201), concerning the sufficient conditions for the existence of the sine transform of d^2f/dx^2 as in (2.201), and its reliance on Theorem 2.15 would apply to the cosine transform and its result (2.202). The same remark also applies to the general results for even derivatives in (2.199), (2.200).

Most of the properties and transforms pairs for the Fourier sine and cosine transforms can be derived, like the above results, in a similar way to those obtained for the exponential Fourier transform of the last two sections. The mathematical grounds for such results can be managed as they are special cases, for odd and even functions, of the Fourier exponential transform, whose very basic results were proved to satisfaction in Theorems 2.14 to 2.18. The most basic operational properties and pairs for the Fourier sine and cosine transforms are listed in Appendices C.3

and C.4, respectively, and we shall select some of them to be proved as exercises.

For lack of space, we may single out here only a few of these properties. The first is to emphasize the usual use of the sine and cosine transforms with even-order derivatives, i.e., to warn against odd-order derivatives. We can show easily, from the first step of the integration by parts toward getting (2.201) (and (2.202)), that the sine transform of the first derivative df/dx (or odd-order derivatives in general) will give an algebraic result, but it is in the Fourier cosine transform space and not the same (desired) Fourier sine transform space,

$$\mathcal{F}_s \left\{ \frac{df}{dx} \right\} = -\lambda F_c(\lambda) \tag{2.203}$$

Similar problem results for the Fourier cosine transform of df/dx (or any odd derivative in general),

$$\mathcal{F}_c \left\{ \frac{df}{dx} \right\} = \lambda F_s(\lambda) - f(0) \tag{2.204}$$

Even though, in general, these illustrations suggest steering away from the sine and cosine transforms when dealing with odd-order derivatives, there are some occasions where we have to face such a situation. In that case we look at our transform space as that of both $F_s(\lambda)$ and $F_c(\lambda)$ and treat them simultaneously in the same problem in the hope of finding their inverse $f(x)$, the desired solution to the problem.

Another important property that we will state here concerns possible jump discontinuities for $f(x)$ and $f'(x)$, where the important operational pairs (2.201) and (2.202) should be modified accordingly. In (2.201), it may happen that the above sufficient conditions for its validity are mildly violated, in which case a modification may be in order. For example, if $f'(x)$ and $f(x)$, instead of being continuous as desired (according to Corollary 1 of Theorem 2.15), have jump discontinuities of size J and J', respectively, at the point $x = c$, then (2.201) is modified to

$$\mathcal{F}_s \left\{ \frac{d^2f}{dx^2} \right\} = -\lambda^2 F_s(\lambda) + \lambda f(0) + \lambda J \cos \lambda c - J' \sin \lambda c, \qquad c > 0$$
$$\tag{2.205}$$

The case for the cosine transform of (2.202) becomes

$$\mathcal{F}_c \left\{ \frac{d^2f}{dx^2} \right\} = -\lambda^2 F_c(\lambda) - f'(0) - \lambda J \sin \lambda c - J' \cos \lambda c, \qquad c > 0$$
$$\tag{2.206}$$

We leave the derivation of these results for the exercises (consult the derivation of (2.121) for Corollary 2 of Theorem 2.15).

Also, for the same class of functions $f(x)$ and $g(x)$ used in Theorem 2.14, which are absolutely integrable, and with respective Fourier sine transforms $F_s(\lambda)$ and $G_s(\lambda)$, we can derive the *convolution theorem*

$$\mathcal{F}_s^{-1}\{F_s(\lambda)G_s(\lambda)\} = \frac{2}{\pi}\int_0^\infty F_s(\lambda)G_s(\lambda)\sin\lambda x\, d\lambda$$

$$= \frac{1}{2}\int_0^\infty [f(|x-\xi|) - f(x+\xi)]g(\xi)\,d\xi \qquad (2.207)$$

A similar convolution theorem can be obtained for the cosine transforms,

$$\mathcal{F}_c^{-1}\{F_c(\lambda)G_c(\lambda)\} = \frac{2}{\pi}\int_0^\infty F_c(\lambda)G_c(\lambda)\cos\lambda x\, d\lambda$$

$$= \frac{1}{2}\int_0^\infty [f(|x-\xi|) + f(x+\xi)]g(\xi)\,d\xi \qquad (2.208)$$

The simple derivation of (2.207)–(2.209) is very similar to that of the Fourier convolution Theorem 2.18 in (2.128), where it makes use of the trigonometric identities for the product of cosine and sine functions. We leave these derivations for the exercises. We note that these convolution theorem type results are not as easy to work with as that of the *exponential* Fourier transform convolution Theorem 2.18 in (2.128).

Also *Parseval's equality* can be derived for both the sine and cosine transforms,

$$\int_0^\infty F_s(\lambda)\overline{G_s(\lambda)}\,d\lambda = \int_0^\infty f(x)\overline{g(x)}\,dx \qquad (2.209)$$

$$\int_0^\infty F_c(\lambda)\overline{G_c(\lambda)}\,d\lambda = \int_0^\infty f(x)\overline{g(x)}\,dx \qquad (2.210)$$

which are left for the exercises.

2.5.2 Applications to Boundary Value Problems on Semi-Infinite Domain

EXAMPLE 2.22a: A Semi-Infinite Heated Bar with One End at Fixed Temperature

To illustrate the use of the sine transform, we will solve for $u(x,t)$, the temperature distribution in a semi-infinite rod, as the solution to the following boundary and initial value problem:

$$\frac{\partial^2 u}{\partial x^2} = \frac{\partial u}{\partial t}, \qquad 0 < x < \infty,\ t > 0 \qquad (E.1)$$

$$u(0,t) = 0, \qquad t > 0 \tag{E.2}$$

$$u(x,0) = f(x), \qquad 0 < x < \infty \tag{E.3}$$

For the modeling of the heat equation (E.1), see Appendix B.2.

We solved this same problem in the first section of this Chapter (see (2.42)–(2.50)) by using the Laplace transform to algebraize the time derivative $\partial u/\partial t$ in the heat equation (E.1). Here we notice that the (even-order) spatial derivative $\partial^2 u/\partial x^2$ and the boundary condition $u(0,t) = 0$ make this problem compatible with the Fourier sine transform as indicated in (2.201). We let $U(\lambda,t) = \mathcal{F}_s\{u(x,t)\}$ be the Fourier sine transform of $u(x,t)$,

$$\mathcal{F}_s\{u(x,t)\} = U(\lambda,t) = \int_0^\infty u(x,t)\sin \lambda x\, dx,$$

then transform both sides of (E.1) and use (2.201) to obtain

$$-\lambda^2 U(\lambda,t) + \lambda u(0,t) = \frac{\partial U(\lambda,t)}{\partial t}$$

$$-\lambda^2 U(\lambda,t) = \frac{\partial U(\lambda,t)}{\partial t} \tag{E.4}$$

after using the boundary condition $u(0,t) = 0$. The solution to (E.4) is

$$U(\lambda,t) = A(\lambda)e^{-\lambda^2 t} \tag{E.5}$$

where

$$A(\lambda) = U(\lambda,0) = \int_0^\infty u(x,0)\sin \lambda x\, dx$$

$$= \int_0^\infty f(x)\sin \lambda x\, dx = F(\lambda)$$

Hence,

$$U(\lambda,t) = F(\lambda)e^{-\lambda^2 t} \tag{E.6}$$

Now we use the inverse sine transform (2.197) to obtain

$$u(x,t) = \frac{2}{\pi} \int_0^\infty F(\lambda)e^{-\lambda^2 t}\sin \lambda x\, d\lambda \tag{E.7}$$

and when we express $F(\lambda)$ as the Fourier sine transforms of the initial temperature $f(x)$, the solution $u(x,t)$ is a double Fourier sine integral, i.e.,

$$u(x,t) = \frac{2}{\pi} \int_0^\infty \int_0^\infty f(y)\sin \lambda y\, dy\, e^{-\lambda^2 t}\sin \lambda x\, d\lambda \tag{E.8}$$

The present Fourier sine transform method for this problem is to be compared to the method of separation of variables for the same problem in Example 1.1 of Chapter 1. The present method is much more clear, direct, and complete. It is in the sense that we known that the problem is suitable for the Fourier sine transform. So we applied this transform and obtained the final result in its same setting. In contrast, even if we can proceed with the separation of variables method for a number of steps, we still have to appeal to the "new setting" of the Fourier sine transform to help represent the nonhomogeneous condition $u(x, 0) = f(x)$, a problem intractable by the product (separation of variables) method.

In the case of the exponential transform, as we did for the infinite bar on $(-\infty, \infty)$ in Example 2.18, we applied the Fourier convolution Theorem 2.18 to the product of the Fourier transforms, like the one in (E.6), to have our final solution $u(x, t)$ as a single integral. Thus the convolution theorem saved us great deal in giving the final solution as a single integral instead of the above double integral in (E.8) for our present solution. We may, of course, use the not-so-simple convolution result (2.207) to reduce the double integral in (E.8) to a single integral. Sometimes for a particular $f(x)$ with its transform $F(\lambda)$, we may be fortunate to find the inverse sine transform of $F(\lambda)e^{-\lambda^2 t}$ in (E.6) as an entry in the tables of Fourier transforms. Short of this, we often resort to numerical computation of the double integral in (E.8) or its not-so-simple single integral equivalent via the convolution result (2.207). This obviously is a demanding job.

Fortunately, with the availability of the fast Fourier transform algorithm, the complexity of these computations is reduced drastically. However, we must also know how to prepare for using such a fast algorithm, a subject which we shall discuss in much detail in Chapter 4 and in particular for the FFT in Section 4.1.3. These preparations are in the sense that we should be aware of the errors involved in approximating our (present infinite) integrals by finite discrete sums; the latter is the the form acceptable for the efficient FFT.

EXAMPLE 2.22b: A Semi-Infinite Heated Bar with Insulated End

The boundary condition at $x = 0$ for the above boundary and initial value problem of the heated semi-infinite bar may vary in many ways according to the physical situation. One such variation makes the problem compatible with the Fourier cosine transform. This is the case when, instead of the fixed zero temperature at the end $x = 0$ in (E.2) above, we have an insulated end $x = 0$. Such a boundary condition is described as $\partial u(0, t)/\partial x = 0$; that is, there is no temperature gradient for the heat to flow at $x = 0$. The compatibility of the heat equation and this boundary condition with Fourier cosine transform is obvious when we consult the cosine transform of d^2f/dx^2 in (2.202) and, in particular, its requirement of having $f'(0)$, the value of the derivative of the transformed function at $x = 0$. Since the method of solving this problem is an exact parallel to that of the above problem with the end at zero temperature, we shall leave it as a simple exercise (see Exercise 2.69a and b).

EXAMPLE 2.23: The Potential Distribution in the Upper Half-Plane

(a) *The case of fixed potential at the edge $y = 0$— a Dirichlet problem.* In Example 2.21 of Section 2.2 on the Fourier exponential transform, we solved the boundary value problem for the potential distribution $u(x,y)$ in the upper half-plane, where the potential is given as $u(x,0) = f(x)$ on the (edge) real line, as shown in Fig. 2.14,

$$\frac{\partial^2 u}{\partial x^2} + \frac{\partial^2 u}{\partial y^2} = 0, \qquad -\infty < x < \infty, \ 0 < y < \infty \tag{E.1}$$

$$u(x,0) = f(x), \qquad -\infty < x < \infty \tag{E.2}$$

and with conditions on $f(x)$ sufficient to justify the necessary Fourier transform operations. There we used the Fourier exponential transform to algebraize the derivative $\partial^2 u/\partial u^2$, as defined on the real line $-\infty < x < \infty$. Then we remarked that with the semi-infinite domain $0 < y < \infty$ in y, an even derivative $\partial^2 u/\partial y^2$ in y, and a given value of $u(x,y)$ at the boundary $y = 0$ as $u(x,0) = f(x)$, the problem is also compatible with the Fourier sine transform for its variation in the y direction. The idea here is to follow the Fourier exponential transform for $\partial^2 u/\partial x^2$ as in (E.4) of Example 2.21,

$$-\lambda^2 U(\lambda,y) + \frac{\partial^2 U(\lambda,y)}{\partial y^2} = 0 \tag{E.3}$$

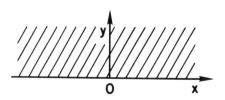

FIG. 2.14 Illustration for Example 2.23.

with a Fourier sine transform

$$\mathcal{F}_s\{U(\lambda,y)\} = \tilde{U}(\lambda,\alpha) \equiv \int_0^\infty U(\lambda,y)\sin \alpha y \, dy \qquad (E.4)$$

to completely algebraize the Laplace equation in $\tilde{U}(\lambda,\alpha) = \mathcal{F}_s\mathcal{F}_e\{u\}$ (the (double) Fourier exponential transform (on x) then Fourier sine transform (on y)),

$$-\lambda^2 \tilde{U}(\lambda,\alpha) - \alpha^2 \tilde{U}(\lambda,\alpha) + \alpha U(\lambda,0) = 0$$

$$\tilde{U}(\lambda,\alpha) = \frac{\alpha}{\lambda^2 + \alpha^2} U(\lambda,0) = \frac{\alpha}{\lambda^2 + \alpha^2} F(\lambda) \qquad (E.5)$$

What remains now is to find the inverse of this (combination) double Fourier transform in (E.5). What we suggest is to use the inverse exponential transform on the above product with respect to λ, where the Fourier convolution Theorem 2.18 can be used to an advantage to transform $\tilde{U}(\lambda,\alpha)$ back to the x space as $\bar{U}(x,\alpha)$. Then we consult the Fourier sine transform pairs to transform $\bar{U}(x,\alpha)$ back from the α-space to the y-space to get $u(x,y)$. We leave these details for an exercise (see Exercise 2.70a)).

The other and possibly more direct, alternative, as far as the Fourier sine transform is concerned, is just to take the Fourier sine transform of the partial differential equation (E.1) on $0 < y < \infty$. We let

$$U_s(x,\alpha) = \mathcal{F}_s\{u(x,y)\} = \int_0^\infty u(x,y))\sin \alpha y \, dy$$

and use it on (E.1) in consultation with (2.202) to have

$$\frac{\partial^2 U_s(x,\alpha)}{\partial x^2} - \alpha^2 U_s(x,\alpha) + \alpha u(x,0) = 0$$

$$\frac{\partial^2 U_s}{\partial x^2}(x,\alpha) - \alpha^2 U_s(x,\alpha) = -\alpha f(x) \qquad (E.6)$$

after involving the boundary condition (E.2). The result is a nonhomogeneous differential equation in x with $-\alpha f(x)$ as the nonhomogeneous term. This can be solved by the method of variation of parameters or its equivalent, the use of the Green's function (see Appendix A). Once (E.6) is solved for $U_s(x,\alpha)$, which is in terms of an integral involving the general (nonhomogeneous) term $f(x)$, we take its inverse sine transform (in α) to arrive at the solution $u(x,y)$. Not surprisingly, this solution is a double integral for general $f(x)$ of (E.3). This is to be compared with a double Fourier exponential and sine transform to obtain $u(x,y)$ from $U(\lambda,\alpha)$ in (E.5). On the side of the latter choice we may keep in mind the fast Fourier transform, but it is now in two dimensions.

We shall leave finding the solution $U_s(x,\alpha)$ of (E.6) and then $u(x,y)$, the inverse sine transform of $U_s(x,\alpha)$, as an exercise which is supplemented by leading steps (see Exercise 2.70b).

(b) *The case of insulated edge $y = 0$— a Neumann problem.* It should be clear that this problem of the potential in the upper half-plane (E.1) would be suitable (in the y direction) for the cosine transform when the edge ($y = 0$) is insulated, i.e., with the boundary condition

$$\frac{\partial u(x,0)}{\partial y} = 0 \tag{E.7}$$

and we shall leave it for an exercise (see Exercise 2.71).

Other practical variations of the boundary condition at $y = 0$, for problems similar to those above on the semi-infinite domain, will necessitate establishing modified versions of the Fourier sine and cosine transforms or having a combination of both of them. We leave the subject of such extensions to the exercises, and we refer the interested reader to other references on the subject, such as Churchill (1972) and Sneddon (1972).

2.6 Higher-Dimensional Fourier Transforms

The definition of the (single) Fourier exponential, sine, and cosine transforms of Sections 2.2–2.5 can be extended to higher dimensions, i.e., to multiple Fourier transforms of functions of two or more variables. For simplicity of notation we will write a function of two variables $f(x,y)$, defined on the infinite xy-plane, as $f(x_1,x_2)$ in two-dimensional space (plane). This allows the extension to functions of n variables $f(x_1,x_2,\ldots,x_n)$ or $f(\mathbf{x})$, where $\mathbf{x}$ is a vector in n-dimensional space. For physical applications in two or three dimensions we may return to the preferable notation $f(x,y)$ and $f(x,y,z)$ or just $f(\mathbf{r})$, $\mathbf{r} = \mathbf{i}x + \mathbf{j}y + \mathbf{k}z$. The development of these higher-dimensional Fourier transforms and their operational pairs will depend to a great extent on the theory and tools that we have developed for the one-dimensional Fourier transform of Section 2.2. The applications will easily extend to higher-dimensional problems such as the temperature distribution in the infinite plane, or space, and the Laplace equation in space.

The main idea behind the derivation of the basic tools for the multiple Fourier transforms is to apply the Fourier transform (2.64) with respect to one variable at a time, considering the rest of the variables as constants at that step. For the case of functions of two variables $f(x_1,x_2)$, $-\infty < x_1 < \infty$, $-\infty < x_2 < \infty$, we Fourier transform with respect to x_1, leaving x_2 as a constant parameter for now, to have

$$\mathcal{F}_{(1)}(\lambda_1,x_2) = \int_{-\infty}^{\infty} e^{-i\lambda_1 x_1} f(x_1,x_2)\,dx_1 \equiv \mathcal{F}_{(1)}\{f\} \qquad (2.211)$$

We use the notation $\mathcal{F}_{(1)}$ (and $F_{(1)}(\lambda_1,x_2)$ only for now) of Sneddon (1972) to indicate a one-dimensional Fourier transform with respect to the first variable x_1. $\mathcal{F}_{(2)}$ denotes a double Fourier transform with respect to the first two variables, and so on.

The double Fourier transform $\mathcal{F}_{(2)}\{f\}$ of $f(x_1,x_2)$ results from "formally" repeating the Fourier transformation on $F_{(1)}(\lambda_1,x_2)$ in (2.211) with respect to the second variable x_2 to have

$$F_{(2)}(\lambda_1,\lambda_2) = \mathcal{F}_{(2)}\{f\} = \int_{-\infty}^{\infty} e^{-i\lambda_2 x_2} \int_{-\infty}^{\infty} e^{-i\lambda_1 x_1} f(x_1,x_2)\,dx_1\,dx_2$$

$$= \int_{-\infty}^{\infty} \int_{-\infty}^{\infty} e^{-i(\lambda_1 x_1 + \lambda_2 x_2)} f(x_1,x_2)\,dx_1\,dx_2$$

$$(2.212)$$

The steps of (2.211), (2.112) can be repeated n times in succession to write $\mathcal{F}_{(n)}(\lambda_1,\lambda_2,\ldots\lambda_n) = \mathcal{F}_{(n)}\{f\}$, the n-dimensional Fourier transform of $f(x_1,x_2,\ldots,x_n)$.

We shall adopt the notation that, once we have completed the multiple Fourier transform with respect to the n variables x_1, x_2, ..., x_n, we use $\mathcal{F}_{(n)}\{f\} = F_{(n)}(\lambda_1,\lambda_2,\ldots,\lambda_n)$. Since it will be clear in the applications, we shall drop the subscript in the transform itself and write it as $F(\lambda_1,\lambda_2,\ldots,\lambda_n)$. The indexed symbol $F_n(\lambda)$ will be reserved for the J_n-Hankel transform of $f(r)$, which we shall cover in Section 2.7.

For the first Fourier transform $\mathcal{F}_{(1)}(\lambda_1,x_2)$ of $f(x_1,x_2)$ with respect to x_1 (leaving x_2 fixed) in (2.211), it is understandable that we assume $f(x_1,x_2)$, as a function of x_1, to satisfy the conditions of the Fourier integral Theorem 2.14 in (2.85) or (2.86). This means that $f(x_1,x_2)$ is absolutely integrable in $x_1 \in (-\infty,\infty)$ (i.e., $\int_{-\infty}^{\infty} |f(x_1,x_2)|\,dx_1 < \infty$) and is piecewise smooth as a function of $x_1 \in (-\infty,\infty)$ for every fixed value of x_2. The same analysis and conditions are then applied to the second variable x_2. These conditions can be repeated on each of the n variables to cover the multiple Fourier transform $F(\lambda_1,\lambda_2,\ldots,\lambda_n)$ of the function of n variables $f(x_1,x_2,\ldots,x_n)$. For our present two-dimensional Fourier transform we must, then, assume $f(x_1,x_2)$ to satisfy the conditions of the Fourier integral Theorem 2.14 of being absolutely integrable (i.e., $\int_{-\infty}^{\infty} |f(x_1,x_2)|\,dx_2 < \infty$) and piecewise smooth with respect to $x_2 \in (-\infty,\infty)$ for every fixed value of x_1.

With these notes in the direction of the existence of the multiple Fourier transform and its inverse, along with the two formal repeated Fourier transformations in (2.211), (2.212), we may write the double exponential Fourier transform as

$$F(\lambda_1, \lambda_2) = \mathcal{F}_{(2)}\{f\} = \int_{-\infty}^{\infty} \int_{-\infty}^{\infty} e^{-i\lambda_1 x_1 - i\lambda_2 x_2} f(x_1, x_2)\, dx_1\, dx_2 \qquad (2.212)$$

with its inverse

$$\mathcal{F}_{(2)}^{-1}\{f\} = f(x_1, x_2) = \frac{1}{4\pi^2} \int_{-\infty}^{\infty} \int_{-\infty}^{\infty} e^{i\lambda_1 x_1 + i\lambda_2 x_2} F(\lambda_1, \lambda_2)\, d\lambda_1\, d\lambda_2$$

$$(2.213)$$

The same method will be followed in developing the Fourier series for functions of several variables based on the Fourier series representation of $f(x)$ that we shall develop in Chapter 3. In general, we can easily define the Fourier transform of $f(\mathbf{x})$, $\mathbf{x} = (x_1, x_2, \ldots, x_n)$, $\boldsymbol{\lambda} = (\lambda_1, \lambda_2, \ldots, \lambda_n)$ in n dimensions as

$$\mathcal{F}_{(n)}\{f\} = F_{(n)}(\boldsymbol{\lambda}) = \int_{-\infty}^{\infty} \overset{\cdots\cdots}{} \int_{-\infty}^{\infty} e^{-i\boldsymbol{\lambda}\cdot\mathbf{x}} f(\mathbf{x})\, dx_1\, dx_2 \cdots dx_n \qquad (2.214)$$

and its inverse

$$\mathcal{F}_{(n)}^{-1}\{f\} = f(\mathbf{x}) = \frac{1}{(2\pi)^n} \int_{-\infty}^{\infty} \overset{\cdots\cdots}{} \int_{-\infty}^{\infty} e^{i\boldsymbol{\lambda}\cdot\mathbf{x}} F(\boldsymbol{\lambda})\, d\lambda_1\, d\lambda_2 \cdots d\lambda_n$$

$$(2.215)$$

Formulas (2.212) and (2.213) and thus (2.214) and (2.215), can be written in trigonometric form parallel to the form (2.86) of Theorem 2.14.

These formulas and other operational pairs for translation, derivative, convolution, etc. can be developed following the same steps we used for the one-dimensional Fourier transform in Section 2.2.

For example, if $f(x_1, x_2)$ is an *even function* in both x_1, and x_2, then its double Fourier transform (2.212) becomes a *double Fourier cosine transform* of it on the first quadrant $0 < x_1 < \infty$, $0 < x_2 < \infty$,

$$\mathcal{F}_{(2)}\{f\} = F_c(\lambda_1, \lambda_2) = \int_0^{\infty} \int_0^{\infty} f(x_1, x_2)\cos\lambda_1 x_1 \cos\lambda_2 x_2\, dx_1\, dx_2 \quad (2.216)$$

$$\mathcal{F}_{(2)}^{-1}\{F_c\} = f(x_1, x_2)$$

$$= \frac{4}{\pi^2} \int_0^{\infty} \int_0^{\infty} F_c(\lambda_1, \lambda_2)\cos\lambda_1 x_1 \cos\lambda_2 x_2\, d\lambda_1\, d\lambda_2 \qquad (2.217)$$

The same can be done for the *double Fourier sine transform* of $f(x_1, x_2)$ as *odd function* in both x_1 and x_2, $-\infty < x_1 < \infty$, $-\infty < x_2 < \infty$.

With the above note, we shall leave the derivation of the following and other basic properties of the double and higher-dimensional Fourier transforms as exercises (see Exercise 2.76). To help with such derivations, we remind the reader of the proper conditions, mentioned above in the spirit of Theorem 2.14, on such functions to ensure the existence of their multiple Fourier transforms and inverses.

1. Translation property

$$\mathcal{F}_{(n)}\{f(\mathbf{x} - \mathbf{a})\} = e^{-i\boldsymbol{\lambda}\cdot\mathbf{a}}\mathcal{F}_{(n)}\{f\} \tag{2.218}$$

where $\mathbf{a}$ is a constant vector in n dimensions, and f here stands for $f(\mathbf{x})$.

2. $\mathcal{F}_{(n)}\left\{\dfrac{\partial f}{\partial x_k}\right\} = i\lambda_k \mathcal{F}_{(n)}\{f\}$, \hfill (2.219)

Clearly the integration by parts method of deriving this result would need the vanishing of f as a function of x_k as $x_k \to \mp\infty$. This can be assumed if we impose conditions like those of Theorem 2.14. For the case of a second partial derivative we have

$$\mathcal{F}_{(n)}\left\{\frac{\partial^2 f}{\partial x_k \partial x_l}\right\} = -\lambda_k \lambda_l \mathcal{F}_{(n)}\{f\} \tag{2.220}$$

For example, the double Fourier transform of the Laplacian $\partial^2 f/\partial x_1^2 + \partial^2 f/\partial x_2^2$ of $f(x_1, x_2)$ is

$$\mathcal{F}_{(2)}\left\{\frac{\partial^2 f}{\partial x_1^2} + \frac{\partial^2 f}{\partial x_2^2}\right\} = [(-i\lambda_1)^2 + (-i\lambda_2)^2]\mathcal{F}_{(2)}\{f\}$$

$$= -(\lambda_1^2 + \lambda_2^2)F(\lambda_1, \lambda_2)$$

$$= -|\lambda^2|F(\lambda_1, \lambda_2) \tag{2.221}$$

and in general

$$\mathcal{F}_{(n)}\left\{\sum_{k=1}^{n}\frac{\partial^2 f}{\partial x_k^2}\right\} = -|\lambda^2|\mathcal{F}_{(n)}\{f\}, \qquad |\lambda|^2 = \sum_{k=1}^{n}\lambda_k^2 \tag{2.222}$$

3. For $F(\lambda) = \mathcal{F}_{(n)}\{f\}$ and $G(\lambda) = \mathcal{F}_{(n)}\{g\}$, we have the *convolution theorem* for the n-dimensional Fourier transforms,

$$\mathcal{F}_{(n)}\{f * g\} = \mathcal{F}_{(n)}\{f\} \cdot \mathcal{F}_{(n)}\{g\} \tag{2.223}$$

where the convolution product $f * g$ of $f(\mathbf{x})$ and $g(\mathbf{x})$ is defined in a similar way to that of (2.127) for Theorem 2.18, as an n-dimensional convolution integral,

$$(f * g)(\mathbf{x}) = \underbrace{\int\int \cdots}_{n} \int (\mathbf{x} - \mathbf{u})g(\mathbf{u})\, du_1\, du_2 \cdots du_n \qquad (2.224)$$

The statement of the convolution Theorem (2.223) for the special case of a two-dimensional Fourier transform is

$$(f * g)(\mathbf{x}) = \int_{-\infty}^{\infty} \int_{-\infty}^{\infty} f(x_1 - u_1, x_2 - u_2)g(u_1, u_2)\, du_1\, du_2$$

$$= \frac{1}{(2\pi)^2} \int_{-\infty}^{\infty} \int_{-\infty}^{\infty} e^{i(\lambda_1 x_1 + \lambda_2 x_2)} F(\lambda_1, \lambda_2) G(\lambda_1, \lambda_2)\, d\lambda_1\, d\lambda_2$$

$$(2.225)$$

It should be easy to show that this convolution product, as in the case of that in (2.127), is commutative: $f * g = g * f$.

The *Parseval equality* follows easily as

$$\underbrace{\int\int \cdots}_{n} \int f(\mathbf{x})\overline{g(\mathbf{x})}\, dx_1\, dx_2 \cdots dx_n$$

$$= \frac{1}{(2\pi)^n} \underbrace{\int\int \cdots}_{n} \int F(\lambda)\overline{G(\lambda)}\, d\lambda_1\, d\lambda_2 \cdots d\lambda_n \qquad (2.226)$$

Next we will illustrate the use of the double Fourier transform for solving problems in two dimensions, such as the temperature distribution in an infinite plate. For such physical problems we will return to the usual notation $f(x,y)$, instead of $f(x_1, x_2)$, for functions of two variables x and y.

EXAMPLE 2.24: Heat Diffusion in an Infinite Plate

In this application we consider the following boundary and initial value problem in $u(x,y,t)$, the temperature distribution in an infinite plate with initial temperature given as $u(x,y,0) = f(x,y)$, $-\infty < x < \infty$, $-\infty < y < \infty$:

$$\frac{\partial^2 u}{\partial x^2} + \frac{\partial^2 u}{\partial y^2} = \frac{\partial u}{\partial t}, \qquad -\infty < x < \infty,\ -\infty < y < \infty,\ t > 0 \qquad (\text{E.1})$$

$$u(x,y,0) = f(x,y), \qquad -\infty < x < \infty,\ -\infty < y < \infty \qquad (\text{E.2})$$

The heat equation (E.1) is derived in Appendix B.2.

Let the double Fourier transform of $u(x,y,t)$ be

$$U(\lambda_1, \lambda_2, t) = \int_{-\infty}^{\infty} \int_{-\infty}^{\infty} e^{-i\lambda_1 x - i\lambda_2 y} u(x,y,t)\, dx\, dy \qquad (\text{E.3})$$

If we apply this transform to (E.1) and use (2.221), which amounts to integrating twice by parts with respect to x and y for $\partial^2 u/\partial x^2$ and $\partial^2 u/\partial y^2$, respectively, we obtain

$$-\lambda_1^2 U(\lambda_1, \lambda_2, t) - \lambda_2^2 U(\lambda_1, \lambda_2, t) = \frac{\partial U(\lambda_1, \lambda_2, t)}{\lambda t} \tag{E.4}$$

$$\frac{\partial U}{\partial t} + (\lambda_1^2 + \lambda_2^2)U = 0 \tag{E.5}$$

We note in the derivation of (E.4) or (2.221) that we need $f(x,y)$, $\partial f/\partial x$, and $\partial f/\partial y$ to vanish as $x, y \to \mp\infty$, a condition that can be secured by imposing their absolute integrability (of Theorem 2.14) over the xy-plane.

To solve the differential equation (E.5) in the unknown double Fourier transform $U(\lambda_1, \lambda_2, t)$, we need an initial condition, which can be obtained from the double Fourier transform of the initial condition $u(x, t, 0) = f(x, y)$ in (E.2):

$$U(\lambda_1, \lambda_2, 0) = \int_{-\infty}^{\infty} \int_{-\infty}^{\infty} e^{-i\lambda_1 x - \lambda_2 y} f(x, y)\, dx\, dy$$

$$= F(\lambda_1, \lambda_2) \tag{E.6}$$

The solution for (E.5) with its initial condition (E.6) is

$$U(\lambda_1, \lambda_2, t) = F(\lambda_1, \lambda_2)e^{-(\lambda_1^2 + \lambda_2^2)t} \tag{E.7}$$

Now we use the inverse double Fourier transform (2.213) to arrive at the solution to the boundary and initial value problem of (E.1) and (E.2):

$$u(x, y, t) = \frac{1}{(2\pi)^2} \int_{-\infty}^{\infty} \int_{-\infty}^{\infty} e^{i(\lambda_1 x + \lambda_2 y)} F(\lambda_1, \lambda_2)e^{-(\lambda_1^2 + \lambda_2^2)t}\, d\lambda_1\, d\lambda_2 \tag{E.8}$$

This gives the solution $u(x, y, t)$, in terms of $f(x, y)$ the initial temperature, as a four-dimensional integral when we consider $F(\lambda_1, \lambda_2)$ in (E.6) for (E.8) as a double integral. To reduce this four-dimensional integral to a possible double integral, we appeal to the convolution theorem in two dimensions as in (2.225).

To use this convolution theorem for (E.8), we need the inverse transform of $e^{-(\lambda_1^2 + \lambda_2^2)t}$. If we extend the result

$$\mathcal{F}^{-1}\left\{e^{-\lambda^2 t}\right\} = \frac{1}{\sqrt{4\pi t}} e^{-x^2/4t} \tag{E.10}$$

to two dimensions, we obtain

$$
\mathcal{F}_{(2)}^{-1}\left\{e^{-(\lambda_1^2+\lambda_2^2)t}\right\} = \frac{1}{\sqrt{4\pi t}}e^{-x^2/4t}\frac{1}{\sqrt{4\pi t}}\cdot e^{-y^2/4t}
$$
$$
= \frac{1}{4\pi t}e^{-(x^2+y^2)/4t}
$$

(E.11)

which will be justified in Section 2.6.1 as the double Fourier transform pair in (2.227).

Now we use (E.11) in (E.8) and apply the convolution theorem (2.225), letting $G(\lambda_1, \lambda_2) = e^{-(\lambda_1^2+\lambda_2^2)t}$, to obtain

$$
u(x,y,t) = \frac{1}{4\pi t}\int_{-\infty}^{\infty}\int_{-\infty}^{\infty}f(\xi,\eta)e^{-((x-\xi)^2+(y-\eta)^2)/(4t)}\,d\xi\,d\eta
$$

(E.12)

For higher-dimensional problems such answers can be neatly written using vector notation like that already used in (2.214), (2.215), (2.224), and (2.226). For the solution (E.12) above we have

$$
u(\mathbf{r},t) = \int_{E_2}f(\rho)\exp\left(\frac{-|\mathbf{r}-\rho|^2}{4t}\right)d\rho, \qquad \mathbf{r} = \mathbf{i}x + \mathbf{j}y, \quad \rho = \mathbf{i}\xi + \mathbf{j}\eta
$$

(E.13)

where the (double) integration is done over E_2, the *two-dimensional* plane.

Problems in three and higher dimensions can be treated in the same way, where the multiple Fourier transform and its inverse (2.214), (2.215) can be employed, along with their basic operational pairs in (2.218) to (2.226). For example, we can consider the same temperature distribution $u(x,y,z;t)$ or $u(\mathbf{r},t)$, $\mathbf{r} = \mathbf{i}x + \mathbf{i}y + \mathbf{k}z$, in an infinite solid (3 space) E_3, with initial condition $u(x,y,z;0) = f(x,y,z)$, $-\infty < x < \infty$, $-\infty < y < \infty$, $-\infty < z < \infty$. For this problem we employ the triple Fourier transform in a similar way to the above example in the plane, to obtain the same type of result as in (E.13),

$$
u(\mathbf{r},t) = \frac{1}{(4\pi t)^{3/2}}\int_{E_3}f(\rho)\exp\left(\frac{-|\mathbf{r}-\rho|}{4t}\right)d\rho
$$

(E.4)

$$
\mathbf{r} = \mathbf{i}x + \mathbf{j}y + \mathbf{k}z, \qquad \rho = \mathbf{i}\xi + \mathbf{j}\eta + \mathbf{k}\zeta
$$

and where the (triple) integration is done over the three-dimensional space E_3.

More on the higher-dimensional Fourier transforms can be found in Sneddon (1972).

2.6.1 Relation Between the Hankel Transform and the Multiple Fourier Transform—Circular Symmetry

In Example 2.24 we have already borrowed a double Fourier transform pair from (E.11)

$$\mathcal{F}_{(2)}\left\{\frac{1}{4\pi t}e^{-(x^2+y^2)/4t}\right\} = e^{-(\lambda_1^2+\lambda_2^2)t} \tag{2.227}$$

This was needed for use of the convolution theorem in (E.12), which made our solution $u(x,y,t)$ as a double integral instead of the four-dimensional Fourier integral of (E.8)—a great saving! The question now is how (2.227) was obtained as an "apparent" extension of the known one-dimensional Fourier transform pair of (E.10). There is a formal way to justify this extension by doing what we did in establishing (or defining!) the double Fourier transform $\mathcal{F}_{(2)}\{f\}$ in (2.212), that is, applying two Fourier transforms in succession, one for $g(x) = (1/\sqrt{4\pi t})e^{-x^2/4t}$ and the second for $h(y) = (1/\sqrt{4\pi t})e^{-y^2/4t}$. However, a better established way is to recognize $f(x,y) = (1/4\pi t)e^{-(x^2+y^2)/4t}$ as a circularly symmetric function in x and y; that is, $f(x,y) = f(\sqrt{x^2+y^2}) = f(|x|)$. This suggests the use of polar coordinates, $x = r\cos\theta$, $y = \sin\theta$, to reduce this function to $f(|x|) = f(r)$, a function of the one (radial) variable r. Indeed, we can take advantage of the circular symmetry of this class of functions, where (as we shall prove next) for such functions $f(\sqrt{x^2+y^2})$, their double Fourier transform $\mathcal{F}_{(2)}\{f\}$ reduces to a one-dimensional J_0-Hankel transform. With such an important relation, the double Fourier pair in (E.11) can be established easily by writing it as a single J_0-Hankel transform, then merely employing the definition (series expansion) of the J_0 Bessel function, which we leave for Exercise 2.77a (see also Exercise 2.76d). This result can be extended to functions of n variables with circular symmetry, that is, $f(|x|)$, $|x|^2 = \sum_{i=1}^{n} x_i^2$, $n = 2, 3, \ldots$, where the n-dimensional Fourier transform $\mathcal{F}_{(n)}$ of $f(|x|)$ becomes a one-dimensional $J_{(n/2)-1}$-Hankel Transform of $f(r)$, $r^2 = \sum_{i=1}^{n} x_i^2$, $n = 2, 3, \ldots$. The Hankel transforms and their applications will be the subject of Section 2.7.

In our applications we will be interested in the case of the J_0-Hankel transform for the circular symmetry in the two spatial variables x and y, like $u(x,y,z) = u(r,z)$, and in the $J_{1/2}$ case for the spherical (or radial) symmetry in the $(n = 3)$ three variables x, y, and z, like $u(x,y,z) = u(\rho)$, $\rho^2 = x^2 + y^2 + z^2$.

Of course, such a relation between the n-dimensional Fourier transform of $f(|x|) = f(r)$ and the $J_{(n/2)-1}$-Hankel transform of $f(r)$ can be used to develop many of the properties of the Hankel transform. We do so by depending on the theory that we have already established for the

Fourier transform in Section 2.2, most of which can easily be extended to the higher-dimensional case of this section. An important example is establishing the existence of the J_0-Hankel transform in (2.236) and even its inverse in (2.237) as we relate it to a double Fourier transform and from there to the Fourier integral Theorem 2.14. This result (2.236), (2.237) will be extended to (2.238), (2.239) for the $J_{(n/2)-1}$-Hankel transform equivalence to the n-dimensional Fourier transform of circularly symmetric functions $f(|\mathbf{x}|)$. We must stress, though, that such a method is clearly limited to $J_{(n/2)-1}$–that is, J_ν with order ν as zero or half a positive integer, since $n = 2, 3, 4, \ldots$. Understandably, the general ν case requires a different mode of proof, and we shall state it in Section 2.7 as Theorem 2.24 without proof.

There are many applications in which circular symmetry is important. For example, the potential (or steady-state temperature) distribution $u(x,y,z)$ in a thick infinite slab, with its two faces $z = a$ and $z = -a$ kept at circularly symmetric potentials $u(x,y,a) = f(r)$ and $u(x,y,-a) = g(r)$, is more simply described by $u(r,z)$, $0 \le r < \infty$, $-a < z < a$, a circularly symmetric potential in the thick infinite slab. Another example is the steady-state temperature $u(x,y,z)$ in a thick infinite slab with a circular hole, where heat enters at a constant rate and the two surfaces at $z = a$ and $z = -a$ are kept at constant temperatures T_1 and T_2, respectively. Here, again, the temperature $u(x,y,z)$ in this domain is circularly symmetric and we represent it by $u(r,z)$, $0 \le r < \infty$, $-a < z < a$. For all such circularly symmetric problems we should replace Cartesian coordinates x, y, z by the two cylindrical coordinates r, z. Thus we have such problems in two variables r and z instead of the three variables x, y, and z. However, as we change to cylindrical coordinates, the Laplace equation

$$\frac{\partial^2 u}{\partial x^2} + \frac{\partial^2 u}{\partial y^2} + \frac{\partial^2 u}{\partial z^2} = 0 \tag{2.228}$$

which governs the potential (or steady-state temperature) distribution in a solid free of a source, will change to one

$$\frac{\partial^2 u}{\partial r^2} + \frac{1}{r} \frac{\partial u}{\partial r} + \frac{\partial^2 u}{\partial z^2} = 0 \tag{2.229}$$

which has a variable coefficient. But we have already shown in (1.24) (of Example 1.6 in Chapter 1) that the J_0-Hankel transform is compatible with $\partial^2 u/\partial r^2 + (1/r)\partial u/\partial r$, the variable-coefficient part of this Laplace differential operator. This means that if we let $U(\lambda,z)$ be the J_0-Hankel transform of $u(r,z)$, that is, $U_0(\lambda,z) = \mathcal{H}_0\{u(r,z)\}$, then the J_0-Hankel transform algebraizes $\partial^2 u/\partial r^2 + (1/r)\partial u/\partial r$,

$$\mathcal{H}_0 \left\{ \frac{\partial^2 u}{\partial r^2} + \frac{1}{r}\frac{\partial u}{\partial r} + \frac{\partial^2 u}{\partial z^2} \right\} = -\lambda^2 U_0(\lambda, z) + \frac{\partial^2 U_0}{\partial z^2} \qquad (2.230)$$

The J_ν-Hankel transform, its existence with its inverse, and its important properties and operations necessary for handling these and other applications will be the subject of Section 2.7. But first we will prove the important relation between the two-dimensional Fourier transform of functions $f(x,y)$, with circular symmetry, and the J_0-Hankel transform.

2.6.2 The Double Fourier Transform of Functions with Circular Symmetry—The J_0-Hankel Transform

Here we will illustrate the reduction of the double Fourier transform of functions with circular symmetry to a single J_0-Hankel transform. Consider the double Fourier transform

$$\mathcal{F}_{(2)}\{f\} = F(\lambda_1, \lambda_2) = \int_{-\infty}^{\infty}\int_{-\infty}^{\infty} e^{-i\lambda_1 x - i\lambda_2 y} f(x,y)\, dx\, dy \qquad \begin{matrix}(2.231)\\(2.212a)\end{matrix}$$

and its inverse

$$\mathcal{F}_{(2)}^{-1}\{F\} = f(x,y) = \frac{1}{4\pi^2}\int_{-\infty}^{\infty}\int_{-\infty}^{\infty} e^{i\lambda_1 x + i\lambda_2 y} F(\lambda_1, \lambda_2)\, d\lambda_1\, d\lambda_2 \qquad \begin{matrix}(2.232)\\(2.213a)\end{matrix}$$

When they both possess circular symmetry we can write them $F(\lambda)$ and $f(r)$, respectively, where

$$\lambda_1^2 + \lambda_2^2 = \lambda^2 \qquad (2.233)$$

and

$$x^2 + y^2 = r^2 \qquad (2.234)$$

We will show that with such symmetry the double Fourier transforms in (2.231) and (2.232) will reduce to the J_0-Hankel transform $2\pi F_0(\lambda)$ and its inverse $f(r)$ [(2.236), (2.237)], respectively. This is done with the help of the integral representation of $J_0(x)$,

$$J_0(x) = \frac{1}{2\pi}\int_0^{2\pi} e^{-ix\cos(\theta - \alpha)}\, d\theta \qquad (2.235)$$

With circular symmetry for $f(r)$, we let $x = r\cos\theta$, $y = r\sin\theta$, $\lambda_1 = \lambda\cos\alpha$, and $\lambda_2 = \lambda\sin\alpha$ in (2.231) and note the new limits for the new variables, $0 \le \theta \le 2\pi$, $0 \le r < \infty$; and $dxdy = rdrd\theta$,

$$\mathcal{F}_{(2)}\{f\} = F(\lambda_1, \lambda_2) = \int_0^{\infty}\int_0^{2\pi} f(r)e^{-i\lambda r\cos(\theta - \alpha)} r\, dr\, d\theta$$

$$\mathcal{F}_{(2)} \{f\} = 2\pi \int_0^\infty rf(r)J_0(\lambda r)\, dr = 2\pi \mathcal{H}_0 \{f\} = 2\pi F_0(\lambda) \qquad (2.236)$$

$$F_0(\lambda) = \mathcal{H}_0 \{f\} = \int_0^\infty rf(r)J_0(\lambda r)\, dr$$

$$F_0(\lambda) = \frac{1}{2\pi} F(\lambda_1, \lambda_2) = \frac{1}{2\pi} \mathcal{F}_{(2)} \{f\} \qquad (2.236)$$

This shows that the double Fourier transform of the circularly symmetric $f(r)$ is also circularly symmetric and that it is reduced to a J_0-Hankel transform $F_0(\lambda)$, where $\mathcal{F}_{(2)} \{f\} = F(\lambda_1, \lambda_2) = 2\pi F_0(\lambda)$ as in (2.236). We note here, in terms of the notation that we have been using, that the form of the double Fourier transform and its inverse in (2.231) and (2.232) is not symmetric, since we have a factor of $1/4\pi^2$ for the inverse transform in (2.232) and of 1 for (2.231). Had we used the symmetric form with factor $1/2\pi$ for both the transform and its inverse in (2.231), (2.232), as in Sneddon (1972) for example, we would have $\mathcal{F}_{(2)} \{f\} = F_0(\lambda)$ or $\mathcal{F}_{(2)} \{f\} = \mathcal{H}_0 \{f\}$ for circularly symmetric $f(r)$. Following the same steps for $f(x,y) = f(r)$, we can prove the J_0-Hankel transform inversion formula (2.237) for the circularly symmetric $F(\lambda_1, \lambda_2) = F(\lambda)$,

$$f(r) = \frac{1}{2\pi} \mathcal{H}_0^{-1} \{F\} = \frac{1}{2\pi} \mathcal{H}_0^{-1} \{2\pi F_0\} = \int_0^\infty \lambda F_0(\lambda)J_0(\lambda r)\, d\lambda$$

$$f(r) = \int_0^\infty \lambda F_0(\lambda)J_0(\lambda r)\, d\lambda \qquad (2.237)$$

where we used $F_0(\lambda) = (1/2\pi)F(\lambda)$ as in (2.236).

As we mentioned before, this result (2.236) can easily be extended to functions $f(|\mathbf{x}|) = f(r)$ of n variables with circular symmetry, where their n-dimensional Fourier transform $\mathcal{F}_{(n)} \{f\}$ results in a circularly symmetric $F(|\lambda|)$, which reduced to a one-dimensional $J_{(n/2)-1}$-Hankel transform. The exact result for $\nu = (n/2) - 1$, $n = 2, 3, \ldots$ is

$$\lambda^\nu \mathcal{F}_{(n)} \{f\} = \lambda^\nu F(\lambda) = \lambda^\nu (2\pi)^\nu F_\nu(\lambda)$$

$$= (2\pi)^\nu \int_0^\infty rr^\nu f(r)J_\nu(\lambda r)\, dr,$$

$$\nu = \frac{n}{2} - 1, \quad n = 2, 3, \ldots$$

$$\lambda^\nu F_\nu(\lambda) = \mathcal{H}_\nu \{r^\nu f(r)\} = \int_0^\infty rr^\nu f(r)J_\nu(\lambda r)\, dr,$$

$$\nu = \frac{n}{2} - 1, \quad n = 2, 3, \ldots, \qquad (2.238)$$

For the two-dimensional Fourier transform ($n = 2$), this clearly reduces to (2.236). We shall leave the proof of the general result (2.238) as an exercise after supplying the essential results necessary for its derivation like the integral representation of $J_{(n/2)-1}(r)$ (see Exercise 2.81).

The same method can be applied to establish the inverse $J_{(n/2)-1}$-Hankel transform as

$$r^\nu f(r) = \mathcal{H}_\nu^{-1}\{\lambda^\nu F_\nu(\lambda)\} = \int_0^\infty \lambda \lambda^\nu F_\nu(\lambda) J_\nu(\lambda r)\, d\lambda,$$

$$\nu = \frac{n}{2} - 1, \quad n = 2, 3, \ldots, \tag{2.239}$$

We will consider (2.238) and (2.239) as the established J_ν-Hankel transform and its inverse, based on the theory of the (multiple) Fourier transform, but, of course, its validity for now is only for $\nu = (n/2) - 1$, a zero or positive half-integer, since our derivation was based on $n = 2, 3, \ldots$ for the two- and higher-dimensional Fourier transforms. The existence of the J_ν-Hankel transform and its inverse for general $\nu > -1/2$ will be stated without proof as Theorem 2.24 in Section 2.7.

As to our applications of the Hankel transform in the next section, we will be concerned with circular symmetry in the xy-plane; thus, we will be working mainly with the special case of the J_0-Hankel transforms (2.236), (2.237).

Besides establishing the existence of the "practical" $J_{(n/2)-1}$-Hankel transform and its inverse, the relation of the multiple Fourier transform of functions with circular symmetry to the $J_{(n/2)-1}$-Hankel transform has other advantages. An important property is that the multiple Fourier transforms supply the very much needed convolution theorem for the Hankel transforms. This is in the sense that the applicability of the convolution theorems that we are familiar with, like Theorems 2.11 and 2.18 for the Laplace and Fourier transforms respectively, depends principally on the particular character of these transforms' kernels being exponential functions. When these exponential kernels are multiplied, their arguments add, which is the basic reason behind the translation property of these transforms that contributed to the derivation of their relatively simple (or applicable) convolution theorems. This, unfortunately, is not the case for the J_ν-Hankel transform with its, obviously, nonexponential Bessel function kernel. So, unless a more general concept of translation for the desired convolution product is adopted, which we will discuss briefly in the next section, there is no convolution theorem for the Hankel transforms in our familiar sense. But in solving problems we are going to have some products of functions in the transform space, which have to be inverted back to the physical space. The way that is available now is to resort to, for example, the J_0-Hankel transform representation as a double Fourier transform (of functions with circular symmetry) and then use the double Fourier (integral) convolution theorem (2.225), as we shall present next in (2.240). This approach forces us to go to two-dimensional Fourier integrals, but the presence of the fast Fourier transform algorithm has alleviated some of the cost of these lengthy computations.

2.6.3 A Double Fourier Transform Convolution Theorem for the J_0-Hankel Transform

Let $F_0(\lambda)$ and $G_0(\lambda)$ be the J_0-Hankel transforms of $f(r)$ and $g(r)$, respectively. If we use (2.236), with (2.233) and (2.234), we can establish a "hybrid" double Fourier-J_0-Hankel transform convolution theorem:

$$\mathcal{H}_0 \left\{ \iint_{-\infty}^{\infty} f\left(\sqrt{\xi^2 + \eta^2}\right) g\left(\sqrt{(x - \xi)^2 + (y - \eta)^2}\right) d\xi\, d\eta \right\}$$

$$= 2\pi F_0(\lambda) G_0(\lambda) \tag{2.240}$$

Such concepts and tools have been used in a variety of problems in optics, where the required circular symmetry exists. Next we will discuss the Hankel transform and illustrate some of its applications.

2.7 The Hankel (Bessel) Transforms

In Chapter 1 we introduced the concept of compatible transforms and listed the Hankel transform (in Table 1.1) among the most commonly used transforms, possibly after the Laplace and Fourier transforms. In the last sections we discussed and illustrated the use of the Laplace and Fourier transforms, which are compatible with constant-coefficient differential operators (or equations). The Bessel functions and their properties are discussed in detail in Appendix A.5. Also, a collection of basic pairs of the Hankel transform is found in Appendix C.5, and those for the finite Hankel transforms are in C.11 and C.12.

In Chapter 1, we also illustrated in Example 1.6 that the J_n-Hankel transform is compatible with the Bessel differential operator—in the sense that it algebraizes the variable-coefficient part of the Bessel differential equation

$$\mathcal{H}_n \left\{ \frac{d^2u}{dr^2} + \frac{1}{r} \frac{du}{dr} - \frac{n^2}{r^2} u \right\} = -\lambda^2 U_n(\lambda) \tag{1.24}$$

where

$$U_n(\lambda) = \mathcal{H}_n\{u(r)\}$$

We also noticed that these "compatibility" concepts are related to the exponential kernels of the Laplace and Fourier transforms versus the (nonexponential) Bessel kernel of the Hankel transform. We mention this as it represents a drawback to the Hankel transform, since it lacks the "usual" translation property of the Laplace and Fourier transforms. This translation property is a consequence of the exponents (of the kernels) adding for the product of the Laplace and Fourier exponential kernels, while, unfortunately, this is not the case for the Bessel function kernel of the Hankel transform. As we mentioned in the last section, this is reflected most clearly in the absence of a practical convolution theorem for the Hankel transform. For the analytical computations, or proofs, concerning the inverse transform of products of J_0-Hankel transforms we would have to resort to the equivalent convolution theorem (2.240) of the double Fourier transform of functions with circular symmetry $f(\sqrt{x^2 + y^2})$.

We will present here the Hankel transform and a theorem for the existence of this transform and its inverse. Then we will develop a few of the basic transform properties that are necessary for the number of applications that follow. Since such properties of the J_ν-Hankel transform depend principally on the properties of the Bessel function kernel J_ν, we leave a number of them as exercises, for which the properties of the Bessel functions in Appendix A.5 can be consulted or we supply the necessary ones that are not in Appendix A.5.

As in the last section, we define the J_ν-Hankel transform of $f(r)$, $0 < r < \infty$, of (2.238) as

$$\mathcal{H}_\nu\{f\} = F_\nu(\lambda) = \int_0^\infty rf(r)J_\nu(\lambda r)\,dr, \qquad \nu = \frac{n}{2} - 1, \ n = 2, 3, \ldots$$

$$(2.241)$$

For the same special case of $\nu = (n/2) - 1$, $n = 2, 3, \ldots$, we showed in (2.239), depending on the theory of the multiple Fourier transforms of radially symmetric functions, that the inverse of this $J_{(n/2)-1}$-Hankel transform is symmetric,

$$\mathcal{H}_\nu^{-1}\{F_\nu\} = f(r) = \int_0^\infty \lambda F_\nu(\lambda)J_\nu(\lambda r)\,d\lambda, \qquad \nu = \frac{n}{2} - 1, \ n = 2, 3, \ldots$$

$$(2.242)$$

It turns out that these results (2.241), (2.242) are valid for general $\nu > -\frac{1}{2}$ for the class of functions $f(r)$, $0 < r < \infty$, such that $\sqrt{r}f(r)$ is piecewise continuous and absolutely integrable on $(0, \infty)$, which we shall state precisely as Theorem 2.24.

THEOREM 2.24: The Existence of the J_ν-Hankel Transform and Its Inverse, $\nu > -\frac{1}{2}$

Let $f(r)$ be defined on $(0, \infty)$ such that $\sqrt{r}f(r)$ is sectionally continuous and absolutely integrable on $(0, \infty)$; then for $\nu > -\frac{1}{2}$ the Hankel transform $\mathcal{H}_\nu\{f\}$,

$$\mathcal{H}_\nu\{f\} = F_\nu(\lambda) = \int_0^\infty rf(r)J_\nu)(\lambda r)\,dr, \qquad \nu > -\frac{1}{2} \tag{2.243}$$

exists, and its inverse is

$$\mathcal{H}_\nu^{-1}\{F_\nu\} = \int_0^\infty \lambda F_\nu(\lambda)J_\nu(\lambda r)\,d\lambda = \frac{1}{2}[f(r+) + f(r-)] \qquad \square \tag{2.244}$$

The proof is long and, as expected, involves many of the properties of Bessel functions; the interested reader can find it in Sneddon (1972, p. 309). For our treatment we are content with the special case $\nu = (n/2) - 1$, $n = 2, 3, \ldots$, that we were able to treat in (2.241), (2.242) via the double Fourier transform theory.

It is of interest to note the parallel between Theorem 2.24 for the J_ν-Hankel transform and the Fourier integral Theorem 2.14. This is especially in regard to the same type of conditions on the transformed function, and the convergence of the integral transform representation to the average value $\frac{1}{2}[f(r_0+) + f(r_0-)]$ at r_0 where the function $f(r)$ has a jump discontinuity. This similarity extends to the validity of a *Riemann–Lebesgue lemma* for the Hankel transform, like that of Theorem 2.13 in (2.91) for the Fourier transform, which we shall state next.

Riemann–Lebesgue Lemma for the J_ν-Hankel Transform

For $f(r)$, $a < r < b$ ($b \geq a \geq 0$), if $\sqrt{r}f(r)$ is piecewise continuous and absolutely integrable in (a,b), then as $\lambda \to \infty$

$$\int_a^b rf(r)J_\nu(\lambda r)\,dr = o(\lambda^{-1/2}), \qquad \nu > -\frac{1}{2}, \ (b \geq a \geq 0) \tag{2.245}$$

This, of course, means that

$$\lim_{\lambda \to \infty} \int_a^b rf(r)J_\nu(\lambda r)\,dr = 0, \qquad \nu > -\frac{1}{2}, \ (b \geq a \geq 0) \tag{2.245a}$$

as the expected parallel to Theorem 2.13. $\qquad \square$

Also a *Parseval equality* exists for the J_ν-Hankel transform, which we state next.

The Parseval Equality for J_ν-Hankel Transform

If $f(r)$ and $g(r)$ satisfy the conditions of the Hankel transform existence Theorem 2.24, and if $F_\nu(\lambda)$ and $G_\nu(\lambda)$ are their respective Hankel transforms, of order $\nu > -\frac{1}{2}$, then

$$\int_0^\infty rf(r)g(r)\,dr = \int_0^\infty \lambda F_\nu(\lambda)G_\nu(\lambda)\,d\lambda \qquad \square \ \ (2.246)$$

We will be satisfied here with a formal proof as

$$\int_0^\infty \lambda F_\nu(\lambda)G_\nu(\lambda)\,d\lambda = \int_0^\infty \lambda F_\nu(\lambda)\left[\int_0^\infty rg(r)J_\nu(\lambda r)\,dr\right]d\lambda$$

$$= \int_0^\infty rg(r)\left[\int_0^\infty \lambda F_\nu(\lambda)J_\nu(\lambda r)\,d\lambda\right]dr$$

$$= \int_0^\infty rg(r)f(r)\,dr, \qquad \nu > -\frac{1}{2} \qquad (2.246)$$

after exchanging the double integrals and using the inverse Hankel transform (2.244) on the last inner integral, where $f(r)$ is assumed continuous.

It seems the right place here to have a convolution theorem for the Hankel transform, yet it is absent in almost all references. The most that one finds is what we had in (2.240), which is done via the double Fourier transform and which means that we have to evaluate a double integral in the Fourier space instead of the desired single integral in the Hankel transform space.

If we insist on a convolution theorem in the Hankel transform's own setting, we must accept a "generalized translation" concept for the Hankel transforms with their (nonexponential) Bessel kernel. This is instead of the usual translation associated with the convolution product of the Laplace and Fourier transforms, which is a consequence of their exponential kernels. We shall leave the topic of the generalized translation (or transformation) to Exercise 2.83, where we will also formally derive a convolution theorem and a Parseval equality for the Hankel transform in its own setting without resorting to the multiple Fourier integrals of circularly symmetric functions. (See Jerri, 1992: Chap. 7 in Marks, vol. 2, 1992).

Next we will present some of the basic properties of the J_ν-Hankel transform (2.243); they clearly depend on the properties of $J_\nu(r)$, the Bessel function of the first kind of order ν, which is the bounded solution as $x \to 0$ of the Bessel differential equation,

$$\frac{d^2u}{dr^2} + \frac{1}{r}\frac{du}{dr} + \left(1 - \frac{\nu^2}{r^2}\right)u(r) = 0 \qquad (2.247)$$

The second solution $Y_\nu(r)$ to this equation is called the *Bessel function of the second kind* of order ν, which diverges as $x \to 0$. Both $J_\nu(r)$ and $Y_\nu(r)$ are bounded as $r \to \infty$. The properties of $J_\nu(r)$ are obtained mostly through the direct application of its infinite series expansion about $r = 0$,

$$J_\nu(r) = \sum_{k=0}^{\infty} \frac{(-1)^k (\tfrac{1}{2}r)^{\nu+2k}}{k!\Gamma(\nu + k + 1)} \tag{2.248}$$

and are listed (with some of them derived) in Appendix A.5. The series representation (2.248) of the Bessel function $J_\nu(r)$ is derived by using the method of Frobenius (see Appendix A.4 for this method and A.5 for these and other special functions of interest).

One of the most important properties of the J_ν-Hankel transform is its compatibility with the above Bessel differential equation; i.e., it reduces the (variable coefficients) differential operator on $u(r)$ in the left side of (2.247) to an algebraic operation on $U(\lambda)$ as $-\lambda^2 U_\nu(\lambda)$, where $U_\nu(\lambda) = \mathcal{H}_\nu\{u\}$. This was shown formally in (1.24) as the final result of Example 1.6,

$$\mathcal{H}_n \left\{ \frac{d^2u}{dr^2} + \frac{1}{r}\frac{du}{dr} - \frac{n^2}{r^2}u \right\} = -\lambda^2 U_n(\lambda) \tag{1.24}$$
$$\tag{2.249}$$

In the two integrations by parts performed in Example 1.6 to produce this result, we were satisfied with assuming the sufficient conditions that would guarantee the vanishing of the resulting integrated terms at the limit points of integration $r = 0$ and $r = \infty$. These appeared as the following in (E.1), (E.2) of Example 1.6:

$$\lim_{r \to 0} ru'(r)J_n(\lambda r) = 0, \qquad \lim_{r \to \infty} ru'(r)J_n(\lambda r) = 0 \tag{2.250}$$

and

$$\lim_{r \to 0} ru(r)\frac{d}{dr}J_n(\lambda r) = 0, \qquad \lim_{r \to \infty} ru(r)\frac{d}{dr}J_n(\lambda r) = 0 \tag{2.251}$$

Then we mentioned that $J_n(r)$ and its derivative dJ_n/dr are known to be bounded as $r \to 0$ and $r \to \infty$, which made the above conditions satisfied if we just have $u(r)$ and its derivative $u'(r)$ such that

$$\lim_{r \to 0} ru'(r) = 0, \qquad \lim_{r \to \infty} ru'(r) = 0 \tag{2.250a}$$

$$\lim_{r \to 0} ru(r) = 0, \qquad \lim_{r \to \infty} ru(r) = 0 \tag{2.251a}$$

However, it turns out, after looking closely at the series (2.248) defining $J_\nu(r)$ and other established properties of $J_\nu(r)$, that the above conditions can be modified to

$$\lim_{r \to 0} r^{\nu+1}\frac{du}{dr} = 0, \qquad \lim_{r \to \infty} r^{1/2}\frac{du}{dr} = 0 \tag{2.250b}$$

$$\lim_{r \to 0} r^\nu u(r) = 0, \qquad \lim_{r \to \infty} r^{1/2} u(r) = 0 \qquad\qquad (2.251b)$$

The first of these conditions can be obtained from its corresponding one in (2.250) after looking at the behavior of $J_\nu(r)$ near $r = 0$ through its series expansion (2.248), where the dominating term, for $\nu > 0$ for example, is proportional to r^ν. The last of the conditions, (2.251b), is valid for all $u(r)$ that have a Hankel transform according to Theorem 2.24. A good feeling for this result may be obtained from the Riemann–Lebesgue lemma in (2.245). The rest of these conditions are left as exercises for the interested reader.

As to other basic properties of the Hankel transform, we definitely don't expect a translation property because of the nonexponential character of the Bessel function kernel $-J_\nu(r)$ (see Exercise 2.83b, where we had to introduce the new concept of generalized transformation for this purpose). However, there is a *scaling property* (see Exercise 2.84a)

$$\mathcal{H}_\nu \{f(ar)\} = \frac{1}{a^2} F_\nu \left(\frac{\lambda}{a} \right), \qquad a > 0 \qquad\qquad (2.252)$$

where we note that the scaling factor here is $1/a^2$ and not $1/a$ as in the case of the Laplace and Fourier transforms. The simple reason for this, as we shall see next, is due to the weight function r in the definition of the Hankel transform (2.243). To derive (2.252),

$$\mathcal{H}_\nu \{f(ar)\} = \int_0^\infty rf(ar) J_\nu(\lambda r) \, dr$$

$$= \frac{1}{a^2} \int_0^\infty raf(ar) J_\nu \left(\frac{\lambda ar}{a} \right) d(ar)$$

$$= \frac{1}{a^2} \int_0^\infty uf(u) J_\nu \left(\frac{\lambda}{a} u \right) du$$

$$= \frac{1}{a^2} F_\nu \left(\frac{\lambda}{a} \right)$$

after letting $u = ar$ in the last integral and using the definition of the J_ν-Hankel transform (2.243) on $f(u)$.

Other pairs of the Hankel transform are left for the exercises, where the necessary properties of the Bessel functions either will be provided or can be found in Appendix A.5.

An exhaustive listing of the Hankel transform pairs may be found in Erdelyi and coauthors (1954, Vol. 2) and in Ditkin and Prudnikov (1965). We warn again about the different notations used in different references. For example, even for this completely symmetric Hankel transform and its inverse in (2.243), (2.244), Ditkin and Prudnikov (1965) and Erdelyi and coauthors (1954) distribute the weight functions r and λ in the transform and its inverse as $\sqrt{r\lambda}$ in both, and they write $F(\lambda, \nu)$ for our $F_\nu(\lambda)$,

$$F_\nu(\lambda) \equiv F(\lambda, \nu) = \int_0^\infty \sqrt{r\lambda} f(r) J_\nu(\lambda r) dr \qquad (2.243a)$$

and

$$f(r) = \int_0^\infty \sqrt{r\lambda} F_\nu(\lambda, \nu) J_\nu(\lambda r) d\lambda \qquad (2.244a)$$

So we have to be careful when consulting such very essential references and make sure to have the modification made from the reference's definition to the definition used in the analysis or applications at hand. As a very simple example, the above definition of the Hankel transform (2.243a) in Ditkin and Prudnikov would result in a different scaling property,

$$\mathcal{H}_\nu \{f(ar)\} = \frac{1}{a} F\left(\frac{\lambda}{a}, \nu\right)$$

compared to ours (2.243), which gave

$$\mathcal{H}_\nu \{f(ar)\} = \frac{1}{a^2} F_\nu\left(\frac{\lambda}{a}\right)$$

in (2.252).

2.7.1 Applications of the Hankel Transforms

Next we will illustrate the application of the Hankel transforms to boundary value problems. Since most of these applications deal with circularly symmetric problems, we shall be working mainly with the J_0-Hankel transforms.

EXAMPLE 2.25: Steady-State Temperature Distribution in a Thick Infinite Slab

Our first application of the Hankel transforms is to solve for the steady-state temperature in a thick infinite slab with the two faces at $z = a$ and $z = -a$ being kept at (circularly) symmetric temperatures given by $f(r)$ and $g(r)$, respectively. For these symmetric conditions we realize that the temperature in the domain of this problem is clearly circularly symmetric, and also it is independent of time (steady state). So we denote the temperature by $u(r, z)$, $0 \le r < \infty$, $-a < z < a$, to be solved for in the following boundary value problem, which is associated with the Laplace equation (E.1) for such steady-state temperature $u(r,z)$,

$$\nabla^2 u \equiv \frac{\partial^2 u}{\partial r^2} + \frac{1}{r}\frac{\partial u}{\partial r} + \frac{\partial^2 u}{\partial z^2} = 0, \qquad 0 < r < \infty, \quad -a < z < a, \qquad \text{(E.1)}$$

$$u(r,a) = f(r), \qquad 0 < r < \infty \tag{E.2}$$

$$u(r,-a) = g(r), \qquad 0 < r < \infty \tag{E.3}$$

For algebraizing the (r) differential operator part $\partial^2 u/\partial r^2 + (1/r)\partial u/\partial r$ of equation (E.1), we use the J_0-Hankel transform (2.243) along with its operational pair (2.249) to have a differential equation in the transform $U_0(\lambda,z) = \mathcal{H}_0\{u(r,z)\}$,

$$-\lambda^2 U_0(\lambda,z) + \frac{d^2 U_0}{dz^2} = 0 \tag{E.4}$$

The solution to this problem is of the following suitable form (among other forms!):

$$U_0(\lambda,z) = A(\lambda)\sinh\lambda(a+z) + B(\lambda)\sinh\lambda(a-z) \tag{E.5}$$

To find $A(\lambda)$ and $B(\lambda)$ we take the Hankel transform of the boundary conditions (E.1), (E.2) to have

$$\mathcal{H}_0\{u(r,a)\} = F_0(\lambda) = A(\lambda)\sinh 2\lambda a$$

$$A(\lambda) = \frac{F_0(\lambda)}{\sinh 2\lambda a} = F_0(\lambda)\operatorname{cosech} 2a\lambda \tag{E.6}$$

and from (E.2) we have

$$\mathcal{H}_0\{u(r,-a)\} = G_0(\lambda) = B(\lambda)\sinh 2a\lambda$$

$$B(\lambda) = G_0(\lambda)\operatorname{cosech} 2a\lambda \tag{E.7}$$

Thus from (E.5) to (E.7), we have the transformed solution as

$$U_0(\lambda,z) = \operatorname{cosech} 2a\lambda[F_0(\lambda)\sinh\lambda(a+z) + G_0(\lambda)\sinh(a-z)] \tag{E.8}$$

where $F_0(\lambda)$ and $G_0(\lambda)$ are the respective J_0-Hankel transforms of $f(r)$ and $g(r)$. The main problem now is to find the solution $u(r,z)$ as the inverse J_0-Hankel transform of the two terms in (E.8), where each term is a product of three Hankel transforms. This should show the need for a convolution theorem. Even for the special case of the same potentials $f(r) = g(r)$ at both surfaces, the solution in the Hankel transform space is still a product,

$$U_0(\lambda,z) = F_0(\lambda)\frac{\cosh\lambda z}{\cosh\lambda a} \tag{E.9}$$

This is still intractable as far as transforming it with a (Hankel) convolution product type of inverse to have the final solution $u(r,z)$ in terms of the boundary function $f(r)$. The only way (E.9) can be inverted is as a one-piece function; otherwise we have to resort to a convolution theorem via the double Fourier transform like (2.240), as we alluded to in Section 2.6.1.

EXAMPLE 2.26: The Vibration of a Large Membrane with Small Symmetric Initial Deformations

This example deals with small transverse vibration of very large membrane, where its initial displacement and velocity cause a circularly symmetric deformation. Because of the symmetry of these initial conditions, the resulting displacement is expected to be also circularly symmetric, and we take it as $u(r,t)$, which satisfies the following initial value problem associated with the wave equation in $u(r,t)$:

$$\frac{\partial^2 u}{\partial r^2} + \frac{1}{r}\frac{\partial u}{\partial r} = \frac{1}{c^2}\frac{\partial^2 u}{\partial t^2}, \qquad 0 < r < \infty, \ t > 0 \tag{E.1}$$

$$u(r,0) = f(r), \qquad 0 < r < \infty \tag{E.2}$$

$$\frac{\partial u}{\partial t}(r,0) = g(r), \qquad 0 < r < \infty \tag{E.3}$$

We let $U_0(\lambda,t)$ be the J_0-Hankel transform of $u(r,t)$ and $F_0(\lambda)$, $G_0(\lambda)$ be the respective Hankel transforms of $f(r)$ and $g(r)$. The J_0-Hankel transform of the wave equation (E.1), after using (2.249), becomes

$$-\lambda^2 U_0(\lambda,t) = \frac{1}{c^2}\frac{d^2 U_0(\lambda,t)}{dt^2}$$

whose solution is

$$U_0(\lambda,t) = A(\lambda)\cos\lambda ct + B(\lambda)\sin\lambda ct \tag{E.4}$$

If we use the initial conditions (E.1) and then (E.2) on (E.4), we have

$$U_0(\lambda,0) = F_0(\lambda) = A(\lambda) \tag{E.5}$$

$$\frac{dU_0(\lambda,0)}{dt} = G_0(\lambda) = cB(\lambda) \tag{E.6}$$

With (E.5) and (E.6) used in (E.4) we have the solution to the transformed problem,

$$U_0(\lambda,t) = F_0(\lambda)\cos\lambda ct + G_0(\lambda)\frac{\sin\lambda ct}{\lambda c} \tag{E.7}$$

What remains now is to find the inverse Hankel transform of these two terms of products of Hankel transforms in (E.7), which is not such a simple problem in the absence of a convolution theorem. However, for particular given initial conditions these products may become a particular function whose inverse transform we can find in the known Hankel transform tables [see Erdelyi and coauthors, (1954b, Vol. 2) or Ditkin and Prudnikov (1965) as two well-known references with exhaustive Hankel and other transform tables]. Otherwise, we just have to struggle with evaluating this inverse Hankel transform as an integral involving the Bessel function, which depends a great deal on our familiarity with the analysis of Bessel functions. After exhausting the search in the tables, one should not underestimate what may be found in a treatise like that of Watson (1966) on the Bessel functions. An example of a very special case that is tractable for a final solution $u(r,z)$, via the Hankel transform tables, is that with the initial displacement $u(r,0) = f(r) = c/\sqrt{1 + r^2/a^2}$, and with zero initial velocity, which we shall leave for an exercise [see Exercise 2.88; also see Exercise 2.87a and Example 2.29 for problems that end with a closed-form (known, or familiar) solution].

EXAMPLE 2.27: The Electrified Unit Disc in Space

In this example we consider the problem of the electrified unit disc in the xy-plane with center at the origin. The potential is held constant as u_0 on the unit disc, and where outside the disc the potential is symmetric with respect to the xy-plane: $\partial u(r,0)/\partial z = 0$, $1 < r < \infty$. Understandably, the potential in space is circularly symmetric, and we represent it by $u(r,z)$. It then must satisfy the following Laplace equation and its *mixed* boundary conditions:

$$\frac{\partial^2 u}{\partial r^2} + \frac{1}{r}\frac{\partial u}{\partial r} + \frac{\partial^2 u}{\partial z^2} = 0, \qquad 0 < r < \infty, \ 0 < z < \infty \tag{E.1}$$

$$u(r,0) = u_0, \qquad 0 \le r < 1 \tag{E.2}$$

$$\frac{\partial u}{\partial z}(r,0) = 0, \qquad 1 < r < \infty \tag{E.3}$$

The mixed boundary conditions are in the sense of the function being given on one part of the boundary ($z = 0$), i.e., for $0 \le r < 1$, while the derivative $\partial u(r,0)/\partial z$ is given on the rest of that boundary for $1 < r < \infty$. Clearly, the r part of the Laplace equation (E.1) is compatible with the J_0-Hankel transform. So we let

$$U_0(\lambda,z) = \int_0^\infty rJ_0(\lambda r)u(r,z)\,dr \tag{E.4}$$

and use (2.249) on (E.1) to obtain

$$\lambda^2 U_0(\lambda,z) = \frac{\partial^2 U_0(\lambda,z)}{\partial z^2} \qquad (E.5)$$

The bounded solution of (E.1) is

$$U_0(\lambda,z) = A(\lambda)e^{-\lambda z} \qquad (E.6)$$

In order to find $A(\lambda)$, we need the condition at $z = 0$ to be transformed; unfortunately, it is given partly as $u(r,0)$ in (E.2) and partly as $u_z(r,0)$ in (E.3), which is not suitable for (E.4). The way around this is to ignore finding $A(\lambda)$ for the moment and write the solution as the inverse Hankel transform of $U_0(\lambda,z)$ in (E.6),

$$u(r,z) = \int_0^\infty \lambda J_0(\lambda r) A(\lambda)e^{-\lambda z} \, d\lambda \qquad (E.7)$$

Now we may apply the first part (E.2) of the mixed boundary condition on (E.7) to obtain

$$u(r,0) = \int_0^\infty \lambda J_0(\lambda r) A(\lambda) d\lambda = u_0, \qquad 0 \le r < 1 \qquad (E.8)$$

where $A(\lambda)$, the function we seek to determine for (E.7), is now involved under an integral in (E.8), which is an *integral equation* in $A(\lambda)$ valid only for $0 \le r < 1$.

The other part of the boundary condition (E.3) on (E.7) gives

$$\frac{\partial u(r,0)}{\partial z} = 0 = \int_0^\infty -\lambda^2 J_0(\lambda r) A(\lambda) d\lambda, \qquad 1 < r < \infty \qquad (E.9)$$

which is another integral equation in $A(\lambda)$ valid only for $1 < r < \infty$. Hence, $A(\lambda)$ is involved in the *dual integral equations* (E.8) and (E.9) that cover all values of r, $0 < r < \infty$. To find $A(\lambda)$ we must consult recent methods of solving dual integral equations. Fortunately, this problem can be solved using the wealth of known integrals involving Bessel functions such as

$$\int_0^\infty J_0(\lambda r) \frac{\sin \lambda}{\lambda} \, d\lambda = \frac{\pi}{2}, \qquad 0 \le r < 1 \qquad (E.10)$$

and

$$\int_0^\infty J_0(\lambda r) \sin \lambda \, d\lambda = 0, \qquad r > 1 \qquad (E.11)$$

We immediately realize after comparing (E.8) with (E.10) and (E.9) with (E.11) that the solution to the dual integral equations (E.8) and (E.9) above is

$$A(\lambda) = \frac{2u_0}{\pi} \frac{\sin \lambda}{\lambda^2} \qquad (E.12)$$

Substituting $A(\lambda)$ from (E.12) in (E.7), the final solution becomes

$$u(r,z) = \frac{2u_0}{\pi} \int_0^\infty J_0(\lambda r) e^{-\lambda z} \frac{\sin \lambda}{\lambda} \, d\lambda \qquad (E.13)$$

It is easy to verify by differentiating (E.13) directly that it is the solution to (E.1). The mixed boundary conditions (E.2), (E.3) can also be verified since

$$u(r,0) = \frac{2u_0}{\pi} \int_0^\infty J_0(\lambda r) \frac{\sin \lambda}{\lambda} \, d\lambda$$

$$= \frac{2u_0}{\pi} \frac{\pi}{2} = u_0, \qquad 0 \le r < 1$$

after using (E.10). Also,

$$\frac{\partial u}{\partial z}(r,0) = \frac{2u_0}{\pi} \int_0^\infty J_0(\lambda r) \sin \lambda \, d\lambda = 0, \qquad 1 < r < \infty$$

after using (E.11).

We may remark here that had we attempted to use the sine or cosine transforms on $\partial^2 u/\partial z^2$ we would still have the difficulty of not having, for all values of r, either u or $\partial u/\partial z$ at $z = 0$. However, other problems with a combination of both (E.2) and (E.3), given each for all values of $0 \le r < \infty$, may be amenable to a combined sine-cosine transform.

EXAMPLE 2.28: Temperature in the Half-Space Due to a Steadily Heated Disc

In this example we consider the temperature distribution in the upper half-space due to heat entering the space at a constant rate q through a circular disc of radius a in the plane $z = 0$, and where the rest of the surface $z = 0$ is insulated. We solve here for the steady state temperature $u(r,z)$, which satisfies the following boundary value problem,

$$\frac{\partial^2 u}{\partial r^2} + \frac{1}{r}\frac{\partial u}{\partial r} + \frac{\partial^2 u}{\partial z^2} = 0, \qquad 0 < r < \infty, \ 0 < z < \infty \qquad (E.1)$$

$$-\kappa \frac{\partial u}{\partial z}(r,0) = \begin{cases} \dfrac{q}{\pi a^2}, & 0 \le r < a \\[2mm] 0, & a < r < \infty \end{cases} \equiv \frac{q}{\pi a^2} p_a(r) \qquad (E.2)$$

where κ is the conductivity of the body and $q/\pi a^2$ is the steady heat flux entering the body through the disc of area πa^2.

We note in (E.2) that we are given the gradient of the temperature on the whole boundary $z = 0$, which is a Neumann boundary condition, compared to the mixed boundary condition in the last Example 2.27. If we let $U_0(\lambda,z) = \mathcal{H}_0\{u(r,z)\}$, Hankel transform (E.1), and use (2.249), we have

$$-\lambda^2 U_0(\lambda,z) + \frac{d^2 U(\lambda,z)}{dz^2} = 0 \tag{E.3}$$

The Hankel transform of the boundary condition (E.2) gives

$$-\kappa \frac{dU_0(\lambda,0)}{dz} = q \frac{J_1(\lambda a)}{\pi \lambda a} \tag{E.4}$$

after using the J_0-Hankel transform pair $\mathcal{H}_0\{p_a(r)\} = aJ_1(\lambda a)/\lambda$, which can be derived easily (see Exercise 2.82a(iii)).

The solution to (E.3) that remains finite as $z \to \infty$ is

$$U_0(\lambda,z) = A(\lambda)e^{-\lambda z} \tag{E.5}$$

and to satisfy the boundary condition (E.4), we have

$$\lambda \kappa A(\lambda) = \frac{qJ_1(\lambda a)}{\pi \lambda a}, \qquad A(\lambda) = \frac{q}{\kappa a \pi} \frac{J_1(\lambda a)}{\lambda^2} \tag{E.6}$$

With this $A(\lambda)$ we have

$$U_0(\lambda,z) = \frac{q}{\pi a \kappa} \frac{J_1(\lambda a)}{\lambda^2} e^{-\lambda z} \tag{E.7}$$

The final solution $u(r,z)$ is the inverse Hankel transform of this $U_0(\lambda,z)$,

$$
\begin{aligned}
u(r,z) &= \frac{q}{\pi a \kappa} \int_0^\infty \lambda \frac{J_1(\lambda a)}{\lambda^2} e^{-\lambda z} J_0(\lambda r)\,d\lambda \\
&= \frac{q}{\pi a \kappa} \int_0^\infty \frac{J_1(\lambda a)}{\lambda} e^{-\lambda z} J_0(\lambda r)\,d\lambda, \qquad z \geq 0,\ r \geq 0
\end{aligned} \tag{E.8}
$$

as a reasonable integral representation of the temperature $u(r,z)$ in the half-space. See also Exercises 2.86 and 2.87.

The next example illustrates an application of the J_0-Hankel transforms with boundary conditions, of a very particular form of function, that results in a simple closed-form solution. The boundary value problem is associated with the *biharmonic equation*, which is often seen in the theory of elasticity,[†]

$$\nabla^2(\nabla^2 u) = \nabla^4 u = 0 \tag{2.253}$$

For $u(r,z)$ we can use (2.249) to write

$$\mathcal{H}_0\{\nabla^2 u\} = \frac{d^2 U_0(\lambda,z)}{dz^2} - \lambda^2 U_0(\lambda,z) \equiv \left[\frac{d^2}{dz^2} - \lambda^2\right] U_0(\lambda,z)$$

So if we let $\nabla^2 u = v(r,z)$, we have

$$\mathcal{H}_0\{\nabla^2 u\} = \mathcal{H}_0\{v\} = V_0(\lambda,z)$$

and

$$\mathcal{H}_0\{\nabla^4 u\} = \mathcal{H}_0\{\nabla^2 v\} = \frac{d^2 V_0(\lambda,z)}{dz^2} - \lambda^2 V_0(\lambda,z)$$

$$= \left[\frac{d^2}{dz^2} - \lambda^2\right] V_0(\lambda,z) = \left[\frac{d^2}{dz^2} - \lambda^2\right] \mathcal{H}_0\{\nabla^2 u\}$$

$$= \left[\frac{d^2}{dz^2} - \lambda^2\right]\left[\frac{d^2}{dz^2} - \lambda^2\right] U_0(\lambda,z)$$

$$\mathcal{H}_0\{\nabla^4 u\} = \left[\frac{d^2}{dz^2} - \lambda^2\right]^2 U_0(\lambda,z) \tag{2.254}$$

This means that the J_0-Hankel transform reduces the radial part of the biharmonic operator in (2.253) to an algebraic one in (2.254) by simple algebraizing that part of the Laplace operator in succession.

EXAMPLE 2.29: The Biharmonic Equation

We consider here the following boundary value problem in $u(r,z)$, which is associated with the biharmonic equation:

$$\nabla^4 u(r,z) = 0, \qquad z > 0, \ r \geq 0 \tag{E.1}$$

and the simple boundary conditions,

$$u(r,0) = f(r), \ r \geq 0 \tag{E.2}$$

$$\frac{\partial u(r,0)}{\partial z} = 0, \ r \geq 0 \qquad u(r,z) \to 0 \text{ as } r \to \infty \tag{E.3}$$

[†]For a related fourth-order differential equation, see the detailed problem of the displacement of an elastic beam in Exercise 2.75.

We let $U_0(\lambda,z) = \mathcal{H}_0 \{u(r,z)\}$ and use the J_0-Hankel transform on (E.1), with the help of the pair we just developed in (2.254), to obtain

$$\left[\frac{d^2}{dz^2} - \lambda^2 \right]^2 U_0(\lambda,z) = 0 \tag{E.4}$$

The bounded solution to this equation (note the repeated roots!) is of the form

$$U_0(\lambda,z) = A(\lambda)e^{-\lambda z} + zB(\lambda)e^{-\lambda z} \tag{E.5}$$

To satisfy the boundary condition (E.3) we must have $dU_0(\lambda,0)/dz = 0$, where

$$\frac{dU_0}{dz}(\lambda,z) = -\lambda A(\lambda)e^{-\lambda z} + B(\lambda)e^{-\lambda z} - \lambda z B(\lambda)e^{-\lambda z} \tag{E.6}$$

$$\frac{dU_0}{dz}(\lambda,0) = 0 = -\lambda A(\lambda) + B(\lambda), \qquad B(\lambda) = \lambda A(\lambda) \tag{E.7}$$

with this result $B(\lambda) = \lambda A(\lambda)$ in (E.5) we have

$$U_0(\lambda,z) = A(\lambda)[1 + \lambda z]e^{-\lambda z} \tag{E.8}$$

The coefficient $A(\lambda)$ is found by Hankel transforming the boundary condition (E.2) and using $U_0(\lambda,0) = A(\lambda)$ from (E.8)

$$\mathcal{H}_0 \{u(r,0)\} = \mathcal{H}_0 \{f(r)\} = F_0(\lambda) = U_0(\lambda,0) = A(\lambda)$$

$$A(\lambda) = F_0(\lambda)$$

Thus the final solution in the transform space is

$$U_0(\lambda,z) = F_0(\lambda)(1 + \lambda z)e^{-\lambda z} \tag{E.9}$$

As we mentioned above, we will use this example to illustrate having a final solution $u(r,z)$, to the boundary value problem, in a simple closed form. What this really means is to look at the available J_0-Hankel transform pairs to find one that has the last two factors in (E.9), and adjust our $f(r)$ in (E.2) to give the exact factor $F_0(\lambda)$ of (E.9) found in the tables. So with the very particular choice of $f(r) = C(a^2 + r^2)^{-1/2}$, we find its J_0-Hankel transform $F_0(\lambda) = Ce^{-a\lambda}/\lambda$, and the inverse J_0-Hankel transform of the resulting (adjusted!) particular form of $U_0(\lambda,z)$ in (E.9)

$$U_0(\lambda,z) = C\frac{e^{-a\lambda}}{\lambda}(1 + \lambda z)e^{-\lambda z} \tag{E.10}$$

is the exact solution of our problem

$$u(r,z) = C\frac{r^2 + (z + a)(2z + a)}{[r^2 + (z + a^2)]^{3/2}} \tag{E.11}$$

See also the answers to Exercises 2.87 and 2.88 for examples (of seldom found!) solutions in a closed form like what we have here in (E.11).

2.8 Laplace Transform Inversion[†]

In Section 2.1 we introduced the Laplace transform and applied it to solving ordinary and partial differential equations with proper auxiliary conditions as well as for solving Volterra integral equations with difference kernels. In all the examples presented we used pairs that we developed or those available in Laplace transform tables. In general, and for more advanced problems, we will need to evaluate the inverse Laplace transform directly, since we may not find the required pair for the problem at hand. Also, this direct method may allow an alternative for the representation of the solution which may be more suitable than others for the given problem and the range of its variables. In addition, the Laplace transform inversion formula will guide us toward one of the methods for the numerical inversion of the Laplace transform, which we shall allude to briefly at the end of this section.

This section is considered optional and may be skipped by those without a background in the basic elements of functions of complex variables.

The formula (2.261) that we had for the Laplace transform inverse in (2.6) will be derived shortly with the aid of the inverse Fourier transform with a complex argument, which we shall present next.

2.8.1 Fourier Transform in the Complex Plane

The Fourier transform

$$F(\lambda) = \int_{-\infty}^{\infty} e^{-i\lambda x} f(x)\, dx \qquad (2.255)$$

can be defined for complex argument $\zeta = \lambda + i\mu$ by considering the Fourier transform of $e^{\mu x} f(x)$,

$$\mathcal{F}\{e^{\mu x} f(x)\}(\lambda) = \int_{-\infty}^{\infty} e^{-i\lambda x} e^{\mu x} f(x)\, dx$$

[†] Optional section.

$$= \int_{-\infty}^{\infty} e^{-i(\lambda + i\mu)x} f(x)\,dx$$

$$= F(\lambda + i\mu) = F(\zeta) \tag{2.256}$$

as a Fourier transform of $f(x)$ with the complex argument $\zeta = \lambda + i\mu$.

In simple terms, a function $F(\zeta)$ is termed *analytic* (or *regular*) if it possesses all its derivatives $d^n F/d\zeta^n$, $n = 0, 1, 2, \ldots$. If the function $F(\zeta)$ is not analytic at a point ζ_0 it is termed *singular* at ζ_0. An example is $g(\zeta) = 1/(\zeta^2 - 1)(\zeta^2 + 4)$, which is analytic everywhere except at the four singular points $\zeta = \mp 1, \mp 2i$.

We may remark here that if $f(x)$ in (2.256) is such that $e^{\mu_1 x} f(x)$ is absolutely integrable on $(-\infty, \infty)$, then its Fourier transform $F(\zeta)$ is analytic for $\mu < \mu_1$. This can be validated by differentiating $F(\zeta)$ in (2.256) with respect to ζ as many times as desired. Such differentiation is allowed inside the integral of (2.256) according to Theorem 2.6, since the integrand $e^{-i\lambda x} e^{\mu x} f(x)$ is absolutely integrable when $e^{\mu x} f(x)$ is,

$$\left| e^{-i\lambda x} e^{\mu x} f(x) \right| = \left| e^{-i\lambda x} \right| \left| e^{\mu x} f(x) \right| = \left| e^{\mu x} f(x) \right|$$

From (2.256) and the inverse Fourier transform formula (2.65), we can immediately write

$$e^{\mu x} f(x) = \frac{1}{2\pi} \int_{-\infty}^{\infty} e^{i\lambda x} F(\zeta)\,d\lambda$$

$$f(x) = \frac{1}{2\pi} \int_{-\infty}^{\infty} e^{i\lambda x} e^{-\mu x} F(\zeta)\,d\lambda$$

$$= \frac{1}{2\pi} \int_{-\infty}^{\infty} e^{i(\lambda + i\mu)x} F(\zeta)\,d\lambda$$

$$= \frac{1}{2\pi} \int_{-\infty}^{\infty} e^{i\zeta x} F(\zeta)\,d\lambda = \frac{1}{2\pi} \int_{-\infty + i\mu}^{\infty + i\mu} e^{i\zeta x} F(\zeta)\,d\zeta$$

$$= \lim_{L \to \infty} \int_{-L + i\mu}^{L + i\mu} e^{i\zeta x} F(\zeta)\,d\zeta, \qquad \zeta = \lambda + i\mu \tag{2.257}$$

which is the inverse of the Fourier transform $F(\zeta)$ with the complex argument $\zeta = \lambda + i\mu$. We note here that the (real) limits of integration $-L, L$ for λ in the top integral are now the complex $-L + i\mu$, $L + i\mu$ for the complex variable of integration $\zeta = \lambda + i\mu$ in (2.257).

Operational pairs for $F(\zeta)$ can be derived in the same way as we derived the corresponding ones for $F(\lambda)$, the Fourier transform with real argument λ. For example,

$$\int_{-\infty}^{\infty} e^{-i\zeta x} \frac{df}{dx}\,dx = i\zeta F(\zeta) \tag{2.258}$$

$$\int_{-\infty}^{\infty} e^{-i\zeta x}(-ixf(x))\,dx = \frac{d}{d\zeta}F(\zeta) \tag{2.259}$$

$$\int_{-\infty}^{\infty} e^{-i\zeta x}f(x-b)\,dx = e^{-ib\zeta}F(\zeta) \tag{2.260}$$

2.8.2 The Laplace Transform Inversion Formula

We will now derive the Laplace transform inversion formula

$$f(x) = \mathcal{L}^{-1}\{F\} = \frac{1}{2\pi i}\lim_{L\to\infty}\int_{\gamma-iL}^{\gamma+iL} e^{zx}F(z)\,dz, \qquad \gamma > \text{Real}\{z_i\} \tag{2.261}$$

where $\{z_i\}$ are the singularities of $F(z)$. We will first relate the Laplace transform to the Fourier transform of causal functions ($f(x) \equiv 0, x < 0$). Then we extend the definitions, of both transforms, to complex variables and use the Fourier inversion formula (2.65) to arrive at a Laplace transform inverse in (2.265). Since we will be relating the Laplace and Fourier transforms for the same function $f(x)$, to avoid confusion we will adopt the temporary notation $F_L(\lambda)$ and $F_F(\lambda)$ for the Laplace and Fourier transforms of $f(x)$, respectively.

For the causal function $f(x)$, it is clear that the Fourier transform is related to the Laplace transform by

$$F_F(\lambda) = \int_0^{\infty} e^{-i\lambda x}f(x)\,dx = \int_0^{\infty} e^{-(i\lambda)x}f(x)\,dx$$

$$= F_L(i\lambda) \tag{2.262}$$

and, in general, if we extend the definition of both transforms to the complex argument $\zeta = \lambda + i\mu$, we have

$$F_F(\zeta) = \int_0^{\infty} e^{-i\zeta x}f(x)\,dx = F_L(i\zeta) \tag{2.263}$$

To obtain $f(x)$ from (2.263) we use the Fourier transform inversion formula (2.257) for functions $F(\zeta)$ of complex variables,

$$f(x) = \frac{1}{2\pi}\lim_{L\to\infty}\int_{-L+i\mu}^{L+i\mu} F_F(\zeta)e^{i\zeta x}\,d\zeta$$

$$= \frac{1}{2\pi}\lim_{L\to\infty}\int_{-L+i\mu}^{L+i\mu} F_L(i\zeta)e^{i\zeta x}\,d\zeta \tag{2.264}$$

where we note that we now have complex limits $-L+i\mu$, $L+i\mu$ for the complex variable $\zeta = \lambda + i\mu$ of the integral.

To arrive at (2.261) we let $\sigma = i\zeta$ in (2.264)

$$f(x) = \frac{1}{2\pi i} \lim_{L \to \infty} \int_{-\mu - iL}^{-\mu + iL} e^{\sigma x} F_L(\sigma) d\sigma \qquad (2.265)$$
$$(2.261)$$

where $-\mu$ is the real part of $\sigma = i\zeta = -\mu + i\lambda$, which can be designated as γ, $\sigma = \gamma + i\lambda$. This is the Laplace transform inversion formula (2.261) as we drop the subscript L (for the Laplace transform) in F_L. The work starting from around equation (2.262) to (2.265) to obtain (2.261) (or (2.265)) may stand as proof of a theorem establishing the inverse Laplace transform (2.261).

The integral in (2.265) is along a line to the right of any singularity of $F(\sigma)$, since, as we remarked for (2.256), the Fourier transform $F_F(\zeta) = F_L(i\zeta) = F(\sigma)$ is analytic for $\mu < \mu_1$ or, in other words, $\gamma = -\mu > -\mu_1$. For example, the Laplace transform of e^{2x} is $1/(\lambda - 2)$ and its analytic continuation to the complex plane is $1/(\sigma - 2)$, which is analytic for $\gamma > 2$ since $e^{2x}e^{\mu_1 x}$ is absolutely integrable for $\mu < \mu_1 = -2$ or $\gamma = -\mu > 2$. Of course, in this simple example this can be seen clearly since $\gamma > 2$ put us to the right of the only singularity at $\sigma_0 = 2$ of the function $F(\sigma) = 1/(\sigma - 2)$.

2.8.3 The Numerical Inversion of the Laplace Transform

In the above analysis for the derivation of the Laplace transform inversion formula (2.261) (or (2.265)) we relied on the symmetric inversion (2.257) of the Fourier transform. The analytic computation of (2.261) is done with the help of complex contour integration, and we leave its illustrations for the interested reader. It may be appropriate at this point to indicate how (2.261) can be evaluated numerically. The idea here will center again on relating to the Fourier transform, where the latter can be approximated via its efficient algorithm, the fast Fourier transform (FFT), which is discussed and illustrated with much detail in Chapter 4. In other words, we will backtrack our development above. To give a brief indication of the general process, we take $\sigma = \gamma + i\lambda$ at a fixed value of γ in (2.261) or (2.265), drop the infinite limits (as is the case in numerical approximations), and call the resulting approximate inverse $f_L(x)$,

$$f_L(x) = \frac{1}{2\pi i} \int_{\gamma - iL}^{\gamma + iL} e^{(\gamma + i\lambda)x} F_L(\gamma + i\lambda) d(\gamma + i\lambda), \qquad \sigma = \gamma + i\lambda$$

$$= \frac{e^{\gamma x}}{2\pi} \int_{-L}^{L} e^{i\lambda x} F_L(\gamma + i\lambda) d\lambda, \qquad F(s) = \mathcal{L}\{f\} \qquad (2.266)$$

where γ is more than the real part of any singularity of $F(\sigma)$. The above integral is a finite-limit (or bandlimited) Fourier integral which can be approximated via the FFT algorithm. However, as we shall see in Chapter 4, a number of errors are incurred in such approximations. With such errors being recognized, we must worry about their magnification by the above factor $e^{\gamma x}$ when γ is positive, keeping in mind that we are finding an approximation to the causal function $f(x)$ on the infinite interval $(0, \infty)$, where x can grow large! It turns out that such a magnification of the error is indeed a serious problem when $\gamma > 0$. Certain remedies for taming such a wild error are available, but they are beyond the scope of this text.

However, we must note that the sought inverse Laplace transform $f_L(x)$ in (2.266) is usually taken as a causal function, which has a jump discontinuity at $x = 0$ in the interior of its domain $(-\infty, \infty)$ as it relates to an inverse Fourier transform according to (2.266). Because of this jump discontinuity, the numerical approximation of its Fourier integral in (2.266) will suffer from a severe error near the discontinuity that may amount to about 9% of the size of the jump there, which is called the Gibbs phenomenon. This error will be discussed in detail in Section 4.1.6G, where it is very hard to get rid of short of changing the sought function by smoothing over its jump discontinuity. With such an error we should anticipate a very serious problem for functions associated with $\gamma > 0$, where the factor $e^{\gamma x}$ in (2.266) will represent a (formidable) magnification factor for this error even for x not very far from zero. The interested reader may find an illustration of this point in Fig. 4.35d, which represents the discrete Fourier transform approximation of the Fourier integral in (2.266) as an approximate representation of

$$h(x) = \begin{cases} e^{-x}, & 0 < x < \infty \\ 0, & x < 0 \end{cases}$$

The Fourier transform of this function is known to be $F(\lambda) = 1/(1 + i\lambda)$, which makes $F(\gamma + i\lambda) = 1/(1 + i\lambda)$ in (2.266) or $F(s) = 1/s$, which corresponds to $f(x) = 1$, and where γ is taken as one. Now it is left to multiply the graph in Fig. 4.35d by $e^{\gamma t} = e^t$ to see how much persisting error we have above the exact value of the constant $f(x) = 1$.

2.8.4 Applications

In the following we will have an example to illustrate the application of the Fourier transform with complex argument and the inversion formula (2.261) for the Laplace transform.

EXAMPLE 2.30: Differential Equations with Variable Coefficients

To illustrate the need for presenting the Fourier transform with complex argument, in addition to its relation to the Laplace transform, we will attempt to solve the following differential equation with variable coefficients, which is another form of Bessel differential equation:

$$x \frac{d^2u}{dx^2} + \frac{1}{2} \frac{du}{dx} - xu = 0 \tag{E.1}$$

We use the Fourier transform with complex argument as in (2.256)

$$U(\zeta) = \int_{-\infty}^{\infty} e^{-i\zeta x} u(x)\, dx \tag{E.2}$$

on (E.1), employing (2.258) twice and then (2.259) on the first term, (2.258) on the second, and (2.259) on the third term of (E.1) to obtain

$$i \frac{d}{d\zeta}[-\zeta^2 U(\zeta)] + \tfrac{1}{2}[i\zeta U(\zeta)] - i \frac{d}{d\zeta} U(\zeta) = 0$$

$$(1 + \zeta^2) \frac{dU}{d\zeta} + \tfrac{3}{2}\zeta U(\zeta) = 0 \tag{E.3}$$

This is a separable differential equation whose solution $U(\zeta)$ is a function of a complex variable,

$$U(\zeta) = C(1 + \zeta^2)^{-3/4} \tag{E.4}$$

which is a *multiple-valued* function. Its inverse Fourier transform (2.257) is

$$u(x) = \frac{C}{2\pi} \int_{-\infty}^{\infty} \frac{e^{i\zeta x}}{(1 + \zeta^2)^{3/4}} \, d\lambda \tag{E.5}$$

This integral is usually evaluated with the aid of complex contour integration. But since we are not assuming that the reader has an in-depth background in this subject, we shall consider such details as optional, and we refer the reader to Weinberger (1965) as a very clear, detailed, and self-contained reference on these computations.

EXAMPLE 2.31: A Laplace Transform Inverse

Here we will find the inverse Laplace transform of $1/s\sqrt{s+1}$. Instead of using (2.261) and the involved contour integration, we note that it can be done by using the convolution theorem (2.35) since $1/s = \mathcal{L}\{1\}$ and, according to (2.15) and (2.36),

$$\frac{1}{\sqrt{s+1}} = \mathcal{L}\left\{\frac{e^{-x}}{\sqrt{\pi x}}\right\} \tag{E.1}$$

Hence, according to the convolution theorem (2.35),

$$\mathcal{L}^{-1}\left\{\frac{1}{s\sqrt{s+1}}\right\} = \int_0^x 1 \frac{e^{-\tau}}{\sqrt{\pi\tau}} \, d\tau = \frac{2}{\sqrt{\pi}} \int_0^{\sqrt{x}} e^{-y^2} \, dy = \operatorname{erf}(\sqrt{x}) \tag{E.2}$$

after letting $\tau = y^2$ and using the definition of the error function in (2.39). The use of complex contour integration for finding this result, via the inverse Laplace transform formula (2.261), is left for the interest of the reader.

2.9 Other Important Integral Transforms

2.9.1 Hilbert Transform

The Hilbert transform of the function $f(x)$ defined on $(-\infty, \infty)$ is

$$F(\lambda) = \mathcal{H}\{f\} = \frac{1}{\pi} P \int_{-\infty}^{\infty} \frac{f(x)\,dx}{x-\lambda} \tag{2.267}$$

where P refers to the *Cauchy principal value* of the improper integral or

$$P \int_{-\infty}^{\infty} \equiv \fint_{-\infty}^{\infty} = \lim_{\epsilon \to 0+} \left[\int_{-\infty}^{\lambda-\epsilon} + \int_{\lambda+\epsilon}^{\infty}\right]$$

The inverse of the Hilbert transform is

$$f(x) = \mathcal{H}^{-1}\{F(\lambda)\} = \frac{1}{\pi} P \int_{-\infty}^{\infty} \frac{F(\lambda)}{x-\lambda} \, d\lambda \tag{2.268}$$

which can be derived with the help of the Fourier integral formula.

In Exercises 2.94 we will illustrate the use of the Hilbert transform for solving an integral equation which appears in the form (2.267).

2.9.2 Mellin Transform

The Mellin transform of the function $f(x)$ defined on $(0, \infty)$ is

$$F(\lambda) = \mathcal{M}\{f\} = \int_0^{\infty} x^{\lambda-1} f(x) \, dx \tag{2.269}$$

and its inverse is

$$f(x) = \mathcal{M}^{-1}\{F\} = \frac{1}{2\pi i} \lim_{L\to\infty} \int_{\gamma-iL}^{\gamma+iL} x^{-\lambda} F(\lambda)\, d\lambda \tag{2.270}$$

It is striking how the form of the inverse Mellin transform is similar to that of Laplace transform (2.261). This becomes clear when we show that the Mellin transform (2.269) reduces to the Laplace transform after substituting $x = e^{-t}$ in the integral (2.269):

$$\int_0^\infty (e^{-t})^{\lambda-1} f(e^{-t}) e^{-t}(-dt) = -\int_0^\infty e^{-\lambda t} f(e^{-t})\, dt$$

The Mellin transform is compatible with the differential operator $x^n\, d^n f/dx^n$,

$$\mathcal{M}\left\{ x^n \frac{d^n f}{dx^n} \right\} = (-1)^n \lambda(\lambda+1)\cdots(\lambda+n-1)F(\lambda) \tag{2.271}$$

which can be proved by n times integration by parts. Also, we should note that the variable-coefficient differential operator $x^n\, d^n f/dx^n$, which is found in Cauchy–Euler's equation

$$a_0 x^n \frac{d^n y}{dx^n} + a_1 x^{n-1}\frac{d^{n-1}y}{dx^{n-1}} + \cdots + a_{n-1}x\frac{dy}{dx} + a_n y = F(x) \tag{2.272}$$

can be reduced easily to a constant-coefficient one by letting $x = e^t$,

$$x^n \frac{d^n f}{dx^n}(x) = (e^t)^n \frac{d^n f}{(e^t)^n dt^n}(e^t) = \frac{d^n f(e^t)}{dt^n}$$

This explains again the origin of the relation between the operational properties of the Mellin and Laplace transforms.

The convolution theorem for the Mellin transform is

$$\mathcal{M}\left\{ \int_0^\infty f_1(\xi) f_2\left(\frac{x}{\xi}\right) \frac{d\xi}{\xi} \right\} = F_1(\lambda) F_2(\lambda) \tag{2.273}$$

or (with such particular convolution product $f_1 * f_2$)

$$\mathcal{M}\{f_1 * f_2\}(x) = F_1(\lambda) F_2(\lambda)$$

Applications of (2.273) are found in algebraizing integral equations which appear in the form of this particular (Mellin transform) convolution product.

2.9.3 The z-Transform and the Laplace Transform

In this chapter we presented the Fourier integral transform, and in Chapter 3 we will present their finite transform versions as the Fourier exponential or sine-cosine series expansion. In Chapter 4 we will present the discrete Fourier transforms (DFT) in (4.3), (4.4), defined on a finite set of points N, which are only an approximation to the Fourier integral transforms or the finite transforms. However, in their own unique discrete setting, we will attempt using the discrete Fourier transforms in Section 4.1.5 to solve boundary value problems, defined on finite discrete points, which are associated with *difference equations* instead of the differential equations of this or the next chapter.

In regard to problems defined on a discrete domain, there are the discretized, or the naturally discrete, initial value problems in $\{u_n\}_{n=0}^{\infty}$ defined on the infinite sequence $\{0, 1, 2, 3, \ldots\}$, i.e., $n \geq 0$. This motivates the possiblity of a discrete analog to the Laplace transform that would be compatible with such discrete initial value problems. Indeed, this is the case with the following z-transform of the sequence $\{u_n\}_{n=0}^{\infty}$:

$$Z\{u_n\} = U(z) = \sum_{n=0}^{\infty} u_n z^{-n}, \qquad z \in C \tag{2.274}$$

where C stands for the set of complex numebrs. We will denote sequences by lowercase letters and their z-transforms by corresponding uppercase letters. The Z notation denotes the operation of applying the z-transform. The transform variable z is now a complex variable which may take any value in the complex plane. In this regard, the z-transform is different and not "as discrete" as the discrete Fourier transforms, in which both the original variable (which we called n) and the transform variable (which we called k) are integers. The fact that the z-transform takes us into the complex plane results in the fact that the inversion of the z-transform (that is, given $U(z)$, find $\{u_n\}$) becomes rather complicated and requires the techniques of contour integration. Nevertheless, by building a table of z-transforms of various sequences and by collecting properties of the z-transform, it will be possible to do the inverse transform by a lookup procedure. Toward this end, we will do an example and leave the few needed pairs for Section 4.1.5 as exercises (see Briggs and Jerri (1985) and the z-transform table in Appendix C.17).

EXAMPLE 2.32:

Find the z-transform of the step sequence

$$u_n = \begin{cases} 0, & n < 0 \\ 1, & n \geq 0 \end{cases} \tag{E.1}$$

By the definition (2.274) we have

$$Z\{u_n\} = \sum_{n=0}^{\infty} u_n z^{-n} = \sum_{n=0}^{\infty} z^{-n} = \sum_{n=0}^{\infty} \left(\frac{1}{z}\right)^n$$

At this point, we appeal to a result which will be used often in all that follows. It is really the extension of the convergence theorem for the geometric series to the complex plane. If $g(z)$ is an expression involving a complex variable z, then the infinite series

$$P(z) = \sum_{n=0}^{\infty} (g(z))^n$$

converges for all z for which $|g(z)| < 1$, and it converges to

$$P(z) = \frac{1}{1 - g(z)}$$

(Compare this to the result for a real number α that $\sum_{n=0}^{\infty} \alpha^n = 1/(1-\alpha)$ for $|\alpha| < 1$.) In the present example, $g(z) = 1/z$. Hence,

$$Z\{u_n\} = \frac{1}{1 - 1/z} = \frac{z}{z-1} \quad \text{for } |z| > 1 \tag{E.2}$$

The following z-transform pairs will be needed for our simple illustrations in Section 4.1.5, and we leave them for the exercises:

$$u_n = \delta_{n,0} = \begin{cases} 1, & n = 0 \\ 0, & n > 1 \end{cases} \qquad U(z) = 1 \tag{2.275}$$

$$u_n = \begin{cases} 0, & n = 0 \\ a^{n-1}, & n \geq 1 \end{cases}, \ a \in C, \quad U(z) = \frac{1}{z-a}, \ |z| > |a| \tag{2.276}$$

$$u_n = \begin{cases} 0 & \text{if } n = 0 \text{ or } n \text{ is odd} \\ (-1)^{(n/2)+1}a^{n-2} & \text{if } n \text{ is even} \end{cases}$$

$$U(z) = \frac{1}{z^2 + a^2}, \ |z| > |a| \tag{2.277}$$

$$u_n = \begin{cases} 0, & n \text{ even} \\ (-1)^{(n-1)/2}a^{n-1}, & n \text{ odd} \end{cases} \quad U(z) = \frac{z}{1+z^2}, \ |z| > |a| \tag{2.278}$$

$$u_n = \sin n\theta, \qquad U(z) = \frac{\sin \theta}{z - 2\cos \theta + z^{-1}}, \ |z| > 1 \tag{2.278}$$

$$u_n = \cos n\theta, \qquad U(z) = \frac{z - \cos \theta}{z - 2\cos \theta + z^{-1}}, \ |z| > 1 \tag{2.279}$$

As expected, the important property of the z-transform for solving difference equations with initial conditions is that of algebraizing the difference equation and at the same time involving the given initial conditions—in other words, the compatibility of the z-transform (2.274) with initial value problems associated with difference equations. Such important z-transform operational pairs are derived in Example 4.5, then used for solving an initial value problem and another describing a problem in combinatorics (Example 4.6).

Relevant References for Chapter 2

The basic references used in this chapter are Ditkin and Prudnikov (1965), Weinberger (1965), Powers (1987), Budak and Fomin (1973), Goldberg (1961), Churchill (1972), Sneddon (1972), Davies (1985), Papoulis (1977, 1968), Jury (1964), Jerri (1985, 1977a) and Briggs and Jerri (1990). Recent books on Fourier analysis are those of Walker (1988) and Körner (1988). For a tutorial review of recent attempts using the modified iterative method for solving nonlinear problems, see our paper (1991). For the most recent general reference on the sampling theorems, see Marks (1991, 1992) Volumes I and II, and in particular, our Chapter 7 in Volume II on the generalization and error analysis of the sampling theorem (1992).

Tables of the transforms: The Laplace transforms may be found in Roberts and Kaufman (1966), which are organized according to the transformed functions as well as their inverses. The Laplace and Fourier transforms are in Erdelyi et al. (1954a), Ditkin and Prudnikov (1965), and Abramowitz and Stegun (1964). The Hankel transforms are in Erdelyi et al. (1954b) and Ditkin and Prudnikov (1965), and other transforms may be found in Erdelyi et al. (1954b). More tables of transforms may be found in the bibliography, in particular Oberhettinger (1971, 1973a,b) and Doetsch (1950, 1951).

Exercises

Section 2.1   Laplace Transforms

2.1. Verify that all the functions in Fig. 2.1a–d and Fig. 2.2 satisfy the conditions of Theorem 2.1, of being sectionally continuous and of exponential order, for their Laplace transform to exist.

2.2. (a) Classify the following functions according to whether they are continuous, piecewise continuous, piecewise smooth, or smooth on the indicated interval.

  (i) $f(x) = x^2 \sin 3x, \; -\pi < x < \pi$

  (ii) $f(x) = 3|x|^{1/2} + x^2, \; -3 < x < 5$

  (iii) $f(x) = \begin{cases} 2, & 0 < x < b \\ -2, & -b < x < 0 \end{cases}, f(x + 2b) = f(x)$

  (iv) $f(x) = \begin{cases} 3|x|^{1/2}, & -1 < x < 1 \\ 4, & 1 \le x < 5 \end{cases}$

  (v) $f(x) = x^2 p_a(x)$, where $p_a(x)$ is the gate function of Fig. 2.2d as defined in (1.31).

  (vi) $f(x) = 1/\sqrt{x}, \; 0 < x < \infty$

  (vii) $f(x) = (\cos x)/x^2, \; 0 \le x < \infty$

(b) Determine whether or not the given function $f(x)$ is of exponential order (as $x \to \infty$), that is, $|f(x)| \le Me^{\alpha x}$ for $x > K$, find suitable M and K, and give the value of α.

  (i) $f(x) = x^2 \sin 3x$

  (ii) $f(x) = \sin(3e^{x^2})$

  (iii) $f(x) = |x| + \sinh 3x$

  (iv) $f(x) = x^3 e^{-2x} \cos x$

  (v) $f(x) = xe^{x^2} \cos(3e^{x^2})$

  (vi) $f(x) = e^{-x^3} \cos 2x$

ANS. (a) (i) Continuous (ii) Continuous (iii) Piecewise smooth (iv) Piecewise continuous (v) Piecewise continuous (vi) Not piecewise continuous (vii) Not piecewise continuous

  (b) (i) $\alpha > 0$ (ii) $\alpha > 0$ (iii) $\alpha > 3$ (iv) $\alpha > -2$ (v) Not of exponential order (vi) $\alpha > 0$.

2.3. (a) Show that $f(x) = |x|^{1/3}$ is continuous but not piecewise smooth on $(-2, 2)$.

(b) Show whether or not $f(x) = (x^2 - 2x)/(x - 2)$ is piecewise continuous on $(-\infty, \infty)$. Find the point (or points) of discontinuity if any!

ANS.   (b) $x = 2$

2.4. (a) Use Theorem 2.1 to test for the sure existence of the Laplace transform for each of the following functions.

   (i) $f(x) = \begin{cases} 3|x|^{1/2}, & -1 < x < 1 \\ 4, & 1 \le x < \infty \end{cases}$

   (ii) $f(x) = (\cos x)/x^2,\ 0 \le x < \infty$

   (iii) $f(x) = 1/\sqrt{x} + \sin(3e^{x^2}),\ 0 \le x < \infty$

   Hint: See part (b) of this exercise.

   (iv) $f(x) = x^2 \sin 3x + |x|^{1/3},\ 0 \le x < \infty$ (See Exercise 2.3a.)

(b) Consider the two functions $f(x) = x^{-1/2},\ 0 < x < \infty$, and $g(x) = x^{-3},\ 0 < x < \infty$. Show that both functions don't satisfy the sufficient conditions of Theorem 2.1 for the existence of their Laplace transforms.

(c) See which one of the two functions in part (b) may still have a Laplace transform.

   Hint: Consult their direct Laplace transform evaluation via the gamma function in (2.15), where it is valid for $f(x) = x^{\nu} = x^{-1/2},\ \nu = -1/2 > -1$. See also Exercise 6b.

ANS.   (a)   (i), (iii) satisfy Theorem 2.1.

   (c) $\mathcal{L}\{x^{-1/2}\} = \sqrt{\pi/s},\ s > 0$, is obtained via the gamma function value in (2.15) which is valid for $\nu = -\frac{1}{2} > -1$ of $f(x) = x^{-1/2}$, but it diverges in the case of $f(x) = x^{-3}$ with $\nu = -3$. See the gamma function figure in Exercise 6d.

2.5. (a) Prove the basic (linear) properties of the Laplace transform in (2.7) and (2.8),

$$\mathcal{L}\{kf(x)\} = kF(s), \qquad k \text{ constant} \tag{2.7}$$

$$\mathcal{L}\{c_1 f_1(x) + c_2 f_2(x)\} = c_1 F_1(s) + c_2 F_2(x),$$

$$c_1, c_2 \text{ constants} \tag{2.8}$$

Hint: Apply the basic (linear) property of integration. Also note that the result in (2.7) is only a special case of (2.8).

(b) Let $f_1(x) = 1$ and $f_2(x) = \sin x$, find their convolution product as $(f_1 * f_2)(x)$, then as $(f_2 * f_1)(x)$ to verify that $(f_1 * f_2)(x) = (f_2 * f_1)(x)$.

2.6. (a) Prove (2.17).

Hint: Use integration by parts for the integral in (2.16), letting $u = x^\nu$ and $dv = e^{-x}dx$.

(b) Use the relation (2.18)

$$\Gamma(\nu)\Gamma(1 - \nu) = \frac{\pi}{\sin \pi \nu}, \qquad 0 < \nu < 1 \tag{E.1}$$

to show that $\Gamma(1/2) = \sqrt{\pi}$; then show that the Laplace transform of $f(x) = x^{-1/2}$, $0 < x < \infty$, does exist as $F(s) = \sqrt{\pi/s}$, $s > 0$.

Hint: See (2.15) and Exercise 1.8c in Chapter 1. Also you may see Exercise 2.58h for a proof of the important result in (E.1).

(c) Use the known integral $\int_{-\infty}^{\infty} e^{-y^2}\, dy = \sqrt{\pi}$ to evaluate $\Gamma(1/2)$.

Hint: Let $x = y^2$.

(d) Use the definition (2.16) for $\Gamma(\nu)$, the gamma function, and its basic properties in (2.17), (2.18) to verify the main features of the graph of $\Gamma(\nu)$ in Fig. 2.15. In particular, note how $\Gamma(\nu)$

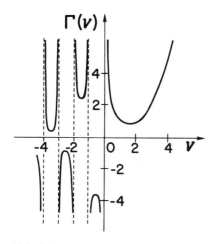

FIG. 2.15 Gamma function $\Gamma(\nu)$.

diverges as ν approaches the nonpositive integers $\nu = 0, -1, -2, -3, \ldots$. You also have $\Gamma(1/2) = \sqrt{\pi}$ and $\Gamma(n + 1) = n!$, $n = 0, 1, 2, 3, \ldots$.

2.7. (a) Prove the following scaling property of the Laplace transform:

$$\mathcal{L}\{f(ax)\} = \frac{1}{a} F\left(\frac{s}{a}\right), \quad \text{where } F(s) = \mathcal{L}\{f\} \tag{E.1}$$

Hint: Make the change of variable $y = ax$ in the resulting Laplace integral.

(b) Prove the *shifting property* of the Laplace transform

$$\mathcal{L}\{e^{-ax}f(x)\} = F(s + a), \quad s + a > \gamma \tag{E.2}$$

where γ is such that $f(x)$ grows at most like $e^{\gamma x}$ as $x \to \infty$.

(c) Use the shifting property in (b) to evaluate the following Laplace transform pair:

$$\mathcal{L}\{e^{-ax} \cos bx\} = \frac{s + a}{(s + a)^2 + b^2}$$

Hint: See the Laplace transform of $\cos at$ in (2.14).

(d) Prove the following complementary result to the shifting property in (E.2); that is, for $f(x - a)$ which vanishes identically for $0 < x < a$, we have for the shifting of the original causal function $f(x)$

$$\mathcal{L}\left\{g(x) = \begin{cases} 0, & 0 < x < a \\ f(x - a), & x > a \end{cases}\right\} = e^{-as} F(s) \tag{E.3}$$

Hint: Use $g(x)$ in the Laplace integral (2.1), noting that the limits of the integral become a to ∞; thus make the change of variable $y = x - a$ to have the limits of the Laplace transform as $y = 0$ to ∞.

(e) Use (E.3) above to find the Laplace transform of the *shifted* causal sine wave in Fig. 2.1c,

$$h(t) = \begin{cases} 0, & t < \dfrac{\pi}{4} \\ \sin\left(t - \dfrac{\pi}{4}\right), & t > \dfrac{\pi}{4} \end{cases}$$

Hint: Use (E.3) with (2.13).

(f) Find the Laplace transform of the *delayed* causal sine wave of Fig. 2.1b,

$$g(t) = \begin{cases} 0, & t \le \dfrac{\pi}{3} \\ \\ \sin t, & t > \dfrac{\pi}{3} \end{cases}$$

Hint: Note that (E.3) is not so useful in this *delayed* causal wave compared to the above *shifted* causal wave of part (d); why?

(g) Find the Laplace transform of the shifted causal cosine wave

$$f(t) = \begin{cases} 0, & t < \dfrac{\pi}{4} \\ \\ \cos\left(t - \dfrac{\pi}{4}\right), & t > \dfrac{\pi}{4} \end{cases}$$

Hint: Use (E.3) with (2.14).

2.8. (a) For Table 2.1 of the Laplace transform pairs prove the following Laplace transfer (operational) pair:

$$\mathcal{L}\left\{ \int_0^x f(\xi)\,d\xi \right\} = \frac{1}{s} F(s) \tag{E.1}$$

Hint: Let $g(x) = \int_0^x f(\xi)\,d\xi$, where clearly $g(0) = 0$, and $dg/dx = f(x)$. Then use $\mathcal{L}\{g(x)\} = G(s)$ with $\mathcal{L}\{f(x)\} = F(s) = \mathcal{L}\{dg/dx\} = sG(s) - f(0) = sG(s)$. This result (E.1) can also be proved by using the Laplace convolution Theorem 2.11 in (2.35) with $f_1(x) = 1$ and $f_2(x) = f(x)$, noting that $f_1(x - \xi)$ of the convolution integral is the constant 1, $f_1(x - \xi) = 1$.

(b) Use the result in (E.1) to find the Laplace transform of $f(t) = \sin at$ from the knowledge of the Laplace transform of $\cos at$:

$$\mathcal{L}\{\cos at\} = \frac{s}{s^2 + a^2}$$

Hint: $\sin at = a \int_0^t \cos at\, dt$

(c) Use the result in (E.1) and knowing that $\mathcal{L}\{1/x^{1/2}\} = \sqrt{\pi/s}$ to find $\mathcal{L}\{x^{1/2}\}$. Check your answer by resorting to (2.15) via the gamma function.

ANS. (c) $F(s) = \frac{1}{2}\sqrt{\pi}/s^{3/2}$

2.9. *The Laplace transform of periodic functions.* Let $f(t)$ be periodic with period T and piecewise continuous on the interval T. Show that

$$\mathcal{L}\{f(t)\} = \frac{\int_0^T e^{-st}f(t)\,dt}{1 - e^{-sT}} \tag{E.1}$$

Hint: Note that for this periodic function $f(t)$,

$$\mathcal{L}\{f(t)\} = \int_0^\infty e^{-st}f(t)\,dt = \sum_{n=0}^\infty \int_{nT}^{(n+1)T} e^{-st}f(t)\,dt \tag{E.2}$$

where $\int_T^{2T} e^{-st}f(t)\,dt = e^{-sT}\int_0^T e^{-sx}f(x)\,dx$ after the change of variable $t = T + x$. Also watch for the resulting infinite *geometric* series

$$\sum_{n=0}^\infty e^{-nsT} = \frac{1}{1 - e^{-sT}} \tag{E.3}$$

2.10. (a) Prove that for the Laplace transform $F(s)$ to exist, it is necessary that it vanishes as s approaches infinity; that is, $\lim_{s\to\infty} F(s) = 0$.

Hint: Take the limit inside the integral of (2.1), after consulting a theorem like the Lebesgue convergence Theorem 2.10 for allowing the interchange of the limit process with the integral.

(b) Verify the result in (a) with all the Laplace transforms you had up to now, as in (2.11)–(2.15), as well as the entries in part (a) of Table 2.1 of the Laplace transform pairs.

(c) Prove that

$$\lim_{t\to 0} f(t) = \lim_{s\to\infty} sF(s)$$

Hint: Take the Laplace transform of $g(t) = f'(t)$, which is $G(s) = sF(s) - f(0+)$, and use the result of (a) for $\lim_{s\to\infty} G(s) = 0$.

2.11. Consider the following infinite series representation for the Bessel function $J_n(x)$:

$$J_n(x) = \sum_{k=0}^\infty \frac{(-1)^k}{k!(k+n)!}\left(\frac{x}{2}\right)^{2k+n} \tag{E.1}$$

For more details of the Bessel functions, see Appendix A.5 and its exercises.

(a)   Use this series to show that $J_0(0) = 1$, $J_1(0) = 0$, and

$$\frac{dJ_0(x)}{dx} = -J_1(x) \tag{E.2}$$

(b)   Use the integral representation of the Bessel function of order zero,

$$J_0(x) = \frac{1}{\pi} \int_0^\pi \cos(x \sin t)\, dt \tag{E.3}$$

to show that

$$\mathcal{L}\{J_0(t)\} = \frac{1}{\sqrt{1 + s^2}} \tag{E.4}$$

Hint: Use (2.1), interchange the two integrals, then use (2.13) for the inner Laplace transform of the sine function. The resulting single integral can still be simplified to give the answer in (E.4). You may also need

$$\int_0^\infty \frac{d\mu}{u^2 + \alpha^2} = \frac{\pi}{2\alpha} \tag{E.5}$$

(c)   Use the Bessel differential equation

$$t^2 J_0'' + t J_0' + t^2 J_0 = 0 \tag{E.6}$$

to find the Laplace transform of $J_0(t)$ as in (E.4). *Hint:* See Example 2.9, and we suggest here finding the arbitrary constant of integration by another method. Divide the Bessel differential equation (E.6) by t; Laplace transform it using (2.19), (2.2), and (2.21); and let $Y(s) = \mathcal{L}\{J_0(t)\}$, noting that $J_0(0) = 1$, $J_0'(0) = -J_1(0) = 0$ from part (a) above or (A.4.15) in Appendix A.4. After solving the resulting first-order differential equation in $Y(s)$, you need to find the arbitrary constant by using $1 = \lim_{t \to 0} J_0(t) = \lim_{s \to \infty} s Y(s)$ property proved in Exercise (2.10c).

(d)   Show that

$$\mathcal{L}\{J_1(t)\} = 1 - \frac{s}{\sqrt{1 + s^2}} \tag{E.7}$$

Hint: Use the result in (E.2) with (2.21) and the Laplace transform of $J_0(t)$ in (E.4).

2.12. (a) Show that

$$\mathcal{L}\{\text{erf}(\sqrt{t})\} = \frac{1}{s\sqrt{s+1}} \tag{E.1}$$

Hint: For $\text{erf}(\sqrt{t}) = (2/\sqrt{\pi})\int_0^{\sqrt{t}} e^{-u^2} du$, write the Taylor series for e^{-u^2}, integrate, Laplace transform, then recognize the resulting series as the binomial expansion of $(1 + 1/s)^{-1/2}$. Another way is to interchange the inner integral of $\text{erf}(\sqrt{t})$ with the Laplace integral, *and be sure to cover the same two-dimensional domain* $0 \le u \le \sqrt{t},\ 0 \le t < \infty$.

2.13. Show that

$$\mathcal{L}\left\{\int_0^t \frac{\sin u}{u}\, du\right\} = \frac{1}{s}\tan^{-1}\frac{1}{s} \tag{E.1}$$

Hint: For $f(t) = \int_0^t ((\sin u)/u)\, du$, we have $tf'(t) = \sin t$ by the fundamental theorem of calculus. Now Laplace transform this $tf'(t) = \sin t$, using (2.2) and (2.21), to have

$$-\frac{d}{ds}[sF(s) - f(0)] = \frac{1}{1+s^2} \tag{E.2}$$

and for the constant of integration of equation (E.2), use $0 = \lim_{t\to 0} f(t) = \lim_{s\to\infty} sF(s)$, the property derived in Exercise 2.10c.

2.14. (a) Use the Euler identity

$$e^{\mp ibt} = \cos bt \mp i\sin bt \tag{E.1}$$

without resorting to integration by parts, to find the Laplace transform of the following functions:

(i) $e^{-at}\cos bt$

(ii) $e^{-at}\sin bt$

(b) Use the definite integral

$$\int_0^\infty e^{-zt}\, dt = \frac{1}{z}, \qquad \text{Re}\, z > 0,\ z = x + iy \tag{E.2}$$

only to evaluate the Laplace transform of the functions in part (a).

Hint: For the complex exponential function inside the integral $e^{-zt} = e^{-(x+iy)t} = e^{-xt}e^{-iyt}$, use the Euler identity for e^{-iyt}, and for the right side $1/z$ of (E.2) write $1/z = \bar{z}/\bar{z}z = (x - iy)/(x^2 +$

y^2). Then equate the real and imaginary parts of the two sides of (E.2).

(c) Check your answer in part (a) or (b) by another method.

Hint: Apply the shifting property (2.36) to (2.13) and (2.14).

ANS. (a) or (b) (i) $\dfrac{s+a}{(s+a)^2+b^2}$ (ii) $\dfrac{b}{(s+a)^2+b^2}$

2.15. (a) Find the inverse Laplace transform of

(i) $F(s) = \dfrac{1}{s^2+9} + \dfrac{1}{s(s+1)}$.

(ii) $F(s) = \dfrac{1}{s^2(s+1)}$. *Hint*: Write the partial fraction:

$$\frac{1}{s^2(s+1)} = \frac{1}{s^2} - \frac{1}{s} + \frac{1}{s+1}$$

(iii) $F(s) = \dfrac{1}{(s-1)^{5/2}}$

(b) Use the convolution theorem (2.35) to find $\mathcal{L}^{-1}\{1/s(s+1)\}$.

ANS. (a) (i) $\tfrac{1}{3}\sin 3s + 1 - e^{-x}$

(ii) $x - 1 + e^{-x}$

(iii) $(1/\Gamma\,(\tfrac{5}{2}))x^{3/2}e^x = (4/3\sqrt{\pi})x^{3/2}e^x$

(b) $1 - e^{-x}$

2.16. (a) Determine whether or not the following problem is compatible with the Laplace transform method. If yes, then solve it.

$$\frac{d^2y}{dt^2} - 5\frac{dy}{dt} + 4y = -8e^{2t}, \qquad 0 < t < \infty \tag{E.1}$$

$$y(0) = 4 \tag{E.2}$$

$$y'(0) = 8 \tag{E.3}$$

(b) Verify that the answer in (a) satisfies the initial value problem (E.1)–(E.3).

ANS. (a)   Yes, since it is a second-order differential equation on $(0, \infty)$ where the two initial conditions $y(0)$ and $y'(0)$ are given for (2.19), and it is also with constant coefficients, $y(t) = 4e^{2t}$.

2.17. Use the Laplace transform to solve for $x(t)$ and $y(t)$ in the system of two differential equations and initial conditions (see Example 2.8).

(a) $\dfrac{dx}{dt} = 4x - 2y, \qquad 0 < t < \infty$

$\dfrac{dy}{dt} = 5x + 2y, \qquad 0 < t < \infty$

$x(0) = 2, \ y(0) = -2$

Hint: Let $\mathcal{L}\{x(t)\} = X(s)$, $\mathcal{L}\{y(t)\} = Y(s)$, Laplace transform (E.1) and (E.2), and solve for $X(s)$ and $Y(s)$ in the two resulting simultaneous (algebraic) equations.

(b) $\dfrac{d^2x}{dt^2} = y + \sin t, \qquad 0 < t < \infty$

$\dfrac{d^2y}{dt^2} = -\dfrac{dx}{dt} + \cos t, \qquad 0 < t < \infty$

$x(0) = 1, \qquad x'(0) = 0$

$y(0) = -1, \qquad y'(0) = -1$

(c) $\dfrac{dy}{dt} - x = te^t, \qquad t > 0$ \hfill (E.1)

$\dfrac{dx}{dt} + y = (t + 1)e^t, \qquad t > 0$ \hfill (E.2)

$x(0) = 1$ \hfill (E.3)

$y(0) = 0$ \hfill (E.4)

ANS. (a) $x(t) = 2e^{3t}[\sin 3t + \cos 3t]$

$y(t) = -2e^{3t}[\cos 3t - 2\sin 3t]$

(b) $x(t) = \cos t, \ y(t) = -\cos t - \sin t$

(c) $y(t) = te^t, \ x(t) = e^t$

2.18. Consider the problem of Example 1.1 in Chapter 1:

$$\frac{\partial u}{\partial t} = \frac{\partial^2 u}{\partial x^2}, \qquad 0 < x < \infty, \; t > 0 \tag{E.1}$$

$$u(0,t) = 0, \qquad t > 0 \tag{E.2}$$

$$u(x,0) = f(x), \qquad 0 < x < \infty \tag{E.3}$$

(a) Determine the part of the problem that is compatible with the Laplace transform. Then Laplace transform it.

(b) Attempt to solve the problem using the Laplace transform to determine the feasibility of this method. Consult Example 2.10.

Hint: You may need to consult Appendix A.1 to solve the resulting nonhomogeneous differential equation in $U(x,s) = \mathcal{L}\{u(x,t)\}$ of part (a). Then consult the Laplace tables to find the inverse Laplace transform of $U(x,s)$.

ANS. (a) $sU(x,s) - f(x) = \dfrac{d^2U(x,s)}{dx^2}, \qquad U(0,s) = 0$

(b) See the discussion in Example 2.10 regarding the more feasible Fourier sine transforms of Section 2.5.

2.19. Find the Laplace transform of the following functions, or expressions in $u(x)$.

(a) $x^{1/2}, \; 0 < x < \infty$

(b) $x - \displaystyle\int_0^x (x-t)^2 u(t)\,dt$

(c) $\displaystyle\int_0^x e^{2(x-t)} \frac{du}{dt}\,dt, \; u(0) = 0$

ANS. (a) $\dfrac{\sqrt{\pi}}{2s^{3/2}}$ (b) $\dfrac{1}{s^2} - \dfrac{2}{s^3} U(s)$ (c) $\dfrac{s}{s-2} U(s)$

2.20. Find the inverse Laplace transform $f(x) = \mathcal{L}^{-1}\{F(s)\}$ of the following functions.

(a) $\dfrac{1}{s - (\lambda + 1)}$

(b) $\dfrac{1}{(s-3)^2+5}$

(c) $\dfrac{1}{(s-3)^2}+\dfrac{1}{s-2}+\dfrac{1}{s}$

(d) $\dfrac{s}{(s^2+1)^2}$

(e) $\dfrac{G(s)}{s-(\lambda+1)}$, $G(s)=\mathcal{L}\{g(x)\}$

(f) $\dfrac{F(s)}{\sqrt{s}}$. *Hint*: Use the convolution theorem.

(g) $\sqrt{s}F(s)$

   Hint: Write $\sqrt{s}F(s)=s(F(s)/\sqrt{s})$; then use (2.2) and the result in (f).

(h) We know that the Laplace transform $sF(s)$ in the Laplace transform space corresponds to the derivative df/dx as in (2.2), while $(1/s)F(s)$ corresponds to the integral $\int_0^x f(x)\,dx$ as in Exercise 2.8a. What would you make of the inverse Laplace transform of $s^{1/2}F(s)$ in part (g) and $(1/s^{1/2})F(s)$ of part (f) above?

ANS. (a) $e^{(\lambda+1)x}$

   (b) $e^{3x}\dfrac{\sin\sqrt{5}x}{\sqrt{5}}$

   (c) $xe^{3x}+e^{2x}+1$

   (d) $\tfrac{1}{2}x\sin x$

   (e) $\displaystyle\int_0^x e^{(\lambda+1)(x-t)}g(t)\,dt$

   (f) $\dfrac{1}{\sqrt{\pi}}\displaystyle\int_0^x \dfrac{f(t)}{\sqrt{x-t}}\,dt$

   (g) $\dfrac{d}{dx}\left[\dfrac{1}{\sqrt{\pi}}\displaystyle\int_0^x \dfrac{f(t)}{\sqrt{x-t}}\,dt\right]$

   (h) The answer in part (g) would represent a *fractional derivative* (of $f(x)$) of order $\tfrac{1}{2}$, while the answer in part (f) represents a *fractional integral* (of $f(x)$) of order $\frac{1}{2}$. (See Exercise 1.8d.)

2.21. (a) Find the Laplace transform of the following Abel's integral equation:

$$1 = \int_0^x \frac{1}{\sqrt{x - \lambda}} u(\lambda) d\lambda$$

(b) Solve for $u(x)$ and verify your answer.

ANS. (a) $1/s = \sqrt{\pi/s}\, U(s)$, $U(s) = \mathcal{L}\{u\}$

(b) $u(x) = 1/\pi\sqrt{x}$

2.22. (a) Find the representation of the following Volterra integral equation in the Laplace transform space:

$$u(x) = x^2 \int_0^x (x - t)u(t)dt$$

(b) Use the Laplace transform to solve the Volterra integral equation

$$u(x) = g(x) + \lambda \int_0^x u(t)dt$$

ANS. (a) $U(s) = (-1)^2 \dfrac{d^2}{ds^2}\left[\dfrac{U(s)}{s^2}\right]$

(b) $U(s) = sG(s)/(s - \lambda)$; write it as

$$U(s) = \frac{s - \lambda + \lambda}{s - \lambda} G(s) = G(s) + \frac{\lambda}{s - \lambda} G(s)$$

Then

$$u(x) = g(x) + \lambda \int_0^x e^{\lambda(x-t)} g(t)dt$$

2.23. Use the Laplace transform method to solve the following *integrodifferential equation*:

$$\frac{d^2u}{dx^2} - 2\frac{du}{dx} + u(x)$$

$$= \cos x - 2\int_0^x \cos(x - t)\frac{d^2u}{dt^2}dt - 2\int_0^x \sin(x - t)\frac{du}{dt}dt$$

$$u(0) = 0, \qquad u'(0) = 0$$

ANS. $U(s) = s/(1 + s^2)^2$, $u(x) = \frac{1}{2}x \sin x$

2.24. Find the Laplace transform of

$$\frac{e^{-a^2/4t}}{\sqrt{\pi t}}, \qquad a > 0$$

Hint: Use $\int_0^\infty e^{(-\alpha^2 x^2 - \beta^2/x^2)} dx = \sqrt{\pi}/2\alpha e^{-2\alpha\beta}$.

ANS. $F(s) = e^{-a\sqrt{s}}/\sqrt{s}$

2.25. Solve the general Abel problem in (2.62).

Hint: Watch for the $s^{1-\alpha}F(s)$ in the Laplace transform of (2.62), where we cannot use an inverse for $s^{1-\alpha}$ since $1 - \alpha > 0$ for the given condition $0 < \alpha < 1$. Instead, write it as

$$s^{1-\alpha}F(s) = s\left[\frac{1}{s^\alpha}F(s)\right] \qquad (E.1)$$

and use the convolution theorem for $(1/s^\alpha)F(s)$ with $F_1(s) = 1/s^\alpha$, $F_2(s) = F(s)$; then use the operation pair (2.2) for the effect of the s factor outside the brackets in (E.1).

ANS. $U(s) = (s^{1-\alpha}/\Gamma(1 - \alpha))F(s)$, $0 < \alpha < 1$. For the solution $u(x)$, see (2.63).

2.26. Consider the temperature distribution $u(r,t)$ in a very long cylinder (with diffusivity k) of radius a with zero initial temperature when the surface is kept at a constant temperature u_0.

(a) Set up the initial boundary value problem.

(b) Use the Laplace transform to reduce the partial differential equation in (a) to an ordinary differential equation in $U(r,s)$ and solve it.

(c) Use the following Laplace transform pair to find the inverse Laplace transform of the solution $U(r,s)$ in part (b). Then find the solution to the initial and boundary value problem in part (a).

$$\mathcal{L}\left\{1 - \frac{2}{a}\sum_{n=1}^{\infty}\frac{e^{-k\alpha_n^2 t}J_0(r\alpha_n)}{\alpha_n J_1(a\alpha_n)}\right\} = \frac{I_0(r\sqrt{s/k})}{sI_0(a\sqrt{s/k})} \qquad (E.1)$$

where $\{a\alpha_n\}$ are the zeros of $J_0(a\alpha) = 0$ and $I_0(x) \equiv J_0(ix)$, $i = \sqrt{-1}$, the modified Bessel function of order zero (see Appendix A.5 and its exercises).

(d) Use the result $u(r,t)$ in part (c) to verify the initial condition $u(r,0) = 0$ and the boundary condition $u(a,t) = u_0$. Also verify that $u(r,t)$ satisfies the heat equation given in (E.4) of the answer to part (a).

Hint: For the boundary condition $u(a,t) = u_0$, consult the Bessel orthogonal series for $f(r) = 1$ on $(0,a)$. (See (3.100), (3.99) in Section 3.3 and Example 3.14, in particular (E.8) and (E.10).) For satisfying the heat equation, remember that $J_0(\lambda r)$ is a solution to the Bessel differential equation

$$\frac{d^2 J_0}{dr^2} + \frac{1}{r}\frac{dJ_0}{dr} + \lambda^2 J_0 = 0 \tag{E.2}$$

which implies that $I_0(\lambda r) \equiv J_0(i\lambda r)$ is a solution of the *modified Bessel differential equation*

$$\frac{d^2 I_0}{dr^2} + \frac{1}{r}\frac{dI_0}{dr} - \lambda^2 I_0 = 0 \tag{E.3}$$

that we needed for solving the Laplace-transformed equation in $U(r,s)$ of the answer of part (b) in (E.7).

(e) Consult Example 3.14 in Section 3.3 for the same problem. Explain why we have to use the modified Bessel function $I_0(\lambda r)$ here, whereas we don't in Example 3.14.

ANS. (a) $\dfrac{\partial^2 u}{\partial r^2} + \dfrac{1}{r}\dfrac{\partial u}{\partial r} = \dfrac{1}{k}\dfrac{\partial u}{\partial t}$, $\qquad 0 \le r < a,\ t > 0$ $\qquad$ (E.4)

$u(r,0) = 0, \qquad 0 \le r < a$ $\hfill$ (E.5)

$u(a,t) = u_0, \qquad t > 0$ $\hfill$ (E.6)

(b) For $U(r,s) = \mathcal{L}\{u(r,t)\}$,

$$\frac{d^2 U}{dr^2} + \frac{1}{r}\frac{dU}{dr} - \frac{s}{k}U = 0, \qquad 0 \le r < a \tag{E.7}$$

$$U(a,s) = \mathcal{L}\{u_0\} = \frac{u_0}{s}$$

Solution: $\qquad U(r,s) = \dfrac{u_0}{s}\dfrac{I_0(r\sqrt{s/k})}{I_0(a\sqrt{s/k})}$ $\hfill$ (E.8)

(c) $\quad u(r,t) = u_0 \left[1 - \dfrac{2}{a} \sum_{n=1}^{\infty} e^{-k\alpha_n^2 t} \dfrac{J_0(r\alpha_n)}{\alpha_n J_1(a\alpha_n)} \right]$ $\qquad$ (E.9)

(e) See the $-(s/k)U$ term in (E.7), which came from Laplace transforming $(1/k)\partial u/\partial t$ in (E.4) and which forced us to consult (E.3) instead of the usual Bessel equation (E.2).

2.27. Attempt the method of separation of variables to solve Exercise 2.26, having in mind the hint given in part (d) for consulting the Bessel–Fourier series expansion (3.100), (3.99). Explain why we will need $J_0(\lambda_k r)$ and we don't run across $I_0(\lambda_k r)$.

ANS. The boundedness of the solution in time will introduce a $-\lambda^2$ factor on the right-hand side of the heat equation (E.4) instead of the positive s factor of Laplace transforming $(1/k)\partial u/\partial t$ in part (b).

2.28. *Double Laplace Transforms.* In Section 2.6 we introduce the Fourier transform of a function of two variables $f(x,y)$, $-\infty < x < \infty$, $-\infty < y < \infty$. The *double* Laplace transform, denoted by $\mathcal{L}_{(2)}$, of $f(x,y)$ on the quadrant $0 < x < \infty$, $0 < y < \infty$ can be defined in a similar way,

$$\mathcal{L}_{(2)}\{f\} = \int_0^{\infty} \int_0^{\infty} f(x,y)e^{-px-qy}\, dx\, dy = F(p,q) \qquad (E.1)$$

and many of its operational properties can be derived in the same way as we did for the (single) Laplace transform of this section. Use the definition of the double Laplace transform in (E.1) to show that

$$\mathcal{L}_{(2)}\{f(x+y)\} = \frac{F(p) - F(q)}{p - q} \qquad (E.2)$$

where $F(p) = \mathcal{L}\{f(x)\}$ and $F(q) = \mathcal{L}\{f(y)\}$.

Hint: Let $y + x = \xi$, $y - x = \eta$, $dx\, dy = \tfrac{1}{2} d\xi\, d\eta$ and note that the first-quadrant domain of integration in x and y will be the rotated, by $\pi/4$, quadrant in ξ and η for the new double integral, with limits of this (iterated) integration $\eta = -\xi$ to ξ and then $\xi = 0$ to ∞.

2.29. With the notation of the double Laplace transform, show that

(a) $\mathcal{L}_{(2)}\left\{\dfrac{\partial u(x,y)}{\partial x}\right\} = pU(p,q) - U_0(q)$ (E.3)

where $U(p,q) = \mathcal{L}_{(2)}\{u(x,y)\}$ and $U_0(q) = \mathcal{L}\{u(0,y)\}$;

$\mathcal{L}_{(2)}\left\{\dfrac{\partial^2 u}{\partial x^2}\right\} = p^2 U(p,q) - pU_0(q) - U_1(q)$ (E.4)

where $U(p,q) = \mathcal{L}_{(2)}\{u(x,y)\}$, $U_0(q) = \mathcal{L}\{u(0,y)\}$, and $U_1(q) = \mathcal{L}\{\partial u(0,y)/\partial x\}$.

(b) $\mathcal{L}_{(2)}\left\{\dfrac{\partial u}{\partial y}\right\} = qU(p,q) - U_2(p)$, $U_2(p) = \mathcal{L}\{u(x,0)\}$ (E.5)

$\mathcal{L}_{(2)}\left\{\dfrac{\partial^2 u}{\partial y^2}\right\} = q^2 U(p,q) - qU_2(p) - U_3(p)$, (E.6)

where $U_3(p) = \mathcal{L}\left\{\dfrac{\partial u(x,0)}{\partial y}\right\}$

2.30. (a) Use the double Laplace transform, with its basic operational properties in Problem 2.29, to solve the following boundary value problem in $u(x,y)$ on the first quadrant, which is associated with a first-order partial differential equation,

$$\frac{\partial u}{\partial x} = \frac{\partial u}{\partial y}, \qquad 0 < x < \infty, \ 0 < y < \infty \qquad (E.1)$$

$$u(x,0) = f(x), \qquad 0 < x < \infty \qquad (E.2)$$

Hint: See the need for a condition $u(0,y)$ in this problem in order to apply the double Laplace transform pair (E.3) of Exercise 2.29 that will algebraize $\partial u/\partial x$ in (E.1).

ANS. (a) Let $u(0,y) = g(y)$ to have $U(p,q) = (G(q) - F(p))/(p - q)$, which is an analytic function on the domain for $\text{Real}\,p > \alpha$, $\text{Real}\,q > \beta$; we must have $G(q) = F(p)$ when $p = q$, $u(x,y) = \mathcal{L}_{(2)}^{-1}\{(F(q) - F(p))/(p - q)\}$.

2.31. Use the single Laplace transform on $\partial u/\partial x$ to solve Exercise 2.30. Compare your results for the two methods.

2.32. Consider the boundary and initial value problem for $u(x,t)$, the displacement of a semi-infinite string:

$$\frac{\partial^2 u}{\partial x^2} = \frac{1}{c^2}\frac{\partial^2 u}{\partial t^2}, \qquad 0 < x < \infty, \ t > 0 \qquad (E.3)$$

$$u(x,0) = f(x), \qquad 0 < x < \infty \tag{E.4}$$

$$\frac{\partial u}{\partial t}(x,0) = g(x), \qquad 0 < x < \infty \tag{E.5}$$

$$u(0,t) = 0, \qquad t > 0 \tag{E.6}$$

(a) Discuss the possibility of using the double Laplace transform to generate the solution obtained by the D'Alembert method in (E.14) of Example 2.19 of Section 2.3 (see Davies, 1985; Sneddon, 1972).

(b) Use the single Laplace transform on $\partial^2 u/\partial t^2$ to solve the problem and compare your results to the answer in part (a) and the D'Alembert solution in Example 2.19.

(c) Consult Section 2.5 for using the Fourier sine transform on $\partial^2 u/\partial x^2$, and solve the problem. Compare the result to the answers in parts (a) and (b) and the D'Alembert solution (see Example 2.19 of the vibrating string in Section 2.3).

Section 2.2   Fourier (Exponential) Transforms

2.33. (a) For the following functions, find the Fourier (exponential) transforms, as defined in (2.64).

(i) $f(x) \equiv p_a(x) = \begin{cases} 1, & |x| < a \\ 0, & |x| > a \end{cases}$

(ii) $f(x) \equiv q_b(x) = \begin{cases} 1 - \dfrac{|x|}{b}, & |x| < b \\ 0, & |x| > b \end{cases}$

(iii) $f(x) = e^{-b|x|}, \qquad b > 0$

(iv) $f(x) \equiv \mathrm{sgn}\, x = \begin{cases} -1, & x < 0 \\ 1, & x > 0 \end{cases}$

Hint: Note that the function is odd; then capitalize on the result of part (i) by involving the symmetry between the Fourier transform and its inverse.

(v) $f(x) = \begin{cases} 1 - x^2, & |x| < 1 \\ 0, & |x| > 1 \end{cases}$

(b)   Check the answers given here for part (a) against what you find in the Fourier exponential transform table in Appendix C.2 and in the basic reference Fourier transform tables in Erdelyi et al. (1954, Vol. 1), Ditkin and Prudnikov (1965), Oberhettinger (1973), and Abramowitz and Stegun (1964). The purpose is to note the variations in the definition of Fourier transform among the different references and be ready to make the minor adjustment when quoting results from different references and transform tables. See the note regarding this around the different definitions in (2.66) to (2.71). Also note that you may not find your desired result in the Fourier exponential tables, but it still may be under the Fourier cosine or sine transform tables, depending on the transformed function $f(x)$ being even or odd, respectively.

(c)   Use the Fourier transform of the gate function in (i) of part (a) to formally validate the integral in (2.88).

Hint: Treat this integral as a special case of the Fourier integral representation of $p_A(x)$ for the particular value of $x = 0$, where you get $p_A(0) = 1$, since $p_A(x)$ is continuous at $x = 0$. You may also see (E.3) of Example 2.15.

(d)   Show that all the functions in part (a) are sectionally smooth.

ANS.   (a)   (i) $F(\lambda) = 2 \dfrac{\sin a\lambda}{\lambda}$

(ii) $F(\lambda) = \dfrac{1}{b} \left[\dfrac{2\sin(b\lambda/2)}{\lambda} \right]^2$

(iii) $F(\lambda) = \dfrac{2b}{b^2 + \lambda^2}$

(iv) $F(\lambda) = \dfrac{2}{i\lambda}$, $i = \sqrt{-1}$

(v) $\dfrac{4}{\lambda^3}(\sin\lambda - \lambda\cos\lambda)$

2.34.   Find the (trigonometric) Fourier cosine and sine transforms $A(\lambda)$ and $B(\lambda)$ in (2.65a), (2.65b) for the functions in part (a) of Exercise 2.33. From your results here, verify that $\pi[A(\lambda) - iB(\lambda)]$ gives the same answer $F(\lambda)$, in Exercise 2.33a, thus verifying this important result that led to the (complex) exponential Fourier transform representation $F(\lambda)$ in (2.64).

2.35. (a) Show that the function

$$f(x) = |x|^{-1/2}(1+x^2)^{-1}, \qquad -\infty < x < \infty \tag{E.1}$$

is absolutely integrable but is not square integrable on $(-\infty, \infty)$; that is, $f(x) \in L_1(-\infty, \infty)$, but $f(x) \notin L_2(-\infty, \infty)$.

(b) Show that the sequence $f_n(x) = e^{-x^2/n^2}$ is *uniformly convergent* to $f(x) \equiv 0$ (as defined in (2.82) for all x) but is not *convergent in the mean* to zero (as defined in (2.83)).

Hint: For showing that the sequence does not converge in the mean to zero, you will need the identity $\int_0^\infty e^{-x^2} dx = \sqrt{\pi}$.

(c) Show that the sequence $f_n(x) = \sqrt{2nx}\, e^{-\frac{1}{2}nx^2}$ converges to $f(x) \equiv 0$ on the interval $[0,1]$, while it does not converge in the mean to $f(x) \equiv 0$.

Hint: For the last part see that $\int_0^1 |f_n(x) - 0|^2 dx = \int_0^1 2nx e^{-nx^2} dx = (1 - e^{-n})$, which approaches 1 (and not zero) as $n \to \infty$.

2.36. (a) Prove Schwarz's inequality (2.76) by considering the following:

(i) Let

$$A = \int_a^b |f|^2 dx, \qquad B = \left| \int_a^b f(x)g(x) dx \right|,$$

$$C = \int_a^b |g|^2(x) dx \tag{E.1}$$

and note that the proof of the Schwarz inequality (2.76) amounts to showing that

$$B \le \sqrt{AC} = \sqrt{A}\sqrt{C} \tag{E.2}$$

as will be proved in the following steps toward (E.5). Consider the case in which at least one of A and B is not zero; the case of $A = B = 0$ is clear. Let $A > 0$ and consider the quadratic expression in λ

$$|\lambda f(x) + g(x)|^2 = \lambda^2 f^2(x) + 2\lambda f(x)g(x) + g^2(x) \ge 0 \tag{E.3}$$

(ii) Integrate the above expression on (a,b) with respect to x to produce A, B, and C in it,

$$A\lambda^2 + 2B\lambda + C \ge 0 \tag{E.4}$$

For this expression to have distinct roots λ_1, and λ_2, we must have $B^2 > AC$, which implies

$$A\lambda^2 + 2B\lambda + C = A(\lambda - \lambda_1)(\lambda - \lambda_2) < 0 \qquad (E.5)$$

for $\lambda_1 < \lambda < \lambda_2$, $A > 0$. This is contradictory to (E.4), hence $B^2 < AC$ as desired in (E.2).

(b) Use another method. See Exercise 3.8 in Chapter 3, or see Weinberger (1965, p. 83).

(c) Extend Schwarz's inequality (2.76) to the infinite interval $(-\infty, \infty)$ as in (2.77).

Hint: (i) For $f(x)$, $g(x) \in L_2(a,b)$ show that $f(x)g(x) \in L_1(a,b)$.

(ii) Then for $f(x)$ and $g(x)$, which are assumed in $L_2(-\infty, \infty)$, for each $\epsilon > 0$ we can find $A(\epsilon)$ such that

$$\int_a^b f^2(x)\,dx < \epsilon, \qquad \int_a^b g^2(x)\,dx < \epsilon \qquad (E.1)$$

for large enough $A(\epsilon) < a < b$ (or $a < b \leq -A(\epsilon)$), as these functions must die out outside $(-A(\epsilon), A(\epsilon))$ for very large $A(\epsilon)$. On such an interval (a,b), $b > a > A(\epsilon)$ for example, use the Schwarz inequality (2.76) to show that

$$\int_a^b |f(x)||g(x)|\,dx < \epsilon, \qquad \begin{array}{l} b > a > A(\epsilon) \\ (\text{or } a < b < -A(\epsilon)) \end{array} \qquad (E.2)$$

Then

$$\int_{-\infty}^{\infty} |f(x)||g(x)|\,dx$$

$$= \lim_{A(\epsilon) \to \infty} \left[\int_{-\infty}^{-A(\epsilon)} |f(x)||g(x)|\,dx \right.$$

$$+ \int_{-A(\epsilon)}^{A(\epsilon)} |f(x)||g(x)|\,dx$$

$$\left. + \int_{A(\epsilon)}^{\infty} |f(x)||g(x)|\,dx \right]$$

$$= 0 + \int_{-\infty}^{\infty} |f(x)||g(x)| \, dx + 0$$

$$\geq \int_{-\infty}^{\infty} f(x)g(x) \, dx \qquad \text{(E.3)}$$

after using (E.2) for the vanishing two integrals. Thus

$$\lim_{\substack{a \to -\infty \\ b \to \infty}} \int_{a}^{b} f(x)g(x) \, dx$$

exists. Now take the limit of (2.76) as $a \to -\infty$, $b \to \infty$, whereby all three infinite integrals exist to give the Schwarz inequality (2.77) on the infinite interval $(-\infty, \infty)$

(d) (i) Establish (2.78), Schwarz's inequality with the weight function $\rho(x) > 0$.

Hint: Replace $f(x)$ and $g(x)$ in (2.77) by $\sqrt{\rho(x)}f(x)$ and $\sqrt{\rho(x)}g(x)$, respectively.

(ii) Generalize (2.76) to involve a weight function $\rho(x) > 0$ as was done above in (i) for (2.78).

(e) Prove the triangle inequality (2.79).

Hint: See the proof done in (3.42), (3.43) of Chapter 3. Also, you do it here by writing

$$|f + g|^2 = (f + g)\overline{(f + g)} = |f|^2 + \overline{f}g + \overline{g}f + |g|^2 \qquad \text{(E.1)}$$

and then using Schwarz's inequality (2.77) on the integrals of both $\overline{f}g$ and $\overline{g}f$ like

$$\left| \int_{-\infty}^{\infty} f(x)g(x) \, dx \right| \leq \left[\int_{-\infty}^{\infty} |f(x)|^2 \, dx \int_{-\infty}^{\infty} |g(x)|^2 \, dx \right]^{1/2} \qquad \text{(E.2)}$$

Finally, integrate (E.1) and use (E.2) to have a complete square for the right side of the inequality in (2.79).

2.37. (a) Prove the version of the Riemann–Lebesgue lemma in Theorem 2.13 by the following method.

(i) Subdivide the interval $[a,b]$ into N subintervals, where the piecewise smooth function $f(x)$ and its derivative df/dx are continuous on each of these subintervals

$[a_i, a_{i+1}]$, $i = 0, 1, 2, \ldots, N - 1$, $a_0 = a$, $a_N = b$. Of course, we have $f(a_i+)$, $f'(a_i+)$, and $f(a_{i+1}-)$, $f'(a_{i+1}-)$ are finite for $f(x)$ piecewise smooth.

(ii)   Replace the integral in (2.91) by the sum of the N integrals of the above N subintervals

$$\int_a^b f(x) \sin \lambda x \, dx = \sum_{i=0}^{N-1} \int_{a_i}^{a_{i+1}} f(x) \sin \lambda x \, dx \qquad \text{(E.1)}$$

Then what is left is to show that each of these integrals vanishes as $\lambda \to \infty$, that is,

$$\lim_{\lambda \to \infty} \int_{a_i}^{a_{i+1}} f(x) \sin \lambda x \, dx = 0 \qquad i = 0, 1, 2, \ldots, N - 1$$

$$\text{(E.2)}$$

(iii)   To show (E.2) do one integration by parts, realizing that it is allowed on each of the closed subintervals $[a_i, a_{i+1}]$, $i = 0, 1, 2, \ldots, N - 1$, since $f(x)$ and $f'(x)$ are continuous there,

$$\int_{a_i}^{a_{i+1}} f(s) \sin \lambda x \, dx = \frac{f(x) \cos \lambda x}{\lambda} \Bigg|_{x = a_i +}^{a_{i+1}-}$$

$$- \frac{1}{\lambda} \int_{a_i}^{a_{i+1}} f'(x) \cos \lambda x \, dx \qquad \text{(E.3)}$$

and

$$\left| \int_{a_i}^{a_{i+1}} f(x) \sin \lambda x \, dx \right| \leq \frac{2M_1}{\lambda} + \frac{M_2(a_{i+1} - a_i)}{\lambda}$$

$$\text{(E.4)}$$

since $f(x)$, $f'(x)$ are bounded on $[a, b]$; that is, $|f(x)| \leq M_1$, $|f'(x)| < M_2$ everywhere on $[a, b]$. Taking the limit as $\lambda \to \infty$ in (E.4) would give (E.2) and hence our result in (2.91)

$$\lim_{\lambda \to \infty} \int_a^b f(x) \sin \lambda x \, dx$$

$$= \sum_{i=0}^{N-1} \lim_{\lambda \to \infty} \int_{a_i}^{a_{i+1}} f(x) \sin \lambda x \, dx = 0 \qquad \text{(E.5)}$$

(b) Use the following "usual" version of the Riemann–Lebesgue lemma for Fourier integrals:

If $f(x)$ is absolutely integrable on $(-\infty, \infty)$, then

$$\lim_{\lambda \to \infty} \int_{-\infty}^{\infty} e^{-i\lambda x} f(x)\, dx = 0$$

to verify its version in (2.91) of Theorem 2.13 as well as its other case where $\sin \lambda x$ is replaced by $\cos \lambda x$.

Hint: Use the Euler identity $e^{-i\lambda x} = \cos \lambda x - i \sin \lambda x$ and the fact that if a complex number vanishes then both its real and imaginary parts must vanish.

(c) Attempt to prove Theorem 2.13 with $f(x)$ piecewise continuous on $[a, b]$.

Hint: See Tyn Myint-U, pp. 98–99 (1973).

2.38. Consider the Riemann–Lebesgue lemma as Theorem 2.13 for the vanishing of the Fourier transform $F(\lambda)$ as $\lambda \to \infty$ in (2.91). Also consider the proofs of its two versions in Exercise 2.37a and b for the transformed function $f(x)$ being sectionally smooth on $(-\infty, \infty)$ in 2.37a and $f(x)$ absolutely integrable in 2.37b.

(a) For Exercise 2.37a, check the functions in Exercise 2.33a to show that they are sectionally smooth, and look at the answers to see that indeed $\lim_{\lambda \to \infty} F(\lambda) = 0$ for all these examples.

Hint: See Exercise 2.33d for all these $f(x)$ being sectionally smooth on $(-\infty, \infty)$.

(b) Test for those functions in Exercise 2.33a that are absolutely integral on $(-\infty, \infty)$ to allow the usual version of the Riemann–Lebesgue lemma in Exercise 2.37b. Check their respective Fourier transforms to validate the result that $F(\lambda)$ vanishes as $\lambda \to \infty$.

(c) Compare the Riemann–Lebesgue lemma with its analog for the Laplace transform, and compare the sufficient condition or (conditions) for the two cases.

ANS. (b) 2.33a (i), (ii), (iii), (v).

(c) $\lim_{s\to\infty} F(s) = 0$ in Exercise 2.10a, and see the sufficient conditions for the existence of the Laplace transform $F(s)$ in Theorem 2.1.

2.39. (a) Check the functions in Exercises 2.33a for the conditions of the Fourier integral Theorem 2.14, which guarantees the existence of their Fourier transform $F(\lambda)$ as well as its inverse $f(x)$.

(b) What would you do about the cases that don't satisfy Theorem 2.14, especially when you have already shown a result for their Fourier transform?

ANS. (a) See Answer of Exercise 2.38b.

(b) Search for a theorem that ensures the existence of what you found, i.e., only $F(\lambda)$, and not the existence of both $F(\lambda)$ and its inverse $f(x)$ as is (so generously) given by Theorem 2.14.

2.40. (a) Check the functions in Exercise 2.33a for being odd and even functions to have, respectively, Fourier sine and cosine transforms, so that Corollaries 1 and 2 of Theorem 2.14 are applied to them. Note that in the tables of Fourier transforms, these odd and even functions are listed under *separate* tables, namely the Fourier sine and cosine transforms, respectively.

(b) For the same functions in Exercise 2.33a, and their Fourier transforms in the answers, search the references on Fourier transform tables given for Exercise 2.33b to verify your answer, having in mind three probable different notations in different references. (This is the same as Exercise 2.33b.)

2.41. (a) Prove (2.111), the shifting property in the physical x-space,

$$\mathcal{F}\{f(x-x_0)\} = e^{-ix_0\lambda}F(\lambda) \qquad \text{(E.1)}$$
$$\text{(2.111)}$$

(b) Prove (2.112), the shifting property in the Fourier λ-space,

$$\mathcal{F}\{e^{icx}\} = F(\lambda - c) \qquad \text{(E.2)}$$
$$\text{(2.112)}$$

(c) Show that

(i) $\mathcal{F}\{f(x)\cos ax\} = \tfrac{1}{2}[F(\lambda+a) + F(\lambda-a)]$

where $F(\lambda) = \mathcal{F}\{f(x)\}$

(ii) $\mathcal{F}\{f(x)\sin ax\} = \frac{1}{2}[F(\lambda + a) - F(\lambda - a)]$.

Hint: Use the definition of Fourier transform and the Euler identities

$$\cos x = \frac{e^{ix} + e^{-ix}}{2}, \qquad \sin x = \frac{e^{ix} - e^{-ix}}{2i}$$

(d)   Prove that $\mathcal{F}\{i \int_{\mp\infty}^{x} g(\xi)d\xi\} = G(\lambda)/\lambda$.

Hint: Let the integral be $f(x)$, $g(x) = df/dx$, and consult (2.114).

(e)   Prove that the Fourier transform of the complex conjugate $\overline{f(x)}$ of $f(x)$ is $\overline{F(-\lambda)}$.

Hint: use the fact that

$$\overline{\int f(x)g(x)\,dx} = \int \overline{f(x)}\,\overline{g(x)}\,dx$$

2.42. (a)   Prove the result (2.117) for the Fourier transform of second derivatives taking into account the justification of your steps as done for (2.114) in Theorem 2.15 and as stated for the general nth-derivative case in Corollary 1 of Theorem 2.15.

(b)   Prove the result (2.126) as an extension or combination of Corollary 1 of Theorem 2.16 in (2.125) and Corollary 1 of Theorem 2.15 in (2.118).

Hint: Use (2.122) for $\mathcal{F}\{(-ix)^n g(x)\} = d^n G(\lambda)/d\lambda^n$, where $g(x) = d^k f/dx^k$, and via (2.118) where we have $G(\lambda) = (i\lambda)^k F(\lambda)$.

2.43. (a)   Prove that the Fourier convolution product of (2.127) is commutative, distributive over addition, and associated as indicated in (2.132)–(2.134).

2.44. Prove the two results in (E.1) and (E.2) if $h(x) = f(x) * g(x)$

(a) $f(x + a) * g(x + a) = h(x + 2a)$ (E.1)

Hint: Write

$$[f(x + a)] * [g(x + a)] = \int_{-\infty}^{\infty} f(x + a - \xi)g(\xi + a)d\xi$$

then make the substitution $t = \xi + a$ inside the integral.

(b) $f(bx) * g(bx) = h(bx)/|b|$.   (E.2)

2.45. (a) Compute the convolution product $p_a(x) * p_a(x)$ of the gate function $p_a(x)$ of Exercise 2.33a(i). Write the result in terms of the triangular function $q_b(x)$ of Exercise 2.33a(ii), observing that the support $(-2a, 2a)$ of $q_b(x)$ is double the $(-a, a)$ of $p_a(x)$, that is, $b = 2a$.

Hint: For the convolution integral

$$(p_a * p_a)(x) \equiv \int_{-\infty}^{\infty} p_a(t) p_a(x - t) dt$$

$$= \int_{-\infty}^{\infty} p_a(t) p_a(t - x) dt$$

it is helpful to graph $p_a(t)$ and the translated $p_a(t - x)$ to find the integral of their product for selected values of x; then plot these values for $(p_a * p_a)(x)$.

(b) Find the Fourier transform of the convolution product $p_a(x) * p_a(x)$ in (a) by using the result of Exercise 2.33a(i) and the convolution theorem.

(c) Do part (b) using the result of Exercise 2.33a(ii) to verify the answer.

(d) Find the triple self-convolution product $(p_a * p_a * p_a)(x)$. Show that its support is $(-3a, 3a)$; i.e., it vanishes identically for $|x| > 3a$. See Exercise 2.62c for the reference to the answer in Ditkin and Prudnikov (1965, p. 178). See Table C.2.

(e) The self-convolution product $q_{2a}(x) = (p_a * p_a)(x)$ in part (b) is continuous. Show that it is sectionally smooth by showing that its derivative has three jump discontinuities, and locate them.

(f) Show that the resulting function in part (d) is continuous, and even though it may look to the eye as smooth, it is not! Determine the derivatives of this function that are not continuous, and locate the position and size of their jump discontinuities.

ANS. (a) $p_a(x) * p_a(x) \equiv q_{2a}(x) = \begin{cases} 2a - |x|, & |x| < 2a \\ 0, & |x| > 2a \end{cases}$

(b) $F(\lambda) = \left(\dfrac{2 \sin a\lambda}{\lambda} \right)^2$

(d)   See the reference given (p. 178, #7.67) and Table C.2.

(e)   The first derivative of $q_{2a}(x)$ has three jump discontinuities located at $x = 0$ and $x = \mp a$.

(f)   The second derivative of the resulting function has
four discontinuities at $x = \mp a$ and $x = \mp 3a$.

2.46.   Find the Fourier transform of

$$f(x) = e^{-\beta x^2}$$

Hint: Write the Fourier integral, complete the square for $\beta t^2 - i\omega t$
in the exponent, then use the result

$$\int_{-\infty}^{\infty} e^{-y^2}\, dy = \sqrt{\pi}$$

Watch for the change of variables of integration; you will have
limits of integration as complex numbers. However, via the use of
a (rectangular) contour integral we can justify the $-\infty$ to ∞ limits
for the resulting integral. (You may also try $\alpha = i\beta$ in the second
entry of Table 2.2.)

ANS. $F(\lambda) = \sqrt{\pi/\beta}\, e^{-\lambda^2/4\beta}$

2.47.   (a)   Find the Fourier transform of $J_0(x)$, the Bessel function of
the first kind of order zero.

Hint: Use the integral representation

$$J_{2n}(x) = \frac{1}{\pi} \int_{-\pi}^{\pi} \cos 2n\theta\, e^{ix\sin\theta}\, d\theta$$

and let $n = 0$, $\omega = \sin\theta$, where $d\theta = d\omega/\sqrt{1-\omega^2}$.

(b)   Find the Fourier transform of $J_1(x)$.

Hint: Use

$$j_{2n+1}(x) = \frac{1}{\pi} \int_{-\pi}^{\pi} \sin(2n+1)\theta\, e^{ix\sin\theta}\, d\theta$$

and see part (a).

ANS.   (a) $F(\lambda) = \mathcal{F}\{J_0\} = \begin{cases} \dfrac{2}{\sqrt{1-\lambda^2}}, & |\lambda| < 1 \\[2mm] 0, & |\lambda| > 1 \end{cases}$

(b) $F(\lambda) = \mathcal{F}\{J_1\} = \begin{cases} -\dfrac{2i\lambda}{\sqrt{1-\lambda^2}}, & |\lambda| < 1 \\ 0, & |\lambda| > 1 \end{cases}$

2.48.   (a)   Find the Fourier transform of the following integral equations in $u(x)$:

(i) $g(x) = \displaystyle\int_{-\infty}^{\infty} \frac{\sin a(x-\xi)}{x-\xi} u(\xi)\,d\xi$

(ii) $u(x) = g(x) + \displaystyle\int_{-\infty}^{\infty} k(x-\xi)u(\xi)\,d\xi$

ANS.   (a)   (i) $G(\lambda) = \begin{cases} \pi U(\lambda), & |\lambda| \le a \\ 0, & |\lambda| > a \end{cases} = \pi U(\lambda)p_a(\lambda)$

$G(\lambda) = \mathcal{F}\{g\}, \qquad U(\lambda) = \mathcal{F}\{u\}$

(ii) $U(\lambda) = G(\lambda) + K(\lambda)U(\lambda), \qquad K(\lambda) = \mathcal{F}\{k(x)\}.$

Section 2.3   Boundary and Initial Value Problems—Solutions by (Exponential) Fourier Transforms

2.49.   (a)   Solve for the potential distribution $u(x,y)$ in the upper half-plane (free of charge: $\nabla^2 u = 0$) where the potential is given on the x-axis as $u(x,0) = e^{-|x|}$, $-\infty < x < \infty$,

$$\frac{\partial^2 u}{\partial x^2} + \frac{\partial^2 u}{\partial y^2} = 0, \qquad -\infty < x < \infty,\ y > 0$$

$$u(x,0) = e^{-|x|}, \qquad -\infty < x < \infty$$

Hint: The Fourier transform is compatible with $\partial^2 u/\partial x^2$. (Why?) You may also need the pair $\mathcal{F}\{e^{-a|y|}\} = 2a/(a^2 + \lambda^2)$, $\mathrm{Re}(a) > 0$. See Example 2.21.

(b)   Explain why the Laplace transform is not compatible with this problem. What about the Fourier sine transform?

(c)   Solve problem (a) when the potential at the edge $y = 0$ is given as

$$u(x,0) = \begin{cases} -u_0, & x < 0 \\ u_0, & x > 0 \end{cases}$$

(d)   Consider the vibrating infinite string of Example 2.19. Use the change of variables $\xi = x - ct$ and $\eta = x + ct$ to show that the wave equation in (E.1) becomes $\partial^2 u / \partial \xi \, \partial \eta = 0$. Integrate and use the initial conditions (E.2), (E.3) to establish its D'Alembert solution in (E.14).

ANS. (a) $u(x,y) = \dfrac{y}{\pi} \displaystyle\int_{-\infty}^{\infty} \dfrac{e^{-|\xi|} d\xi}{y^2 + (x - \xi)^2}$

(b) For $\partial^2 u / \partial y^2$ on $(0, \infty)$ here we need $\partial u(x,0) / \partial y$ in addition to the given $u(x,0) = f(x)$ for the Laplace transform. The Fourier sine transform is compatible with $\partial^2 u / \partial y^2$, $0 < y < \infty$, since we are given $u(x,0)$, which is needed in (1.16) or (2.201).

(c) $u(x,y) = \dfrac{2u_0}{\pi} \tan^{-1} \dfrac{x}{y}$

2.50. Consider the potential distribution in the upper half-plane as in Example 2.21, where the potential was given at the boundary $y = 0$, which is termed a *Dirichlet problem*. Now consider the problem with the *Neumann-type condition*, where the (normal) derivative of the potential $\partial u / \partial y$ is given on the boundary $y = 0$,

$$\frac{\partial^2 u}{\partial x^2} + \frac{\partial^2 u}{\partial y^2} = 0, \qquad -\infty < x < \infty, \ 0 < y < \infty \qquad \text{(E.1)}$$

$$\frac{\partial u}{\partial y}(x,0) = g(x), \qquad -\infty < x < \infty \qquad \text{(E.2)}$$

$u(x,y)$ is bounded in the domain, and u and $\partial u / \partial x$ vanish as $|x| \to \infty$. Use the change of variable $v(x,y) = \partial u(x,y) / \partial y$, where $u(x,y) = \int_a^y v(x,\eta) d\eta$, to reduce the above (Neumann) problem (E.1), (E.2) to the (Dirichlet) problem of Example 2.21 in $v(x,y)$ (where the value, and not the derivative, of the new function $v(x,0) = g(x)$ is given on the boundary $y = 0$ as needed in (1.16) for the Fourier sine transform on $\partial^2 u / \partial y^2$ in Example 2.21).

ANS. The boundary value problem in $v(x,y)$ is

$$\frac{\partial^2 v}{\partial x^2} + \frac{\partial^2 v}{\partial y^2} = 0, \qquad -\infty < x < \infty, \ 0 < y < \infty$$

$$v(x,0) = g(x), \qquad -\infty < x < \infty$$

with

$$v(x,y) = \frac{y}{\pi} \int_{-\infty}^{\infty} \frac{g(\xi)d\xi}{(\xi-x)^2 + y^2}$$

from Example 2.21 and

$$u(x,y) = \int_a^y v(x,\eta)d\eta = \frac{1}{2\pi} \int_{-\infty}^{\infty} g(\xi) \ln \left[\frac{(\xi-x)^2 + y^2}{(\xi-x)^2 + a^2} \right] d\xi$$

2.51. (a) Consider a very long cylinder $(-\infty < z < \infty)$ of radius a with its axis along the z-axis. Solve for the potential distribution $u(x,z)$ due to having a symmetric part around the origin with length b, which is kept at a constant potential u_0 ($u = u_0$, $-b < z < b$), the rest of the surface being grounded ($u = 0$, $|z| > b$). Solve for $u(r,z)$ (independent of θ; why?) in the following boundary value problem:

$$\nabla^2 u \equiv \frac{1}{r} \frac{\partial}{\partial r} \left(r \frac{\partial u}{\partial r} \right) + \frac{\partial^2 u}{\partial z^2} = 0, \qquad 0 < r < a, \qquad \text{(E.1)}$$
$$-\infty < z < \infty$$

$$u(a,z) = \begin{cases} u_0, & |z| < b \\ 0, & |z| > b \end{cases} \qquad \text{(E.2)}$$

Hint: Use Fourier transform on $\partial^2 u/\partial z^2$, employing the result of Exercise 2.33a(i) for the Fourier transform of $u(a,z)$; then consult Exercise 2.26d for the resulting modified Bessel differential equation,

$$\frac{d^2}{dr^2} U(r,\lambda) + \frac{1}{r} \frac{d}{dr} U(r,\lambda) - \lambda^2 U(r,\lambda) = 0 \qquad \text{(E.3)}$$

$$U(a,\lambda) = 2u_0 \frac{\sin b\lambda}{\lambda} \qquad \text{(E.4)}$$

where $U(r,\lambda) = \mathcal{F}\{u(r,z)\}$.

(b) Here $I_0(\lambda r)$ is one of the modified Bessel equation's (E.3) solutions which is bounded at $r = 0$. What about solving for the same problem outside the cylinder?

Hint: Check $I_0(x)$ for divergence as $x \to \infty$ and search for the second solution $K_0(x)$, which is bounded as $x \to \infty$. See Appendix A.5.

(c) Do the problem in part (a) for the case where, instead of $u = u_0$ constant on $(-b,b)$, we have

$$u(a,z) = \begin{cases} -u_0, & -b < z < 0 \\ u_0, & 0 < z < b \\ 0, & |z| > b \end{cases}$$

Hint: Note that the solution $u(r,z)$ would be an odd function in z.

ANS. (a) $u(r,z) = \dfrac{2u_0}{\pi} \displaystyle\int_0^\infty \dfrac{\sin b\lambda}{\lambda} \dfrac{I_0(\lambda r)}{I_0(\lambda a)} \cos \lambda z \, d\lambda$

where $I_0(x)$ is the modified Bessel function of the first kind of order zero. See the hint and answer of Exercise 2.24b and properties of Bessel functions in Appendix A.5. Consequently, note that the solution to problem (E.3), (E.4) is

$$\mathcal{F}\{u\} = U(r,\lambda) = 2u_0 \frac{\sin b\lambda}{\lambda} \frac{I_0(\lambda r)}{I_0(\lambda a)},$$

$$-\infty < \lambda < \infty$$

(b) $u(r,z) = \dfrac{2u_0}{\pi} \displaystyle\int_0^\infty \dfrac{\sin \lambda b}{\lambda} \dfrac{K_0(\lambda r)}{K_0(\lambda a)} \cos \lambda z \, d\lambda$

(c) $u(r,z) = \dfrac{2u_0}{\pi} \displaystyle\int_0^\infty \dfrac{1 - \cos \lambda b}{\lambda} \dfrac{I_0(\lambda r)}{I_0(\lambda a)} \sin \lambda z \, d\lambda$

2.52. Solve the following boundary and initial value problem associated with the diffusion equation for neutrons in an infinite medium $(-\infty < x < \infty)$:

$$\frac{\partial u}{\partial \tau} = \frac{\partial^2 u}{\partial x^2} + \delta(x)\delta(\tau), \qquad -\infty < x < \infty, \ \tau > 0 \qquad \text{(E.1)}$$

where $u(x,\tau)$ is the number of neutrons per unit volume per unit time. The initial condition is an impulse source at time $\tau = 0$,

$$u(x,0) = \delta(x) \qquad \text{(E.2)}$$

where $\delta(x)$ is the Dirac delta function (2.150a). We also note the impulse-type source $\delta(x)\delta(\tau)$ in (E.1) as the nonhomogeneous term of the above diffusion equation. Also, we impose the necessary condition

$$\lim_{|x| \to \infty} u(x, \tau) = 0 \qquad \text{(E.3)}$$

Hint: $\mathcal{F}_e\{\delta(x)\} = 1$ and $\mathcal{F}_e\{e^{-ax^2}\} = \sqrt{\pi/a}\, e^{-\lambda^2/4a}$, $a > 0$.

ANS. $u(x, \tau) = \dfrac{1}{2\sqrt{\pi\tau}} e^{-x^2/4\tau}$

2.53. Solve the following integral equations in $u(x)$:

(a) $\displaystyle\int_{-\infty}^{\infty} \frac{u(\xi)\,d\xi}{(x - \xi)^2 + a^2} = \frac{1}{x^2 + b^2}, \qquad 0 < a < b$

Hint: Use the Fourier convolution theorem and remember that $\mathcal{F}_e\{1/(x^2 + b^2)\} = (\pi/b)e^{-b\lambda}$:

(b) $\displaystyle\int_{-\infty}^{\infty} u(x - \xi)u(\xi)\,d\xi = e^{-x^2}$

Hint: $\mathcal{F}_e\{e^{-ax^2}\} = \sqrt{\pi/a}\, e^{-\lambda^2/4a}$.

ANS. (a) $u(x) = a(b - a)/b\pi [x^2 + (b - a)^2]$

(b) $u(x) = (4/\pi)^{1/4} e^{-2x^2}$

2.54. Solve for the temperature distribution in the infinite bar of Example 2.18 when the initial temperature is given as

(a) $u(x, 0) = p_a(x) = \begin{cases} 1, & |x| < a \\ 0, & |x| > a \end{cases}$

Attempt to write your answer in the simplest form.

Hint: See (E.13) of Example 2.18, use the commutivity of the convolution product, and remember the definition of the error function $\text{erf}(x)$ in (2.39).

(b) The initial temperature is given as constant T_0,

$$u(x, 0) = T_0, \qquad -\infty < x < \infty$$

ANS. (a) $u(x, t) = \dfrac{1}{2}\left[\text{erf}\left(\dfrac{x + a}{2\sqrt{t}}\right) - \text{erf}\left(\dfrac{x - a}{2\sqrt{t}}\right)\right]$

(b) $u(x,t) = T_0 \operatorname{erf}\left(\dfrac{x}{2\sqrt{t}}\right)$

2.55. Solve for the potential distribution $u(x,y)$ in an infinite strip of width a, where the upper-plane surface at $y = a$ is kept at a potential $u(x,a) = f(x)$, while the lower surface at $y = 0$ is grounded.

ANS. $u(x,y) = \dfrac{1}{\pi}\displaystyle\int_0^\infty \int_{-\infty}^\infty f(\xi)\dfrac{\sinh \lambda y}{\sinh \lambda a}\cos \lambda(\xi - x)\,d\xi\,d\lambda$

2.56. Solve for the temperature distribution $u(x,t)$ in a semi-infinite thin bar $(0 < x < \infty)$, where its end $x = 0$ is insulated, i.e., $\partial u(0,t)/\partial x = 0$, and the initial temperature is given as

$u(x,0) = f(x) = e^{-ax^2}, \qquad 0 < x < \infty$

Hint: Because we have zero derivative $\partial u(0,t)/\partial x$ at $x = 0$, we can extend the $u(x,t)$ as an even function in x on $(-\infty,\infty)$, then use the exponential Fourier transform. Also, you may see Section 2.5 for the (natural) use of the cosine Fourier transform.

ANS. $u(x,t) = \dfrac{1}{\sqrt{1+4at}}\,e^{-ax^2/(1+4at)}$

Section 2.4 Signals and Linear Systems

2.57. (a) For the signal $f(t)$ in (2.137) and its bandlimited case $f_a(t)$ in (2.139), show that they are related as follows:

$$f_a(t) = \int_{-\infty}^\infty f(\tau)\frac{\sin a(t - \tau)}{\pi(t - \tau)}\,d\tau \qquad\qquad \text{(E.1)}$$

Hint: Use the convolution theorem in (2.128) for (2.139) and invoke the Fourier transform of $p_a(\omega)$ in (2.140). Also see the derivation of (E.1) in (2.183).

(b) Show that, as expected,

$\lim_{a\to\infty} f_a(t) = f(t)$

which is also to show that

$\lim_{a\to\infty}\dfrac{\sin a(t - \tau)}{\pi(t - \tau)}$

behaves like the Dirac delta function $\delta(t - \tau)$.

Hint: Use (E.1) and recognize the definition of the Dirac delta function

$$\delta(t - \tau) \equiv \lim_{a \to \infty} \frac{\sin a(t - \tau)}{\pi(t - \tau)}$$

after justifying taking the limit as $a \to \infty$ inside the integral of (E.1).

(c) Use the result in part (b) for the definition of the Dirac delta function as a limit of a Fourier transform

$$\delta(t - \tau) = \lim_{a \to \infty} \frac{\sin a(t - \tau)}{\pi(t - \tau)}$$

to show

(i) $\delta(t) = \mathcal{F}^{-1}\{1\}$

(ii) $\delta(t - b) = \mathcal{F}^{-1}\left\{e^{ib\omega}\right\}$

$$\textit{Hint:} \qquad \frac{\sin at}{\pi t} = \frac{1}{2\pi} \frac{2 \sin at}{t} = \frac{1}{2\pi} \int_{-a}^{a} e^{i\omega t} \, d\omega$$

$$= \frac{1}{2\pi} \int_{-\infty}^{\infty} p_a(\omega) e^{i\omega t} \, d\omega$$

$$\lim_{a \to \infty} \frac{\sin at}{\pi t} = \delta(t) = \frac{1}{2\pi} \int_{-\infty}^{\infty} \lim_{a \to \infty} p_a(\omega) e^{i\omega t} \, d\omega$$

$$= \frac{1}{2\pi} \int_{-\infty}^{\infty} e^{i\omega t} \, d\omega = \mathcal{F}^{-1}\{1\}$$

(d) Verify that the other definition of the Dirac delta function $\delta(t)$ in (2.161) has the inverse Fourier transform of 1.

Hint: $\mathcal{F}\left\{e^{-ax^2}\right\} = \sqrt{\pi/a}\, e^{-\lambda^2/4a}$.

(e) Compare the sampling series expansion (2.184) for the bandlimited signal $f_a(t)$ in (2.139) of part (a) and the integral representation for the same $f_a(t)$ in (E.1) of part (a). What is the main difference?

(f) Inside the integral (E.1) of part (e), can we use $f_a(\tau)$ instead of $f(\tau)$? Also, for the sampling series (2.184) of part (e), can we replace $f_a(n\pi/a)$ by $f(n\pi/a)$?

ANS. (e) The main difference is that we have a not necessarily bandlimited function $f(\tau)$ inside the integral, but we have *samples* of a (must be!) bandlimited function $f_a(n\pi/a)$ inside the sampling series.

(f) We can replace $f(\tau)$ by $f_a(\tau)$ inside the integral, which amounts to passing our bandlimited signal again through the same ideal low-pass filter. We cannot replace $f_a(n\pi/a)$ by $f(n\pi/a)$ in the series; if we do there is an *aliasing error* (see (2.186) and more detailed discussion in Chapter 4).

2.58. (a) In deriving the sampling expression (2.184) for the bandlimited signal $f_a(t)$ we used the pulse train (2.185) for $f_a(\tau)$ in the convolution integral (2.183) representing $f_a(t)$. Instead, use the pulse train for $(\sin a(t - \tau))/\pi(t - \tau)$ in the integral (2.183) to derive the sampling expansion (2.184).

Hint: The needed interchange of the infinite sum and the integral is justifiable for the class of functions considered.

(b) The divergent Fourier series

$$\frac{1}{T} \sum_{n=-\infty}^{\infty} e^{i2\pi nt/T} \equiv \sum_{k=-\infty}^{\infty} \delta(t + kT) \tag{E.1}$$

defines the impulse train in (2.156) and Fig. 2.7. Use 20 to 40 terms in the partial sum of the Fourier series on the left-hand side to verify the impulse train in Fig. 2.7. For future reference, see also what was done for the impulse train (3.199) in Fig. 3.13 of the Bessel–Fourier series, and its associated Bessel sampling expansion.

(c) Let $f_a(t)$ and $g_a(t)$ be functions bandlimited to $(-a,a)$; i.e., their Fourier transforms $F(\omega)$ and $G(\omega) \equiv 0$ for $|\omega| > a$ as in (2.138). Prove the following discretization of their convolution product:

$$(f_a * g_a)(t) \equiv \int_{-\infty}^{\infty} f_a(\tau)g_a(t - \tau)\,d\tau$$

$$= \frac{1}{2a} \sum_{n=-\infty}^{\infty} f_a\left(\frac{n\pi}{a}\right) g_a\left(t - \frac{n\pi}{a}\right) \tag{E.1}$$

Hint: Use the pulse train (2.185) for $f_a(\tau)$ above and follow the same steps that lead to the sampling expansion in (2.184). For the Fourier series-integrals method of establishing (E.1) and other related results see Exercise 3.31 in Chapter 3.

(d)   Prove that

$$\int_{-\infty}^{\infty} \frac{\sin(at - n\pi)}{at - n\pi} \frac{\sin(at - m\pi)}{at - m\pi}\, dt = \begin{cases} 0, & m \neq n \\ \dfrac{\pi}{a}, & m = n \end{cases}$$

(E.2)

which means that these sampling functions of (2.184) are *orthogonal on the infinite interval* $(-\infty, \infty)$.

Hint: Use the convolution integral (2.183), and note that the input is bandlimited to the bandlimit a of the ideal low-pass filter.

(e)   Prove the more general result of (E.2) as the following *Hardy's integral*:

$$\int_{-\infty}^{\infty} \frac{\sin a(y - x)}{y - x} \frac{\sin b(y - z)}{y - z}\, dy = \frac{\pi \sin b(z - x)}{z - x},$$

$$0 < b < a$$

(E.3)

then establish its special case in (E.2).

Hint: Note the fact that $b < a$, and follow the hint in part (a). You can also use the Parseval equality from (2.128) with $x = 0$.

(f)   Attempt a parallel to the Hardy integral (E.2) for the J_0-Hankel transform.

Hint: See Exercise 3.61d in Section 3.6, where such a result is developed in a parallel simple way.

(g)   As a different application of the sampling expansion (2.184), consider the following $f_{\pi/2}(x)$, the bandlimited Fourier integral representation of (the combined) gamma functions of x:

$$f_{\pi/2}(x) = \frac{\pi\Gamma(\alpha - 1)}{2^{\alpha - 2}\Gamma((\alpha + x)/2)\Gamma((\alpha - x)/2)}$$

$$= \int_{-\pi/2}^{\pi/2} (\cos t)^{\alpha - 2} e^{ixt}\, dt, \qquad t > 1$$

(E.4)

Derive the sampling expansion for the function $f_{\pi/2}(x)$ in terms of its sampler $f_{\pi/2}(n)$.

(h) Use the general result in part (g) to prove the important result of the gamma function in (2.18).

$$\frac{1}{\Gamma(x/2)\Gamma(1-x/2)} = \frac{\sin(\pi/2)x}{(\pi/2)x} \tag{E.5}$$

Hint: Set $\alpha = 2$ in the result of part (g).

ANS. (f) $\dfrac{1}{\Gamma((a+x)/2)\Gamma((a-x)/2)}$

$$= \sum_{n=-\infty}^{\infty} \frac{1}{\Gamma((a+2n)/2)\Gamma((a-2n)/2)} \frac{\sin((\pi/2)x - n\pi)}{(\pi/2)x - n\pi}$$

For a more general result, see Jerri and Glasser (1980).

2.59. (a) For a real signal $g(t)$, show that in general its Fourier transform $H(\omega)$ has real and imaginary parts $H_R(\omega)$ and $H_I(\omega)$, $H(\omega) = H_R(\omega) + iH_I(\omega)$.

Hint: Use the Euler identity $e^{-i\omega t} = \cos\omega t - i\sin\omega t$.

(b) With the result in part (a) show that the Fourier transform of real and even function $g(t)$ is a real function $G(\omega)$ and that if $g(t)$ is real and odd, then its transform $G(\omega)$ is pure imaginary.

(c) Consider the causal function $h(t)$ in (2.175) and its Fourier transform $H(\omega)$ in (2.176). Graph the real and imaginary parts $H_R(\omega)$ and $H_I(\omega)$ of $H(\omega) = H_R(\omega) + iH_I(\omega)$ and its amplitude $A(\omega) = |H(\omega)|$.

(d) For a real function $f(t)$ and its transform $F(\omega)$, show that if $F(\omega)$ is truncated to the half-line as $2F(\omega)u_0(\omega)$, where $u_0(\omega)$ is the unit step function, then $f_I(t)$, the imaginary part of $f(t)$, is the Hilbert transform of $f(t)$,

$$f_I(t) = \mathcal{H}\{f\} \equiv \frac{1}{\pi}P\int_{-\infty}^{\infty} \frac{f(\tau)}{t - \tau}\, d\tau$$

as defined in (2.267).

Hint: $u_0(\omega) = 1 + \operatorname{sgn}\omega$, $\mathcal{H}\{1/\pi t\} = -i\operatorname{sgn}\omega$.

(e) (i) Find the impulse response when the system function of a low-pass filter is $\phi_2(x) = (\phi_1 * \phi_1)(x) = (p_a * p_a)(x)$.

(ii) Write the integral representation for the output of an input $f(t)$ to the system of part (i).

Hint: See (2.170), (2.172) and a more parallel development from (2.182) to (2.184) for that of $H(\omega) = p_a(\omega)$.

ANS. (e) (i) $h(t) = \left(\dfrac{\sin at}{\pi t} \right)^2$

(ii) $f_{2a}(t) = \displaystyle\int_{-\infty}^{\infty} \left[\dfrac{\sin a(t - \tau)}{\pi(t - \tau)} \right]^2 f(\tau)\,d\tau$

2.60. This exercise illustrates the integral representation, via the Green's function as in (2.145), of a boundary value problem associated with a nonhomogeneous differential equation like (2.143). Consider the following boundary value problem in $u(x)$, $0 < x < 1$:

$$\frac{d^2u}{dx^2} + b^2u = f(x), \qquad b \neq 0, \ 0 < x < 1 \tag{E.1}$$

$$u(0) = 0 \tag{E.2}$$

$$u(1) = 0 \tag{E.3}$$

(a) Show that $v_1(x) = \sin bx$ and $v_2(x) = \sin b(1 - x)$ are two (linearly independent) solutions of the *homogeneous equation associated* with (E.1),

$$\frac{d^2v}{dx^2} + b^2v = 0, \qquad b \neq 0, \ 0 < x < 1 \tag{E.4}$$

and that $v_1(x)$ satisfies the first boundary condition (E.2) while $v_2(x)$ satisfies the second boundary condition (E.3). Note this "branching" of the solutions in satisfying the two boundary conditions.

(b) Consider the Green's function with its "two branches" definition, as a function of x on $0 \le x \le \xi$ and another function (branch) on $\xi \le x \le 1$,

$$G(x,\xi) = \begin{cases} \dfrac{\sin b(1 - \xi)\sin bx}{b \sin b}, & 0 \le x \le \xi \\[3mm] \dfrac{\sin b\xi \, \sin b(1 - x)}{b \sin b}, & \xi \le x \le 1 \end{cases} \tag{E.5}$$

Show that, in contrast to each of $v_1(x)$ and $v_2(x)$, which satisfy only one of the boundary conditions ((E.2) for $v_1(x)$, (E.3) for $v_2(x)$), the Green's function in (E.5) satisfies both boundary conditions (E.2) and (E.3).

Hint: Use the first branch in (E.5) (with its domain $0 \le x < \xi$) for the boundary condition (E.2) at $x = 0$ and the second branch (with its proper domain $\xi \le x \le 1$) for the boundary condition (E.3) at $x = 1$.

(c) Show that the Green's function in (E.5) satisfies the homogeneous differential equation (E.4).

(d) For future reference, it is beneficial to show in this case (which is true in general for self-adjoint differential operator in (E.1)) that the Green's function is symmetric, i.e., $G(x,\xi) = G(\xi,x)$.

Hint: Be *extremely careful* about how the domain of each branch of $G(x,\xi)$ changes (or switches) as you write $G(\xi,x)$.

(e) The integral representation for the solution of the (nonhomogeneous) boundary value problem (E.1)–(E.3), via the Green's function (E.5), is

$$u(x) = \int_0^1 G(x,\xi)f(\xi)d\xi \tag{E.6}$$

Verify this statement; i.e., show that $u(x)$ in (E.6) satisfies the differential equation (E.1) and the two boundary conditions (E.2), (E.3).

Hint: (i) For the boundary conditions (E.2), (E.3), apply them on x inside the integral (E.6) and consult part (b) for the fact that the Green's function satisfies both of them.

(ii) For satisfying the differential equation, we must write the integral as two parts for $0 \le \xi \le x$ and $x \le \xi \le 1$ since the Green's function is defined differently (as two branches of (E.5)) there. Then on each integral we need to apply the generalization of the fundamental theorem of calculus as the following *Leibnitz rule*:

$$\frac{d}{dx} \int_{\alpha(x)}^{\beta(x)} F(x,\xi)d\xi = \int_{\alpha(x)}^{\beta(x)} \frac{\partial F(x,\xi)}{\partial x} d\xi$$

$$+ F(x, \beta(x)) \frac{d\beta}{dx} - F(x, \alpha(x)) \frac{d\alpha}{dx} \qquad \text{(E.7)}$$

(f) Consider the problem in (E.1)–(E.3), except that $b = 0$ in (E.1). Show that the integral representation for the solution of such a boundary value problem is

$$u(x) = \int_0^1 G(x, \xi) f(\xi) d\xi \qquad \text{(E.8)}$$

where

$$G(x, \xi) = \begin{cases} x(1 - \xi), & 0 \le x \le \xi \\ \xi(1 - x), & \xi \le x \le l \end{cases} \qquad \text{(E.9)}$$

Hint: For more details on the Green's function and similar other problems, see Chapter 4 in the author's book on introductory integral equations (1985).

2.61. In statistics, the *distribution function* $F(x)$ is defined as the probability that the random variables $X \le x$ for every real number x. For continuous x, the *probability density* $p(x)$ is related to $F(x)$ as follows:

$$F(x) = \int_{-\infty}^x p(t) dt \qquad \text{(E.1)}$$

The role of the Fourier integral representation appears when the *characteristic function* $f(t)$, of a distribution function $F(x)$, is defined as the Fourier transform of its probability density $p(x)$,

$$f(t) = \int_{-\infty}^{\infty} e^{ixt} p(x) dx \qquad \text{(E.2)}$$

(a) Show that $\lim_{x \to \infty} F(x) = 1$, and $dF/dx = p(x)$.

(b) Show that $\mathcal{F}\{F(x)\} = if(t)/t$.

Hint: See (E.1) and the Fourier transform of $\int_{-\infty}^x f(\xi) d\xi$ in Table 2.2, noting the variation in the definitions (E.2) and (2.65).

(c) For the normal distribution, the probability density is

$$p(x) = \frac{1}{\sigma \sqrt{2\pi}} \exp\left(-\frac{(x - m)^2}{2\sigma^2}\right)$$

where σ^2 is the variance. Find the characteristic function $f(t)$.

ANS. (c) $f(t) = \exp\left(itm - \tfrac{1}{2}\sigma t^2\right)$

2.62. Consider the hill function of order $R + 1$, $\phi_{R+1}(a(R + 1), \omega)$ as defined in (2.190), which as we have indicated, is the Fourier transform of $[(2\sin at)/t]^{R+1}$. Consult Exercise 2.45a, b, and d for the special cases of $R = 0$, 1, and 2, respectively.

(a) Use the definition (2.190) and its above transform to prove that $\phi_{R+1}(a(R + 1), \omega)$ is an even function.

Hint: Note that the transform $[(2\sin at)/t]^{R+1}$ is an even function, which results in its Fourier integral being a real cosine integral.

(b) For $R = 0$ and 1, verify that the integral of the function $\phi_1(a, \omega)$ and of $\phi_2(2a, \omega)$, on their respective domains $(-a, a)$ and $(-2a, 2a)$, is exactly equal to the area of one rectangle for $\phi_1(a, \omega)$ with height at the center of its domain and two rectangles with heights at the centers of the two subdomains of $\phi_2(2a, \omega)$. Consult Exercise 2.46a for showing that the support of $\phi_2(2a, \omega)$ is $2a$, double that of $\phi_1(a, \omega) = p_a(\omega)$. For ϕ_R generalization, see Jerri (1982c, 1992).

(c) The exact explicit form of $\phi_{R+1}(a(R + 1), \omega)$ is given in Ditkin and Prudnikov (1965, pp. 178–179) for each of the $(R + 1)/2$ and $(R + 2)/2$ subintervals of $[0, a(R + 1)]$ for R odd and R even, respectively (see part b). From such expressions verify that ϕ_{R+1} is an even function (part a) in ω, a polynomial of degree R with continuous derivatives up to $R - 1$, and vanishes identically outside the interval $[-a(R+1), a(R + 1)]$. See Table C.2 in Appendix C.

(d) Write the Fourier integral representation of the hill function of order $R + 1$ as given in (2.190). See the hint in part (a).

(e) Prove the result in (2.192). You may consult Exercises 3.30d and 3.34a.

ANS. (d) $\phi_{R+1}(a(R + 1), \omega) = \dfrac{1}{2\pi} \displaystyle\int_{-\infty}^{\infty} e^{-i\omega t} \left[\dfrac{2\sin at}{t}\right]^{R+1} dt$

Section 2.5 Fourier Sine and Cosine Transforms

2.63. (a) Verify the Fourier cosine transform and its inverse (2.99), (2.100) for the function

$$f(x) = \begin{cases} 1, & 0 \le x < a \\ \frac{1}{2}, & x = a \\ 0, & x > a \end{cases}$$

(b) To what value does the Fourier cosine transform representation in part (a) converge at $x = \frac{2}{3}a$, a, and $12a$?

ANS. (a) $F_c(\lambda) = (\sin \lambda a)/\lambda$; use it in (2.100), write $\sin \lambda a \times \cos \lambda x = \frac{1}{2}[\sin \lambda(a+x) + \sin \lambda(a-x)]$, perform the two integrals, noting that they are of the same form as the integral of $F_c(\lambda)$, and then use a graph to combine the two results to arrive at your sectionally continuous function $f(x)$.

(b) $1, \frac{1}{2}, 0$

2.64. (a) Find the Fourier sine and cosine transforms of $f(x) = e^{-bx}$, $b > 0, x > 0$.

(b) With the Fourier transform of $f(x)$ in part (a) write the Fourier sine representation $f_s(x)$ and the Fourier cosine representation $f_c(x)$. Determine to what value each of these two representations converges at $x = 0, a, b$, and $-2b$.

(c) Prove the Parseval equality for the Fourier sine and cosine transforms in (2.209) and (2.210), respectively.

ANS. (a) $F_s(\lambda) = \dfrac{\lambda}{b^2 + \lambda^2}$

$F_c(\lambda) = \dfrac{b}{b^2 + \lambda^2}$

(b) At $x = 0$, the Fourier sine representation $f_s(x)$ converges to $\frac{1}{2}[f_s(0+) + f_s(0-)] = \frac{1}{2}[1 - 1] = 0$ and the Fourier cosine representation $f_c(x)$ converges to $\frac{1}{2}[f_c(0+) + f_c(0-)] = 1$. This difference is due to the odd extension of the sine representation at $x = 0$

versus the even extension of the cosine representation at $x = 0$. Thus there is a jump discontinuity at $x = 0$ for the (odd) sine representation, while the (even) cosine representation is continuous at $x = 0$. At $x = a$, $b > 0$, the function e^{-bx} is continuous and both representations converge to the same value e^{-ba} and e^{-b^2} at $x = a$ and b, respectively. For $x = -2a$, $f_s(-2a) = -e^{2ab}$, $f_c(-2a) = e^{2ab}$ because of the odd and even extensions of these two respective Fourier representations.

2.65. (a) Find the Fourier cosine transform of

$$f(x) = \begin{cases} \cos x, & 0 < x < a \\ 0, & x > a \end{cases} \tag{E.1}$$

(b) Solve for $f(x)$ in the following integral equation:

$$\int_0^\infty f(x) \cos \lambda x \, dx = K(\lambda) = \frac{1}{\sqrt{\lambda}} \tag{E.2}$$

Hint: Find the inverse Fourier cosine transform of $K(\lambda)$. See the table in Appendix C.4.

(c) Solve the integral equation in $u(x)$

$$\int_0^\infty u(x) \sin \lambda x \, dx = \begin{cases} 1 - \lambda, & 0 \le \lambda < 1 \\ 0, & \lambda > 1 \end{cases}$$

Hint: Find the inverse sine transform.

ANS. (a) $F_c(\lambda) = \dfrac{1}{2} \left[\dfrac{\sin(1 - \lambda)a}{1 - \lambda} + \dfrac{\sin(1 + \lambda)a}{1 + \lambda} \right]$

(b) $f(x) = \sqrt{\dfrac{2}{\pi x}}$

(c) $u(x) = \dfrac{2(x - \sin s)}{\pi x^2}$

2.66. Consider the function

$$f(x) = \begin{cases} 1, & 0 < x < 1 \\ 0, & x > 1 \end{cases}$$

(a) Find the Fourier sine and cosine transforms of this function as given in (2.193) and (2.194), respectively. It is important to consult the usual forms (2.193a) and (2.194a) in case we resort to the tables, where this form is used most often.

(b) Find the Fourier cosine transform of the above function, from knowing the Fourier exponential transform of the gate function

$$p_a(x) = \begin{cases} 1, & |x| < a \\ 0, & |x| > a \end{cases} \tag{E.1}$$

$$\mathcal{F}_e\{p_a(x)\} = \frac{2\sin\lambda a}{\lambda} \tag{E.2}$$

Hint: Extend the function as an even function.

(c) Use the result in Exercise 2.64a, to find the Fourier (exponential) transform of

$$f(x) = e^{-|x|}, \qquad -\infty < x < \infty$$

Hint: This is an even function, its $F_e(\lambda) = 2F_c(\lambda)$.

ANS. (a) $F_s(\lambda) = \dfrac{1 - \cos\lambda}{\lambda}$; see $\dfrac{\sqrt{2}}{\pi}\dfrac{1 - \cos\lambda}{\lambda}$ for (2.193a) type definition.

$F_c(\lambda) = \dfrac{\sin\lambda}{\lambda}$; see $\dfrac{\sqrt{2}}{\pi}\dfrac{\sin\lambda}{\lambda}$ for (2.194a) type definition.

(b) $F_e(\lambda) = \dfrac{2}{1 + \lambda^2}$

2.67. Consider the following boundary value problem describing the potential distribution $u(x,y)$ in the first quadrant with the lower edge being grounded and the left edge $x = 0$ kept at a given potential as specified in the boundary condition (E.3); see Fig. 2.16,

$$\frac{\partial^2 u}{\partial x^2} + \frac{\partial^2 u}{\partial y^2} = 0, \qquad x > 0, \, y > 0 \tag{E.1}$$

$$u(x,0) = 0, \qquad x > 0 \tag{E.2}$$

$$u(0,y) = \begin{cases} 1, & 0 < y < 1 \\ 0, & y > 1 \end{cases} \tag{E.3}$$

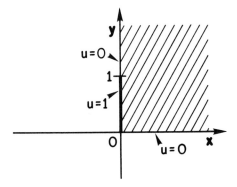

FIG. 2.16 Illustration for Exercise 2.67.

(assume $\nabla u \to 0$ as $x^2 + y^2 \to \infty$). (E.4)

(a) Find a compatible transform and solve the problem.

(b) Is there another transform compatible with the problem? If so, use it in succession with the one you used in part (a) to completely algebraize the problem; then find the solution and compare with the answer in (a).

Hint: Avoid the Laplace transforms, since for the second order derivatives here you are not given the function and its derivative *independently* at a boundary $x = 0$ or $y = 0$.

ANS. (a) The compatible transform is a Fourier sine transform for $\partial^2 u / \partial y^2$.

$$u(x,y) = \frac{2}{\pi} \int_0^\infty \frac{1 - \cos \lambda}{\lambda} e^{-\lambda x} \sin \lambda y \, d\lambda \qquad (E.5)$$

(b) A sine transform on $\partial^2 u / \partial x^2$, thus applying a *double Fourier sine* transform. With $U_2(\lambda, \mu) = \mathcal{F}_s^2 \{u(x,y)\}$, the final algebraic equation in $U_2(\lambda, \mu)$ is

$$-(\lambda^2 + \mu^2)U_2(\lambda, \mu) + \lambda U(x = 0, \mu) + \mu U(\lambda, y = 0),$$

$$-(\lambda^2 + \mu^2)U_2(\lambda, \mu) = -\frac{\lambda}{\mu}(1 - \cos \mu) + 0$$

$$U_2(\lambda,\mu) = \frac{\lambda}{\mu} \frac{(1 - \cos\mu)}{\lambda^2 + \mu^2} \tag{E.6}$$

$$u(x,y) = \frac{4}{\pi^2} \int_0^\infty \int_0^\infty \frac{\lambda(1 - \cos\mu)}{\mu(\lambda^2 + \mu^2)}$$

$$\times \sin\lambda x \sin\mu y \, d\lambda \, d\mu \tag{E.7}$$

$$= \frac{2}{\pi} \int_0^\infty e^{-\mu x} \frac{(1 - \cos\mu)}{\mu} \sin\mu y \, d\mu \tag{E.8}$$

where the inverse Fourier sine transform $e^{-\mu x} = \mathcal{F}_s^{-1}\{\lambda/(\lambda^2 + \mu^2)\}$ was used to reduce the double Fourier integral in (E.7) to the final single Fourier integral in (E.8).

2.68. (a) Solve Exercise 2.67 when the edge $y = 0$ is *insulated*, i.e., $\partial u(x,0)/\partial y = 0$, instead of being grounded.

(b) Attempt another transform to algebraize $\partial^2 u/\partial x^2$ in the Laplace equation. If there is one, try it with the cosine transform used in part (a) to completely algebraize the Laplace equation. Find the solution and compare it to the one found in part (a).

(c) Consider Exercise 2.50, and attempt a Fourier cosine transform for $\partial^2 u/\partial y^2$ in addition to the Fourier exponential transform for $\partial^2 u/\partial x^2$. Solve and compare your answer.

ANS. (a) $u(x,y) = \dfrac{2}{\pi} \displaystyle\int_0^\infty \frac{\sin\lambda}{\lambda} e^{-\lambda x} \cos\lambda y \, d\lambda$

2.69. (a) Solve for the temperature distribution $u(x,t)$ in a semi-infinite bar with zero initial temperature and where the heat flow $\partial u(0,t)/\partial x$ at the end $x = 0$ is given by $g(t)$; i.e., solve the boundary and initial value problem in $u(x,t)$,

$$\frac{\partial^2 u}{\partial x^2} = \frac{\partial u}{\partial t}, \qquad 0 < x < \infty, \ t > 0 \tag{E.1}$$

$$u(x,0) = 0, \qquad 0 < x < \infty \tag{E.2}$$

$$\frac{\partial u(0,t)}{\partial x} = g(t), \qquad 0 < t < \infty \tag{E.3}$$

Hint: Use the Fourier cosine transform to reduce the problem to a first-order nonhomogeneous differential equation in $U(\lambda,t) = \mathcal{F}_c\{u(x,t)\}$. Then use the Laplace transform to completely algebraize the problem in $\tilde{U}(\lambda,s) = \mathcal{L}\{U(\lambda,t)\}$, assuming that $g(t)$ in (E.3) is Laplace transformable (you may choose to use the integrating factor, instead of the Laplace transform, to solve for $U(\lambda,t)$). Also, you may need the cosine transform pair

$$\mathcal{F}_c\{e^{-ax^2}\} = \frac{1}{2}\sqrt{\frac{\pi}{a}}\,e^{-\lambda^2/4a}, \qquad a > 0.$$

(b) Solve the above problem for initial temperature given by $u(x,0) = e^{-x}$, $0 < x < \infty$, and where the end $x = 0$ is being insulated. Verify your answer.

ANS. (a) $\displaystyle u(x,t) = \frac{-1}{\sqrt{\pi}}\int_0^t \frac{g(\tau)}{\sqrt{t-\tau}}e^{-x^2/4(t-\tau)}\,d\tau$

(b) $\displaystyle u(x,t) = \frac{2}{\pi}\int_0^\infty \frac{1}{1+\lambda^2}\cos\lambda x \cdot e^{-\lambda^2 t}\,d\lambda$

2.70. (a) Consider the potential distribution in the upper half-plane of Example 2.23, where the potential is given at the edge $x = 0$. Use the Fourier exponential transform and the Fourier sine transform, as suggested in (E.3)–(E.5) of Example 2.23, to completely algebraize the potential equation (E.1) in $\tilde{U}(\lambda,\alpha) = \mathcal{F}_s\mathcal{F}_e\{u\}$.

(b) For the problem in part (a), use the Fourier sine transform only to algebraize $\partial^2 u/\partial y^2$ in the Laplace equation, as indicated in (E.6) of Example 2.23. Then solve the nonhomogeneous equation (E.6) in $U_s(x,\alpha) = \mathcal{F}_s\{u\}$, and finally obtain the final solution as the inverse sine transform $u(x,y) = \mathcal{F}_s^{-1}\{U_s(x,\alpha)\}$.

Hint: For the nonhomogeneous differential equation, see Appendix A.2.

2.71. Consider the problem in Exercise 2.70 except that the edge $y = 0$ is being insulated, i.e., $\partial u(x,0)/\partial y = 0$.

(a) Use a repeated Fourier exponential transform and then the Fourier cosine transform to solve the problem.

(b) Use the Fourier cosine transform only to algebraize $\partial^2 u/\partial y^2$ in the potential equation (E.1) of Example 2.23, solve the resulting nonhomogeneous differential equation in $U_c(x,\alpha) =$

$\mathcal{F}_c\{u(x,y)\}$, then find the inverse Fourier cosine transform for the final solution $u(x,y)$.

Hint: For the nonhomogeneous differential equation, see Appendix A.2.

2.72. Consider Exercise 2.68 if the edge $y = 0$ is with the boundary condition

$$u(x,0) + h\frac{\partial u}{\partial y}(x,0) = 0 \qquad (E.1)$$

(a) Try the general method of Section 1.6 to find the transform with kernel $K(y,\lambda)$ that algebraizes $\partial^2/\partial y^2$ and at the same time involves the present boundary condition (E.1) at $y = 0$. The desired kernel $K(y,\lambda)$ must be a solution of the following boundary value problem:

$$\frac{d^2K}{dy^2}(y,\lambda) = -\lambda^2 K(y,\lambda) \qquad (E.2)$$

$$\frac{dK}{dy}(0,\lambda) + \frac{1}{h}K(0,\lambda) = 0 \qquad (E.3)$$

where the solution $K(y,\lambda) = A\cos\lambda y + B\sin\lambda y$ of (E.2) that satisfies (E.3) is

$$K(y,\lambda) = \lambda\cos\lambda y - \frac{1}{h}\sin\lambda y \qquad (E.4)$$

(b) Use this kernel for the desired compatible transform:

$$T\{f\} = F(\lambda) = \int_0^\infty \left(\lambda\cos\lambda y - \frac{1}{h}\sin\lambda y\right)f(y)\,dy \qquad (E.5)$$

and verify that its inverse is

$$T^{-1}\{F\} = f(y) = \frac{2}{\pi}\int_0^\infty F(\lambda)\frac{\lambda\cos\lambda y - (1/h)\sin\lambda y}{1/h^2 + \lambda^2}\,d\lambda \qquad (E.6)$$

(c) Prove the important operational property of this new Fourier transform (E.5)

$$T\left\{\frac{d^2f}{dy^2}\right\} = -\lambda^2 F(\lambda) - \lambda\left[f'(0) + \frac{1}{h}f(0)\right] \qquad (E.7)$$

(d) Use this new transform to solve the potential problem of Exercise 2.68 with the above boundary condition (E.1) at $y = 0$.

ANS. (d) $u(x,y) = \dfrac{2}{\pi} \displaystyle\int_0^\infty \left[\sin\lambda + \dfrac{\cos\lambda - 1}{\lambda h} \right]$

$\qquad\qquad \times e^{-\lambda x} \dfrac{\lambda\cos\lambda - (1/h)\sin\lambda y}{1/h^2 + \lambda^2} \, d\lambda$

2.73. (a) In Exercise 2.18, use the Laplace and the Fourier sine transforms to completely algebraize equation (E.1); then find the final solution.

Hint: Let $\bar{U}(\lambda,s) = \mathcal{F}_s\{\mathcal{L}[u(x,t)]\} = \mathcal{F}_s\{U(x,s)\}$, and note that we have what we need for the sine transform of $\partial^2 u/\partial x^2$ as required in (2.201); i.e., we have $U(0,s) = 0$.

(b) Explain the difficulty in finding the final solution in Exercise 2.18 and the advantage of the method in part (a).

(c) Find the solution of Exercise 2.18 via the Laplace transform.

Hint: For the resulting nonhomogeneous differential equation, see Appendix A.2.

(d) Use the Fourier sine transform only to solve Exercise 2.18.

Hint: See that it is done in Example 2.22a.

ANS. (a) The final (transformed) algebraic equation in $\bar{U}(\lambda,s)$ $\equiv \mathcal{F}_s\mathcal{L}\{u\}$ is

$$s\bar{U}(\lambda,s) - F_s(\lambda) = -\lambda^2 \bar{U}(\lambda,s) \qquad\qquad \text{(E.1)}$$

$$\bar{U}(\lambda,s) = \frac{F_s(\lambda)}{\lambda^2 + s} \qquad\qquad\qquad \text{(E.2)}$$

whose combined Laplace–Fourier sine transforms inverse is the final solution $u(x,t)$,

$$u(x,t) = \frac{2}{\pi} \int_0^\infty \left[\int_0^\infty f(\xi)\sin\lambda\xi\,d\xi \right] e^{-\lambda^2 t} \sin\lambda x\,d\lambda$$
$$\text{(E.3)}$$

(b) The Laplace-transformed equation $d^2 U/dx^2 - sU(x,s)$ $= -f(x)$, $U(0,s) = 0$ is nonhomogeneous, which needs more work to establish its solution. This shows the advantage of using the Fourier sine transform.

(c) See the answer of Example 2.22a.

2.74. Consider the heat equation problem in Exercise 2.18 with everything the same except that the edge $x = 0$ is insulated; i.e., $\partial u(0,t)/\partial x = 0$ instead of being at fixed temperature zero.

(a) Use a Fourier transform to solve the problem.

   Hint: See Example 2.22b for using the Fourier cosine transform on $\partial^2 u/\partial x^2$ since you are given the derivative at $x = 0$, $\partial u(0,t)/\partial x = 0$, that you need for (2.202).

(b) Use the Laplace transform and a Fourier transform to completely algebraize the heat equation (E.1) in Exercise 2.18, when the edge $x = 0$ is insulated as in part (a) above.

   Hint: Use the Fourier cosine transform on $\partial^2 u/\partial x^2$ since you have $\partial u(0,t)/\partial x = 0$, which is needed for (2.202).

ANS. (a) See $u(x,t)$ in the answer of part (b).

   (b) The final (transformed) algebraic equation in $\bar{U}(\lambda,s)$ $= \mathcal{F}_c \mathcal{L}\{u\}$ is

   $$sU(\lambda,s) - F_c(\lambda) = -\lambda^2 U(\lambda,s) \tag{E.1}$$

   $$\bar{U}(\lambda,s) = \frac{F_c(\lambda)}{s + \lambda^2} \tag{E.2}$$

   whose (double Laplace–Fourier cosine transform) inverse is

   $$u(x,t) = \frac{2}{\pi} \int_0^\infty \left[\int_0^\infty f(\xi) \cos \xi \lambda \, d\xi \right] e^{-\lambda^2 t} \cos \lambda x \, dx$$

2.75. Consider the following boundary value problem that is associated with a fourth-order differential equation in $y(x)$, *the deflection of an infinite elastic beam* like a railroad track. We will consider $y(x)$, $0 < x < \infty$, because of the symmetry of the problem about $x = 0$, where we describe this symmetry by the boundary condition $y'(0)$ at $x = 0$ where the load is applied. We state the problem in dimensionless form,

$$\frac{d^4 y}{dx^4} = -y(x), \qquad 0 < x < \infty \tag{E.1}$$

$$y'(0) = 0 \tag{E.2}$$

$$y(\infty) = 0 \tag{E.3}$$

$$y'''(0) = 1 \tag{E.4}$$

$$y'''(\infty) = 0 \tag{E.5}$$

where the boundary condition in (E.4) is a consequence of the static equilibrium condition (in physical dimensions),

$$2 \int_0^\infty ky(x)\,dx = P \tag{E.6}$$

where k is the spring constant of the foundation per unit length of the beam, P is the load, and with the dimensionless form we have $k/EI = P/2EI = 1$, where EI is the flexural rigidity.

(a) Find the compatible transform to algebraize the boundary value problem, and solve for the transform of $y(x)$.

(b) Solve the problem by finding the inverse of the resulting transform in part (a).

(c) In this format of the deflection of the railroad track, the *Wasiutynski condition* for the spacing S of the rail cross ties $S \leq \pi/4\beta$, where $\beta = (k/4EI)^{1/4}$, *the characteristic of the system.* Attempt to predict this condition via the use of the sampling expansion (2.184), where S would stand for the samples spacing π/a. Show that this condition corresponds to assuming that the displacement $y(x)$ dies out essentially (close to identically zero) for $|\lambda| > 2\sqrt{2}$; i.e., $y(x)$ is bandlimited with bandlimit $a = 2\sqrt{2}$. Consult the physical dimensions, where $\lambda\xi_0 = \lambda(E/k)^{1/4} = 2\sqrt{2}$.

(d) The assumption in part (c) of essentially bandlimited $Y_c(\lambda)$ carries with it an *aliasing error.* Show that the maximum on this aliasing error is about 1.4% at $x = 0$.

Hint: Consult (2.187) for an upper bound on the aliasing error.

ANS. (a) This problem is compatible with the Fourier cosine transform, since we can show, after four integrations by parts like those leading to (2.202) and (2.200), that

$$\mathcal{F}_c\left\{\frac{d^4y}{dx^4}\right\} = \int_0^\infty \frac{d^4y}{dx^4} \cos \lambda x \, dx$$

$$= \lambda^4 Y_c(\lambda) + \lambda^2 y'(0) - y'''(0) \tag{E.9}$$

where we have used conditions (E.3) and (E.5) in the process, and where the remaining conditions (E.2), (E.4) are needed for the cosine transform operator (E.9). The transformed equation in $Y(\lambda)$ is

$$\lambda^4 Y_c(\lambda) - 1 = Y_c(\lambda)$$

$$Y_c(\lambda) = \frac{1}{1 + \lambda^4}$$

(b) $y(x) = \mathcal{F}_c^{-1}\left\{\frac{1}{1 + \lambda^4}\right\} = e^{-x/\sqrt{2}} \sin\left(\frac{\pi}{4} + \frac{x}{\sqrt{2}}\right)$

which can be found in the tables (see Erdelyi et al., 1954, Vol. 1).

(d) $|\epsilon_A(x)| \leq \dfrac{4}{\pi} \displaystyle\int_{2\sqrt{2}}^{\infty} \dfrac{1}{1 + \lambda^4}\, d\lambda \sim 0.0192$

which is about 1.4% of $y(0) = \sqrt{2}/2$ in $y(x)$ of part (b).

Section 2.6 Higher-Dimensional Fourier Transforms

2.76. (a) Let $F_{(2)}(\lambda_1, \lambda_2)$ and $G_{(2)}(\lambda_1, \lambda_2)$ be the double Fourier transforms of $f(x,y)$ and $g(x,y)$, respectively, and let a, b, c, and d be constants. Prove the following very basic properties that extend from the (single) Fourier transforms:

   (i) $\mathcal{F}_{(2)}\{af(x,y) + bg(x,y)\} = aF_{(2)}(\lambda_1, \lambda_2) + bG_{(2)}(\lambda_1, \lambda_2)$

   (ii) $\mathcal{F}_{(2)}\{f(x - c, y - d)\} = e^{i(c\lambda_1 + d\lambda_2)}F_{(2)}(\lambda_1, \lambda_2)$

   (iii) $\mathcal{F}_{(2)}\{f(cx, dy)\} = \dfrac{1}{cd} F_{(2)}(\lambda_1, \lambda_2)$

   (iv) $\mathcal{F}_{(2)}\{e^{-i(cx + dy)}\} = F_{(2)}(\lambda_1 - c, \lambda_2 - d)$

(b) Prove results similar to those in part (a) for the triple Fourier transform.

(c) Prove the following separation property for the double Fourier transforms: If $f(x,y)$ is separable in x and y, i.e., $f(x,y) = g(x)h(y)$, then its double Fourier transform is separable, and

$$\mathcal{F}_{(2)}\{f(x,y)\} = \mathcal{F}\{g(x)\}\,\mathcal{F}\{h(y)\} = G(\lambda_1)H(\lambda_2)$$

(d) Take advantage of this separation property to find the double Fourier transform of the following functions:

   (i) $f(x,y) = e^{-\pi(x^2 + y^2)}$, $-\infty < x < \infty, \ -\infty < y < \infty$

Hint: $\mathcal{F}\left\{e^{-ax^2}\right\} = \sqrt{\pi/a}\,e^{-\lambda^2/4a}$.

(ii) $f(x,y) = \begin{cases} 1, & -a < x < a, \ -b < y < b \\ 0, & \text{otherwise} \end{cases}$

(iii) $f(x,y) = e^{-x^2-|y|}$, $-\infty < x < \infty, \ -\infty < y < \infty$

(e) Extend the separation property of the double Fourier transform to the triple Fourier transform and prove it.

ANS. (d) (i) $F_{(2)}(\lambda_1, \lambda_2) = e^{-\lambda_1^2/4\pi}e^{-\lambda_2^2/4\pi} = e^{-(\lambda_1^2+\lambda_2^2)/4\pi}$

(ii) $F_{(2)}(\lambda_1, \lambda_2) = 4\,\dfrac{\sin a\lambda_1}{\lambda_1}\dfrac{\sin b\lambda_2}{\lambda_2}$

(iii) $F_{(2)}(\lambda_1, \lambda_2) = 2\sqrt{\pi}e^{-\lambda_1^2/4} \cdot \dfrac{1}{1 + \lambda_2^2}$

2.77. (a) Establish the double Fourier transform pair in (2.227) that was needed for Example 2.24 as (E.11), the extension to two dimensions of the (single) Fourier transform pair in (E.10).

Hint: We recognize that the function to be transformed in (2.227) is *circularly symmetric*, which according to (2.236) becomes a single J_0-Hankel transform. Now use the series expansion for the kernel $J_0(\lambda r)$ of this transform, recognizing the integral defining the gamma (or factorial) function, and the power series expansion of e^{-x^2}.

(b) Show that

$$\mathcal{F}_{(2)}\left\{ \frac{e^{-k\sqrt{x^2+y^2}}}{\sqrt{x^2+y^2}} \right\} = \frac{1}{\sqrt{k^2+\lambda^2}}, \qquad k > 0$$

Hint: Recognize this as a circularly symmetric function whose double Fourier transform according to (2.236) becomes a single J_0-Hankel transform. Also note that $\int_0^\infty e^{-kr}J_0(\lambda r)\,dr = 1/\sqrt{k^2+\lambda^2}$, a result that we had in Exercise 2.11b as the Laplace transform of $J_0(\lambda r)$.

ANS. (a) $f_0\left(\sqrt{x^2+y^2}\right) = f_0(r) = \displaystyle\int_0^\infty \lambda J_0(\lambda r)F_0(\lambda)\,d\lambda,$

$$F_0(\lambda) = F\left(\sqrt{\lambda_1^2 + \lambda_2^2}\right)$$

$$\mathcal{H}_0^{-1}\left\{e^{-\sqrt{\lambda_1^2+\lambda_2^2}}\right\} = \int_0^\infty \lambda e^{-\lambda^2 t} J_0(\lambda r)\,d\lambda = f_0(r)$$

$$f_0(r) = \sum_{n=0}^\infty \frac{(-\frac{1}{4}r^2)^n}{n!n!} \int_0^\infty \lambda^{2n+1} e^{-\lambda^2 t}\,d\lambda$$

$$= \frac{1}{2t}\sum_{n=0}^\infty \frac{1}{n!}\left(-\frac{r^2}{4t}\right)^n = \frac{1}{2t}e^{-r^2/4t}$$

$$= \frac{1}{2t}e^{-(x^2+y^2)/4t} = 2\pi\frac{1}{4\pi t}e^{-(x^2+y^2)/4t}$$

2.78. Prove that the J_0-Hankel transform of $f(r) = e^{-\pi r^2}$ is

$$F_0(\lambda) = \frac{1}{2\pi}e^{-\lambda^2/4}$$

Hint: $f(r) = e^{-\pi(x^2+y^2)}$ is circularly symmetric, so we can use its equivalence to a double Fourier transform as in (2.236), then rely on the separation property in Exercise 2.76c and the Fourier transform of e^{-x^2} as given in Exercise 2.76d (i).

2.79. Consider the potential distribution $u(x,y,z)$ in the upper half-space ($z > 0$), where the xy-plane base is kept at a potential given by

$$u(x,y,0) = e^{-x^2+y^2}, \qquad -\infty < x < \infty, \ -\infty < y < \infty$$

(a) Solve the problem using the double Fourier transform.

(b) Since the problem is circularly symmetric, consult the J_0-Hankel transform of Section 2.7 for solving it.

2.80. In Section 2.4 we introduced the hill function of order $R + 1$, $\phi_{R+1}(a(R + 1), \omega)$, in (2.190), where its special case of order one is the gate function

$$\phi_1(a, \omega) = p_a(\omega) = \begin{cases} 1, & |\omega| < a \\ 0, & |\omega| > a \end{cases} \tag{E.1}$$

(a) Extend the definition of the hill function of order one, $\phi_1(x)$, to two dimensions, $\phi_1(x,y)$, then find its double Fourier transform.

(b) Extend the definition of the hill function of order two, $\phi_2(x) = \phi_1 * \phi_1$, to two dimensions as $\phi_2(x,y)$, then find its double Fourier transform.

ANS. (a) $\phi_1(x,y) = \begin{cases} 1, & |x| < 1 \text{ and } |y| < 1 \\ 0, & |x| > 1 \text{ and } |y| > 1 \end{cases}$

$$\Phi_1(\lambda, \mu) = \mathcal{F}_{(2)}\{\phi_1(x,y)\} = 4\frac{\sin \lambda}{\lambda}\frac{\sin \mu}{\mu}$$

(b) $\Phi_2(\lambda, \mu) = \mathcal{F}_{(2)}\{\phi_2(x,y)\} = \left[4\frac{\sin \lambda}{\lambda}\frac{\sin \mu}{\mu}\right]^2$

2.81. Derive the result (2.238) of reducing the radially symmetric n-dimensional Fourier transform $\mathcal{F}_{(n)}\{f(|\mathbf{x}|)\}$ to a one-dimensional $J_{(n/2)-1}$-Hankel transform for $n = 2, 3, \ldots$.

Hint: See the derivation of (2.236) for $n = 2$, then use the general relation

$$dx_2 dx_3 \cdots dx_n = \frac{r^{n-2} 2(\pi)^{n/2-1/2}}{\Gamma(\frac{1}{2}n - \frac{1}{2})}$$

and

$$\int_0^\pi \sin^{n-2}\theta\, e^{i\lambda r \cos\theta}\, d\theta = \frac{2^{n/2-1}\pi^{1/2}\Gamma(\frac{1}{2}n - \frac{1}{2})}{(\lambda r)^{n/2-1}} J_{(n/2)-1}(\lambda r)$$

See Sneddon (1972, pp. 79–82).

Section 2.7 Hankel Transforms

The Bessel functions and their properties that are needed for this section are covered in Appendix A.5 and are often supplied here as hints.

2.82. (a) Consult Theorem 2.24 for the existence of the J_0-Hankel transforms of the following functions and then find the transforms.

(i) $f(r) = 1/r,\ 0 < r < \infty$

(ii) $f(r) = e^{-ar}/r,\ 0 < r < \infty,\ a > 0$

Hint: Consult the Laplace transform of $J_0(\lambda t)$; $\mathcal{L}\{J_0(\lambda t)\} = 1/\sqrt{\lambda^2 + s^2}$.

(iii) $f(r) = p_a(r) = \begin{cases} 1, & 0 < r < a \\ 0, & r > a \end{cases}$

Hint: Use $d(x^n J_n(x))/dx = x^n J_{n-1}$ for $n = 1$ and note that $J_1(0) = 0$.

(b) For the Hankel transforms of the functions in part (a), illustrate that they must satisfy the Riemann–Lebesgue lemma of the Hankel transform as stated in (2.245) and as in (2.245a) of their vanishing as $\lambda \to \infty$.

(c) Since the Hankel transform $F(\lambda)$ and its inverse $f(r)$ are symmetric, do we expect $F(\lambda)$ to satisfy Theorem 2.24 for the existence of its (inverse) transform? If so, verify that the functions $f(r)$ in part (a) satisfy the Rieman–Lebesgue lemma.

ANS. (a)   (i) $F(\lambda) = 1/\lambda$

(ii) $F(\lambda) = 1/\sqrt{a^2 + \lambda^2}$

(iii) $F(\lambda) = aJ_1(\lambda a)/\lambda$

(b) They all satisfy (2.245), and very clearly (2.245a), where they vanish as $\lambda \to \infty$.

(c) All $F(\lambda)$ in part (a) do satisfy Theorem 2.24. Thus, as expected, their transforms $f(r)$ satisfy the Riemann–Lebesgue lemma and all $f(r)$ in part (a) vanish as $r \to \infty$.

2.83. (a) Let $F_n(\lambda)$ and $G_n(\lambda)$ be the J_n-Hankel transforms of $f(r)$ and $g(r)$, respectively. Prove the Parseval equality

$$\int_0^\infty rf(r)g(r)\,dr = \int_0^\infty \lambda F_n(\lambda)F_n(\lambda)G_n(\lambda)\,d\lambda \qquad (E.1)$$

Hint: Write the representation (2.243) for $F_n(\lambda)$ and then interchange the resulting two integrals, using (2.244) last for $g(r)$.

(b) For the Hankel transform in part (a) define the *generalized translation* $f(r\theta x)$ of $f(r)$ by x as

$$f(r\theta x) \equiv \int_0^\infty \lambda F_n(\lambda)J_n(\lambda r)J_n(\lambda x)\,d\lambda \qquad (E.2)$$

Then define the J_n-Hankel convolution product $f * g$ of f and g as

$$(f * g)(x) \equiv \int \tau f(\tau \theta x) g(\tau) d\tau \tag{E.3}$$

(i) Prove that this convolution product is commutative, i.e., $f * g = g * f$.

(ii) Prove the following J_n-Hankel transform convolution theorem:

$$\mathcal{H}_n \{f * g\} = F_n(\lambda) G_n(\lambda) \tag{E.4}$$

Hint: See the proof of the Parseval equality (E.1) in part (a).

2.84. Prove that

(a) $\mathcal{H}_n \{f(ar)\} = \dfrac{1}{a^2} F_n \left(\dfrac{\lambda}{a} \right)$

(b) $\mathcal{H}_n \left\{ \dfrac{f(r)}{r} \right\} = \dfrac{\lambda}{2n} [F_{n-1}(\lambda) - F_{n+1}(\lambda)]$

Hint: Use the identity $J_\nu(x) = (x/2\nu)[J_{\nu-1}(x) - J_{\nu+1}(x)]$ in Appendix A.5.

2.85. (a) Solve the following boundary value problem in $u(r,z)$:

$$\frac{\partial^2 u}{\partial r^2} + \frac{1}{r} \frac{\partial u}{\partial r} + \frac{\partial^2 u}{\partial z^2} = 0, \qquad 0 \le r < \infty, 0 < z < \infty \tag{E.1}$$

$$\frac{\partial u}{\partial z}(r,0) = \begin{cases} -c, & r < a \\ 0, & r > a \end{cases} \tag{E.2}$$

Hint: Use the J_0-Hankel transform, and consult the Hankel transform pair in Exercise 2.82(iii).

(b) Is there a possibility of applying another transform to fully algebraize the potential equation (E.1)? If so, use it and solve the problem.

Hint: Try the Fourier cosine transform in the z direction, and remember that $\mathcal{F}_c \{e^{-az}\} = a/(\mu^2 + a^2) = F_c(\mu)$.

(c) Use the separation-of-variables method of Section 1.1 to solve the problem in (a).

ANS. (a) $u(r,z) = \dfrac{c}{a} \displaystyle\int_0^\infty \dfrac{J_1(\lambda a)}{\lambda} e^{-\lambda z} J_0(\lambda r)\,d\lambda$

(b) Same as in (a)

2.86. Let $u(r,z)$ be the steady-state temperature distribution in a semi-infinite solid ($z > 0$, $r > 0$), or a half-space, with a *steady source* of *heat* $q(r)$ that varies with the radial distance r only, and where the face $z = 0$ is kept at zero temperature, i.e.,

$$\dfrac{\partial^2 u}{\partial r^2} + \dfrac{1}{r}\dfrac{\partial u}{\partial r} + \dfrac{\partial^2 u}{\partial r} = -q(r), \qquad 0 < r < \infty,\ 0 < z < \infty \quad \text{(E.1)}$$

$$u(r,0) = 0, \qquad 0 < r < \infty \quad \text{(E.2)}$$

(a) Solve for $u(r,z)$.

Hint: Use J_0-Hankel transform on (E.1) to obtain

$$\dfrac{d^2 U_0}{dz^2} - \lambda^2 U_0(\lambda,z) = -Q(\lambda) \quad \text{(E.3)}$$

where $U_0(\lambda,z) = \mathcal{H}_0\{u(r,z)\}$ and $Q(\lambda) = \mathcal{H}_0\{q(r)\}$, and note that $Q(\lambda)$ is constant for the differential equation (E.3), whose solution is

$$U_0(\lambda,z) = B(\lambda)e^{-\lambda z} + \dfrac{Q(\lambda)}{\lambda^2} \quad \text{(E.4)}$$

The transform of the boundary condition (E.2) is $U_0(\lambda,0) = 0$ and, used in (E.4), gives $B(\lambda) = -Q(\lambda)/\lambda^2$.

(b) Attempt another transform besides the above J_0-Hankel transform to fully algebraize (E.1).

(c) Attempt the method of separation of variables (Section 1.1 in Chapter 1) or a variation of it to solve this problem.

ANS. (a) $u(r,z) = \displaystyle\int_0^\infty Q(\lambda)\dfrac{1 - e^{-\lambda z}}{\lambda} J_0(\lambda r)\,d\lambda$

2.87. Consider Exercise 2.86 without a heat source ($q = 0$) and where the face ($z = 0$) is kept at a temperature that varies with r only as $u(r,0) = 1/\sqrt{1 + r^2}$:

$$\dfrac{\partial^2 u}{\partial r^2} + \dfrac{1}{r}\dfrac{\partial u}{\partial r} + \dfrac{\partial^2 u}{\partial z^2} = 0, \qquad 0 < r < \infty,\ 0 < z < \infty \quad \text{(E.1)}$$

$$u(r,0) = \frac{1}{\sqrt{1+r^2}} \qquad\qquad (E.2)$$

(a) Solve the problem.

   Hint: Consult the J_0-Hankel transform pair in Exercise 2.82(ii) for the Hankel transform of the boundary condition (E.2).

(b) Consider solving for the potential distribution $u(r,z)$ in the half-space, where the potential on the base $(z = 0)$ has the same radial dependence as in (E.2).

(c) Is the problem in (a) or (b) compatible with a Fourier transform? Which one? If so, would you prefer it over the Hankel transform? Attempt to solve the problem in (a) via this Fourier transform.

(d) If in part (c) you found a compatible Fourier transform, use the Hankel and this Fourier transform to completely algebraize the Laplace equation in (E.1). Find the solution $u(r,z)$ and compare it to the answer in part (a) or (c).

(e) Use separation of variables to solve the problem in (a).

ANS. (a) $u(r,z) = 1/\sqrt{(z+1)^2 + r^2}$

   (b) This is the same problem as the steady-state temperature distribution of part (a).

   (c) The Fourier sine transform for $\partial^2 u/\partial z^2$, since we are given the potential at $z = 0$ in (E.2).

   (d) See (a) and (c).

2.88. In Example 2.26 consider the vibration of the circularly symmetric large membrane with initial displacement given by $u(r,0) = c/\sqrt{1+r^2/a^2}$ and with zero initial velocity. Establish the solution as $u(r,t) = ca\,\mathrm{Re}[r^2 + (a + ict)^2]^{-1/2}$, $i = \sqrt{-1}$, then verify it.

Hint: $\mathcal{H}_0\{1/\sqrt{1+r^2/a^2}\} = ae^{-\lambda a}/\lambda$, and remember the perfect symmetry of the Hankel transform and its inverse. You may see Exercise 1.6 for elements of complex numbers to find the real part (Re) of a complex-valued function.

ANS. $u(r,t) = ca[(a^2 + r^2 - c^2t^2)^2 + 4a^2c^2t^2]^{-1/4} \cos\phi,$

where $\tan 2\phi = \dfrac{2act}{(a^2 + r^2 - c^2t^2)}$

2.89. Consider the electrostatic potential distribution $u(r,z)$ between two grounded infinite parallel plates located at $z = \mp a$, which is due to a point charge at the origin. This potential $u(r,z)$ satisfies the Laplace equation except at the origin, where it has a *singular behavior like* $q/\sqrt{r^2 + z^2}$. So we assume $u(r,z) = q/\sqrt{r^2 + z^2} + v(r,z)$, where $v(r,z)$ satisfies the Laplace equation in space,

$$\frac{\partial^2 v}{\partial r^2} + \frac{1}{r}\frac{\partial v}{\partial r} + \frac{\partial^2 v}{\partial z^2} = 0, \qquad 0 \le r < \infty, \; -a < z < a \qquad (\text{E.1})$$

and the boundary conditions of *grounded surfaces* at $z = \mp a$

$$u(r, \mp a) = v(r, \mp a) + \frac{q}{\sqrt{r^2 + a^2}} = 0$$

$$v(r, \mp a) = -\frac{q}{\sqrt{r^2 + a^2}}, \qquad 0 \le r < \infty \qquad (\text{E.2})$$

(a) Solve for the boundary value problem in $v(r,z)$, and thus $u(r,z)$.

(b) Use the separation-of-variables method or a variation on it to solve the problem.

Hint: Attempt this product method for $v(r,z)$.

ANS. $u(r,z) = \dfrac{q}{\sqrt{r^2 + z^2}} - q \displaystyle\int_0^\infty \dfrac{\cosh \lambda z}{\cosh \lambda a} e^{-\lambda a} J_0(\lambda r)\, d\lambda$

2.90. Consider $Y_n(r)$, the Bessel function of order n of *the second kind* that we present in Appendix A.5, and define the *Weber transform* $f(r)$ on $(0, \infty)$ as

$$W_\nu\{f\} = F_\nu(\lambda) = \int_0^\infty rf(r)Z_\nu(\lambda r)\, dr \qquad (\text{E.1})$$

with

$$Z_\nu(\lambda r) = J_\nu(\lambda r)Y_\nu(\lambda a) - Y_\nu(\lambda r)J_\nu(\lambda a) \qquad (\text{E.2})$$

where, clearly, $Z_\nu(\lambda a) = 0$. The inverse is

$$W_\nu^{-1}\{F_\nu(\lambda)\} = f(r) \equiv \int_0^\infty \lambda F_\nu(\lambda) \frac{Z_\nu(\lambda r)\, d\lambda}{J_\nu^2(\lambda a) + Y_\nu^2(\lambda a)}$$

(a) Show that this Weber transform algebraizes the Bessel differential operator as

$$W_\nu \left\{ \frac{d^2u}{dr^2} + \frac{1}{r}\frac{du}{dr} - \frac{\nu^2}{r^2}u \right\} = -\lambda^2 U_\nu(\lambda) - \frac{2}{\pi}u(a) \quad \text{(E.3)}$$

(b) Write the boundary value problem describing the steady-state temperature distribution in a thick infinite slab of thickness $2b$, through which there is a cylindrical hole of radius a. The plane faces at $z = \mp b$ are kept at constant temperature T_0, while the cylindrical surface $r = a$ is kept at temperature zero.

(c) Use the Weber transform to solve the boundary value problem in part (b).

(d) Consider the problem in part (b) for an infinite slab in space, where the wall (surface) of the infinite circular hole (of radius a) is kept at fixed temperature T_1 and the rest of the infinite solid started initially at temperature T_0. Find the solution $u(r,t)$, since the problem is clearly independent of z.

ANS. (b) $\quad \dfrac{\partial^2 u}{\partial r^2} + \dfrac{1}{r}\dfrac{\partial u}{\partial r} + \dfrac{\partial^2 u}{\partial z^2} = 0, \qquad 0 \le r < \infty, \; -b < z < b$

$u(r, \mp b) = T_0, \qquad 0 \le r < \infty$

$u(a,z) = 0, \qquad -b < z < b$

(c) $\quad U_0(\lambda, z) = -\dfrac{2T_0}{\pi \lambda^2}\left[1 - \dfrac{\cosh \lambda z}{\cosh \lambda b} \right]$

$u(r,z) = -\dfrac{2T_0}{\pi} \displaystyle\int_0^\infty \dfrac{1}{\lambda}\left[1 - \dfrac{\cosh \lambda z}{\cosh \lambda b} \right]$

$\qquad \times \dfrac{Z_0(\lambda r)\,d\lambda}{J_0^2(\lambda a) + Y_0^2(\lambda a)}, \qquad 0 \le r < \infty, \; -b < z < b$

(d) $\quad \dfrac{\partial^2 u}{\partial r^2} + \dfrac{1}{r}\dfrac{\partial u}{\partial r} = \dfrac{1}{k}\dfrac{\partial u}{\partial t}$

$u(r,0) = T_0, \qquad a < r < \infty$

$u(a,t) = T_1, \qquad t > 0$

$u(r,t) = T_0 + 2\dfrac{T_1 - T_0}{\pi} \displaystyle\int_0^\infty \dfrac{1}{\lambda}\dfrac{(1 - e^{-k\lambda^2 t})Z_0(\lambda r)}{J_0^2(\lambda a) + Y_0^2(\lambda a)}\,d\lambda$

2.91. Consider the *Laplacian* in cylindrical coordinates (r,θ,z)

$$\nabla^2 = \Delta_3 = \frac{\partial^2}{\partial r^2} + \frac{1}{r}\frac{\partial}{\partial r} + \frac{1}{r^2}\frac{\partial^2}{\partial\theta^2} + \frac{\partial^2}{\partial z^2}$$

(a) Give its special case for axial symmetry Δ_a.

(b) Assume a separable solution $u(r,\theta,z) = v(r,z)e^{i\nu\theta}$, and show that

$$\Delta_3 u(r,\theta,z) = \left(\frac{\partial^2}{\partial r^2} + \frac{1}{r}\frac{\partial}{\partial r} - \frac{\nu^2}{r^2} + \frac{\partial^2}{\partial z^2} \right) v(r,z)e^{i\nu\theta}$$

(c) Find the J_ν-Hankel transform of $\Delta_3(v(r,z)e^{i\nu\theta})$.

(d) Consider the biharmonic operation $\nabla^4 = \Delta_3^2 = \Delta_3\Delta_3$, used often in the theory of elasticity. Use repeated application of the result in part (c) to find the J_ν-Hankel transform of the biharmonic operator Δ_3^2 on $u(r,\theta,z)$;i.e., find $\mathcal{H}_\nu\{\Delta_3^2(v(r,z)e^{i\nu\theta})\}$.

Hint: See the derivation of $\mathcal{H}_0\{\Delta_3^2\}$ in (2.254) that was used for the biharmonic equation in Example 2.29.

ANS. (a) $\nabla^2 = \Delta_a = \dfrac{\partial^2}{\partial r^2} + \dfrac{1}{r}\dfrac{\partial}{\partial r},\qquad \Delta_a \equiv \Delta_2$

(c) $\left(\dfrac{d^2}{dz^2} - \lambda^2 \right) V_\nu(\lambda,z)e^{i\nu\theta},\qquad V_\nu(\lambda,z) = \mathcal{H}_\nu\{v(r,z)\}$

(d) $\left(\dfrac{d^2}{dz^2} - \lambda^2 \right)^2 V_\nu(\lambda,z)e^{i\nu\theta},\qquad V_\nu(\lambda,z) = \mathcal{H}_\nu\{v(r,z)\}$

2.92. Consider the problem of free small vibrations of an infinite elastic plate with circularly symmetric initial displacement and zero initial velocity. The displacement here is clearly circularly symmetric, and we denote it by $u(r,t)$. It is known that $u(r,t)$ is governed by the following equation involving the (axially symmetric) biharmonic operator ∇^4 (see Exercise 2.91 and Example 2.29):

$$b^2\left(\frac{\partial^2}{\partial r^2} + \frac{1}{r}\frac{\partial}{\partial r} \right)^2 u + \frac{\partial^2 u}{\partial t^2} = 0, \qquad 0 < r < \infty,\ t > 0 \qquad \text{(E.1)}$$

where b is the ratio of the flexural rigidity of the plate and its mass per unit area.

(a) Set up the initial value problem in $u(r,t)$.

(b) Use the J_0-Hankel transform to find $U_0(\lambda,t)$, the transform of $u(r,t)$, and then invert it to find $u(r,t)$.

Hint: See Exercise 2.91, and also Example 2.29.

(c) Attempt the method of separation of variables (of Section 1.1) to solve the problem.

ANS. (a) (E.1) with the initial conditions

$$u(r,0) = f(r), \qquad 0 < r < \infty \qquad (E.2)$$

$$u_t(r,0) = 0, \qquad 0 < r < \infty \qquad (E.3)$$

(b) $U_0(\lambda,t) = F_0(\lambda)\cos(bt\lambda^2)$

$$u(r,t) = \int_0^\infty \lambda F_0(\lambda)\cos(bt\lambda^2)J_0(\lambda r)d\lambda$$

Section 2.8 Laplace Transform Inversion

2.93. (a) Consider the Laplace transform $G(s) = 1/(s + 1)$, $s > \gamma > -1$, and equation (2.266) of its possible numerical inversion via the Fourier integral, along with the related discussion at the end of Section 2.8.3. Show that such an approximated numerical inverse is given by the data in Fig. 4.35d (dots) for the exact inverse of

$$g(t) = \begin{cases} 0, & t < 0 \\ e^{-t}, & t > 0 \end{cases}$$

Hint: Use $\gamma = 0$ in (2.266) and follow the discussions and the example given at the end of Section 2.8.3.

(b) Discuss the error involved in part (a) and how it would compare to that of the example of $G(s) = 1/s$ in Section 2.8.3.

(c) Do parts (a) and (b) for the Laplace transform

$$G(s) = \frac{1}{s - 1}, \qquad s > \gamma > 1$$

Hint: Let $\gamma = 2$.

ANS. (a)   See Fig. 4.35d for the approximate answer (dots) and the exact answer (solid line).

   (b)   The error is what shows in Fig. 4.35d without magnification, since the factor in (2.266) is $e^{\gamma t} = 1$ for $\gamma = 0$. This error is a *Gibbs phenomenon* in the Fourier analysis approximation of functions with jump discontinuities, here at $t = 0$.

   (c)   The factor $e^{\gamma t} = e^{2t}$, which is a (major) magnification factor to multiply Fig. 4.35d (dots) by to have the approximate numerical inverse of $G(s) = 1/(s-1)$—a formidable error!—where we know the exact answer as

$$g(t) = \begin{cases} 0, & t < 0 \\ e^t, & t > 0 \end{cases}$$

Section 2.9   Other Important Transforms

2.94. (a)   Solve the following (singular) integral equation in $u(x)$:

$$P \int_{-\infty}^{\infty} \frac{u(x)}{x - \lambda} \, dx = \sin \lambda \tag{E.1}$$

Hint: Note that the integral is in the form of Hilbert transform of $u(x)$ as in (2.267). Consult the Hilbert transform tables to find

$$P \int_{-\infty}^{\infty} \frac{\cos bx \, dx}{x - \lambda} = -\pi \sin bx \tag{E.2}$$

and use it to solve (E.1)

ANS. (a) $u(x) = (-1/\pi) \cos x$

2.95. (a)   Find the Mellin transform of e^{-ax}, $a > 0$.

   (b)   Use the result in part (a) to find the Mellin transform of the (singular) integral equation in $u(x)$,

$$u(x) = e^{-ax} + \int_0^{\infty} e^{-x/\xi} u(\xi) \frac{d\xi}{\xi}, \qquad a > 0$$

Hint: Note that the integral is in the Mellin convolution form (2.273) with $f_1(\xi) = u(\xi)$ and $f_2(x/\xi) = e^{-x/\xi}$.

ANS.   (a) $\mathcal{M}\{e^{-ax}\} = \Gamma(\lambda)/a^{\lambda}$, $a > 0$.

(b) $U(\lambda) = \mathcal{M}\{u(x)\} = \Gamma(\lambda)/a^{\lambda} + \Gamma(\lambda)U(\lambda)$.

3

FINITE TRANSFORMS

Fourier Series and Coefficients

In Chapter 1 we introduced the general finite transform of $u(x)$ on the interval I as

$$U(\lambda_m) = \int_I \rho(x)\overline{K(\lambda_m,x)}u(x)\,dx \tag{3.1}$$

where $\{K(\lambda_m,x)\}$ is an orthogonal set on I, and $\overline{K(\lambda_m,x)}$ is the complex conjugate of $K(\lambda_m,x)$. The inverse of this transform is the Fourier series of $u(x)$ on I, i.e.,

$$u(x) = \sum_m \frac{U(\lambda_m)K(\lambda_m,x)}{N(\lambda_m)}, \qquad x \in I = (a,b) \tag{3.2}$$

where $N(\lambda_m)$ is the *norm square* of $K(\lambda_m,x)$,

$$N(\lambda_m) = \|K(\lambda_m,x)\|^2 \equiv \int_a^b \rho(x)|K(\lambda_m,x)|^2\,dx \tag{3.3}$$

In Section 1.5, we showed how to find the compatible transforms for boundary value problems associated with second-order differential equations (1.87)–(1.88). The essence of the results was that the desired kernel must satisfy a similar (homogeneous) boundary value problem (1.75)–(1.77). In Section 1.6, we illustrated, among other transforms, how to find compatible finite sine and Hankel transforms in Examples 1.10 and 1.11, respectively. The latter was done by using the self-adjoint form L_s

329

of the second-order operator L_2 in the general problem (1.90)–(1.92) and finding the kernel in (1.96), (1.99)–(1.100), (1.102).

In this chapter, we will introduce more finite transforms with emphasis on their applications to solving boundary value problems and their use for signal representations in connection with the sampling expansion, as already done in Section 2.4 of Chapter 2 for the bandlimited functions of the exponential Fourier transform (2.138). The spacings required by such sampling expansions will be vital for the derivation of the discrete Fourier and other transforms of Chapter 4.

In the last Section 3.7 of this chapter we will try to touch on the difficulty such (linear) finite transforms (or the integral transforms of Chapter 2) will face in handling nonlinear problems. The notion is illustrated by a way of an example. This is related to a very specialized analytic iterative method for nonlinear mechanical vibration problems which involves the Laplace transform and requires very detailed Laplace transform pairs. It is to be followed by a brief summary of a recent numerical (modified) iterative method for a variety of nonlinear boundary value problems that involves the Fourier and other transforms. The latter is done again to point out the difficulty of handling nonlinear problems with the (linear) Fourier, or other more general, transforms.

Consider the orthogonal set of functions $\{K(\lambda_n,x)\}$, with respect to the weight function $\rho(x)$, on the interval (a,b),

$$\int_a^b \rho(x)K(\lambda_n,x)\overline{K(\lambda_m,x)}\,dx = N(\lambda_n)\delta_{m,n} \tag{3.4}$$

where $\delta_{m,n}$ is the *Kronecker delta*,

$$\delta_{m,n} = \begin{cases} 0, & m \neq n \\ 1, & m = n \end{cases} \tag{3.5}$$

and $N(\lambda_n)$ is the *norm square* of $K(\lambda_n,x)$,

$$N(\lambda_n) = \int_a^b \rho(x)|K(\lambda_n,x)|^2\,dx \equiv \|K(\lambda_n,-)\|^2 \tag{3.3}$$

The basic differential equations and their solutions of such important special functions $K(\lambda,x)$ will be covered in some detail in Appendix A. Also, the basic properties of such special functions are listed in Appendix A.5, and many of them can be proved as basic exercises. For more on special functions, the well-known references (Erdelyi, 1953a, 1953b; Abramowitz and Stegun, 1964) can be consulted.

We define the *general finite transform* of $f(x)$ on (a,b) as

$$F(\lambda_n) = \int_a^b \rho(x)\overline{K(\lambda_n,x)}f(x)\,dx \tag{3.6}$$

Here we refer to the interval $(a,b)\rho$ to indicate the weight function $\rho(x)$ used, but we will assume it clearly understood from the integral. From the Fourier series expansion (1.36), (1.37), or (1.67)–(1.68), the inverse transform $f(x)$ of $F(\lambda_n)$ is the Fourier series representation of $f(x)$ on (a,b), in terms of the same orthogonal set (3.1); that is,

$$f(x) = \sum c_n K(\lambda_n,x) = \sum \frac{F(\lambda_n)K(\lambda_n,x)}{N(\lambda_n)} \tag{3.7}$$

with Fourier coefficients $c_n = F(\lambda_n)/N(\lambda_n)$. The limits of this infinite Fourier series depend on the particular orthogonal set $\{K(\lambda_n,x)\}$.

When we use an *orthonormal* set (i.e., $N(\lambda_n) = 1$), as in (1.85) or (1.38), our finite transform $F(\lambda_n)$ becomes exactly the Fourier coefficient c_n.

We should point out that the finite transform methods do not solve problems that are intractable by the Fourier series method; instead, they facilitate arriving directly at the solution. This is especially true when we have nonhomogeneous problems.

Our first examples of finite transforms include the finite sine and cosine transforms in Section 3.2. These are followed by the finite exponential Fourier transform in Section 3.3, the finite Hankel transform in Section 3.4, and the classical orthogonal polynomial transforms like the Legendre, Tchebychev, Laguerre, and Hermite transforms in Section 3.5.

The finite transform (3.6) is a Fourier coefficient to its inverse $f(x)$, the general Fourier series, or orthogonal expansion of $f(x)$ in terms of the orthonormal set ($N(\lambda_n) = 1$ in (3.7)) of functions $\{K(\lambda_n,x)\}$ on (a,b). The first question that comes to mind concerns the convergence of this Fourier series for it to exist as representing $f(x)$ on (a,b). Furthermore, this series and also the integral of the coefficients represent, as transforms, our very basic operational tools. This means that they will be exposed to varied mathematical operations like differentiation, integration, and the limit process. In general, it is known that these operations are not allowed, as we saw in the case of the Fourier transforms in the last chapter, unless the series or the series after the particular operation has a certain "good enough" (or satisfactory) convergence. Thus to put our finite transform, and in particular its inverse, as the infinite Fourier series on sound mathematical grounds, we must study and know what type of convergence the Fourier series have for representing a particular class of functions $f(x)$ on (a,b). For this reason, we shall, before introducing the various finite transforms and their applications, cover the topic of Fourier series and the orthogonal expansion in Section 3.1. The emphasis will be on illustrating the most basic types of convergence in Section 3.1.1, while the proofs are left to Section 3.1.3.

We will start by introducing the (complex) exponential Fourier series and its equivalent, the familiar trigonometric sine-cosine series, which are for periodic functions. Then we will state very carefully the familiar types of convergence and illustrate them for a good number of examples of Fourier series expansion in Section 3.1.1.

Such elementary convergence theorems are done for *piecewise smooth functions*, or *square integrable functions*, on (a,b). The complete detailed proofs of these theorems for the Fourier trigonometric sine-cosine series are done in Section 3.1.3 after we introduce the basic elements for the convergence of the general infinite series in Section 3.1.2.

3.1   Fourier (Trigonometric) Series and General Orthogonal Expansion

---

The Exponential Fourier Series and the Trigonometric Fourier Series

In this section we will present the (complex) exponential Fourier series expansion of $f(x)$ on the symmetric interval $(-l,l)$,

$$f(x) = \sum_{n=-\infty}^{\infty} c_n e^{in\pi x/l}, \qquad -l < x < l \tag{3.8}$$

with c_n as its Fourier coefficients,

$$c_n = \frac{1}{2l} \int_{-l}^{l} f(x) e^{-in\pi x/l}\, dx \tag{3.9}$$

Parallel to what we did for the Fourier exponential transforms in (2.64), (2.65)—i.e., showed that they are equivalent to the Fourier sine-cosine transforms (2.64a), (2.65a), (2.65b)—this *exponential* Fourier series (3.8) is also equivalent to the following (usual) form of the (*trigonometric*) Fourier *sine-cosine* series,

$$f(x) = a_0 + \sum_{n=1}^{\infty} a_n \cos\frac{n\pi x}{l} + b_n \sin\frac{n\pi x}{l}, \qquad -l < x < l \tag{3.10}$$

$$a_0 = \frac{1}{2l} \int_{-l}^{l} f(x)\, dx \tag{3.11}$$

$$a_n = \frac{1}{l} \int_{-l}^{l} f(x) \cos \frac{n\pi x}{l} \, dx \qquad (3.12)$$

$$b_n = \frac{1}{l} \int_{-l}^{l} f(x) \sin \frac{n\pi x}{l} \, dx \qquad (3.13)$$

Note that in some texts a_0 in (3.10) is replaced by $\frac{1}{2}a_0$, where a common formula for a_n, $n = 0, 1, 2, \ldots$, is given as in (3.12) for all $a_0, a_1, a_2, \ldots$. The equivalence of the above two representations (3.10) and (3.8) can easily be shown when we use the following *Euler identities* for $\cos(n\pi x/l)$ and $\sin(n\pi x/l)$,

$$\cos \theta = \frac{1}{2} [e^{i\theta} + e^{-i\theta}]$$

$$\sin \theta = \frac{1}{2i} [e^{i\theta} - e^{-i\theta}] = -\frac{i}{2} [e^{i\theta} - e^{-i\theta}] \qquad (3.14)$$

(also note that $e^{\mp i\theta} = \cos \theta \mp i \sin \theta$) in the Fourier sine-cosine series (3.10). Then we group the coefficients of $e^{in\pi x/l}$ and $e^{-in\pi x/l}$ to have

$$f(x) = a_0 + \frac{1}{2} \sum_{n=1}^{\infty} a_n (e^{in\pi x/l} + e^{-in\pi x/l}) - ib_n(e^{in\pi x/l} - e^{-in\pi x/l})$$

$$= a_0 + \sum_{n=1}^{\infty} \frac{1}{2}(a_n - ib_n)e^{in\pi x/l} + \frac{1}{2}(a_n + ib_n)e^{-in\pi x/l}$$

$$= c_0 + \sum_{n=1}^{\infty} c_n e^{in\pi x/l} + c_{-n}e^{-in\pi x/l}$$

After defining $c_0 = a_0$, $c_n = \frac{1}{2}(a_n - ib_n)$, and $c_{-n} = \frac{1}{2}(a_n + ib_n)$, and writing the "dummy" variable $-n$ in the last sum as m, we have

$$f(x) = c_0 + \sum_{n=1}^{\infty} c_n e^{in\pi x/l} + \sum_{m=-\infty}^{m=1} c_m e^{im\pi x/l}$$

and if we combine the two series we arrive at the desired result (3.8),

$$f(x) = \sum_{n=-\infty}^{\infty} c_n e^{in\pi x/l} \qquad (3.8)$$

The coefficients c_n can easily be seen in the form (3.9),

$$c_n = \frac{1}{2l} \int_{-l}^{l} f(x) e^{-in\pi x/l} \, dx \qquad (3.9)$$

since for n a positive integer,

$$c_n = \tfrac{1}{2}(a_n - ib_n) = \frac{1}{2l} \int_{-l}^{l} f(x) \left[\cos \frac{n\pi x}{l} - i \sin \frac{n\pi x}{l} \right] dx$$

$$= \frac{1}{2l} \int_{-l}^{l} f(x)e^{-in\pi x/l} dx$$

and the same can be shown for c_{-n} with $-n$ as a negative integer.

The Fourier exponential series and its coefficients (3.8), (3.9) can also be written in the following form:

$$f(x) = \sum_{n=-\infty}^{\infty} c_n e^{-in\pi x/l} \tag{3.8a}$$

$$c_n = \frac{1}{l} \int_{-l}^{l} f(x)e^{in\pi x/l} dx \tag{3.9a}$$

with a simple change of the dummy variable n to $-n$ in (3.8) and then (3.9). This is also done to reassert the form of our general definition of the finite transform (3.6) with the kernel $\overline{K(\lambda_n, x)} = \bar{e}^{in\pi x/l} = e^{in\pi x/l}$. Of course, (3.8), (3.9) and (3.8a), (3.9a) are equivalent and we are free to work with either one. This brings up another reason for mentioning the form (3.8a), (3.9a), which is its possible very usual use in some specific fields, where we should be ready for it for our own convenience.

Even and Odd Functions

Just as we did for the Fourier sine-cosine transforms, it should be clear from the symmetric integrals of the Fourier coefficients (3.10)–(3.12) that the special cases of odd and even functions $f_o(x)$, $f_e(x)$ on $(-l, l)$ have a sine and cosine Fourier series representation, respectively,

$$f_o(x) = \sum_{n=1}^{\infty} b_n \sin \frac{n\pi x}{l}, \qquad -l < x < l \tag{3.15}$$

with

$$b_n = \frac{2}{l} \int_{0}^{l} f(x) \sin \frac{n\pi x}{l} dx \tag{3.16}$$

and

$$f_e(x) = a_0 + \sum_{n=1}^{\infty} a_n \cos \frac{n\pi x}{l}, \qquad -l < x < l \tag{3.17}$$

with

$$a_0 = \frac{1}{l} \int_{0}^{l} f(x) dx \tag{3.18}$$

$$a_n = \frac{2}{l} \int_0^l f(x) \cos \frac{n\pi x}{l} \, dx \qquad (3.19)$$

If we compare the above two series (3.15), (3.17), we notice that they add up to just the sine-cosine series (3.10) for $f(x)$ on $(-l,l)$, which, of course, is not necessarily even or odd. This suggests that any function $f(x)$ on $(-l,l)$ can be written as the sum of odd and even functions: $f(x) = f_e(x) + f_o(x)$ on $(-l,l)$. It can be proved by writing

$$f(x) = \tfrac{1}{2}[f(x) + f(-x)] + \tfrac{1}{2}[f(x) - f(-x)] = f_e(x) + f_o(x)$$

where the first term is an even function and the second term is an odd one. The latter can be verified by changing x to $-x$ in each of these two terms in brackets to see that the first bracket is unchanged, while the second one picks up a minus sign.

In this line we may look closely at the Fourier sine-cosine series (3.10) for $f(x)$, to note that the coefficient a_0 represents the average value f_{av} of $f(x)$ on the interval $(-l,l)$,

$$a_0 = f_{av} = \frac{1}{2l} \int_{-l}^l f(x) \, dx$$

This may be interpreted, in favor of the Fourier series representation, to say that just one term of the infinite series (3.10) gives us the series' best approximation possible for the wave $f(x)$, namely its average value over its domain $(-l,l)$.

Periodic Functions—Inheritance from Their Fourier (Trigonometric) Series Representation

We note that either one of the above exponential (3.8) or trigonometric (3.10) Fourier series represents $f(x)$ on $(-l,l)$ and also extends it as a periodic function $\bar{f}(x)$ on $(-\infty, \infty)$ with period $2l$; that is, $f(x \mp 2l) = f(x)$. This is seen when we substitute $x \mp 2l$ for x in the series (3.10) (or (3.8)) where we know that the sine and cosine functions are periodic with period $2l$; that is,

$$\sin \frac{n\pi}{l}(x \mp 2l) = \sin\left(\frac{n\pi x}{l} \mp 2n\pi\right) = \sin \frac{n\pi x}{l}$$

and

$$\cos \frac{n\pi}{l}(x \mp 2l) = \cos\left(\frac{n\pi x}{l} \mp 2n\pi\right) = \cos \frac{n\pi x}{l}$$

$$f(x \mp 2l) = a_0 + \frac{1}{2}\sum_{n=1}^{\infty} a_n \cos \frac{n\pi x}{l}(x \mp 2l) + b_n \sin \frac{n\pi}{l}(x \mp 2l)$$

$$= a_0 + \frac{1}{2}\sum_{n=1}^{\infty} a_n \cos \frac{n\pi x}{l} + b_n \sin \frac{n\pi x}{l} = f(x),$$

$f(x \mp 2l) = f(x)$. It can also easily be shown that such $f(x)$ is periodic with period $2ml$, $m = 1, 2, 3, 4, \ldots$. This is a formal presentation of the exponential—and the more familiar trigonometric sine-cosine—Fourier series of periodic functions.

Of course, we will give a number of illustrations of these series representations for different functions. But, more important is the mathematical basis for the convergence of those series to certain class of such functions. The latter will be presented in Section 3.1.1 in the form of Theorems 3.1 and 3.4 for *pointwise convergence* and *convergence in the mean* to the class of piecewise smooth (or square integrable) functions on $(-l,l)$. The *uniform convergence* Theorem 3.2 is given for piecewise smooth functions and the condition of no jump between the end points: $f(-l) = f(l)$. These theorems will be followed by three very detailed Examples, 3.2, 3.3, and 3.4, of computing Fourier series and their Fourier coefficients as presented above in (3.10)–(3.13). The discussion following these computations will illustrate the application of the convergence theorems (Theorems 3.1–3.4) to the Fourier series representation of functions in Examples 3.2–3.4. We will also illustrate in Example 3.4A how the Nth partial sum of the Fourier series approximates the represented function, along with a brief discussion of the truncation error resulting from such approximation. In Example 3.4B we will illustrate the numerical computations for Fourier integrals (transforms), including the Fourier coefficients' integrals. Detailed analyses of the main errors in such Fourier analysis computations are the subject of Chapter 4, in particular Sections 4.1.6 and 4.1.7.

Before we present the convergence theorems and their applications in Section 3.1.1, we feel it instructive to present some basic graphical illustrations of the Fourier series representation of periodic functions. The complete detailed proofs of the convergence theorems will be delayed until Section 3.1.3, after we cover the basic elements of the convergence for general infinite series in Section 3.1.2.

Figure 3.1a shows the (odd) function $f(x) = x$, $-l < x < l$, and its periodic extension $\bar{f}(x)$,[†] with period $2l$. Note that the function is continuous on $(-l,l)$, while its periodic extension, which is forced on it by its Fourier trigonometric series (3.10), has jump discontinuities of size $J = 2l$ at all (end) points $x_m = \mp(2m + 1)l$, $m = 0, 1, 2, 3, \ldots$. Of course, such periodic extensions of the trigonometric series representation can be used to an advantage if we are in need of replicas of $f(x)$, $-l < x < l$, which

---

[†] The bar on $\bar{f}$ for this periodic extension of $f(x)$ should not be confused with the thick bar on $\bar{f}$, used primarily in Chapter 4, for the superposition of all the translations by nT of the (in general) not necessarily periodic function f.

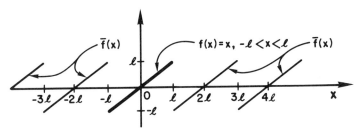

FIG. 3.1a   Periodic extension of the *odd* function $f(x) = x$, $-l < x < l$, period $2l$.

is the case in the analysis of electrical signals and circuits. Figure 3.1b shows the (even) function $f(x) = x^2$, $-l < x < l$, and its periodic extension $\bar{f}(x)$, with period $2l$, which is continuous on $(-\infty, \infty)$. We summarize that the periodic extension for odd functions (with $f(l) \neq 0$) results in jumps at the end points $\mp(2m + 1)l$ of all its periodic intervals, which makes its periodic extension $\bar{f}(x)$ piecewise continuous. On the other hand, the periodic extension for even functions is a continuous periodic extension at the end points $\mp(2m + 1)l$ of all its periodic intervals. Thus the periodic extension $\bar{f}(x)$ of the even (continuous) function in Fig. 3.1b, as given by the Fourier series (3.10), is continuous on the whole real line.

As we shall show soon, and as we did in the case of the Fourier integral representation in Chapter 2, the Fourier series will make the compromise to converge to a value at the middle of the jumps, besides possible other jumps in the interior, of the points of discontinuity $x_m = \mp(2m + 1)l$. Such values assigned by the Fourier series are, of course, not defined values for our function $f(x)$. The reason is that $f(x)$ is defined only at the interior of the basic period $(-l, l)$ and the interior of the (extended) periods

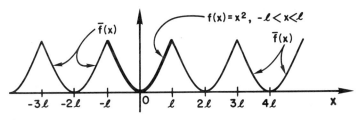

FIG. 3.1b   Periodic extension of the *even* function $f(x) = x^2$, $-l < x < l$, period $2l$.

of its periodic extension $\bar{f}(x)$. Clearly, there is no such problem for the periodic extension of our particular example of the continuous even function $f(x) = x^2$, $-l < x < l$, in Figure 3.1b. Our particular examples here happed to be both continuous on $(-l,l)$, and our conclusion was for jump discontinuities at the ends $\pm l$ of the period for the odd functions and continuity for the even function. Of course, in general, such even and odd functions may be piecewise continuous on $(-l,l)$. In this case these jump discontinuities will be reproduced in the periodic extension of both the odd and even functions (see also Figs. (3.6), (3.7), (3.5)).

In the spirit of Chapter 4, where we will discuss in detail the error involved in the practical application of the Fourier series and integrals, such jump discontinuities will represent a "headache" for our computations. For example, while computing the Fourier series (3.10) of $f(x)$, we can only evaluate and add a finite number of terms of the Fourier series, which we term the Nth partial sum $S_N(x)$

$$S_N(x) = a_0 + \sum_{n=1}^{N} a_n \cos \frac{n\pi x}{l} + b_n \sin \frac{n\pi x}{l} \qquad (3.20)$$

In this case, we must admit an error $\epsilon_N = |f(x) - S_N(x)|$, termed the *truncation error*, in our approximation of $f(x)$ by the finite partial sum $S_N(x)$ and not the mathematically exact (infinite) Fourier series (3.10). We will state a result concerning the "stubbornness" of this error at x being very close to the point of the jump discontinuities of $\bar{f}(x)$, which may be a surprise to some (see Figures 3.5c,d)! But it should not be surprising when we realize that it reflects the difference between the reality of the actual numerical computations for a somewhat arbitrary but bounded function, in terms of a *finite* number of very smooth trigonometric functions, and the mathematically exact (but ideal!) analytical expressions in terms of the very smooth trigonometric functions, in this case an *infinite* Fourier series. This result says that *if J is the size of the jump at $x = l$, i.e., $J = f(l+) - f(l-)$, then the maximum truncation error ϵ_N at a point x very close to l, termed the "Gibbs phenomenon," is about 9% of the jump size J, regardless of how large a (finite) number of terms N we take for the partial sum $S_N(x)$.* This means that, for example, in our illustration of the odd function $f(x) = x$ in Fig. 3.1a, the partial sum of sine functions $S_N(x)$ overshoots in the small neighborhood of $x = l$ before it comes down to its value $S_N(l) = 0$ at $x = l$, as seen in Fig. 3.1c. On the other hand, for the even function $f(x) = x^2$, we don't expect such extreme difficulty for its partial sum of cosine functions $S_N(x)$. The situation, of course, will be different when we want to speak of a Fourier series for the derivative $f'(x) = 2x$, $-l < x < l$, of this even function x^2, which will have a

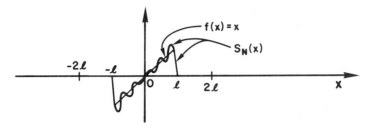

FIG. 3.1c   Fourier series approximation $S_N(x)$ of $\overline{f}(x)$, $-l < x < l$, in part (a).

jump discontinuity at $x = \mp(2m + 1)l$, $m = 0, 1, 2, 3, \ldots$ Here its expected Fourier approximation $dS_N(x)/dx$ will suffer a Gibbs phenomenon at these points just like that of $f(x) = x$, $-l < x < l$, in Fig. 3.1a. The Gibbs phenomenon will be discussed and illustrated in Section 4.1.6G of Chapter 4.

Fourier Series of f(x) on the General Interval (a, b)

Both the Fourier exponential and sine and cosine series (3.6) and (3.10) are written for a function on the symmetric interval $(-a, a)$. For a wave $f(x)$ on the general interval (a, b), we can derive the following Fourier series representation on (a, b):

$$f(x) = a_0 + \sum_{n=1}^{\infty} a_n \cos \frac{2\pi n}{b - a} (x - \tfrac{1}{2}(a + b))$$

$$+ b_n \sin \frac{2\pi n}{b - a} (x - \tfrac{1}{2}(a + b)), \qquad a < x < b \tag{3.21}$$

$$a_0 = \frac{1}{b - a} \int_a^b f(x)\, dx \tag{3.22}$$

$$a_n = \frac{2}{b - a} \int_a^b f(x) \cos \frac{2\pi n}{b - a} (x - \tfrac{1}{2}(a + b)) \tag{3.23}$$

$$b_n = \frac{2}{b - a} \int_a^b f(x) \sin \frac{2\pi n}{b - a} (x - \tfrac{1}{2}(a + b)) \tag{3.24}$$

This is done, as shown in Fig. 3.2, by translating our y coordinates to the middle of the interval (a, b); that is, let $x' = x - \tfrac{1}{2}(a + b)$ to have our function now $f(x) = g(x')$ on a symmetric interval in x' as $x' \in (-c, c) = (-(b - a)/2, (b - a)/2)$. However, for our transforming back to x we have

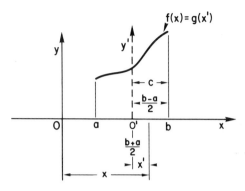

FIG. 3.2 $f(x)$, $a < x < b$, in the translated coordinates as $f(x) = g(x')$, $-c < x' < c$.

its limits from $x = a$ to $x = b$,

$$f(x) = g(x') = a_0 + \sum_{n=1}^{\infty} a_n \cos \frac{n\pi x'}{c} + b_n \sin \frac{n\pi x'}{c},$$

$$c = \frac{b-a}{2}, \quad -c < x' < c$$

$$f(x) = a_0 + \sum_{n=1}^{\infty} \left[a_n \cos \frac{2n\pi}{b-a}(x - \tfrac{1}{2}(a+b)) \right.$$

$$\left. + b_n \sin \frac{2n\pi}{b-a}(x - \tfrac{1}{2}(a+b)) \right], \quad a < x < b \tag{3.21}$$

$$a_0 = \frac{1}{2c} \int_{-c}^{c} g(x')\,dx' = \frac{1}{b-a} \int_{a}^{b} f(x)\,dx \tag{3.22}$$

$$a_n = \frac{1}{c} \int_{-c}^{c} g(x') \cos \frac{n\pi x'}{c}\,dx'$$

$$= \frac{2}{b-a} \int_{a}^{b} f(x) \cos \frac{2n\pi}{b-a}(x - \tfrac{1}{2}(a+b)) \tag{3.23}$$

$$b_n = \frac{1}{c} \int_{-c}^{c} g(x') \sin \frac{n\pi x'}{c}\,dx'$$

$$= \frac{2}{b-a} \int_{a}^{b} f(x) \sin \frac{2n\pi}{b-a}(x - \tfrac{1}{2}(a+b)) \tag{3.24}$$

after noting again that the limits $-c$ to c in x' go back as a to b in x.

Now that we have presented just the forms of the (usual) Fourier sine and cosine series and the exponential series representation for periodic functions, there remain the two questions regarding this association of periodic functions and Fourier series. The first and most important question is what kind of functions $f(x)$, with period $2l$ (in (3.10) or (3.8)), are deserving of the Fourier series representation? The second question is, how does the quality of the convergence of the Fourier series depend on the properties of the represented functions? The second question concerns the condition under which the convergence of this series allows certain operations. For example, what are the conditions on $f(x)$ of this series for the convergent series to allow its differentiation or integration? The answer to the first question, as we may suspect from the last chapter's Fourier integral representation, lies, for a simple theory, in limiting our functions to being piecewise smooth on $(-l,l)$. Here, the series will still assign its own value at the middle of a jump discontinuity, where the function is not even defined. Furthermore, even if the function is continuous on $(-l,l)$, its periodic extension may not be defined at the end points $\mp(2m + 1)l$. Thus, if we limit our functions even more to being continuous with piecewise smooth derivative, and $f(-l+) = f(l-)$, then the series and the function coincide. This should stand as a motivation behind the conditions on $f(x)$ for having a uniform convergent Fourier series, as we shall state in Theorem 3.2.

All these conditions on the periodic functions $f(x)$ will soon be carefully stated as Theorems 3.1 to 3.4 and illustrated in Section 3.1.1. The theorems will then be proved in Section 3.1.3A on orthogonal expansions and Fourier series, after covering the basic elements of infinite series in Section 3.1.2.

The next question would be how the Fourier coefficients were derived. The answer is dependent on the orthogonality property of the trigonometric functions $\{1, \sin(n\pi x/l), \cos(n\pi x/l)\}_{n=1}^{\infty}$. The orthogonality of functions (1.66) was discussed as general property of the regular Sturm–Liouville problem in Example 1.9 and was then used for deriving (1.68) the Fourier coefficients of the general orthogonal expansion (1.67). To show here, for example, that the sine functions of the trigonometric Fourier series (3.10) are orthogonal on $(-l,l)$, we must show that for $n \neq m$ the following integral of their product vanishes:

$$\int_{-l}^{l} \phi_n(x)\phi_m(x)\,dx = \int_{-l}^{l} \sin\frac{n\pi x}{l} \sin\frac{m\pi x}{l}\,dx = 0, \qquad n \neq m$$

This was assigned as (a simple) Exercise 1.12b in Section 1.3 of Chapter 1.

The easier way of showing the orthogonality of the trigonometric functions of the Fourier series (3.10) is to recognize both $\sin(n\pi x/l)$ and $\cos(n\pi x/l)$ as eigenfunctions of the special case of the (regular) Sturm–Liouville problem (1.63)–(1.65) with $\rho = 1$, $r = 1$, $q = 0$, and the parameter λ is replaced by $-\lambda^2$ as we will illustrate in the following example for the problem defined on $(0,l)$. The case for the symmetric interval $(-a,a)$ or the nonsymmetrical interval (a,b) follows readily.

EXAMPLE 3.1: Sturm–Liouville Problem as Source of Orthogonal Functions

(a) We can show that $\sin(n\pi x/l)$ are solutions of the boundary value problem in $y(x)$,

$$\frac{d^2y}{dx^2} + \lambda^2 y = 0, \qquad 0 < x < l \tag{E.1}$$

$$y(0) = 0 \tag{E.2}$$

$$y(l) = 0 \tag{E.3}$$

where

$$y(x) = c_1 \cos \lambda x + c_2 \sin \lambda_x \tag{E.4}$$

$c_1 = 0$ to satisfy (E.2), and λ has to be discretized to $\lambda_n = n\pi/l$ to satisfy (E.3),

$$y_n = c_2 \sin \frac{n\pi x}{l} \tag{E.5}$$

(b) The same can be done for $\cos(n\pi x/l)$ as solutions to the boundary value problem:

$$\frac{d^2y}{dx^2} + \lambda^2 y = 0, \qquad 0 < x < l \tag{E.6}$$

$$y'(0) = 0 \tag{E.7}$$

$$y'(l) = 0 \tag{E.8}$$

(c) For the boundary value problem,

$$\frac{d^2y}{dx^2} + \lambda^2 y = 0, \qquad 0 < x < l \tag{E.9}$$

$$y'(0) = 0 \tag{E.10}$$

$$y(l) = 0 \tag{E.11}$$

we find the solutions as

$$y_n(x) = \cos\left(\frac{2n + 1}{2}\right)\frac{\pi x}{l} \qquad (E.12)$$

which must be orthogonal on $(0,l)$, and also on $(-l,l)$!

The last, but very important, question concerns the operations which are permissible in this Fourier (orthogonal) series representation of the given function ((3.10) or (3.8)) if we are to involve it in our analysis. For example, we may want to show that this function is a solution of a differential equation. The question is, can we dare differentiate this series, i.e., enter the differentiation operation inside the infinite series, and how would that affect its "precious" convergence? This, as we shall see, will depend on how good the represented function is and, to be simple, if the function itself is good enough to allow one or more differentiations. The other operation would be to integrate the infinite series term by term. Again, we feel it necessary, as we saw for the Fourier transforms, to present such conditions as very precise theorems. We will prove the most basic of theorems in Section 3.1.3. We will do these proofs after introducing in Section 3.1.2 the basic elements of the convergence of infinite series in general. Then our Fourier series and the general orthogonal expansion are taken as special cases.

3.1.1 Convergence of the Fourier Series

Parallel to the Fourier integral theorem, of the Fourier transforms, we will state here the fundamental theorem on the (*pointwise*) *convergence* of Fourier series (Theorem 3.1) for piecewise smooth functions. Like the proof of the Fourier integral formula (Theorem 2.14), the proof of this theorem is a bit long; however, we have developed all the necessary tools for it. In order to present illustrations for the Fourier series, along with its convergence, we shall state this important theorem and concentrate on its application in the next few examples. We will also state here the *uniform convergence* Theorems 3.2, 3.2a for continuous and piecewise smooth functions and the (practical) *convergence in the mean* Theorems 3.4, 3.4a for piecewise continuous functions (or the more general class of square integrable functions). These theorems will be illustrated in the following examples; their complete proofs will be delayed until after the next section, when we cover the basic elements of convergence of series in general and the Fourier series and orthogonal expansion as particular cases. To name one basic reason for delaying these proofs, it is the need for the orthogonality property of the elements (bases) of the Fourier series expansion and its resulting Bessel inequality (3.60) for approximating piecewise continuous (or square integrable) functions $f(x)$ on the finite interval (a,b).

THEOREM 3.1: Pointwise Convergence of Fourier Series

If $f(x)$ is a piecewise smooth function on the interval $(-l,l)$, its Fourier series converges at each point x of this interval, and

$$a_0 + \sum_{k=1}^{\infty} a_k \cos \frac{k\pi x}{l} + b_k \sin \frac{k\pi x}{l} = \frac{f(x+) + f(x-)}{2} \tag{3.25}$$

for x in the interior of the interval, i.e., $x \in (-l,l)$. For the end points $x = \mp l$, the series converges to $(f(-l+) + f(l-))/2$. Clearly, when $f(x)$ is continuous for all $x \in (-l,l)$, the series (3.25) will converge to $(f(x) + f(x))/2 = f(x)$—i.e., it converges uniformly to $f(x)$—on $(-l,l)$. □

A more general version of the above (more familiar) pointwise convergence is the following Theorem 3.1a, which gives this convergence for *absolutely integrable* functions on $(-l,l)$:

THEOREM 3.1a: Pointwise Convergence of Fourier Series
(Another Form)

If $f(x)$ is absolutely integrable on $(-l,l)$, i.e., $\int_{-l}^{l} |f(x)| dx < \infty$, its trigonometric Fourier series (3.10) converges at points where $f(x)$ is differentiable. If the function $f(x)$ is has a jump discontinuity at a point x_0, then the series converges to the average of the left- and right-hand limits of $f(x)$, i.e., to $(f(x_0+) + f(x_0-))/2$. □

EXAMPLE 3.2:

The Fourier series for the function

$$f(x) = \begin{cases} 2, & -\pi < x < \pi/2 \\ -1, & \pi/2 < x < \pi \end{cases}$$

(as in Fig. 3.3) converges pointwise to $f(x)$ for $-\pi < x < \pi/2$ and $\pi/2 < x < \pi$ since the function is absolutely integrable on $(-\pi, \pi)$,

$$\int_{-\pi}^{\pi} |f(x)| dx = \int_{-\pi}^{\pi/2} 2 \, dx + \int_{\pi/2}^{\pi} 1 \, dx = 3\pi + \frac{\pi}{2} = \frac{7\pi}{2}$$

The function has a jump discontinuity at $x = \pi/2$, and the series converges to $(2 - 1)/2 = 1/2$ (the dot in Fig. 3.3). Note that the value 1/2 given

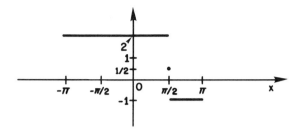

FIG. 3.3 The Fourier series and the function $f(x)$ are the same on $(-\pi, \pi)$ except at $x = \pi/2$, where $f(x)$ has a jump discontinuity.

by the series at $x = \pi/2$ here is not a value of $f(x)$, since $f(x)$ is not defined at $x = \pi/2$. The series converges to the value of $f(x) = 2$ for $-\pi < x < \pi/2$ and $f(x) = -1$ for $\pi/2 < x < \pi$, where the function is obviously continuous. The same conclusions can be arrived at from the familiar version of Theorem 3.1, where the function here is piecewise smooth on $(-\pi, \pi)$.

Theorem 3.1 is for piecewise smooth functions. To show the dependence of a better quality of convergence—uniform convergence for all $x \in [-l, l]$—on the quality of the functions represented, we will state Theorem 3.2 on uniform convergence of Fourier series. The proof will need Bessel's inequality (3.60), which we shall develop soon for the general orthogonal series expansion.

THEOREM 3.2: Uniform Convergence of Fourier Series

If a function $f(x)$

(i)   is continuous and piecewise smooth on an interval $[-l, l]$, and
(ii) assumes equal values at the end points of this interval, i.e., $f(-l) = f(l)$,

then its Fourier series is uniformly convergent on the interval $[-l, l]$, and

$$a_0 + \sum_{k=1}^{\infty} a_k \cos \frac{k\pi x}{l} + b_k \sin \frac{k\pi x}{l} = f(x) \tag{3.25}$$

at each x in $[-l, l]$. □

The complete proof is done in Example 3.6. For completeness, we present another very close version of this uniform convergence theorem as Theorem 3.2a.

THEOREM 3.2a: Uniform Convergence of Fourier Series
                           (Another Form)

If $f(x)$ is

(i)   continuous for $-l \leq x \leq l$,
(ii) $f(l) = f(-l)$, and
(iii) $\int_{-l}^{l} f'^2(x) dx < \infty$,

then the trigonometric Fourier series converges uniformly to $f(x)$. $\square$

In Theorem 3.2 we note how the continuity of $f(x)$ on $[-l,l]$ and the equality of the values of $f(x)$ at the end points $x = \mp l$ were added to the only condition, of piecewise smooth $f(x)$ on $[-l,l]$, of the pointwise convergence Theorem 3.1 in order to have uniform convergence.

This improvement in the kind of convergence may raise the question of how good it can be, or how fast the series will converge, if we are to have even better-quality functions than those used for Theorem 3.2. In Theorem 3.2 we offered only $f(x)$ as continuous, where its first derivative $f'(x)$ is piecewise continuous ($f(x)$ piecewise smooth). What if we offer $f(x)$ and all its derivatives up to $d^m f/dx^m$ (m large) as continuous on $[a,b]$? The answer, as we may intuitively suspect and will state next in Theorem 3.3, is that the uniform convergence of the Fourier series, of such very smooth functions (plus similar conditions to those of Theorem 3.2 at the end points), will be fast and depends on the degree of smoothness m. It will turn out, as we will see in the following theorem, that its Fourier coefficients a_k and b_k will behave like $1/k^{m+1}$ for k large. This means that the terms in the series (3.10) will die out fast, and we can safely truncate it at some large $k = K$, where the truncation error will be small depending on the degree of smoothness m.

THEOREM 3.3:   Fast Uniform Convergence of Fourier Series for
                           Very Smooth Functions

If

(i) $f(x)$ and all its derivatives up to order m (i.e., $f(x)$, $f'(x)$, ..., $f^{(m)}(x)$, $m \geq 0$), are continuous on an interval $[a,b]$,

(ii) these derivatives assume equal values at the ends of the interval $[-l,l]$, i.e.,

$$f(-l) = f(l), \qquad f'(-l) = f'(l), \ldots, f^{(m)}(-l) = f^{(m)}(l) \qquad (3.26)$$

and

(iii) the derivative $f^{(m+1)}(x)$ is piecewise continuous on the interval $[-l,l]$, then

(i) The coefficients a_k and b_k of the Fourier series (3.10) for $f(x)$ satisfy the (asymptotic) relations

$$a_k = o\left(\frac{1}{k^{m+1}}\right) \quad \text{and} \quad b_k = o\left(\frac{1}{k^{m+1}}\right) \qquad \text{as } k \to \infty \qquad (3.27)$$

by the asymptotic notation with small o, $a_k = o\left(1/k^{m+1}\right)$, we mean that $\lim_{k\to\infty}\left(a_k/(1/k^{m+1})\right) = 0$.

Moreover,

(ii) the series

$$\sum_{k=1}^{\infty} k^n(|a_k| + |b_k|), \qquad n = 0, 1, 2, \ldots, m \qquad (3.28)$$

are convergent. $\qquad\qquad\qquad\qquad\qquad\qquad\qquad\qquad\qquad\qquad\square$

The complete proof is done in Example 3.7. Theorem 3.3 will be proved in detail right after the proof of the uniform convergence Theorem 3.2, and after we have developed tools of the general orthogonal expansion including Bessel's inequality in (3.62), (3.61).

Speaking of Bessel's inequality and the orthogonal expansion, another more practical type of convergence is used there, namely the *convergence in the mean* to *square integrable functions* $f(x)$ on the finite interval $(-l,l)$, i.e., $f(x) \in L_2(-l,l)$:

$$\int_{-l}^{l} |f(x)|^2\, dx < \infty \qquad (3.29)$$

We note that this class of functions is larger than the class of *piecewise continuous functions* $P(-l,l)$; the latter has bounded functions $f(x) \in P(-l,l)$ which are clearly square integrable on our given finite interval $(-l,l)$. Here we given an example of a function $f(x) = 1/x^{1/3}$, $0 < x < \pi$, which is not piecewise continuous on $[0,\pi]$ since it diverges at $x = 0$; however, it is square integrable on $(0,\pi)$,

$$\int_0^\pi \left(\frac{1}{x^{1/3}}\right)^2 dx = \int_0^\pi x^{-2/3}\, dx = 3x^{1/3}\,\Big|_{x=0}^{\pi} = 3\pi^{1/3}$$

which shows that square integrable functions on a finite interval are not necessarily piecewise continuous on that interval.

The following two versions of the convergence in the mean Theorems 3.4, 3.4a are respectively for the piecewise continuous functions $f(x) \in P(-l,l)$ and the more general and wider class of square integrable functions $f(x) \in L_2(-l,l)$. The latter is the most popular version in Fourier analysis, but we will prove Theorem 3.4 since it has conditions parallel to those of Theorem 3.1, as the very basic and elementary theorem on convergence of Fourier series.

THEOREM 3.4:   Convergence in the Mean of Fourier Series to Piecewise Continuous Functions

If $f(x) \in P(-l,l)$—i.e., it is piecewise continuous on $(-l,l)$—then its Fourier series (3.10) converges in the mean to $f(x)$; with the nth partial sum $f_n(x) \equiv S_n(x)$ of the Fourier series (3.10) of $f(x)$,

$$f_n(x) \equiv S_n(x) = a_0 + \sum_{k=1}^{n} a_k \cos \frac{k \pi x}{l} + b_k \sin \frac{k \pi x}{l}$$

we have

$$\lim_{n \to \infty} \int_{-l}^{l} [f(x) - f_n(x)]^2 \, dx = 0 \tag{3.30}$$

or in other words, given an arbitrary $\epsilon > 0$, there is an integer n_0 such that

$$\int_{-l}^{l} [f(x) - f_n(x)]^2 \, dx < \epsilon \quad \text{for } n > n_0 \qquad \square \tag{3.31}$$

The proof is done in Theorem 3.10 to show the *completeness* of the trigonometric (sine-cosine) Fourier basis in the series (3.10) as a special case of the orthogonal functions $K(\lambda_m, x)$ of the general orthogonal expansion (3.2). It will become very clear from Theorem 3.7 that the completeness is equivalent to the above *convergence in the mean* and to *the allowance of integrating term by term* of such convergent-in-the-mean Fourier series to square integrable functions $f(x) \in L_2(a,b)$.

Next we state the other stronger version of convergence in the mean of the Fourier series (3.10) to the larger class of square integrable functions $f(x) \in L_2(-l,l)$.

THEOREM 3.4a: Converges in the Mean of Fourier Series to Square Integrable Functions

If $f(x)$ is square integrable on $(-l,l)$—i.e., $\int_{-l}^{l} |f(x)|^2 dx < \infty$—then its Fourier trigonometric series (3.10) converges in the mean to $f(x)$ and we write

$$f(x) \sim a_0 + \sum_{k=1}^{\infty} a_k \cos \frac{k\pi x}{l} + b_k \sin \frac{k\pi x}{l}, \qquad -l < x < l \qquad \square \quad (3.32)$$

The sign $\sim$ is used because the series may not coincide with $f(x)$ at a set of a finite number of separate (zero-width) points. This can be seen from the condition of the square integrability of $f(x)$, where such an integral is unaffected even if $f(x)$ is defined with removable discontinuities at a set of separate points, or if it is defined to be infinite at such points. This, to put it simply, is because such an infinite value of $f(x)$ is given with zero width, hence it contributes nothing to the integral.

More details of defining such functions and the practical meaning of this convergence in the mean will be given in the prelude to proving the equivalent statement of the completeness of the trigonometric basis $\{1, \sin(k\pi x/l), \cos(k\pi x/l)\}_{k=1}^{\infty}$ of (3.32) on $(-l,l)$. This is also done in the much larger scope of the general orthogonal expansion of $f(x)$,

$$f(x) = \sum_{n=1}^{\infty} c_n \phi_n(x), \qquad a < x < b \qquad (3.33)$$

in terms of the general orthogonal set $\{\phi_n(x)\}$ on (a,b). The real general result for the class of square integrable functions, $f(x) \in L_2(a,b)$, is that this very general orthogonal expansion—compared to just using trigonometric functions in (3.32)—converges in the mean to $f(x) \in L_2(a,b)$.[†]

Clearly, in Example 3.2, the series converges in the mean, according to Theorem 3.4 because the function

$$f(x) = \begin{cases} 2, & -\pi < x < \pi/2 \\ -1, & \pi/2 < x < \pi \end{cases}$$

[†] This symbol $L_2(a,b)$ should not be confused with the second order differential operator L_2 that we used, primarily, in Chapter 1.

is piecewise continuous on $(-\pi, \pi)$, and according to Theorem 3.4a because it is square integrable there,

$$\int_{-\infty}^{\infty} |f(x)|^2 dx = \int_{-\pi}^{\pi/2} 4 dx + \int_{\pi/2}^{\pi} 1 dx = 6\pi + \frac{\pi}{2} = \frac{13\pi}{2}$$

As we mentioned before, after having stated Theorems 3.1–3.4 on the different types of convergence of the Fourier series (3.10), we will now present a number of illustrations of this Fourier series expansion with reference to illustrating the applications and the results of these convergence theorems. In addition, we will touch on the practical side of the computations of Fourier series and integrals, the subject of Chapter 4. This is done by computing the nth partial sum $S_n(x)$ of the series to see with detailed graphs how it approximates $f(x)$ and how it will have trouble near the jump discontinuities! The Fourier integrals had to be approximated by a discrete sum for its numerical computation.

Illustration of the Fourier Trigonometric Series and Its Convergence

EXAMPLE 3.3:

Consider the functions in parts (a)–(c) for their Fourier series, periodic extension, and discussion of Fourier series convergence:

(a) $f(x) = x, \qquad -\pi < x < \pi$ (E.1)

which is the function in Fig. 3.1a with $l = \pi$.

The Fourier sine-cosine series (3.10) for $f(x)$ on the interval $(-\pi, \pi)$ is

$$f(x) \sim a_0 + \sum_{n=1}^{\infty} a_n \cos nx + b_n \sin nx, \qquad -\pi < x < \pi \qquad \text{(E.2)}$$

$$a_0 = \frac{1}{2\pi} \int_{-\pi}^{\pi} f(x) \, dx \qquad \text{(E.3)}$$

$$a_n = \frac{1}{\pi} \int_{-\pi}^{\pi} f(x) \cos nx \, dx, \qquad n = 0, 1, 2, \ldots \qquad \text{(E.4)}$$

$$b_n = \frac{1}{\pi} \int_{-\pi}^{\pi} f(x) \sin nx \, dx, \qquad n = 1, 2, \ldots \qquad \text{(E.5)}$$

We note that $f(x) = x$ is an *odd* function on the symmetric interval $(-\pi, \pi)$ (see Fig. 3.4a); hence the integrals for a_0, a_n in (E.3) and

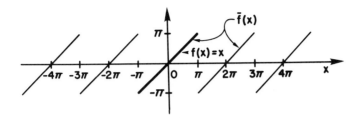

FIG. 3.4a   Periodic extension $\bar{f}(x)$ of (the *odd* function) $f(x) = x$, $-\pi < x < \pi$, period 2π.

(E.4) must vanish, and we are left with b_n to evaluate,

$$b_n = \frac{2}{\pi} \int_0^\pi x \sin nx \, dx = \frac{2}{\pi} \left[\frac{\sin nx}{n^2} - \frac{x \cos nx}{n} \right]_0^\pi$$

$$= \frac{2}{\pi} \left[\left\{ \frac{\sin n\pi}{n^2} - \frac{\pi \cos n\pi}{n} \right\} - \left\{ \frac{\sin 0}{n^2} - 0 \right\} \right]$$

$$= -\frac{2}{n} \cos n\pi = -\frac{2}{n} (-1)^n = \frac{2}{n} (-1)^{n+1} \qquad \text{(E.6)}$$

where we used one integration by parts for evaluating the integral.

With b_n in (E.6) and $a_0 = a_n = 0$, the Fourier series representation (E.2) for $f(x) = x$, $-\pi < x < \pi$, becomes a *sine series*:

$$x \sim \sum_{n=1}^\infty \frac{2}{n} (-1)^{n+1} \sin nx$$

$$x \sim 2 [\sin x - \tfrac{1}{2} \sin 2x + \tfrac{1}{3} \sin 3x + \cdots], \qquad -\pi < x < \pi \qquad \text{(E.7)}$$

The (sine) series representation (E.7) for the function in (E.1) would extend $f(x) = x$ as periodic with period 2π as shown in Fig. 3.4a.

We note that series (E.7) gives the same answer to the problem of representing the function

$$f(x) = x, \qquad 0 < x < \pi \qquad \text{(E.8)}$$

but assuming that this function is first extended as an odd function

$$f(x) = x, \qquad -\pi < x < \pi \qquad \text{(E.1)}$$

which is its natural extension as we had it in (E.1). However, the same function of (E.8) would require a cosine series if it was extended (against its natural tendency!) as an even function,

$$f_{even}(x) = |x| = \begin{cases} x, & 0 < x < \pi \\ -x, & -\pi < x < 0 \end{cases} \tag{E.9}$$

since in this case b_n of (E.5) would vanish, as we will show in part (b) of this example (Fig. 3.4b).

In relation to these two very special cases of odd (E.1) and even (E.9) extensions of $f(x)$ of (E.8), other extensions like

$$f(x) = \begin{cases} x, & 1 < x < \pi \\ 1, & -\pi < x < 0 \end{cases} \tag{E.10}$$

as illustrated in Fig. 3.4c of part (c) of this example, would require a general sine-cosine series (E.2), as there is no obvious reason for the vanishing of either coefficients a_n or b_n. This will be illustrated in part (c).

The periodic extension of the function $f(x)$ in (E.1) is clearly piecewise smooth with jump discontinuities, of size $J = 2\pi$, at all end points $x = \mp(2n + 1)\pi$, $n = 0, 1, 2, \ldots$. The Fourier series, according to Theorem 3.1, converges to the periodic extension $\bar{f}(x)$, where it is continuous; i.e., it coincides with the solid straight lines. However, the series assumes the value of zero (like $(f(\pi-) + f(\pi+))/2 = 0$) at all the jump discontinuities $x_n = \mp(2n + 1)\pi$, $n = 0, 1, 2, 3, \ldots$. Such values, as assigned by the series, are not equal to values of the periodic extension $\bar{f}(x)$ of $f(x)$ in (E.1), since $\bar{f}(x)$ is not even defined at these points of jump discontinuities $x_n = \mp(2n + 1)\pi$, $n = 0, 1, 2, 3, \ldots$. Because of the difference between the Fourier series values and the periodic extension $\bar{f}(x)$, uniform convergence here is out of the question. If we consult Theorem 3.2 on uniform convergence, we see that this function does not satisfy its condition of equality at the end points of the period, i.e., $f(-\pi) = -\pi \neq f(\pi) = \pi$. Having the other conditions satisfied, this one condition $f(-\pi) \neq f(\pi)$ is the one that prevented $\bar{f}(x)$ from being continuous to have uniform convergence. As we shall see in part (b) for even functions, this condition is automatically satisfied, and so if the function is continuous in the interior, its periodic extension $\bar{f}(x)$ will be continuous and its Fourier series converges uniformly.

As for the convergence in the mean of the Fourier series (E.2) to the (odd) function in (E.1), it is clearly valid by Theorem 3.4, since the function is piecewise continuous on $(-\pi, \pi)$ and is square integrable there,

$$\int_{-\pi}^{\pi} x^2\, dx = \frac{x^3}{3}\bigg|_{-\pi}^{\pi} = \frac{2\pi^3}{3}$$

It can be said that convergence in the mean is insensitive to mathematically "undesirable" jump discontinuities or even infinite "pulses." It does have a great support in applications, since, for example, for signals with our function as the current $i(t)$ we usually assume that the wave on $(-\pi, \pi)$ has only finite energy, i.e., $\int_{-\pi}^{\pi} i^2(t)\, dt < \infty$, regardless of whether there are impulses at separate points of time. In quantum mechanics, where our function is the wave function $\phi(x)$ for a particle in a rectangular box, $-\pi < x < \pi$, the physical interpretation is that $|\phi(x)|^2 \Delta x$ is the probability of finding the particle in the interval Δx. Thus, the probability being 1 of finding this "imprisoned" particle between the two walls $x = \mp\pi$ of this box requires that $\int_{-\pi}^{\pi} |\phi(x)|^2\, dx = 1 < \infty$, again a practical condition that suits the taste of the convergence in-the-mean Theorem 3.4a.

(b) The Fourier series expansion for the function (see Fig. 3.4b)

$$f(x) = |x| = \begin{cases} x, & 0 < x < \pi \\ -x, & -\pi < x < 0 \end{cases} \tag{E.11}$$

as we mentioned at the end of part (a), is for an *even* function on a symmetric interval $(-\pi, \pi)$ where b_n vanishes and

$$a_0 = \frac{1}{2\pi} \int_{-\pi}^{\pi} |x|\, dx = \frac{2}{2\pi} \int_{0}^{\pi} x\, dx = \frac{1}{\pi} \frac{x^2}{2}\bigg|_{0}^{\pi}$$

$$= \frac{1}{\pi} \cdot \frac{\pi^2}{2} = \frac{\pi}{2} \tag{E.12}$$

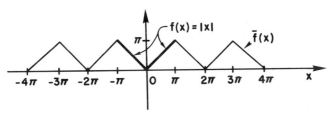

FIG. 3.4b Periodic extension $\bar{f}(x)$ of (the *even* function) $f(x) = |x|$, period 2π.

$$a_n = \frac{1}{\pi} \int_{-\pi}^{\pi} |x| \cos nx \, dx = \frac{2}{\pi} \int_{0}^{\pi} x \cos nx \, dx$$

$$= \frac{2}{\pi} \left[\frac{1}{n^2} \cos nx + \frac{x}{n} \sin nx \right] \Big|_{x=0}^{\pi}$$

$$= \frac{2}{\pi} \left[\left\{ \frac{1}{n^2} \cos n\pi + \frac{\pi}{n} \sin \pi \right\} - \left\{ \frac{1}{n^2} + 0 \right\} \right]$$

$$= \frac{2}{\pi} \left[\frac{1}{n^2} (-1)^n + \frac{\pi}{n} \sin 0 - \frac{1}{n^2} \right] = \frac{2}{\pi n^2} [(-1)^n - 1] \qquad \text{(E.13)}$$

So, if we use $b_n = 0$ with a_0 of (E.12) and a_n of (E.13), the Fourier series for $f(x)$ in (E.11) would be the following *cosine series*:

$$|x| \sim \frac{\pi}{2} + \sum_{n=1}^{\infty} \frac{2}{\pi n^2} [(-1)^n - 1] \cos nx$$

$$= \frac{\pi}{2} - \frac{4}{\pi} \left[\cos x + \frac{1}{9} \cos 3x + \frac{1}{25} \cos 5x + \cdots \right] \qquad \text{(E.14)}$$

The periodic extension for $f(x)$ of (E.11) as represented by this cosine series is continuous with period 2π as illustrated in Fig. 3.4b.

In contrast to the odd function in part (a), whose periodic extension (Fig. 3.4a) has jump discontinuities, the periodic extension of the present even function is continuous. Thus its Fourier series (E.14) converges uniformly to its periodic extension $\bar{f}(x)$, where the sign $\sim$ in (E.14) should be replaced by an equality sign. We see that the conditions of Theorem 3.2 (of piecewise smooth $f(x)$ and $f(-\pi) = f(\pi) = \pi$) are satisfied. For its other version, Theorem 3.2a, the conditions are also satisfied, including

$$\int_{-\pi}^{\pi} [f'(x)]^2 \, dx = \int_{-\pi}^{0} (-1)^2 \, dx + \int_{0}^{\pi} 1^2 \, dx = \pi + \pi = 2\pi$$

being finite.

Convergence in the mean follows from Theorem 3.4, since the function is piecewise continuous and is square integrable on $(-\pi, \pi)$ (for its other version Theorem 3.4a),

$$\int_{-\pi}^{\pi} |f(x)|^2 \, dx = 2 \int_{0}^{\pi} x^2 \, dx = \frac{2\pi^3}{3}$$

(c)   The function

$$f(x) = \begin{cases} x, & 0 < x < 2 \\ 1, & -2 < x < 0 \end{cases} \qquad \text{(E.15)}$$

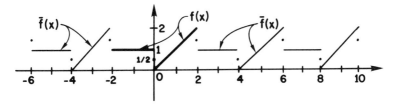

FIG. 3.4c Periodic extension of $\overline{f}(x)$ of

$$f(x) = \begin{cases} x, & 1 < x < 2 \\ 0, & -2 < x < 0 \end{cases}, \qquad \text{period } 4$$

as shown in Fig. 3.4c is neither even nor odd, so for its Fourier series (E.2) we must evaluate all the Fourier coefficients a_0, a_n, and b_n as in (E.3)–(E.5),

$$a_n = \frac{1}{2} \int_{-2}^{2} f(x) \cos \frac{n\pi x}{2} \, dx$$

$$= \frac{1}{2} \left[\int_{-2}^{0} 1 \cos \frac{n\pi x}{2} \, dx + \int_{0}^{2} x \cos \frac{n\pi x}{2} \, dx \right]$$

$$= \frac{1}{2} \left[\left\{ \frac{\sin(n\pi x/2)}{n\pi/2} \right\} \Big|_{x=-2}^{0} \right.$$

$$\left. + \left\{ \frac{1}{(n\pi/2)^2} \cos \frac{n\pi}{2} x + \frac{x}{(n\pi/2)} \sin \frac{n\pi}{2} x \right\} \Big|_{x=0}^{2} \right]$$

$$= \frac{1}{2} \left[\{0\} + \left\{ \frac{4}{n^2\pi^2} \cos n\pi + \frac{4}{n\pi} \sin n\pi - \left(\frac{4}{n^2\pi^2} \cos 0 + 0 \right) \right\} \right]$$

$$= \frac{1}{2} \left[\frac{4}{n^2\pi^2} (-1)^n + 0 - \frac{4}{n^2\pi^2} \right] = \frac{2}{n^2\pi^2} [(-1)^n - 1] \qquad \text{(E.16)}$$

$$a_0 = \frac{1}{4} \int_{-2}^{2} f(x) \, dx = \frac{1}{4} \left[\int_{-2}^{0} 1 \, dx + \int_{0}^{2} x \, dx \right]$$

$$= \frac{1}{4} \left[x \Big|_{-2}^{0} + \frac{x^2}{2} \Big|_{0}^{2} \right] = \frac{1}{4} \left[2 + \frac{4}{2} \right] = 1 \qquad \text{(E.17)}$$

Note how a_0 represents the average value 1 of the function in (E.15) on the interval $(-2, 2)$, as is also clear from Fig. 3.4c.

$$b_n = \frac{1}{2} \int_{-2}^{2} f(x) \sin \frac{n\pi x}{2} \, dx$$

$$= \frac{1}{2} \left[\int_{-2}^{0} \sin \frac{n\pi x}{2} \, dx + \int_{0}^{2} x \sin \frac{n\pi x}{2} \, dx \right]$$

$$= \frac{1}{2} \left[-\frac{\cos(n\pi x/2)}{n\pi/2} \Big|_{x=-2}^{0} \right.$$

$$\left. + \left\{ \frac{1}{(n\pi/2)^2} \sin \frac{n\pi}{2} x - \frac{x}{(n\pi/2)} \cos \frac{n\pi}{2} x \right\} \Big|_{x=0}^{2} \right]$$

$$= \frac{1}{2} \left[-\frac{2}{n\pi} + \frac{2}{n\pi} \cos(-n\pi) + \left\{ 0 - \frac{4}{n\pi} \cos n\pi - 0 \right\} \right]$$

$$= \frac{1}{n\pi} [-1 + (-1)^n - 2(-1)^n] = -\frac{1}{n\pi} [1 + (-1)^n] \qquad \text{(E.18)}$$

With the values of a_n, a_0, and b_n as in (E.16), (E.17), and (E.18), respectively, the Fourier (sine-cosine) series of $f(x)$ in (E.15) becomes

$$f(x) \sim 1 + \sum_{n=1}^{\infty} \left[\frac{2}{\pi^2 n^2} \{(-1)^n - 1\} \cos \frac{n\pi x}{2} \right.$$

$$\left. - \frac{1}{\pi n} [1 + (-1)^n] \sin \frac{n\pi x}{2} \right], \qquad -2 < x < 2 \qquad \text{(E.19)}$$

The periodic extension $\bar{f}(x)$ of $f(x)$ in (E.15) is with period 4 as illustrated in Fig. 3.4c. The periodic extension $\bar{f}$ here is sectionally smooth with jump discontinuities at $\mp 2n$, $n = 0, 1, 2, \ldots$, where the Fourier series converges to 1/2 at the points of discontinuity $x = \mp 4n$, $n = 0, 1, 2, 3, \ldots$ (e.g., at $x = 0$, it assumes the value $(f(0-) + f(0+))/2 = (1 + 0)/2 = \frac{1}{2}$) while it converges to 3/2 at the points of discontinuity $x = \mp 2(2n + 1)$, $n = 0, 1, 2, 3, \ldots$ (e.g., at $x = -2$, it assumes the value $(f(-2-) + f(-2+))/2 = (2 + 1)/2 = 3/2$. The Fourier series values at the jump discontinuities of $f(x)$ are indicated in Fig. 3.4c as dots.

The Fourier series (E.19) converges to the function of (E.15) in the mean on $(-2, 2)$, since it is piecewise continuous there for Theorem 3.4 to apply. Also, it is square integrable on the interval for the other version Theorem 3.4a to apply,

$$\int_{-2}^{2} |f(x)|^2 \, dx = \int_{-2}^{0} 1 \, dx + \int_{0}^{2} x^2 \, dx = 2 + \frac{8}{3} = \frac{14}{3}$$

In all the above three parts of Example 3.3, we may be concerned with problems of initial and boundary values with the function in its physical interval $(0, l)$, such as a heated rod of length l, a vibrating string of length l, or an electrified rectangular plate of width l, where the potential distribution is studied for its x-dependence. For these applications to appeal to the Fourier series representation of such functions on $(0, l)$, they must submit to the Fourier trigonometric series (3.8) natural property of extending them periodically with period $2l$. Such exact replicas or periodic extensions are a welcome advantage in the eyes of the signals or circuits analyst, as the wave $f(t)$ on $(-l, l)$ is being automatically duplicated in time with its period of $2l$.

In the following we will also discuss and illustrate the practical side of the numerical computation (approximation) of the (of course finite sum) Fourier series of such functions, after making sure that the series converges! We will take a function like that of part (a) in Example 3.3 to illustrate in some details how the Nth partial sum $S_N(x)$ of its Fourier series approximates the function. This is done with due attention given to this approximation near the jump discontinuities.

EXAMPLE 3.4:   Numerical Computations of Fourier Series and
Integrals Fourier Series

Fourier Series   In Example 3.3 we discussed how the Fourier series represents a periodic function with period $2a$,

$$f(x) \sim a_0 + \sum_{n=1}^{\infty} a_n \cos \frac{n\pi x}{a} + b_n \sin \frac{n\pi x}{a}, \qquad -a < x < a \qquad \text{(E.1)}$$

$$a_0 = \frac{1}{2a} \int_{-a}^{a} f(x) \, dx \qquad \text{(E.2)}$$

$$a_n = \frac{1}{a} \int_{-a}^{a} f(x) \cos \frac{n\pi x}{a} \, dx, \qquad n = 1, 2, 3, \ldots \qquad \text{(E.3)}$$

$$b_n = \frac{1}{a} \int_{-a}^{a} f(x) \sin \frac{n\pi x}{a} \, dx, \qquad n = 1, 2, 3, \ldots \qquad \text{(E.4)}$$

where for sectionally smooth periodic extension $\bar{f}(x)$, the series in (E.1) converges at x_0 to $\frac{1}{2}[\bar{f}(x_0+) + \bar{f}(x_0-)]$.

It is clear that even if we assume that we know the coefficients a_0, a_n, and b_n from evaluating their integrals (E.2), (E.3), and (E.4), respectively, we still cannot compute $f(x)$ from (E.1) exactly, since we cannot take infinitely many terms in the series in (E.1). So we must be satisfied with a finite N-term partial sum $S_N(x)$ as an approximation to $f(x)$ and hope that the *(truncation) error* $\epsilon_N(x) = |f(x) - S_N(x)|$ becomes negligible as we increase N. In Example 3.4A we will give graphic illustrations of how this error (often) decreases with increasing N. More detailed analysis of such error, along with suggested remedies for reducing it, is a main topic of Chapter 4, especially Section 4.1.6E.

For the solution of the boundary value problems where we employ Fourier series—for example, in the case of the temperature distribution $u(x,t)$, $0 < x < a$, $t > 0$, as we shall see in Section 3.2. (Example 3.10)—we also have an infinite series like

$$u(x,t) = \sum_{n=1}^{\infty} b_n e^{-(n^2\pi^2/a^2)kt} \sin\frac{n\pi x}{a}, \qquad 0 < x < a, \ k > 0 \qquad \text{(E.5)}$$

which must be truncated to a partial sum $S_N(x,t)$,

$$S_N(x,t) = \sum_{n=1}^{N} b_n e^{-(n^2\pi^2/a^2)kt} \sin\frac{n\pi x}{a}, \qquad k > 0 \qquad \text{(E.6)}$$

to approximate $u(x,t)$ with a (time-dependent) error $\epsilon_N(x,t) = |u(x,t) - S_N(x,t)|$. In this case we may look at (E.6) as a truncated Fourier series with time-dependent coefficients $b_n(t)$,

$$b_n(t) = b_n e^{-(n^2\pi^2/a^2)kt} \qquad \text{(E.7)}$$

The same is done for $u(x,t)$ the displacement of a finite string, and $u(x,y)$, the potential distribution in a rectangle.

Fourier Integrals Here we must have a word regarding the numerical computations even of the (infinite limits) Fourier integrals (of Sections 2.2—2.5 in Chapter 2). This is very relevant to the theme of this book, as it encompasses the (infinite) integral transforms, the finite (Fourier series) and coefficients transforms as well as the discrete transforms. The latter is our vehicle to its fast algorithm, the fast Fourier transform (FFT). Thus, when it comes to numerical computations of the Fourier representation and we think of the FFT, we must realize that the Fourier integrals and the Fourier series are both in the same ball game of having to go through the process of discretization and finite summation. This is the format that can be presented to the FFT. Both of these operations involve errors, in our approximation of the Fourier integrals or series, and we must be very careful about them. We shall discuss these errors in detail in Chapter 4, particularly Sections 4.1.6 and 4.1.7. This is, of course, unless those elegant (infinite) Fourier integral representations can be found exactly in terms of known (elementary!) functions, which is very often not the case and is a type of luck we cannot count on at all.

For functions defined on the infinite interval $(-\infty, \infty)$ we use (infinite) Fourier integrals,

$$f(x) \sim \int_0^\infty [A(\lambda)\cos \lambda x + B(\lambda)\sin \lambda x]\, d\lambda, \qquad -\infty < x < \infty \qquad \text{(E.8)}$$

$$A(\lambda) = \frac{1}{\pi} \int_{-\infty}^\infty f(x)\cos \lambda x\, dx \qquad \text{(E.9)}$$

$$B(\lambda) = \frac{1}{\pi} \int_{-\infty}^\infty f(x)\sin \lambda x\, dx \qquad \text{(E.10)}$$

where, again, for the same practical consideration we must truncate the infinite integral (E.8) to a finite limit L, instead of ∞, to approximate $f(x)$ by $f(x,L)$,

$$f(x,L) = \int_0^L [A(\lambda)\cos \lambda x + B(\lambda)\sin \lambda x]\, d\lambda \qquad \text{(E.11)}$$

with a truncation error $\epsilon_L(x) = |f(x) - f(x,L)|$. This, of course, is on the assumption that we can compute $A(\lambda)$ and $B(\lambda)$ exactly from (E.9) and (E.10), respectively. However, in general, we have to compute (E.9) and (E.10) numerically by truncating their infinite integrals to finite integrals, of the type in (E.11), which results in approximations to $A(\lambda)$ and $B(\lambda)$.

Another very important point for computing the (approximate) Fourier integral in (E.11), in comparison to the Fourier series, is that the integrand here is a continuous function of λ which has to be discretized to become a function of $n\Delta\lambda$; then we sum over $n = 1$ to $n = N$ to approximate the integral (E.11) as

$$\sum_{n=1}^N [A(n\Delta\lambda)\cos nx\Delta\lambda + B(n\Delta\lambda)\sin nx\Delta\lambda]\frac{L}{N} \qquad \text{(E.12)}$$

where $\Delta\lambda = (L - 0)/N = L/N$. This means that the computation of the infinite Fourier integral (E.8) will suffer from two errors, namely those of (1) *truncating its infinite limit to a finite limit* as $f(x,L)$ in (E.11) and (2) *discretizing (sampling) this last integral to a (finite) sum* as in (E.12), as we shall illustrate in Example 3.4B. In contrast, the Fourier series is already a sum, so there is no discretization but a truncation of its infinite limit as $S_N(x)$ in (E.6).

Considering these two errors of *truncation* in (E.11) and *discretization* in (E.12), for approximating the Fourier integrals, and the truncation error of the Fourier series in (E.6), we must realize the importance of increasing N for minimizing these errors, This would mean very lengthy computations for the integrals $A(\lambda)$, $B(\lambda)$ as in (E.9), (E.10) and for the final integral (E.11) of $f(x)$, which makes the Fourier method of analyzing problems very lengthy and expensive. Fortunately, a very fast algorithm for computing these finite trigonometric sums was developed about 25 years ago which reduced the number of computations with a decisive advantage, especially where we need it as N becomes very large. This algorithm is called fast Fourier transform (FFT) and is available with most computers. In essence, this algorithm capitalizes on the periodicity of the (periodic) trigonometric functions involved to come up with a very fast computation scheme. With such an efficient and appealing algorithm, we should also caution that it is very important to know how to prepare our input data for it and minimize the error before we call for its very fast computing. Since we don't intend to indulge in these details, which we leave for Sections 4.1.3, 4.1.6, and 4.1.7, we illustrate here the direct computations without the use of the FFT.

EXAMPLE 3.4A: Truncated Fourier Series

(a) Consider the function (see Example 3.3a, Fig. 3.4a, and Fig. 3.5a)

$$f(x) = 3x, \qquad -2 < x < 2$$

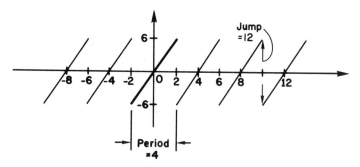

FIG. 3.5a Periodic extension of $f(x) = 3x$, $-2 < x < 2$, $f(x \mp 4n) = f(x)$, $n = 0, 1, 2, \ldots$, period 4.

This function is an odd function which has a Fourier sine series representation

$$f(x) \sim \sum_{n=1}^{\infty} b_n \sin \frac{n\pi x}{2} \tag{E.13}$$

$$b_n = \frac{2}{2} \int_0^2 3x \sin \frac{n\pi x}{2} \, dx$$

$$= \frac{6}{2} \left[\left\{ \frac{4}{n^2 \pi^2} \sin \frac{n\pi x}{2} - \frac{2}{n\pi} \cos \frac{n\pi x}{2} \right\} \Big|_{x=0}^{2} \right]$$

$$= 3 \left[0 - \frac{4}{n\pi} \cos n\pi - 0 + 0 \right] = -\frac{12}{n\pi} \cos n\pi$$

$$= \frac{12}{n} (-1)^{n+1} \tag{E.14}$$

$$f(x) \sim \frac{12}{\pi} \sum_{n=1}^{\infty} \frac{(-1)^{n+1}}{n} \sin \frac{n\pi x}{2} \tag{E.15}$$

as we showed in Example 3.3a.

To compute for $f(x)$ we must truncate the series to N terms,

$$S_N(x) = \frac{12}{\pi} \sum_{n=1}^{N} \frac{(-1)^{n+1}}{n} \sin \frac{n\pi x}{2} \tag{E.16}$$

Figure 3.5b illustrates how one, three, and five terms of the partial sum $S_N(x)$ in (E.16) are not such good approximations to the function $f(x) = 3x$ on the interval $(0, 2)$. Increasing N to 10, 15, and then 20 in Fig. 3.5c improved the approximation close to $x = 0$ (far away from where a jump discontinuity is at $x = 2.0$) but definitely not for the x values close to $x = 2.0$.

When N was increased to 100, as in Fig. 3.5d, the approximation became very satisfactory for x far from where the jump is at $x = 2.0$ but was still bad close to $x = 2.0$. Even when N was increased to 500, as in Fig. 3.5e, the error for x values near 2.0 still persisted. Such an error can be explained in terms of the continuity of the periodic extension of Fig. 3.5a near $x = 0$, and, by contrast, it has a large jump discontinuity of $|f(2+) - f(2-)| = |-6 - 6| = 12$ at $x = 2.0$ as seen in Fig. 3.5a. It is known that such an error near jump discontinuities will persist regardless of how large N is, and it is termed the *Gibbs phenomenon*. A way to alleviate it would be effectively to smooth this edge at $x = 2.0$, which means that we are

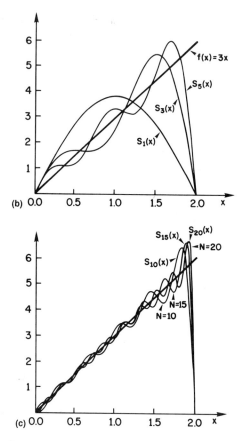

FIG. 3.5b,c (b) $S_N(x)$ approximation of $f(x) = 3x$, $0 < x < 2$, for $N = 1, 3, 5$. (c) $S_N(x)$ approximation of $f(x) = 3x$, $0 < x < 2$, for $N = 10, 15, 20$.

changing the function to suit the Fourier series inherited smoothness of its sine and cosine bases of approximation.

(b)   Consider the other function

$$f(x) = \begin{cases} 1, & -4 < x < 0 \\ \sin x & 0 < x < 4 \end{cases}$$

This function, as illustrated in Fig. 3.6a, is neither odd nor even,

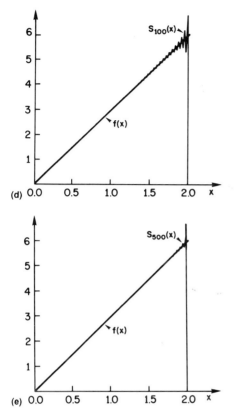

FIG. 3.5d,e   (d) $S_N(x)$ approximation of $f(x) = 3x$, $N = 100$—the obvious "Gibbs phenomenon" near the jump discontinuity at $x = 2$. (e) $S_N(x)$ approximation of $f(x) = 3x$, $N = 500$, and still a Gibbs phenomenon near $x = 2.0$.

and so it has the following Fourier sine-cosine series:

$$f(x) \sim a_0 + \sum_{n=1}^{\infty} a_n \cos \frac{n\pi x}{4} + b_n \sin \frac{n\pi x}{4}, \qquad -4 < x < 4 \quad \text{(E.17)}$$

$$a_0 = \frac{5 - \cos 4}{8} \qquad\qquad\qquad\qquad \text{(E.18)}$$

$$a_n = \frac{4[1 - (-1)^n \cos 4]}{16 - n^2 \pi^2}; \qquad\qquad\qquad \text{(E.19)}$$

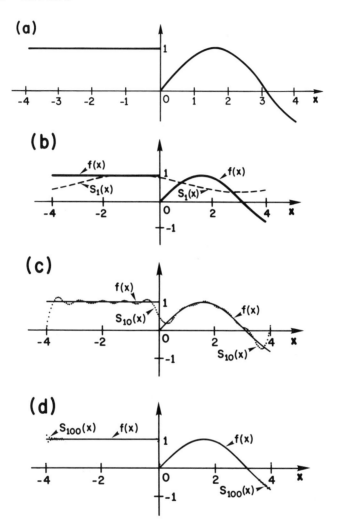

FIG. 3.6 (a) $f(x) = \begin{cases} 1, & -4 < x < 0 \\ \sin x, & 0 < x < 4 \end{cases}$ (b) $S_1(x)$ approximation of $f(x)$ in part (a). (c) $S_{10}(x)$ approximation of $f(x)$ in part (a). (d) $S_{100}(x)$ approximation of $f(x)$ in part (a).

$$b_n \frac{(-1)^n - 1}{n\pi} + \frac{(-1)^n n\pi \sin 4}{16 - n^2\pi^2} \tag{E.20}$$

The partial sum $S_N(x)$ that we consider here for approximating $f(x)$ is

$$S_N(x) = a_0 + \sum_{n=1}^{N} a_n \cos \frac{n\pi x}{4} + b_n \sin \frac{n\pi x}{4} \tag{E.21}$$

Figure 3.6b illustrates the approximation of $S_1(x)$, which in this case has three terms associated with a_0, a_1, and b_1 in (E.21). Figure 3.6c shows the approximation with $N = 10$, and we notice the bad approximation around $x = 0$, ∓ 4, where the function has jump discontinuities and hence a Gibbs phenomenon. In Fig. 3.6d we see that this error (Gibbs phenomenon) near the jump discontinuities is still apparent even when we increase the number of the terms in the series (E.21) to $N = 100$. We may observe that this error persists and is about 9% of the size of the jump discontinuity, no matter how large we take N. This will be the subject of Section 4.1.6G in Chapter 4, where we will show analytically the reason for this 9% and discuss a possible remedy for alleviating this problem.

The next example will illustrate the introduction of jump discontinuities when a naturally periodic function is expanded on an interval $(-a, a)$ with a width $2a$ that is not equal to its natural period. In this example (part c) for $f(x) = \sin x$, we purposely take the interval $(-4, 4)$ and not the natural one $(-\pi, \pi)$ (for its period of 2π) to show the resulting jump discontinuities of the periodic extension $f(x)$ of the trigonometric Fourier series expansion of $f(x) = \sin x$ on $(-4, 4)$. This is done here because in practice we may have a good feeling for a periodic wave, but to write its Fourier series representation we must know the exact period $2a$ to use in the Fourier series. Often we have to make an estimate of this period, and the error in such an estimate will result in those jump discontinuities, as we shall see in the following example.

(c) Consider the function

$$f(x) = \sin x, \qquad -4 < x < 4 \tag{E.22}$$

This function is its own Fourier series of one term: $f(x) = \sin x$, $-\pi < x < \pi$, if we are to expand it on its natural period $(-\pi, \pi)$. However, let us assume that we had estimated its period as 8 instead of 2π, so that we expand it on $(-4, 4)$ and we expect this part of it (Fig. 3.7a) on $(-4, 4)$ to be repeated periodically with period 8. As an odd function, it is represented by the following Fourier sine

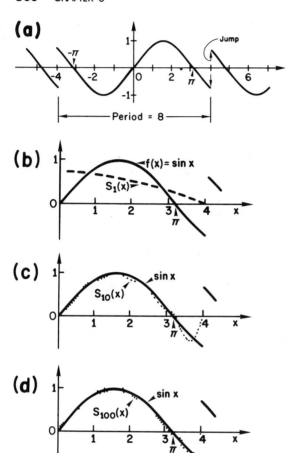

FIG. 3.7 (a) $f(x) = \sin x$, $-4 < x < 4$ period 8. (b–d) Fourier series approximations of $f(x) = \sin x$, $-4 < x < 4$: (b) $S_1(x)$ approximation of $f(x) = \sin x$, $-4 < x < 4$, of part (a); (c) $S_{10}(x)$ approximation of $f(x) = \sin x$, $-4 < x < 4$, of part (a); (d) $S_{100}(x)$ approximation of $f(x) = \sin x$, $-4 < x < 4$, of part (a). Note the different scale in part (a).

series:

$$f(x) \sim \sum_{n=1}^{\infty} b_n \sin \frac{n\pi x}{4}, \qquad -4 < x < 4 \qquad (E.23)$$

$$b_n = \frac{2}{4} \int_0^4 \sin x \sin \frac{n\pi x}{4} \, dx$$

$$= 2\pi(\sin 4) \frac{n \cos n\pi}{16 - n^2 \pi^2} \qquad (E.24)$$

$$S_n(x) = 2\pi \sin 4 \sum_{n=1}^{N} (-1)^n \frac{n}{16 - n^2 \pi^2} \sin \frac{n\pi x}{4} \qquad (E.25)$$

Figure 3.7a illustrates the function on $(-4, 4)$, where the jump discontinuity should be noted at $x = \mp 4$ and where we expect it at $x = \mp 4(2n + 1)$ for the periodic extension of this function. With this jump discontinuity, we also expect the truncation error (Gibbs phenomenon) to persist near $x = \mp 4(2n + 1)$. On the other hand, we expect a good approximation by a reasonable number of terms N of the series for such a smooth function in the interior of $(-4, 4)$. Figure 3.7b shows the poor approximation of one term of the sine series (E.25). Figure 3.7c shows the approximation with $N = 10$ terms, which clearly suffers near the jump discontinuity at $x = 4$. When N was increased to 100, as illustrated in Fig. 3.7d, the approximation improves dramatically away from $x = 4$ but the error is still very noticeable close to $x = 4$.

Direct Numerical Computations of Fourier Integrals

The following example will illustrate the difficulties we may expect in direct numerical computations of (infinite) Fourier transforms. To name a few of these, we first have to truncate the infinite interval $(-\infty, \infty)$ of the Fourier transform to a finite interval $(-L, L)$. Second, we have to discretize the resulting finite-limit integral to have it as an infinite sum, whence it will immediately behave like a Fourier series, which has a periodic structure instead of the (generally) nonperiodic form of a transform on $(-\infty, \infty)$. Finally, this infinite sum must be truncated.

The following computations should be considered very elementary (or primitive) compared to what is done nowadays via the discrete Fourier transform (DFT), the vehicle to its efficient algorithm, the fast Fourier transform (FFT), as we shall see in Chapter 4. The following Example 3.4B will also be redone as Example 4.7 in Section 4.1.7 of Chapter 4 for illustrating the use of the discrete Fourier transforms.

EXAMPLE 3.4B: Numerical Computations of Fourier Transforms (Integrals)

(a) Consider the function defined on $(-\infty, \infty)$, as illustrated in Fig. 3.8a.

$$f(x) = \begin{cases} 0, & -\infty < x < 0 \\ e^{-x}, & 0 < x < \infty \end{cases} \tag{E.26}$$

The Fourier integral representation of this (causal) function (Fig. 3.8a) is

$$f(x) \sim \int_0^\infty [A(\lambda)\cos \lambda x + B(\lambda)\sin \lambda x]\,d\lambda, \qquad -\infty < x < \infty \tag{E.27}$$

where $A(\lambda)$ and $B(\lambda)$ are known in a simple form,

$$A(\lambda) = \frac{1}{\pi} \int_{-\infty}^\infty f(x)\cos \lambda x\,dx$$

$$= \frac{1}{\pi} \int_0^\infty e^{-x} \cos \lambda x\,dx = \frac{1}{\pi}\frac{1}{1+\lambda^2} \tag{E.28}$$

$$B(\lambda) = \frac{1}{\pi} \int_{-\infty}^\infty f(x)\sin \lambda x\,dx$$

$$= \frac{1}{\pi} \int_0^\infty e^{-x} \sin \lambda x\,dx = \frac{1}{\pi}\frac{\lambda}{1+\lambda^2} \tag{E.29}$$

So the final (infinite) integral representation (E.27) of $f(x)$ in (E.26) is

$$f(x) \sim \frac{1}{\pi} \int_0^\infty \left[\frac{1}{1+\lambda^2} \cos \lambda x + \frac{\lambda}{1+\lambda^2} \sin \lambda x \right] d\lambda \tag{E.30}$$

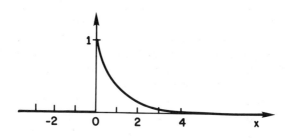

FIG. 3.8a   (a) $f(x) = \begin{cases} 0, & -\infty < x < 0 \\ e^{-x}, & 0 < x < \infty \end{cases}$.

which is to be computed. This is an infinite integral which we must first *truncate* to a finite-limit integral

$$f(x,L) = \frac{1}{\pi} \int_0^L \left[\frac{1}{1 + \lambda^2} \cos \lambda x + \frac{\lambda}{1 + \lambda^2} \sin \lambda x \right] d\lambda \qquad \text{(E.31)}$$

as we have indicated earlier. Then we have to approximate this finite integral (E.31) by a sum. For such a sum we have to discretize the above integrand by choosing $\Delta\lambda = (L - 0)/N$, $\lambda_n = n\Delta\lambda$, where N is the number of terms in the resulting sum,

$$S_N(x,L) = \frac{1}{\pi} \sum_{n=1}^{N} \left[\frac{1}{(n\Delta\lambda)^2 + 1} \cos nx\Delta\lambda \right.$$

$$\left. + \frac{n\Delta\lambda}{(n\Delta\lambda)^2 + 1} \sin nx\Delta\lambda \right] \Delta\lambda \qquad \text{(E.32)}$$

Of course, even x has to be discretized (sampled), but this can be left at the end for the final desired number of x values.

Now we must discuss deciding on a reasonable value of the truncation limit L and of the discretization number N. The number L should be chosen in such a way that the value of the integrand in (E.31) at $x = L$ becomes very small. For the sake of illustration, and not necessarily accuracy, we look here for a value of L such that the integrand decreases to less than 10%. A look at the slowest-varying term $\lambda/(1 + \lambda^2)$ in the integrand of (E.31) indicates that $L = 20$ will suffice for (E.31). Now we turn to a reasonable N for approximating the finite integral (E.31) by the finite sum (E.32). Here, we should pay attention to the presence of the oscillating functions $\cos \lambda x$ and $\sin \lambda x$ in the integrand of (E.31), which are to be approximated by a finite number of rectangles as in (E.32). In this case we leave N as the parameter in our illustrations to emphasize this dependence. For example, if we take $N = 10$ with $L = 20$ we have $\Delta\lambda = 20/10 = 2$, which is very coarse for the variation of the integrand in (E.31). The result of these computations is illustrated in Fig. 3.8b, where we find a very bad approximation. This is a clear disagreement which should not surprise us. For example, the periodic extension shown in Fig. 3.8b is the result of the periodicity of the trigonometric functions $\cos nx\Delta\lambda = \cos 2nx$ and $\sin nx\Delta\lambda = \sin 2nx$ in (E.32), which are periodic with period π. Hence (E.32) stands as a truncated Fourier sine-cosine series that approximates the part of the function $(f(x), -\infty < x < \infty)$ that is given on $(-\pi/2, \pi/2)$, only then renders it (with its truncation error) periodic with period π. With this note,

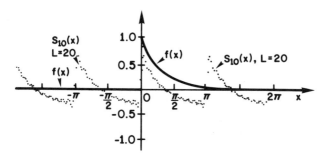

FIG. 3.8b $f(x)$ approximation by $S_N(x)$, $-\infty < x < \infty$; $N = 10$, $L = 20$. Note the periodicity of $S_N(x)$ with (the small) period of π.

then for Fig. 3.8b it only makes sense to compare the approximation of (E.32) to only that part of $f(x)$ on $(-\pi/2, \pi/2)$.

By increasing N to $N = 100$ we have $\Delta\lambda = L/N = 20/100 = 0.2$, which is a much more suitable increment for the integral in (E.31) and its approximation sum in (E.32). Also, the trigonometric functions in (E.32) will now have a period of 10π instead of π, where (E.32) approximates $f(x)$ on the much larger interval $(-5\pi, 5\pi)$. We also have enough distance of 5π for e^{-x} to become small and close to the zero value of $f(x)$ at $x = 5\pi$. These results are illustrated in Fig. 3.8c, where the agreement (with $L = 20$ and $N = 100$) looks fairly good. However, we can still notice a large error near the jump discontinuity at $x = 0$. Also somewhat noticeable are some wiggles, which can be attributed to both the truncation (finite L) error and the discretization (sampling finite N) error of (E.32).

As mentioned earlier, we will redo this example as Example 4.7 in Section 4.1.7, where we will use the discrete Fourier transform.

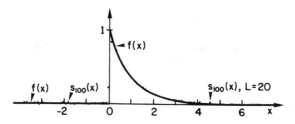

FIG. 3.8c $f(x)$ approximation by $S_N(x)$; $-\infty < x < \infty$, $N = 100$, $L = 20$. Note how $S_N(x)$ here has a much larger period of 10π.

The computations related to Fig. 3.8c will be illustrated in Fig. 4.34d of Example 4.7 with only $N = 32$ and a period of 8 instead of the present needed period of 10π.

(b) The function illustrated in Fig. 3.9a,

$$f(x) = e^{-|x|}, \qquad -\infty < x < \infty \tag{E.33}$$

is an even function which can be represented by the cosine part of the Fourier integral,

$$e^{-|x|} \sim \frac{2}{\pi} \int_0^\infty \frac{1}{1 + \lambda^2} \cos \lambda x \, d\lambda \tag{E.34}$$

Here we note that the

$$A(\lambda) = \frac{2}{\pi} \frac{1}{1 + \lambda^2}$$

decreases faster than $\lambda/(1 + \lambda^2)$ of the sine integral in (E.30), so instead of $L = 20$ we may choose $L = 10$ for the cosine integral in (E.34). The requirement of a fine $\Delta\lambda$ for the cosine part of (E.32) remains the same as in the last example (part a). For $N = 10$ with $L = 10$ we now have a smaller (better) $\Delta\lambda = L/N = 10/10 = 1$ than that of Fig. 3.8b of the last example. This will also increase the period of the sum (E.32) to 2π instead of π. These results are illustrated in Fig. 3.9a.

With $N = 100$ we have $\Delta\lambda = 0.1$ and we should expect a lower discretization error, but of course we may have increased the truncation error by decreasing L from 20 to 10 compared to the last example. These results are shown in Fig. 3.9b, where the clear error at $x = 0$ is due primarily to the truncation of the integral and the

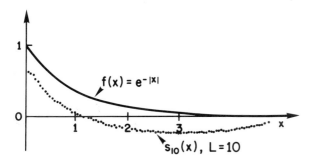

FIG. 3.9a $F(x) = e^{-|x|}$ and its approximation, $S_{10}(x)$, $-\infty < x < \infty$, $N = 10$, $L = 10$.

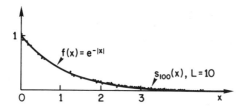

FIG. 3.9b $f(x) = e^{-|x|}$ and its approximation $S_{100}(x)$, $-\infty < x < \infty$, $N = 100$, $L = 10$.

series. We should also note, compared to part (a) of the above example in (E.26), that the present even function $e^{-|x|}$ is continuous at $x = 0$, so there is no jump discontinuity to worry about.

Of course, we should consider the computations of this example as elementary; the main purpose here is to illustrate again, and at a very early stage, the various errors incurred in any practical computations of the Fourier integrals. Thus we should count on our analysis and illustrations for using the discrete Fourier transform of Chapter 4 for such computations. Then we will also be prepared to reap the benefits of using the efficient fast Fourier transform algorithm. The computations for $f(x) = e^{-|x|}$, $-\infty < x < \infty$, in (E.33) will be the very detailed subject of Exercise 4.1 in Chapter 4.

3.1.2[†] Elements of Infinite Series—Convergence Theorems

Our concern here regarding the convergence of the Fourier trigonometric series (3.10)–(3.13), the exponential series (3.8), (3.9), or the general orthogonal expansion (3.33) can be summed up as whether the partial sum $S_N(x)$ of this series

$$S_N(x) = \sum_{k=1}^{N} c_k \phi_k(x) \tag{3.34}$$

converges as $N \to \infty$ and, if so, for what values of x. We will search for at least sufficient conditions for the limit of this sequence $S_N(x)$ to exist.

[†] This Section 3.1.2 and the following Section 3.1.3 are suggested for the graduate course adoption of this book on transform methods. However, it should be within easy reach of aspiring undergraduates.

The sequence $S_N(x)$ is also a special case of the series

$$S_n(x) = \sum_{k=1}^{n} u_k(x) \tag{3.35}$$

where $u_k(x) = c_k \phi_k(x)$, and where we used n instead of N as it is customary to speak of the sequence S_n. We will return in the next section, in (3.51) and thereafter, to using $S_N(x)$, as it is again customary in the orthogonal expansion to speak of the Nth partial sum $S_N(x)$ of (3.34).

We shall state, and sometimes prove, the basic theorems concerning the convergence of the infinite series

$$S(x) = \lim_{n \to \infty} S_n(x) = \lim_{n \to \infty} \sum_{k=1}^{n} u_k(x) \tag{3.36}$$

whose results will be applied to our special cases of the Fourier trigonometric series and the general orthogonal expansion. Of course, the following analysis for the convergence of sequence is not by any means limited to the above nth partial sum type of sequence $S_n(x)$. For this reason it may be instructive to use $f_n(x)$ for the sequence that converges to $f(x)$, i.e., $\lim_{n \to \infty} f_n(x) = f(x)$, but to keep our notation in line with other sources, and since our aim is to apply it to $S_n(x)$ of the series, we shall adhere to using $S_n(x)$. We have already defined in the last section what we mean by convergence of the sequence $S_n(x)$ to $f(x)$ pointwise (for each x), uniform (for all x), and in the mean in (3.30). We will repeat them here along with some important related definitions.

Definition: Pointwise Convergence. The sequence $S_n(x)$ is said to converge to $f(x)$ pointwise if

$$\lim_{n \to \infty} S_n(x) = f(x) \quad \text{at each } x \tag{3.37}$$

If such convergence is valid for all values of x in the interval $[a,b]$, the convergence of $S_n(x)$ to $f(x)$ is termed *uniform convergence*. $\quad\square$

We stress here that uniform convergence means that $f(x) - \lim_{n \to \infty} S_n(x) \equiv 0$ for all $x \in [a,b]$. The uniform convergence, as we have seen in the last section, is very desirable by itself for $f(x)$ and also as a main condition to allowing the interchange of other operations with the limit process of the series. Its shortcoming is that it may limit the class of functions which it represents. This limitation is in the direction of the functions being continuously differentiable or, better yet, infinitely differentiable. This is unlike the following convergence in the mean of $S_n(x)$ to $f(x)$, which would allow a larger (more practical) class of functions and even admits a difference $f(x) - \lim_{n \to \infty} S_n(x)$, which may be very large but only on a (finite) set of *separate points* in the interval $[a,b]$.

Definition: Convergence in the Mean Square. The sequence $S_n(x)$ is said to converge to $f(x)$ in the mean (square) on (a,b) if

$$\lim_{n \to \infty} \int_a^b |f(x) - S_n(x)|^2 \, dx = 0 \qquad \square \quad (3.38)$$

The integral

$$\int_a^b |f(x) - S_n(x)|^2 \, dx \equiv \int_a^b |\epsilon_n(x)|^2 \, dx \qquad (3.39)$$

is called the *mean square deviation*, i.e., error of the function $f(x)$ from the sequence $S_n(x)$ on the interval (a,b). We note that this integral is not affected if $f(x) - S_n(x)$ is a large number at a set of separate points x_i, $i = 1, 2, \ldots, M$, since these values do not contribute anything to the integral. Hence, when we have convergence in the mean of $S_n(x)$ to $f(x)$, we cannot say that $f(x) - \lim_{n \to \infty} S_n(x) \equiv 0$. Instead, we write $f(x) \sim \lim_{n \to \infty} S_n(x)$ or $\lim_{n \to \infty} S_n(x) = f(x)$ *almost everywhere* (in short a.e.) to indicate the possible difference at the set of separate points in the domain of $f(x)$ (see the discussion of integration in the Lebesgue sense versus the familiar integration in the Riemann sense around equation (2.75) in Chapter 2). So for the infinite series (3.36) converging in the mean square to $f(x)$ we write

$$f(x) \sim \lim_{n \to \infty} \sum_{k=1}^{n} u_k(x) \qquad (3.40)$$

or

$$f(x) \sim \sum_{k=1}^{\infty} u_k(x) \qquad (3.40a)$$

or

$$f(x) = \sum_{k=1}^{\infty} u_k(x) \quad \text{a.e. on } (a,b) \qquad (3.40b)$$

As we can see from the definition (3.38) of convergence in the mean (square), the integral will involve the square of the function $f(x)$ considered, i.e., $\int_a^b f^2(x) \, dx$. As we indicated in the discussion related to Theorem 3.4a, this convergence is compatible with functions that are square integrable on (a,b) or in $L_2(a,b)$:

$$\int_a^b |f(x)|^2 dx < \infty$$

We stress here again that this integral, *in the sense of Lebesgue*, allows $f(x)$ to be "discontinuous" in that it can have large values at a separate finite number of points in the interval (a,b). For this class of functions we will present next two very important tools, namely the Schwarz inequality in (3.41) and the triangle inequality in (3.42).

The Schwarz Inequality

For two functions $f(x)$ and $g(x)$ defined and square integrable on (a,b) (i.e., $f,g \in L_2(a,b)$) the *Schwarz inequality* states that

$$\left| \int_a^b f(x)g(x)\,dx \right|^2 \le \int_a^b |f(x)|^2\,dx \int_a^b |g(x)|^2\,dx \tag{3.41}$$

or

$$\left| \int_a^b f(x)g(x)\,dx \right| \le \sqrt{\int_a^b |f(x)|^2\,dx} \sqrt{\int_a^b |g(x)|^2\,dx}$$

The proof was outlined in full detail in Exercise 2.36a,b, and it was extended in (2.78) to functions on the infinite interval $(-\infty,\infty)$ with weight function $\rho(x)$ in Exercise 2.36c,d.

Another tool that can be derived easily from the Schwarz inequality is the *triangle inequality*

$$\left\{ \int_a^b |f(x) + g(x)|^2\,dx^{1/2} \right\} \le \left\{ \int_a^b |f(x)|^2 \right\}^{1/2} + \left\{ \int_a^b |g(x)|^2\,dx \right\}^{1/2} \tag{3.42}$$

This can be proved by using

$$
\begin{aligned}
|f(x) + g(x)|^2 &= (f(x) + g(x))(\overline{f(x) + g(x)}) \\
&= (f(x) + g(x))(\overline{f(x)} + \overline{g(x)}) \\
&= |f(x)|^2 + f(x)\overline{g(x)} + g(x)\overline{f(x)} + |g(x)|^2
\end{aligned} \tag{3.43}
$$

in the above integral to have

$$\int_a^b |f(x) + g(x)|^2\,dx$$

$$= \int_a^b [|f(x)|^2 + f(x)\overline{g(x)} + g(x)\overline{f(x)} + |g(x)|^2\,dx$$

$$= \int_a^b |f(x)|^2\, dx + \int_a^b f(x)\overline{g(x)}\, dx + \int_a^b g(x)\overline{f(x)}\, dx$$

$$+ \int_a^b |g(x)|^2\, dx$$

$$\leq \int_a^b |f(x)|^2\, dx + 2 \left\{ \int_a^b |f(x)|^2\, dx \int_a^b |g(x)|^2\, dx \right\}^{1/2}$$

$$+ \int_a^b |g(x)|^2\, dx$$

$$\int_a^b |f(x) + g(x)|^2\, dx \leq \left[\left\{ \int_a^b |f(x)|^2\, dx \right\}^{1/2} + \left\{ \int_a^b |g(x)|^2\, dx \right\}^{1/2} \right]^2$$

$$\left\{ \int_a^b |f(x) + g(x)|^2\, dx \right\}^{1/2}$$

$$\leq \left\{ \int_a^b |f(x)|^2\, dx \right\}^{1/2} + \left\{ \int_a^b |g(x)|^2\, dx \right\}^{1/2} \tag{3.42}$$

after using the Schwarz inequality (3.41) on each of the two middle integrals of $g(x)\overline{f(x)}$ and $f(x)\overline{g(x)}$ and realizing the resulting complete square on the right side of the inequality.

Also, we will show next, by the simple use of the Schwarz inequality, *that a function which is square integrable on the finite integral (a,b) is absolutely integrable on (a,b)*:

$$\int_a^b |f(x)|\, dx \leq \int_a^b 1 f(x)\, dx \leq \left[\int_a^b 1\, dx \right]^{1/2} \left[\int_a^b |f(x)|^2\, dx \right]^{1/2}$$

$$\leq (b-a)^{1/2} \int_a^b |f(x)|^2\, dx$$

$$\int_a^b |f(x)|\, dx < \infty$$

since $(b-a)^{1/2}$ is finite and $f(x)$ is square integrable on (a,b).

We shall now state and prove a theorem concerning the integration of sequences (or series) that are convergent in the mean—in other words, allowing the interchange of an integration operation with the limit process of the convergence of the sequence. For convergent series, this means allowing the integration to be entered inside the infinite sum, i.e., *to integrate it term by term*. Moreover, we find that such an integration operation even improves the convergence of the series. This is in the sense that while the series converges (only) in the mean, its integrated resulting new series converges uniformly.

THEOREM 3.5:   Term-by-Term Integration of a Series Convergent in the Mean

Let $S_n(x)$ be a sequence that converges in the mean to $f(x)$ on an interval (a,b); then

(i)   For any x_0 and x belonging to the interval (a,b), we have

$$\lim_{n \to \infty} \int_{x_0}^{x} S_n(x)\,dx = \int_{x_0}^{x} f(x)\,dx \tag{3.44}$$

or, in other words, for the series $S_n(x) = \sum_{k=1}^{n} u_k(x)$,

$$\int_{x_0}^{x} f(x)\,dx = \int_{x_0}^{x} \sum_{k=1}^{\infty} u_k(x)\,dx = \sum_{k=1}^{\infty} \int_{x_0}^{x} u_k(x)\,dx$$

$$= \lim_{n \to \infty} \int_{x_0}^{x} \left[\sum_{k=1}^{n} u_k(x)\,dx \right]$$

$$= \lim_{n \to \infty} \sum_{k=1}^{n} \int_{x_0}^{x} u_k(x)\,dx \tag{3.45}$$

(ii)   Moreover, for any $x_0 \in [a,b]$, the resulting integrated sequence (or series) $\int_{x_0}^{x} S_n(x)\,dx$ converges uniformly to $\int_{x_0}^{x} f(x)\,dx$ on $[a,b]$. $\square$

Proof: We have, by the hypothesis of the convergence in the mean of $S_n(x)$ to $f(x)$,

$$\lim_{n \to \infty} \int_{a}^{b} |f(x) - S_n(x)|^2\,dx = 0 \tag{E.1}$$

To prove (3.44) in part (i), and consequently part (ii), we consider only the difference of the integrals in (3.44) without the limit,

$$\int_{x_0}^{x} [f(x) - S_n(x)]\,dx \tag{E.2}$$

We then show, with the help of the Schwarz inequality (3.41) for the subinterval $(x_0,x) \subseteq [a,b]$, that the limit of the integral in (E.2) as $n \to \infty$

converges for all $x \in [a, b]$,

$$\int_{x_0}^{x} 1[f(x) - S_n(x)] dx \le \left\{ \int_{x_0}^{x} 1^2 dx \right\}^{1/2} \left\{ \int_{x_0}^{x} |f(x) - S_n(x)|^2 dx \right\}^{1/2}$$

(E.3)

But both integrands in the above two integrals are positive and are taken on the subinterval $(x_0, x) \subseteq [a, b]$; thus they are bounded by their respective integration on the whole interval $[a, b]$,

$$\int_{x_0}^{x} [f(x) - S_n(x)] dx \le \left\{ \int_{a}^{b} 1^2 dx \right\}^{1/2} \left\{ \int_{a}^{b} |f(x) - S_n(x)|^2 dx \right\}^{1/2}$$

$$= (b - a)^{1/2} \left\{ \int_{a}^{b} |f(x) - S_n(x)|^2 dx \right\}^{1/2}$$

(E.4)

where this upper bound is independent of x. Thus by taking the limit of (E.4) as $n \to \infty$ and using our assumption (E.1) of $S_n(x)$ convergent in the mean to $f(x)$, we have the desired results of (i) and (ii)

$$\lim_{n \to \infty} \int_{x_0}^{x} [f(x) - S_n(x)] dx = (b - a)^{1/2} \lim_{n \to \infty} \int_{a}^{b} |f(x) - S_n(x)|^2 dx$$

$$= 0$$

(E.5)

$$= \int_{x_0}^{x} f(x) dx - \lim_{n \to \infty} \int_{x_0}^{x} S_n(x) dx = 0 \quad \text{for all } x \in [a, b],$$

$$\int_{x_0}^{x} f(x) dx = \lim_{n \to \infty} \int_{x_0}^{x} S_n(x) \quad \text{for all } x \in [a, b]$$

(3.44)

The fact that this result is valid (uniformly) for all $x \in [a, b]$ makes this resulting series, after integration, uniform convergent. For the infinite series $\sum_{k=1}^{\infty} u_k(x)$, this result means that

$$\int_{x_0}^{x} f(x) dx = \lim_{n \to \infty} \int_{x_0}^{x} S_n(x) dx = \lim_{n \to \infty} \int_{x_0}^{x} \left[\sum_{k=1}^{n} u_k(x) \right] dx$$

$$= \lim_{n \to \infty} \sum_{k=1}^{n} \int_{x_0}^{x} u_k(x) dx$$

$$= \sum_{k=1}^{\infty} \int_{x_0}^{x} u_k(x) dx = \int_{x_0}^{x} f(x) dx = \int_{x_0}^{x} \sum_{k=1}^{\infty} [u_k(x)] dx$$

(E.6)

as allowing the integration term by term in the convergent-in-the-mean type of series in the last line. ∎

By mentioning that the sequence $S_n(x)$ can be the partial sum $S_n(x) = \sum_{k=1}^{n} u_k(x)$ of the infinite series, we have already covered the allowance of integrating term by term of the infinite series $\sum_{k=1}^{\infty} u_k(x)$ that is convergent in the mean to $f(x)$. To stress this property (which we shall use often in Fourier series) of square integrable functions, we state it as a corollary to the above theorem.

Corollary: If the infinite series $\sum_{k=1}^{\infty} u_k(x)$ with integrable terms converges in the mean to an integrable function $f(x)$, then

$$\int_{x_0}^{x} f(x)\,dx = \int_{x_0}^{x} \left[\sum_{k=1}^{\infty} u_k(x) \right] dx = \sum_{k=1}^{\infty} \int_{x_0}^{x} u_k(x)\,dx \qquad \square \quad (3.45)$$

In the above theorem, or corollary, we saw the "smoothing" effect of the integration of a convergent-in-the-mean series to make it uniformly convergent. On the other hand, the differentiation of a convergent sequence may worsen the convergence of the resulting series, so before we differentiate we should prepare for a "good convergent" series to end up at least with some convergence for the resulting differentiated series. For example, the series

$$\sum_{n=1}^{\infty} \frac{(-1)^n}{n} x^n$$

is convergent at $x = 1$, as an alternating series with decaying coefficients $|c_n| = |(-1)^n/n| = 1/n$. However, after differentiation

$$\frac{d}{dx} \sum_{n=1}^{\infty} \frac{(-1)^n}{n} x^n$$

and if we allow the interchange of the differentiation and infinite summation, we have

$$\sum_{n=1}^{\infty} \frac{(-1)^n}{n} nx^{n-1} = \sum_{n=1}^{\infty} (-1)^n x^{n+1}$$

which is obviously not convergent at $x = 1$, as the differentiation had taken away the decaying effect of the coefficient $|c_n| = 1/n$, which is now $|d_n| = |-1| = 1$.

To put a warning in perspective, we state the following theorem concerning differentiating a sequence (or series) $S_n(x)$ that is convergent in the mean. The proof will parallel to some extent that of Theorem 3.1. Then we will state a corollary that deals with allowing the differentiation term by term of infinite series (of continuously differentiable terms).

THEOREM 3.6: Differentiation of a Convergent in the Mean
(Continuously Differentiable) Sequence (or Series)

If for a continuously differentiable sequence $S_n(x)$ (or series) $S_n(x) = \sum_{k=1}^{n} c_k u_k(x)$ (i.e., $dS_n(x)/dx$ is continuous), we have the following:

(i) convergence in the mean to a function $f(x)$ on (a,b).
(ii) the sequence of its derivatives $S_n'(x)$ is also convergent in the mean to a continuous function $g(x)$ on (a,b); i.e., $g(x) \sim \lim_{n \to \infty} S_n'(x)$. Then the function $f(x)$ is differentiable on (a,b), and

$$\frac{df}{dx} = g(x) \sim \lim_{n \to \infty} S_n'(x) = \sum_{k=1}^{\infty} c_k u_k'(x) \quad \text{on } (a,b) \qquad \square \quad (3.46)$$

Proof: To show that $f(x)$ is differentiable, it is sufficient to show that it satisfies *the fundamental theorem of calculus*:

$$f(x) - f(x_0) = \int_{x_0}^{x} f'(x) \, dx \tag{E.1}$$

This we will attempt to prove by showing the vanishing of the new sequence

$$a_n(x) = \left| S_n(x) - S_n(x_0) - \int_{x_0}^{x} f'(x) \, dx \right| \tag{E.2}$$

as $n \to \infty$.

$S_n(x)$ is assumed differentiable, so we can use the fundamental theorem of calculus to write

$$S_n(x) - S_n(x_0) = \int_{x_0}^{x} S_n'(x) \, dx \tag{E.3}$$

If we use this result for $S_n(x) - S_n(x_0)$ in (E.2) with $g(x)$ instead of $f'(x)$ for now, along with Schwarz's inequality, we have

$$\left| S_n(x) - S_n(x_0) - \int_{x_0}^{x} g(x) \, dx \right| = \left| \int_{x_0}^{x} [S_n'(x) - g(x)] \, dx \right|$$

$$\leq \left\{ \int_{x_0}^{x} 1 \, dx \right\}^{1/2} \left\{ \int_{x_0}^{x} |S_n'(x) - g(x)|^2 \, dx \right\}^{1/2} \tag{E.4}$$

The two integrands above are positive and bounded since $S_n(x)$ is assumed continuously differentiable, thus $S_n'(x)$ is continuous on $[a,b]$; also, $g(x) = \lim_{n \to \infty} S_n'(x)$ is assumed continuous in condition (ii). So the two integrals on the right are finite on the subinterval $(x_0, x) \subseteq (a,b)$, which makes their product bounded by the integration of each over the larger interval (a,b),

$$\left| S_n(x) - S_n(x_0) - \int_{x_0}^{x} g(x)\,dx \right| \leq (b-a)^{1/2} \left\{ \int_{a}^{b} |S_n'(x) - g(x)|^2\,dx \right\}^{1/2}$$

(E.5)

By having, from condition (ii), $S_n'(x)$ convergent in the mean to $g(x)$, we can take the limit of (E.5) as $n \to \infty$ to have our desired result

$$\lim_{n \to \infty} \left| S_n(x) - S_n(x_0) - \int_{x_0}^{x} g(x)\,dx \right| = 0$$

$$f(x) - f(x_0) = \int_{x_0}^{x} g(x)\,dx$$

which, according to the fundamental theorem of calculus, defines $g(x)$ as $f'(x)$, $f(x) - f(x_0) = \int_{x_0}^{x} f'(x)\,dx$ to complete our proof. ∎

The following corollary is the statement of *differentiating an infinite series term by term*, whose proof we leave as an exercise.

Corollary:

(i) If the infinite series

$$\sum_{k=1}^{\infty} u_k(x) = f(x)$$

(3.47)

of continuously differentiable terms $u_k(x)$—i.e., $u_k'(x)$ are continuous on (a,b)—is convergent in the mean on (a,b), and

(ii) the resulting series

$$\sum_{k=1}^{\infty} u_k'(x)$$

(3.48)

obtained by differentiating the series (3.47) of $f(x)$ term by term, is convergent in the mean to a continuous function $g(x)$,

then the infinite series (3.47) of $f(x)$ is differentiable on (a,b) and $g(x) = df/dx$,

$$\sum_{k=1}^{\infty} u_k'(x) = \frac{df}{dx}$$

(3.49)

that is,

$$\frac{d}{dx} \sum_{k=1}^{\infty} u_k(x) = \sum_{k=1}^{\infty} \frac{du_k(x)}{dx} \qquad\qquad \Box \ (3.50)$$

Next, we have a few words about the possible relations between the three basic types of convergence of a sequence $S_n(x)$: *pointwise convergence* for each $x \in (a,b)$, uniform *convergence* for all $x \in [a,b]$, the *convergence in the mean* on $[a,b]$, except for a finite number of separate points there where the series may even diverge. We will now illustrate such relations between these three basic types of convergence by examples or full proofs, some of which are left for the exercises.

EXAMPLE 3.5:   Relations Between the Three Basic Types of
Convergence

(a)   Uniform convergence implies convergence in the mean, but not the converse.

(b)   Pointwise convergence does not imply convergence in the mean.

(a)   We prove the first part of uniform convergence:

$$\lim_{n \to \infty} S_n(x) = f(x) \quad \text{for all } x \in [a,b] \tag{E.1}$$

implying convergence in the mean:

$$\lim_{n \to \infty} \int_a^b |f(x) - S_n(x)|^2 \, dx = 0 \tag{E.2}$$

and we leave the discussion of the converse in part (a), that convergence in the mean does not imply uniform convergence, for Exercise 3.9.

Proof of part (a):

(i)   From the definition of uniform convergence in (E.1) we write

$$|S_n(x) - f(x)| < \epsilon_1 = \sqrt{\frac{\epsilon}{b-a}}$$

for all $n > N(\epsilon)$ and all $x \in [b,a]$ \qquad\qquad (E.3)

where we note that $N(\epsilon)$ is independent of x for uniform convergence.

If we square both sides and integrate from a to b we have

$$\int_a^b |f(x) - S_n(x)|^2 dx < \frac{\epsilon}{b-a} \int_a^b 1\,dx = \frac{\epsilon(b-a)}{(b-a)} = \epsilon$$

(E.4)

as valid for all $n > N(\epsilon)$ and all $x \in [a,b]$, which is the ϵ and $N(\epsilon)$ language of defining convergence in the mean as desired in (E.2). This shows that (i) uniform convergence implies convergence in the mean. A comment on the converse not being true follows next.

(ii) Convergence in the mean of $S_n(x)$ to $f(x)$ does not necessarily imply uniform convergence of $S_n(x)$ to $f(x)$. This is the subject of Exercise 3.9, which will also show an example of a sequence convergent in the mean on $[a,b]$ but divergent at each point of the interval by not attaining a unique limit. ■

(b) This is an example to show that pointwise convergence of $S_n(x)$ to $f(x)$ at each point of $[a,b]$ does not imply that $S_n(x)$ converges in the mean to $f(x)$ on $[a,b]$.

The sequence $c_n(x) = \sqrt{2n}x\,e^{-1/2nx^2}$ converges to zero on the interval $(0,1]$ as $n \to \infty$, as can easily be shown with the use of L'Hôpital rule (for $c_n(x)$ as a function of the variable n approaching infinity):

$$\lim_{n\to\infty} c_n(x) = \lim_{n\to\infty} \frac{2^{1/2}n^{1/2}x^{1/2}}{e^{nx^2/2}} = \lim_{n\to\infty} \frac{2^{1/2}x^{1/2}\tfrac{1}{2}n^{-1/2}}{\tfrac{1}{2}x^2 e^{nx^2/2}}$$

$$= 2^{1/2}x^{-3/2}\lim_{n\to\infty}\frac{1}{\sqrt{n}e^{nx^2/2}} = 0 \equiv f(x)$$

for all $x \in (0,1]$.

However, this sequence does not converge in the mean to this $f(x) \equiv 0$ on $(0,1]$, since

$$\int_0^1 |c_n(x) - 0|^2 dx = \int_0^1 2nxe^{-nx^2}dx = 1 - e^{-n}$$

and

$$\lim_{n\to\infty}\int_0^1 |c_n(x) - 0|^2 dx = \lim_{n\to\infty}(1 - e^{-n}) = 1 \neq 0$$

that is, $c_n(x)$ does not converge in the mean to $f(x) = 0$ on $(0,1]$. Thus, in general, pointwise convergence does not imply convergence in the mean.

We have covered, in the above section, the basic elements of infinite sequences (and series) that are necessary for our development of convergence for Fourier series and orthogonal expansions, as a special case of infinite series. In particular, we covered the most familiar types of convergence, namely *pointwise convergence, uniform convergence,* and *convergence in the sense of the mean square* of the sequence $S_n(x) = \sum_{k=1}^{n} u_k$, of the nth partial sum of the series. Also, we have shown how such convergence of the series is affected (often for the better) by integrating it term by term (Theorem 3.5 and its corollary) and may be worse after differentiating it term by term (Theorem 3.6 and its corollary).

We now turn to our main concern, the series of the orthogonal expansion

$$S_N(x) = \sum_{k=1}^{N} c_k \phi_k(x) \tag{3.51}$$

of (3.34) as a special type of series, where the above basic theorems will be applied to it. We will then specialize to the familiar orthogonal expansion of the trigonometric Fourier sine-cosine series (3.10)–(3.13), where

$$\{\phi_k(x)\} = \left\{ 1, \cos\frac{k\pi x}{a}, \sin\frac{k\pi x}{a} \right\}_{k=1}^{\infty}$$

or its equivalent, the Fourier (complex) exponential series in (3.8), (3.9) with $\{\phi_k(x)\} \equiv \{e^{ik\pi x/l}\}_{k=1}^{\infty}$.

3.1.3 The Orthogonal Expansions—Bessel's Inequality and Fourier Series

In Section 1.5.2A of Chapter 1, we proved in Example 1.9 that the solutions $\{\phi_m(x)\}$ of the (regular) Sturm–Liouville problem (1.63)–(1.65) are orthogonal on the interval $(a,b)_\rho$ with respect to a weight function $\rho(x) > 0$;

$$\int_a^b \rho(x)\phi_n(x)\overline{\phi_m(x)}\,dx = 0, \qquad n \neq m \tag{3.52}$$
$$\tag{1.66}$$

and we discussed the orthogonal or general Fourier series, expansion of $f(x)$ on (a,b),

$$f(x) = \sum_{n=1}^{\infty} c_n \phi_n(x), \qquad a < x < b \tag{3.53}$$
$$\tag{1.67}$$

Then, assuming that the convergence of this series allows integration term by term, which we are about to justify, we used the orthogonality property (1.66) of $\{\phi_n(x)\}$ to derive the following form for the Fourier coefficients:

$$c_n = \frac{\int_a^b \rho(x)f(x)\overline{\phi_n(x)}\,dx}{\int_a^b \rho(x)|\phi_n|^2\,dx} \tag{3.54}$$
$$\tag{1.68}$$

It is the main purpose of this section to show that (i) for square integrable functions $f(x)$ on $(a,b)_\rho$,

$$\int_a^b \rho(x)|f(x)|^2\,dx < \infty \tag{3.55}$$

and (ii) the set of orthogonal functions $\{\phi_n(x)\}$ being *complete* on the indicated interval $(a,b)_\rho$, the above-mentioned term-by-term integration of the orthogonal series is valid.

By *complete orthogonal set* on $(a,b)_\rho$ we mean that its orthogonal series (3.53) converges in the mean for every piecewise continuous function $f(x) \in P(a,b)_\rho$ (or square integrable function $f(x) \in L_2(a,b)_\rho$). We term the set $\{\phi_n(x)\}$ *closed* if every piecewise continuous function on $(a,b)_\rho$ that is orthogonal to all $\{\phi_n(x)\}$ is the zero function on $(a,b)_\rho$ (of course, here we mean the zero on (a,b) except for a set of finite separate points). We will soon prove in Theorem 3.10 that the (orthogonal) trigonometric set of functions $\{1, \cos(n\pi x/a),\ \sin(n\pi x/a)\}_{n=1}^\infty$ is complete and closed on $(-a,a)$. This makes it also the natural time to speak in terms of the convergence in the mean of the series (3.53) to a square integrable function. It is clear that a piecewise continuous function $f(x) \in P(a,b)$ is *square integrable*; i.e., $f(x) \in L_2(a,b)$, as $f(x)$ is bounded on (a,b). Moreover, in the development of the convergence in the mean of the orthogonal series we will obtain important tools of analysis like *Bessel's inequality* (3.62), which will give us some idea about the truncation error of the partial sum $S_N(x)$ of the series

$$S_N(x) = \sum_{n=1}^N c_n \phi_n(x) \tag{3.51}$$

Also, as we have mentioned before, the Bessel inequality will help us in the proof of Theorem 3.2 and Theorem 3.3(ii) for uniform convergence. Another tool will be *Parseval's equality* (3.63), among the many applications of which is the evaluation of infinite series like $\sum_{n=1}^\infty 1/n^2 = \pi^2/6$ and others like it. We may mention that in introductory calculus we know by the p-test, for example, that series $\sum_{n=1}^\infty \frac{1}{n^2}$ is convergent, but we learned of no way to obtain its limit as $\pi^2/6$.

The following development, then, is for the expansion of square integrable functions, or piecewise continuous functions on (a,b), in terms of a complete orthogonal set $\{\phi_n(x)\}$, whose usual source is the Sturm–Liouville boundary value problem (1.63)–(1.65). This includes the *regular* Sturm–Liouville problem, covered in Section 1.5.2B, and an important *singular* case of Sturm–Liouville problem. The latter is associated with differential equations of variable coefficients that have a removable singularity (regular singular equations—see (3.75)–(3.77)) at a point or points (usually of interest to us) in the domain of the problem. A number of such equations and their solutions are discussed in Appendix A. An example is the Bessel differential equation (1.6) with regular singularity at an end point $x = 0$. In Section 1.5.2B we considered the regular case of the Sturm–Liouville problem to have a chance of introducing and illustrating the orthogonality of its solutions, without indulging too much, at that stage, in varied differential equations like those with regular singularity.

We will postpone the illustration of the orthogonal solutions of the Sturm–Liouville problem, including the regular singular one, to Section 3.1.3B, until we have developed the Bessel inequality (3.61), (3.62) and the Parseval equality (3.63), (3.65) of the general orthogonal expansion for square integrable functions on (a,b).

Consider the partial sum

$$S_n(x) = \sum_{n=1}^{N} c_n \phi_n(x) \tag{3.51}$$

As in the last section, we say that $S_N(x)$ *approximates* $f(x)$ *in the mean* on (a,b) if E_N, the *mean square deviation*,

$$E_N = \int_a^b \rho(x)|f(x) - S_N(x)|^2 \, dx \tag{3.56}$$

is minimum, and that $S_N(x)$ *converges to* $f(x)$ *in the mean* if $\lim_{N \to \infty} E_N = 0$. An orthogonal set $\{\phi_n(x)\}$ which gives convergence in the mean for every square integrable function $f(x)$ on the indicated interval is called a *complete set*.

Next we shall show that the mean square deviation E_N is indeed minimum if we choose the coefficient c_n in the orthogonal expansion (3.51) as the following b_n of (3.54):

$$b_n = \frac{\int_a^b \rho(x)f(x)\phi_n(x)\,dx}{\int_a^b \rho(x)\phi_n^2(x)\,dx} \tag{3.57}$$

which is the form of the Fourier coefficient. The treatment here is for real-valued functions. For complex-valued functions, $\phi_n(x)$ should be replaced by its conjugate $\overline{\phi_n(x)}$, and $\phi_n^2(x)$ above by $|\phi_n(x)|^2$, as given in (3.54). If we expand the series for $S_N(x)$ in (3.56), we write

$$
\begin{aligned}
E_N &= \int_a^b \rho[f - (c_1\phi_1 + c_2\phi_2 + \cdots + c_N\phi_N)]^2\, dx \\
&= \int_a^b \rho f^2\, dx - 2\int_a^b \rho f(x)[c_1\phi_1 + c_2\phi_2 + \cdots + c_N\phi_N]\, dx \\
&\quad + \int_a^b \rho[c_1\phi_1 + c_2\phi_2 + \cdots + c_N\phi_N][c_1\phi_1 + \cdots + c_N\phi_N]\, dx \quad (3.58)
\end{aligned}
$$

where with the use of (3.57) we note in the middle integral

$$
\begin{aligned}
\int_a^b \rho(x)f(x)c_j\phi_j(x)\, dx &= c_j \int_a^b \rho(x)f(x)\phi_j(x)\, dx \\
&= c_j b_j \int_a^b \rho(x)\phi_j^2(x)\, dx, \qquad j = 1, 2, 3, \ldots
\end{aligned}
$$

after using only the definition of b_j in (3.57) for the above integral. Note that we have not yet committed c_j of (3.51) to the form of Fourier coefficient as in (3.57). The last integral in (3.58) involves products of ϕ_1 and ϕ_j, which are orthogonal. Hence we have

$$
\int_a^b \rho(x)\phi_i(x)\phi_j(x)\, dx = \begin{cases} \displaystyle\int_a^b \phi_j^2(x)\, dx, & i = j \\ 0, & i \neq j \end{cases}
$$

which makes every term in that integral vanish except when $i = j$ and gives us

$$
E_N = \int_a^b \rho f^2\, dx - 2\sum_{n=1}^N c_n b_n \int_a^b \rho\phi_n^2\, dx + \sum_{n=1}^N c_n^2 \int_a^b \rho\phi_n^2\, dx \quad (3.58)
$$

If we combine the terms under summation and complete the square we obtain

$$
\begin{aligned}
E_N &= \int_a^b \rho f^2\, dx + \sum_{n=1}^N \int_a^b \rho\phi_n^2\, dx \left[c_n - \frac{\int_a^b \rho f \phi_n\, dx}{\int_a^b \rho\phi_n^2\, dx} \right]^2 \\
&\quad - \sum_{n=1}^N \frac{\left[\int_a^b \rho f \phi_n\, dx \right]^2}{\int_a^b \rho\phi_n^2\, dx} \quad (3.59)
\end{aligned}
$$

With b_n given by (3.57), the middle term of the right-hand side of (3.59) becomes

$$\sum_{n=1}^{N} \int_a^b \rho \phi_n^2 \, dx \, [c_n - b_n]^2$$

This term, which is positive, vanishes if we choose $c_n = b_n$ as in (3.57), which is the first time we commit the coefficient c_n, in the approximation (3.51), to a Fourier coefficient form b_n of (3.57). Thus

$$E_N = \int_a^b \rho f^2 \, dx - \sum_{n=1}^{N} c_n^2 \int_a^b \rho \phi_n^2 \, dx \qquad (3.60)$$

which is termed the *Bessel identity* that gives the minimum E_N.

From (3.56) it is clear that E_N is nonnegative and hence (3.60) gives *Bessel's inequality*.

$$\sum_{n=1}^{N} c_n^2 \int_a^b \rho \phi_n^2 \, dx \le \int_a^b \rho f^2 \, dx \qquad (3.61)$$

For the special case of the trigonometric Fourier series (3.10)–(3.13), Bessel's inequality is

$$2a_0^2 + \sum_{n=1}^{N} (a_n^2 + b_n^2) \le \frac{1}{l} \int_{-l}^{l} f^2(x) \, dx \qquad (3.61a)$$

Now if $f(x)$ is square integrable, the above sum on the left in (3.61), as a sequence S_n, is bounded by the finite integral on the right. So if we take the limit as $N \to \infty$ of (3.61) we obtain a very useful Bessel inequality,

$$\sum_{n=1}^{\infty} c_n^2 \int_a^b \rho(x) \phi_n^2(x) \, dx \le \int_a^b \rho(x) f^2(x) \, dx \qquad (3.62)$$

and in the case of the trigonometric Fourier series (3.10)–(3.13) we have

$$2a_0^2 + \sum_{n=1}^{\infty} (a_n^2 + b_n^2) \le \frac{1}{l} \int_{-l}^{l} f^2(x) \, dx \qquad (3.62b)$$

These are the types of Bessel inequalities, (3.62) or (3.61), that we shall need, soon after this development, for the proof of the uniform convergence of Fourier (trigonometric) series (Theorem 3.2) in Example 3.6. The relation of better convergence to the smoothness of the function expanded (Theorem 3.3) is presented in Example 3.7. Also, the Bessel inequality is instrumental in the proof of the completeness of the trigonometric functions (basis) of Fourier series in Theorem 3.10, as well as the completeness of more general orthogonal functions like the Legendre polynomials in Example 3.8.

If in addition we have that the orthogonal set $\{\phi_n(x)\}$ in (3.60) is complete, it means that the orthogonal expansion (3.51) of this set on $[a,b]$ converges in the mean to each square integrable function $f(x)$ on $[a,b]$. Thus for E_N of (3.56), $\lim_{N\to\infty} E_N = 0$, where its use in (3.60) results in the Bessel inequalities (3.61), (3.62) becoming the *Parseval equality*,

$$\sum_{n=1}^{\infty} c_n^2 \int_a^b \rho(x)\phi_n^2(x)\,dx = \int_a^b \rho(x)f^2(x)\,dx \tag{3.63}$$

In other words, with a complete orthogonal set, the Bessel inequalities become Parseval's equality for the square integrable functions $f(x)$. The Parseval equality (3.63) for the trigonometric Fourier series (3.10)–(3.13) becomes

$$2a_0^2 + \sum_{n=1}^{\infty} (a_n^2 + b_n^2) = \frac{1}{l} \int_{-l}^{l} f^2(x)\,dx \tag{3.63b}$$

Now if $f(x)$ is square integrable and $\{\phi_n(x)\}$ is a complete orthogonal set, we may use the Parseval equality (3.63) in (3.60) to obtain

$$E_N = \int_a^b \rho f^2\,dx - \sum_{n=1}^{N} c_n^2 \int_a^b \rho\phi_n^2\,dx$$

$$= \sum_{n=1}^{\infty} c_n^2 \int_a^b \rho\phi_n^2\,dx - \sum_{n=1}^{N} c_n^2 \int_a^b \rho\phi_n^2\,dx = \sum_{N+1}^{\infty} c_n^2 \int_a^b \rho\phi_n^2\,dx$$

$$E_N = \sum_{N+1}^{\infty} c_n^2 \int_a^b \rho\phi_n^2\,dx \tag{3.64}$$

as the *least square deviation*, obtained with the help of the completeness property of the orthogonal set $\{\phi_n\}$ that gave us the Parseval equality. This least square deviation error is in contrast to that of the mean square deviation in (3.56), which is not optimal, since it has not yet felt the advantage of the orthogonality and completeness property of the particular orthogonal basis $\{\phi_n(x)\}$ that gave Parseval's equality (3.63).

The following generalization of Parseval's equality (3.63), to the product $f(x)g(x)$ of two square integrable (real-valued) functions instead of just $f^2(x)$ as in (3.63), can be obtained easily from (3.63) and is left for an exercise (see Exercise 1.18a).

If $f(x)$ and $g(x)$ are square integrable on (a,b), then

$$\int_a^b \rho(x)f(x)g(x)\,dx = \sum_{n=1}^{\infty} c_n d_n \int_a^b \rho\phi_n^2\,dx \qquad (3.65)$$

$$= \sum_{n=1}^{\infty} c_n d_n \|\phi_n\|^2$$

where c_n and d_n are the respective orthogonal expansion (Fourier) coefficients for $f(x)$ and $g(x)$ as defined in (3.54), and $\|\phi_n\|^2$ is the *norm square* of $\phi_n(x)$ on (a,b), defined as $\|\phi_n\|^2 \equiv \int_a^b \rho(x)_n^2(x)\,dx$. In case these $f(x)$ and $g(x)$ are complex valued, the above generalized Parseval's equality takes the form

$$\int_a^b \rho(x)f(x)\overline{g(x)}\,dx = \sum_{n=1}^{\infty} c_n \overline{d_n} \|\phi_n\|^2 \qquad (3.65a)$$

where $\overline{g(x)}$ and $\overline{d_n}$ stand for the complex conjugates of $g(x)$ and d_n.

The detailed illustrations we had in Section 3.1.1 were for the very special case of the trigonometric (sine-cosine) Fourier series expansion. If we are to give any illustration of the orthogonal expansion away from the usual trigonometric series, the simplest example would be with the Legendre polynomials $P_n(x)$, which are solutions of the Legendre equation (3.84) (see also (3.89)–(3.94) and Appendix A.5).

We can easily verify that the Legendre polynomials,

$$P_0(x) = 1, \qquad P_1(x) = x, \qquad P_2(x) = \tfrac{3}{2}x^2 - \tfrac{1}{2}, \qquad P_3(x) = \tfrac{5}{2}x^3 - \tfrac{3}{2}x$$

are orthogonal on $(-1,1)$. For example,

$$\int_{-1}^1 P_0(x)P_1(x)\,dx = \int_{-1}^1 x\,dx = \left.\frac{x^2}{2}\right|_{-1}^1 = 0$$

For the general case of $P_n(x)$ we use the *Rodrigues formula* (see (3.89))

$$P_n(x) = \frac{1}{2^n n!}\frac{d^n}{dx^n}(x^2 - 1)^n, \qquad n = 0, 1, 2, \ldots$$

like

$$P_2(x) = \frac{1}{8}\frac{d^2}{dx^2}(x^2 - 1)^2 = \frac{1}{8}\frac{d^2}{dx^2}(x^4 - 2x^2 + 1)$$

$$= \frac{1}{8}(12x^2 - 4) = \frac{3}{2}x^2 - \frac{1}{2}$$

Also, it is easy to write the first four terms of the (Legendre polynomials) orthogonal series expansion of a (very special) simple function like $f(x) =$

x^3 (which is square integrable on $(-1,1)$ since $\int_{-1}^{1} x^6 \, dx = \frac{2}{7}$) in terms of the above four polynomials:

$$x^3 = c_0 P_0 + c_1 P_1 + c_2 P_2 + c_3 P_3$$

where, for example,

$$c_1 = \frac{\int_{-1}^{1} x^3 x \, dx}{\int_{-1}^{1} x^2 \, dx} = \frac{(x^5/5)\big|_{-1}^{1}}{(x^3/3)\big|_{-1}^{1}} = \frac{3}{5}, \qquad \text{etc.}$$

It is left as an exercise to find $c_3 = 3$ and $c_0 = c_1 = c_4 = c_5 = 0$. With these values we can write $x^3 = \frac{2}{5} P_3(x) + \frac{3}{5} P_1(x)$ as a finite (Legendre polynomials) orthogonal expansion of this particular function $f(x) = x^3$ on $(-1,1)$. This example is related to Example 3.8, where we prove the completeness property of the orthogonal (infinite) set of Legendre polynomials $\{P_n(x)\}_{n=0}^{\infty}$ on $(-1,1)$. On the other hand, $f(x) = x^{1/2}$ is also square integrable on $(-1,1)$ since $\int_{-1}^{1} x \, dx = 0$, but it is obvious that we cannot write an orthogonal series expansion for it in terms of the above four Legendre polynomials, which shows that they are *not* complete. Indeed, the infinite set of the Legendre polynomials $\{P_n(x)\}_{n=0}^{\infty}$ is complete, as we shall prove in Example 3.8. Thus we need an infinite (Legendre) orthogonal series expansion (see (3.91), (3.92), (3.90)) in terms of $P_n(x)$ for representing $f(x) = x^{1/2}$ on $(-1,1)$,

$$x^{1/2} = \sum_{n=0}^{\infty} c_n P_n(x), \qquad -1 < x < 1,$$

$$c_n = \frac{2n+1}{2} \int_{-1}^{1} x^{1/2} P_n(x) \, dx$$

For the coefficients c_n we used norm square $\|P_n\|^2 = 2/(2n+1)$, as needed in (3.57), and which can be derived with the help of the Rodrigues formula (3.89) for computing $P_n(x)$ (see Exercises A.28 and A.27 in Appendix A).

We may emphasize and summarize our preceding results, which center around the convergence in the mean of the orthogonal expansion of the complete orthogonal set to square integrable functions $f(x)$ on (a,b), as the following Theorem 3.7 on convergence in the mean.

THEOREM 3.7: Convergence in the mean and its equivalents—Parseval's equality and integration term by term of the orthogonal series

For the square integrable functions $f(x)$ and $g(x)$ (with respect to the weight function $\rho(x)$ on (a,b)) and the orthogonal set $\{\phi_n(x)\}$ on the same interval (a,b), the following three results are equivalent:

(i) The last result for the validity of Parseval's equality (3.63) is equivalent to the orthogonal set $\{\phi_n(x)\}$ being complete.

(ii) Result (i) is also equivalent to the convergence of the orthogonal expansion in the mean to the square integrable functions $f(x)$ on (a,b).

(iii) The statement of the generalization of the Parseval equality,

$$\int_a^b \rho(x) f(x) g(x)\, dx = \sum_{n=1}^{\infty} c_n d_n \| \phi_n \|^2 \tag{3.65}$$

is equivalent to allowing the term-by-term integration of the (convergent in the mean) orthogonal series of square integrable functions,

$$g(x) = \sum_{n=1}^{\infty} d_n \phi_n(x), \qquad a < x < b \qquad \square$$

Proof:

(a) The first two parts of the theorem are clear, and even part (iii) is seen clearly as we write the orthogonal expansion of $g(x)$ inside the above integral and use the definition of the Fourier coefficients c_n of $f(x)$,

$$\int_a^b \rho(x) f(x) \left[\sum_{n=1}^{\infty} d_n \phi_n(x) \right] dx = \sum_{n=1}^{\infty} d_n \left[\int_a^b \rho(x) f(x) \phi_n(x)\, dx \right]$$

$$= \sum_{n=1}^{\infty} d_n c_n \| \phi_n \|^2 \qquad \blacksquare \ (3.66)$$

As we mentioned above, a set $\{\phi_n(x)\}$ is also termed *closed* on the interval (a,b) if every piecewise continuous function, i.e., $f(x) \in P(a,b)$ (generally square integrable, i.e., $g(x) \in L_2(a,b)$), which is orthogonal to all the members of the orthogonal set $\phi_n(x)$ is the zero element of the set $P(a,b)$ (or $L_2(a,b)$). This means that such an $f(x)$ would, if continuous, be identically zero, $f(x) \equiv 0$, and has value at the middle of its jump discontinuity if it is piecewise continuous. So this function can have nonzero values only at a finite number of points of the interval (a,b), as one member of the piecewise continuous (or square integrable) function set. We leave it as an exercise (see Exercise 3.10) to show that if the set $\{\phi_n(x)\}$ is orthogonal and complete, then it is closed. With the help of this result we can show that if two functions $f(x)$ and $g(x)$ have the same Fourier series (of complete orthogonal functions $\{\phi_n(x)\}$ on (a,b)), then these functions coincide as a member of the piecewise continuous set $P(a,b)$. This means, from the definition of these sets, that the two functions $f(x)$ and $g(x)$ may differ only at a finite number of separate points of the interval (a,b).

A. Proofs of the Convergence Theorem for Fourier Series

The Uniform Convergence of Trigonometric Fourier series

Before we set out to prove that the set (basis) of the trigonometric series

$$\{\phi_n(x)\}_{n=1}^{\infty} = \left\{ 1, \sin \frac{\pi x}{l}, \sin \frac{2\pi x}{l}, \cos \frac{2\pi x}{l}, \ldots, \sin \frac{k\pi x}{l}, \cos \frac{k\pi x}{l}, \ldots \right\}$$

(3.67)

is complete, we will first prove in Example 3.6 that its partial sum $S_n(x)$, the trigonometric polynomial,

$$S_n(x) = a_0 + \sum_{k=1}^{n} a_k \cos \frac{k\pi x}{l} + b_k \sin \frac{k\pi x}{l}$$

(3.68)

converges uniformly to a continuous piecewise smooth function $f(x)$ on $[-l,l]$ which assumes equal values at the end points of the interval, i.e., $f(-l) = f(l)$, as stated in Theorem 3.2. We have delayed the proof of this theorem until we established the Bessel inequality (3.61), (3.62), as a consequence of the property of orthogonal functions on $[-l,l]$, which is the case for our present trigonometric set. The last condition $f(-l) = f(l)$ clearly makes $\bar{f}$ the periodic extension, with period $2l$, continuous.

We still need some preliminaries before we start the proof of Theorem 3.2 for uniform convergence in Example 3.6, the proof of Theorem 3.3 for accelerated convergence in Example 3.7, and Theorem 3.10 for the completeness of the trigonometric set in (3.67).

The way we termed the nth partial sum of the trigonometric functions $S_n(x)$ as a "trigonometric polynomial" is understood when we use the Euler identities

$$\cos \frac{k\pi x}{l} = e^{ik\pi x/l} + e^{-ik\pi x/l} = \tfrac{1}{2} \left[z^k + \omega^k \right]$$

where

$$z = e^{i\pi x/l} > 0, \qquad \omega = e^{-i\pi x/l}$$

and

$$\sin \frac{k\pi x}{l} = \frac{1}{2i} \left[e^{ik\pi x/l} - e^{-ik\pi x/l} \right] = \frac{1}{2i} \left[z^k - \omega^k \right]$$

in (3.68), then collect the z^k terms and ω^k terms to have

$$S_n = a_0 + \frac{1}{2} \sum_{k=1}^{n} a_k [z^k + \omega^k] - i b_k [z^k - \omega^k]$$

$$= a_0 + \sum_{k=1}^{n} \frac{1}{2} [a_k - i b_k] z^k + \frac{1}{2} [a_k + i b_k] \omega^k$$

$$z = e^{i \pi k/l}, \qquad \omega = e^{-i \pi k/l} \tag{3.69}$$

This parallels the very usual approximation of a continuous function on $[a,b]$ by an algebraic polynomial $Q_m(x)$,

$$Q_m = A_0 + A_1 x + A_2 x^2 + \cdots + A_m x^m = \sum_{k=1}^{m} A_k x^k \tag{3.70}$$

The approximation by the nth partial sum of Legendre polynomials $P_n(x)$ on $[-1, 1]$,

$$S_n(x) = \sum_{k=1}^{n} c_k P_k(x) \tag{3.71}$$

can be looked at as a special case of the algebraic polynomial approximation (3.70), since $P_k(x)$ is a polynomial of degree k, i.e.,

$$P_k(x) = \sum_{i=1}^{k} \gamma_i x^i$$

and

$$S_n(x) = \sum_{k=1}^{n} c_k \left[\sum_{i=1}^{k} \gamma_i x^i \right] = \sum_{j=1}^{n} \beta_j x^j$$

after collecting the coefficients of x^j as β_j.

There is even a very strong theorem for approximating a continuous function $f(x)$ on $[a,b]$ by an algebraic polynomial $Q_m(x)$, which is the following Weierstrass polynomial approximation Theorem 3.8. We will state the theorem here and leave its proof to an exercise with complete detailed steps (see Exercise 3.11).

THEOREM 3.8: Weierstrass Polynomial Approximation of
Continuous Functions

If $f(x)$ is continuous on an interval $[a,b]$, then given $\epsilon > 0$, there exists
an algebraic polynomial $Q_m(x)$,[†]

$$Q_m(x) = \sum_{k=1}^{m} A_k x^k \tag{3.72}$$

such that

$$|f(x) - Q_m(x)| < \epsilon \quad \text{for } x \in [a,b] \qquad \square$$

We expect the proof of this theorem to be a reasonable exercise because it shares the idea of the proof of uniform convergence
(Theorem 3.2), Theorem 3.3, and the proof of Theorem 3.10 for the
completeness of the trigonometric set in (3.67) on $[-l,l]$. The idea is
that the assumed function $f(x)$ is replaced by $g(x)$, a very close function
with just the right quality, and that $g(x)$ approximates $f(x)$ in the mean
on $[-l,l]$. For pointwise convergence, the replacement $g(x)$ is made continuous around a jump discontinuity x_0 of $f(x)$ by defining it as a straight
line in the neighborhood $(x_0 - \delta; x_0 + \delta)$, which will be done in complete
detail in the proof of the completeness of the trigonometric set (basis)
in Theorem 3.10. For $f(x)$ just continuous on $[-l,l]$, $f(-l) = f(l)$, the
replacement $g(x)$ is made continuous and piecewise smooth for the proof
of uniform convergence.

The other reason for having the Weierstrass polynomial approximation theorem here is that it will be needed in Example 3.8 for proving
that the Legendre polynomials' set (basis) $\{P_0, P_1, P_2, \ldots\}$ is complete
on $[-l,l]$. This follows the same process, in the sense that we will
need the uniform convergence of the trigonometric polynomials (3.68) (or
(3.69)) for proving the completeness of the trigonometric set (of (3.67))
in Theorem 3.10.

Theorem 3.8 is a very important theoretical tool for the existence of
a polynomial that approximates any continuous function $f(x)$ to any pre-
assigned accuracy $\epsilon > 0$. But on the practical side, it does not say much
about how to construct such a polynomial. This question was raised
and answered by Tchebychev in the way that among all polynomials he

---

[†] The symbol $Q_m(x)$ used here for a *polynomial* of degree m should be differentiated from $Q_m(x)$, the *Legendre function of the second kind*, in (A.5.35)–(A.5.36) of Appendix A, which is an infinite series.

proved the existence and uniqueness of a polynomial, called the *Tcheby-chev polynomial*, that gives the least error (deviation) in its approximation of continuous functions on a finite interval $[a,b]$. The Tchebychev polynomials and other important polynomials and special functions are discussed in Appendix A.

In line with the replacement of the given function $f(x)$ by a better one $g(x)$ on $[-l,l]$, we will state the following basic theorem, which is essential to the proof of the uniform convergence Theorem 3.2 that will follow it in Example 3.6. The proof of this Theorem 3.9 will exemplify how such a replacement of $f(x)$ by $g(x)$ is done, but we feel it will be much in our way of getting to the uniform convergence proof and the completeness proof, where this replacement is done in complete detail. Thus we leave the proof of the present theorem to an exercise and advise looking the proof of the completeness in Theorem 3.10 and how the replacement is carefully done.

BASIC THEOREM 3.9:

For every continuous function $f(x)$ defined on the internal $[a,b]$, and for every $\epsilon > 0$, there exists a continuous *piecewise smooth* function $g(x)$ defined on this interval such that

$$|f(x) - g(x)| < \frac{\epsilon}{2} \quad \text{for all } x \in [a,b]$$

and

$$g(a) = f(a), \qquad g(b) = f(b) \qquad\qquad\qquad \square$$

We have $\epsilon/2$ above and not ϵ, since for a corresponding neighborhood of width $\delta(\epsilon)$ we may use its half-width $\frac{1}{2}\delta(\epsilon)$ with $\epsilon/2$.

EXAMPLE 3.6: Proof of the Uniform Convergence Theorem 3.2 for the Fourier Trigonometric Series

In this example we are to prove Theorem 3.2: "If a continuous and *piecewise smooth* function $f(x)$ defined on the interval $[-l,l]$ assumes equal values at the end points $x = l$ of this interval, i.e., $f(-l) = f(l)$, then its Fourier series

$$a_0 + \sum_{k=1}^{\infty} a_k \cos \frac{k\pi x}{l} + b_k \sin \frac{k\pi x}{l} \tag{E.1}$$

with coefficients

$$a_0 = \frac{1}{2l} \int_{-l}^{l} f(x)\,dx \tag{E.2}$$

$$a_k = \frac{1}{l} \int_{-l}^{l} f(x) \cos \frac{k\pi x}{l} \, dx \tag{E.3}$$

$$b_k = \frac{1}{l} \int_{-l}^{l} f(x) \sin \frac{k\pi x}{l} \, dx \tag{E.4}$$

converges uniformly on the interval $[-l,l]$ and the series is equal to $f(x)$ at each point of this interval."

Proof: The proof here will make clear the need for Bessel's inequality (3.62), the condition $f(-l) = f(l)$, and the condition that the derivative $f'(x)$ is piecewise continuous on $[-l,l]$, which gives us $f'(x)$ being square integrable on this interval, i.e., $\int_{-l}^{l} f'^2(x) \, dx < \infty$. The proof depends on an elementary theorem for the uniform convergence of infinite series $\sum_{k=1}^{\infty} u_k(x)$ in general. This is the dominance theorem; that is, if all elements of the series $\{u_k(x)\}$ are bounded by constant sequences M_k on the whole interval $[a,b]$, i.e., $|u_k(x)| < M_k$, $k = 1, 2, 3, \ldots$, and if this (dominant) series $\sum_{k=1}^{\infty} M_k$ converges, then the series $\sum_{k=1}^{\infty} u_k(x)$ converges uniformly on $[a,b]$.

In our case we have $\{u_k(x)\} = \{a_0, a_k \cos(k\pi x/l), b_k \sin(k\pi x/l)\}$ defined on $[-l,l]$ and it is clear that they are dominated by $\{M_k\} = \{|a_0|, |a_k|, |b_k|\}$ since $|a_0| = |a_0|$, $|a_k \cos(k\pi x/l)| \leq |a_k|$, and $|b_k \sin(k\pi x/l)| \leq |b_k|$. So the Fourier series (E.1) is now dominated by the series

$$|a_0| + \sum_{k=1}^{\infty} |a_k| + |b_k| \tag{E.5}$$

which we want to show is convergent. To this end we are set to show that this series (E.5) is dominated by a convergent series via some integration by parts and the "deserving" Bessel inequality for the coefficients α_0, α_k, and β_k of the following Fourier series (E.6)–(E.9) for the square integrable $f'(x)$ on $[-l,l]$:

$$f'(x) \sim \alpha_0 + \sum_{k=1}^{\infty} \alpha_n \cos \frac{k\pi x}{l} + \beta_k \sin \frac{k\pi x}{l} \tag{E.6}$$

$$\alpha_0 = \frac{1}{2l} \int_{-l}^{l} f'(x) \, dx \tag{E.7}$$

$$\alpha_k = \frac{1}{l} \int_{-l}^{l} f'(x) \cos \frac{k\pi x}{l} \, dx \tag{E.8}$$

$$\beta_k = \frac{1}{l} \int_{-l}^{l} f'(x) \sin \frac{k\pi x}{l} \, dx \tag{E.9}$$

This means that we are to show from the integration by parts that

(i) $\quad |a_k| + |b_k| \le \dfrac{l}{\pi} \left(\dfrac{|\alpha_k|}{k} + \dfrac{|\beta_k|}{k} \right), \qquad k = 1, 2, \ldots$ (E.10)

and with the help of the Bessel inequality for $f'(x)$ that

(ii) $\quad \dfrac{\sum_{k=1}^{\infty} |\alpha_k|}{k} + \dfrac{|\beta_k|}{k} \le \sum_{k=1}^{\infty} \dfrac{1}{2} (\alpha_k^2 + \beta_k^2) + \dfrac{1}{k^2}$ (E.11)

is convergent, since $\sum_{k=1}^{\infty} 1/k^2$ converges by the p-test ($p = 2$) and the series $\sum_{k=1}^{\infty} \alpha_k^2 + \beta_k^2$ is convergent by the Bessel inequality (3.62) for the square integrable $f'(x)$ on $[-l, l]$,

$$2\alpha_0^2 + \sum_{k=1}^{\infty} \alpha_k^2 + \beta_k^2 \le \dfrac{1}{l} \int_{-l}^{l} f'^2(x)\, dx$$

Thus if we have (i) and (ii) as in (E.10), (E.11) we can conclude that the series

$$|a_0| + \sum_{k=1}^{\infty} |a_k| + |b_k|$$ (E.5)

is convergent, but this series dominates our Fourier series (E.1) for all $x \in [-l, l]$, so the Fourier series (E.1) converges uniformly on $[-l, l]$. What remains is to prove the results (i) and (ii) of (E.10), (E.11) that we need now to conclude the result in (E.5). For (i), as we mentioned above, we consider the Fourier series of $f'(x)$ on $[-l, l]$ as in (E.6)–(E.9). To involve these α_k, β_k with a_k, b_k in a relation like (E.10), we consider the Fourier integrals for the coefficients a_k, b_k and purposely nominate $f(x)$ for a $u(x)$, in one integration by parts, to result in a Fourier coefficient of $du/dx = f'(x)$ in the first resulting integral, i.e.,

$$a_k = \dfrac{1}{l} \int_{-l}^{l} f(x) \cos \dfrac{k\pi x}{l}\, dx = \dfrac{1}{l} \dfrac{l}{k\pi} f(x) \sin \dfrac{k\pi x}{l} \bigg|_{x=-l}^{l}$$

$$- \dfrac{1}{l} \dfrac{l}{k\pi} \int_{-l}^{l} f'(x) \sin \dfrac{k\pi x}{l}\, dx$$

$$a_k = -\dfrac{l}{\pi} \dfrac{\alpha_k}{k}$$ (E.13)

since $\sin(k\pi x/l)$ vanishes at $x = \mp l$, and we have used the definition of α_k as in (E.8).

The same can be done for b_k, except we have to watch for $\cos(k\pi x/l)$ not vanishing at $x = \mp l$, and they have the values $\cos k\pi = (-1)^k$. To have a result similar to (E.13), we will see the need for the assumption of $f(-l) = f(l)$, which will make the term involving the nonzero $\cos k\pi$ vanish.

$$b_k = \frac{1}{l} \int_{-l}^{l} f(x) \sin \frac{k\pi x}{l} \, dx = -\frac{1}{l} \frac{l}{k\pi} f(x) \cos \frac{k\pi x}{k} \bigg|_{x=-l}^{x=l}$$

$$+ \frac{1}{l} \frac{l}{k\pi} \int_{-l}^{l} f'(x) \cos \frac{k\pi x}{l} \, dx$$

$$= \frac{-1}{k\pi} \cos k\pi [f(l) - f(-l)] + \frac{l}{\pi} \frac{\alpha_k}{k} \tag{E.14}$$

$$b_k = \frac{l}{\pi} \frac{\alpha_k}{k} \tag{E.15}$$

after using the important assumption (for uniform convergence) that $f(x)$ have the same value at the end points, i.e., $f(-l) = f(l)$.

At this stage we may point out that these Fourier coefficients a_k, b_k represent the main topic of this chapter, as the finite cosine and sine transforms, which we shall discuss and use in Section 3.2. In this respect, some applications may need to have $f(l) - f(-l) \neq 0$; for example, we may know that the temperature at the end $x = l$ is different from that at $x = -l$. The result (E.14) with jump discontinuity would fit such a situation and should be kept as a generalization to the finite sine transform when $f(l) \neq f(-l)$.

From the results (E.13) and (E.15) we have (i) in (E.10),

$$|a_k| + |b_k| = \left| -\frac{l}{\pi} \frac{\alpha_k}{k} \right| + \left| \frac{l}{\pi} \frac{\beta_k}{k} \right| \le \frac{l}{\pi} \left(\frac{|\alpha_k|}{k} + \frac{|\beta_k|}{k} \right) \tag{E.16}$$

Now we show part (ii), where the series of the above sequence is dominated by a convergent series, with the help of some simple identities and the Bessel inequality for $f'(x)$ on $[-l, l]$, which says that the following series is convergent:

$$2\alpha_0^2 + \sum_{k=1}^{\infty} \alpha_k^2 + \beta_k^2 \le \frac{1}{l} \int_{-l}^{l} f'^2(x) \, dx \tag{E.17}$$

We want to show that

$$\frac{|\alpha_k|}{k} + \frac{|\beta_k|}{k} \le \frac{1}{2} (\alpha_k^2 + \beta_k^2) + \frac{1}{k^2} \tag{E.18}$$

For this we can write

$$0 \le \left(|\alpha_k| - \frac{1}{k} \right)^2 = \alpha_k^2 - 2 \frac{|\alpha_k|}{k} + \frac{1}{k^2} \tag{E.19}$$

$$0 \le \left(|\beta_k| - \frac{1}{k} \right)^2 = \beta_k^2 - 2\frac{|\beta_k|}{k} + \frac{1}{k^2} \tag{E.20}$$

whence

$$\frac{|\alpha_k|}{k} \le \frac{1}{2}\left(\alpha_k^2 + \frac{1}{k^2} \right), \qquad \frac{|\beta_k|}{k} \le \frac{1}{2}\left(\beta^2 k + \frac{1}{k^2} \right) \tag{E.21}$$

to have (E.18) after adding these two inequalities, which (along with (E.16)) proves part (ii) in (E.11). Thus the proof of Theorem 3.2 for uniform convergence of the trigonometric Fourier series in (E.1) is completed. ∎

Our next example is to prove Theorem 3.3. In Theorem 3.2, just a bit of "goodness" for $f(x)$ in being continuous piecewise smooth on $[-l,l]$ and $f(l) = f(-l)$, we captured uniform convergence for the Fourier trigonometric series. In Theorem 3.3 we add more goodness of the functions $f(x)$ being continuously differentiable to order m, i.e., very smooth for large m, to see how much we can reap of fast convergence for the Fourier series.

EXAMPLE 3.7: Proof of Theorem 3.3 for Faster Uniform Convergence of Fourier Trigonometric Series of Very Smooth Functions

We are to prove Theorem 3.3, that if

(i) $f(x)$ and all its derivatives up to the mth order $f^{(m)}(x)$, $m \ge 0$, are continuous on the interval $[-l,l]$,

(ii)   they assume equal values at the end points $x = \mp l$, i.e.,

$$f(-l) = f(l), \qquad f'(-l) = f'(l), \ldots, f^{(m)}(-l) = f^{(m)}(l) \tag{E.1}$$

and

(iii)   the derivative $f(x)^{(m-1)}$ is piecewise continuous on $[-l,l]$,

then we have the following:

(i)   The Fourier coefficients,

$$a_k = \frac{1}{l} \int_{-l}^{l} f(x) \cos \frac{k\pi x}{l}\, dx \tag{E.2}$$

$$b_k = \frac{1}{l} \int_{-l}^{l} f(x) \sin \frac{k\pi x}{l} \, dx \tag{E.3}$$

of $f(x)$ satisfy the asymptotic relations

$$a_k = o\left(\frac{1}{k^{m+1}}\right), \quad b_k = o\left(\frac{1}{k^{m+1}}\right) \quad \text{as } k \to \infty \tag{E.4}$$

(ii)   Moreover, the series:

$$\sum_{k=1}^{\infty} k^n \left[|a_k| + |b_k| \right], \quad n = 0, 1, 2, \ldots, m \tag{E.5}$$

are convergent (strong convergence).

Here we use (small) o in the *asymptotic notation* to mean that $\lim_{k\to\infty} a_k/(1/k^{m+1}) = 0$ With the above assumptions, we shall integrate by parts in (E.2), as done for Theorem 3.2 in Example 3.6, starting by letting $u(x) = f'(x)$, then for the second integration again letting $u(x) = f''(x)$, and repeating this process $m + 1$ times to obtain, depending on $m + 1$ even or odd,

$$a_k = \begin{cases} \pm \left(\dfrac{l}{k\pi}\right)^{m+1} \dfrac{1}{l} \displaystyle\int_{-l}^{l} f^{m+1}(x) \cos \dfrac{k\pi x}{l} \, dx \\[2ex] \quad = \left(\dfrac{l}{k\pi}\right)^{m+1} A_k, \quad m+1 \text{ even} \\[3ex] \pm \left(\dfrac{l}{k\pi}\right)^{m+1} \dfrac{1}{l} \displaystyle\int_{-l}^{l} f^{m+1}(x) \sin \dfrac{k\pi x}{l} \, dx \\[2ex] \quad = \left(\dfrac{l}{k\pi}\right)^{m+1} B_k, \quad m+1 \text{ odd} \end{cases} \tag{E.6}$$

For this result we realize that the boundary conditions (E.1), as was the case in proving Theorem 3.2, make all the integrated terms of the integrations by parts vanish. In (E.6), A_k and B_k are the respective Fourier cosine and sine coefficients $f^{(m+1)}(x)$. Also, the negative sign is for $m = 0, 1, 4, 5, \ldots$ and the positive sign is for $m = 2, 3, 6, 7, \ldots$.

The same method can be used for b_k to have

$$
b_k = \begin{cases} \pm \left(\dfrac{l}{k\pi} \right)^{m+1} \dfrac{1}{l} \displaystyle\int_{-l}^{l} f^{(m+1)}(x) \sin \dfrac{k\pi x}{l} \, dx \\[4mm] \quad = \pm \left(\dfrac{l}{k\pi} \right)^{m+1} B_k, \qquad m+1 \text{ odd} \\[4mm] \pm \left(\dfrac{l}{k\pi} \right)^{m+1} \dfrac{1}{l} \displaystyle\int_{-l}^{l} f^{(m+1)}(x) \cos \dfrac{k\pi x}{l} \, dx \\[4mm] \quad = \left(\dfrac{l}{k\pi} \right)^{m+1} A_k, \qquad m+1 \text{ even} \end{cases}
$$
(E.7)

where the $-$ and $+$ signs go, as in (E.6), with $m = 0,\ 1,\ 4,\ 5,\ \ldots$ and $m = 2,\ 3,\ 6,\ 7,\ \ldots$ respectively.

From (E.6) and (E.7) we can write

$$
|a_k| + |b_k| = \left(\frac{l}{\pi} \right)^{m+1} \left[\frac{|A_k|}{k^{m+1}} + \frac{|B_k|}{k^{m+1}} \right]
$$
(E.8)

But A_k and B_k, as the Fourier coefficients for the piecewise continuous function $g(x) = d^{m+1}f(x)/dx^{m+1}$, both vanish as $k \to \infty$, which is a necessary condition for the convergence of its Fourier series. Thus the relation (E.8) would mean that

$$
\lim_{k \to \infty} \frac{a_k}{1/k^{m+1}} = \left(\frac{l}{\pi} \right)^{m+1} \lim_{k \to \infty} |A_k| = 0
$$

$$
\lim_{k \to \infty} \frac{b_k}{1/k^{m+1}} = \left(\frac{l}{\pi} \right)^{m+1} \lim_{k \to \infty} |B_k| = 0
$$

which is what we mean by the asymptotic notation (for large k) in (E.4) of part (i) of the theorem:

$$
a_k = o\left(\frac{1}{k^{m+1}} \right) \quad \text{and} \quad b_k = o\left(\frac{1}{k^{m+1}} \right), \qquad k \text{ very large}
$$
(E.4)

To prove part (ii) of the theorem in (E.5), we have from (E.8) that

$$
k^m [|a_k| + |b_k|] = \left(\frac{l}{\pi} \right)^{m+1} \left[\frac{|A_k|}{k} + \frac{|B_k|}{k} \right]
$$

$$
\leq \frac{1}{2} \left(\frac{l}{\pi} \right)^{m+1} \left[|A_k|^2 + |B_k|^2 + \frac{2}{k^2} \right]
$$
(E.9)

since, as we did in the proof of Theorem 3.2 in Example 3.6 for (E.18) (see also (E.19)–(E.21)), we can show easily that

$$\frac{|A_k|}{k} \le \frac{1}{2}\left(|A_k|^2 + \frac{1}{k^2}\right), \qquad \frac{|B_k|}{k} \le \frac{1}{2}\left(|B_k|^2 + \frac{1}{k^2}\right) \tag{E.10}$$

From (E.9) we see that the series (E.5), which we are to prove convergent, is dominated by the sum of two series

$$\sum_{k=1}^{\infty} k^m(|a_k| + |b_k|) \le \frac{1}{2}\left(\frac{l}{\pi}\right)^{m+1}\sum_{k=1}^{\infty}|A_k|^2 + |B_k|^2 + \left(\frac{l}{\pi}\right)^{m+1}\sum_{k=1}^{\infty}\frac{1}{k^2} \tag{E.11}$$

the last of which is clearly convergent by the p-test. The first series on the right is convergent when we appeal to the following Bessel inequality for sectionally continuous $f^{(m+1)}(x)$ on $[-l,l]$,

$$2A_0^2 + \sum_{k=1}^{\infty}(|A_k|^2 + |B_k|^2) \le \frac{1}{l}\int_{-l}^{l}|f(x)^{(m+1)}|^2\,dx \tag{E.12}$$

Thus from (E.11) and (E.12) we prove that *all* the series in (E.5) are convergent with k^n, $n = 0, 1, 2, \ldots, m$, since we are proving them in (E.11) as convergent for the largest n, namely m, which completes the proof of Theorem 3.3.

The Completeness of the Set of (Orthogonal) Trigonometric Functions

It is now instructive to show by direct computations that the orthogonal set $\{1, \sin(n\pi x/l), \cos(n\pi x/l)\}_{n=1}^{\infty}$ of the trigonometric Fourier series (3.10)–(3.13) is complete on $(-l,l)$. It will soon become very clear that this completeness property is equivalent to the convergence in the mean of the Fourier series. Moreover, both of these properties are in turn equivalent to the validity of the Parseval equality (3.65).

THEOREM 3.10: Completeness of the Trigonometric Functions of
Fourier Series—Convergence in the Mean

The trigonometric set (system)

$$1, \cos\frac{\pi x}{l}, \sin\frac{\pi x}{l}, \cos\frac{2\pi x}{k}, \sin\frac{2\pi x}{l}, \ldots, \cos\frac{k\pi x}{l}, \sin\frac{k\pi x}{l}, \ldots$$

is complete. $\qquad\qquad\qquad\qquad\qquad\qquad\qquad\square$

Proof: We are to prove that the nth partial sum $S_n(x)$ (or $f_n(x)$)

$$f_n(x) \equiv S_n(x) = a_0 + \sum_{k=1}^{\infty} a_k \cos \frac{k\pi x}{l} + b_n \sin \frac{k\pi x}{l} \qquad \text{(E.1)}$$

as a best approximation to the function $f(x)$, *converges in the mean* for every piecewise continuous (or square integrable) $f(x)$ defined on $[-l,l]$, i.e., to show that

$$\lim_{n \to \infty} \int_{-\infty}^{\infty} |f(x) - f_n(x)|^2 \, dx = 0 \qquad \text{(E.2)}$$

or in other words, given an arbitrary $\epsilon > 0$, we have

$$\int_{-l}^{l} |f(x) - f_n(x)|^2 \, dx < \epsilon \quad \text{for } n > n_0 \qquad \text{(E.3)}$$

for every piecewise continuous functions on $[-l,l]$. Of course, we have the trigonometric set being orthogonal, and by the use of this orthogonality and the Bessel identity in (3.60), the integral

$$\int_{-\infty}^{\infty} |f(x) - f_n(x)|^2 \, dx = \int_{-l}^{l} f^2(x) \, dx - l \left\{ 2a_0^2 + \sum_{k=1}^{n} a_k^2 + b_k^2 \right\} \qquad \text{(E.4)}$$

represents the *minimum of the mean square deviation*.

The proof will center around considering a Fourier series for a continuous function $g(x)$ which coincides with $f(x)$ except at the jump discontinuities, where it assumes a straight line in a small neighborhood around these points and where it has equal values at the end points of the period $g(-l) = g(l)$, as shown in Fig. 3.10. The Fourier series for this

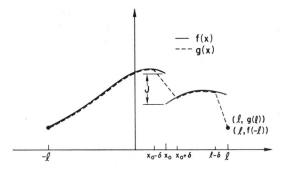

FIG. 3.10 $f(x)$ with jump discontinuity at x_0, and its continuous replacement $g(x)$ with $g(-l) = g(l) = f(-l)$.

continuous function is uniformly convergent, so its minimum mean square deviation can be made as small as desired, $\epsilon > 0$ for $n > n_0$, i.e.,

$$\int_{-l}^{l} |g(x) - g_n(x)|^2 \, dx < \epsilon, \qquad n > n_0 \tag{E.5}$$

We will show that $g_n(x)$ approximates $f(x)$ on $(-l,l)$ within the same mean square deviation ϵ, i.e.,

$$\int_{-l}^{l} |f(x) - g_n(x)|^2 \, dx < \epsilon \tag{E.6}$$

Then we will combine results (E.4), (E.6) to show the desired result in (E.3). A piecewise continuous function may have a finite number of jump discontinuities, but to simplify our proof we will treat it with only one jump discontinuity at x_0 in the interior $(-l,l)$ with jump size $J = f(x_0+) - f(x_0-)$ as shown in Fig. 3.10. There is no loss of generality in doing this, as we can treat the rest of the jump discontinuities in the same way.

We may mention here again that the Fourier series approximation $S_n(x)$ will suffer from the Gibbs phenomenon at the jump discontinuities, where for even very large n, $S_n(x)$ will overshoot the function $f(x)$ near x_0 by about 9% of the size of the jump J at x_0 (see Figs. 3.5b–d and 4.30). The essence of the present proof will also be needed in Section 4.1.6G of Chapter 4 (see Figs. 4.29 and 4.30), when we attempt to find a way to reduce this Gibbs phenomenon at an interior point, or an end point for the periodic extension of $f(x)$ (generated by its (periodic) Fourier series representation). This is done by replacing $f(x)$ by the same type of continuous function $g(x)$, which coincides with $f(x)$ where $f(x)$ is continuous (solid line in Fig. 3.10) and where $g(x)$ (dotted line in Fig. 3.10) is represented by a straight line in a small neighborhood $(x_0 - \delta, x_0 + \delta)$, near the interior point x_0 to make it continuous on $(-l,l)$. This is also done at the end point $x = l$ if $f(-l) \neq f(l)$, where $g(x)$ is a straight line on the left side neighborhood of $x = l$ from the point $(l - \delta, g(l - \delta))$ or $(l - \delta, f(l - \delta))$ to the point $(l, g(l))$ or $((l, f(-l)))$. The $g(x)$ here is continuous and has a continuous periodic extension, where the Fourier series converges to it uniformly according to Theorem 3.1. However, it is not $f(x)$ but a "smoothed approximation" to $f(x)$. As we shall see when we discuss and illustrate the Gibbs phenomenon in full detail in Section 4.1.6G, it turns out that the overshooting near the jump discontinuity x_0 for $f(x)$ will be much reduced for its approximate replacement, the continuous $g(x)$ (see Figs. 4.30 and 4.29). With this note, we have a $g(x)$ coinciding with $f(x)$ except for the small neighborhood

$(x_0 - \delta, x_0 + \delta)$ and the half-neighborhood $(l - \delta, l)$, where δ is very small, and $g(l) = g(-l) \neq f(l)$.

We have $f(x)$ bounded, and therefore so is $g(x)$, so we take $f(x) \leq M$, $g(x) \leq M$ on $[-l, l]$. We will show now for the constructed $g(x)$ with sufficiently small δ that

$$\int_{-l}^{l} |f(x) - g(x)|^2 \, dx < \frac{\epsilon}{4} \quad \text{for arbitrary } \epsilon > 0,$$

$$\int_{-l}^{l} |f(x) - g(x)|^2 \, dx = \int_{-l}^{x_0 - \delta} |f(x) - g(x)|^2 \, dx + \int_{x_0 - \delta}^{x_0 + \delta} |f(x) - g(x)|^2 \, dx$$

$$+ \int_{x_0 + \delta}^{l - \delta} |f(x) - g(x)|^2 \, dx + \int_{l - \delta}^{l} |f(x) - g(x)|^2 \, dx$$

$$= 0 + \int_{x_0 - \delta}^{x_0 + \delta} |f(x) - g(x)|^2 \, dx + 0 + \int_{l - \delta}^{l} |f(x) - g(x)|^2 \, dx$$

$$\leq \int_{x_0 - \delta}^{x_0 + \delta} [|F(x)| + |g(x)|]^2 \, dx + \int_{l - \delta}^{l} [|f(x)| + |g(x)|]^2 \, dx$$

$$\leq (2M)^2 2\delta + (2M)^2 \delta = 8M^2 \delta + 4M^2 \delta = 12M^2 \delta = \frac{\epsilon}{4} \quad (E.7)$$

for an arbitrary δ, provided that δ is sufficiently small. In the last integrals we used the bound M on both $f(x)$ and $g(x)$, i.e.,

$$\{|f(x)| + |g(x)|\}^2 \leq \{M + M\}^2 = (2M)^2 = 4M^2$$

and we note that the second integral is with δ support (width) while the first one is with 2δ support (width). The function $g(x)$ is continuous on $[-l, l]$ and has equal values $g(-l) = g(l)$ at the end points $\mp l$; thus by Theorem 3.2, the Fourier trigonometric series converges to it on $[-l, l]$ uniformly. Here we use $g_n(x)$ instead of $S_n(x)$ for the nth partial sum and α_n, β_n for the Fourier coefficients to differentiate them from the nth partial sum (or trigonometric polynomial) $f_n(x)$ of (E.1) and its Fourier coefficients a_k, b_k. The uniform convergence of the Fourier series of $g(x)$ means that its partial sum,

$$g_n(x) = \alpha_0 + \sum_{k=1}^{n} \alpha_n \cos \frac{k \pi x}{l} + \beta_n \sin \frac{k \pi x}{l} \quad (E.8)$$

can approximate $g(x)$ on $[a, b]$ with any arbitrary $\epsilon_1 > 0$ (for convenience we take $\epsilon_1 = \sqrt{\epsilon/8l}$),

$$|g(x) - g_n(x)| < \epsilon_1 = \sqrt{\frac{\epsilon}{8l}} \quad \text{for all } x \in [a, b] \quad (E.9)$$

With this bound on $|g(x) - g_n(x)|$ we can have a bound of $\epsilon/4$ on the following integral:

$$\int_{-l}^{l} [g(x) - g_n(x)]^2 dx < \int_{-l}^{l} \frac{\epsilon}{8l} dx = \frac{\epsilon}{8l} 2l = \frac{\epsilon}{4} \qquad (E.10)$$

With the help of (E.7) and (E.10) we will show here that $g_n(x)$ approximates $f(x)$ with a mean square deviation less than ϵ. If we invoke the inequality $(a + b)^2 \leq 2a^2 + 2b^2$ for the integrand in (E.10) we obtain

$$\int_{-l}^{l} |f(x) - g_n(x)|^2 dx = \int_{-l}^{l} |(f(x) - g(x)) + (g(x) - g_n(x))|^2 dx$$

$$\leq 2 \int_{-l}^{l} |f(x) - g(x)|^2 dx + 2 \int_{-l}^{l} |g(x) - g_n(x)|^2 dx$$

$$< 2\frac{\epsilon}{4} + 2\frac{\epsilon}{4} = \epsilon$$

$$\int_{-l}^{l} |f(x) - g_n(x)|^2 dx < \epsilon \qquad (E.6)$$

after using (E.7) and (E.10) in the last two integrals, respectively. This means that $g_n(x)$ approximates $f(x)$ with a mean square error less than ϵ, but it is known that $f_n(x)$ of (E.1) gives the minimum of such error, hence we conclude it must have this ϵ as a bound, and we have

$$\int_{-l}^{l} |f(x) - f_n(x)|^2 dx < \epsilon \qquad (E.11)$$

With the Bessel identity (E.4) of (3.60) we can write this as

$$E_n = \int_{-l}^{l} |f(x) - f_n(x)|^2 dx = \int_{-l}^{l} f^2(x) - \left\{ 2a_0^2 + \sum_{k=1}^{n} a_k^2 + b_k^2 \right\} < \epsilon \qquad (E.12)$$

But the Bessel identity (3.60) is valid for all n, which makes it clear that we can have (E.12) for $n > n_0$,

$$\int_{-l}^{l} |f(x) - f_n(x)|^2 dx < \epsilon, \qquad n > n_0 \qquad (E.12)$$

which is the same thing as

$$\lim_{n \to \infty} \int_{-l}^{l} |f(x) - f_n(x)|^2 dx = 0$$

and the proof is concluded for the completeness of the trigonometric *Fourier basis* on $[-l, l]$. ∎

Before we prove the completeness of the Legendre polynomials in the next Example 3.8, we may mention the completeness of the complex exponential functions $\{e^{ik\pi x/l}\}_{k=-\infty}^{\infty}$ on $(-l,l)$. This, with the above development, should be straightforward, as these functions can be written in terms of the set of the trigonometric functions through the Euler identities (3.14). These are the identities used at the beginning of this chapter to show the same result in the other direction, namely that the trigonometric Fourier series (3.10) reduces to the exponential Fourier series (3.8).

Next we show that the complete orthogonal trigonometric system is closed; i.e., the only member that can be orthogonal to all its members is the zero function (when it is continuous). The proof is simple as the function, if continuous, has a Fourier representation, but since it is orthogonal to all the trigonometric functions, then all a_0, a_n, b_n are zero. Thus this continuous function is represented by a Fourier series that converges to it uniformly as $f(x) \equiv 0$, $x \in [-l,l]$. This is for the assumption of $f(x)$ continuous, and we shall consider the case of piecewise continuous $f(x)$. If such $f(x)$ is orthogonal to all members of the trigonometric system, then again all its coefficients are zero. But still this Fourier series converges to the middle of the jump discontinuity, which means it can have nonzero values only at such a finite number of points of discontinuity in the interval $(-l,l)$. But this element being zero on $(-l,l)$ except for such a separate finite set of points is allowed from the essence of the definition of what we mean by a closed set. This also reminds us, as we have illustrated in a number of examples, that the piecewise continuous function is uniquely specified by its Fourier trigonometric series representation except possibly at a finite number of separate points of jump discontinuities in the interval $(-l,l)$. This is so as the above zero element is allowed to be different from zero, but only at such a separate set of points.

The next example is to prove the completeness of the Legendre polynomials $P_n(x)$ on $(-l,l)$.

EXAMPLE 3.8: The Completeness of Legendre Polynomials

In the Theorem 3.10, we proved in full detail the completeness of the trigonometric functions $\{1, \cos(k\pi x/l), \sin(k\pi x/l)\}_{k=1}^{\infty}$ on $(-l,l)$. We note that these functions are solutions of a very simple differential equation with constant coefficients, $d^2y/dx^2 + \lambda^2 y = 0$, and also that they naturally pride themselves on being periodic with period $2l$. This resulted in the Fourier series, of a sectionally continuous $f(x)$ on $(-l,l)$, automatically extending $f(x)$ on the whole real line with period $2l$. In practice, as in solving for vibrations of a circular membrane, the hydrogen atom, or the potential equation in a sphere, we will face more sophisticated differential equations with variable coefficients. As we discussed earlier, and will discuss further in Appendix A, these equations may be singular at the end points of their domain. The Legendre polynomials arise in such situations, and of course they are nowhere periodic. We mention this since we are going to start our proof for their completeness in parallel with the preceding proof of Theorem 3.10 for the (periodic!) trigonometric functions. In the present case we also consider any piecewise continuous (or square integrable) function $f(x)$ on $(-l,l)$, to prove that the orthogonal expansion of Legendre polynomials

$$\sum_{k=1}^{\infty} c_k P_k(x) \tag{E.1}$$

converges in the mean to $f(x)$. As we did in the proof of Theorem 3.10, we find a replacement $g(x)$ for $f(x)$ so that they coincide where $f(x)$ is continuous, and $g(x)$ is defined as a straight line in $(x_0 - \delta, x_0 + \delta)$, a small neighborhood around the jump discontinuity at x_0. However, we don't need to assume that $g(-l) = g(l)$; this was done in Theorem 3.10 to satisfy the special characteristic of the Fourier trigonometric series, which gives a periodic extension to $g(x)$, and $g(-l) = g(l)$ is necessary for having such an extension be continuous. Here we seldom talk about a general orthogonal expansion, like the present Legendre one or the Bessel one in (3.97) beyond the basic interval (a,b) of its orthogonality. However, very recent research, concerning the Bessel orthogonal series (3.97), for example, seems to indicate the necessity for looking at such a (nonperiodic) representation of $f(x)$ beyond its basic interval of expansion $(0,b)$. We will discuss this topic briefly in Section 4.1 to indicate the necessity of looking at the behavior of series beyond the interval $(0,b)$ of the Bessel series expansion. For a replacement $g(x)$ on $(-l,l)$ we can show, as we did for Theorem 3.10, that for any $\epsilon > 0$,

$$\int_{-l}^{l} |f(x) - g(x)|^2 \, dx < \frac{\epsilon}{4} \tag{E.2}$$

From the Weierstrass polynomial approximation Theorem 3.8, this continuous function $f(x)$ also has an algebraic polynomial of degree m,

$$Q_m(x) = A_0 + A_1 x + A_2 x^2 + \cdots + A_m x^m \tag{E.3}$$

that approximates it within an arbitrary small $\epsilon_1 = \frac{1}{2}\sqrt{\epsilon/2}$

$$|g(x) - Q_m(x)| < \frac{1}{2}\sqrt{\frac{\epsilon}{2}} \tag{E.4}$$

which gives

$$\int_{-1}^{1} |g(x) - Q_m(x)|^2 \, dx < \int_{-1}^{1} \frac{1}{4} \frac{\epsilon}{2} = \frac{\epsilon}{4} \tag{E.5}$$

We have shown a few examples of the Legendre polynomials $P_0(x) = 1$, $P_1(x) = x$, $P_2(x) = \frac{3}{2}x^2 - \frac{1}{2}$, and in general, for n a nonnegative integer, we have the Rodrigues formula

$$P_n(x) = \frac{1}{2^n n!} \frac{d^n}{dx^n} (x^2 - 1)^n, \qquad n = 0, 1, 2, \ldots \tag{E.6}$$

which are solutions of the Legendre differential equation (3.84). Here we can easily say that each one is a linear combination of powers of x to the highest degree n of $P_n(x)$. In the same way, we can show that x^j, $j = 0, 1, 2, \ldots$, can be written as a linear combination of the Legendre polynomials up to $P_j(x)$,

$$x^j = \sum_{k=1}^{j} b_k P_k(x) \tag{E.7}$$

If we write this for each power term in the algebraic polynomial $Q_m(x)$ of (E.3) we have

$$Q_m(x) = \sum_{j=0}^{m} A_j \sum_{k=1}^{j} b_k P_k(x) = \sum_{i=1}^{m} \alpha_i P_i(x) \tag{E.8}$$

We can easily verify the inequality $(a + b)^2 \leq 2a^2 + 2b^2$, which we use with (E.2) and (E.5) to arrive at the approximation of $Q_m(x)$ to $f(x)$

$$|f(x) - Q_m(x)|^2 = |\{f(x) - g(x)\} + \{g(x) - Q_m(x)\}|^2$$

$$\leq 2|f(x) - g(x)|^2 + |g(x) - Q_m(x)|^2$$

$$\int_{-1}^{1} |f(x) - Q_m(x)|^2 \, dx \leq 2 \int_{-1}^{1} |f(x) - g(x)|^2 \, dx$$

$$+ 2 \int_{-1}^{1} |g(x) - Q_m(x)|^2 \, dx$$

$$< 2\frac{\epsilon}{4} + 2\frac{\epsilon}{4} = \epsilon$$

$$\int_{-1}^{1} |f(x) - Q_m(x)|^2 \, dx < \epsilon \tag{E.9}$$

where (E.2) and (E.5) are used in the above two integrals on the right of the inequality.

Now let us consider the orthogonal Legendre polynomial expansion of $f(x)$ on $(-1, 1)$

$$f(x) = \sum_{k=0}^{\infty} c_k P_k(x) \tag{E.10}$$

with its Fourier coefficients

$$c_k = \frac{1}{\|P_k\|^2} \int_{-1}^{1} f(x) P_k(x) dx \tag{E.11}$$

As we indicated earlier, the norm square $\|P_k\|^2$ can be calculated with the help of Rodrigues formula (E.6) (see (3.89)) and integrations by parts to give (see Exercises A.28 and A.27)

$$\|P_k\|^2 = \int_{-1}^{1} P_k^2 \, dx = \frac{2}{2k+1} \tag{E.12}$$

Hence c_k of (E.11) can be written as

$$c_k = \frac{2k+1}{2} \int_{-1}^{1} f(x) P_k(x) dx \tag{E.13}$$

This expansion (E.10), of course, deserves the property that its mth partial sum $f_m(x)$

$$f_m(x) = \sum_{k=0}^{m} c_k P_k(x), \tag{E.14}$$

which is a polynomial of degree m, gives the minimum square deviation to $f(x)$. So

$$\int_{-1}^{1} |f(x) - f_m(x)|^2 \, dx < \epsilon \tag{E.15}$$

since the arbitrary polynomial $Q_m(x)$ in (E.9) cannot compete with this special polynomial $f_m(x)$ for the least square deviation associated with the orthogonal expansions nature of $f_m(x)$ in (E.14). We will show explicitly that (E.15) with $f_n(x)$ is valid for all $n \geq m$, which can be attained via the Bessel identity (3.60) of the orthogonal expansion in (E.14)

$$E_m \equiv \int_{-1}^{1} |f(x) - f_m(x)|^2 \, dx = \int_{-1}^{1} f^2(x) dx - \sum_{k=1}^{m} c_k^2 \|P_k\|^2 < \epsilon \tag{E.16}$$

But the Bessel identity (E.16) of (3.60) is valid for all m, so we take it for $n \geq m$ and we have

$$\int_{-1}^{1} |f(x) - f_n(x)|^2 \, dx < \epsilon \quad \text{for all } n \geq m \tag{E.17}$$

which is what we mean by the Legendre series expansion converging to each $f(x)$ that is piecewise continuous (or square integrable) on $(-1,1)$ This concludes the proof that the (orthogonal) Legendre polynomial set $\{P_n(x)\}_{n=0}^{\infty}$ is complete on $(-1,1)$.

Attempts to Accelerate the Convergence of Fourier Series

Theorem 3.3 is the realization (instigated by Theorem 3.2 for uniform convergence) that better functions, i.e., smoother functions, deserve a rapidly convergent series. The question now arises of whether we can do something to the series of a function $f(x)$ to make it rapidly convergent— i.e., with coefficients of order $o(1/k^m)$ and m high. The answer here lies in whether the function $f(x)$ happens to be a sum of two functions $f(x) = \phi(x) + g(x)$, with the series of $\phi(x)$ known to us from experience or tables, and converges very slowly with coefficients β_n of say, order $o(1/n)$. So if we feel that the coefficients α_n of the given function $f(x)$ are also of order $o(1/n)$, we may attempt to compute the Fourier series for the difference $g(x) = f(x) - \phi(x)$. The hope here is that the slow $o(1/n)$ behavior of the coefficients of $f(x)$ and $\phi(x)$ cancels out, and we end up with coefficients γ_n for $g(x)$ that decay more rapidly than $o(1/n)$. Example 3.9 will illustrate such attempts. Of course, $f(x)$ may not satisfy the condition $f(l) = f(-l)$ at the end points, necessary for uniform convergence, and it would be a welcome if $\phi(x)$ has the same values there: $\phi(l) = f(l)$, $\phi(-l) = f(-l)$. In this case our rapidly convergent series for the function $g(x) = f(x) - \phi(x)$ will also satisfy $g(l) = g(-l)$ as the welcome property of uniform convergence.

The general discussion above is illustrated in the following more realistic example, where we happened to know a function $\phi(x)$ with such a slowly convergent sine series.

EXAMPLE 3.9: Improving the Convergence of Fourier Series

We are given the following Fourier sine series that represents a function $f(x)$ on $(-\pi, \pi)$:

$$f(x) \sim \sum_{n=1}^{\infty} (-1)^n \frac{n^3}{n^4 + 1} \sin nx, \quad -\pi < x < \pi \tag{E.1}$$

The function $f(x)$ is known to us through this series, and to compute it we must compute the series. In doing so, we realize that this series is slowly convergent as its coefficients are such that $|b_n| = n^3/(n^4 + 1) = o(1/n)$; i.e., it varies like $1/n$ for large n. For different values of $x \in (-\pi, \pi)$ we shall need to consider very many terms of this series if we are to hope for any reasonable accuracy of our computations. The idea now is that can we separate (or subtract) from $n^3/(n^4 + 1)$ the slowly varying $1/n$, and hopefully what is left is a rapidly varying term. This happened to be the case since $n^3/(n^4 + 1) - 1/n = 1/(n^5 + n)$, a rapidly varying coefficient. The key here is whether we know the function $\phi(x)$ for the slowly varying coefficient $o(1/n)$. This happened to be the case, since we know from Example 3.3 that

$$\frac{x}{2} = \sum_{n=1}^{\infty} \frac{(-1)^{n+1}}{n} \sin nx, \qquad -\pi < x < \pi \tag{E.2}$$

so

$$f(x) = \sum_{n=1}^{\infty} (-1)^n \frac{n^3}{n^4 + 1} \sin nx$$

$$= -\sum_{n=1}^{\infty} \frac{(-1)^{n+1}}{n} \sin nx + \sum_{n=1}^{\infty} \frac{(-1)^n}{n^5 + n} \sin nx$$

$$f(x) = -\frac{x}{2} + \sum_{n=1}^{\infty} \frac{(-1)^n}{n^5 + n} \sin nx$$

that is, our unknown function $f(x)$ is computed via a much more rapidly convergent series with coefficients $\gamma_n = (-1)^n/(n^5 + n)$ instead of $\alpha_n = (-1)^n n^3/(n^5 + n)$ given in (E.1).

B. The Regular and (Regular) Singular Sturm–Liouville Problems—General Finite Transforms

We would like to cover here in detail the important case of the Sturm–Liouville boundary value problem with regular (removable) singularity, but space in this section limits us to defining this problem. The very detailed treatment is covered in Appendixes A.4 and A.5. The differential equation in $u(x)$

$$\frac{d^2u}{dx^2} + q_1(x)\frac{du}{dx} + q_2(x)u = 0, \qquad a < x < b \tag{3.75}$$

is said to be *regular* if $q_1(x)$ and $q_2(x)$ are continuous on $[a,b]$. A Sturm–Liouville problem in (3.81) is called *singular* when the coefficients $r(x)$ or $\rho(x)$ vanish at one end or both ends of a finite interval $[a,b]$, or if the differential equation is defined on a semi-infinite or infinite domain. The singularity at x_0 of the equation (3.75) is called *removable* if

$$\lim_{x \to x_0} (x - x_0)q_1(x) \quad \text{exists} \tag{3.76}$$

and

$$\lim_{x \to x_0} (x - x_0)^2 q_2(x) \quad \text{exists} \tag{3.77}$$

We will soon discuss the latter class of differential equations with removable singularity, or what is termed *regular singular equations*; we do this since a number of our applications and transforms are solutions to such "singular" Sturm–Liouville problem. The first example that comes to mind is the problem associated with Bessel differential equation,

$$\frac{d^2u}{dx^2} + \frac{1}{x}\frac{du}{dx} + \left(\lambda^2 - \frac{\nu^2}{x^2} \right)u(x) = 0, \qquad 0 < x < 1 \tag{3.78}$$

$$u(0) = 0 \tag{3.79}$$

$$u(1) = 0 \tag{3.80}$$

where the differential equation has a removable singularity at $x = 0$ since

$$\lim_{x \to 0} x \cdot \frac{1}{x} = 1 \quad \text{and} \quad \lim_{x \to 0} x^2 \left(\lambda^2 - \frac{n^2}{x^2} \right) = -n^2$$

The Completeness of the Eigenfunctions of the Regular Sturm–Liouville Problem

The first practical question now is whether the set of orthogonal solutions to the regular Sturm–Liouville problem (1.78)–(1.80), which we discussed in Section 1.5.2B of Chapter 1, is complete. The answer is in the affirmative, as we shall state without proof (see Weinberger, 1965) in the following Theorem 3.11.

THEOREM 3.11:

For the Sturm–Liouville problem (1.63)–(1.65) with the same regular differential equation but with the special case of the boundary values $u(0) = u(1) = 0$ (note here λ is replaced by $-\lambda^2$).)

$$\frac{d}{dx}\left[r(x)\frac{du}{dx} \right] - q(x)u(x) + \lambda^2\rho(x)u(x) = 0, \qquad 0 < x < 1 \tag{3.81}$$

$$u(0) = 0$$

$$u(1) = 0 \tag{3.82}$$

where $r(x)$, $r'(x)$, $q(x)$, and $\rho(x)$ are continuous, and $r(x) > 0$, $\rho(x) > 0$, and $q(x)$ nonnegative for $0 \leq x \leq 1$, the eigenfunctions (continuously differentiable) $\{u_n(x)\}$, as solutions to (3.81), (3.82), constitute a complete orthogonal set. The orthogonality of $\{u_n(x)\}$ on $(0, 1)$ is seen as a special case of the general regular Sturm–Liouville problem (1.63)–(1.65), which we stated and proved in Example 1.9 of Chapter 1. $\qquad\square$

Singular Sturm–Liouville Problem with Removable Singularity—(Regular) Singularity Problem

The second question concerns the fact that many functions in physical applications, like the Bessel functions, are not solutions of the above regular Sturm–Liouville problem. In particular, the regularity conditions may be satisfied in the interior of the interval (a,b) but not at one or both of the end points. For example, in the case of the *Bessel differential equation*,

$$x^2 \frac{d^2u}{dx^2} + x\frac{du}{dx} + (\lambda^2 x^2 - \nu^2)u = 0, \qquad 0 < x < 1 \tag{3.83}$$

and its self-adjoint form

$$\frac{d}{dx}\left(x\frac{du}{dx}\right) + \left(\lambda^2 x - \frac{\nu^2}{x}\right)u = 0, \qquad 0 < x < 1 \tag{3.83a}$$

is with $r(x) = x$, $\rho(x) = x$, which approach zero toward the end point $x = 0$, but $q(x) = \nu^2/x$ becomes infinite as x approaches the same end point $x = 0$. But, as we showed above for (3.78), the point $x = 0$ is a removable singularity. In cases like this of $q(x)$, the explicit boundary condition at $x = 0$ in Theorem 3.11 is replaced by a boundedness on the solutions there.

Another important example where $r(x)$ approaches zero at both ends is that of the *Legendre differential equation*,

$$(1-x^2)\frac{d^2u}{dx^2} - 2x\frac{du}{dx} + n(n+1)u = 0, \qquad -1 < x < 1 \tag{3.84}$$

whose solution is the Legendre polynomial $P_n(x)$ as given in (E.6) of Example 3.8, where we proved the completeness of $\{P_n(x)\}_{n=0}^{\infty}$ on $(-1, 1)$. The self-adjoint form of the Legendre equation (3.84) is

$$\frac{d}{dx}\left[(1-x^2)\frac{du}{dx}\right] + n(n+1)u, \qquad -1 < x < 1 \tag{3.84a}$$

Here we have $r(x) = 1 - x^2$, $q(x) = 0$, $\rho(x) = 1$ and $n(n+1)$, where $r(x)$ violates the regularity condition by vanishing at both end points

$x = \mp 1$. In Appendix A.5 we will use power series expansion about (the nonsingular point) $x_0 = 0$, as illustrated in Appendix A.3, to arrive at the Legendre polynomial $P_n(x)$ as the bounded solution to the Legendre equation (3.84).

As for the existence of solutions to the above Bessel equation (3.83) it is with a regular singularity, and solutions do exist. The *method of Frobenius* is used to construct these solutions, which will be treated and illustrated in detail in Appendix A.4 and for the Bessel functions in Appendix A.5.

Consider the (regular) singular Sturm–Liouville problem associated with the Bessel differential equation,

$$\frac{d}{dx}\left(x \frac{du}{dx}\right) + \left(\lambda^2 x - \frac{\nu^2}{x}\right) u = 0, \qquad 0 < x < 1 \tag{3.83a}$$

with the boundary condition at $x = 1$,

$$u(1) = 0 \tag{3.85}$$

and with boundedness condition at $x = 0$ of u staying bounded and $x\, du/dx \to 0$ as $x \to 0$. It can be shown, using arguments beyond the space allowed in this book (see Weinberger, 1965) that the solutions are a complete orthogonal set on $(0, 1)$. These are the set of Bessel functions $\{J_\nu(j_{\nu,m}x)\}_{m=1}^{\infty}$ of the first kind of order ν. Here $j_{\nu,m}$, $m = 1, 2, \ldots$, are the zeros of the Bessel function $J_\nu(x)$, which is the consequence of satisfying the boundary condition (3.85)

$$u(1) = 0 = J_\nu(j_{\nu,m}), \qquad m = 1, 2, 3, \ldots \tag{3.86}$$

Of course, this result can be extended, with simple scaling, to the interval $(0, a)$, whence $u_m(x) = J_\nu((j_{\nu,m}/a)x)$.

In the direction covering the orthogonality of the eigenfunctions of a (regular) singular Sturm–Liouville problem like the Bessel one in (3.83a), we state the following theorem and leave its proof for the interested reader.

THEOREM 3.12:

Consider the Sturm–Liouville problem

$$\frac{d}{dx}\left[r(x) \frac{du}{dx}\right] - q(x)u(x) + \lambda^2 \rho(x)u(x) = 0, \qquad a < x < b \tag{3.87}$$

with the boundary condition

$$u(b) = 0 \tag{3.88}$$

and $r(a) = 0$, $r(x) > 0$ for $a < x \leq b$, but $q(x)$, $\rho(x)$ satisfy the same conditions of the regular Sturm–Liouville problem as being continuous and $\rho(x) > 0$ on $a \leq x \leq b$. Also, the solution $u(x)$ and its derivative $u'(x)$ have finite limits as $x \to a_+$. Then the eigenfunctions, if they exist, are orthogonal on (a, b). □

The Legendre Polynomials' Orthogonal Expansion and Their Finite Legendre Transforms

Among many other important special functions in Appendix A.5, the bounded solution to Legendre equation (3.84), (3.84a) is the Legendre polynomial $P_n(x)$ of degree n. This is established with power series expansion about $x = 0$ (a regular point of Legendre equation (3.83)), as a polynomial of degree n, with its Rodrigues formula,

$$P_n(x) = \frac{1}{2^n n!} \frac{d^n}{dx^n} (x^2 - 1)^n, \qquad n = 0, 1, 2, \ldots \tag{3.89}$$

(which is derived in Exercise A.26 of Appendix A.5). This can be used easily to show that the Legendre polynomials $P_n(x)$ are orthogonal on $(-1, 1)$, i.e.,

$$\int_{-1}^{1} P_n(x) P_m(x) \, dx = 0, \qquad n \neq m$$

and that the normsquare (see Exercises A.28 and A.27),

$$\|P_n\|^2 \equiv \int_{-1}^{1} P_n^2(x) \, dx = \frac{2}{2n + 1} \tag{3.90}$$

With this orthogonality property and the completeness we proved in Example 3.10 for these Legendre polynomials, we can easily write the formal orthogonal expansion for sectionally continuous (or square integrable) functions $f(x)$ on $(-1, 1)$,

$$f(x) = \sum_{n=0}^{\infty} c_n P_n(x), \qquad -1 < x < 1 \tag{3.91}$$

$$c_n = \frac{2n + 1}{2} \int_{-1}^{1} f(x) P_n(x) \, dx \tag{3.92}$$

It is the integral above that we shall use as the *finite Legendre transform* of $f(x)$ in Section 3.5.1,

$$F(n) = \int_{-1}^{1} f(x) P_n(x) \, dx \tag{3.93}$$

We then define its inverse transform $f(x)$ via the Legendre orthogonal expansion (3.91) as

$$f(x) = \sum_{n=0}^{\infty} \frac{2n+1}{2} F(n) P_n(x) \tag{3.94}$$

This finite Legendre transform and its inverse (3.93), (3.94) is a special case of the general finite transform of the next section defined by the integral in the numerator of (3.54) (for real-valued $\phi_n(x)$),

$$F(n) = \int_a^b \rho(x) f(x) \phi_n(x) \, dx \tag{3.95}$$

and its inverse $f(x)$ as given by (3.53) with $c_n = \frac{F(n)}{\|\phi_n\|^2}$,

$$f(x) = \sum_{n=1}^{\infty} \frac{F(n)}{\|\phi_n\|^2} \phi_n(x) \tag{3.96}$$

Along with other such general finite transforms, besides the Fourier trigonometric ones, in the rest of the chapter we will emphasize the operational properties of such transforms. These are used to facilitate finding the solution of their compatible differential equations in parallel to what we did with the Fourier, Laplace, Hankel, and other transforms in Chapter 2. As to the convergence of the orthogonal series (3.94), for example, as the inverse finite Legendre transform, we will concentrate on the convergence in the mean for the piecewise continuous (or square integrable) functions $f(x)$ on $(-1, 1)$. In Section 4.2.1B,D we will cover the discrete version of the above Legendre transform as given in (its exact form) (4.159), (4.160), where each of the new discrete transforms and its inverse is given only as a sequence of n sample values.

The Bessel Functions' Orthogonal Expansion and Their Finite Hankel (Bessel) Transforms

We already indicated that the Bessel functions $\{J_\nu(j_{\nu,m} x)\}$ constitute a complete orthogonal set on the interval $(0, 1)$ with respect to the weight function $\rho(x) = x$ and where $\{j_{\nu,m}\}_{m=1}^{\infty}$ are the zeros of $J_\nu(x)$ (see Table A.1 in Appendix A.5 for 20 of these zeros for $\nu = 0, 1, 2, 3, 4, 5$).

For $f(x)$ sectionally continuous on $(0, 1)$, its formal orthogonal expansion in terms of the Bessel functions $J_\nu(j_{\nu,m} x)$ is

$$f(x) = \sum_{m=1}^{\infty} c_m J_\nu(j_{\nu,m} x), \qquad 0 < x < 1 \tag{3.97}$$

Multiplying both sides of (3.95) by $xJ_\nu(j_{\nu,k}x)$ and then integrating from 0 to 1, we obtain

$$\int_0^1 xJ_\nu(j_{\nu,k}x)f(x)\,dx = \sum_{m=1}^\infty c_m \int_0^1 xJ_\nu(j_{\nu,k}x)J_\nu(j_{\nu,m}x)\,dx$$

The above operation of integrating the infinite series (3.97) term by term is allowed by Theorem 3.5 if we assume that $f(x)$ is square integrable on the interval $(0,1)$ with respect to the weight function $\rho(x) = x$. And from the orthogonality of the Bessel functions, the above infinite series becomes one term,

$$\int_0^1 xJ_\nu(j_{\nu,k}x)f(x)\,dx = c_k \int_0^1 xJ_\nu^2(j_{\nu,k}x)\,dx$$

Hence

$$c_k = \frac{\int_0^1 xJ_\nu(j_{\nu,k}x)f(x)\,dx}{\int_0^1 xJ_\nu^2(j_{\nu,k}x)\,dx} = \frac{\int_0^1 xJ_\nu(j_{\nu,k}x)f(x)\,dx}{\|J_\nu\|^2} \tag{3.98}$$

Computations of the integral in the denominator can be done easily with the aid of the properties of the Bessel function that are established in Appendix A.5. The integral in the numerator, of course, depends on the given function $f(x)$.

As we remarked above concerning the finite Legendre transforms (3.93), (3.94), it is also the integral in the numerator of (3.98) that will defined the *finite Hankel transform* in Section 3.4,

$$F_\nu(j_{\nu,m}) = \int_0^1 xJ_\nu(j_{\nu,m}x)f(x)\,dx \tag{3.99}$$

with its *inverse* $f(x)$ defined via the Bessel orthogonal series (3.97) as

$$f(x) = \sum_{m=1}^\infty \frac{F_\nu(j_{\nu,m})J_\nu(j_{\nu,m}x)}{\int_0^1 xJ_\nu^2(j_{\nu,m}x)\,dx} \tag{3.100}$$

Again we shall emphasize the operational properties of the finite Hankel (Bessel) transform (3.99) for the purpose of facilitating the solution of problems associated with Bessel differential equations. We will also attempt to find a discrete version of these finite Hankel transforms (4.192) in Section 4.3 of Chapter 4. As we remarked for the finite Legendre transforms (3.93) and (3.94), the above finite Hankel transform and its inverse (3.98), (3.99) are a special case of the general finite transform (3.95), (3.96), which we shall cover in detail in the following sections.

The Nonperiodicity of the Orthogonal Expansion (in General)

As indicated very briefly in Example 3.8, the Bessel orthogonal expansion (3.97), (3.98), as a general Fourier series expansion and away from the periodic Fourier trigonometric series, is clearly not periodic. But recent research indicates the need to look at the Bessel series representation (3.97) of $f(x)$ beyond its interval of expansion $(0,1)$, where, in contrast to the Fourier trigonometric series representation, we will not find periodic replicas of $f(x)$. The research concerns the Bessel orthogonal series (3.97) or (3.100), for example, and its needed "approximate" relation to the (infinite) Hankel (Bessel) transform (2.242). This is in the hope of finding efficient algorithms for such general orthogonal series or transforms. Also, it is to find possible estimates for a number of errors incurred in our practical computations of such infinite series or integrals. The general orthogonal series or transforms, or their finite series approximations for the computation, are in general not periodic. In order to do the analysis for such (nonperiodic) general Bessel orthogonal expansions, in parallel to what was done for the (periodic) Fourier (trigonometric) series, we have to look at the function's representation beyond its interval of expansion $(0,1)$. We shall discuss this new topic briefly in Chapter 4 and point out the necessity for finding an estimate for the practical errors of computing the infinite series or integral as a finite sum, which is what our computing machines can do. In so doing we will be compelled to define a *"generalized" type of translation*, and a generalized *"repetition" or periodicity*, which seems to be in the right direction for giving the first estimate of one of the important errors.

In the last section, we discussed three types of convergence for the special case of Fourier trigonometric series, but to present a simple theory for the general orthogonal expansion, we shall consider mainly the convergence in the mean. This is guaranteed by Theorem 3.7 when we work with piecewise continuous (or square integrable) functions. Such a class of functions is practical and not so limited to work with in the analysis of signals, in physics, or in other boundary and initial value problems.

Finite Transforms—The Orthogonal Expansion Coefficients (Transform) and Series Expansion (Inverse Transform)

We conclude this section by saying that we have covered in detail with precise statements, and most of the time, proofs of the convergence for the Fourier trigonometric series and the orthogonal series expansion. This is important, as the orthogonal series (3.96) represents the inverse of the finite transform (3.95), the main topic of this chapter on the finite transforms,

$$F(n) = \int_a^b \rho(x)f(x)\phi_n(x)\,dx \qquad (3.95)$$

and its inverse $f(x)$ as given by (3.53),

$$f(x) = \sum_{n=1}^{\infty} \frac{F(n)}{\|\phi_n\|^2}\phi_n(x) \qquad (3.96)$$

We should recall here that in this section we covered the pointwise convergence, uniform convergence, better convergence, and convergence in the mean of the Fourier trigonometric series, while we were limited by the space to only the analysis and proof of convergence in the mean for the general orthogonal expansion (3.96).

In Sections 3.2–3.5 we will present the most important special cases of this general finite transform $F(n)$ of (3.95). This includes the *Fourier sine* and *cosine transforms* in Section 3.2, the *Fourier exponential transforms* in Section 3.3, the *Hankel (Bessel) transforms* in Section 3.4, and the classical orthogonal polynomials transforms in Section 3.5. The latter is illustrated for its important special cases of the Legendre, Laguerre, Hermite, and Tchebychev polynomial transforms.

The emphasis in the presentations of all these finite transforms is on their very basic operational properties, which are necessary for illustrating the application of these transforms in solving boundary value problems, most often on a finite domain. We will also use these transforms for their simpler representation of signals in the transform space. A case in point is a generalized sampling expansion related to finite transforms with general kernel away from the usual trigonometric or exponential kernels of the Fourier transforms.

3.2 Fourier Sine and Cosine Transforms

We introduce the *finite sine transform* of $f(x)$, which is defined on the finite interval $(0, \pi)$ as

$$F_s(n) = \int_0^\pi \sin nx\, f(x)\,dx \equiv \mathfrak{f}_s\{f\} \qquad (3.101)$$

with its inverse as the Fourier sine series of $f(x)$,

$$f(x) = \frac{2}{\pi}\sum_{n=1}^{\infty} F_s(n)\sin nx, \qquad 0 < x < \pi \qquad (3.102)$$

The generalization of this and other finite transforms to $f(x)$ on $(0,a)$ needs a simple scaling of the interval $(0,\pi)$, and we leave it for an exercise (see Exercise 3.14). The *finite cosine transform* of $f(x)$, which is defined on $(0,\pi)$, is

$$F_c(n) = \int_0^\pi \cos nx f(x)\,dx \equiv \mathfrak{f}_c\{f\} \tag{3.103}$$

and its inverse is the Fourier cosine series of $f(x)$,

$$f(x) = \frac{F_c(0)}{\pi} + \frac{2}{\pi}\sum_{n=1}^\infty F_c(n)\cos nx, \qquad 0 < x < \pi \tag{3.104}$$

If we compare both (3.101), (3.102) and (3.103), (3.104) with the general orthogonal expansion (3.6), (3.7),

$$F(\lambda_n) = \int_a^b \rho(x)\overline{K(\lambda_n,x}f(x)\,dx \tag{3.6}$$

$$f(x) = \sum c_n K(\lambda_n,x) = \sum \frac{F(\lambda_n)K(\lambda_n,x)}{N(\lambda_n)} \tag{3.7}$$

with Fourier coefficients $c_n = F(\lambda_n)/N(\lambda_n)$, and $N(\lambda_n)$ is the *norm square* of $K(\lambda_n,x)$,

$$N(\lambda_n) = \int_a^b \rho(x)|K(\lambda_n,x)|^2\,dx \equiv \|K(\lambda_n,-)\|^2 \tag{3.3}$$

we note that $c_0 = 0$, $c_n = (2/\pi)F_s(n)$ for the sine transform and $c_0 = (1/\pi)F_c(0)$, $c_n = (2/\pi)F_c(n)$ for the cosine transform. This can be verified by evaluating $N(\lambda_n)$ in (3.3) to find $N(n) = \pi/2$, $n \neq 0$, for both the sine and cosine transforms and $N(0) = \pi$ for the cosine transform (see Exercise 3.14c).

As we mentioned earlier, the convergence of the Fourier sine series (3.102) and the Fourier cosine series (3.104), as the inverse finite sine and cosine transforms, respectively, was the very detailed subject of Section 3.1.

We have already illustrated (in Example 1.4 of Chapter 1; in Section 2.5 for both the infinite sine and cosine transforms in (2.201), (2.202); and in Example 1.10 for a finite sine transform) that the sine and cosine transforms are compatible with even-order derivatives. This result was established by using integration by parts. Following the same method, we can show after performing two integrations by parts that the finite Fourier sine and cosine transforms algebraize d^2f/dx^2 on $(0,\pi)$ in the following forms, respectively:

$$\int_0^{\pi} \sin nx \frac{d^2f}{dx^2} \, dx = n \left\{ f(0) - (-1)^n f(\pi) \right\} - n^2 F_s(n) \tag{3.105}$$

and

$$\int_0^{\pi} \cos nx \frac{d^2f}{dx^2} \, dx = (-1)^n f'(\pi) - f'(0) - n^2 F_c(n) \tag{3.106}$$

This says that the finite sine transform requires the value of the function at the two end points 0 and π, while the finite cosine transform requires the derivative at the same two end points.

Sometimes we may be given, in addition to $f'(0) = 0$, a linear combination of $f(x)$ and its derivative at the other end point $x = a$; for example,

$$f'(a) + hf(a) = 0 \tag{3.107}$$

where h is a constant, which will be illustrated in Example 3.11. In this case we can define a finite cosine transform

$$F_c(\lambda_n) = \int_0^a \cos \lambda_n x f(x) \, dx \tag{3.108}$$

where due to (3.107) the set $\{\lambda_n\}$ consists of the zeros of

$$\tan \lambda_n a = \frac{h}{\lambda_n}, \qquad n = 1, 2, \ldots \tag{3.109}$$

which are usually found by numerical methods. The set $\{\cos \lambda_n x\}$ is still orthogonal on $(0, a)$, since it consists of solutions of a Sturm–Liouville problem. The norm square $N(\lambda_n)$ for this orthogonal set can be evaluated as

$$N(\lambda_n) = \int_0^a \cos^2 \lambda_n x \, dx = \frac{a(\lambda_n^2 + h^2) + h}{2(\lambda_n^2 + h^2)} \tag{3.110}$$

The inverse transform $f(x)$ of (3.108) is the Fourier series

$$f(x) = 2 \sum_{n=1}^{\infty} \frac{(\lambda_n^2 + h^2) F_c(\lambda_n)}{a(\lambda_n^2 + h^2) + h} \cos \lambda_n x \tag{3.111}$$

(see Exercise 3.15).

EXAMPLE 3.10: Diffusion in a finite domain

The following problem describes the temperature distribution in a rod of length π with both ends at zero temperature and with initial temperature

$f(x)$.

$$\frac{\partial u}{\partial t} - \frac{\partial^2 u}{\partial x^2} = 0, \qquad 0 < x < \pi, \ t > 0 \tag{E.1}$$

$$u(0,t) = 0, \qquad t > 0 \tag{E.2}$$

$$u(\pi,t) = 0, \qquad t > 0 \tag{E.3}$$

$$u(x,0) = f(x), \qquad 0 < x < \pi \tag{E.4}$$

This problem is the same as that of Example 1.1 or Example 2.22a, except that here the rod is of finite length π. As in these examples, this problem can also be solved by first using the method of separation of variables, which we illustrated in Example 1.1, which may solve for the homogeneous equations (E.1)–(E.3). With this method we must then appeal to Fourier analysis to resolve the problem of the remaining nonhomogeneous condition (E.4). In this case we look for a Fourier series (instead of the infinite integrals of Examples 1.1 and 2.22a) representation of $f(x)$ on $(0,\pi)$ in (E.4) in terms of (orthogonal) functions that have already satisfied conditions (E.1)–(E.3) as functions of x (see Exercise 3.19a). Thus we may say that the separation-of-variables method combined with Fourier analysis does this and other similar problems, which we shall do here with Fourier analysis only, i.e., via a finite sine transform and its inverse. This makes the present transform method *more direct*, in the sense that we start and end the problem in the setting of the transform and its inverse, without appealing to another method that depends primarily on, in advance, guessing the form of the solution as being separable.

We will illustrate here the use of the finite Fourier sine transform for the second derivative with respect to x, since we have $u(0,t)$ and $u(\pi,t)$ in (E.2), (E.3) as required by (3.105). Hence, we say that the finite sine transform is compatible with the present problem. We let

$$U_s(n,t) = \int_0^\pi u(x,t) \sin nx \, dx \tag{E.5}$$

and apply it to (E.1); then we use (3.105) and (E.2), (E.3) to obtain

$$\frac{\partial U_s(n,t)}{\partial t} = -n^2 U_s(n,t) \tag{E.6}$$

We note that some authors use the total derivative $dU_s(n,t)/dt$ instead of $\partial U_s(n,t)/\partial t$ in the transformed space; we prefer the latter, which we use most often in this book.

This is a first-order (homogeneous) partial differential equation in $U_s(n,t)$, with the solution

$$U_s(n,t) = A(n)e^{-n^2 t} \tag{E.7}$$

To find $A(n)$ we use the transform of the nonhomogeneous condition (E.4) to obtain

$$U_s(n,0) = A(n) = \int_0^\pi u(x,0)\sin nx\, dx = \int_0^\pi f(x)\sin nx\, dx = F_s(n)$$

(E.8)

Hence,

$$U_s(n,t) = F_s(n)e^{-n^2 t}$$

(E.9)

The solution $u(x,t)$ is obtained as the inverse sine transform (3.102) of $U_s(n,t)$; that is,

$$u(x,t) = \frac{2}{\pi}\sum_{n=1}^\infty F_s(n)e^{-n^2 t}\sin nx$$

(E.10)

where

$$F_s(n) = \int_0^\pi f(x)\sin nx\, dx$$

(E.8)

We recall here that in case we need to evaluate this solution numerically via the fast Fourier transform, we will have a double Fourier transform to evaluate for (E.8)–(E.10) in the event that we have a general $f(x)$ that did not result in $F_s(n)$ of (E.8) as a simple function in a closed form. We leave it as an exercise to obtain the same result (E.10) by the method of separation of variables (see Exercise 3.19a).

We must also add that the transform method, besides being more direct than the separation-of-variables method, can also handle nonhomogeneous boundary conditions via (3.105), (3.106). The separation-of-variables method needs some advanced planning in reducing the problem to equivalence with another problem or problems with homogeneous conditions except for one that is left for the Fourier analysis representation. These homogeneous equations are in general not separable, but as we mentioned earlier we can find some sufficient conditions that will inform us of such separability (see Exercise 1.4). After the equations are separated and the Fourier analysis is applied, we have a result for one problem. This is to be added to the solutions of the others to make up the solution of their sought equivalent (physical) problem. A familiar simple example is that in which the two ends $x = 0$ and $x = \pi$ of the rod in Example 3.10 are kept at temperatures T_1 and T_2, respectively— i.e., $u(0,t) = T_1$ and $u(\pi,t) = T_2$—whence the problem will have three nonhomogeneous conditions, a very heavy burden for the method of separation of variables! The way around that is to consider the solution $u(x,t) = w(x,t) + v(x)$ as a sum of a *transient* solution $w(x,t)$, where

$\lim_{t \to \infty} w(x,t) = 0$, and a steady-state solution $v(x)$. This will result in a simple solution for

$$v(x) = \frac{T_2 - T_1}{\pi} x + T_1$$

where $w(x,t)$ satisfies three homogeneous equations like (E.1)–(E.3) and only one nonhomogeneous equation of the form (E.4). Thus $w(x,t)$ can be solved for as in Example 3.10, and then $v(x)$ is added to it to make the final (physical) solution for the temperature distribution $u(x,t) = w(x,t) + v(x)$. The details of solving this problem are left as an exercise (see Exercise 3.19b).

EXAMPLE 3.11: Radiation at the boundary

This example illustrates the use of the more general finite cosine transform of (3.108). Consider the boundary value problem which describes the heat conduction in a bar of length a with radiation occurring at the end $x = a$; the end $x = 0$ is insulated and the initial temperature is constant u_0; that is,

$$\frac{\partial u}{\partial t} = \frac{\partial^2 u}{\partial x^2}, \qquad 0 < x < a, \ t > 0 \tag{E.1}$$

$$\frac{\partial u(0,t)}{\partial x} = 0, \qquad t > 0 \tag{E.2}$$

$$\frac{\partial u(a,t)}{\partial x} + hu(a,t) = 0, \qquad t > 0, \ h = \text{constant} \tag{E.3}$$

$$u(x,0) = u_0, \qquad 0 < x < a \tag{E.4}$$

Condition (E.3) describes the radiation at the end $x = a$, and (E.2) describes the insulated end $x = 0$. Let

$$U_c(\lambda_n, t) = \int_0^a u(x,t) \cos \lambda_n x \, dx \tag{E.5}$$

$$\tan a \lambda_n = \frac{h}{\lambda_n} \tag{E.6}$$

If we apply the transform (E.5) to (E.1)—perform two integrations by parts and use (E.2), (E.3), and (E.6)—we obtain

$$\frac{\partial U_c(\lambda_n, t)}{\partial t} = -\lambda_n^2 U_c(\lambda_n, t) \tag{E.7}$$

The bounded solution to (E.7) is

$$U_c(\lambda_n, t) = A(\lambda_n) e^{-\lambda_n^2 t} \tag{E.8}$$

To find $A(\lambda_n)$ we take the transform of (E.4) so that

$$A(\lambda_n) = U_c(\lambda_n, 0) = \int_0^a u(x, 0) \cos \lambda_n x \, dx = \int_0^a u_0 \cos \lambda_n x \, dx \quad \text{(E.8)}$$

$$= u_0 \frac{\sin a \lambda_n}{\lambda_n}$$

$$U_c(\lambda_n, 0) = \frac{u_0 \sin a \lambda_n}{\lambda_n} e^{-\lambda_n^2 t} \quad \text{(E.9)}$$

The solution $u(x,t)$ is the inverse transform (3.111) of (E.9):

$$u(x, t) = 2u_0 \sum_{n=1}^{\infty} \frac{\lambda_n^2 + h^2}{a(\lambda_n^2 + h^2) + h} \frac{\sin a \lambda_n}{\lambda_n} e^{-\lambda_n^2 t} \cos \lambda_n x \quad \text{(E.10)}$$

For the use of separation of variables on this problem, see Exercise 3.21a.

3.3 Fourier (Exponential) Transforms

The finite sine and cosine transforms are special cases of the finite Fourier exponential transform

$$F(n) = \int_{-\pi}^{\pi} e^{-inx} f(x) \, dx \quad \text{(3.112)}$$

$$f(x) = \frac{1}{2\pi} \sum_{n=-\infty}^{\infty} F(n) e^{inx}, \quad -\pi < x < \pi \quad \text{(3.113)}$$

when $f(x)$ is odd or even, respectively. Clearly, this finite Fourier transform pair is a special case of the general finite transform pair (3.6), (3.7) that corresponds to $K(\lambda_n, x) = e^{inx}$, $a = -\pi$ and $b = \pi$. The convergence for the (inverse transform) Fourier series was discussed in detail, with ample illustrations, in Section 3.1.1, and with the supporting basic theorems of convergence and their proofs done later in Section 3.1.3.

This transform is used for problems defined on a symmetric finite interval such as $(-\pi, \pi)$ and for all orders of derivatives. This, of course, contrasts with the finite sine and cosine transforms which handle functions on $(0, \pi)$ and are applied to even-order derivatives only as in (3.105), (3.106). For example, the finite Fourier transform of df/dx is

$$\int_{-\pi}^{\pi} e^{-inx} \frac{df}{dx} \, dx = e^{-inx} f(x) \Big|_{-\pi}^{\pi} + \int_{-\pi}^{\pi} (in) e^{-inx} f(x) \, dx$$

$$= (-1)^n \{f(\pi) - f(-\pi)\} + inF(n) \tag{3.114}$$

This also takes care of a possible "jump" condition $f(\pi) - f(-\pi)$ between the two boundary values. Without such a jump, i.e., for periodic boundary conditions $f(\pi) = f(-\pi)$, the transform (3.114) of df/dx becomes $inF(n)$. We must mention that the generality in (3.114) above, of having the jump $f(\pi) - f(-\pi)$, may be used to an advantage. This is in the sense that we may be given the difference between the temperatures at the end points $u(\pi, t) - u(-\pi, t) = T_1$, and not the usual temperatures for the separate ends $u(\pi, t) = T_2$ and $u(-\pi, t) = T_1$ independently.

The following example illustrates the use of the finite exponential Fourier transform for solving a boundary value problem on $(-\pi, \pi)$ that is associated with a first-order partial differential equation.

EXAMPLE 3.12:

To solve the following boundary value problem:

$$\frac{\partial u}{\partial y} + \frac{\partial u}{\partial x} = 0, \qquad -\pi < x < \pi, \ 0 < y < \infty \tag{E.1}$$

$$u(-\pi, y) = u(\pi, t), \qquad 0 < y < \infty \tag{E.2}$$

$$u(x, 0) = f(x), \qquad -\pi < x < \pi \tag{E.3}$$

we first apply the finite Fourier transform (3.112),

$$U(n, y) = \int_{-\pi}^{\pi} e^{-inx} u(x, y) \, dx \tag{E.4}$$

to (E.1) and use (3.114) and (E.2) to obtain

$$\frac{\partial U(n, y)}{\partial y} + inU(n, y) = 0 \tag{E.5}$$

The solution to (E.5) is

$$U(n, y) = A(n) e^{-iny}$$

where

$$U(n, 0) = \int_{-\pi}^{\pi} e^{-inx} u(x, 0) \, dx = \int_{-\pi}^{\pi} e^{-inx} f(x) \, dx = F(n) = A(n) \tag{E.6}$$

Hence,

$$U(n, y) = F(n) e^{-iny} \tag{E.7}$$

and its inverse, obtained through (3.113), is

$$u(x,y) = \frac{1}{2\pi} \sum_{n=-\infty}^{\infty} F(n) e^{-iny} e^{inx} \tag{E.8}$$

The finite Fourier transform for functions defined on $(-a,a)$ is obtained from (3.102), (3.103) through a simple change of scale:

$$F\left(\frac{n\pi}{a}\right) = \int_{-a}^{a} \exp\left(-\frac{in\pi}{a}x\right) f(x)\,dx \tag{3.115}$$

$$f(x) = \frac{1}{2a} \sum_{n=-\infty}^{\infty} F\left(\frac{in\pi}{a}\right) \exp\left(\frac{n\pi}{a}x\right), \qquad -a < x < a \tag{3.116}$$

As in the case of the Fourier exponential transforms, we can replace i by $-i$ in (3.115), (3.116) to have an equivalent representation that is preferred in applications in some fields of study

$$F\left(\frac{n\pi}{a}\right) = \int_{-a}^{a} \exp\left(\frac{in\pi x}{a}\right) f(x)\,dx \tag{3.115a}$$

$$f(x) = \frac{1}{2a} \sum_{n=-\infty}^{\infty} F\left(\frac{n\pi}{a}\right) \exp\left(-\frac{in\pi x}{a}\right) \tag{3.116a}$$

We note again that either version (3.115), (3.116) or (3.115a), (3.116a) of the finite Fourier exponential transforms is a special case of the general orthogonal expansion (finite transform) (3.6), (3.7) with the kernel $K(\lambda_n,x)$ as a (complex) exponential function $e^{in\pi x/a}$ (or $e^{-in\pi x/a}$).

3.3.1 The Finite Fourier Exponential Transform and the Sampling Expansion

It is clear from (3.116) that this series expansion for $f(x)$ on $(-a,a)$ is periodic with period $2a$. With regard to the important sampling expansion (2.184) that we covered in Section 2.4.2, we may consider $\{F(n\pi/a)\}$ of (3.115) as samples of the *bandlimited* function

$$F_a(\lambda) = \int_{-a}^{a} e^{-i\lambda x} f(x)\,dx \tag{3.117}$$

at $\lambda_n = n\pi/a$. However, we must stress that in this case $f(x)$ is assumed to vanish identically beyond $(-a,a)$; hence, it is not allowed the periodic extension beyond $(-a,a)$ that is so generously offered by the Fourier series (3.116). To be careful, we must remember the presence of the gate function $p_a(x)$ for $p_a(x)f(x)$ in the Fourier integral,

$$F_a(\lambda) = \int_{-\infty}^{\infty} p_a(x)f(x)e^{-i\lambda x}\,dx = \int_{-a}^{a} f(x)e^{-i\lambda x}\,dx \qquad (3.118)$$

that defined $F_a(\lambda)$ as bandlimited to $(-a,a)$. Here $p_a(x)$ is the gate function,

$$p_a(x) = \begin{cases} 1, & |x| < a \\ 0, & |x| > a \end{cases} \qquad (3.119)$$

Next we will use all these and other Fourier analysis tools to derive the sampling expansion of bandlimited signals, where we will use the Fourier analysis notations appropriate in the field of signals and systems.

The Sampling Expansion (The Shannon[†] Sampling Theorem)
Here we shall adhere to the notation of the Fourier representation, used in Section 2.4.2, for the analysis of bandlimited signal $f_a(t)$,

$$f_a(t) = \int_{-a}^{a} e^{i\omega t} F(\omega)\,d\omega \qquad \begin{matrix}(2.138)\\(3.120)\end{matrix}$$

where $F(\omega)$ is its spectrum. In Section 2.4.2, and without the advantage of having the Fourier (series) analysis that we have developed in this chapter thus far, we used the impulse train as a (shortcut) tool for developing the sampling series expansion (2.184)

$$f_a(t) = \sum_{n=-\infty}^{\infty} f_a\left(\frac{n\pi}{a}\right) \frac{\sin(at - n\pi)}{at - n\pi} \qquad \begin{matrix}(2.184)\\(3.121)\end{matrix}$$

for this bandlimited signal $f_a(t)$. We also hinted, then, at the clear possibility of using the Fourier series representation of $F(\omega)$ on $(-a,a)$ inside the integral of $f_a(t)$ in (3.120). But we decided to delay such a natural mode of derivation of (3.121) until this section. Clearly, we expect conditions on the function $F(\omega)$ that reflect a reasonable convergence for its Fourier series representation on $(-a,a)$ in order to allow certain operations on such series. As we shall see, the most obvious of such operations is the exchange of the integral of (3.120) with the infinite sum

---

[†]This theorem is also attributed to Whittakers Kotelńikov, Someya and Raáb (see Shannon (1949), Jerri 1992, 1977a, and Marks 1991, 1992).

of the Fourier series of $F(\omega)$. This, as we know from Theorem 3.7, is allowed for $F(\omega)$ square integrable on $(-a,a)$, that is, $\int_{-a}^{a} |F(\omega)|^2 d\omega < \infty$. So the sampling theorem interpolation for a bandlimited signal $f_a(t)$ in terms of its samples $f_a(n\pi/a)$ states that for the bandlimited signal (3.120) with *square integrable* $F(\omega)$ on $(-a,a)$ we have

$$f_a(t) = \sum_{n=-\infty}^{\infty} f_a\left(\frac{n\pi}{a}\right) \frac{\sin a(t - n\pi/a)}{a(t - n\pi/a)} \tag{3.121}$$

One of the simplest proofs starts with writing the Fourier series expansion (3.116a) for $F(\omega)$ of (3.120) in terms of the orthogonal set of functions $\{e^{-(in\pi/a)\omega}\}$ on the interval $(-a,a)$, that is,

$$F(\omega) = \sum_{n=-\infty}^{\infty} c_n \exp\left(-\frac{in\pi}{a}\omega\right) \tag{3.122}$$

According to (3.115a), the Fourier coefficients c_n are obtained as

$$c_n = \frac{1}{2a} \int_{-a}^{a} F(\omega) \exp\left(\frac{in\pi}{a}\omega\right) d\omega = \frac{1}{2a} f_a\left(\frac{n\pi}{a}\right) \tag{3.123}$$

after consulting (3.120) for recognizing the above integral as $f_a(n\pi/a)$. The Fourier series (3.122) then becomes a

$$F(\omega) = \frac{1}{2a} \sum_{n=-\infty}^{\infty} f_a\left(\frac{n\pi}{a}\right) \exp\left(-\frac{in\pi}{a}\omega\right) \tag{3.124}$$

To obtain $f_a(t)$ as in (3.120), we multiply (3.124) by $e^{i\omega t}$ and then integrate both sides from $-a$ to a:

$$f_a(t) = \int_{-a}^{a} F(\omega) e^{i\omega t} d\omega$$

$$= \frac{1}{2a} \int_{-a}^{a} e^{i\omega t} \sum_{n=-\infty}^{\infty} f_a\left(\frac{n\pi}{a}\right) \exp\left(-\frac{in\pi}{a}\omega\right) d\omega$$

The assumption that $F(\omega)$ is square integrable is sufficient for exchanging the integration with the infinite summation, according to Theorem 3.7 in (3.65), (3.66); thus we obtain

$$f_a(t) = \frac{1}{2a} \sum_{n=-\infty}^{\infty} f_a\left(\frac{n\pi}{a}\right) \int_{-a}^{a} e^{i\omega(t-n\pi/a)} d\omega$$

$$= \sum_{n=-\infty}^{\infty} f_a\left(\frac{n\pi}{a}\right) \frac{\sin a(t - n\pi/a)}{a(t - n\pi/a)}$$

after doing the simple integration as done in (1.32).

We should mention that a number of methods are available for deriving the above sampling expansion. The most obvious, among those that use Fourier analysis, is just to write the Fourier series expansion on $(-a,a)$ for $e^{i\omega t}$ instead of $F(\omega)$ inside the Fourier integral of (3.120), then integrate term by term. The latter operation is allowed, since $e^{i\omega t}$ is clearly square integrable on $(-a,a)$. Other methods include the use of the Parseval equality (3.65), which can also be extended to allow relaxing the condition of square integrability on $F(\omega)$ to just absolute integrability. For example, the Bessel function $J_0(\pi t)$ has a bandlimited function representation

$$J_0(\pi t) = \frac{1}{\pi} \int_{-\pi}^{\pi} \frac{e^{ixt}}{\sqrt{\pi^2 - x^2}} \tag{3.125}$$

with a Fourier spectrum $F(x) = 1/\sqrt{\pi^2 - x^2}$ which is absolutely integrable and not square integrable since

$$\frac{1}{\pi} \int_{-\pi}^{\pi} \frac{1}{\sqrt{\pi^2 - x^2}} = J_0(0) = 1$$

while

$$\int_{-\pi}^{\pi} \frac{1}{\pi^2 - x^2} \, dx = \frac{1}{\pi^3} \tanh^{-1} \frac{x}{\pi} \Big|_{x=-\pi}^{\pi} = \infty$$

So such an extension of the sampling expansion is very useful for interpolating signal $f(t)$ whose spectrum $F(x)$ is not necessarily square integrable.

One of the most powerful methods for developing the sampling expansion (3.121) and most of its extensions is through the use of complex contour integration, which is considered optional in our treatment, but which we shall include, for interested readers, among the above methods in Exercise 3.29. The extension of the sampling expansion to functions of more than one variable follows in a simple way after consulting the multiple Fourier integrals in Section 2.6 (see Exercise 3.26b).

We should stress here that the importance of the sampling expansion lies not only in interpolating signals but also in specifying the required spacing π/a in terms of the bandlimit a. Such a spacing requirement plays an important role in developing the discrete Fourier transforms, whose efficient algorithm is the useful fast Fourier transform (FFT), as we shall cover in Section 4.1. As we recognize the importance of the sample spacing π/a, we may mention that the sampling expansion (3.121) for $f_a(t)$ can be extended to use the samples $f(2n\pi/a)$ of the function and its derivative $f'(2n\pi//a)$ with the clear advantage of doubling the sample spacing to $2\pi/a$ instead of π/a in (3.121). The derivation of this result is left for an exercise (see Exercise 3.27b).

As discussed in Section 2.4.2, two obvious errors may be involved in the practical application of the sampling expansion (3.121) of signals, namely those of the truncation error ϵ_N in (2.185),

$$\epsilon_N(t) = \left| f_a(t) - \sum_{n=-N}^{N} f_a\left(\frac{n\pi}{a}\right) \frac{\sin a(t - n\pi/a)}{a(t - n\pi/a)} \right| \qquad (2.185)$$

and the aliasing error ϵ_A in (2.186),

$$\epsilon_A = \left| f(t) - \sum_{n=-\infty}^{\infty} f\left(\frac{n\pi}{a}\right) \frac{\sin(at - n\pi)}{at - n\pi} \right| \qquad (2.186)$$

The truncation error ϵ_N is due to including only a finite number $2N + 1$ of samples instead of the infinite number required by the sampling series (3.121). The aliasing error is due to the uncertainty of knowing exactly the bandlimit a of the required (bandlimited) signal $f_a(t)$. Upper bounds or estimates for such truncation and aliasing errors are of utmost importance in signal analysis, where developing better and tighter error bounds is an ongoing research field. We will attempt the derivation of simple bounds for these error in the exercises (see Exercise 3.28, 3.29).

In Section 2.4.2 we also alluded to the possibility of sampling for time-limited signals of (2.188) with a stern warning in terms of observing the uncertainty principle for such a situation!

Another extension of the sampling expansion is to signals represented by more general finite limit (bandlimited!) transforms with kernel $K(x,t)$ like the Bessel function $J_n(xt)$, and away from the trigonometric kernel of the Fourier transforms. We shall present this expansion as the generalized sampling theorem of Kramer and Weiss (in Section 3.6) after we cover the general finite (Sturm–Liouville type) transforms in the next sections. For coverage of all aspects of the sampling expansions see Jerri (1977a); and for the most recent general reference see Marks (1991), (1992).

3.4 Hankel (Bessel) Transforms

We present here the finite J_n-Hankel transform as another example of the Sturm–Liouville type transform (3.6), (3.7).

$$\mathsf{h}_n\{f\} = F_n(\lambda_k) = \int_0^a r J_n(\lambda_k r) f(r)\, dr \qquad (3.126)$$

$$J_n(\lambda_k a) = 0, \qquad k = 1, 2, \dots. \tag{3.127}$$

where we use h_n for the present finite J_n-Hankel transform to differentiate it from the $\mathcal{H}_n$ we used for the infinite J_n-Hankel integral transform of Section 2.7.

Here $\{\lambda_k a\} = \{j_{n,k}\}$ are the zeros of $J_n(x)$. Twenty such zeros for J_n $n = 0, 1, 2, 3, 4, 5$, are presented in Table A.1 of Appendix A.5. The inverse is obtained from (3.7) as

$$f(r) = \frac{2}{a^2} \sum_{k=1}^{\infty} \frac{F(\lambda_k) J_n(\lambda_k r)}{J_{n+1}^2(a\lambda_k)} \tag{3.128}$$

after evaluating $N(\lambda_k)$,

$$N(\lambda_k) = \int_0^a r J_n^2(\lambda_k r)\, dr = \frac{a^2}{2}\, [J_{n+1}(a\lambda_k)]^2 \tag{3.129}$$

and with the help of some properties of the Bessel functions [see Exercise 3.35a, Exercise A.25d in Appendix A.5 and (A.5.18)],

$$J_n'(x) = -J_{n+1}(x) + \frac{n}{x} J_n(x) \tag{3.130}$$

The convergence of the Bessel–Fourier series (3.128), as the inverse finite J_n-Hankel transform here, was the subject of general discussion in Section 3.1.3B concerning the convergence of the orthogonal series expansion in terms of solutions of the Sturm–Liouville problem. In particular, we relied on such convergence for the derivation of the finite J_ν-Hankel transform (3.99),

$$F_\nu(j_{\nu,m}) = \int_0^1 x J_\nu(j_{\nu,m} x) f(x)\, dx \tag{3.99}$$

from its inverse

$$f(x) = \sum_{m=1}^{\infty} \frac{F_\nu(j_{\nu,m}) J_\nu(j_{\nu,m} x)}{\int_0^1 x J_\nu^2(j_{\nu,m} x)\, dx} \tag{3.100}$$

after consulting Theorem 3.5 for the convergence in the mean of this series for the square integrable function $f(x)$ on $(0, 1)$. With that, Theorem 3.5 assured the validity of integrating this series term by term, as was done in the step leading to (3.98) and the above finite Hankel transform in (3.99). As we mentioned before, $J_n(x)$ is one of the two solutions of Bessel differential equation (3.83a) which is bounded at $x = 0$. Hence, the expansion (3.128) is good for problems inside a disc of radius a, that is, $0 \le r \le a$. An attempt at a discrete algorithm to approximate the finite

Hankel transforms (3.126), (3.128) and the Hankel transforms of Section 2.7 will be presented very briefly in Section 4.3 with some numerical illustrations.

Following the steps used in Example 1.6 for the derivation of the (infinite limit) J_n-Hankel transform pair in (1.24), we can easily show, after two integrations by parts and using (3.127), that the finite Hankel transform is compatible with the following main part of the Bessel differential equation. In other words,

$$h_n \left\{ \frac{1}{r} \frac{d}{dr} (rf'(r)) - \frac{n^2}{r^2} f(r) \right\} = -\lambda_k^2 F_n(\lambda_k) + a\lambda_k J_{n+1}(a\lambda_k) f(a)$$

$$(3.131)$$

since

$$\int_0^a r \left\{ \frac{1}{r} \frac{d}{dr} (rf(r)) - \frac{n^2}{r^2} \right\} J_n(\lambda_k r) \, dr$$

$$= -\lambda_k^2 F_n(\lambda_k) - a\lambda_k J_n'(a\lambda_k) f(a)$$

$$= -\lambda_k^2 F_n(\lambda_k) + a\lambda_k J_{n+1}(a\lambda_k) f(a) \qquad (3.132)$$

where we employed property (3.130) for the case $J_n(a\lambda_k) = 0$ in (3.127), i.e.,

$$J_n'(a\lambda_k) = -J_{n+1}(a\lambda_k) \qquad (3.130a)$$

From (3.131) we note that the finite Hankel transform (3.126) associated with condition (3.127) requires $f(a)$, the value of the transformed function $f(x)$ at the end point $x = a$. As is the case with the finite cosine transform, we may consider a modification of this Hankel transform to handle boundary conditions at $r = a$ like $f'(a)$, or the general case that involves a linear combination of the function and its derivative; for example,

$$f'(a) + hf(a) = 0. \qquad (3.133)$$

as we shall illustrate in Section 3.4.1 and Example 3.14.

EXAMPLE 3.13: Temperature distribution in a long cylinder

As an example, we consider the temperature distribution $u(r,t)$ in a long cylinder $(-\infty < z < \infty$—no variation in the z direction) with radius one and with constant surface temperature u_1 and zero initial temperature.

$$\frac{1}{\kappa} \frac{\partial u}{\partial t} = \frac{\partial^2 u}{\partial r^2} + \frac{1}{r} \frac{\partial u}{\partial r}, \qquad 0 \le r < 1, \ t > 0 \qquad (E.1)$$

$$u(1,t) = u_1, \qquad t > 0 \qquad (E.2)$$

$$u(r,0) = 0, \qquad 0 \le r < 1 \qquad (E.3)$$

where κ is the diffusivity.

We note here that we can use both the finite Hankel transform on $u(r,-)$ in the variable r or the Laplace transform on $u(-,t)$ in the variable t, since they are compatible with the differential operators in r and t, respectively, and since the boundary conditions required by the respective transforms, i.e., $u(1,t)$ and $u(r,0)$, are given. Indeed, we may even use both transforms successively to algebraize completely the diffusion equation (E.1), which we shall leave for an exercise (see Exercise 3.40). Also note that for this long cylinder, we assumed that the solution is independent of z where equation (E.1) corresponds to the case $n = 0$ in (3.132). Hence, we will deal with the finite J_0-Hankel transform $H_0(\lambda_k)$ associated with $J_0(x\lambda_k)$, where $\{\lambda_k\} \equiv \{\lambda_{0,k}\}$ are the zeros of $J_0(\lambda_k) = 0$.

Let the finite J_0-Hankel transform of $u(r,t)$ be

$$U_0(\lambda_k,t) = \int_0^1 rJ_0(\lambda_k r)u(r,t)dr \equiv h_0\{u\} \tag{E.4}$$

We shall assume that the notation $h_0\{u\}$ is understood from the context of this section to be for the finite J_0-Hankel transform, and it is not to be confused with the notation $\mathcal{H}_0\{u\}$ used for the J_0-Hankel transform (of functions on $(0,\infty)$) in Section 2.7. Hence, if we take the Hankel transform of (E.1) and use (3.132) and (E.2), we obtain

$$\frac{1}{\kappa}\frac{\partial U_0}{\partial t}(\lambda_k,t) = -\lambda_k^2 U_0(\lambda_k,t) + u_1\lambda_k J_1(\lambda_k) \tag{E.5}$$

a nonhomogeneous differential equation in t whose solution can be obtained by using *the integrating factor* to have

$$U_0(\lambda_k,t) = \frac{u_1}{\lambda_k}J_1(\lambda_k)\left\{1 - e^{-\kappa\lambda_k^2 t}\right\} \tag{E.6}$$

where we have used the fact that $U_0(\lambda_k,0) = 0$ as the Hankel transform of (E.3).

To find the solution $u(r,t)$ we take the inverse J_0-Hankel transform of (E.6) using (3.128) and obtain

$$u(r,t) = 2u_1 \sum_{k=1}^{\infty} \frac{J_1(\lambda_k)}{\lambda_k}\frac{\{1 - e^{-\kappa\lambda_k^2 t}\}}{J_1^2(\lambda_k)}J_0(\lambda_k r) \tag{E.7}$$

This series solution can be further simplified when we realize that the Fourier Bessel series (3.128) of $f(r) = 1$ on $(0,1)$ is

$$1 = 2 \sum_{k=1}^{\infty} \frac{J_1^{-1}(\lambda_k)}{\lambda_k}J_0(\lambda_k r) \tag{E.8}$$

which can be established with the help of the identity

$$\frac{d}{dx}[x^n J_n(ax)] = ax^n J_{n-1}(ax) \tag{E.9}$$

Hence, (E.7) becomes

$$u(r,t) = u_1 - 2u_1 \sum_{k+1}^{\infty} e^{-\kappa \lambda_k^2 t} \frac{J_0(\lambda_k r)}{\lambda_k J_1(\lambda_k)}, \quad J_0(\lambda_k) = 0 \tag{E.10}$$

The problem of the above example is associated with a *Dirichlet condition* where the value of the temperature is given on the boundary $r = 1$. In case the gradient $\partial u / \partial r$ is given on the boundary, such as $\partial u(1,t)/\partial r = 0$, we have a *Neumann condition*. For such a problem we can use the same finite Hankel transform (3.126), except that it will be tied with the condition $J_n'(\lambda_k') = 0$ instead of $J_n(\lambda_k) = 0$, where here $\{\lambda_k'\}$ are the zeros of $J_n'(x)$ (see Exercise 3.37, 3.39d).

3.4.1 Another Finite Hankel Transform

Just as in the case of the finite Fourier cosine transform (3.108)–(3.111), which involved the general boundary condition $f'(a) + hf(a) = 0$ at the end point $x = a$ as in (3.107), the finite Hankel transform can be modified to handle such a condition at $r = a$. According to the (regular singular) Sturm–Liouville problem, the Bessel function $\{J_n(\lambda_k r)\}$, where $\{\lambda_k a\}$ are the zeros of the equation

$$\lambda J_n'(\lambda a) + h J_n(\lambda a) = 0 \tag{3.134}$$

can be shown as orthogonal on the interval $(0, a)$. Theorem 3.12 addressed this question of orthogonality for the special boundary condition related to $J_n(\lambda a) = 0$. Equation (3.134) is the result of the more general boundary condition (on $J_n(\lambda r)$),

$$f'(a) + hf(a) = 0 \tag{3.133}$$

that describes *Newton's law of convection* or cooling at the surface $r = a$. This suggests defining another type of finite Hankel transform:

$$F_n(\lambda_k) = \int_0^a r J_n(\lambda_k r) f(r) \, dr \tag{3.135}$$

$$\lambda_k J_n'(\lambda_k a) + h J_n(\lambda_k a) = 0, \qquad k = 1, 2, \ldots \tag{3.136}$$

where the $\{\lambda_k\}$ are determined from (3.134).

Indeed, we can show, with the help of the Bessel function properties (see Exercise 3.38c), that the norm square of these functions is

$$\|J_n(\lambda_k)\|^2 \equiv \int_0^a rJ_n^2(\lambda_k r)\,dr = \frac{[(\lambda_k^2 + h^2)a^2 - n^2]J_n^2(a\lambda_k)}{2\lambda_k^2} \tag{3.137}$$

According to the generalized Fourier series expansion (3.7), the inverse transform is

$$f(r) = 2\sum_{k=1}^{\infty} \frac{F_n(\lambda_k)\lambda_k^2 J_n(\lambda_k r)}{[(\lambda_k^2 + h^2)a^2 - n^2]J_n^2(a\lambda_k)} \tag{3.138}$$

after using (3.137).

Again, we can follow the same steps we used for the integral (infinite limit) Hankel transform to show that this finite Hankel transform is compatible with the Bessel differential equation and gives

$$h_n\left\{\frac{1}{r}\frac{d}{dr}(rf'(r)) - \frac{n^2}{r^2}f(r)\right\}$$

$$= -\lambda_k^2 F_n(\lambda_k) - \frac{a\lambda_k J_n'(a\lambda_k)}{h}\left\{\frac{df}{dr}(a) + hf(a)\right\} \tag{3.139}$$

where we can clearly see how the condition $df(a)/dr + hf(a)$ of (3.133) is involved in this transformation. We illustrate the use of this Hankel transform in the following example.

EXAMPLE 3.14: Convective Cooling of a Long Cylinder

The finite Hankel transform defined by (3.135), (3.136), and (3.138) is suitable for the temperature distribution $u(r,t)$ in a long cylinder (take $-\infty < z < \infty$), which is cooling by radiating heat to the outside medium according to Newton's law of cooling (3.133). Here we take the initial temperature to be a constant u_1.

If we let the radius of the cylinder be a, Newton's law of cooling (with the outside medium at temperature zero) will state that

$$\frac{\partial u}{\partial r}(a,t) = -hu(a,t) \tag{E.1}$$

where h is a constant.

The initial and boundary value problem becomes

$$\frac{1}{\kappa}\frac{\partial u}{\partial t} = \frac{\partial^2 u}{\partial r^2} + \frac{1}{r}\frac{\partial u}{\partial r}, \qquad 0 \le r < a,\ t > 0 \tag{E.2}$$

$$\frac{\partial u}{\partial r}(a,t) + hu(a,t) = 0, \qquad t > 0 \tag{E.3}$$

$$u(r,0) = u_1, \qquad 0 < r < a \tag{E.4}$$

Let the J_0-Hankel transform of $u(r,t)$ be

$$U_0(\lambda_k,t) = \int_0^a ru(r,t)J_0(\lambda_k r)\,dr \tag{E.5}$$

where $\{a\lambda_k\}$ are the zeros of

$$\lambda J_0'(\lambda a) + hJ_0(\lambda a) = 0 \tag{E.6}$$

We take the Hankel transform of (E.2) and use (3.139) to obtain

$$\frac{1}{\kappa}\frac{\partial U_0}{\partial t}(\lambda_k,t) = -\lambda_k^2 U_0(\lambda_k,t) \tag{E.7}$$

where the last term on the right side of (3.139) vanishes because of the boundary condition (E.3). The solution to (E.7) is

$$U_0(\lambda_k,t) = A(\lambda_k)e^{-\kappa\lambda_k^2 t} \tag{E.8}$$

where $A(\lambda_k)$ is determined from the transformed initial condition (E.4)

$$U_0(\lambda_k,0) = \int_0^a rJ_0(\lambda_k r)u(r,0)\,dr$$

$$= u_1 \int_0^a rJ_0(\lambda_k r)\,dr = \frac{u_1 aJ_1(a\lambda_k)}{\lambda_k} = A(\lambda_k) \tag{E.9}$$

after using (E.4) and the fact that $h_0\{1\} = aJ_1(a\lambda_k)/\lambda_k$ (which is left as an exercise).
Hence,

$$U_0(\lambda_k,t) = \frac{u_1 aJ_1(a\lambda_k)}{\lambda_k}e^{-\kappa\lambda_k^2 t} \tag{E.10}$$

To obtain the solution $u(r,t)$ we use the inverse Hankel transform (3.138) with $F_0(\lambda_k) = U_0(\lambda_k,t)$ of (E.10); that is,

$$u(r,t) = \frac{2u_1}{a}\sum_{k=1}^{\infty}\frac{\lambda_k J_1(a\lambda_k)e^{-\kappa\lambda_k^2 t}J_0(\lambda_k r)}{(\lambda_k^2 + h^2)J_0^2(a\lambda_k)} \tag{E.11}$$

The subject of other types of Hankel transforms will be covered in the exercises. These include some variation of boundary conditions.

3.5 Classical Orthogonal Polynomial Transforms

As another example of the general finite transform (3.95), (3.96) (or (3.6), (3.7)), we introduce transforms with polynomial kernels $\{\phi_n(x)\} = \{p_n(x)\}$. These classical orthogonal polynomials,

$$p_n(x) = k_n x^n + k'_n x^{n-1} + \cdots \tag{3.140}$$

are orthogonal on the interval (a,b), with respect to the weight function $\rho(x)$, so that

$$\int_a^b \rho(x) p_n^2(x)\,dx = h_n \tag{3.141}$$

The orthogonal polynomial expansion of $f(x)$ on the interval (a,b) is

$$f(x) = \sum_{j=0}^{\infty} \frac{1}{h_j} F(j) p_j(x) \tag{3.142}$$

which represents the inverse transform of the classical orthogonal polynomial transform,

$$F(j) = \int_a^b \rho(x) f(x) p_j(x)\,dx \tag{3.143}$$

Two of the particular cases we will cover here are those of the finite *Legendre* and *Laguerre* transforms with the Legendre polynomials $P_n(x)$ and the Laguerre polynomial $L_n(x)$ as kernels, respectively. For these important polynomials and other special functions, see Appendix A.5. These two examples represent two distinct cases of the general finite transforms in the sense that the Legendre polynomials $P_n(x)$ are orthogonal on the *finite* interval $(-1, 1)$, while the Laguerre polynomials $L_n(x)$ are orthogonal on the *infinite* interval $(0, \infty)$ with respect to $\rho(x) = e^{-x}$; i.e.,

$$\int_{-1}^{1} P_n(x) P_m(x)\,dx = \begin{cases} \dfrac{2}{2n+1}, & n = m \\ 0, & n \ne m \end{cases} \tag{3.144}$$

$$\int_0^{\infty} e^{-x} L_n(x) L_m(x)\,dx = \begin{cases} 1, & n = m \\ 0, & n \ne m \end{cases} \tag{3.145}$$

This should explain the objection to calling transforms (3.142), (3.143) finite, to avoid confusing them with just finite limits of integration, and why some authors call them instead *Sturm–Liouville type transforms*. They are indeed the Fourier coefficients (3.94) of the Fourier series (3.95), depending on which one we call the transform, when we simply write the general orthogonal expansion (3.95), (3.94) of $f(x)$. This was the case with the discrete Fourier transforms when we called either one a transform in Chapter 1. However, in the present case, the transform is the Fourier coefficient $F(j)$, since in this chapter we are speaking of finite *integral* transforms.

The discrete algorithms for computing these (finite) orthogonal polynomial transforms are discussed in Section 4.2.

3.5.1 Legendre Transforms

The Legendre transform of $f(x)$, defined on $(-1, 1)$, is

$$F(j) = \int_{-1}^{1} f(x)P_j(x)\,dx \equiv \mathfrak{l}\{f\} \tag{3.146}$$

where l stands for the finite Legendre transform operator. The inverse of this Legendre transform is the *Legendre Fourier series*

$$f(x) = \sum_{j=0}^{\infty} \frac{1}{h_j} F(j)P_j(x) = \frac{1}{2}\sum_{j=0}^{\infty}(2j + 1)F(j)P_j(x) \equiv \mathfrak{l}^{-1}\{F\} \tag{3.147}$$

The convergence of this and other orthogonal expansions series was discussed in Section 3.1.3, where Example 3.8 covered the proof of the completeness of these Legendre polynomials on $(-1, 1)$. The Legendre polynomial $P_j(x)$ is a solution of the Legendre differential equation

$$\frac{d}{dx}\left\{(1-x^2)\frac{du}{dx}\right\} + j(j + 1)u(x) = 0 \tag{3.148}$$

The simplest expression for $P_n(x)$ is obtained through its Rodrigues formula:

$$P_j(x) = \frac{1}{2^j j!} \frac{d^j(x^2 - 1)^j}{dx^j} \tag{3.149}$$

As an example, we use (3.149) to show that $P_0(x) = 1$, $P_1(x) = x$, and $P_2(x) = \frac{1}{2}(3x^2 - 1)$.

As we have done for the Hankel transform, we can show that the finite Legendre transform (3.146) is compatible with the differential operator of the Legendre differential equation type; i.e.,

$$\mathfrak{l}\left\{\frac{d}{dx}(1-x^2)\frac{du}{dx}(x)\right\} = -j(j + 1)U(j) \tag{3.150}$$

where $U(j) = l\{u(x)\}$.

This is done by resorting to the general Sturm–Liouville transform results in Chapter 1 (Section 1.6 with its Examples 1.9–1.11), or by the usual method of integrating by parts twice, and with the help of the fact that $P_n(x)$ satisfies (3.148) as we saw in the case of the Hankel transform

$$\int_{-1}^{1} \frac{d}{dx}\left\{(1-x^2)\frac{du}{dx}\right\}P_j(x)\,dx$$

$$= (1-x^2)\frac{du}{dx}P_j(x)\Big|_{-1}^{1} - \int_{-1}^{1}\frac{du}{dx}(1-x^2)\frac{dP_j(x)}{dx}\,dx \qquad (3.151)$$

Since du/dx and $P_j(x)$ are finite at $x = \mp 1$, the first term vanishes. Now we apply another integration by parts on the remaining integral of (3.151) and make use of (3.148). We then obtain

$$\int_{-1}^{1} \frac{d}{dx}\left\{(1-x^2)\frac{du}{dx}\right\}P_j(x)\,dx$$

$$= -\left[u(x)(1-x^2)\frac{dP_j(x)}{dx}\Big|_{-1}^{1} - \int_{-1}^{1}\left\{u(x)\frac{d}{dx}\left\{(1-x^2)\frac{dP_j(x)}{dx}\right\}\right\}\right]$$

$$= 0 + \int_{-1}^{1} u(x)\{-j(j+1)P_j(x)\}\,dx$$

$$= -j(j+1)\int_{-1}^{1} u(x)P_j(x)\,dx = -j(j+1)F(j) \qquad (3.152)$$

after noting that the first term vanishes at $x = \mp 1$ and using the definition (3.146) for our transform $F(j)$.

Before giving an example illustrating the use of the finite Legendre transform, we present a typical application of the Legendre differential equation as it appears, for example, in the process of the method of separation of variables.

When we discuss potential distribution in three-dimensional space (free of charge) we know that the potential $v(x,y,z)$ inside the domain of interest D is governed by the Laplace equation

$$\nabla^2 v = \frac{\partial^2 v}{\partial x^2} + \frac{\partial^2 v}{\partial y^2} + \frac{\partial^2 v}{\partial z^2} = 0, \qquad (x,y,z) \in D \qquad (3.153)$$

However, the shape of the boundary β of the domain D (see Fig. 3.11) may necessitate the use of suitable coordinates to simplify the description of the boundary. This is essential if one is to apply separation of variables. For example, if we are to study the potential distribution outside a sphere of radius a and centered at the origin, it is necessary to

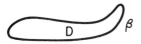

FIG. 3.11   A domain D and its boundary β.

use spherical coordinates (ρ, ϕ, θ) (see Fig. 3.12). The boundary as the surface of the sphere can be described simply by the (separable!) constant value of one of the spherical coordinates instead of the (nonseparable) Cartesian coordinates description of the same sphere: $x^2 + y^2 + z^2 = a^2$. This is especially accommodating to the method of separation of variables and allows boundary conditions to separate to conditions such that each of the independent variables is a constant. In this case, for example, a constant potential v_0 on the sphere surface $v(a, \phi, \theta) = v_0$ is a boundary condition associated with the constant value of $\rho = a$. This replaces (the nonseparable!) $v(x, y, \sqrt{a^2 - x^2 - y^2})$ if we are to stay with Cartesian coordinates. In spherical coordinates, (ρ, ϕ, θ), the Laplace equation (3.153) for $v(\rho, \phi, \theta)$, after tedious computations, becomes

$$\nabla^2 v \equiv \frac{1}{\rho^2}\left[\frac{\partial}{\partial \rho}\left(\rho^2 \frac{\partial v}{\partial \rho}\right) + \frac{1}{\sin\theta}\frac{\partial}{\partial \theta}\left(\sin\theta \frac{\partial v}{\partial \theta}\right) + \frac{1}{\sin^2\theta}\frac{\partial^2 v}{\partial \phi^2}\right] = 0,$$

$$0 \leq r < \infty, \ 0 < \theta < \pi, \ 0 < \phi < 2\pi \qquad (3.154)$$

In case the potential is independent of the angle ϕ as $v(\rho, \theta)$, and if we write $\sin^2\theta = 1 - \cos^2\theta = 1 - x^2$ where $x = \cos\theta$, we obtain

$$\frac{1}{\rho^2}\left[\frac{\partial}{\partial \rho}\left(\rho^2 \frac{\partial v}{\partial \rho}\right) + \frac{\partial}{\partial x}(1 - x^2)\frac{\partial v}{\partial x}\right] = 0, \qquad 0 \leq \rho < \infty, \ -1 < x < 1$$

$$(3.155)$$

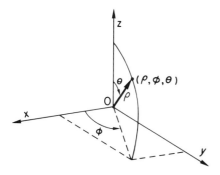

FIG. 3.12   Spherical coordinates (ρ, ϕ, θ).

It is now clear that the Legendre differential operator

$$\frac{d}{dx}(1-x^2)\frac{du}{dx}$$

of the Legendre equation (3.148) will appear after separation of the variables $v(\rho,x) = R(\rho)u(x)$. However, it is not yet clear how the arbitrary constant of the separation of variables became $-j(j+1)$ as in (3.148). Such a choice serves to satisfy the reasonable periodic condition $v(\rho, \cos 0) = v(\rho, \cos 2\pi)$ and it also terminates the infinite series representation of the solution of $u(x)$ to j terms, resulting in $P_j(x)$ as a finite sum power series solution to the singular Legendre differential equation (3.148). The derivation of the power series solution of the Legendre equation is given in Appendix A.5.2. With such a choice, we can easily show that $\rho^j P_j(\cos\theta)$ and $\rho^{-(j+1)}P_j(\cos\theta)$ are the two solutions of the partial differential equation (3.155), after recognizing that the resulting differential equation in $R(\rho)$ is of Cauchy–Euler type with solution of the form $R(\rho) = \rho^\gamma$ as we shall show next in Example 3.15.

We could continue this method of separation of variables to a conclusion. However, we shall instead illustrate (its equivalent!) the use of the finite Legendre transforms to solve for the potential distribution associated with a spherical boundary.

EXAMPLE 3.15:   Potential distribution inside a sphere

In this example, we will solve for the potential $v(\rho, \cos\theta)$ inside a unit sphere where the potential on the surface of the sphere is given as $v(1, \cos\theta) = f(\cos\theta)$.

We let $x = \cos\theta$ to use (3.155) for Laplace equation

$$\frac{1}{\rho^2}\left[\frac{\partial}{\partial\rho}\left(\rho^2\frac{\partial v}{\partial\rho}\right) + \frac{\partial}{\partial x}(1-x^2)\frac{\partial v}{\partial x}\right] = 0, \qquad 0 \le \rho < 1,\ -1 < x < 1$$

$$(E.1)$$

where the boundary condition becomes

$$v(1,x) = f(x), \qquad -1 < x < 1 \tag{E.2}$$

We let $V(\rho,j)$ and $F(j)$ be the finite Legendre transforms of $v(\rho,x)$ and $f(x)$, respectively, then take the Legendre transform (3.146) of (E.1) and use the algebraization property (3.150) to have

$$\frac{1}{\rho^2}\left[\frac{d}{d\rho}\left(\rho^2\frac{dV}{d\rho}(\rho,j)\right) - j(j+1)V(\rho,j)\right] = 0 \tag{E.3}$$

This equation, after performing the first derivatives, can be recognized as a Cauchy–Euler equation with solution of the form $R(\rho) = \rho^\gamma$ (see

Exercise 3.48) that results in two solutions $A(j)\rho^j$ and $B(j)\rho^{-(j+1)}$. We choose the first since it is bounded at $\rho = 0$ to suit our problem. Thus we can simply verify that

$$V(\rho,j) = \rho^j F(j) \tag{E.4}$$

is a solution to (E.3), where $A(j) = F(j)$ is obtained from the Legendre transform of the boundary condition (E.2),

$$l\{v(1,x)\} = V(1,j) = l\{f(x)\} = F(j) \tag{E.5}$$

To find the solution to the original problem (E.1), (E.2), we take the inverse Legendre transform (3.147) of $V(\rho,j)$ to obtain

$$v(\rho,\cos\theta) = \frac{1}{2}\sum_{j=0}^{\infty}(2j + 1)F(j)\rho^j P_j(\cos\theta) \tag{E.6}$$

This method, of course depends on the availability of the Legendre transform $F(j)$ in a closed form from, possibly, the existing tables of finite Legendre transforms. Otherwise, we may have to evaluate $F(j)$ numerically. In this respect, the discrete version of the Legendre transforms will be of value, which we shall discuss in Section 4.2.1D and illustrate with numerical results (for this example) in Example 4.9 of Section 4.2.3B.

3.5.2 Laguerre Transform

As we indicated earlier, the Laguerre transform $F(j)$ of the function $f(x)$ on $(0,\infty)$,

$$F(j) = \int_0^{\infty} e^{-x}L_j(x)f(x)dx \equiv L_G\{f\} \tag{3.156}$$

is different from the rest of the finite transforms in the sense that the kernel $L_j(x)$ is orthogonal on the semi-infinite interval $(0,\infty)$ as in (3.145), instead of on the usual finite interval of the typical Fourier (or orthogonal expansion) coefficients. Here we used L_G for the finite Laguerre transform operator to differentiate it from the L_j of the present Laguerre polynomial and the l of the above finite Legendre transforms operator in (3.146).

The simplest form for the (orthogonal) Laguerre polynomial $L_j(x)$ is obtained through its Rodrigues formula,

$$L_j(x) = \frac{e^x}{j!}\frac{d^j}{dx^j}(x^j e^{-x}) = j!\sum_{k=0}^{j}\frac{(-1)^k x^k}{(k!)^2(j-k)!} \tag{3.157}$$

The Laguerre polynomial $L_j(x)$ is a solution to the *Laguerre differential equation* (see Appendix A.5.3),

$$x\frac{d^2u}{dx^2} + (1-x)\frac{du}{dx} + ju = 0 \tag{3.158}$$

The set $\{L_j(x)\}$ is orthogonal on the semi-infinite interval with respect to the weight function $\rho(x) = e^{-x}$, that is,

$$\int_0^\infty e^{-x}L_n(x)L_m(x)dx = \begin{cases} 0, & n \neq m \\ 1, & n = m \end{cases}$$

The inverse Laguerre transform is

$$f(x) = \sum_{j=0}^\infty F(j)L_j(x) \equiv L_G^{-1}\{F\} \tag{3.159}$$

For a sufficiently smooth function $u(x)$ and with growth slower than e^x as $x \to \infty$, we can show by using two integrations by parts that the Laguerre transform (3.156) is compatible with the differential part of the Laguerre differential equation (3.158), i.e.,

$$L_G\left\{x\frac{d^2u}{dx^2} + (1-x)\frac{du}{dx}\right\} = -jU(j) \tag{3.160}$$

where $U(j) = L_G\{u(x)\}$.

To do the integration by parts, or the Sturm–Liouville analysis, for establishing (3.160) it is advisable to write (3.158) in its self-adjoint form:

$$\frac{d}{dx}\left[xe^{-x}\frac{dL_j}{dx}\right] = -je^{-x}L_j(x) \tag{3.161}$$

after multiplying (3.158), with $u(x) = L_j(x)$, by e^{-x}. The discrete version of this Laguerre transform will be discussed in Section 4.2.1C.

To do the integration by parts, or the Sturm–Liouville analysis, for establishing (3.160) it is advisable to write (3.158) in its self-adjoint form:

$$\frac{d}{dx}\left[xe^{-x}\frac{dL_j}{dx}\right] = -je^{-x}L_j(x) \tag{3.161}$$

after multiplying (3.158), with $u(x) = L_j(x)$, by e^{-x}. The discrete version of this Laguerre transform will be discussed in Section 4.2.1C.

3.5.3 Hermite Transforms

Another example of classical orthogonal polynomial transforms (3.142), (3.143) is the Hermite transform of the function $f(x)$, which is defined on the infinite interval $(-\infty, \infty)$,

$$F(j) = \int_{-\infty}^{\infty} e^{-x^2} H_j(x) f(x) \, dx \equiv h\{f\}$$
(3.162)

where $H_j(x) = p_j(x)$ of (3.143) is the Hermite polynomial, which is a solution of the Hermite equation (see Appendix A.5.3)

$$\frac{d^2u}{dx^2} - 2x\frac{du}{dx} + 2ju = 0$$
(3.163)

The simplest form for the Hermite polynomial $H_j(x)$ is obtained, like other classical orthogonal polynomials, through its Rodrigues formula

$$H_j(x) = (-1)^j e^{x^2} \frac{d^j}{dx^j}(e^{-x^2}) = j! \sum_{k=0}^{[j/2]} \frac{(-1)^k (2x)^{j-2k}}{k!(j-2k)!}$$
(3.164)

where $[j/2]$ is the greatest integer $\leq j/2$.

The set $\{H_j(x)\}$ is *orthogonal on the infinite interval* $(-\infty, \infty)$ with the weight function e^{-x^2}; i.e.,

$$\int_{-\infty}^{\infty} e^{-x^2} H_n(x) H_m(x) \, dx = \begin{cases} \sqrt{\pi} 2^n n!, & n = m \\ 0, & n \neq m \end{cases}$$
(3.165)

Attention should be paid to the different definitions of $H_j(x)$ in different references. Using the orthogonality property and the norm square in (3.165), we can show that the inverse Hermite transform of (3.162) is

$$f(x) = \sum_{j=0}^{\infty} \frac{2^{-j}}{\sqrt{\pi} j!} F(j) H_j(x) \equiv h^{-1}\{F\}$$
(3.166)

Following the same analysis as in the Laguerre transform, we can show, for sufficiently smooth functions $u(x)$ and with growth slower than e^{x^2} as $x \to \infty$, that the Hermite transform (3.162) is compatible with the differential part of the Hermite differential equation (3.163); i.e.,

$$h\left\{\frac{d^2u}{dx^2} - 2x\frac{du}{dx}\right\} = -2jU(j)$$
(3.167)

where $U(j) = h\{u(x)\}$.

3.5.4 Tchebychev Transforms

The Tchebychev transform is another one of the special classical orthogonal polynomial transforms (3.146), (3.147) with the Tchebychev polynomial

$T_j(x)$ as its kernel, and where $T_j(x)$ is a solution of the Tchebychev differential equation

$$(1-x^2)\frac{d^2u}{dx^2} - x\frac{du}{dx} + j^2u = 0 \tag{3.168}$$

The Tchebychev polynomials $T_j(x)$ have the following very useful relation to the trigonometric cosine function:

$$T_j(x) = \cos(j \cos^{-1}x) \tag{3.169}$$

which have been used to a great advantage in numerical analysis (see Exercise 3.54c).

Also, as expected, the set $\{T_j(x)\}$ is orthogonal on the interval $(-1, 1)$ with respect to the weight function $\rho(x) = (1-x^2)^{-1/2}$ so that

$$\int_{-1}^{1}(1-x^2)^{-1/2}T_n(x)T_m(x)\,dx = \begin{cases} \tfrac{1}{2}\pi, & m = n \neq 0 \\ 0, & m \neq n \\ \pi, & m = n = 0 \end{cases} \tag{3.170}$$

The finite Tchebychev transform of $f(x)$ on $(-1, 1)$ is

$$F(j) = \int_{-1}^{1}(1-x^2)^{-1/2}T_j(x)f(x)\,dx \equiv t\,\{f\} \tag{3.171}$$

and we can show through the orthogonality property (3.170) that its inverse is

$$f(x) = \frac{1}{\pi}F(0) + \frac{2}{\pi}\sum_{j=1}^{\infty}F(j)T_j(x) \equiv t^{-1}\{F\} \tag{3.172}$$

As we did in the case of the Legendre transform (3.146), we can either take this transform (3.171) as a special case of the Sturm–Liouville transform of Section 1.6 or use two integrations by parts to show that it is compatible with the differential operator part: $(1-x^2)d^2u/dx^2 - x\,du/dx$ of (3.168); i.e.,

$$t\left\{(1-x^2)\frac{d^2u}{dx^2} - x\frac{du}{dx}\right\}$$

$$= \int_{-1}^{1}(1-x^2)^{-1/2}\left[(1-x^2)\frac{d^2u}{dx^2} - x\frac{du}{dx}\right]T_j(x)\,dx$$

$$= -j^2F(j) \tag{3.173}$$

The applications of the above classical orthogonal polynomials transforms to solving boundary value problems are left for the exercises. The use of such finite transforms and the finite Hankel transform of the last section will be apparent in relation to the generalized sampling expansion of the following section.

The discrete versions (4.159) of the finite classical orthogonal polynomials transform of this section are introduced in Section 4.2. Their basic properties and applications are discussed briefly and then illustrated for the discrete Laguerre and Legendre transforms in Sections 4.2.1 and 4.2.3.

3.6 The Generalized Sampling Expansion

In Sections 2.4.2 and 3.3 we presented the sampling expansion (2.184) or (3.121) for the finite limit Fourier transform (2.138) as a representation of bandlimited signals. Here we will present a generalization of the sampling expansion for functions represented by general integral transforms with kernel $K(x,t)$, for example, as a solution of the Sturm–Liouville problem (see Section 1.5.2A).

Consider $f(t)$ as the following K-transform of $F(x)$:

$$f(t) = \int_I \rho(x)K(x,t)F(x)\,dx \tag{3.174}$$

where $F(x)$ is square integrable on the interval I and $\{K(x,t_n)\}$ is a complete orthogonal set on the interval I with respect to a weight function $\rho(x)$. The generalized (or Kramer–Weiss)[†] sampling theorem gives the following sampling expansion for $f(t)$ of (3.174).

$$f(t) = \lim_{N \to \infty} \sum_{|n| \leq N} f(t_n)S_n(t) \tag{3.175}$$

where

$$S_n(t) = S(t,t_n) = \frac{\int_I \rho(x)K(x,t)\overline{K(x,t_n)}\,dx}{\int_I \rho(x)|K(x,t_n)|^2\,dx} \tag{3.176}$$

is the *sampling function*. Here $\overline{K(x,t)}$ is the complex conjugate of $K(x,t)$. As in the proof of the sampling expansion (3.121), the simplest proof is to write the orthogonal expansion for $F(x)$ in (3.174) in terms of $\overline{K(x,t_n)}$;

[†] See Jerrri (1992) for some history of this important development as well as extensions and error analysis. Also, see Kramer (1959) and Weiss (1957).

i.e.,

$$F(x) = \sum_{n=1}^{\infty} c_n \overline{K(x,t_n)} \tag{3.177}$$

$$c_n = \frac{\int_I \rho(x)F(x)K(x,t_n)\,dx}{\int_I \rho(x)|K(x,t_n)|^2\,dx} = \frac{f(t_n)}{\int_I \rho(x)|K(x,t_n)|^2\,dx} \tag{3.178}$$

after using (3.174) to give the value of the integral as $f(t_n)$. Here $f(t_n)$ is easily recognized as the finite K-transform of $F(x)$, $x \in I$. Then multiply both sides of (3.177) by $\rho(x)K(x,t)$ and formally integrate term by term to obtain

$$\int_I \rho(x)K(x,t)F(x)\,dx = f(t) = \sum_{n=1}^{\infty} \frac{f(t_n)\int_I \rho(x)K(x,t)\overline{K(x,t_n)}\,dx}{\int_I \rho(x)|K(x,t_n)|^2\,dx}$$

$$= \sum_{n=1}^{\infty} f(t_n)S_n(t) \tag{3.175}–(3.176)$$

after using (3.174) for $f(t)$ and (3.176) for the sampling function $S_n(t)$, which we sometimes write $S(t,t_n)$. The term-by-term integration of the above orthogonal expansion series is justified according to Theorem 3.10, since $F(x)$ is square integrable on the interval I. We indicate that the same proof can be followed when $K(x,t)$ of (3.174) is expanded in terms of the same orthogonal functions $K(x,t_n)$. However, the shortest proof is to use Parseval's equality (3.65) (see Exercise 3.56a) for the integral in (3.174) with the Fourier coefficients c_n of (3.178) and $S_n(t)$ of (3.176) for $F(x)$ and $K(x,t)$, respectively. It is clear that the sampling expansion (2.184), (3.121) is a special case of (3.175), (3.176) corresponding to $K(x,t) = e^{ixt}$. For a comprehensive treatment of the generalized sampling theorem, and in particular its (recent) error analysis, see the author's Chapter 7 in Marks (1992) on the subject (Jerri, 1992). Also see Jerri (1977) for the first comprehensive coverage of all aspects of the sampling theorems.

As mentioned above, the conditions on the kernel $K(x,t)$ in (3.174) for this theorem are exhibited by the solutions of the Sturm–Liouville problem, which we will illustrate for the case of $K(x,t) = J_m(xt)$, the Bessel function of the first kind of order m. In this case the transform (3.174) becomes the following finite limit J_m-Hankel (or Bessel) transform:

$$f(t) = \int_0^1 xJ_m(xt)F(x)\,dx \tag{3.179}$$

The sampling functions $S_n(t)$ of (3.176) is

$$S_n(t) = S(t, t_{m,n}) = \frac{\int_0^1 x J_m(xt) J_m(x t_{m,n}) \, dx}{\int_0^1 x [J_m(xt)]^2 \, dx}$$

$$= \frac{2 t_{m,n} J_m(t)}{(t_{m,n}^2 - t^2) J_{m+1}(t_{m,n})} \qquad (3.180)$$

where the $\{t_{m,n} \equiv j_{m,n}\}$ are the zeros of the Bessel function J_m; that is, $J_m(t_{m,n}) = 0$, $n = 1, 2, \ldots$. Here familiar properties of the Bessel functions are used (see Exercise 3.56b,c and Appendix A.5) to evaluate the integrals of (3.180). The final sampling series (3.175) for the finite limit Hankel transform becomes

$$f(t) = \sum_{n=1}^{\infty} f(t_{m,n}) \frac{2 t_{m,n} J_m(t)}{(t_{m,n}^2 - t^2) J_{m+1}(t_{m,n})}, \qquad J_m(t_{m,n}) = 0, \ n = 1, 2, \ldots$$

$$(3.181)$$

We note here that the weight function $\rho(x) = x$ is introduced explicitly in (3.179) instead of being implicit in the product $K(x,t) F(x)$. Another illustration of this generalized sampling function $S_n(t)$ of (3.176) for other finite limit transform is the subject of Exercise 3.57.

A simple comparison of the sampling expansion of bandlimited functions (3.121) and the above Bessel-type sampling expansion (3.181) indicates a very basic and relevant difference in the form of their sampling functions. In the usual (or Fourier) sampling expansion (3.121), the sampling function

$$S_n(t) = S\left(t - \frac{n\pi}{a}\right) = \frac{\sin a(t - n\pi/a)}{a(t - n\pi/a)}$$

is time-invariant; that is, it depends on the difference $t - t_n = t - n\pi/a$ of the two times t and $t_n = n\pi/a$. This is a simple translation by $t_n = n\pi/a$ of the Fourier transform $(\sin at)/at$ of a gate function $(a/2) p_a(\omega)$,

$$p_a(\omega) = \begin{cases} 1, & |\omega| < a \\ 0, & |\omega| > a \end{cases} \qquad (3.182)$$

It is the result of multiplying $p_a(\omega)$ by $e^{-(in\pi/a)\omega}$ in the Fourier integral leading to the final sampling series (3.121). We stress that this important simple shift or translation property for such a sampling function, or other functions such as $f(t - \tau)$, is the result of multiplying their corresponding Fourier transform by a complex exponential function $e^{-i\omega t}$, which is of the same form as the kernel of the (Fourier) transform.

In contrast, we observe that the sampling function

$$S_n(t) = S(t, t_{m,n}) = \frac{2t_{m,n}J_m(t)}{(t_{m,n}^2 - t^2)J_{m+1}(t_{m,n})},$$

$$J_m(t_{m,n}) = 0, \quad n = 1, 2, \ldots \tag{3.180a}$$

of the Bessel sampling series (3.181) depends on t and t_n in a complicated way and is far from being time-invariant. However, this sampling function, as seen from the Bessel integral leading to it in (3.180), is also the result of multiplying the gate function $p_1(x)$ by $J_m(xt_{m,n})$ before taking its J_m-Hankel transform. But, contrasting the sampling function of (3.121), it does lack the well-known translation property that is associated with the exponential kernel-type transforms.

If we are to give a physical interpretation of these Bessel, or other general, sampling series expansions of functions, we can draw a parallel to the Fourier (and Laplace) transforms. This is in the sense that we can call $S(t, t_n)$ in (3.180a) a "generalized translation": $S(t\theta t_n)$ of $S(t)$ by $t = t_n$, where $S(t)$ is the J_m-Hankel transform of the gate function $p_1(x)$. For example, in the case of the J_0-Hankel transform we can show that $\mathcal{H}_0\{p_1(x)\} = S(t) = J_1(t)/t$, while its above generalized translation is

$$\mathcal{H}_0\{p_1(x)J_0(xt_{0,n})\} = S(t, t_{0,n}) = \frac{2t_{0,n}J_0(t)}{(t_{0,n}^2 - t^2)J_1(t_{0,n})} \equiv S(t\theta t_{0,n})$$

The generalized translation, for *nonexponential* kernel transforms, along with an associated convolution theorem and other basic properties are the subjects of the following section.

3.6.1 Generalized Translation and Convolution Products

Consider the κ-transform

$$f(t) = \int \rho(\omega)K(\omega, t)F(\omega)\,d\omega \equiv \kappa\{f\} \tag{3.183}$$

with its Fourier-type (symmetric) inverse

$$F(\omega) = \int \rho(t)\overline{K(\omega, t)}f(t)\,dt \equiv \kappa^{-1}\{f\} \tag{3.184}$$

The limits of integration here are left for the specific integral transform. We may mention here that our most familiar examples of such symmetric transforms are those where the kernel is a function of the product ωt, that is, $K(\omega, t) = K(\omega t)$. For this special case we have the Fourier and J_m-Hankel transforms with kernels $K(\omega t)$ as $e^{i\omega t}$ and $J_m(\omega t)$, respectively.

A. Generalized Translation

As illustrated above, for the generalized sampling expansion (3.181), when we deal with transforms of nonexponential kernels, we don't expect the usual translation or "shift" property that we are so accustomed to in dealing with the Laplace and Fourier transforms. This shift property was very important in defining the convolution products that are essential for developing the convolution Theorems 2.11 and 2.18 in (2.35) and (2.127) for the Laplace and Fourier transforms, respectively. Thus, in considering a convolution product for such transforms (3.183), (3.184), which are, in general, not with exponential kernels, and to give the first physical interpretation of the generalized sampling expansion (3.175) or (3.181), it was necessary to introduce a compatible transformation that we called a *"generalized translation"* (Jerri, 1972). The τ-generalized translation $f(t\theta\tau)$ of $f(t)$ is defined for such transforms (3.183), (3.184) by

$$f(t\theta\tau) \equiv \int \rho(\omega)F(\omega)K(\omega,t)\overline{K(\omega,\tau)}\,d\omega \qquad (3.185)$$

This definition is not surprising, since the translation for the Fourier transform is also the result of multiplying the transformed function by an exponential function which is of the same type as the transform kernel. We have already indicated that the form $S(t,t_n)$, of the sampling function (3.180) in the Bessel sampling series (3.181), represents a generalized translation which can now be written as $S(t\theta t_n)$. To illustrate this idea, we can show that the J_0-Hankel transform of the gate function $p_a(\omega)$ is $S(t) = aJ_1(at)/t$. To find the τ-generalized translation $S(t,\tau) \equiv S(t\theta\tau) = aJ_1(a(t\theta\tau))/t\theta\tau$ of this function $S(t)$, we compute the J_0-Hankel transform of $J_0(\omega\tau)p_a(\omega)$,

$$\frac{aJ_1(a(t\theta\tau))}{t\theta\tau} = \int_0^a \omega J_0(\omega t)J_0(\omega\tau)\,d\omega$$

$$= \frac{atJ_1(at)J_0(a\tau) - a\tau J_1(a\tau)J_0(at)}{t^2 - \tau^2} \qquad (3.185a)$$

$$\equiv S(t\theta\tau)$$

as can be found with the aid of the Bessel function properties in Appendix A.5 (see Exercises A.23—A.25 in Appendix A.5 and Exercise 3.56b).

B. Generalized Convolution Product

Let $f(t)$ and $g(t)$ be the integral transforms of $F(\omega)$ and $G(\omega)$, respectively; we define the *convolution product* $(f * g)(t)$ of f and g

by

$$(f * g)(t) = \int \rho(\tau) g(\tau) f(t\theta\tau) d\tau = \int \rho(\omega) F(\omega) G(\omega) K(\omega, t) d\omega$$

(3.186)

It is clear that this convolution product is commutative, $f * g = g * f$, and it can be shown to be associative too, that is, $f * (g * h) = (f * g) * h$.

It should be easy now to state formally a *convolution theorem* for the general transforms (3.183), (3.184),

$$\kappa \{f * g\}(t) = F(\omega) G(\omega)$$

(3.187)

and we leave its proof as an exercise (see Exercise 3.61c).

C. Generalized Hill Functions

Now we are in a position to define the generalized hill function $\psi_{R+1}(\omega)$, paralleling that of $\phi_{R+1}(\omega)$ in (2.190), to be the R-fold convolution product (3.186) of the gate function $\psi_1(\omega) = p_I(\omega)$ of (3.182) so that

$$\psi_{R+1}(\omega) = (\psi_1 * \overset{R}{\cdots} * \psi_1)(\omega) = \int \rho(x) \psi_1(x) \psi_R(\omega \theta x) dx$$

$$= \int \rho(t) [\xi_1(t)]^{R+1} K(t, \omega) dt$$

(3.188)

where $\xi_1(t) \equiv \xi_1(I, t)$ is the κ-transform of $\psi_1(\omega)$, which is bandlimited to I.

One should note here that the general translation for constructing the new hill function (3.188) is not easy to perform; however, there is a method for computing the regular hill functions $\phi_{R+1}(\omega)$ (B-splines) of (2.190), which can be extended easily to express the ψ_{R+1}, defined on (a, b), as a series in the orthogonal set $\{K(\omega, t_n)\}$. As seen here,

$$\psi_{R+1}(\omega) = \sum \frac{[\xi(t_n)]^{R+1} K(\omega, t_n)}{\|K(\cdot, t_n)\|_2^2}$$

(3.189)

Here $\|K(\cdot, t_n)\|^2 = \int \rho(\omega) |K(\omega, t_n)|^2 d\omega$. A close look at the factor $[\xi(t_n)]^{R+1}$ inside the series would indicate its role in making the series self-truncating. For example, in the case of the J_0-Hankel transform,

$$\xi_1(t_n) = \frac{J_1(t_n)}{t_n}, \qquad t_n = j_{0,n}; \text{ the zeros of } J_0(x)$$

As we indicated in Section 2.4.3, the usual hill function $\phi_{R+1}(a(R + 1), \omega)$ vanishes identically outside the wider interval $(-a(R + 1), a(R + 1))$ instead of $(-a, a)$ of the gate function $\phi_1(a, \omega)$. Parallel with this, it is reasonable to expect that the bandlimit J, associated with $\psi_{R+1}(\omega)$, is larger than I of $\psi_1(\omega)$, which should be investigated for each individual transform. In the meantime, we write $J(I,R)$ to indicate this dependence, where $J = (R + 1)I$ for the Fourier transform, and it also turns out to be the case for the J_0-Hankel transform. Illustrations of this and the preceding concepts will be for the J_0-Hankel (Bessel) transform,

$$H_0(\omega) = \int_0^\infty th_0(t)J_0(t\omega)\,dt \tag{3.190}$$

and its inverse $h_0(t)$,

$$h_0(t) = \int_0^\infty \omega H_0(\omega)J_0(\omega t)\,d\omega \tag{3.191}$$

which we shall leave for the exercises; see Exercise 3.62 for the domain $(0, 2a)$ of $\psi_2(2a, \omega)$ compared to $(0, a)$ for $\psi_1(a, \omega)$.

Another important property is that of the discretization of the convolution product (3.186).

D. Discretization of the Convolution Product

The discretization of the convolution product (3.186) is valid when f and g are bandlimited to $(0,a)$, that is, $F(\omega) = G(\omega) \equiv 0$, $|\omega| > a$; then

$$\int_0^\infty \rho(\tau)f(\tau)g(t\theta\tau)\,d\tau = \sum_n \frac{f_a(t_n)g_a(t\theta t_n)}{\|K(\cdot,t_n)\|_2^2}$$

$$= \int_0^a \rho(\omega)F(\omega)G(\omega)K(\omega,t)\,d\omega. \tag{3.192}$$

The proof parallels that of the generalized sampling theorem (3.174), (3.176), which we leave for an exercise (see Exercise 3.63). The special case of $K(\omega,t) = J_0(\omega t)$ and $G(\omega) = \psi_1(\omega) = p_a(\omega)$ gives the Bessel sampling expansion for $f(t)$,

$$f(t) = \sum_{n=1}^\infty f(t_{0,n}) \frac{aJ_1(t\theta t_{0,n})}{t\theta t_{0,n}} \frac{1}{(a^2/2)J_1^2(j_{0,n})}$$

$$= \sum_{n=1}^\infty f(t_{0,n}) \frac{2t_{0,n}J_0(at)}{a(t_{0,n}^2 - t^2)J_1(j_{0,n})} \tag{3.193}$$

as a particular case of (3.181) with $m = 0$.

3.6.2 Impulse Train for Bessel Orthogonal Series Expansion for a (New) Bessel-Type Poisson Summation Formula

In Section 2.4.2 we saw how the impulse train (2.156) in Fig. 2.7 was so fundamental in the analysis of systems and signals. In particular, this train of equally spaced (or periodic) Dirac delta functions (impulses) is useful for deriving the samples or pulse train (2.157) in Fig. 2.8 and also the impulse response in (2.169) of the system. In addition, with the aid of such an impulse train, we were able to derive the sampling series expansion (2.184) for bandlimited functions, even before we covered the Fourier series expansion in Section 3.3.1. The latter is the usual mode of the simplest direct method of proving this sampling expansion, as we have already shown in deriving the same result as (3.121).

We mention here that in the sequel we may refer to this impulse train as a pulse train (2.156), where a pulse is meant to describe a Dirac delta function (impulse) or its generalization in (3.195) for other integral transforms besides the usual Fourier transform.

In this section we developed the generalized sampling expansion (3.175), (3.176), associated with transform kernels that are not necessarily trigonometric functions of the Fourier transform. This means that we will not expect the familiar shift property for these transforms and, more importantly, the periodicity of their associated orthogonal series expansion. The derivation of the generalized sampling series, using orthogonal series expansion, does not depend on the periodicity; however, any attempt at a physical interpretation of this series will! This situation was faced about two decades ago, when the above "generalized translation" (3.185) concept had to be introduced as the analog of the simple shift (or translation) of the trigonometric kernels of the Fourier transforms.

In Chapter 4 we will discuss in detail the derivation and computations of discrete Fourier transforms (DFTs), with emphasis on the errors incurred in their approximating the (trigonometric) Fourier series of this chapter and the Fourier integrals of the last chapter. Such error analysis must be done, since it is the form of the (finite sum of samples) discrete transforms that is eligible for an efficient computation via the fast Fourier transform (FFT) algorithm. In other words, the FFT is no more than a fast algorithm for computing the discrete Fourier transforms of Chapter 4. A very essential tool of this Fourier analysis approximation, which exposes one of the main errors, namely the *aliasing error*, is the *Poisson summation formula* (4.58). This important formula will be derived with the (essential) aid of the (Fourier-type) impulse train (2.156). Such a derivation depends entirely on the concept of the *periodic extension* of the (Fourier expanded) functions considered, as we shall see in (4.179)–(4.184), where we must use the (periodic) impulse train in (2.156).

This mode of derivation should make it clear that the well-known Poisson sum formula (4.58) is exclusively for (trigonometric) Fourier kernel transforms and orthogonal series. Hence, if we are to attempt a discrete Bessel (Hankel) transform, for example, we shall lack this necessary periodicity property, and in turn the familiar impulse train (2.156) and the well-known Poisson sum formula (4.58).

In Section 4.3 we will present preliminaries that might help in attempting to establish a discrete Hankel transform, which, unfortunately, does not exist at present. Our attempt starts with establishing analogs (or generalizations) of the above necessary tools of analysis, such as a new impulse (or pulse!) train, and a Poisson summation formula associated with the Hankel transform and the Bessel orthogonal series expansion. These two new tools, which we shall present and derive, respectively, in this section and Section 4.3, are fundamental to any attempt at a discrete Hankel transform pair. This will become abundantly clear after we see the derivation in (4.179)–(4.184) of the familiar Poisson formula (4.58) in Section 4.1, which will also be redone at the start of Section 4.3.

Next we will present the impulse train associated with the J_0-Bessel orthogonal series expansion as shown in Fig. 3.13, which will be used for deriving the (new) *Bessel-type Poisson summation formula* (4.190) in Section 4.3. To further clarify the method of this derivation, we will do it after presenting two methods for deriving the Poisson sum formula (4.58). The first method is clearly dependent on the periodicity of the Fourier series, which offers no hope for our development. The second uses the

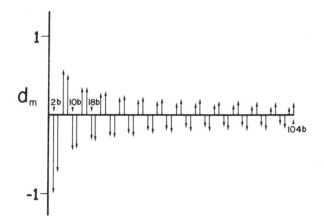

FIG. 3.13   The *impulse train* of the J_0-Bessel (Fourier) series, its spacings at $2mb$, and decreasing amplitudes $|d_m|$, $m = 0, 1, 2, \ldots, 52$ (From Jerri, 1988. Courtesy of Journal of Applicable Analysis.)

pulse train (2.156), which gives a glimpse of hope, if we are to present a generalized version of the pulse train (2.156), which is what we shall do next.

A. Impulse Train for Bessel Orthogonal Series Expansion

We first introduce the "generalized delta function" $\delta(\omega\theta x)$ for the general transforms (3.183), (3.184),

$$\int \rho(t)K(t,\omega)\overline{K(t,x)}\,dt \equiv \delta(\omega\theta x) \tag{3.194}$$

(and where we may refer to $\delta(\omega)$ as an impulse or a pulse).

It is to be noted that because of the weight function $\rho(t)$ used in the definition of the above general transforms, to use this delta function (3.194) to locate the pulses in a pulse train, similar to what is done with $\delta(\omega - x)$ of the special case of Fourier transforms in (2.156), it should be (formally) written as $\delta(\omega\theta x) = \delta(\omega - x)/\rho(\omega)$; i.e.,

$$\int \rho(\omega)F(\omega)\delta(\omega\theta x)\,d\omega = \int \rho(\omega)F(\omega)\frac{\delta(\omega - x)}{\rho(\omega)}\,d\omega = F(x) \tag{3.195}$$

For this preliminary treatment we may sometimes call the analog of the impulse train in (2.156) a pulse train. Next, we introduce the pulse train in a manner similar to that of the Fourier trigonometric series (3.9) or (3.10), but with the sense of the present general translation and, of course, without the usual periodicity of the (trigonometric) Fourier series.

Consider the orthogonal set $\{K(\omega,t_n)\}$ on the interval I; we define the *pulse (or impulse) train*, via the following divergent series at $\{c_m\}$, as

$$X(\omega) = \sum_m d_m\delta(\omega\theta c_m) \equiv \sum_n \frac{K(\omega,t_n)}{\|K(\omega,t_n)\|_2^2} \tag{3.196}$$

where

$$\|K(\omega,t_n)\|_2^2 = \int \rho(\omega)|K(\omega,t_n)|^2\,d\omega \tag{3.197}$$

and where the locations $\{c_m\}$ for the pulses are to be determined for the particular transform of interest. As we shall soon illustrate for the case of the Fourier–Bessel series, and in contrast to the case of the usual trigonometric series, d_m in (3.196) may change sign and in general $|d_m|$ is not uniform. The very important special case of the impulse train of the (usual) trigonometric Fourier analysis is defined via the following divergent Fourier series:

$$\sum_{n=-\infty}^{\infty} \delta(t + nT) = \frac{1}{T}\sum_{n=-\infty}^{\infty} e^{(i2\pi n/T)t} \tag{3.198}$$

in the t-space. A symmetric one, of course, can be written in the ω-frequency space.

In the case of the J_0-Hankel transform, we have the following formal divergent Fourier–Bessel series expansion on $(0,b)$, whose natural (nonperiodic) extension defines an impulse train, $X(\omega)$:

$$X(\omega) = \sum_{m=0}^{\infty} d_m \delta(\omega\theta2mb) = \frac{2}{b^2} \sum_{n=1}^{\infty} \frac{J_0(\omega j_{0,n}/b)}{J_1^2(j_{0,n})}, 0 < \omega < \infty,$$

$$J_0(j_{0,n}) = 0,^\dagger \ n = 1,2,\ldots \tag{3.199}$$

The location of the impulses $\{\omega_m = c_m = 2mb\}$ was verified numerically and can also be supported by using an asymptotic expansion (Watson, 1966, p. 584) with $|d_m|$ decreasing. It is also shown (or verified) that $d_0 = 1$; $|d_1| = 1$; $d_1, d_2, d_5, d_6, \ldots$ are negative; and $d_3, d_4, d_7, d_8, \ldots$ are positive, as illustrated in Fig. 3.13.

As we remarked earlier, this impulse train (3.199) can now be written (formally) in terms of the usual (Fourier-type) delta functions as

$$X(\omega) = \sum_{m=0}^{\infty} d_m \delta(\omega\theta2mb) = d_0 \frac{\delta(\omega)}{\omega} + \sum_{m=1}^{\infty} d_m \frac{\delta(\omega - 2mb)}{\omega}$$

$$= \frac{2}{b^2} \sum_{n=1}^{\infty} \frac{J_0(\omega j_{0,n}/b)}{J_1^2(j_{0,n})} \tag{3.200}$$

Considering the integral of the Fourier coefficients on the right-hand side of (3.200) as 1, we find

$$\int_0^b \omega X(\omega) J_0\left(\frac{\omega j_{0,n}}{b}\right) d\omega = d_0 \int_0^b \omega \frac{\delta(\omega)}{\omega} J_0\left(\frac{\omega j_{0,n}}{b}\right) d\omega = d_0 = 1$$

necessitates assigning $d_0 = 1$, after recognizing that the other impulses with $d_1, d_2, d_3, \ldots$ fall outside the interval of integration $(0,b)$.

$$X(\omega) = \frac{\delta(\omega)}{\omega} + \sum_{m=1}^{\infty} d_m \frac{\delta(\omega - 2mb)}{\omega} = \frac{2}{b^2} \sum_{n=1}^{\infty} \frac{J_0(\omega j_{0,n}/b)}{J_1^2(j_{0,n})} \tag{3.201}$$

If we formally take the Hankel transform of both sides of (3.201) we have

$$x(t) = \int_0^\infty \omega \frac{\delta(\omega)}{\omega} J_0(\omega t) d\omega + \sum_{m=1}^{\infty} d_m \int_0^\infty \omega \frac{\delta(\omega - 2mb)}{\omega} J_0(\omega t) d\omega$$

$\dagger$ Twenty zeros $\{j_{m,n}\}_{n=1}^{20}$ of $J_m(x)$ for $m = 0, 1, 2, 3, 4, 5$ are listed in Table A.1 of Appendix A.5.

$$= \frac{2}{b^2} \sum_{n=1}^{\infty} \frac{\int_0^{\infty} \omega J_0(\omega t) J_0(\omega j_{0,n}/b) \, d\omega}{J_1^2(j_{0,n})}$$

$$x(t) = 1 + \sum_{m=1}^{\infty} d_m J_0(2mbt) = \frac{2}{b^2} \sum_{n=1}^{\infty} \frac{\delta(t \theta j_{0,n}/b)}{J_1^2(j_{0,n})} \tag{3.202}$$

That is, the result is *another impulse train* in the t-space, where the pulses are located at $\{t_n = j_{0,n}/b\}$. We may remark again that for the purpose of locating these pulses we write $\delta(t \theta j_{0,n}/b) = (1/t)\delta(t - j_{0,n}/b)$. Also, if we accept the numerical or asymptotic result that d_1, d_2, d_5, d_6, ... are negative while d_3, d_4, d_7, d_8 ... are positive, we may use (3.202) at $t = 0$, where the right-hand side vanishes, to have a basic relation between these d_n,

$$1 - [|d_1| + |d_2| + |d_5| + |d_6| + \cdots] + d_3 + d_4 + d_7 + d_8 + \cdots = 0 \tag{3.203}$$

It was verified numerically, to very good accuracy, that this equality (3.203) is satisfied when all these terms $|d_m|$ are normalized in terms of $|d_1| = 1$. Figure 3.13 illustrates the locations $\{2mb\}$ and the variation of d_m for $m = 1$ to 52, where we used 40 terms in the series of the right-hand side of (3.201) to generate them. In the same figure we note the relative constancy of the normalized pulses $|d_m/d_1|$, $m \neq 0$, as m becomes large. This means that we can write the impulse train (3.201) in the ω-space as

$$X(\omega) = \frac{\delta(\omega)}{\omega} - \frac{\delta(\omega - 2b)}{\omega} + \sum_{m=2}^{\infty} d_m \frac{\delta(\omega - 2mb)}{\omega} = \frac{2}{b^2} \sum_{n=1}^{\infty} \frac{J_0(\omega j_{0,n}/b)}{(j_{0,n})} \tag{3.204}$$

For future reference we will use $X(\omega)$ for the impulse train of the left-hand side of equation (3.204), while $X_s(\omega)$ will stand for its above Bessel series representation on the right of (3.204).

We may note here how the pulses of $X(\omega)$ repeat at $\{2mb\}$ with equal spacing of $2b$, while the pulses of $x(t)$ in (3.202) repeat at $\{j_{0,n}/b\}$, which are not equally spaced, except asymptotically where $j_{0,n+1} - j_{0,n} \sim \pi$ for large n. This would also mean that, in contrast to the symmetric and equal spacing of the samples of Fourier series in both ω and t, we have here two different spacings for the samples; hence we should expect two versions of the extension of the Poisson sum formula (4.58) for the Hankel transforms. In such a development we shall mainly concentrate on the first version with equal spacing starting with the impulse train $X(\omega)$ in (3.204).

With the present development of the (*new*) *impulse train of the Bessel series* (3.204) as shown in Fig. 3.13, we will wait until Section 4.3.1 to develop the *Poisson summation formula* (4.190) *associated with the Bessel–Fourier series and the Hankel transforms*. We do so after developing the known Poisson sum formula (4.58) of the Fourier series and transforms in order to draw its parallel in the derivation of the new Bessel-type Poisson sum formula (4.190).

3.7[†] A Remark on the Transform Methods and Nonlinear Problems

---

Integral transform methods are used mainly, or exclusively, for solving linear problems. This is evident as in almost all texts on integral transforms, there is not even a reference to nonlinear problems. An exception was a brief mention of one example in Sneddon's old book on Fourier Transforms (1951) which was then deleted in his later treatise on the use of integral transforms (1972). Another exception was Pipe's (1965) treatment of nonlinear initial value problems of mechanical vibrations which was exclusively Laplace transform method, combined with iterations that depended entirely on an exhaustive and specially prepared Laplace transforms tables in order to come up with an "exact" solution. We note that this method, and especially its exhaustive special Laplace transform tables, is limited to the nonhomogeneous term of the differential equation being a trigonometric one like $F(t) = c_1 \cos \omega_1 t + c_2 \sin \omega_2 t$. Thus a little change in $F(t)$ like $F(t) = J_0(t)$ would render such special Laplace tables useless! As an example of some particular attempt of the transform method at nonlinear problems, we present the following nonlinear initial value problem with a cosine driving force. In this example we will illustrate the basic idea behind such an "iterative operational" method. This is followed by a brief remark and references concerning our recent attempt to modify known (numerical) iterative methods for solving nonlinear problems in which the linear part of the differential operator is compatible with transforms such as those discussed in this book.

EXAMPLE 3.16: A Nonlinear Initial Value Problems

Consider the following initial value problem in $y(t)$:

$$\frac{d^2 y}{dt^2} + k^2 y + h y^2 = A \cos \omega t \tag{E.1}$$

[†] Optional section.

$$y(0) = B \tag{E.2}$$

$$y'(0) = 0 \tag{E.3}$$

which describes the displacement $y(t)$ with the restoring force $-(k^2y + hy^2)$, and a driving force of the trigonometric form $F(t) = a\cos\omega t$. Pipe's "iterative operational" method starts by assuming the solution in the form

$$y(t) = y_0(t) + y_1(t) + y_2(t) + \cdots \tag{E.5}$$

whose Laplace transform is

$$Y(s) = Y_0(s) + Y_1(s) + Y_2(s) + \cdots \tag{E.6}$$

where $y_0(t)$ is the first approximation, which is taken as the solution of the corresponding linear problem, i.e., with $h = 0$ in (E.1):

$$\frac{d^2y_0}{dt^2} + k^2y_0 = A\cos\omega t \tag{E.1}'$$

$$y_0(0) = B \tag{E.2}'$$

$$y_0'(0) = 0 \tag{E.3}'$$

Let us first use the Laplace transform to find $y_0(t)$; then we will return to the nonlinear problem (E.1)–(E.3) in an attempt to find $y_1(t)$ of (E.5), the second term in the expansion for the final solution $y(t)$.

To find $y_0(t)$, we apply the Laplace transform on (E.1)$'$, using the appropriate pairs that we developed in Section 2.1.1, to have

$$s^2Y_0(s) - sy_0(0) - y_0'(0) + k^2Y_0(s) = \frac{As}{s^2 + \omega^2} \tag{E.7}$$

$$[s^2 + k^2]Y_0(s) = sB + \frac{As}{s^2 + \omega^2} \tag{E.8}$$

after using (2.19) and (E.2), (E.3) for $\mathcal{L}\{d^2y/dt^2\}$, and (2.14) for $\mathcal{L}\{\cos\omega t\} = s/(s^2 + \omega^2)$.

$$Y_0(s) = \frac{Bs}{s^2 + k^2} + \frac{As}{(s^2 + \omega^2)(s^2 + k^2)} \tag{E.9}$$

If we use partial fractions on the second term of (E.9) we have

$$Y_0(s) = \frac{Bs}{s^2 + k^2} + \frac{As}{(\omega^2 - k^2)(s^2 + k^2)} + \frac{As}{(k^2 - \omega^2)(s^2 + \omega^2)} \tag{E.10}$$

Now we can use the Laplace pair (2.14) to find the Laplace inverse of $Y_0(s)$,

$$y_0(t) = B\cos kt + \frac{A}{\omega^2 - k^2}\cos kt + \frac{A}{k^2 - \omega^2}\cos\omega t,$$

$$= \left[B + \frac{A}{\omega^2 - k^2} \right] \cos kt + \frac{A}{k^2 - \omega^2} \cos \omega t \tag{E.11}$$

With (E.9) as the Laplace transform of the linearized problem, we will return to take the Laplace transform of the nonlinear problem (E.1)–(E.3), where, for the moment, we leave the Laplace transform of the quadratic term y^2 as $\mathcal{L}\{y^2\}$,

$$Y(s) = \frac{Bs}{s^2 + k^2} + \frac{As}{(s^2 + \omega^2)(s^2 + k^2)} - \frac{h}{s^2 + k^2} \mathcal{L}\{y^2(t)\}$$

$$Y(s) = Y_0(s) - \frac{h}{s^2 + k^2} \mathcal{L}\{y^2\} \tag{E.12}$$

The method now moves to find an approximation for $y^2(t)$ and then takes its Laplace transform for (E.12). From the expansion (E.5), we have

$$y^2(t) = [y_0(t) + y_1(t) + y_2(t) + \cdots]^2$$

$$= y_0^2 + 2y_0y_1 + y_1^2 + 2(y_0 + y_1)y_2 + y_2^2 + \cdots \tag{E.13}$$

where we have obtained only its first term $y_0^2(t)$ as in (E.11). If we substitute this in (E.12) and compare the result with (E.6), we have

$$Y(s) = Y_0(s) - \frac{h}{s^2 + k^2} \mathcal{L}\left[y_0^2 + 2y_0y_1 + y_1^2 + 2(y_0 + y_1)y_2 + \cdots \right]$$

$$= Y_0(s) + Y_1(s) + Y_2(s) + \cdots \tag{E.14}$$

We now assign

$$Y_1(s) = -\frac{h}{s^2 + k^2} \mathcal{L}\{y_0^2\} \tag{E.15}$$

$$Y_2(s) = -\frac{h}{s^2 + k^2} \mathcal{L}\{2y_0y_1 + y_1^2\} \tag{E.16}$$

$$Y_3(s) = -\frac{h}{s^2 + k^2} \mathcal{L}\{2(y_0 + y_1)y_2\} \tag{E.17}$$

$$\vdots$$

In (E.15) we may say that, in principle, $Y_1(s)$ is known, since we know $y_0(t)$ in (E.11) (and hence $y_0^2(t)$) as at least three terms. However, the problem is much more involved, as we are supposed to find $\mathcal{L}\{y_0^2(t)\}$, which will involve more new terms from squaring $y_0(t)$ in (E.11). The result now is to be Laplace-transformed and then multiplied by $-h/(s^2 + k^2)$ in (E.15) to find $Y_1(s)$, and finally the inverse Laplace transform of this is computed to obtain $y_1(t)$. This is all for only the second term of the approximation (E.5) for the solution $y(t)$. Such involved Laplace transform pairs need specially prepared tables to cover mainly problems with trigonometric-type driving forces like $F(t) = c_1 \cos \omega_1 t + c_2 \sin \omega_2 t$. We may note again that a little change in $F(t)$ like $F(t) = J_0(t)$ would render many such special tables useless.

We will leave the details of finding $\mathcal{L}\{y_0^2\}$ and then $y_1(t) = \mathcal{L}^{-1}\{Y_1(s)\}$ for the interested reader to consult (Pipes, 1965, pp. 7–12). In principle, the method continues, as once we have $y_0(t)$ and $y_1(t)$ we can use them in (E.16) to find $Y_2(s)$, and hence $y_2(t) = \mathcal{L}^{-1}\{Y_2(s)\}$ as the third term in (E.5) for approximating $y(t)$. This process can be continued for (E.17) to find $Y_3(s)$ and then $y_3(t)$. Even for the three terms y_0, y_1 and y_2 of the approximation (E.5), it is easy to imagine the involvement in the number of terms and the range of Laplace transform pairs needed, especially when the driving force has a few more terms than the one $\cos \omega t$ term (E.1) that we used in this example.

The rest of the typical transform methods involving the transforms we discuss in this book are basically numerical iterative methods. Such an iterative method starts with a first approximation for the nonlinear problem, which is usually the solution to the linearized problem, and then iterates to find the higher-order approximations. Of fundamental importance to such an iterative method is its convergence to the solution. In that regard, we have attempted (1983a, 1984 1987a,b, 1991) a modified iteration scheme that improved on the conditions for such convergence. We shall make the initial basic steps of this method as exercises with the detailed results for their answers found in the references given (see Exercises 3.67 to 3.72). For a general reference to this *modified iterative method* and its various applications, see our recent tutorial paper on the subject (Jerri, 1991, 1992).

Relevant References to Chapter 3

The basic references used in this chapter are Weinberger (1965), Budak and Fomin (1973), Edwards (1967a), Alexits (1961), Powers (1987), Sneddon (1972), Churchill (1972), Tranter (1951), Coddington and Levinson (1955), Papoulis (1968, 1977), Davies (1985), Jerri (1977a, 1983, 1984, 1987a,b, 1988, 1991, 1992). See also the references to the varied exercises and those in the bibliography.

Exercises

Section 3.1   Fourier (Trigonometric) Series and General Orthogonal Expansion

3.1. (a) Use the orthogonality property (3.4) to derive the expression for the Fourier coefficients $c_n = F(\lambda_n)/N(\lambda_n)$ as in (3.7), where $F(\lambda_n)$ is the finite transform of $f(x)$ as given in (3.6).

Hint: See (1.83) for the formal derivation of (3.7), and (3.54) for the completely justified derivation in (3.58) to (3.60).

(b) Prove that the set $\{e^{in\pi x/l}\}_{n=-\infty}^{\infty}$ is orthogonal on $(-l,l)$; then use this property to derive the Fourier exponential series coefficients c_n in (3.9).

(c) Use the method in part (b) to derive the Fourier coefficient c_n in (3.9a) for the form (3.8a) of the Fourier exponential series.

3.2. (a) Show that the Fourier exponential series (3.8) (or (3.8a)) represents a periodic function $f(x)$ with period $2l$.

(b) Show that if $f(x)$ is periodic with period p, then it is periodic with period mp; that is, any m integer multiple of the basic period p.

Hint: $f(x) = f(x+p)$ for all x, as periodic with period p. Let $x = x' + p$, $f(x) = f(x'+p) = f(x') = f(x+p) = f(x'+2p)$, $f(x') = f(x'+2p)$, periodic with period $2p$, and so on.

3.3. (a) In Example 3.1c, verify that the set of solutions in (E.12) satisfies the boundary value problem (E.9)–(E.11).

(b) Solve the following two boundary value problems:

(i) $\dfrac{d^2y}{dx^2} + \lambda^2 y = 0, \qquad 0 < x < b$ (E.1)

$y(0) = 0$ (E.2)

$y'(b) = 0$ (E.3)

(ii) $\dfrac{d^2y}{dx^2} + \lambda^2 y = 0, \qquad 0 < x < b$ (E.4)

$y'(0) = 0$ (E.5)

$y'(b) + hy(b) = 0$ (E.6)

Hint: See Exercise A.4 in Appendix A for some other variations of such boundary value problems.

ANS. (i) $y_n(x) = A \sin(n + \frac{1}{2})\pi x/b$

(ii) $y_n(x) = A \cos \lambda_n x$, where $\{\lambda_n\}$ are the solutions of $\tan \lambda_n b = h/\lambda_n$.

3.4. Consider each of the following functions with their indicated interval (a,b) and period p:

(i) $f(x) = x$, $\quad -1 < x < 1, p = 2$

(ii) $f(x) = \begin{cases} 1, & -1 < x < 0, \\ \sin x & 0 < x < 1, \end{cases} \quad p = 2$

(iii) $f(x) = \sin x$, $\quad -1 < x < 1, p = 2$

(iv) $f(x) = x^2$, $\quad -2 < x < 2, p = 4$

(v) $f(x) = \begin{cases} x, & -\pi < x < 0, \\ 1, & 0 < x < \pi, \end{cases} \quad p = 2\pi$

(vi) $f(x) = |x|$, $\quad -1 < x < 1, p = 2$

(vii) $f(x) = \cos x$, $\quad \pi < x < \pi, p = 2\pi$

(viii) $f(x) = \cos x$, $\quad -2 < x < 2, p = 4$

(Note that the first three functions were used practically in Example 3.4A for most of the computations of this problem.)

(a) Graph the periodic extension (at least three periods) and observe the possible jump discontinuities, as done in Example 3.3.

(b) Write the Fourier series expansion. Give the value to which the corresponding Fourier series of each function converges at $x = 0, x = 1, x = 2.5, x = -7, x = 12, x = 2\pi$.

(c) Which of these functions has its Fourier series converging to it uniformly?

(d) The Fourier series of $f(x) = \cos x$, $-\pi < x < \pi$, in (vii) has one term. Why? What happened in the case of $f(x) = \cos x$, $-2 < x < 2$, in (viii)?

(e) As done in Example 3.4A, compute the Nth partial sum $S_N(x)$ of the Fourier series to approximate its $f(x)$; then graph and compare with $f(x)$. Use the same N values, $N = 1, 3, 5, 10, 15$, and 20, as in Example 3.4A; then vary N, and draw your own conclusions about the truncation error and the possible Gibbs phenomenon in some of these problems. Of course, you need to look at the periodic extension of $f(x)$ for at least three periods. It would be helpful later to look up the more refined error analysis of Fourier series in Exercise 4.42 of Chapter 4.

ANS. (b) For the first three functions in part (iv), the Fourier series converges to:

$$0, \ 1, \ 2.25, \ 1, \ 0, \ (8 - 2\pi)^2$$

at the given points, respectively.

(c) Those without discontinuities in their periodic extension.

(d) On $-\pi < x < \pi$, the function $\cos x$ has its natural period of 2π.

3.5. (a) Write the Fourier series of the following function on the *nonsymmetric* interval $(-2, 3)$ with period 5.

$$f(x) = \begin{cases} 1, & -2 < x < 0, \\ \sin x, & 0 < x < 3 \end{cases} \quad \text{period } p = 5$$

Hint: See (3.21)–(3.24)

(b) Compare the Fourier series with that of the function on the symmetric interval $(-1, 1)$ in part (ii) of Exercise 3.4. Use the Fourier series in part (a) to compute for this function and its periodic extension on the interval $(-6, 10)$.

3.6. Repeat the analysis done for the functions of Exercise 3.4 in part (e) for the truncation error and Gibbs phenomenon of the Fourier series representation in Exercise 3.5 of $f(x)$ on $(-3, 2)$.

3.7. For the two functions in Example 3.4B:

(a) Repeat the numerical computations to approximate their Fourier integral representation starting with the N and L limits in the example. Then vary these limits and draw your own conclusions about reducing the *truncation* and *discretization* errors. It would be helpful (later) to look up the more refined error analysis of Fourier integrals in Exercise 4.1 of Chapter 4.

(b) Repeat the analysis of part (a) for the following functions defined on the indicated (infinite) interval

(i) $g(t) = \dfrac{2}{1 + 4\pi^2 t^2}$, $-\infty < t < \infty$

(ii) $g(t) = \dfrac{l}{1 + 4\pi^2 t^2}$, $0 < t < \infty$

(iii) $g(t) = (1 - t)e^{-|t|}$, $-\infty < t < \infty$

Section 3.1.2   Elements of Infinite Series—Convergence Theorems

3.8. (a) Prove the Schwarz inequality (3.41) (the same problem as Exercise 2.36b of Chapter 2).

(b) Generalize the Schwarz inequality (3.41) to involve a weight function $\rho(x) > 0$ (see Exercise 2.25e in Chapter 2).

(c) Prove the following Schwarz inequality for infinite series:

$$\left| \sum_{n=1}^{\infty} c_n d_n \right|^2 \leq \sum_{n=1}^{\infty} c_n^2 \sum_{n=1}^{\infty} d_n^2 \tag{E.1}$$

by (i) using the inequality,

$$|cd| \leq \frac{1}{2}\left\{ \alpha^2 c^2 + \frac{1}{\alpha^2} d^2 \right\} \quad \text{for } c, d \text{ real numbers, } \alpha > 0 \tag{E.2}$$

and (ii) considering

$$\sum_{n=M+1}^{N} c_n d_n \quad \text{with } \alpha^2 = \frac{\sum_{n=M+1}^{n} d_n^2}{\sum_{n=M+1}^{N} c_n^2} \tag{E.3}$$

to get the Schwarz inequality for such finite sums. For the convergence of the infinite sums to get (E.1), consult Exercise 2.25d in Chapter 2, which dealt with the same idea of finite integrals versus infinite integrals.

3.9. *Convergence in the mean does not imply uniform convergence.* This exercise deals with an example of a sequence $f_n(x)$ that converges in the mean to $f(x) \equiv 0$ on the interval $[0, 1]$, but it diverges on the same interval.

Construct the sequence $f_1(x)$, $f_2(x)$, $\ldots$, $f_n(x)$, $\ldots$ on the interval $[0, 1]$ in the following way.

$$f_1(x) \equiv \begin{cases} 1, & 0 \leq x \leq \tfrac{1}{2} \\ 0, & \tfrac{1}{2} < x \leq 1 \end{cases}, \quad f_2(x) = \begin{cases} 0, & 0 \leq x \leq \tfrac{1}{2} \\ 1, & \tfrac{1}{2} < x \leq 1 \end{cases}$$

Then divide the interval into four subintervals, and define f_3, f_4, f_5, and f_6.

$$f_3(x) = \begin{cases} 1, & 0 \leq x \leq \tfrac{1}{4} \\ 0, & \tfrac{1}{4} < x \leq 1 \end{cases}, \quad f_4(x) = \begin{cases} 0, & 0 \leq x < \tfrac{1}{4} \\ 1, & \tfrac{1}{4} < x \leq \tfrac{1}{2} \\ 0, & \tfrac{1}{2} < x \leq 1 \end{cases}$$

$$f_5(x) = \cdots$$

(a) Write $f_5(x)$ and $f_6(x)$; then subdivide the interval $[0, 1]$ to $2^3 = 8$ subintervals and write f_7 to f_{14}. Continue this for 2^n subintervals with width $1/2^n$ to show that, along with those in (E.1), (E.2), we now have $m = 2(2^n - 1)$ members of the sequence $f_1, f_2, \ldots, f_m$.

(b) Show that the sequence $f_m(x)$, defined as 1 on a subinterval of width $1/2^n$ of the interval $[0, 1]$ and zero elsewhere on $[0, 1]$, is convergent in the mean to $f(x)$, and find this $f(x)$.

Hint: Note that $f_n^2(x) = f_n(x)$ on $[0, 1]$, for all n.

(c) Prove that $\lim_{n \to \infty} f_n(x)$ does not exist because it does not approach a unique limit.

Hint: Watch for the open left end of the inner subintervals that supports $f_n(x) = 1$, and show that for very large N there are two sequences $f_n(x)$ and $f_{n'}(x)$, $n, n' > N$, such that $f_n(x)$ approaches zero while $f_n'(x)$ approaches 1 at the points x.

ANS. (b) $f(x) \equiv 0$.

Section 3.1.3 The Orthogonal Expansion—Bessel's Inequality and Fourier Series

3.10. (a) Prove that if the system $\{\phi_n(x)\}$ is *orthogonal* and *complete* on (a, b) it is *closed*. This is to show that any piecewise continuous function $f(x) \in P$ that is orthogonal to all $\phi_n(x)$, i.e.,

$$\int_a^b f(x)\phi_k(x)\,dx = 0, \qquad k = 1, 2, \ldots \tag{E.1}$$

must vanish identically $(f(x) \equiv 0)$ at all points of its continuity on (a,b).

Hint: Condition (E.1) gives zero Fourier coefficients (3.57) for $f(x)$, and by the completeness of $\{\phi_n(x)\}$ we are allowed the Parseval equality (3.65), which leads us to $\int_a^b f^2(x)\,dx = 0$. To conclude from this that $f(x) \equiv 0$ at its points of continuity x on (a,b), we use a contradiction argument. (See the detailed proof of (E.2) in Exercise B.14a of Appendix B.)

(b) Prove that if two functions $f(x)$ and $g(x)$ are piecewise continuous (or square integrable), they have the same orthogonal expansion with respect to a complete orthogonal system on (a,b), i.e., a closed system. (See part (a)). Then $f(x)$ and $g(x)$ must coincide as elements of $P(a,b)$, the set of piecewise functions; that is, they may differ only at a finite number of points of the interval (a,b).

Hint: Consider the orthogonal expansion for $\psi(x) = g(x) - f(x)$,

$$\psi(x) = f(x) - g(x) = \sum_{k=1}^{\infty} c_k \phi_k(x)$$

$$= \sum_{k=1}^{\infty} a_k \phi_k(x) - \sum_{k=1}^{\infty} b_k \phi_k(x) \tag{E.1}$$

where $c_k = a_k - b_k = 0$ since $a_k = b_k$ by hypothesis. Use this $c_k = 0 = \|\phi_k\|^2 \int_a^b \psi(x)\phi_k(x)$, $k = 1, 2, \ldots$, to conclude that $\psi(x)$ is orthogonal to all members of the (complete) orthogonal set $\{\phi_k(x)\}$, and use the result in part (a) to reach the final conclusion that $\psi(x) \equiv 0$ at all points x on $[a,b]$ where $\psi(x)$ is continuous. Thus $f(x) - g(x) \equiv 0$ when they are continuous, but may differ at a finite number of points of $[a,b]$.

3.11. Prove Theorem 3.8, the Weierstrass polynomial approximation of a continuous function.

Hint: Consider the function $F(x)$ on the larger interval $[-l,l]$ that strictly contains $[a,b]$, and where $F(x) = f(x)$ on $[a,b]$ but $F(x)$ is extended as a straight line on $[-l,a)$ and $(b,l]$ in such a way that $F(-l) = F(l)$—that is, to make $F(x)$ continuous with $F(-l) = F(l)$ for Theorem 3.2 to allow it a uniform convergent Fourier trigonometric series with $S_n(x)$ as in (3.68),

$$|F(x) - S_n(x)| < \frac{\epsilon}{2} \quad \text{for all } x \in [-l,l] \tag{E.1}$$

(a) Use the (absolutely convergent) power series for $\cos(k\pi x/l)$ and $\sin(k\pi x/l)$ in $S_n(x)$ to obtain an infinite power series $\sum_{k=0}^{\infty} A_k x^k = S_n(x)$, which converges uniformly to $S_n(x)$ for all x, and in particular on $[-l,l]$,

$$|S_n(x) - Q_m(x)| < \frac{\epsilon}{2}, \quad x \in [-l,l] \tag{E.2}$$

where $Q_m(x)$ is an mth partial sum (polynomial of degree m) of the infinite power series of $S_n(x)$ in (E.2),

$$Q_m(x) = \sum_{k=0}^{m} A_k x^k \tag{E.3}$$

(b) From (E.1) and (E.2), show that

$$|F(x) - Q_m(x)| < \frac{\epsilon}{2} + \frac{\epsilon}{2} = \epsilon, \quad x \in [-l,l] \tag{E.4}$$

and in particular for $x \in [a,b] \subseteq [-l,l]$, where $F(x) = f(x)$, and (E.4) gives the desired answer

$$|f(x) - Q_m(x)| < \epsilon \quad \text{for } x \in [a,b] \tag{E.5}$$

which says that if $f(x)$ is continuous on the interval $[a,b]$, then it can be approximated to within ϵ by a simple algebraic polynomial like $Q_m(x)$ of (E.3). We note that this $Q_m(x)$ polynomial of degree m should not be confused with $Q_m(x)$, the infinite series (A.5.35), (A.5.36) representing the second solution of Legendre differential equation (A.5.26) in Appendix A.5.

3.12. As in the case of the Bessel differential equation (3.78), show the following.

(a) The *modified* Bessel differential equation

$$x^2 \frac{d^2y}{dx^2} + x \frac{dy}{dx} - x^2 y - n^2 y = 0 \tag{E.1}$$

is also with regular singularity at $x = 0$.

(b)   Indeed, this equation is the Bessel differential equation with only $x^2 y$ replaced by $-x^2 y = (ix)^2 y$. Show that a solution to (E.1) is $J_n(ix) \equiv i^n I_n(x)$, where $I_n(x)$ is the *modified Bessel function of order n*. (See (A.5.22), (A.5.23).)

(c)   Show that the Legendre equation has two regular singularities at $x = 1$ and -1.

3.13. (a)   Write the first few terms of the Legendre–Fourier series (3.91), (3.92) for

$$f(x) = \begin{cases} -1, & -1 < x < 0 \\ 1, & 0 < x < 1 \end{cases} \tag{E.1}$$

(b)   Write the first few terms of the Legendre polynomial expansion (3.91), (3.92) for the function

$$f(x) = \begin{cases} 1, & 0 < x < 1 \\ 0, & -1 < x < 0 \end{cases} \tag{E.2}$$

Hint: For the coefficient c_n in (3.92), use the Rodrigues formula (3.89) for $P_n(x)$, where after one integration it is clear that the $d^{n-1}(x^2 - 1)^n / dx^{n-1}$ would have at least one factor of $(x^2 - 1)$ that vanishes at $x = 1$. The evaluation of this derivative at the other limit of integration $x = 0$ would need a binomial expansion of $(x^2 - 1)^n$; apply the $n - 1$ order differentiation and then substitute $x = 0$ in the final result. Clearly $c_0 = \frac{1}{2}$.

(c)   Use the answer in part (b) with 5 or 10 terms in the Legendre series to illustrate its approximation of the function on $(-1, 1)$.

(d)   Use the result in part (b) to repeat the computations of part (c) for x values on the larger interval $(-3, 3)$ in order to illustrate the *nonperiodic nature* of this general (Legendre) orthogonal expansion.

ANS.   (a) $f(x) = \frac{3}{2} P_1(x) - \frac{7}{8} P_3(x) + \frac{11}{16} P_5(x) + \cdots,$

$-1 < x < 1$ for $f(x)$ of (E.1).

(b) $f(x) = \frac{1}{2} + \sum_{n=0}^{\infty} \frac{(-1)^n(4n+3)(2n)!}{4^{n+1}(n+1)!n!} P_{2n+1}(x),$

$-1 < x < 1$ for $f(x)$ of (E.2).

Section 3.2 Fourier Sine and Cosine Transforms

3.14. (a) Generalize the definition of the finite sine transform, $f(x)$ on $(0, \pi)$ in (3.101), (3.102), to that of $f(x)$ on $(0, a)$.

Hint: Make the simple change of variable $x = (\pi/a)\xi$, where $dx = (\pi/a)d\xi$.

(b) Do the same as in part (a) for the cosine transform (3.103), (3.104) of $f(x)$ on $(0, \pi)$.

(c) Find $N(\lambda_n)$, the norm square of (3.3), for the finite sine and cosine transforms.

Hint: Use $\cos^2 x = \frac{1}{2}(1 + \cos 2x)$.

ANS. (a) $F_s(n) \equiv F_s\left(\frac{n\pi}{a}\right) = \int_0^a f(x)\sin\frac{n\pi x}{a}\, dx$ \hfill (E.1)

$f(x) = \frac{2}{a}\sum_{n=1}^{\infty} F_s\left(\frac{n\pi}{a}\right)\sin\frac{n\pi x}{a}, \qquad 0 < x < a$ \hfill (E.2)

(b) $F_c(s) \equiv F_c\left(\frac{n\pi}{a}\right) = \int_0^a f(x)\cos\frac{n\pi x}{a}\, dx$

$f(x) = \frac{F_c(0)}{a} + \frac{2}{a}\sum_{n=1}^{\infty} F_c\left(\frac{n\pi}{a}\right)\cos\frac{n\pi x}{a},$

$0 < x < a$

(c) For the sine transform $N(\lambda_n) = a/2, \qquad n \neq 0.$

For the cosine transform $N(\lambda_n) = \begin{cases} \dfrac{a}{2}, & n \neq 0 \\ a, & n = 0 \end{cases}$

3.15. (a) Use a graphical method to find a few of the zeros $\{\lambda_n\}$ of equation (3.109).

Hint: Graph $y_1(\lambda) = \tan \lambda a$ and $y_2(\lambda) = h/\lambda$ as functions of λ; then locate where they intersect at $\lambda = \lambda_n$, i.e., where $y_1(\lambda_n) = y_2(\lambda_n) = \tan \lambda_n a = h/\lambda_n.$

(b) Evaluate $N(\lambda_n)$ in (3.110), the norm square of the cosine transform (3.108).

Hint: Use $\cos^2 x = \frac{1}{2}(1 + \cos 2x)$ for the integration; then consult (3.109).

3.16. (a) For the finite sine transform (3.101) show that the norm square $N(\lambda_n)$ of (3.3) is $N(\lambda_n) = \pi/2$ for the sine series (3.102). Then use the orthogonality property on the Fourier sine series (3.102) to derive (3.101).

Hint: For $N(\lambda_n)$ in the first part, see the result in Exercise 3.14c, $N(\lambda_n) = \pi/2$, $n \neq 0$.

(b) Use the same method as in (a) to derive (3.103) from (3.104).

Hint: Note that you will need the result for the norm square $N(\lambda_n)$ which was done in 3.14c as

$$N(\lambda_n) = \begin{cases} \dfrac{\pi}{2}, & n \neq 0 \\ \pi, & n = 0 \end{cases}$$

3.17. (a) Use integration by parts to prove (3.106).

(b) Use Green's identity (1.50) to prove (3.105) and (3.106).

Hint: For (3.105), take $v(x) = \sin nx$, $u(x) = f(x)$ and $L_2 = d^2/dx^2$. Here we have $A_0(x) = 1$, $A_1(x) = 0$. For (3.106), take $v(x) = \cos nx$, ...

3.18. (a) Derive (3.108).

Hint: Use (3.111), multiply by $\cos \lambda_m x$, integrate, and use the orthogonality property of $\{\cos \lambda_n x\}$ on $(0,a)$ with the help of (3.110).

(b) Derive (3.111).

Hint: Consult (3.96), and use (3.110), which was derived in Exercise 3.15b.

3.19. (a) Solve the problem of Example 3.10 by using the method of separations of variables, which was illustrated in Example 1.1.

(b) In Example 3.10 let the two ends $x = 0$ and $x = \pi$ of the bar be kept at constant temperatures T_1 and T_2, respectively; i.e., replace (E.2) and (E.3) by

$$u(0,t) = T_1, \qquad t > 0 \tag{E.2}$$

$$u(\pi,t) = T_2, \qquad t > 0 \tag{E.3}$$

and the rest of the conditions (E.1) and (E.4) are the same. Use the method of separation of variables to solve for $u(x,t)$.

Hint: Write $u(x,t) = \omega(x,t) + v(x)$, where $\omega(x,t)$ is the *transient* temperature (i.e., $\lim_{t\to\infty}\omega(x,t) = 0$), and $v(x)$ is the *steady-state* temperature. The separation of variables will apply to $\omega(x,t)$ in a problem equivalent to the one in Example 3.10.

(c) Use the method of the sine transform of this section to solve the problem of part (b).

(d) Compare the efforts in the two approaches in parts (c) and (b).

(e) Write the answers for the problem in part (a) and the problem in part (c) when the bar is of length l instead of π.

Hint: Make a simple change of scale by letting $x = (\pi/l)\xi$, $0 < \xi < l$.

ANS. (b) $u(x,t) = v(x) + \omega(x,t) = T_1 + \dfrac{T_2 - T_1}{\pi}(x - \pi) + \omega(x,t)$

where $\omega(x,t)$ is the solution of a problem equivalent to that of Example 3.10 except that $f(x)$ is replaced by $g(x) = f(x) - v(x)$.

3.20. (a) In reference to Exercise 3.19a, is the Laplace transform compatible with the problem of Example 3.10? If so, use it to solve the problem.

Hint: If you face a resulting nonhomogeneous differential equation, see Exercises A.6 to A.12 in Appendix A, and consult part (b).

(b) If in part (a) the Laplace transform is compatible with $\partial u/\partial t$, then use both the Laplace and finite sine transforms to completely algebraize the problem, thus avoiding solving a second-order differential equation in $U(x,s) = \mathcal{L}\{u(x,t)\}$. Find the final solution, and compare to the answer in part (a) and the answer of Exercise 3.19a.

Hint: Let $\bar{U}(n,s) = \mathcal{L}\{U(n,t)\}$ where $U(n,t)$ is the finite sine transform of $u(x,t)$.

ANS. (a) The Laplace transform is compatible with $\partial u/\partial t$, since we are given $u(x,0) = U_0$, which is needed for $\mathcal{L}\{\partial u/\partial t\} = sU(x,s) - U(x,0)$ in (2.2),

$$\frac{d^2U}{dx^2} - sU(x,s) = -f(x) \tag{E.1}$$

The solution to this differential equation is better done via the finite sine transform of part (b).

(b) $-n^2\bar{U}(n,s) - s\bar{U}(n,s) = -F_s(n)$

$$\bar{U}(n,s) = \frac{F_s(n)}{s+n^2},$$

$$U(n,t) = \mathcal{L}^{-1}\{\bar{U}(n,s)\} = F_s(n)e^{-n^2t}$$

$$u(x,t) = \frac{2}{\pi}\sum_{n=1}^{\infty}F_s(n)e^{-n^2t}\sin nx$$

which is the same as the answer of Example 3.10.

3.21. (a) As in Exercise 3.19a, use the method of separation of variables to find the solution of Example 3.11.

Hint: Watch for the boundary condition (E.3) and consult (3.111) for your Fourier series analysis of the desired function (where you need a Fourier cosine series).

(b) If the Laplace transform is also compatible with this problem (which part?), would you rather use it? How would you compare the finite cosine transform, the Laplace transform, and the separation-of-variables methods for solving this problem?

ANS. (a) The same answer as for Example 3.11.

(b) The Laplace transform is compatible with $\partial u/\partial t$, since we are given $u(x,0) = u_0$. For the Laplace transform method we have $d^2U(x,s)/dx^2 - sU(x,s) = -u_0$. $U(x,s) = A(s)e^{x\sqrt{s}} + B(s)e^{-x\sqrt{s}} + u_0/s$, and applying (E.2) and (E.3) we have

$$U(x,s) = \frac{u_0}{s}$$

$$- hu_0\frac{(e^{x\sqrt{s}} + e^{-x\sqrt{s}})}{(e^{a\sqrt{s}} - e^{-a\sqrt{s}}) + h(e^{a\sqrt{s}} + e^{-a\sqrt{s}})}$$

whose *inverse Laplace transform* is clearly difficult to find. So it is easier to use the Finite cosine transform or the separation of variables method.

Section 3.3 Fourier (Exponential) Transforms

3.22. (a) Derive (3.112), using (3.113).

Hint: See Exercise 3.16a.

(b) We have proved (3.114) by using integration by parts; now prove it by using Green's identity (1.50).

Hint: See Exercise 3.17b. In (1.50) let $v(x) = e^{-inx}$, $u(x) = f(x)$, and $L = d/dx$. Here $A_0 = 0$, $A_1 = 1$.

3.23. (a) Solve the problem in Example 3.12 when there is a jump of 1 between the two boundary values, i.e., $u(\pi,y) - u(-\pi,y) = 1$ instead of (E.2).

Hint: See (3.114) for the transform of df/dx when there is a jump $J = f(\pi) - f(-\pi)$.

(b) Use the method of separation of variables to solve Example 3.12.

Hint: Apply the separation of variables for (E.1), (E.2); you will need the Fourier exponential series representation of $f(x)$ as in (3.113) to accommodate the nonhomogeneous condition $u(x,0) = f(x)$ in (E.3).

(c) How does separation of variables fare in solving the problem in part (a)?

ANS. (a) $u(x,y) = \dfrac{1}{2\pi} \displaystyle\sum_{n=-\infty}^{\infty} [F(n)e^{-iny}$

$+ \dfrac{i}{n}(-1)^n(1 - e^{-iny})]e^{inx}$

(b) As in (E.8) of Example 3.12.

3.24. (a) Let $f(n)$ and $G(n)$ be the finite exponential transforms of $f(x)$ and $g(x)$, as defined in (3.112), respectively. Prove the

Parseval equality

$$\sum_{n=-\infty}^{\infty} F(n)\overline{G(n)} = 2\pi \int_{-\pi}^{\pi} f(x)\overline{g(x)}dx \qquad (E.1)$$

Hint: Multiply both sides of (3.113) by $\overline{g(x)}$, integrate, interchange the summation and the integral (justify!), and consult (3.112) for defining $\overline{G(n)}$.

(b) Derive the Parseval equality in (E.1) for functions defined on the more general interval $(-a, a)$ instead of $(-\pi, \pi)$; i.e., derive (3.115), (3.116) from (3.112), (3.113), then follow part (a), or use the simple change of variable $x = (\pi/a)\xi$ in the integral of (E.1).

(c) Use the Parseval equality derived in part (b) to prove the sampling expansion (3.121).

Hint: In the integral of (3.121), write the Fourier series for $e^{i\omega t}$ and $F(\omega)$ on $(-a, a)$, and note that their respective Fourier coefficients are

$$2a\, \frac{\sin(at - n\pi)}{(at - n\pi)} \quad \text{and} \quad f_a\left(\frac{n\pi}{a}\right)$$

(d) In the derivation of the sampling expansion (3.121) in part (c) via the Parseval equality, there does not seem to be any worry about justifying the interchange of the integration with the infinite summation, as in part (a). Why?

Hint: Consult Theorem 3.7 and see the equivalence between the validity of the Parseval equality with the interchange of the integration and the infinite sum of the Fourier series of square integrable functions.

(e) In deriving the sampling expansion (3.121) we used the finite exponential Fourier transforms pair (3.116a), (3.115a). Show that we can just as well use the equivalent pair (3.116), (3.115) to arrive at (3.121) with possibly one extra simple change of variable for the index of summation: $n = -m$.

ANS. (b) $\displaystyle\sum_{n=-\infty}^{\infty} F(n)\overline{G(n)} = 2a \int_{-a}^{a} f(x)\overline{g(x)}dx$

(d) The validity of interchanging integration with the infinite summation of the Fourier series expansion, for square integrable functions, is equivalent to the validity of the Parseval equality of the same class of functions as given in Theorem 3.7.

3.25. In all our derivations of the sampling expansion (3.121), via the Fourier series in the text in (3.122)–(3.124) or via the Parseval equality in Exercise 3.24c, we were limited to bandlimited signals $f(t)$ of *square integrable* $F(\omega)$ on $(-a,a)$. Search for a method of justifying the sampling expansion for the bandlimited Bessel function (3.125), which is a Fourier transform of an absolutely integrable, *and not square integrable* $F(x) = 1/\sqrt{\pi^2 - x^2}$ on $(-\pi, \pi)$.

Hint: Consult the following *Hölder inequality*, as an extension of the Schwarz inequality. For $f(x) \in Lp(a,b)$, $g(x) \in Lp'(a,b)$, $1/p + 1/p' = 1$, the *Hölder inequality* is

$$\left| \int_a^b f(x)g(x)\,dx \right| \le \left[\int_a^b |f(x)|^p\,dx \right]^{1/p} \left[\int |g(x)|^{p'}\,dx \right]^{1/p'}$$

where $f \in L_p(a,b)$ means $\int_a^b |f(x)|^p\,dx < \infty$.

Use this inequality on $|J_0(t) - S_N(t)|$ and note that $e^{i\omega t} \in L_\infty(-\pi, \pi)$ for $p = 1$, and $p' = \infty$. $S_N(t)$ is the Nth partial sum of the sampling expansion for $J_0(t)$. There is also a similar extension of the Parseval equality that is valid for the above class of functions with $1/p + 1/p' = 1$ instead of the special case we have been using, which corresponds to $p = p' = 2$. [You may consult also Brown's paper (1968b) and the author's (1973), (1977a).]

3.26. (a) Derive the Fourier exponential series for a function of two variables $f(x,y)$ defined on the rectangle $-a \le x \le a$, $-b \le y \le b$, which is periodic in both direction (periods $2a$ and $2b$, respectively) and is square integrable on the rectangle, i.e.,

$$\int_{-a}^{a} \int_{-b}^{b} |f(x,y)|^2\,dx\,dy < \infty \tag{E.1}$$

Hint: Consult the derivation of the two-dimensional Fourier transform (2.211), (2.212) in Section 2.6, which would go here as follows: We write Fourier series for $f(x,y)$ on $(-a,a)$ *as a function of* x, treating y for the movement as constant, but the Fourier coefficient $C_n(y)$ appears as a function of y. Then we write the Fourier series expansion of this function $C_n(y)$ as a function of y on $(-b,b)$.

(b)   Use the double Fourier series in part (a) to derive the sampling expansion for the function of two variables $f(t_1,t_2)$ with its spectrum $F(\lambda_1,\lambda_2)$ vanishing identically outside its support as the rectangle $-a \leq \lambda_1 \leq b,\ -b \leq \lambda_2 \leq b$.

ANS.   (a) $f(x,y) = \displaystyle\sum_{n=-\infty}^{\infty} \sum_{m=-\infty}^{\infty} c_{n,m} \exp\left[-\left(\frac{in\pi x}{a} + \frac{im\pi y}{b}\right)\right]$

$$c_{n,m} = \frac{1}{4ab} \int_{-a}^{a} \int_{-b}^{b} f(x,y) \exp\left(\frac{in\pi x}{a}\right)$$

$$\times \exp\left(\frac{im\pi y}{b}\right) dx\, dy$$

(b) $f(t_1,t_2) = \displaystyle\sum_{n=-\infty}^{\infty} \sum_{m=-\infty}^{\infty} f\left(\frac{n\pi}{a}, \frac{m\pi}{b}\right) \frac{\sin(at_1 - n\pi)}{at_1 - n\pi}$

$$\times \frac{\sin(bt_2 - m\pi)}{bt_2 - m\pi}$$

3.27.   In (3.120) we have the bandlimited function $f_a(t)$ as the Fourier transform of $p_a(\omega)F(\omega)$, where $p_a(\omega)$ is the gate function, or what we called the hill function $\phi_1(a,\omega)$ of order 1 as a special case (2.191) of the hill function $\phi_{R+1}(a(R+1),\omega)$ in (2.190).

(a)   Show that (try it graphically)

$$P_{2a}(\omega) = \phi_1(2a,\omega) = \phi_2(2a,\omega) + \frac{d}{d\omega}\phi_2(2a,\omega) \qquad (E.1)$$

(b)   Use this representation in (E.1) for $p_a(\omega)$ of (3.120); then write its (exponential) Fourier series expansion to develop the following sampling series expansion with *samples of the function and its derivative*. Also, see Exercise 3.33a,b.

$$f_a(t) = \sum_{n=-\infty}^{\infty} \left[f_a\left(\frac{2n\pi}{a}\right) + \left(t - \frac{2n\pi}{a}\right) f_a'\left(\frac{2n\pi}{a}\right) \right]$$

$$\times \left[\frac{\sin(at - 2n\pi)}{at - 2n\pi}\right]^2 \qquad (E.2)$$

(c) Explain why the spacing of the samples in the sampling series of (E.2) is $2\pi/a$, i.e., twice that required by the sampling series (3.121) with samples of the function $f_a(n\pi/a)$ only.

Hint: For the derivation of (E.2) for $f_a(t)$ we use $p_a(\omega)$ from (E.1), where we then need $\phi_2(a,\omega)$ in (E.1). But $\phi_2(a,\omega) = \phi_1(a/2,\omega) * \phi_1(a/2,\omega)$, where the Fourier series is written on $(-a/2, a/2)$ to relate to $\phi_1(a/2,\omega)$, thus effectively doubling the sampling spacing as we see it in

$$\exp\left(\frac{in\pi}{a/2}\right) = \exp\left(\frac{i2n\pi}{a}\right)$$

instead of $e^{in\pi/a}$ that we used for (3.121). Also, see Exercise 3.33a,b.

3.28. In practice, the signals that we deal with are not necessarily bandlimited (3.120) as required by the sampling expansion (3.121). The *aliasing error*

$$\epsilon_A(t) = |f(t) - f_s(t)| \tag{E.1}$$

is the result of applying the sampling representation $f_s(t)$ as in (3.121) to signals $f(t)$ with samples $f(n\pi/a)$ even when they are not bandlimited, or are bandlimited to different limits beyond $-a$ to a. Assume that the Fourier transform $F(\omega)$ of $f(t)$ is absolutely integrable, i.e., $\int_{-\infty}^{\infty} |F(\omega)| d\omega < \infty$. Attempt to find an estimate of $\epsilon_A(t)$.

Hint: (a) In (E.1) write $f(t)$ as an (infinite limits) Fourier transform of $F(\omega)$, and $f_s(t)$ as $f_a(t)$ in (3.121). Then write $f_a(n\pi/a)$ as a (bandlimited) integral, and take the integration outside the sum. Last, observe the Fourier series of $e^{i\omega t}$ on $(-a,a)$, and use the proper inequality.

(b) You may also consult the Poisson sum formula (4.58) in Chapter 4 to arrive at the answer for part (a). See Brown (1968a) and Weiss (1963).

ANS. $\epsilon_A(t) \leq \dfrac{1}{\pi} \displaystyle\int_{|\omega|>a}^{\infty} |F(\omega)| d\omega$

3.29. Consider the Nth partial sum $S_N(t)$ of the sampling series expansion (3.121), where we define the truncation error as $\epsilon_T = \epsilon_N \equiv$

$|f_a(t) - S_N(t)|$. Use Fourier analysis to give an indication of how the truncation error ϵ_T depends on the truncation limit N.

Hint: Attempt an upper bound on the remainder of the sampling series. The usual method uses complex contour integration; see the authors' tutorial paper (1977a) and the references therein.

ANS. A very basic estimate is a bound on the remainder of the sampling series

$$\epsilon_T \leq M \sum_{|n|>N} \frac{(-1)^n}{at - n\pi}, \quad \text{where } |f_a(t)| \leq M$$

and clearly $|\sin at| \leq 1$. A very rough bound is obtained by considering the leading (first term: $n = N$) of this series,

$$\epsilon_T \leq \frac{2Mat}{a^2 t^2 - N^2 \pi^2}$$

3.30. (a) Write the Fourier series representation for the hill function of order $R + 1$, $\phi_{R+1}(a(R + 1)), \omega)$ on $[-a(R + 1), a(R + 1)]$ as given in (2.190).

(b) Write the Fourier integral representation of $\phi_{R+1}(a(R+1), \omega)$. See Exercise 2.62d in Chapter 2.

(c) What is the difference between the representations of ϕ_{R+1} in parts (a) and (b)? Why?

(d) Prove the self-truncating sampling series in (2.192).

Hint: Instead of considering the Fourier series expansion for $p_a(x)F(x)$, as we did for deriving the usual sampling series in (3.121), consider the Fourier series here for $F(x)[p_{ra}(x) * \phi_m(aq,x)]$, $q + r = 1$, $0 < r < 1$.

(e) This part is designed toward a general proof of the special two cases in Exercise 2.62b of Chapter 2. Prove that $\bar{\phi}_{R+1}$, the average value of ϕ_{R+1} over the interval $[-a(R + 1), a(R + 1)]$, is equal to the average of its $R + 1$ values at the middle of each of the $R + 1$ equal subintervals of $[-a(R + 1), a(R + 1)]$. It is sufficient to consider the case of R where ϕ_{R+1} is defined on $R + 1 = 2m$ equal subintervals. The case of even R will follow in the same way with $R + 1 = 2m + 1$ equal subintervals. Let $\rho_k(\omega)$ be the part of ϕ_{R+1} defined on the two symmetric subintervals $(-2(k + 1)a, -2ka)$ and $(2ka, 2(k +$

1)a). $\rho_k(\mp(2k + 1)a)$ are the values of the hill function ϕ_{R+1} at the middle of two such symmetric kth subintervals. The problem is to prove that

$$\frac{2}{R + 1} \sum_{k=0}^{(R-1)/2} \rho_k(\mp(2k + 1)a) = \tilde{\phi}_{R+1} \equiv \frac{(2a)^{R+1}}{R + 1} \qquad \text{(E.1)}$$

Hint: Use the Fourier series of ϕ_{R+1} in part (a) for $\rho_k(\omega)$ and then take the finite sum in (E.1) and use the following identity:

$$\sum_{k=0}^{(R-1)/2} \cos(2k + 1)\alpha = \frac{\sin(R - 1)\alpha}{2 \sin \alpha} \qquad \text{(E.2)}$$

[see also the author's papers (1977b,a).]

ANS. (a) With $c = a(R + 1)$:

$$\phi_{R+1}(a(R + 1), \omega) = \frac{a_0}{2} + \sum_{n=1}^{\infty} a_n \cos \frac{n\pi\omega}{c}$$

$$a_n = \frac{1}{c} \int_{-c}^{c} \phi_{R+1}(a(R + 1), \omega) \cos \frac{n\pi\omega}{c} \, d\omega$$

$$c = a(R + 1)$$

(b) See the answer to Exercise 2.62d in Chapter 2.

(c) In part (a) we have Fourier series that represents $\phi_{R+1}(a(R + 1), \omega)$ on $(-a(R + 1), a(R + 1))$ and repeats this picture periodically with period $2a(R + 1)$. In part (b) we have a representation for ϕ_{R+1} on the whole real line, where in this case ϕ_{R+1} coincides with that of part (a) on the interval $(-a(R + 1), a(R + 1))$, but it *vanishes identically* outside this interval.

3.31. (a) Let $f_a(x)$ and $g_a(x)$ be bandlimited functions (2.138) to $(-a, a)$; i.e., their Fourier transforms $F(\omega) = G(\omega) \equiv 0$ for $|\omega| > a$. Prove the following *discretization of the Fourier convolution product*:

$$f_a * g_a \equiv \int_{-\infty}^{\infty} f_a(\xi) g_a(x - \xi) \, d\xi$$

$$= \frac{1}{2a} \sum_{n=-\infty}^{\infty} f_a \left(\frac{n\pi}{a} \right) g_a \left(x - \frac{n\pi}{a} \right) \qquad (E.1)$$

Hint: Write the (exponential) Fourier series for either $p_a(\lambda)F(\lambda)$ or $p_a(\lambda)G(\lambda)$; then take the Fourier transform of $p_a(\lambda)F(\lambda)G(\lambda)$ for the convolution integral on the left side of (E.1). You may also use the Parseval equality (3.65).

(b) Use result (E.1) to show that the discretized convolution product is commutative, i.e., $f_a * g_a = g_a * f_a$.

(c) Use result (E.1) in (a) to prove the sampling expansion (3.121) (or (2.184)) for $f_a(t)$.

Hint: Note that $g_a(x) = (2\sin ax)/x$ is bandlimited to a as the Fourier transform of $p_a(\lambda)$.

(d) Use only the summation in (E.1) and the commutativity property in part (b) to establish the sampling expansion (3.121) for $f_a(t)$.

Hint: Take the right side of (E.1) in part (a) with $g_a(x)$ as in part (c); then use the commutativity property of the convolution product to have a series resulting in one term $f_a(x)$, since

$$\frac{\sin n\pi}{n\pi} = 0 \quad \text{for } n \neq 0, \quad \text{and}$$

is 1 at $n = 0$. You may note that this may be the *shortest proof of the sampling expansion* (2.184).

(e) Let $f_a(t)$ and $g_b(t)$ be bandlimited to a and b, respectively, as in (2.138). Derive the discretization of the convolution product for (i) $a > b$ and (ii) $a < b$.

Hint: Write the Fourier transform of the convolution product as

$$\frac{1}{2\pi} \int_{-\infty}^{\infty} e^{i\lambda x} p_a(\lambda)F(\lambda)p_b(\lambda)G(\lambda)$$

$$= \frac{1}{2\pi} \int_{-b}^{b} e^{i\lambda x} F(\lambda)G(\lambda)d\lambda, \qquad a > b \qquad (E.2)$$

$$= \frac{1}{2\pi} \int_{-a}^{a} e^{i\lambda x} F(\lambda)G(\lambda)d\lambda, \qquad a < b \qquad (E.3)$$

Then use Fourier series expansion for $F(\lambda)$ on $(-a,a)$ in (E.2) and for $G(\lambda)$ on $(-b,b)$ in (E.3).

ANS. (e) (i) $h_b(x) \equiv \displaystyle\int_{-\infty}^{\infty} f_a(\xi)g_b(x-\xi)d\xi$

$$= \frac{1}{2a} \sum_{n=-\infty}^{\infty} f_a\left(\frac{n\pi}{a}\right)g_b\left(x - \frac{n\pi}{a}\right) \qquad (E.4)$$

(ii) $h_a(x) \equiv \displaystyle\int_{-\infty}^{\infty} f_a(\xi)g_b(x-\xi)d\xi$

$$= \frac{1}{2b} \sum_{n=-\infty}^{\infty} f_a\left(x - \frac{n\pi}{b}\right)g_b\left(\frac{n\pi}{b}\right) \qquad (E.5)$$

3.32. (a) Derive the sampling expansion (2.138) by

(i) First writing the exponential Fourier series for $p_a(x)$.

(ii) Using the Parseval equality (3.65) with $f_1(x) = p_a(x)$ and $f_2(x) = F(x)$. See also Exercise 3.31c.

(b) Prove (E.1) in Exercise 3.31a for the discretization of the convolution product by using Parseval equality (3.65).

3.33. (a) The sampling expansion (3.121) or (2.138) can be extended to use the samples of the function as well as the samples of its derivative with the obvious advantage of sampling at double the spacings $\{2n\pi/a\}$ instead of $\{n\pi/a\}$ of (3.121) for the same bandlimited function $f_a(t)$. In this case, (3.121) is modified to the following result to be proved:

$$f_a(t) = \sum_{n=-\infty}^{\infty} \left[f_a\left(\frac{2n\pi}{a}\right) + \left(t - \frac{2n\pi}{a}\right)f'\left(\frac{2n\pi}{a}\right) \right]$$

$$\times \left[\frac{\sin a(t - 2n\pi/a)}{a(t - 2n\pi/a)} \right]^2$$

Hint: Consult Exercise 3.27 to find a combination of two functions $F_1(\omega)$ and $F_2(\omega)$ that add up to $p_a(\omega)F(\omega)$ to give the transform $f_a(t)$ of the left side such that the transforms of $F_1(\omega)$ and $F_2(\omega)$ result in the first and second terms of the series (E.1), respectively.

(b) Attempt to write a result that involves the samples of the second derivative too.

Hint: The safest way is to use complex contour integration to construct such series [see the author's tutorial paper (1977a)]. For a Fourier analysis (lengthy) method see Papoulis (1977); better yet, you may be able to find $F_1(\omega)$, $F_2(\omega)$, and $F_3(\omega)$ in parallel to the $F_1(\omega)$ and $F_2(\omega)$ of part (a). Note that the latter hope may be risky; the most assured and accurate result is via complex integration.

3.34. (a) Use the discretization of the convolution product in Exercise 3.31a and the Fourier transform of the hill function ϕ_{R+1} of (2.190) to derive the "self-truncating" sampling series (2.192). See Exercise 3.30d and its hint.

(b) Compare (2.192) with the sampling series (2.138) or (3.121). The physical interpretation is that $f_a(t)$ of (3.121) is bandlimited to $(-a, a)$ as a result of passing the nonbandlimited signal $f(t)$ through an ideal low filter or "window" $p_a(\lambda)$. Determine the shape of such a "window" for (2.192) when $m = 2$ and $m = 3$. Graph both windows. See the Fourier transforms table in Appendix C.2.

ANS. (b) See Exercise 2.62 in Chapter 2, and in particular the reference to Ditkin and Prudnikov (1965), where the exact answer for Exercise 2.62c (and the present problem) is found.

Section 3.4 Hankel (Bessel) Transforms

In Appendix A.5 and its exercises we have most of the properties of the Bessel functions that are needed for this section and other sections involving them, like Sections 2.7, 3.6, and 4.3.

3.35. (a) Derive the result in (3.129).

Hint: See Appendix A.5 for the properties of Bessel functions, in particular equations (A.5.18) and (A.5.17) and Exercise A.25.

(b) Use (3.128) to derive (3.126).

Hint: Note that the set $\{J_n(\lambda_k r)\}_{k=1}^{\infty}$, where $J_n(\lambda_k a) = 0$, is orthogonal on $(0, a)$ with respect to the weight function

$\rho(r) = r$. Of course, you can see (3.128) as a special case of (3.95), (3.96) or (3.1)–(3.3).

(c) Prove (3.131).

Hint: See Example 1.6 in Chapter 1.

3.36. (a) Establish the result (E.8) in Example 3.13.

Hint: Let $n = 1$ in (E.9) of this example to have

$$\int_0^1 rJ_0(r\lambda_k)\,dr = \int_0^1 \frac{1}{\lambda_k}\frac{d}{dr}[rJ_1(r\lambda_k)]\,dr$$

$$= \frac{r}{\lambda_k}J_1(\lambda_k r)\bigg|_{r=0}^1 = \frac{J_1(\lambda_k)}{\lambda_k}$$

(b) Use the method of separation of variables to solve Example 3.13.

Hint: Use $u(r,t) = R(r)T(t)$ for the homogeneous equations (E.1) and (E.3); then appeal to the Bessel–Fourier series expansion to satisfy the nonhomogeneous condition (E.2) with the help of the result in (E.8) of Example 3.13, which is proved in part (a) of this problem.

3.37. Solve the problem in Example 3.13 when the surface of the cylinder is insulated, i.e., $\partial u(1,t)/\partial r = 0$, $t > 0$.

Hint: You will need $\{\lambda_k\} = \{\lambda_{1,k}\}$ as zeros of $J_1(\lambda_{1,k}) = 0$ since this boundary condition requires $J_0'(\lambda) = -J_1(\lambda_{1,k}) = 0$. See Exercise 3.39d.

3.38. (a) Derive the Hankel transform (3.135), (3.136) and its inverse (3.138) that is associated with the boundary conditions (3.133) or (3.136) instead of (3.127).

(b) Establish the result in (3.137). See Exercise 3.39c.

(c) Use the method of separation of variables to solve Example 3.14.

Hint: See Exercise 3.36, where a similar hint is given. Use $u(r,t) = R(r)T(t)$ for the (separable) homogeneous equations (E.2) and (E.3). Then for the nonhomogeneous condition (E.4), appeal to the Bessel–Fourier series (3.138) associated with condition (3.136), which is (E.6) here, and take advantage of the integral (E.9), which was done in Exercise 3.36a.

3.39. (a) Prove that

$$\int_0^a rJ_n(\lambda r)J_n(\mu r)\,dr = \frac{a\mu J_n(\lambda a)J_n'(\mu a) - \lambda J_n(\mu a)J_n'(\lambda a)}{\lambda^2 - \mu^2}$$

(E.1)

Hint: See Exercises 3.35 and 3.36, where we depend on Appendix A.5 and its exercises.

(b) Use L'Hôpital's rule for (E.1) as μ approaches λ to show that

$$\int_0^a rJ_n^2(\lambda r) = \frac{a^2}{2}\left[J_n'^2(\lambda a) + \left(1 - \frac{n^2}{\lambda^2}\right)J_n^2(\lambda a)\right]$$

(E.2)

Hint: As you differentiate the numerator in the right-hand side of (E.1) you will have $J_n''(\lambda a)$, which can be eliminated by recognizing $J_n(x)$ as a solution of the Bessel differential equation

$$x^2 J_n''(x) + x J_n'(x) + (x^2 - n^2)J_n(x) = 0$$

(c) Use the result in (E.2) to derive the norm square in (3.137).

Hint: In (E.2) let $\lambda = \{\lambda_k\}$ be the zeros of (3.136),

$$\lambda_k J_n'(\lambda_k a) + hJ_n(\lambda_k a) = 0 \qquad\qquad \text{(E.3)}$$
$$\qquad\qquad\qquad\qquad\qquad\qquad\qquad\qquad \text{(3.136)}$$

and use this to eliminate $J_n'(\lambda_k a)$ in (E.2).

(d) Derive the norm square of the Bessel series needed in Exercise 3.37 when the surface of the cylinder of Example 3.13 is insulated.

Hint: This is a special case of the boundary condition in (3.136) with $h = 0$ and $n = 0$.

ANS. (d) $\|J_0(\lambda_k)\|^2 = \dfrac{a^2}{2}J_0^2(a\lambda_{1,k})$, $\quad J_1(j_{1,k}) = 0$, $\quad k = 1, 2, \ldots$
For the zeros $\{j_{1,k}\}$, see Table A.1 in Appendix A.5.

3.40. Consider the temperature distribution $u(r,z,t)$ in a cylinder of unit radius and unit thickness, where the initial temperature has radial and z dependence only. The entire surface (lateral, top, and bottom) is kept at temperature zero.

(a) Why did we describe the temperature by $u(r,z,t)$ and not the more general $u(r,\theta,z,t)$?

(b) Write the initial and boundary value problem in $u(r,z,t)$.

(c) Is there more than one transform that is compatible with this problem? If so, use the two transforms for the spatial dependence in r and z (i.e., the boundary value part of the problem) to reduce the heat equation to a first-order differential equation in t. Solve this simple equation; then invert your double transform to find the final solution $u(r,z,t)$.

(d) In part (c), would it be beneficial to use a third transform, i.e., the Laplace transform, to completely algebraize the problem in the three-dimensional transform space?

(e) Does the method of separation of variables apply in this problem? If so, use it to solve the problem and compare your answer to that in part (d).

ANS. (a) Primarily, the initial condition is independent of θ, and also the boundary values are constants.

(b)
$$\frac{\partial^2 u}{\partial r^2} + \frac{1}{r}\frac{\partial u}{\partial r} + \frac{\partial^2 u}{\partial z^2} = \frac{1}{k}\frac{\partial u}{\partial t},$$

$$\begin{cases} 0 < r < 1, & t > 0 \\ 0 < z < 1, & t > 0 \end{cases} \tag{E.1}$$

$$u(r,z,0) = f(r,z), \qquad 0 < r < 1,\ 0 < z < 1 \tag{E.2}$$

$$u(r,0,t) = 0, \qquad 0 < r < 1,\ t > 0 \tag{E.3}$$

$$u(r,1,t) = 0, \qquad 0 < r < 1,\ t > 0 \tag{E.4}$$

$$u(1,z,t) = 0, \qquad 0 < z < 1,\ t > 0 \tag{E.5}$$

(c) Yes, a (*finite*) J_0-Hankel transform for the radial dependence, and a (*finite*) sine transform for the z dependence. Of course, there is also the Laplace transform for the time dependence.

$$U(r,m,t) = f_s\{u(r,z,t)\}$$

$$= 2\int_0^1 u(r,z,t)\sin m\pi z\, dz$$

$$\tilde{U}(\lambda_i, m, t) = h_0 \{U(r, m, t)\}$$

$$= \frac{2}{j_1^2(\lambda_i)} \int_0^1 rU(r, m, t)J_0(\lambda_i r)\,dr,$$

$$J_0(\lambda_i) = 0, \quad i = 1, 2, \ldots$$

$$\tilde{U} = h_0[f_s\{u\}]$$

$$= \frac{4}{J_1^2(\lambda_i)} \int_0^1 \int_0^1 ru(r, z, t)\sin m\pi z J_0(\lambda_i r)\,dr$$

$$\frac{d\tilde{U}}{dt} = -k(\lambda_i^2 + m^2\pi^2)\tilde{U},$$

$$\tilde{U}(\lambda_i, m, 0) = h_0[f_s\{f(r, z)\}]$$

$$\tilde{U}(\lambda_i, m, t) = e^{-k(\lambda_i^2 + m^2\pi^2)}\tilde{U}(\lambda_i, m, 0)$$

$$u(r, z, t) = \sum_{i=1}^{\infty}\sum_{m=1}^{\infty} \frac{4}{J_1^2(\lambda_i^2)}e^{-k(\lambda_i^2 + m^2\pi^2)t}$$

$$\times \tilde{U}(\lambda_i, m, 0)J_0(\lambda_i r)\sin m\pi z$$

(d) Not really, since we have a very simple separable first-order differential equation for the time dependence of the initial value part of the problem.

(e) Yes, to the r and z spatial dependence, since the conditions for them are homogeneous equations. They are also given at constant values of the concerned variables, i.e., $r = 1$ and $z = 0, 1$.

3.41. (a) Solve the problem of Exercise 3.40 for the particular initial condition of constant temperature T_0, i.e., $u(r, z, 0) = T_0$.

(b) What is the steady-state solution?

ANS. (a) See the series solution answer of Exercise 3.40c with the particular *double (finite) Hankel-sine transform*:

$$\tilde{U}(\lambda_i, m, 0) = \frac{4T_0(1 - (-1)^m)}{m\pi \lambda_i J_1(\lambda_i)},$$

$$J_0(\lambda_i) = 0, i = 1, 2, \ldots$$

(b) $\lim_{t \to \infty} u(r, z, t) = 0$

3.42. Consider the vibrating membrane of Appendix B.1 for the displacement $u(r,\theta,t)$ of a *circular membrane* of unit radius whose rim (boundary) is fixed. The vibrations start from rest with initial (vertical) displacement that is antisymmetric with respect to the x axis, i.e., $u(r,\theta,0) = f(r,\theta)$, $f(r,-\theta) = -f(r,\theta)$ as an odd function in θ. For the wave equation in $u(r,\theta,t)$, see (B.1.8).

(a) Set up the boundary and initial value problem in $u(r,\theta,t)$.

(b) What are the compatible transforms for this problem?

   Hint: See Exercise 3.41.

(c) Use the transforms in part (b) to solve the problem.

ANS. (a) $\displaystyle \nabla^2 u = \frac{\partial^2 u}{\partial r^2} + \frac{1}{r}\frac{\partial u}{\partial r} + \frac{1}{r^2}\frac{\partial^2 u}{\partial \theta^2} = \frac{1}{c^2}\frac{\partial^2 u}{\partial t^2},$

$\qquad 0 < r < 1, \ 0 < \theta < 2\pi$ \hfill (E.1)

$\qquad u(1,\theta,t) = 0, \qquad 0 \le \theta < 2\pi, \ t > 0$ \hfill (E.2)

$\qquad$ *(also periodic in θ with period 2π)*

$\qquad u(r,\theta,0) = \begin{cases} f(r,\theta), & 0 < \theta \le \pi \\ -f(r,\theta), & \pi < \theta < 2\pi \end{cases}$ \hfill (E.4)

$\qquad u_t(r,\theta,0) = 0, \qquad 0 \le \theta \le 2\pi, \ 0 \le r \le 1$

(b) The finite J_n-Hankel transform for the radial dependence, and a finite sine transform for the θ dependence.

(c) $\displaystyle u(r,\theta,t) = \sum_{n=1}^{\infty}\sum_{k=1}^{\infty} \bar{U}(\lambda_{n,k},n)J_n(\lambda_{n,k}r)\sin n\theta \cos \lambda_{n,k}ct$

$\bar{U}(\lambda_{n,k},n) = h_n\left[f_s\{u(r,\theta,0)\}\right]$

$\displaystyle = \frac{4}{\pi J_{n+1}^2(\lambda_{n,k})}\int_0^1\int_0^{\pi} rf(r,\theta)J_n(\lambda_{n,k}r)\sin\theta\, dr\, d\theta$

$J_n(\lambda_{n,k}) = 0, \qquad k = 1,2,\ldots$

3.43. Consider the temperature distribution $u(r,t)$ in a hollow long cylinder of inner radius a and outer radius b. The inner and outer surfaces are kept at zero temperature, and the cylinder started with initial temperature $f(r)$.

(a) Write the boundary and initial value problem for $u(r,t)$.

(b) Show that a compatible transform for the radial variation of this problem is with kernel:

$$K(\lambda_n r) \equiv Y_0(\lambda_n a)J_0(\lambda_n r) - J_0(\lambda_n a)Y_0(\lambda_n r)$$

where Y_0 is the *Bessel function of the second kind of order zero* and $\{\lambda_n\}$ are the zeros of $Y_0(a\lambda_n)J_0(b\lambda_n) - J_0(a\lambda_n)Y_0(b\lambda_n) = 0$.

Hint: See Appendix A.5 and its exercises for these Bessel functions.

(c) Use the (combined) Hankel transform

$$U(\lambda_n,t) = \mathcal{H}_c\{u(r,t)\} = \int_a^b rK(\lambda_n r)u(r,t)\,dr$$

to solve the problem.

(d) Use the method of separation of variables to solve the problem.

ANS. (a) $\dfrac{\partial^2 u}{\partial r^2} + \dfrac{1}{r}\dfrac{\partial u}{\partial r} = \dfrac{1}{k}\dfrac{\partial u}{\partial t}, \qquad a < r < b,\ t > 0$

$u(a,t) = 0, \qquad t > 0$

$u(b,t) = 0, \qquad t > 0$

$u(r,0) = f(r), \qquad a < r < b$

Also $|u(r,t)| < M$

(c) $u(r,t) = \displaystyle\sum_{n=1}^{\infty} \dfrac{U(\lambda_n,0)}{\|K(\lambda_n,-)\|} e^{-k\lambda_n^2 t}K(\lambda_n r)$

$U(\lambda_n,0) = \displaystyle\int_a^b rf(r)K(\lambda_n r)\,dr$

$\|K(\lambda_n,-)\|^2 = \displaystyle\int_a^b rK^2(\lambda_n r)\,dr$

Section 3.5 Classical Orthogonal Polynomial Transforms

3.44. (a) Use the orthogonality property (3.144) to derive (3.146) from (3.147).

(b) Use the Rodrigues formula (3.149) to establish the Legendre polynomials $P_0(x)$, $P_1(x)$, $P_2(x)$, and $P_3(x)$. See Appendix A.5.

(c) Illustrate that $P_2(x)$ and $P_3(x)$ are orthogonal on $(-1, 1)$.

3.45. (a) Find the Legendre–Fourier series, as in (3.147), for the following functions:

(i) $f(x) = \sqrt{\dfrac{1-x}{2}}$, $\qquad -1 < x < 1$

(ii) $f(x) = \begin{cases} 0, & -1 < x < 0 \\ 1, & 0 < x < 1 \end{cases}$

Hint: See the table in Appendix C.13.

(iii) $f(x) = \begin{cases} 0, & -1 \le x < 0 \\ \frac{1}{2}, & x = 0 \\ 1, & 0 < x \le 1 \end{cases}$

(iv) $f(x) = |x|$, $\qquad -1 < x < 1$

(v) $f(x) = x^3$, $\qquad -1 < x < 1$

(vi) $f(x) = \dfrac{1}{1 - 2ax + a^2}$, $\qquad -1 < x < 1,\, a > 1$

Hint: See the table in Appendix C.13.

(vii) $f(x) = \begin{cases} 0, & -1 < x < 0 \\ x, & 0 < x < 1 \end{cases}$

(b) To what value does the Legendre–Fourier series converge at

$x = 0,$ $\qquad x = \frac{1}{2},$ $\qquad x = 1,$ $\qquad x = -\frac{2}{3}.$

(c) For which of these functions in part (a) does the series converge uniformly?

(d) Attempt to compute the Legendre–Fourier series representation (found in part a) of each of these functions on the larger interval $(-2, 3)$. This is to verify the absence of the periodicity that we always have for the trigonometric Fourier series in (3.10)–(3.13). Graph your results on $(-2, 3)$.

(e) Compare your computations of the Legendre–Fourier series on $(-1, 1)$ in part (d) with the *discrete* Legendre transform approximation of three of these functions (in part *a*), as done in Exercise 4.32 of Chapter 4. See the detailed answer of Exercise 4.32d.

ANS.   (a)   (ii) $f(x) = \frac{1}{2}P_0(x) + \frac{3}{4}P_1(x) - \frac{7}{16}P_3(x) + \frac{11}{32}P_5(x) - \cdots$

   (iv) $f(x) = \frac{1}{2}P_0(x) + \frac{5}{8}P_2(x) - \frac{9}{48}P_4(x) + \cdots$

   (vi) $f(x) = P_0(x) + aP_1(x) + a^2P_2(x) + a^3P_3(x) + \cdots$

   (vii) $f(x) = \frac{1}{4}P_0(x) + \frac{1}{2}P_1(x) + \frac{5}{16}P_2(x) - \frac{3}{32}P_4(x) + \cdots$

(b)   (i)   To the value of the function.

   (ii)   To $\frac{1}{2}$ at $x = 0$, the rest to the value of the function.

   (iii)   To the value of the function.

   (iv)   To the value of the function.

   (v)   To the value of the function.

   (vi)   To the value of the function.

   (vii)   To the value of the function.

   (viii)   To the value of the function.

(c)   See the answer to part (b).

3.46.   (a)   Use the orthogonality property (3.145) to derive (3.156) from (3.159).

(b)   Establish (3.161) as the self-adjoint form of (3.158).

(c)   Use the method of separation of variables to solve Example 3.15.

3.47.   Consider the problem of the *electrified disc* in Example 2.27 of Section 2.7, where the Hankel transform was used to solve it. As was done in Exercise 1.5 of Chapter 1, consider now the change for this problem to the *oblate spheroidal* coordinate, which we presented for Exercise 1.5 in order to better accommodate the representation of the boundary conditions in (E.5), (E.6) and the resulting differential equation in (E.8) of the same exercise.

(a)   Use the finite Legendre transform $l\{v(\mu, \zeta)\} = V(n, \zeta)$ on the differential operator of (E.8) in the μ variable to reduce it to a simpler differential equation in $V(n, \zeta)$.

(b) Use the Legendre transform of the boundary condition $v(\mu, 0) = u_0$ in (E.5) to show that $V(n, 0) = 0$ for positive n.

Hint: For the Legendre integral of $V(n, 0)$ note that $P_n(x)$ is orthogonal to $1 = P_0(x)$, which gives $V(0, 0) = 2$ and $V(n, 0) = 0$ for $n > 0$.

(c) Since ζ tending to infinity corresponds to $z \rightarrow \mp\infty$, as seen in (E.3) of Exercise 1.5, we can safely assume that the potential $v(\mu, \zeta)$ vanishes as ζ tends to infinity. Show that $V(n, \zeta)$ in part (a) vanishes as $\zeta \rightarrow \infty$ and, together with $V(n, 0) = 0$ for $n > 0$, that $V(n, \zeta)$ also vanishes for $n > 0$.

(d) Use the result in part (c) to show that the transformed equation in part (a) becomes an *exact homogeneous differential equation* in $V(n, \zeta)$, $n > 0$. Then solve it for $V(n, \zeta)$, and use the finite Legendre transform inverse to find $v(\mu, \zeta)$.

Hint: For the arbitrary constant of integration use $V(0, 0) = 2u_0$.

(e) Transform the final resulting $v(\mu, \zeta)$ in part (d) to polar coordinates as $u(r, z)$, and try to compare it with the result obtained via the Hankel transform in Example 2.27 of Section 2.7.

ANS. (a) $\dfrac{d}{d\zeta}\left[(1 + \zeta^2)\dfrac{dV(n, \zeta)}{d\zeta}\right] = n(n + 1)V(n, \zeta)$

(d) $v(\mu, \zeta) = \dfrac{2u_0}{\pi}\cot^{-1}\zeta$

3.48. (a) Use the Legendre transform to solve for the potential distribution inside a hollow sphere of unit radius when the upper half of its surface is kept at a constant potential u_0 and the lower part of the surface is grounded.

Hint: See Example 3.15.

(b) Use separation of variables to solve the problem in part (a).

ANS. (a) $u(\rho, \theta) \equiv v(\rho, \cos\theta) = \dfrac{u_0}{2}\left[1 + \tfrac{3}{2}\rho P_1(\cos\theta)\right.$

$\left. - \tfrac{7}{8}\rho^3 P_3(\cos\theta) + \tfrac{11}{16}\rho^5 P_5(\cos\theta) + \cdots\right],$

$$0 < \rho < 1$$

3.49. (a) Use the Legendre transform to solve for the potential distribution outside the sphere of Example 3.15

(b) Use the Legendre transform to solve for the potential distribution outside the sphere of part (a) in Exercise 3.48.

ANS. (a) $v(\rho, \cos\theta) = \frac{1}{2}\sum_{j=0}^{\infty}(2j+1)F(j)\rho^{-j-1}P_j(\cos\theta)$, $\qquad 1 < \rho$

(b) $v(\rho, \cos\theta) = \dfrac{u_0}{2\rho}\left[1 + \dfrac{3}{2\rho}P_1(\cos\theta)\right.$

$\qquad\qquad\left. -\dfrac{7}{8\rho^3}P_3(\cos\theta) + \tfrac{11}{16}\rho^{-5}P_5(\cos\theta) + \cdots\right]$, $\qquad \rho > 1$

3.50. Consider the potential distribution $u(\rho, \theta, \phi)$ inside the unit sphere when the surface is kept at a potential $u(1, \theta, \phi) = f(\theta, \phi)$, which is antisymmetric with respect to the yz-plane, i.e., an odd function in ϕ.

With $P_n^m(x)$ as the *associated Legendre polynomial* as discussed in Appendix A.5, show that the P_n^m-associated Legendre transform

$$F(j) = \int_{-1}^{1} f(x)P_j^m(x)\,dx \equiv l_m\{f\} \tag{E.1}$$

$$f(x) = \sum_{j=0}^{\infty}\frac{(2j+1)(j-m)!}{2(j+m)!}F(j)P_j^m(x) \equiv l_m^{-1}\{F\} \tag{E.2}$$

is compatible with the angular variation of the Laplace equation.

Hint: You may consult Section 1.5 for the general Sturm–Liouville transforms, of which the present transform is a clear special case.

3.51. Use the associated Legendre transform developed in Exercise 3.50 to solve its boundary value problem of the potential $u(\rho, \theta, \phi)$.

Hint: Note that the solution must be bounded at the origin to choose the r^n dependence, and it is bounded at $\theta = 0$ to choose the bounded $p_n^m(\cos\theta)$ and not $Q_n^m(\cos\theta)$. Also, the problem is periodic in ϕ.

ANS. $u(\rho,\theta,\phi) = v(\rho,\cos\theta,\phi) = u(\rho,x,\phi)$

$$= \frac{1}{2\pi} \sum_{m=1}^{\infty} \sum_{j=0}^{\infty} \frac{(2j+1)(j-m)!}{(j+m)!} F(j,m) P_j^m(x) \sin m\phi$$

where

$$F(j,m) = l\{f_s\{f(\theta,\phi)\}\}$$

$$= \int_{-1}^{1} \int_{0}^{2\pi} f(\theta,\phi) P_j^m(x) \sin m\phi \, dx \, d\phi$$

$x = \cos\theta$.

3.52. Use the separation-of-variables method to solve the problem in Exercise 3.50 and to arrive at the same answer.

3.53. After doing Exercises 3.51 and 3.52, use inspection to show that $u(r,\theta,\phi) = r^2 \sin^2\theta \cos 2\phi$ is a solution to our problem with the particular value of the potential on the surface of the sphere being $u(1,\theta,\phi) = \sin^2\theta \cos 2\phi$.

Hint: The solution must be even in ϕ; also, $\sin^2\theta = 1 - \cos^2\theta = 1 - x^2$, which relates in a very simple way to $P_2^2(x)$, the associated Legendre polynomial $P_j^m(x)$ with $m = 2$ and $j = 2$.

3.54. (a) Use the orthogonality property (3.170) of the Tchebychev polynomials to derive the inverse transform (3.172) of (3.171).

(b) Show that this finite Tchebychev transform in (3.171) is compatible with the differential operator $(1-x^2)d^2/dx^2 - x\,d/dx$ of (3.168); i.e., prove (3.173),

$$\int_{-1}^{1} (1-x^2)^{-1/2} T_n(x) \left[(1-x^2)\frac{d^2u}{dx^2} - x\frac{du}{dx}\right] dx$$

$$= -n^2 F(n)$$

(c) Prove the important representation of the Tchebychev polynomial in (3.169)

$$T_n(x) = \cos(n\cos^{-1}x)$$

$$= x^n - \binom{n}{2}x^n(1-x^2) + \binom{n}{4}x^{n-4}(1-x^2)^2 + \cdots$$

Hint: Let $u = \cos x$, $T_n(x) = \cos nu = $ real part of $e^{inu} = \cos nu + i \sin nu = (\cos u + i \sin u)^n$. Now write the binomial expansion of $(\cos u + i \sin u)^n$ and take its real part. $\binom{n}{j} \equiv \frac{n!}{(n-j)!j!}$

3.55. (a) Show the very close relation between the Tchebychev polynomial expansion of $f(x)$ on $-1 \le x \le 1$, $x = \cos\theta$, and its Fourier cosine series as $F(\theta) = f(\cos\theta)$ on the interval $0 \le \theta \le \pi$.

Hint: Write the Fourier cosine series for $F(\theta)$, and use the important property $T_n(\cos\theta) = \cos n\theta$ that we developed in Exercise 3.54.

(b) Show that the Tchebychev series expansion for the $f(x) = \mathrm{sgn}\,x$ is

$$\mathrm{sgn}\,x = \frac{4}{\pi} \sum_{n=0}^{\infty} \frac{(-1)^n T_{2n+1}(x)}{2n+1}, \qquad -1 < x < 1 \qquad \text{(E.1)}$$

Hint: For evaluating the coefficients of this series see the answer to part (a), where you can evaluate a simple integral of a trigonometric function instead of that of Tchebychev polynomial.

(c) Compute the partial sum of the Tchebychev series in (E.1) of $\mathrm{sgn}\,x$ for $N = 5$ and 10, and make your observation about the presence of the Gibbs phenomenon near the jump discontinuity at $x = 0$ and its absence near the end points $x = \mp 1$, a remarkable property of the Tchebychev polynomial expansion!

ANS. (a) $f(\cos\theta) = F(\theta) = \dfrac{a_0}{2} + \displaystyle\sum_{n=1}^{\infty} a_n \cos n\theta$

$$= \frac{a_0}{2} + \sum_{n=1}^{\infty} a_n T_n(x)$$

$$a_n = \frac{2}{\pi} \int_0^{\pi} f(\cos\theta) \cos n\theta \, d\theta$$

$$= \frac{2}{\pi} \int_{-1}^{1} f(x) T_n(x)(1 - x^2)^{-1/2} \, dx$$

$$a_0 = \frac{1}{\pi} \int_0^{\pi} (1 - x^2)^{-1/2} f(x) \, dx$$

Section 3.6   The Generalized Sampling Theorem

3.56. (a)   Use the Parseval equality in (3.65) to prove the generalized sampling expansion in (3.175), (3.176).

Hint: Let $f(x) = \rho(x)K(x,t)$, $g(x) = F(x)$.

(b)   Use the following properties of the Bessel function $J_n(x)$ to establish (3.180) the sampling function $S(t,t_{m,n})$ of the Bessel-type sampling series (3.181).

$$\int xJ_n^2(ax)\,dx = \frac{x^2}{2}\left[J_n^2(ax) - J_{n-1}(ax)J_{n+1}(ax)\right] \qquad \text{(E.1)}$$

$$\int xJ_n(ax)J_n(bx)\,dx$$

$$= \frac{bxJ_n(ax)J_{n-1}(bx) - axJ_{n-1}(ax)J_n(bx)}{a^2 - b^2} \qquad \text{(E.2)}$$

$$J_{n-1}(x) + J_{n+1}(x) = \frac{2n}{x}J_n(x) \qquad \text{(E.3)}$$

For the derivation of these and other properties of Bessel functions, see Appendix A.5 and its exercises.

(c)   Use the facts that $J_m(xt)$ is a solution of the Bessel differential equation (as in (3.78) for $\lambda = t$ and $\nu = m$), and

$$J_m(at_{m,n}) = 0, \qquad n = 1, 2, \ldots \qquad \text{(E.1)}$$

to derive the sampling function of the Bessel sampling series (3.180) in part (a).

Hint: Follow the same method of proving the orthogonality property of the solutions of the Sturm–Liouville problem as done in Example 1.9. So, write the Bessel differential equations in their self-adjoint form for $J_m(xt)$ and $J_m(xt_{m,n})$ as (E.4) and (E.5), respectively, multiply (E.4) by $J_m(xt_{m,n})$ and (E.5) by $J_m(xt)$, and then subtract and integrate from 0 to 1, remembering to make use of (E.1).

(d)   Attempt any basic expression for a bound on the truncation error of the Bessel sampling series in (3.181).

Hint: See Exercise 3.29, Jerri and Joslin (1982b), and Jerri (1992).

ANS. (d) $\quad \epsilon_T \leq \dfrac{M}{a} \left| \displaystyle\sum_{n>N} \dfrac{1}{(t^2 - t_{m,n}^2)J_{m+1}(t_{m,n})} \right|,$

where we used $|f_a(t)| \leq M$ and clearly $|J_m(at)| \leq 1$.

3.57. Consider $P_t(x)$, the *Legendre function*, which is the generalization, or extension, of the *Legendre polynomial* $P_n(x)$ for t unrestricted. Now consider the Legendre function transform

$$f(t) = \int_{-1}^{1} F(x)P_{t-1/2}(x)\,dx \tag{E.1}$$

(a) Show that for $F(x)$ square integrable on $[-1,1]$, this transform satisfies the conditions of the generalized sampling theorem in (3.174), (3.176).

   Hint: See that $\{K(t_n,x)\} = \{P_n(x)\}$ (for $f(t_n) = f(n + \frac{1}{2})$), is the orthogonal Legendre polynomials set on $[-1,1]$.

(b) Write the sampling series (3.176) for the present Legendre transform in (E.1).

   Hint: For the evaluation of the integral for the sampling function, consult Erdelyi et al. (1953a, p. 170, Eq. 3.12(17)), Campbell (1964), and Jerri (1969a).

(c) Try part (a) for the more general Legendre transform with kernel $P_t^\mu(x)$ as the *associated Legendre function*.

   Hint: Consult the last and first references given in part (b).

(d) Develop the generalized sampling theorem and its expansion that is associated with the kernel $K(x,t) = C_t^\nu(x)$ as the Gegenbauer function.

   Hint: See parts (a) and (b) for references; the answer is in the last reference of the hint to part (b).

ANS. (b) $\quad f(t) = \displaystyle\sum_{n=0}^{\infty} f\left(n + \tfrac{1}{2}\right) S_n(t)$

$$S_n(t) = \frac{2n + 1}{\pi} \frac{\sin \pi (t - n - \frac{1}{2})}{(t - n - \frac{1}{2})(t + n + \frac{1}{2})}$$

(c) In the last reference of the hint to part (b).

3.58. Under certain conditions, the generalized finite-limit integral transform (3.174) of the generalized sampling expansion (3.175), (3.176) may be equivalent to a bandlimited function, and thus we may, instead, use the usual Shannon (Fourier) sampling (2.184). This often happens when the kernel $K(x,t)$ is itself a bandlimited function.

(a) Consider the (bandlimited) representation of the Bessel function,

$$J_n(tx) = \frac{(t/2x)^n}{\sqrt{\pi}\Gamma(n+1/2)} \int_{-x}^{x} e^{i\omega t}(x^2 - \omega^2)^{n-1/2} d\omega \qquad \text{(E.1)}$$

Show that the finite-limit Hankel transform,

$$h(t) = \int_0^1 x J_n(xt) H(x) dx \qquad \text{(E.2)}$$

can be reduced to $t^n f_a(t)$, where $f_a(t)$ is a bandlimited function (with bandlimit $a = 1$). Thus $h(t)/t^n = f_a(t)$ is worthy of the simpler (Fourier) sampling expansion in (2.184).

Hint: Substitute $J_n(xt)$ as in (E.1) in the integral of (E.2), then interchange the order of the two integrals. For the general equivalence problem of the usual (Shannon) and the generalized sampling theorem, see Jerri (1969b) and Campbell (1964).

(b) As in part (a), use the following integral representation of the Legendre function (with t not restricted to half integers);

$$P_{t-1/2}(\cos\theta) = \frac{\sqrt{2}}{2\pi} \int_{-\theta}^{\theta} \frac{e^{i\alpha t}}{\sqrt{\cos\alpha - \cos\theta}} d\alpha \qquad \text{(E.3)}$$

to show that the *finite-limit* Legendre (function) transform

$$f(t) = \int_{-1}^{1} P_{t-1/2}(x)g(x) dx \qquad \text{(E.4)}$$

relates in a very simple way to a bandlimited function with band limit $a = \pi$.

Hint: See the steps suggested by the hint in part (a), where we substitute from (E.3) in the integral of (E.4), then interchange the resulting two integrals.

3.59. *Sampling for infinite-limit transforms.* In all our illustrations of the generalized sampling expansion (3.174)–(3.176) we considered finite-limit integral transforms with $\{K(x,t_n)\}$ being orthogonal on a finite interval. The following exercises deal with illustrations of the generalized sampling expansion where the set $\{K(x,t_n)\}$ is orthogonal on an infinite interval.

(a)   We know that the Laguerre polynomials $L_n(x)$, as presented in Section 3.5, are orthogonal on the semi-infinite interval $(0,\infty)$ with weight function $\rho(x) = e^{-x}$. We consider the Laguerre function $L_\nu(x)$, with $\nu \geq 0$, and the following particular Laguerre (function) transform:

$$f(\nu) = \int_0^\infty e^{-x} L_\nu\left(\frac{\lambda x}{1-\lambda}\right) F(x)\,dx, \qquad -1 < \lambda < 1,\; \nu \geq 0$$

(E.1)

Derive the generalized sampling expansion of this $f(\nu)$.

Hint: See Jerri (1976) and the tutorial paper on the general subject of the sampling expansion (Jerri, 1977a).

(b)   We also know that the Hermite polynomials $H_n(x)$ (of Section 3.5) are orthogonal on the infinite interval $(-\infty,\infty)$ with a weight function $\rho(x) = e^{-x^2}$. The parabolic cylinder (Hermite–Weber) functions $D_n(x)$ are simply related to $H_n(x)$ as

$$H_n(x) = 2^{n/2} e^{x^2} D_n(\sqrt{2}x),$$

which are orthogonal on $(-\infty,\infty)$ with constant weight function $\rho(x) = 1$. Consider the *parabolic cylinder* (Hermite–Weber) function *transform*

$$f(\nu) = \int_0^\infty D_\nu(x) f(x)\,dx$$

and develop its (generalized) sampling expansion.

Hint: See Jerri (1982).

3.60.   (a)   In Exercise 3.27 we developed the sampling expansion for bandlimited functions where both the samples of the signal and its derivative were used to advantage in doubling the sample spacing to $2\pi/a$, where a is the bandlimit. This exercise deals with the possible extension of the (generalized) Bessel sampling series in (3.181) to involve the samples of the finite Hankel transform and its derivative.

Hint: Here we lack the periodicity of the Fourier trigonometric kernel as well as the simple shifting properties as a consequence of the exponential kernel of the Fourier transform. The best lead is to use complex contour integration.

See Jerri (1977a) and the reference therein for the detailed derivation of this and other similar results.

(b) Use the same method to derive the result in Exercise 3.27.

3.61. (a) Show that the generalized translation (3.185a) by $\tau = 0$ for the J_0-Hankel transform leaves the transform unchanged, i.e., $f(t\theta 0) = f(t)$. Verify that for the J_0-Hankel transform of the gate function $p_a(\omega)$.

Hint: Note $\overline{J_0(\omega 0)} = J_0(0) = 1$ for (3.185).

(b) Prove that the convolution product $f * g$ in (3.186) of the generalized transforms is (i) commutative and (ii) associative.

(c) Prove the convolution theorem in (3.187) for the general transforms (3.183), (3.184). For the application of the generalized translation to a time-varying system, see Jerri (1972), (1992).

Hint: See the derivation of the convolution Theorem 2.18 of the (special case) Fourier transforms.

(d) Prove that

$$\int_0^\infty \frac{yJ_1(a(y\theta x))}{y\theta x} \frac{J_1(b(y\theta z))}{y\theta z} \, dy = \frac{J_1(b(z\theta z))}{(z\theta x)},$$

$$0 < b < a$$

where $J_1(a(y\theta x))/y\theta x$ stands for the "generalized" J_0-Hankel type translation of $J_1(ay)/y$.

Hint: Use the Parseval equality associated with the J_0-Hankel transform, recognizing that $J_1(ay)/y$ and $J_1(a(y\theta x))/y\theta x$ are the J_0-Hankel transforms of the gate function $p_a(t)$ and $J_0(xt)p_a(t)$, respectively. For other proofs, see the author's note (1983b). For the application of the generalized translation to time-varying systems, see Jerri (1972), (1977a) and (1992).

3.62. (a) In Exercise 2.62 of Chapter 2 we showed that the (Fourier) self-convolution product of the gate function $p_a(\omega) = \phi_1(\omega)$ is $\phi_2(\omega)$,

$$\phi_2(\omega) = \phi_1 * \phi_1 = \begin{cases} 2a - |\omega|, & |\omega| < 2a \\ 0, & |\omega| > 2a \end{cases} \tag{E.1}$$

Attempt to find $\psi_2(\omega)$, the (J_0-Bessel) self-convolution $\psi_2(\omega) = \psi_1 * \psi_1$ of $p_a(\omega) = \psi_1(\omega)$, as a general (Bessel-type) hill function corresponding to $R = 1$ of the general case in (3.188).

(b) Investigate the domain of $\psi_2(\omega)$ and see if $\psi_2(\omega)$ vanishes identically only outside the larger domain $(0, 2a)$.

Hint: You may resort to writing the J_0-Hankel transform as a double Fourier transform to take advantage of the usual translation property, and thus look at the overlap of the two domains as discs of radius a to arrive at the domain of the convolved gate functions (or circular apertures!) over these discs.

3.63. (a) Prove the discretization of the general convolution product in (3.192).

Hint: See the derivation of the generalized sampling theorem in (3.174)–(3.176).

(b) Use the discretization of the general convolution product in (3.192) to prove the generalized sampling theorem in (3.174)–(3.176).

3.64. (a) In parallel to the divergent Fourier series in (3.198) that defines the *impulse train* in the t-space, develop its analog in the Fourier ω-space.

(b) Verify the presence of the impulse train by attempting to compute numerically the (divergent) series

$$\sum_{n=1}^{\infty} \sin \frac{n\pi}{a} t$$

Try 10, 20, and then 40 terms for the series to see how the pulses develop, and establish their locations.

3.65. Consider the (divergent) Bessel–Fourier series in (3.199) that defines the impulse train for the Bessel–Fourier series analysis.

(a) Use 10, 20, and then 40 terms in the series to verify the Bessel type impulse train in Fig. 3.13. In particular, verify (i) the impulse locations at $\omega_m = 2mb$ and (ii) the variations of their amplitudes $\{d_m\}$.

(b) Use your results for d_m in part (a) to see how they fit the identity in (3.203).

(c) Show that $d_0 = 1$ in (3.200).

Hint: When we integrate on $(0,b)$ for this result after (3.200), the pulses (with amplitudes) d_1, d_2, ... are outside the interval of integration $(0,b)$.

3.66. Use the series

$$1 + \sum_{m=1}^{\infty} d_m J_0(2mbt)$$

with d_m from Exercise 3.65a to verify numerically that this series also defines an impulse train in the transform t-space with the pulses located at $t_n = j_{0,n}/b$. For more details, see Jerri (1988).

Section 3.7† A Remark on the Transform Methods and Nonlinear Problems

3.67. Consider the following boundary value problem for the nonlinear chemical concentration $y(x)$ in a *planar catalyst pellet*:

$$\frac{d^2y}{dx^2} = \phi^2 y^2, \qquad 0 < x < 1$$

$$y'(0) = 0, \qquad y(1) = 1$$

where ϕ is the Theile modulus.

(a) Find a compatible transform that algebraizes d^2y/dx^2, and thus write the (nonlinear) *summation* equation in the transform space.

Hint: Try a finite cosine transform, and consult the first reference given in the hint for the following part (b).

(b) Try an iterative method to solve the resulting representation in part (a) for $\phi = 1$, 5, 10, and 50.

Hint: Watch for the convergence of such iterative method. For a modified iterative method that converges and a detailed answer, see Jerri (1983b), Jerri, Weiland, and Herman (1987a), and the general reference Jerri (1991).

---

†Optional.

3.68. As in Exercise 3.67, consider the same type of problem for the chemical concentration $u(r)$ inside a *circular pellet* of unit radius, assuming the same type of boundary conditions.

(a) Write the boundary value problem in $u(r)$.

(b) Find the compatible transform that algebraizes the radial differential operation.

Hint: Try a finite J_0-Hankel transform and see the second reference given for Exercise 3.67.

(c) As in part (c) of Exercise 3.67, attempt an iterative method to solve the nonlinear summation equation representation in the Hankel transform space.

Hint: See the method and answer in the second reference given for Exercise 3.67.

ANS. (a) $\dfrac{1}{r}\dfrac{d}{dr}\left(r\dfrac{du}{dr}\right) = \phi^2 u^2, \qquad 0 < r < 1$ (E.1)

$u_r(0) = 0$ (E.2)

$u(1) = 1$ (E.3)

(b) See hint for part (c).

3.69. For the same type of problem as in Exercises 3.67 and 3.68, consider the chemical concentration $u(\rho)$ in a spherical pellet of unit radius.

(a) Set up the boundary value problem in $u(\rho)$.

(b) Find a transform that algebraizes the radial differential operation.

Hint: Try the finite transform:

$$\mathcal{R}\{u\} \equiv U(n\pi) = \int_0^1 \rho^2 u(\rho)\,\frac{\sin n\pi\rho}{\rho}\,d\rho$$

$$u(\rho) = 2\sum_{n=1}^{\infty} U(n\pi)\frac{\sin n\pi\rho}{\rho}$$

(c) As in parts (c) of Exercises 3.67 and 3.68, use an iterative method to solve the resulting nonlinear summation equation in part (b).

Hint: For a modified iterative method and the answer, see the second reference given for Exercises 3.67 and 3.68 and the general reference on the method in Jerri (1991).

ANS. (a) $\dfrac{1}{\rho^2}\dfrac{d}{d\rho}\left(\rho^2\dfrac{du}{d\rho}\right) = \phi^2 u^2, \qquad 0 < \rho < 1$

$u_\rho(0) = 0, \qquad u(1) = 1$

(b) See hint for parts (b) and (c).

3.70. Consider the variable coefficient equation in $u(x,t)$,

$$\dfrac{\partial u}{\partial t} - x\dfrac{\partial u}{\partial x} = 0, \qquad -1 < x < 1, \ t > 0$$

$$u(-1,t) = u(1,t), \qquad t > 0$$

$$u(x,0) = \cos \pi x, \qquad -1 < x < 1$$

whose exact solution is known as $u(x,t) = \cos(\pi x e^t)$.

(a) Find a transform that algebraize the spatial differential operator $x\frac{\partial u}{\partial x}$.

Hint: Try the finite exponential Fourier transform in (3.115), (3.116).

(b) Attempt an iterative method to solve for the transform in part (a); then invert to find the solution.

Hint: See Jerri (1984 or 1991) and the references to Exercises 3.67–3.69.

3.71. Consider the (nonlinear) Korteweg-deVries (KdV) equation $u(x,t)$

$$u_t + 6uu_x + u_{xxx} = 0, \qquad -\infty < x < \infty, \ t > 0 \qquad \text{(E.1)}$$

along with its *solitary* initial wave

$$u(x,0) = 2k^2 \operatorname{sech}^2 kx, \qquad -\infty < x < \infty \qquad \text{(E.2)}$$

(a) Apply the exponential Fourier transform to show that it transforms the nonlinear KdV equation in (E.2) to a nonlinear *self-convolution integral equation* in $U(\lambda,t) = \mathcal{F}\{u(x,t)\}$.

(b) Attempt an iterative method to solve for the resulting nonlinear *integrodifferential equation* in $U(\lambda,t)$ in part (a) with $U(\lambda,0) = \mathcal{F}\{u(x,0)\}$ as given in the answer to part (a).

Hint: See the references to Exercises 3.67 to 3.69, Jerri (1984, 1991), and Jerri and Tse (1987b).

ANS. (a) $\dfrac{dU(\lambda,t)}{dt} = i\lambda^3 U(\lambda,t)$

$$- 6i \int_{-\infty}^{\infty} (\lambda - \mu)U(\lambda - \mu,t)U(\mu,t)d\mu$$

$$U(\lambda,0) = 2\pi\lambda \operatorname{csch} \frac{\pi}{2k}\lambda$$

3.72. Consider the (nonlinear) *Benjamin–Ono* (B-O) equation in $u(x,t)$,

$$u_t + 2uu_x + \mathcal{H}\{u_{xx}\} = 0, \qquad -\infty < x < \infty, \; t > 0 \qquad (E.1)$$

where $\mathcal{H}\{f\}$ is the *Hilbert transform* as defined in (2.267).

(a) Show the following relation between the Fourier transform $F(\lambda) = \mathcal{F}\{f\}$ and the Hilbert transform $\mathcal{H}\{f\}$:

$$\mathcal{F}\{\mathcal{H}\{f\}\} = -i \operatorname{sgn} \lambda F(\lambda) \qquad (E.2)$$

which is clearly very vital for any attempt at Fourier transforming the B-O equation in (E.1).

Hint: Note that for $\mathcal{F}\{1/\pi t\} = -i \operatorname{sgn}\omega$ you may also consult Exercise 2.59e. Also, for this problem, Exercise 3.71, the nonlinear Schrödinger equation, and the two-dimensional nonlinear Kadomtsev–Petviashvili (K-P) equation, see the references for Exercise 3.71.

(b) Find the Fourier transform of the B-O equation in $U(\lambda,t) = \{u(x,t)\}$,

(c) See Exercise 3.71 for attempting an iterative method for solving the nonlinear integrodifferential equation in $U(\lambda,t)$ of part (b).

ANS. (a) $\dfrac{dU}{dt} = 2i \int_{-\infty}^{\infty} (\lambda - \mu)U(\lambda - \mu,t)U(\mu,t)d\mu$

$$\lambda^2 i(\operatorname{sgn}\lambda)U(\lambda,t)$$

4

DISCRETE TRANSFORMS

The fast Fourier transform (FFT), which has been studied and applied extensively in the past twenty-five years, is a very efficient algorithm for computing the discrete Fourier transform (DFT). As we shall see shortly, the discrete Fourier transform is only an approximation to the Fourier integral or series that we would like to compute. Having this in mind, our preliminary discussion will center on the main errors involved in such an approximation before appealing to the fast Fourier transform algorithm.

In some instances, where our main interest is in the discrete data and its transformation back and forth, the discrete Fourier transforms play an analogous role for *difference equations* to that of integral transforms for differential equations. Indeed, we can use the algebra of *summation by parts*, instead of integration by parts, to show that the discrete transforms are compatible with difference equations (see Section 4.1.5). However, in most instances we seek the difference equation as a last resort for approximating the differential equation; hence at the end we still face approximating an integral or infinite sum by a finite discrete sum.

In this spirit we will present the z-transform, which is the finite (or discrete) analog of the Laplace transform. Then we will illustrate the application of this z-transform to solving relevant discrete initial value problems.

In the first section we will present the discrete Fourier transform as a finite sum of discrete values, with its basic properties and how it approximates Fourier series and integrals. This is supported by a number of

509

numerical and graphical illustrations. It is followed by a simple treatment establishing the fast Fourier transform algorithm, with emphasis on the significance of its applications. The last part of this section will cover the practical applications of the discrete Fourier transform and its fast algorithm, the FFT, for computing Fourier series and integrals. The emphasis will be on clear illustrations of the error incurred in such approximation. These include the *truncation error* and the *aliasing error*, which will be discussed in full detail and in the direction of finding some remedies to correct for them. Also, we will have a very detailed discussion of the *Gibbs phenomenon* with serious attempts to minimize it.

In parallel with the discrete Fourier transform, we will discuss other classical discrete transforms in the sense that their kernels are the discrete version of the corresponding integral transform's kernels. In Sections 4.2 and 4.3 we will present the discrete orthogonal polynomial transform and briefly mention an attempt at a discrete Hankel transform. Such discrete transforms may be distinguished from others with kernels like the Walsh and Rademacher functions, which are not solutions of the difference equation analog of Sturm–Liouville type differential equations.

Of central importance in the development of the discrete Fourier transforms, for the determination of the required spacing of its discrete values (samples) and for its approximation of Fourier series or integrals, is the *sampling theorem*, which we discussed in detail as (2.184) of Section 2.4.2 and proved it as (3.121) of Section 3.3. In developing the general discrete transforms, we shall utilize the generalized version of this sampling theorem. Such a *generalized sampling theorem* is basically the extension to transforms with general kernels as solutions of the Sturm–Liouville problem, which we presented and proved and illustrated in Section 3.6.

4.1 Discrete Fourier Transforms

As with most of the development over the past twenty-five years that appeared in the electrical engineering literature, we will adapt its notation[†] for the most part in our presentations in this chapter.

Consider the Fourier transform $G(f)$ of $g(t)$,

$$G(f) = \int_{-\infty}^{\infty} g(t)e^{-j2\pi ft}\, dt, \qquad j = \sqrt{-1} \tag{4.1}$$

[†]A detailed treatment may be found in Brigham (1974, 1988), Papoulis (1977), and, with a numerical analysis touch, Hamming (1973).

and its inverse,

$$g(t) = \int_{-\infty}^{\infty} G(f)e^{j2\pi ft}\, df \tag{4.2}$$

where f is the frequency in hertz (cycles per second). In many examples we will use $2\pi f$ for the frequency and t for time. The discrete Fourier transform $\tilde{G}$ and its inverse $\tilde{g}$ are defined respectively as,

$$\tilde{G}_n \equiv \tilde{G}\left(\frac{n}{NT}\right) = \sum_{k=0}^{N-1} \tilde{g}(kT)e^{-j2\pi nk/N} \equiv \tilde{\mathcal{F}}\{\tilde{g}_k\},$$

$$n = 0, 1, \ldots, N-1 \tag{4.3}$$

at its discrete frequency values $f_n = n/NT$,

$$\tilde{g}_k = \tilde{g}(kT) = \frac{1}{N}\sum_{n=0}^{N-1} \tilde{G}\left(\frac{n}{NT}\right) e^{j2\pi nk/N} = \tilde{\mathcal{F}}^{-1}\{\tilde{G}_n\},$$

$$k = 0, 1, \ldots, N-1 \tag{4.4}$$

at its discrete time values $t_k = kT$.

Here we used $\tilde{G}_n$ and $\tilde{g}_n$ instead of G_n and g_n to indicate that the discrete transform is only an approximation of the continuous transform. We will soon verify that $\tilde{G}$ and $\tilde{g}$ are exact discrete transforms of each other. However, it must be stressed that, in general, $\tilde{G}$ and $\tilde{g}$ are not Fourier transforms of each other in the sense of (4.1), (4.2). [The detailed derivation of (4.3), (4.4) is presented in Section 4.1.4.]

As in the case of the exponential transforms, these discrete transforms in (4.3), (4.4) are very symmetric in terms of their kernels and the domain of the summation. Thus some authors may refer to one of them as the transform where the other is its inverse. Here we have already referred to $\tilde{G}_n \equiv \tilde{G}(n/NT)$ in (4.3) as the DFT of $\tilde{g}_k \equiv \tilde{g}(kT)$ in (4.4), where this choice, obviously, defines $\tilde{g}_k$ as the inverse discrete Fourier transform (IDFT) of $\tilde{G}_n \equiv \tilde{G}(n/NT)$. However, and especially in relation to other work, we should feel free to refer to either one as the DFT of the other without worrying about any confusion. Also, for abbreviation, we shall sometimes denote $\tilde{G}(n/NT)$ and $\tilde{g}(kT)$ by $\tilde{G}_n$ and $\tilde{g}_k$, respectively (also, we may use $\tilde{G}(n)$ and $\tilde{g}(n)$, $\tilde{G}$ and $\tilde{g}$).

For the same reason of abbreviation, and in line with the notation of most authors, we shall write the discrete transforms pair (4.3), (4.4) in its following equivalent (more compact) form via the use of $W = e^{j2\pi/N}$, the

the Nth root of unity,

$$\tilde{G}_n = \sum_{k=0}^{N-1} \tilde{g}_k W^{-nk}, \qquad W = e^{j2\pi/N}, \qquad 0 \leq n \leq N-1 \qquad (4.3a)$$

$$\tilde{g}_k = \frac{1}{N} \sum_{n=0}^{N-1} \tilde{G}_n W^{nk}, \qquad 0 \leq k \leq N-1 \qquad (4.4a)$$

(sometimes $W = e^{-j2\pi/N}$ is also used).

These compact forms are needed to simplify the notation when lengthy analytical computations are performed, as in solving difference equations by using these discrete transforms in Section 4.1.5. Another reason for this notation arises in the derivation of the FFT algorithm for these discrete transforms. This is in the sense that writing $W = e^{j2\pi/N}$ as the Nth root of unity should focus attention on the most important property of the kernels $e^{(j2\pi/N)k} = W^k$ of such transforms as being *cyclic* with period N; that is, $(W^k)^N = W^{kN} = (W^N)^k = 1^k = 1$. This is the main property of the *trigonometric (periodic)* type kernels of the Fourier transforms that gave it its present discrete version (4.3), (4.4) and, more important, its efficient FFT algorithm for computing these discrete transforms.

In the following example we will present a very simple illustration of finding the inverse discrete Fourier transform $\tilde{g}_k$ of the constant sequence $\tilde{G}_n = 1$ for $0 \leq n \leq N-1$.

EXAMPLE 4.1: A Discrete Fourier Transform Pair

In this example we will find $\tilde{g}_k$, the inverse DFT of the constant sequence $\tilde{G}_n = 1$ for $0 \leq n \leq N-1$. In (4.4a) above we substitute $\tilde{G}_n = 1$ to have

$$\tilde{g}_k = \mathcal{F}^{-1}\{1\} = \frac{1}{N} \sum_{n=0}^{N-1} 1 \cdot W^{nk} = \frac{1}{N} \sum_{n=0}^{N-1} (W^k)^n \qquad (E.1)$$

$$= \frac{1}{N} \frac{1 - W^{kN}}{1 - W^k} \qquad (E.2)$$

after recognizing the last sum as a finite geometric series. As long as $k \neq 0$, the numerator of this last expression is zero (since $W^{kN} = 1$) and the denominator is nonzero. If $k = 0$, we return to the original sum (E.1) to find

$$\tilde{g}_0 = \frac{1}{N} \sum_{n=0}^{N-1} 1 \cdot W^{n \cdot 0} = \frac{1}{N} \cdot N = 1$$

Therefore,

$$\tilde{g}_k = \tilde{\mathcal{F}}^{-1}\{1\} = \begin{cases} 0 & \text{if } k \neq 0 \\ 1 & \text{if } k = 0 \end{cases} \equiv \delta_{k,0} \tag{E.3}$$

or

$$\tilde{G}_n = \tilde{\mathcal{F}}\{\delta_{k,0}\} = 1, \qquad 0 \leq n \leq N - 1 \tag{E.4}$$

Here $\delta_{k,0}$ is the familiar *Kronecker delta*.

As we attempt to find the inverse DFT of the simple sequence $\tilde{G}_n = n$, $0 \leq n \leq N - 1$, we will quickly realize that we need more tools of the difference calculus. This means that we need the *summation by parts* method to perform the required finite summation of (4.4a), and we shall present and illustrate this in Section 4.1.5.

We should note that in comparison with the exponential Fourier transform pair (4.1), (4.2), the limits of summation of the discrete transforms (4.3), (4.4) are not symmetric around 0. This is done basically for the convenience of electrical engineers, and indeed discrete Fourier transforms can be defined with summation limits n_0 to $n_0 + N - 1$, where n_0 is an integer. This will become clear when we recognize that (4.3) and (4.4) are sums of periodic sequences of N samples; hence we can sum from any integer n_0 to $n_0 + N - 1$ instead of 0 to $N - 1$. This will be clarified further when we prove the discrete orthogonality (4.49) of these transforms in Section 4.1.4A. In particular, we can choose $n_0 = -M$ with $N = 2M + 1$ to have

$$\tilde{G}\left(\frac{n}{(2M+1)T}\right) = \sum_{k=-M}^{M} \tilde{g}(kT)\exp\left(-\frac{j2\pi nk}{2M+1}\right),$$
$$n = -M,\ldots,0,\ldots,M \tag{4.5}$$

$$\tilde{g}(kT) = \frac{1}{2M+1}\sum_{n=-M}^{M}\tilde{G}\left(\frac{n}{(2M+1)T}\right)\exp\left(\frac{j2\pi nk}{2M+1}\right),$$
$$k = -M,\ldots,0,\ldots,M \tag{4.6}$$

with symmetric limits of summation, which is the normal form of the exponential Fourier series or integrals. In our presentation, we can choose either of the above two forms (4.3), (4.4) or (4.5), (4.6) in a way that is suitable for the approximated Fourier integral or series under discussion. Of course, in case of even functions, it is easy to use either of these forms, which is the case for our first illustrations of Figs. 4.1–4.8.

4.1.1   Fourier Integrals, Series, and the Discrete Transforms

We should emphasize that our main concern here is the analysis of how the discrete transforms approximate Fourier integrals or series, then the FFT.

We should mention now that the finite sum (4.3) of the discrete Fourier transform $\tilde{G}(n/NT)$ would have to be multiplied by $\Delta t = T$ for it to approximate the samples $G(n/NT)$ as represented by the Fourier integral (4.1); that is,

$$G\left(\frac{n}{NT}\right) \sim T\tilde{G}\left(\frac{n}{NT}\right)$$

Such adjustment is not necessary for approximating $g(kT)$ of (4.2) by the $\tilde{g}(kT)$ of (4.4) because of the $1/N = (1/NT)T$ factor in (4.4), where the $1/NT$ gives the needed $\Delta f = 1/NT$, and T would adjust the input $\tilde{G}$ to $T\tilde{G}$ that approximates G as we have indicated above,

$$g(kT) \sim \tilde{g}(kT)$$

We note that the samples of the discrete transform $\tilde{G}(n/NT)$ start at $f_0 = 0$ and stop at $f_{N-1} = (N-1)/NT$ and that $\tilde{G}$ is periodic with period of N samples, since the exponential function $e^{-j2\pi nk/N}$ in (4.3) is periodic in n with period of N samples. We say that these periodic samples of $\tilde{G}(n/NT)$ are over the period $(0, N \cdot (1/NT) = (0, 1/T)$. The same can be said about $\tilde{g}(kT)$ in (4.4), whose samples start at $t_0 = 0$ and stop at $t_{N-1} = (N-1)T$; then they repeat with period of N samples over the interval $(0, NT)$. With this periodicity property of N samples for (4.3), (4.4), the symmetric form (4.5), (4.6) can be accommodated by translating the $N/2$ to $N-1$ samples, in the second half of the period of (4.3), (4.4), to the left by one period of N samples.

Such a periodicity property should be stressed form the beginning, since, as we shall see later, it will be the cornerstone in establishing the fast Fourier transform algorithm for computing these discrete Fourier transforms. The periodicity property may also remind us of the Fourier (trigonometric) series as the possible origin of these discrete transforms, hence a starting point for deriving their form (4.3), (4.4). If we recognize and associate these two closely related notions of the periodicity and the Fourier series origin with the discrete transforms, we can also appreciate the difficulty facing such (periodic) discrete transforms (4.3), (4.4) in approximating the (nonperiodic) Fourier integrals (4.1), (4.2), as compared to approximating the (periodic) Fourier series. Such difficulties are also clear from simple comparison of the (infinite) Fourier integral (4.1) and that of its approximation (4.3) as the *discrete finite sum* offered by the discrete Fourier transform. The *discreteness* in (4.3) may again remind us of the periodicity of Fourier series with $\bar{g}(kT)$ as the Fourier coefficients. The *finiteness* of the sum would, of course, introduce the usual truncation error of approximating Fourier integrals or infinite Fourier series by a finite sum.

A simplistic approach to overcoming these two difficulties of the discreteness (or periodicity) and the finiteness (truncation) is to take T very small, for $\bar{g}(kT)$ approximation of $g(t)$, and N very large for the limit NT to approach the point after which $g(t)$ will make very negligible contributions to the infinite integral (4.1) of $G(f)$. Of course, this situation is greatly helped if $G(f)$ is time-limited to b—that is, $g(t)$ vanishes identically for $|t| > b$—where we can pick $TN = b$. However, we would still have the discreteness problem to tackle.

It is true that, with the help of the fast Fourier transform, N can be taken very large, but we are also reminded of the finiteness, or in other words the time limitations set on the fastest of today's computing machines. So to look at it from a mathematical point of view, the Fourier integral (4.1) cannot be computed via the discrete Fourier transform (4.3) without incurring some errors, regardless of how fast an algorithm the discrete transform has. Our main discussion of this section will center on the study of such errors, namely the discreteness (sampling, or periodicity) error and the truncation error.

To start the error analysis discussion, we turn to the Fourier series as the basis for establishing the discrete Fourier transforms.

A. Fourier Series Approximation—Truncation Error

Consider the periodic function $G_s(f)$ on the same interval $(-1/2T, 1/2T)$ of the discrete Fourier transform (4.3) or (4.5). The exponential Fourier series representation of $G_s(f)$ is (see (3.116a) and (3.115a) in Section 3.3)

$$G_s(f) = \sum_{k=-\infty}^{\infty} c_k e^{-j2\pi kTf}, \qquad -\frac{1}{2T} < f < \frac{1}{2T} \tag{4.7}$$

$$c_k = T \int_{-1/2T}^{1/2T} G_s(f) e^{2\pi jkTf} \, df \tag{4.8}$$

Here we use the subscript "s" for $G_s(f)$ to emphasize its Fourier series (periodic) representation with period $1/T$. Indeed, this series' periodic extension $G_s(f)$ can be written as $\bar{G}_s$, the sum of all the translations by n/T (integer multiple of the correct period $1/T$) of $G_s(f)$,

$$\bar{G}_s(f) = \sum_{n=-\infty}^{\infty} G_s\left(f + \frac{n}{T}\right), \qquad -\infty < f < \infty \tag{4.9}$$

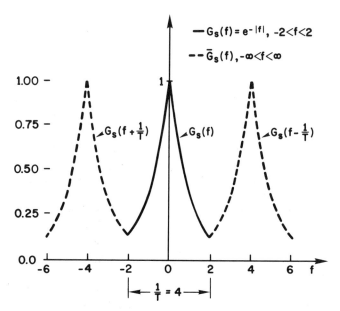

FIG. 4.1α The Fourier series (periodic) representation of $G_s(f)$, $-2 < f < 2$.

which (here) shows exact replicas of (the nontranslated) $G_s(f)$ on $(-1/2T, 1/2T)$. Figure 4.1a illustrates the function

$$G_s(f) = e^{-|f|}, \qquad -\frac{1}{2T} < f < \frac{1}{2T}$$

on its (basic) domain $(-1/2T, 1/2T)$ and two replicas of its periodic extensions $\bar{G}_s(f)$ with period $1/T = 4$.

We stress here that the *periodic extension* $G_s(f)$, $-\infty < f < \infty$, of the Fourier series representation $G_s(f)$ in (4.7) is the same as $\overline{G}_s(f)$, the *superposition of all the translations by n/T of $G_s(f)$* on the right side of (4.9), as long as we know that we have a periodic wave $G(f)$ on $(-1/2T, 1/2T)$ with *exact* period $1/T$. Thus we write it as $G_s(f)$, $\check{G}(f)$, or $\overline{G}_s(f)$. This is very suitable for our first step toward using the discrete Fourier transform in (4.10), where the function at hand $G(f)$ is committed as periodic with period $1/T$ by the fact that its samples $G(n/NT)$ are expressed in terms of the (periodic) Fourier series in (4.10). However, for a general function $G(f)$ to be treated via the same DFT concept, we may not even know its exact period $1/T$, and we have to guess it as $1/T' < 1/T$ to be used in (4.9). The more likely situation is that the function at hand $G(f)$, $-\infty < f < \infty$, is not even periodic (or $1/T = \infty$), which is what we face with functions represented by the Fourier integral. In this situation, and to use the DFT, we are forced to estimate a period $1/T$ to use in (4.10), with a very obvious error! This turns out to be one of the main topics in the error analysis of the DFT approximation of Fourier integrals, namely the *aliasing error*. Such an error can be clearly illustrated when the above concept of the superposition of all the translations by n/T of $G(f)$ is used for this given wave $G(f)$ with (the forced upon it) period $1/T$. In this case, and in contrast to the (clean) periodic extension $\tilde{G}_s(f)$ of (4.9), the superposition denoted here by $\overline{G}(f)$

$$\overline{G}(f) = \sum_{n=-\infty}^{\infty} G\left(f + \frac{n}{T}\right)$$

will show *overlap* between the translations, which makes $\overline{G}(f)$ much different from $G(f)$ on $(-1/2T, 1/2T)$. This overlap is the essence of the *aliasing error*. This is illustrated clearly for the function $G(f) = e^{-|f|}$, $-\infty < f < \infty$, in Fig. 4.6 with a period of 4. For most of our treatment we shall reserve the thick bar $^{-}$ on $\overline{G}$ or $\overline{g}$ for such illustration of the translations by an integer multiple of (an often wrong!) period $1/T$, and this should be differentiated from the short thin bar $^{-}$ used for the periodic extension in (4.9), and the long thin bar $\overline{}$ used for complex conjugation.

Figure 4.1b illustrates the Fourier coefficients c_k of the above function (Fig. 4.1a),

$$c_k = 2T \frac{1 + (-1)^{k+1}e^{-1/2T}}{1 + (2\pi kT)^2}$$

which is, of course, discrete but nonperiodic. From (4.7) (and Fig. 4.1a), we note that a discrete frequency $f = n/NT$ can be chosen for $G_s(f)$ where the Fourier series (4.7) would have the same type of discrete terms,

$$G_s\left(\frac{n}{NT}\right) = \sum_{k=-\infty}^{\infty} c_k e^{-j2\pi kn/N}, \qquad n = 0, 1, 2, \ldots, N - 1 \qquad (4.10)$$

as the discrete transform (4.3), except that it is an infinite instead of a finite series. Hence, given c_k, if we are interested only in the above N samples of $G_s(n/NT)$, the discrete transform (4.3) would be very natural, where we have only a truncation error, which is due to the finite sum of N terms. We give this example to isolate and illustrate the *truncation error* from the other more serious error that we often face in approximating Fourier integrals by the discrete Fourier transforms, namely that due to

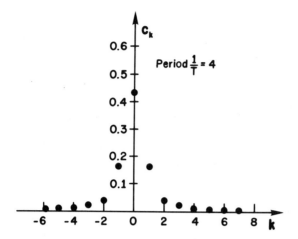

FIG. 4.1b The Fourier coefficients (nonperiodic) c_k of $G_s(f)$ in part (a).

the nonperiodic versus periodic nature of the two systems. Figure 4.1c illustrates the truncation error when we use only 33 terms ($k = -16$ to 16) to approximate the Fourier series (4.7) of $G_s(f)$ in Fig. 4.1a. The truncation error manifests itself as ripples especially around the ends of the period $\mp 1/2T = \mp 2$ (see also Fig. 3.5b–d, for example, where the ripples are very evident near the end point $x = 2$).

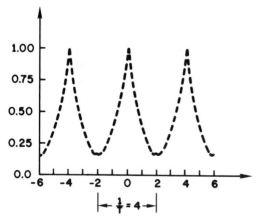

FIG. 4.1c Truncation error (only) for the samples $G_s(n/NT)$ of the partial sum of the Fourier series (4.7) with 33 terms ($k = -16$ to 16).

The absence of a sampling error in approximating the samples of the Fourier series in (4.10) by the discrete transform (4.3) hinges on using the exact Fourier series period $1/T$ when applying the discrete Fourier transform. Use of the wrong period, for example, $1/T' < 1/T$, for the discrete transform would result in a new periodic extension of different frequency because of the overlapping between neighboring translations of the original function $G_s(f)$. This is illustrated in Fig. 4.2a for the periodic extension $\overline{G}_s(f)$ of Fig. 4.1a with period $1/T = 4$, and a lower period $1/T' = 3$. Figure 4.2b shows the effect of this truncation of the period of $G_s(f)$ from 4 to 3 on this function's Fourier coefficients c_k whose

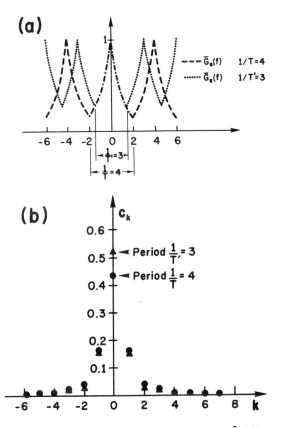

FIG. 4.2 (a) The periodic extension $\tilde{G}_s(f)$ with periods $1/T = 4$ and $1/T' = 3$. (b) The effect of the truncation of the period of $G_s(f)$ on its coefficients c_k.

values are clearly diferrent at $k = 0$. For smooth periodic waves, such a truncation of the period is too abrupt and would cause serious error in its c_k, which we will discuss in Section 4.1.6b.

Such an overlap, as in Fig. 4.2a, is another graphical way to describe the error incurred in the discrete Fourier transform's approximation of functions that do not share its exact periodicity, or that are not even periodic, such as the functions represented by Fourier integrals [on $(-\infty, \infty)$ or $(0, \infty)$] as seen in Figs. 4.3, 4.5, 4.6, and 4.7b. To approximate the latter nonperiodic functions of the Fourier integral representation, by the discrete Fourier transform, is to force upon them the periodicity of the DFT. As a result, we should assume and use a very large period. Thus we speak of the *discretization* (*sampling*) or periodicity being forced upon the Fourier integrals by the DFT. As will soon become clear, this error is also described as an *aliasing error*. Also, the approximation of the Fourier coefficients c_k (4.8) by the discrete transform (4.4) will suffer from the other main error which can be looked at in two equivalent ways: c_k being forced to be periodic, or its integral in (4.8) being discretized. Once this error is admitted for the approximation of c_k, the truncation error may not be such a problem because of the finite limits of the integral.

B.   Fourier Integral Approximation—Sampling and Truncation Errors

With the overlapping of Fig. 4.2a as some measure of the above discretization or sampling error, we are led to the main question of approximating the Fourier integral $G(f)$ in (4.1) by the discrete transform $\tilde{G}(n/NT)$ of (4.3). In this case, no matter what finite period $1/T$ we pick for (4.3), there will always be an overlap between the translations by n/T of the original function $G(f)$. This is, of course, unless $g(t)$ is bandlimited to the period $1/T$; that is, $G(f)$ vanishes identically for $|f| > 1/T$, where there will be no overlapping between the translations. Figure 4.3 illustrates the function

$$G(f) = e^{-|f|}, \qquad -\infty < f < \infty$$

along with two of its translations $G(f \mp 1/T)$, by a period $1/T = 4$, where we notice the overlappings between the three repetitions of the function. Figure 4.4 illustrates the Fourier transform of $G(f)$,

$$g(t) = \frac{2}{1 + (2\pi t)^2}, \qquad -\infty < t < \infty$$

along with two of its translations, $g(t \mp NT)$ by a period $NT = 32 \cdot \frac{1}{4} = 8$. as the period set by the discrete transform $\bar{g}(kT)$ in (4.4) with $N = 32$

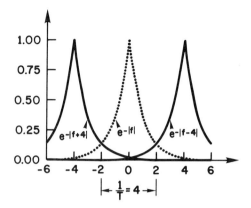

FIG. 4.3 The function $G(f) = e^{-|f|}$, $-\infty < f < \infty$, and its two translations $e^{-|f \mp 4|}$, $-\infty < f < \infty$.

samples. We note here that, due to the large period of 8, the overlapping is much less than that of Fig. 4.3. Figure 4.5 is a magnification of Fig. 4.4 to better illustrate these overlappings. Since $G(f)$ and $g(t)$ must be approximated on the finite intervals $(-1/2T, 1/2T)$ and $(-NT, NT)$ set by the discrete transforms (4.3), (4.4), a compromise must then be found for the size of each of these intervals, depending on the given functions.

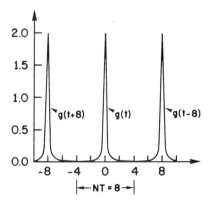

FIG. 4.4 The function $g(t) = 2/[1 + (2\pi t)^2]$, $-\infty < t < \infty$, and its two translations $2/[1 + 4\pi^2(t \mp 8)^2]$, $-\infty < t < \infty$.

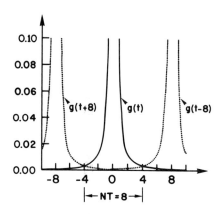

FIG. 4.5 The function $g(t) = 2/(1 + 4\pi^2 t^2)$, $-\infty < t < \infty$, and its two translations $2/[1 + 4\pi^2(t \mp 8)^2]$, $-\infty < t < \infty$ (a magnification of Fig. 4.4).

Such overlappings are, of course, absent in the case of the Fourier series (4.9), where we were able to write its *periodic extension*

$$\overline{G}_s(f) = \sum_{n=-\infty}^{\infty} G_s\left(f + \frac{n}{T}\right) = G_s(f), \quad -\infty < f < \infty \tag{4.9}$$

as the superposition of all the translations by n/T, an integer multiple of its period $1/T$. In comparison with this, the *superposition of all the translations by the same n/T* of the Fourier integral $G(f)$,

$$\overline{G}(f) = \sum_{n=-\infty}^{\infty} G\left(f + \frac{n}{T}\right) \tag{4.11}$$

would obviously not be replicas of the same $G(f)$ on $(-1/2T, 1/2T)$. Figure 4.6 illustrates a clear difference between this superposition $\overline{G}(f)$ with an 81-term sum in (4.11) and the actual function $G(f) = e^{-|f|}$ on the interval $(-1/2T, 1/2T) = (-2, 2)$. As we shall see in Section 4.1.6, it is this difference that actually measures the error that we have called a *sampling, discretization,* or *nonperiodicity* error, and which is often called the *aliasing* error of approximating Fourier integrals by discrete Fourier transforms. To be more specific, it is the N samples of this superposition (4.11), $\overline{G}(n/NT)$, $n = 0, 1, \ldots, N - 1$, on the interval $(-1/2T, 1/2T)$ that define the discrete transform $\check{G}(n/NT)$ of (4.3),

$$\check{G}\left(\frac{n}{NT}\right) \equiv \overline{G}\left(\frac{n}{NT}\right) = \sum_{k=-\infty}^{\infty} G\left(\frac{n}{NT} + \frac{k}{T}\right)$$

$$= G\left(\frac{n}{NT}\right) + \sum_{|k|=1}^{\infty} G\left(\frac{n}{NT} + \frac{k}{T}\right) \qquad (4.11a)$$

Hence we can define the aliasing error ϵ_A for G at these samples as

$$\epsilon_A G = \tilde{G}\left(\frac{n}{NT}\right) - G\left(\frac{n}{NT}\right) = \overline{G}\left(\frac{n}{NT}\right) - G\left(\frac{n}{NT}\right)$$

$$= \sum_{|k|=1}^{\infty} G\left(\frac{n}{NT} + \frac{k}{N}\right), \qquad n = 0, 1, \dots, N-1 \qquad (4.11b)$$

which is the sum of all the translation overlappings with the basic function G on $(-1/2T, 1/2T)$, as may be seen in Fig. 4.6.

The superposition of $\overline{G}(f)$ is usually described as the *folding* of $G(f)$, which means that all the values of the function $G(f)$ on its infinite domain $-\infty < f < \infty$ are being folded in the basic finite interval $(-1/2T, 1/2T)$. This is in the sense that we have $G(f)$ on $(-1/2T, 1/2T)$; then the translation $G(f - 1/T)$ would bring the part of $G(f)$ on $(-3/2T, -1/2T)$ to $(-1/2T, 1/2T)$, as illustrated in Fig. 4.3 with $1/T = 4$. In the same way, $G(f - 2/T)$ would bring the part of $G(f)$ on $(-5/2T, -3/2T)$ to $(-1/2T, 1/2T)$, and so on. The translations $G(f + 1/T)$, $G(f + 2/T)$, ... would bring the parts of $G(f)$ on $(1/2T, 3/2T)$, $(3/2T, 5/2T)$, ... to $(-1/2T, 1/2T)$, as illustrated in Fig. 4.3. It should be understood that once these parts are folded as in (4.11) to make $\overline{G}(f)$, then having $\overline{G}(f)$ on $(-1/2T, 1/2T)$ it is impossible to get back the exact values of $G(f)$; that is, we cannot unfold $G(f)$ out of $\overline{G}(f)$. So from (4.11a) it can be

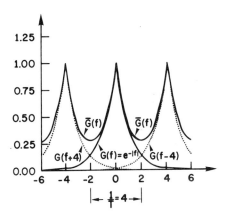

FIG. 4.6 The function $G(f) = e^{-|f|}$, $-\infty < f < \infty$, and $\overline{G}(f)$, the superposition of its 81 translations by $n/T = 4n$.

stated that the discrete transform $\tilde{G}(n/NT)$ approximation of the function $G(n/NT)$ represents the folding of all the sampled values of G on $(-\infty, \infty)$ into $(-1/2T, 1/2T)$, which cannot be unfolded to get the real identity of $G(n/NT)$ on $(-1/2T, 1/2T)$; hence the presence of the aliasing error. This would also mean that given $\tilde{G}(n/NT)$, if we are to ask about the samples of the original function $G((k+N)/NT)$ beyond the interval $(-1/2T, 1/2T)$, the discrete transform, as periodic with period N, has to "alias" and gives us those corresponding ones (modulo mN) in the basic replica on the interval $(-1/2T, 1/2T)$.

An excellent example of the folding of frequencies (aliasing) due to sampling is what we see when we watch a speeding stage coach in movie westerns. As the stage coach starts and then speeds up, we observe that the wheels go faster, then they start to slow down, stop, go backward, then forward, and so on. This is explained in light of the above discussion by the fact that we have only a limited number N of frames per second on the film (the interval $(-1/2T, 1/2T)$), which is the sampled version of the actual scene. So the actual fast rotations (frequencies) of the wheels $(G(n/NT))$, which we are supposed to see, are folded to within the range of the discrete frequencies 0 to $N-1$ of the sampled picture on $(-1/2T, 1/2T)$. Hence any higher speed of rotation beyond $N-1$ is folded to its corresponding value, according to (4.11a), in the range 0 to $N-1$.

In the same manner, we define $\bar{g}(t)$, the superposition of all the translations, by nNT, of the Fourier integral $g(t)$ in (4.2) with its infinite domain $(-\infty < t < \infty)$,

$$\bar{g}(t) = \sum_{n=-\infty}^{\infty} g(t + nNT) \tag{4.12}$$

Figure 4.7a shows a small difference between this superposition $\bar{g}(t)$ and the actual function $g(t) = 2/(1 + (2\pi t)^2)$ on the interval $(-NT/2, NT/2) = (-4, 4)$, where $N = 32$ and $T = \frac{1}{4}$. Figure 4.7b is a magnification of Fig. 4.7a to show more clearly the aliasing (or sampling) error.

As we shall also see in Section 4.1.6, it is the samples of this superposition $\bar{g}(kT)$ that make the discrete Fourier transform $\tilde{g}(kT)$ of (4.4),

$$\tilde{g}(kT) = \bar{g}(kT) = \sum_{n=-\infty}^{\infty} g(kT + nNT) = g(kT) + \sum_{|n|=1}^{\infty} g(kT + nNT) \tag{4.12a}$$

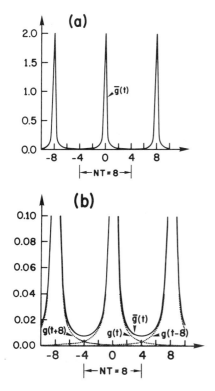

FIG. 4.7 (a) $\bar{g}(t)$, the superposition of 81 translations by $nNT = 8n$ of $g(t)$ in Fig. 4.4. (b) $\bar{g}(t)$, the superposition of 81 translations by $nNT = 8n$ of $g(t)$ in Fig. 4.4. Observe the ordinate scales of parts (a) and (b) (a magnification of part (a)).

Hence the aliasing error $\epsilon_A g$ of the approximation of the samples $g(kT)$ by the discrete transform $\tilde{g}(kT)$ on the interval $(0, NT)$ is

$$\epsilon_A g \equiv \tilde{g}(kT) - g(kT) = \bar{g}(kT) - g(kT)$$

$$= \sum_{|n|=1}^{\infty} g(kT + nNT), \qquad k = 0, 1, \ldots, N-1 \qquad (4.13)$$

Now we turn to a brief discussion of the truncation error and its definite relation to the aliasing error involved in approximating Fourier integrals by discrete transforms. We may state that the reason for the aliasing error of the Fourier integral $G(f)$ is the same as that for the truncation error of its Fourier transform $g(t)$ and vice versa. For example, assume that $g(t)$ is bandlimited to $(-1/2T, 1/2T)$, which means that $G(f)$ vanishes outside $(-1/2T, 1/2T)$; hence there is no overlapping for $\overline{G}(f)$. This means that G is free of aliasing, which can be associated with a previous remark regarding the lack of truncation error for such bandlimited functions $g(t)$.

The other example is of the Fourier coefficients c_k of (4.8), which is aliased but would not suffer a truncation error due to its finite-limit integral. Such lack of truncation error for c_k may be looked at in terms of the absence of aliasing for its transform, the periodic Fourier series $G_s(f)$ in (4.7), (4.9).

As a final note, we may observe that while the truncation error is characterized by ripples, especially close to the end of the period, the aliasing error, as evidenced in Figs. 4.6 and 4.7, shows an "overall" change for the function over the period.

C Discrete Cosine and Sine Transforms

In addition to the common form of the discrete (exponential) Fourier transform (4.3), (4.4) and the symmetric form (4.5), (4.6), we will present the discrete Fourier cosine-sine form.

The Fourier exponential series (4.7), (4.8) is, of course, equivalent to the Fourier sine-cosine series,

$$G_s(f) = \frac{a_0}{2} + \sum_{k=1}^{\infty} [a_k \cos 2\pi k Tf + b_k \sin 2\pi k Tf],$$

$$\frac{1}{2T} < f < \frac{1}{2T}. \text{ (Note the use of } \frac{a_0}{2} \text{ here)} \tag{3.10a}$$

$$a_k = 2T \int_{-1/2T}^{1/2T} G_s(f) \cos 2\pi k Tf \, df \tag{3.12a}$$

$$b_k = 2T \int_{-1/2T}^{1/2T} G_s(f) \sin 2\pi k Tf \, df \tag{3.13a}$$

which is obtained from (4.7), (4.8) after using the *Euler identity*,

$$e^{\mp jx} = \cos x \mp j \sin x, \qquad j = \sqrt{-1} \tag{3.14a}$$

and where

$$c_0 = \frac{a_0}{2}, \qquad c_{\pm k} = \frac{1}{2}(a_k \mp jb_k) \tag{4.14}$$

The discrete Fourier sine-cosine transform can be obtained from the exponential discrete transforms in the same way as we obtained the Fourier cosine-sine series from the exponential one.

Consider $\tilde{G}_n$, the DFT in its form (4.3a),

$$\tilde{G}_n = \sum_{k=0}^{N-1} \tilde{g}_k W^{-nk}, \qquad W = e^{2\pi j/N}, \qquad 0 \le n \le N-1 \qquad (4.3),(4.3a)$$

Matters will be simplified somewhat if N is assumed to be even. Therefore, let $N = 2M$. Rewriting (4.3a) with $N = 2M$ gives

$$\tilde{G}_n = \sum_{k=0}^{2M-1} \tilde{g}_k e^{-j\pi nk/M}$$

$$= \tilde{g}_0 + \sum_{k=1}^{M-1} \tilde{g}_k e^{-j\pi nk/M} + \tilde{g}_M(-1)^n + \sum_{k=M+1}^{2M-1} \tilde{g}_k e^{-j\pi nk/M}$$

In the last sum, define a new index by letting $l = 2M - k$ and recall the periodicity property $\exp(-j\pi n(2M - l))/M = \exp(j\pi nl/M)$

$$\tilde{G}_n = \tilde{g}_0 + \sum_{k=1}^{M-1} \tilde{g}_k e^{-j\pi nk/M} + \tilde{g}_M(-1)^n + \sum_{l=1}^{M-1} \tilde{g}_{2M-l} e^{j\pi nl/M}$$

$$= \tilde{g}_0 + \sum_{k=1}^{M-1} \tilde{g}_k e^{-j\pi nk/M} + \tilde{g}_M(-1)^n + \sum_{k=1}^{M-1} \tilde{g}_{2M-k} e^{j\pi nk/M}$$

(replacing l by k)

$$= \tilde{g}_0 + \sum_{k=1}^{M-1} \left[(\tilde{g}_k + \tilde{g}_{2M-k}) \cos\left(\frac{\pi nk}{M}\right) \right.$$

$$\left. + j(\tilde{g}_{2M-k} - \tilde{g}_k) \sin\left(\frac{\pi nk}{M}\right) \right] + \tilde{g}_M(-1)^n$$

$$\tilde{G}_n \equiv \frac{A_0}{2} + \sum_{k=1}^{M-1} \left[A_k \cos\left(\frac{\pi nk}{M}\right) + B_k \sin\left(\frac{\pi nk}{M}\right) \right] + \frac{A_M(-1)^n}{2},$$

$$0 \le n \le N-1 \qquad (4.15)$$

In this final line, we have defined the *cosine coefficients*, A_k, and the *sine coefficients*, B_k. Using expression (4.4a) for $\tilde{g}_k$, we find that

$$A_k = \tilde{g}_k + \tilde{g}_{2M-k} = \frac{2}{N} \sum_{n=0}^{N-1} \tilde{G}_n \cos\left(\frac{2\pi nk}{N}\right),$$

$$0 \le k \le \frac{N}{2} = M \tag{4.16}$$

$$B_k = j(\tilde{g}_{2M-k} - \tilde{g}_k) = \frac{2}{N} \sum_{n=0}^{N-1} \tilde{G}_n \sin\left(\frac{2\pi nk}{N}\right),$$

$$1 \le k \le \frac{N}{2} - 1 = M - 1 \tag{4.17}$$

Representation (4.15) is referred to as the *sine-cosine form of the DFT*. Given the complex coefficients $\tilde{g}_k$ of a sequence $\tilde{G}_n$, it is always possible to find the sine and cosine coefficients A_k and B_k through the relations

$$A_0 = 2\tilde{g}_0, \qquad A_M = 2\tilde{g}_M$$

$$A_k = \tilde{g}_k + \tilde{g}_{2M-k}, \qquad B_k = j(\tilde{g}_{2M-k} - \tilde{g}_k),$$

$$1 \le k \le M - 1 \tag{4.16a), (4.17a}$$

Notice that for a sequence $\{\tilde{G}_n\}_{n=0}^{N-1}$ there are N complex coefficients or alternatively a total of N sine and cosine coefficients.

We may note that the DFT representation (4.3a) above is *cyclic*; that is, it is periodic with period N, since from (4.3a) we can easily see that $\tilde{G}_0 = \tilde{G}_N$ since $W^N = e^{2\pi jkN} = 1$. Thus if we plan to use this DFT for solving difference equations in Section 4.1.5, it will be for those of (discrete) boundary value problems with *periodic boundary conditions*.

The Discrete Cosine Transforms (DCT)—Even Sequences

Now consider the situation in which $\{\tilde{G}_n\}$ is an *even sequence* on the interval $0 \le n \le 2M$, that is, $\tilde{G}_{2M-n} = \tilde{G}_n$. With this assumption, the cosine and sine coefficients can be simplified when we realize that $\cos(\pi nk/M)$ is an even function of n on $[0,N]$ for all k and $\sin(\pi nk/M)$ an odd function of n on $[0,N]$ for all k. This gives us from (4.16), (4.17)

$$A_k = \frac{2}{N} \sum_{n=0}^{N-1} \tilde{G}_n \cos\left(\frac{2\pi nk}{N}\right)$$

$$= \frac{1}{M} \left\{ \tilde{G}_0 + \sum_{n=1}^{M-1} \tilde{G}_n \cos\left(\frac{\pi nk}{M}\right) + \tilde{G}_M(-1)^k \right.$$

$$\left. + \sum_{n=1}^{M-1} \tilde{G}_{2M-n} \cos\left(\frac{\pi k(2M-n)}{M}\right) \right\}$$

$$= \frac{1}{M} \tilde{G}_0 + \frac{2}{M} \sum_{n=1}^{M-1} \tilde{G}_n \cos\left(\frac{\pi nk}{M}\right) + \frac{1}{M} \tilde{G}_M(-1)^k, \qquad 0 \le k \le M$$

At the same time, all the sine coefficients in (4.17) vanish, that is, $B_k = 0$ for $1 \le k \le M - 1$, due to the evenness of $\{\tilde{G}_n\}$ and the oddness of the $\sin(\pi nk/M)$ terms. Thus we have for an even sequence $\{\tilde{G}_n\}$ on $0 \le n \le N$,

$$\tilde{G}_n = \frac{A_0}{2} + \sum_{k=1}^{M-1} A_k \cos\left(\frac{\pi nk}{M}\right) + \frac{A_M}{2}(-1)^n, \qquad 0 \le n \le M \qquad (4.18)$$

where

$$A_k = \frac{1}{M}\left[\tilde{G}_0 + 2\sum_{n=1}^{M-1} \tilde{G}_n \cos\left(\frac{\pi nk}{M}\right) + \tilde{G}_M(-1)^k\right], \qquad 0 \le k \le M$$
(4.19)

But now notice what has happened. By incorporating the evenness of the sequence $\{\tilde{G}_n\}$ into the DFT, all of the sine terms have disappeared. Furthermore, the original sequence of length N is now used only for $0 \le n \le M = N/2$. Therefore, we may interpret (4.18), (4.19) as a transform pair relating two sequences $\{\tilde{G}_n\}$ and $\{A_k\}$ of length $M + 1$, where M may be any integer. This pair constitutes the *discrete cosine transform* (DCT).

The important property of the DCT pair is that the sequences $\{\tilde{G}_n\}$ and $\{A_k\}$ which it relates are no longer periodic in the same way that the DFT is. However, these sequences do have a special symmetry. If the expression (4.18) for $\tilde{G}_n$ is evaluated outside the interval $0 \le n \le M$, the sequence that is produced is the *even extension* of the sequence $\{\tilde{G}_n\}_{n=0}^M$. In particular, this means that $\{\tilde{G}_n\}$ is an even sequence about $n = 0$ (i.e., $\tilde{G}_n = \tilde{G}_{-n}$) and $\{\tilde{G}_n\}$ is an even sequence about $n = M$ (i.e., $\tilde{G}_{M-n} = \tilde{G}_{M+n}$). Figure 4.8a shows a sequence $\{\tilde{G}_n\}$ defined on $0 \le n \le M$ and its even extension outside this interval. This symmetry about the end points of the interval on which $\{\tilde{G}_n\}$ is defined will be an important property when the DCT is used to solve difference equations associated with boundary conditions as this "evenness" at both ends of the sequence (domain) of their definition. Clearly, the DCT is compatible with the boundary conditions $\tilde{G}_1 - \tilde{G}_{-1} = 0$ and $\tilde{G}_{M+1} - \tilde{G}_{M-1} = 0$, which is the vanishing of the central difference of the sequence $\tilde{G}_n$ at the end points $n = 0$ and $n = M$.

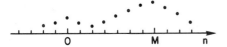

FIG. 4.8a   A sequence $\{\tilde{G}_n\}$ on $0 \le n \le M$ and its even extension.

The Discrete Sine Transform (DST)—Odd Sequences
A similar procedure may be followed to produce the discrete sine transform (DST). We now assume that $\{\tilde{G}_n\}$ is an odd sequence on an interval of length $N = 2M$. This means that $\tilde{G}_n = -\tilde{G}_{2M-n}$. The terms $\cos(\pi nk/M)$ and $\sin(\pi nk/M)$ have the same symmetry as before. Again, we look at the expressions (4.17) and (4.16) for the sine and cosine coefficients in the DFT. We have

$$B_k = \frac{2}{N} \sum_{n=0}^{N-1} \tilde{G}_n \sin\left(\frac{2\pi nk}{N}\right)$$

$$= \frac{1}{M} \sum_{n=1}^{M-1} \tilde{G}_n \sin\left(\frac{\pi nk}{M}\right) + \sum_{n=1}^{M-1} \tilde{G}_{2M-n} \sin\left(\frac{\pi k(2M-n)}{M}\right)$$

$$= \frac{2}{M} \sum_{n=1}^{M-1} \tilde{G}_n \sin\left(\frac{\pi nk}{M}\right) \quad \text{for } 1 \le k \le M - 1$$

At the same time, all of the cosine coefficients in (4.16) vanish—that is, $A_k = 0$ for $0 \le k \le M$—due to the oddness of $\{\tilde{G}_n\}$ and the evenness of the $\cos(\pi nk/M)$ terms. Thus we have for an odd sequence $\{\tilde{G}_n\}$ on $0 \le n \le N$

$$\tilde{G}_n = \sum_{k=1}^{M-1} B_k \sin\left(\frac{\pi nk}{M}\right), \qquad 1 \le n \le M - 1 \tag{4.20}$$

$$B_k = \frac{2}{M} \sum_{n=1}^{M-1} \tilde{G}_n \sin\left(\frac{\pi nk}{M}\right), \qquad 1 \le k \le M - 1 \tag{4.21}$$

As in the case of the DCT, the original sequence $\{\tilde{G}_n\}$ of length N is used only for $1 \le n \le M - 1$. These $M - 1$ values of the sequence $\{\tilde{G}_n\}$ generate $M - 1$ sine coefficients $\{B_k\}$. Equations (4.20), (4.21) comprise a transform pair relating two sequences $\{\tilde{G}_n\}$ and $\{B_k\}$ of length $M - 1$ where M may be any integer. This pair of relations is the *discrete sine transform* (DST).

The sequences $\{\tilde{G}_n\}$ and $\{B_k\}$ defined by the DST pair also have a special symmetry which should be examined. The functions $\sin(\pi nk/M)$, regarded as functions of either n or k, are odd functions about $n = 0$ (or $k = 0$) and $n = M$ (or $k = M$). Also note that $\tilde{G}_0 = \tilde{G}_M = B_0 = B_M = 0$. Therefore, if the expression (4.20) for $\tilde{G}_n$, for example, is evaluated outside the interval $0 \le n \le M$, the sequence that is produced is the *odd extension* of the sequence $\{\tilde{G}_n\}_{n=0}^{M}$. In particular, this means that $\tilde{G}_n = -\tilde{G}_{-n}$ and $\tilde{G}_{M-n} = -\tilde{G}_{M+n}$. Figure 4.8b shows a sequence defined on $0 \le n \le M$ and its odd extension outside this interval. The symmetry of

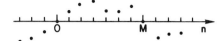

FIG. 4.8b   A sequence on $0 \leq n \leq M$ and its odd extension.

the sequences produced by the DST about the end points will determine when the DST should be chosen to solve difference equations. Clearly, the DST is compatible with the boundary conditions $\tilde{G}_0 = 0$ and $\tilde{G}_M = 0$.

Now some remarks should be made concerning the actual computation of the DCT and DST of sequences. Given a sequence $\{\tilde{G}_n\}_{n=0}^{M}$, its DCT and its DST may be computed using (4.19) and (4.21). Inversely, if the coefficients $\{A_k\}_{k=0}^{M}$ or $\{B_k\}_{k=1}^{M-1}$ are known, then the sequence $\{\tilde{G}_n\}$ may be reconstructed using either (4.18) or (4.20), respectively. In situations in which the sums of (4.18)–(4.21) cannot be evaluated explicitly, the problem must be done numerically. However, as with the DFT, these calculations can be expedited significantly by using modifications of the FFT.

Here we used A_k and B_k, instead of the same $\tilde{g}_k$, for the discrete cosine (4.20) and sine (4.21) transforms, respectively, to distinguish them from the exact Fourier cosine and sine coefficients a_k, b_k of (3.12a)–(3.13a) of $G_s(f)$ on $(-1/2T, 1/2T)$ in (3.10a). To be more specific, we give the relation between these two important sets of transforms, namely the *discrete* Fourier cosine and sine transforms A_k, B_k and their analogs, the *finite* Fourier cosine and sine transforms a_k, b_k:

$$A_k = a_k + \sum_{m=1}^{\infty} (a_{2Mm-k} + a_{2Mm+k}), \tag{4.22}$$

$$B_k = b_k + \sum_{m=1}^{\infty} (-b_{2Mm-k} + b_{2Mm+k}) \tag{4.23}$$

with the special case

$$A_0 = a_0 + 2 \sum_{m=1}^{\infty} a_{2Mm} \tag{4.24}$$

These relations can be derived along the lines of $\overline{G}$ in (4.11), (4.11a), the superposition of all the translations, by n/T, of G on $(-1/2T, 1/2T)$, which we shall discuss in complete detail in Section 4.1.4. Hence, in parallel with the development of the aliasing error in (4.11b), we can recognize the infinite sums in (4.22), (4.23), and (4.24) as the *aliasing errors* for approximating the Fourier coefficients a_k, b_k, and a_0 of (3.12a),

(3.13a) by the above discrete cosine-sine transforms, A_k, B_k, and A_0, respectively. Figure 4.8c illustrates the aliasing of $a_k = c_k$, the Fourier (cosine) coefficients of the (even) periodic function,

$$G_s(f) = e^{-|f|}, \qquad -\frac{1}{2T} < f < \frac{1}{2T}$$

of Fig. 4.1, with $2M = 32$ and with an 81-term partial sum for the series in (4.22). The aliasing error in Fig. 4.8c is small, which is a consequence of this particular example of fast-decaying Fourier coefficients as seen in Fig. 4.1c and where we have made the good decision to take enough of them ($M = 16$).

Before we start the analysis for deriving the discrete transforms (4.3), (4.4); (4.5), (4.6); or (4.15)–(4.17), we will give more illustrations with emphasis on how to deal with the expected complex-valued transforms.

4.1.2 Computing for Complex-Valued Functions[†]

Before we indulge in the development of the discrete Fourier transform and its properties, or its efficient computational algorithm the fast Fourier transform, we will inquire into the kinds of functions that are involved in these computations. We may agree that our input function $g(t)$ for the Fourier integral (4.1) is usually a real-valued function. However, we see immediately that the integral involves a complex-valued exponential functions,

$$e^{-j2\pi ft} = \cos 2\pi ft - j \sin 2\pi ft$$

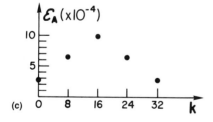

FIG. 4.8c   The aliasing error for the Fourier cosine coefficients

$$a_k = 2T \frac{1 + (-1)^{k+1} e^{-1/2T}}{1 + (2\pi kT)^2}$$

of the function $G_s(f) = e^{-|f|}$, $-1/2T < f < 1/2T$, in (a). $T = \frac{1}{4}$.

[†] For a brief review of complex numbers, see Exercise 1.6.

with a real part $\cos 2\pi ft$ and a pure imaginary part $-\sin 2\pi ft$ multiplied by the imaginary number $j = \sqrt{-1}$. Since computing machines deal with real numbers, the way to compute this complex exponential function is to evaluate its real part $\cos 2\pi ft$ and set it aside, and then to evaluate its pure imaginary part $-\sin 2\pi ft$, to be multiplied later by $j = \sqrt{-1}$ and added to the real part $\cos 2\pi ft$. For $z = x + iy$, a complex number with x and y as its real and imaginary parts, respectively, we know that any complex-valued function $f(z)$ can be written as the sum of a real part $\mathrm{Re}(x,y)$ and an imaginary part $\mathrm{Im}(x,y)$, (or $f(z) = f_r + jf_i$)

$$f(z) = \mathrm{Re}(x,y) + j\,\mathrm{Im}(x,y)$$

Our first example is $e^{2\pi jft} = e^z$ with the exponent $z = j(2\pi ft) = jy$ as a pure imaginary number,

$$e^{jy} = \cos y + j\sin y$$

with $\mathrm{Re}(x,y) = \cos y$ and $\mathrm{Im}(x,y) = \sin y$. Another example is

$$f(z) = z^2 = (x + jy)^2 = x^2 - y^2 + j2xy$$

with $\mathrm{Re}(x,y) = x^2 - y^2$ and $\mathrm{Im}(x,y) = 2xy$.

Even if $g(t)$ in (4.1) is a real-valued function, it is going to be multiplied by the complex-valued function $e^{-j2\pi ft}$ and then integrated, and we expect the result $G(f)$ to be in general a complex-valued function. The same is true of the inverse Fourier transform (4.2), the discrete exponential transforms (4.3) and (4.4), and the exponential Fourier series (4.7) and (4.8). So it is instructive to know, before any computations, what types of input and output function we have, so that we can direct separate computations for the real part (Re) and the imaginary part (Im). We follow this until the end then we combine the two numbers as a complex number $\mathrm{Re} + j\,\mathrm{Im}$. In our graphical illustrations, we shall graph the real and imaginary parts Re and Im separately. Sometimes it is instructive to graph the absolute value of the function (see Fig. 4.37 and 4.39, the answers to Exercises 4.11c, and 4.11d, respectively.)

$$|f(z)| = |\mathrm{Re}(x,y) + j\,\mathrm{Im}(x,y)| = \sqrt{\mathrm{Re}^2(x,y) + \mathrm{Im}^2(x,y)}$$

Since the Fourier exponential transforms (4.1), (4.2) have symmetric limits $(-\infty, \infty)$ around the origin, we can benefit form the fact that such an integral vanishes if its integrand is an odd function, and it is twice the integral from 0 to ∞ if the integrand is an even function. Consider the real and imaginary parts of the Fourier exponential transform of a function $g(t)$, given by

$$G(f) = \int_{-\infty}^{\infty} g(t)e^{-j2\pi ft}\,dt = \int_{-\infty}^{\infty} g(t)[\cos 2\pi ft - j\sin 2\pi ft]\,dt$$

$$= \int_{-\infty}^{\infty} g(t)\cos 2\pi ft\,dt - j\int_{-\infty}^{\infty} g(t)\sin 2\pi ft\,dt$$

$$\equiv G_r(f) + jG_i(f) \tag{4.25}$$

where G_r and G_i represent, respectively the real and imaginary parts of the complex function G. For $g(t)$ as an even function, we note that in the first integral of (4.25), as the real part of $G(f)$, the integrand $g(t)\cos 2\pi ft$ is an even function, which allows us to write the integral as $2\int_0^{\infty} g(t)\cos 2\pi ft$. In the second integral, the integrand $g(t)\sin 2\pi ft$ is an odd function, so the integral vanishes. This leaves us with one integral for $G(f)$ as a real function,

$$G(f) = 2\int_0^{\infty} g(t)\cos 2\pi ft\,dt \tag{4.26}$$

which we can show is an even function. Hence it can be directly approximated by the discrete cosine ($B_k = 0$) transform (4.19). The result of this example demonstrates that the Fourier exponential transform of a real and even function $g(t)$ is a real and even function $G(f)$. Figures 4.1 to 4.7 all involve such real even functions.

For $g(t)$ a real function, we can show from (4.25) that its transform is a complex function

$$G(f) = \text{Re}(f) + j\,\text{Im}(f) \equiv G_r(f) + jG_i(f)$$

where clearly the first integral of (4.25) is even in f and the second integral is odd in f. The case of $g(t)$ as a real odd function would result in the first integral of (4.25) vanishing and the second interval becoming $2\int_0^{\infty} g(t)\sin 2\pi ft\,dt$,

$$G(f) = -2j\int_0^{\infty} g(t)\sin 2\pi ft\,dt \tag{4.27}$$

a pure imaginary function which is also odd in f.

In the following we show that the transform of a complex function $g(t) = g_r(t) + jg_i(t)$, where g_r and g_i represent respectively the real and imaginary parts of the complex function g, is also a complex function $G(f) = G_r + jG_i$.

From (4.1) we have

$$G(f) = \int_{-\infty}^{\infty} g(t)e^{-j2\pi ft}\,dt = \int_{-\infty}^{\infty} [g_r(t) + jg_i(t)][\cos 2\pi ft - j\sin 2\pi ft]\,dt$$

Now we multiply inside the integral, using $j^2 = -1$, and collect the real and imaginary parts separately to obtain

$$G(f) = \int_{-\infty}^{\infty} [g_r(t)\cos 2\pi ft + g_i(t)\sin 2\pi ft]\, dt$$

$$-j \int_{-\infty}^{\infty} [g_r(t)\sin 2\pi ft - g_i(t)\cos 2\pi ft]\, dt$$

$$= G_r(f) + jG_i(f) \tag{4.28}$$

which is a complex function, where $G_i(f)$ is the negative of the second integral. In another special case where $g(t)$ is pure imaginary $(g_r(t) = 0)$ and odd, we see from (4.28) that

$$G(f) = \int_{-\infty}^{\infty} g_i(t)\sin 2\pi ft\, dt + j \int_{-\infty}^{\infty} g_i(t)\cos 2\pi ft\, dt$$

$$= 2 \int_{0}^{\infty} g_i(t)\sin 2\pi ft\, dt$$

as a real odd function.

For our analysis we may need to show that both the real and imaginary parts of an odd (even) complex function are odd (even) functions. For example, we have

$$g(-t) = g_r(-t) + jg_i(-t)$$

and

$$-g(t) = -g_r(t) - jg_i(t)$$

but if the function $g(t)$ is odd, then $g(-t) = -g(t)$,

$$g(-t) = g_r(-t) + jg_i(-t) = -g_r(t) - jg_i(t) = -g(t)$$

But since two complex numbers are equal if, and only if, their respective real and imaginary parts are equal, the last equation gives

$$g_r(-t) = -g_r(t)$$

$$g_i(-t) = -g_i(t)$$

We can then conclude that both g_r and g_i are odd functions when g is an odd function. The same can be shown for the even complex functions. Table 4.1 summarizes the effect of Fourier transformation on various complex and real functions. Since the Fourier exponential transform is symmetric, we note that the properties in Table 4.1 are symmetric for the inverse transform.

TABLE 4.1 Fourier Exponential Transforms of Real and Complex Functions

$g(t)$	$G(f)$
1. $g = g_r$ (real)	$G = G_r$ even $+ jG_i$ odd
2. $g = jg_i$ (imaginary)	$G = G_r$ odd $+ jG_i$ even
3. $g = g_r$ even $+ jg_i$ odd	$G = G_r$ (real)
4. $g = g_r$ odd $+ jg_i$ even	$G = jG_i$ (imaginary)
5. $g = g_r$ even	$G = G_r$ even
6. $g = g_r$ odd	$G = jG_i$ odd
7. $g = jg_i$ even	$G = jG_i$ even
8. $g = jg_i$ odd	$G = G_r$ odd
9. $g = (g_r + jg_i)$ even	$G = (G_r + jG_i)$ even
10. $g = (g_r + jg_i)$ odd	$G = (G_r + jG_i)$ odd

Our first illustration of computing Fourier transforms, in Section 4.1.4, is for the real *causal* function (Fig. 4.9a),

$$g(t) = \begin{cases} e^{-t}, & t > 0 \\ 0, & t < 0 \end{cases} \tag{4.29}$$

whose Fourier integral can be evaluated as

$$G(f) = \frac{1 - j2\pi f}{(2\pi f)^2 + 1} = \frac{1}{(2\pi f)^2 + 1} - j\frac{2\pi f}{(2\pi f)^2 + 1} = G_r + jG_i \tag{4.30}$$

as a complex function whose real part $G_r(f) = 1/(1 + (2\pi f)^2)$ is an even function (Fig. 4.9b) and whose imaginary part $G_i(f) = -2\pi f/(1 + (2\pi f)^2)$ is an odd function (Fig. 4.9c) as indicated in the first entry of Table 4.1. The third row in Table 4.1 shows the symmetry of the same property; that is, the inverse Fourier exponential transform of a real function $G(f)$ is a complex function $g(t)$ with an even real part g_r and an odd imaginary part g_i.

In Fig. 4.9 we illustrate the causal function $g(t)$ of (4.29) and separate the real $G_r(f)$ part from the imaginary $G_i(f)$ part of its Fourier transform $G(f)$ of (4.30). In Fig. 4.9d we illustrate the absolute value $|G(f)|$ of the transform,

$$|G(f)| = \left| \frac{1}{1 + (2\pi f)^2} - j\frac{2\pi f}{1 + (2\pi f)^2} \right|$$

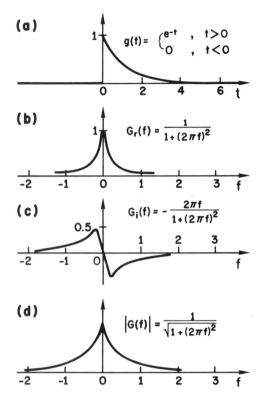

FIG. 4.9 The causal function $g(t)$ with real $G_r(f)$ and imaginary $G_i(f)$ parts of its Fourier transform and the absolute value of the transform $|G(f)|$.

$$= \sqrt{\left[\frac{1}{1+(2\pi f)^2}\right]^2 + \left[\frac{-2\pi f}{1+(2\pi f)^2}\right]^2}$$

$$= \sqrt{\frac{1+(2\pi f)^2}{[1+(2\pi f)^2]^2}} = \frac{1}{\sqrt{1+(2\pi f)^2}} \tag{4.31}$$

To further illustrate the development of the discrete Fourier transform (and its aliasing (sampling) and truncation error in approximating Fourier integral and series) we will consider a more general example than that of Fig. 4.1. In the following example we use the (4.3), (4.4) form of the discrete transforms.

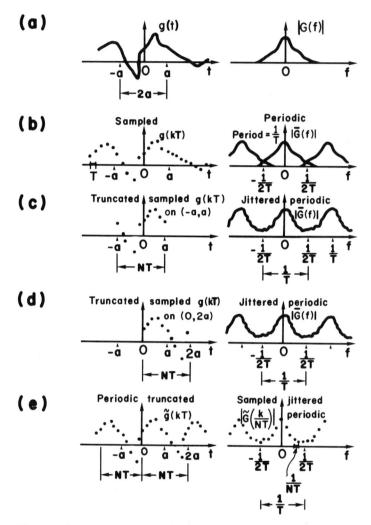

FIG. 4.10 The process of approximating the Fourier integrals in part (a) as discrete Fourier transforms (DFTs) in part (e).

Consider the real function $g(t)$ in Fig. 4.10a. Since its Fourier transform (4.3) is a complex-valued functions, as indicated above (see Table 4.1), we graph its absolute value $|G(f)|$. So although $|G(f)|$ appears to be even in the figure, it does not have to be even in our analysis, where we shall refer, in general, to $G(f)$. We could have used a special case of a real even function $g(t)$ (Figs. 4.3, 4.4) to give a real Fourier transform (see Table 4.1), as we shall do later in Fig. 4.13, but we preferred a general-type function here. The discrete transform starts with sampling $g(t)$ as in Fig. 4.10b, with sampling spacing T which can be taken as a Fourier coefficient $g(kT)$. This results in forcing $G(f)$ to a new shape [precisely $\overline{G}(f)$ of (4.11)] that is to be periodic with period $1/T$. Because of this sampling of $g(t)$ and the resulting periodicity of $\overline{G}$, we see that the actual picture G is overlapping with its forced upon periodic extensions, hence the aliasing error, and very clearly in their superposition $\overline{G}(f)$ of (4.11) as illustrated in Fig. 4.6. We may inject here the role of the sampling theorem, which says that if the function $g(t)$ had been bandlimited to $1/2T$—that is, $G(f)$ vanishes identically outside $(-1/2T, 1/2T)$—then this sampling $g(kT)$ of $g(t)$ is natural. This is in the sense that the $g(kT)$, $k = 0, 1, 2, \ldots$, are the Fourier coefficients of $G(f)$; hence we have the inherited natural periodic extension of G with no overlap or "aliasing." So this step (Fig. 4.10b) involves no error when dealing with $g_{1/2T}(t)$ as a function bandlimited to $1/2T$, or the Fourier series expansion $G_s(f)$ of $G(f)$ as defined on $(-1/2T, 1/2T)$. Another way to describe the overlap aliasing error is in terms of misuse of the sampling theorem. This means that we are either sampling for nonbandlimited functions or assuming the wrong bandlimit, i.e., using the wrong sample spacing T.

Since the goal for a discrete transform is to end up with a finite sum, the next step is to truncate the original infinite sequence $g(kT)$ to N terms as in Fig. 4.10c. This would bring the usual Fourier series truncation error for $G(f)$, which, as we know, manifests itself in ripples especially close to the ends of the period, $-1/2T$ and $1/2T$. This truncation is usually done by multiplying the sequence $g(kT)$ by the gate function

$$
p_a(t) = \begin{cases} 1, & -a < t < a \\ 0, & |t| > a \end{cases}
$$

with $a = NT/2$, as it would show in Fig. 4.10c. The samples $g(kT)$ on $(-a, a)$ are the ones suitable for the symmetric discrete form in (4.6). However, and as we mentioned earlier, because of the popular form (4.4) and the fact that the original fast Fourier transform was established for it, we have to translate the samples on $(-a, 0)$ to the interval $(a, 2a)$ as seen in Fig. 4.10d. We note that the gate function is very abrupt in truncating the sequence, and as we shall see in Section 4.1.6, other truncation functions (*windows*; see Figs. 4.31, 4.33) with a smoother drop can be tried to minimize the ripples of the truncation error for $G(f)$ around $\mp 1/2T$. At this stage we have the N-term sequence $\{g(kT)\}_{k=0}^{N-1}$, with a periodic but still continuous $\overline{G}(f)$ on $(-1/2T, 1/2T)$, and this should show the effect of *aliasing (overlap)* and *truncation (ripples)* as compared with Fig. 4.10a. The final step is then to discretize (to sample) $\overline{G}(f)$. The question concerns the sample spacing Δf and the effect on the already completed finite sequence $g(kT)$ of Fig. 4.10d. This can be accomplished when we follow the step of Fig. 4.10b in the opposite direction, with very careful attention to the sampling theorem requirement for sample spacing. Since we don't want to disturb the sequence $g(kT)$, which is truncated at $t = NT$, we think of periodically extending it with period NT (Fig. 4.10e). This means roughly a sampling of $\overline{G}$ in Fig. 4.10d, as a function $\overline{G}(f)$ time-limited to t of the interval $(0, NT)$, which can be sampled with $\Delta f = 1/NT$ as in Fig. 4.10e. This description of arriving at the discrete transform from the Fourier transform could be supported with basic Fourier analysis, which we will revisit in Section 4.1.3 after developing the basic property of discrete transforms, namely discrete orthogonality. However, we must add that the resulting discrete transforms $\tilde{g}$ and $\tilde{G}$ are not obviously Fourier transforms of each other in the sense of (4.1) and (4.2).

A.   Sampling on the Symmetric Interval

Usually, the finite exponential transform is defined on a symmetric interval $(-a, a)$, but here we truncated $g(nT)$ from 0 to NT instead of $-NT/2$ to $NT/2$. The reason may lie in the tradition of electrical engineers, who see the real time function $g(t)$ defined on $(0, \infty)$ rather than $(-\infty, \infty)$. Consequently, even if the Fourier transform $G(f)$ is defined on $(-\infty, \infty)$, it still is not very convenient for them to work with negative frequencies. In Fig. 4.10e, we have the discrete transform $\tilde{G}(n/NT)$ as a sequence, $f_n = -1/2T$ to $1/2T$ with its periodic extensions. However, its expression (4.3) is taken from $f_n = 0$ to $1/T$. This again, because of tradition, would cost us the inconvenience of working with the values from $f_n = 1/2T$ to $f_n = 1/T$, as the exact replica of the actual part of the negative frequencies on the negative side $-1/2T$ to 0. The question of how to deal with the arbitrary function defined on $(-a, a)$ is well illustrated in the modification done for $g(kT)$ of Fig. 4.10c to that of Fig. 4.10d, where we actually translated the part on $(-a, 0)$ to $(a, 2a)$ by a complete period $2a = NT$.

At this point we should make an important remark concerning the truncation error for $\tilde{g}(kT)$ as the inverse DFT of $\tilde{G}$, and where we expect this error around the end points $t = \mp a$ in Fig. 4.10c or 4.10e, which are equivalent to the one point $t = a$ in Fig. 4.10d. This is so as not to mistake $t = 2a$ in Fig. 4.10d as an end point for the domain of $\tilde{g}(kT)$; it is indeed equivalent to the middle point $t = 0$ in Fig. 4.10c, and there should be little trouble there for a continuous $g(t)$ on $(-a, a)$. However, this warning becomes essential when there is a jump discontinuity, for example, at $t = 0$ as in the case of the important class of causal functions we shall consider next in (4.29) and Fig. 4.11. In Fig. 4.11c we expect a truncation error around the end points $t = \mp a$ and a "stubborn" Gibbs phenomenon (see Section 4.1.6G) around $t = a$. Equivalently, when we look at Fig. 4.11d we should recognize the error around $t = 0, 2a$ as a Gibbs phenomenon, and that around $t = a$ as the truncation error.

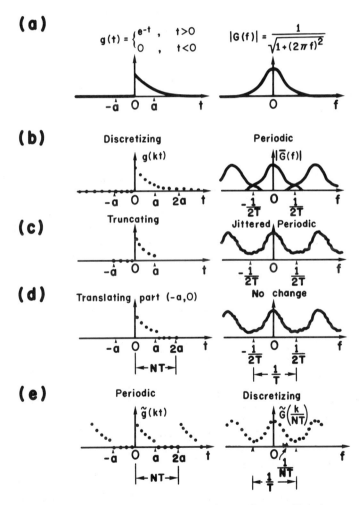

FIG. 4.11   Graphical derivation of the discrete Fourier transform pair.

The next example is the causal functions of Fig. 4.11,

$$g(t) = \begin{cases} e^{-t}, & t > 0 \\ 0, & t < 0 \end{cases} \tag{4.29}$$

with its transform

$$G(f) = \frac{1}{1 + j2\pi f} = \frac{1 - j2\pi f}{(1 - j2\pi f)(1 + j2\pi f)} = \frac{1 - j2\pi f}{1 + (2\pi f)^2}$$

$$= \frac{1}{1 + (2\pi f)^2} + j\frac{-2\pi f}{1 + (2\pi f)^2} = \operatorname{Re} G + j \operatorname{Im} G \tag{4.30}$$

$$\equiv G_r + jG_i$$

as a complex-valued function whose real and imaginary parts are both defined on the whole infinite interval $(-\infty, \infty)$.

Figure 4.11 is another illustration, like that of Fig. 4.10, for the discrete Fourier transform approximation of this specific function $g(t)$ (Fig. 4.11) and the absolute value of its transform

$$|G(f)| = \sqrt{\left[\frac{1}{1 + (2\pi f)^2}\right]^2 + \left[\frac{-2\pi f}{1 + (2\pi f)^2}\right]^2} = \frac{1}{\sqrt{1 + (2\pi f)^2}} \tag{4.31}$$

To clarify further the translation necessary for the nonsymmetric form of the discrete transform (4.3), (4.4), we consider this causal function and its transforms of Fig. 4.9 and prepare it as an input for such form. Let us assume that we had decided on the symmetric intervals $(-a, a)$ for $g(t)$ and $(-b, b)$ for its transforms, as illustrated in the first column of Fig. 4.12. The second column of Fig. 4.12 illustrates the necessary translations that have these transforms ready as inputs for the (4.3), (4.4) form.

We should also remember that the outputs of (4.3), (4.4) take the second-column form, which has to be changed back to the first-column form as its actual graph.

Of course, if we are to compute for $g(t)$ in (4.4), then we have to prepare the real and imaginary parts of $G(f)$ as in Fig. 4.12b', c' and combine them for the input G in (4.4).

We reiterate our comments regarding what we should expect as a truncation error around the end points of the symmetric interval (e.g., $t = \mp a$ in Fig. 4.12a and $f = \mp b$ in Fig. 4.12c, d) and the Gibbs phenomenon especially around an interior point such as $t = 0$ in Fig. 4.12a for the causal function $g(t)$. As we shall see in Example 4.7 and Fig. 4.35, when we receive the DFT output, it is listed as in the second column of Fig. 4.12, and we advise that it should be returned to its natural symmetric format

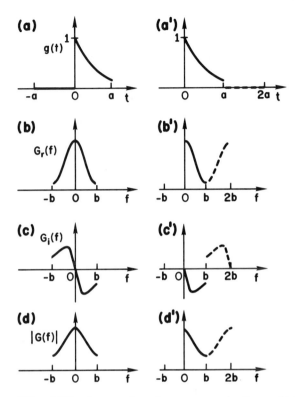

FIG. 4.12 Inputs for the symmetric form (4.5), (4.6) in (a)–(d) and the nonsymmetric form (4.3), (4.4) in (a′)–(d′) of the discrete Fourier transform.

in the first column of Fig. 4.12 before any attempt to look at or interpret the errors in the DFT approximation of our functions. We do that because we notice that sometimes the seeming end points in the second column are easily confused with the actual end points of the domain in the first column. This is what we shall follow in our detailed Example 4.7 and its Fig. 4.35 for the functions of Figs. 4.11 and 4.12 (compare Figs. 4.35c and 4.35d). Otherwise, if we accept the second column format of Fig. 4.12, we should interpret the errors around $t = a$ in Fig. 4.12a′ and $f = b$ in Fig. 4.12b′, c′, d′ as truncation errors, which can be helped by an optimal truncation window, as discussed in detail in Section 4.1.6I. On the other hand, we interpret the error around $t = 0$ and $t = 2a$ in Fig. 4.12a′ as a Gibbs phenomenon, which is due to the jump discontinuity at the interior point $t = 0$ of the interval $(-a, a)$, a very stubborn error which we discuss in detail with suggested remedies in Section 4.1.6G (see Fig. 4.30). This error must be watched especially when we work with causal functions with (often) a jump discontinuity at $t = 0$. Also, this situation is evident if we are to use the DFT for the numerical inversion of Laplace transforms, where we expect the output $g(t)$ as a causal function.

Another place where we have a Gibbs phenomenon in Fig. 4.12 is at $f = \mp b$ for $G_i(f)$ in Fig. 4.12c (or equivalently at $f = b$ in Fig. 4.12c'), which is due to the jump discontinuity in the periodic extension of (the odd function) $G_i(f)$. There is no Gibbs phenomenon in Fig. 4.12b for $G_r(f)$ and Fig. 4.12d for $|G(f)|$ because there are no jump discontinuities, since both functions are continuous on $(-b, b)$, and as even functions their periodic extension is also continuous.

In Fig. 4.13 we consider the real even function $e^{-|t|}$

$$g(t) = e^{-|t|}, \qquad -\infty < t < \infty \tag{4.32}$$

with its transform

$$G(f) = \frac{2}{1 + (2\pi f)^2}, \qquad -\infty < f < \infty \tag{4.33}$$

which is also a real even function. We may note that as in this example, when either $g(t)$ or $G(f)$ is a real even function, the part of $\bar{g}$ or $\bar{G}$ on $(-N/2, 0)$, $(0, N/2)$ and its periodic extension on $(N/2, N)$ are easily related, and hence the equivalence of the (4.3), (4.4) and (4.5), (4.6) forms. However, in the general case, as we have indicated for Figs. 4.10–4.12, we have to translate the part $(-N/2, 0)$ by N to the right to get its needed replica on $(N/2, N)$ for the (4.3), (4.4) form. Since the DFT or the FFT output is presented in this form, we advise that the data on $(N/2, N)$ be translated back to its natural position $(-N/2, 0)$, where we can see the original picture of the output wave in its more natural setting. This would allow a clearer Fourier analysis interpretation with its approximation errors, even though it may represent an extra step to the electrical engineer.

B. Preparing for the Computer Program of the FFT Algorithm

In the previous discussions, we presented the three different forms of the discrete Fourier transform pairs; the typical form (4.3), (4.4); the symmetric form (4.5), (4.6); and the cosine-sine form (4.15)–(4.17). In the next section we will discuss the FFT algorithm for computing these discrete transforms, which is available on most computers. The very familiar program for the FFT is designated as FFTCC: [†] the fast Fourier transform

---

[†] The up-to-date versions are FFTCF, FFTCB in the IMSL Math.-library, Ed. 1.1.

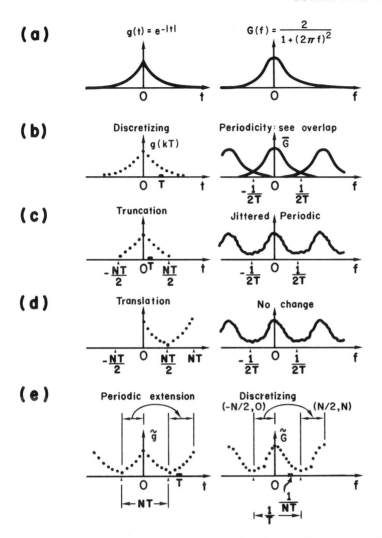

FIG. 4.13 Graphical development of the discrete Fourier transforms. Approximation of $g(t) = e^{-|t|}$, $-\infty < t < \infty$, and its (even) transform $G(f)$, $-\infty < f < \infty$.

for complex-valued sequence computation. The sine-cosine form is also available.

As we shall see, this complex form of the FFTCC is still a little different from either of the forms (4.3), (4.4) and (4.5), (4.6) that we have presented. In the following, we will discuss the adjustments necessary for the usual form (4.3), (4.4) to suit the FFTCC discrete form.

The FFTCC routine uses the following form of the (complex) discrete Fourier transform:

$$X_{k+1} = \sum_{j=0}^{N-1} A_{j+1} e^{2\pi ijk/N}, \qquad k = 0, 1, 2, \ldots, N-1, \ i = \sqrt{-1} \qquad (4.4b)$$

to compute the (complex) sequence X_{k+1} as the discrete Fourier transform of the (complex) sequence A_{j+1}. We will first adjust the notation in the above form (4.4b) to avoid any confusion with what we already presented for (4.3)–(4.6). This means that we should stay with $j = \sqrt{-1}$ instead of $i = \sqrt{-1}$ in the above form, and we use l instead of j for the above index of summation,

$$X_{k+1} = \sum_{l=0}^{N-1} A_{l+1} e^{2\pi jlk/N}, \qquad k = 0, 1, 2, \ldots, N-1, \ j = \sqrt{-1} \qquad (4.4c)$$

This form can be adopted, after dividing by N, for the (4.4) form to compute

$$\tilde{g}(kT) = \frac{1}{N} X_{k+1}, \qquad k = 0, 1, 2, \ldots, N-1$$

the discrete Fourier transform of

$$\tilde{G}\left(\frac{l}{NT}\right) = A_{l+1}, \qquad l = 0, 1, 2, \ldots, N-1$$

The computer program for the FFTCC uses $X(K+1)$, $K = 0$, 1, 2, ..., $N-1$ and $A(L+1)$, $L = 0$, 1, 2, ..., $N-1$ for the complex sequences X_{k+1} and A_{l+1} respectively. Thus we use the index l, instead of the usual n for $\tilde{G}(n/NT)$, to avoid confusing $n \equiv N$, the index in the program, with the summation limit $N-1$.

The complex sequences A_{l+1} and X_{k+1} in (4.4b) are input and output, respectively, as ordered pairs

$$A_{l+1} = \left(\tilde{G}_r\left(\frac{l}{NT}\right), \tilde{G}_i\left(\frac{l}{NT}\right)\right) \equiv \tilde{G}_r\left(\frac{l}{NT}\right) + j\tilde{G}_i\left(\frac{l}{NT}\right),$$
$$l = 0, 1, 2, \ldots, N-1$$

$$X_{k+1} = N(\tilde{g}_r(kT), \tilde{g}_i(kT)) \equiv N[\tilde{g}_r(kT) + j\tilde{g}_i(kT)],$$
$$k = 0, 1, 2, \ldots, N-1$$

The same form (4.4c) can be used, after some careful adjustments, to compute $\tilde{G}(n/NT)$, the discrete transform of $\tilde{g}(kT)$ in (4.3). If we compare (4.3) and (4.4) we note that the main difference is in the complex conjugation (j is changed to $-j$) of the complex exponential kernel $e^{-2\pi jkn/N}$ in (4.3). This means that by taking the complex conjugate of both sides of (4.3), we can have the form (4.4),

$$\overline{\tilde{G}\left(\frac{l}{NT}\right)} = \sum_{k=0}^{N-1} \tilde{g}(kT)e^{-j2\pi lk/N}$$

$$= \sum_{k=0}^{N-1} \overline{\tilde{g}(kT)}e^{j2\pi lk/N}, \qquad l = 0, 1, 2, \ldots, N-1 \qquad (4.4d)$$

after using the fact that the complex conjugate of the sum/product is the sum/product of the complex conjugates. This form is now ready for the IMSL form (4.4c) with the input

$$A_{k+1} = \overline{\tilde{g}(kT)} = \overline{(\tilde{g}_r(kT), \tilde{g}_i(kT))} = (\tilde{g}_r(kT), -\tilde{g}_i(kT))$$

as the complex conjugate of $\tilde{g}(kT)$, the input to (4.3). The output is also the complex conjugate of the output sought in (4.3),

$$X_{l+1} = \overline{\tilde{G}\left(\frac{l}{NT}\right)} = \overline{\left(\tilde{G}_r\left(\frac{l}{NT}\right), \tilde{G}_i\left(\frac{l}{NT}\right)\right)}$$

$$= \left(\tilde{G}_r\left(\frac{l}{NT}\right), -\tilde{G}_i\left(\frac{l}{NT}\right)\right)$$

This means that to use the IMSL form (4.4b) to compute $\tilde{G}(l/NT)$ of (4.3), we must input the complex conjugate $\overline{\tilde{g}(kT)}$ of $\tilde{g}(kT)$ (by merely changing the sign of its imaginary part $\tilde{g}_i(kT)$ to $-\tilde{g}_i(kT)$) to obtain $\overline{\tilde{G}(l/NT)}$, which can then be complex-conjugated to have the desired $\tilde{G}(l/NT) = \overline{\overline{\tilde{G}(l/NT)}}$ of (4.3).

We may add that the FFTCC[†] program takes any positive integer number N of terms for the discrete transform (4.4c). This is a great improvement over other usual programs for the fast Fourier transform, since they are limited to $N = 2^n$. The latter is usually done to simplify the presentation of the development of the FFT algorithm, which we shall adopt for our discussion of the FFT in the next section.

4.1.3 The Fast Fourier Transform

The basic properties, considered here and mainly in the next section, for the discrete Fourier transforms are done with the most familiar forms

---

[†] See footnote on p. 544

(4.3), (4.4) or (4.3a), (4.4a) of these transforms,

$$\tilde{G}_n \equiv \tilde{G}\left(\frac{n}{NT}\right) = \sum_{k=0}^{N-1} \tilde{g}(kT)e^{-j2\pi nk/N}, \qquad n = 0, 1, \ldots, N-1$$

$$f_n = n\Delta f = n\frac{1}{NT} \tag{4.3}$$

$$\tilde{g}_k \equiv \tilde{g}(kT) = \frac{1}{N}\sum_{n=0}^{N-1} \tilde{G}\left(\frac{n}{NT}\right) e^{j2\pi nk/N}, \qquad k = 0, 1, \ldots, N-1$$

$$t_k = k\Delta t = kT \tag{4.4}$$

We will show that such results can be modified to the (4.5), (4.6) symmetric form or the (4.19)–(4.21) cosine-sine form. In the last section, we discussed the adjustments necessary for the above forms to suit the discrete form used in the most familiar computer program (FFTCC) for the fast Fourier transform in the IMSL library. (See also the FFT computer programs in Brigham, 1988.)

To prepare for the eventual efficient algorithm for computing these transforms, i.e., the fast Fourier transforms, let us observe the number of multiplications and additions (operations) necessary for computing $\tilde{G}(n/NT)$ for $n = 0, \ldots, N-1$. We call N the order of the discrete transform, and we note that in the case of $n = 0$ we don't need multiplication inside the sum, since the exponential kernel is 1. This is the same for the first value $\tilde{g}(0)$ of (4.4). To prepare for the total direct computations we will assume that the necessary $(N-1)^2$ values of the discrete kernel $e^{-j2\pi kn/N}$, for the relevant $n, k = 1, \ldots, N-1$, are already computed and stored. It is clear that for each $n = 0, \ldots, N-1$ we will need N additions after the $N-1$ multiplications, a total of N^2 additions. Since additions are considered to be very simple operations on computing machines (as compared to multiplications) we shall concern ourselves with the $(N-1)^2$ multiplications as the required number of operations for the direct computation of the discrete transform $\tilde{G}(n/NT)$ of order N. The same can be shown for the $(N-1)^2$ operations for $\tilde{g}(kT)$ of order N as in (4.4).

In our present treatment, we will focus on the few particular properties of the discrete transforms that will contribute to establishing the fast Fourier transform, and its final result of reducing this number of $(N-1)^2$ direct operations to $(N/2)\log_2 N$. This is in parallel with Papoulis's (1977) development.

The most important property of the discrete transforms (4.3), (4.4) that led to the fast computation algorithm, the FFT, is the *cyclic property* of the discrete kernel $e^{-j2\pi nk/N}$, which repeats the value $e^{-j2\pi m/N}$ whenever

$nk = lN + m$ (l, m, n, N integers)

$$e^{-j2\pi nk/N} = e^{-j2\pi(lN+m)/N} = e^{-j2\pi l}e^{-j2\pi m/N} = e^{-j2\pi m/N} \qquad (4.34)$$

That is, it is periodic with period N. Also it has the simple value of 1 whenever nk is an integer multiple of N. Intuitively, one should take advantage of the repetitions as nk ranges from 1 to $(N-1)^2$, simply indicating the locations of these repetitions and signaling the machine to avoid their repeated computation. The fast Fourier transform goes much beyond this in capitalizing on such repetitions to the utmost.

As a starting point of the FFT algorithm, we will show next *how a discrete transform $\tilde{G}(n/2MT)$ in (4.3) of order $N = 2M$ is written as a simple combination of two discrete transforms, each of order M.* Hence, roughly, the $(2M-1)^2$ operations required for $\tilde{G}(n/2MT)$ are reduced to $2(M-1)^2$ operations, a saving by a factor of about a half.

We will use the following simple odd-even property of the discrete kernel $e^{-j2\pi nk/2M}$:

$$e^{-j2\pi nk/2M} = \begin{cases} e^{-j2\pi n2l/2M} = e^{-j2\pi nl/M}, & k = 2l, \ l = 0, 1, 2, \ldots, M-1 \\ e^{-j2\pi n(2l+1)/2M} = e^{-j2\pi nl/M}e^{-j\pi n/M}, & k = 2l+1, \\ l = 0, 1, 2, \ldots, M-1 \end{cases}$$

$$(4.35)$$

We will assume that the $(M-1)^2$ values of $e^{-j2\pi nl/M}$ and the $(M-1)$ values of $e^{-j\pi n/M}$ are already computed and stored. So, we need $(M-1)^2 + (M-1)$ different computations to cover the, seemingly, $(2M-1)^2$ computations necessary for $e^{-j2\pi nk/2M}$, $n, k = 0, 1, 2, \ldots, 2M-1$. Now we will prove the following basic result for establishing the FFT algorithm:

$$\tilde{G}\left(\frac{n}{2MT}\right) = \tilde{F}\left(\frac{n}{MT}\right) + e^{-j\pi n/M}\tilde{H}\left(\frac{n}{MT}\right) \qquad (4.36)$$

where $\tilde{F}$ and $\tilde{H}$ are the order M discrete transforms respectively of the even samples $\tilde{g}(2lT) \equiv \tilde{f}(lT)$ and the odd samples $\tilde{g}((2l+1)T) \equiv \tilde{h}(lT)$ of $\tilde{g}(kT)$ in (4.4), which are, of course, of order M. To prove (4.36) we write

$$\tilde{G}\left(\frac{n}{2MT}\right) = \sum_{k=0}^{2M-1} \tilde{g}(kT)e^{-j2\pi nk/2M}$$

$$= \sum_{k=0,2,\ldots,2M-2} \tilde{g}(kT)e^{-j2\pi nk/2M}$$

$$+ \sum_{k=1,3,\ldots,2M-1} \tilde{g}(kT)e^{-j2\pi nk/2M} \qquad (4.37)$$

after separating the even and odd terms of the summation, as the first and second sums, respectively. Now we make a simple change of variables $k = 2l$ and $k = (2r + 1)$ in the first and second sums respectively to have

$$\bar{G}\left(\frac{n}{2MT}\right) = \sum_{l=0}^{M-1} \bar{g}(2lT)e^{-j2\pi 2ln/2M} + \sum_{r=0}^{M-1} \bar{g}((2r+1)T)e^{-j2\pi(2r+1)n/2M}$$

$$= \sum_{l=0}^{M-1} \bar{g}(2lT)e^{-j2\pi ln/M} + e^{-j\pi n/M}\sum_{r=0}^{M-1} \bar{g}((2r+1)T)e^{-j2\pi rn/M}$$

$$= \sum_{l=0}^{M-1} \tilde{g}(lT)e^{-j2\pi ln/M} + e^{-j\pi n/M}\sum_{r=0}^{M-1} \tilde{h}(rT)e^{-j2\pi rn/M}$$

$$= \tilde{F}\left(\frac{n}{MT}\right) + e^{-j\pi n/M}\tilde{H}\left(\frac{n}{MT}\right)$$

after defining $\tilde{f}(lT) \equiv \bar{g}(2lT)$, $\tilde{h}(rT) \equiv \bar{g}((2r+1)T)$, and using (4.3) for $\tilde{F}$ and $\tilde{H}$, which as we shall show next will be effectively of order $N = M$ instead of the $2M$ for $\bar{G}$.

Now we can see the advantage of (4.36), in that as $\bar{G}(n/2MT)$ goes through its period of $2M$ sample values, both $\tilde{F}$ and $\tilde{H}$ will go through two periods of their M samples, where the second period is a repetition, except for the factor $e^{-j\pi n/M}$ which will change by only a minus sign for $n = M + l > M$,

$$e^{-j\pi(M+l)/M} = e^{-j\pi}e^{-j\pi l/M} = -e^{-j\pi l/M}$$

and which we will count as one operation, since it is different from addition. So now with only M samples for each of $\tilde{F}$ and $\tilde{G}$, in (4.36) we require $2(M-1)^2$ operations (multiplications) for computing them, plus $(M-1)$ multiplications for the factor $e^{-j\pi n/N}$ and one subtraction operation for $n = M + l > M$. This is a total of operations

$$2(M-1)^2 + M - 1 + 1 = 2(M-1)^2 + M \tag{4.38}$$

(mainly multiplications), instead of the $(2M-1)^2$ multiplications, required for the direct (4.3) computation of $\bar{G}(n/2MT)$ without the use of (4.36). Hence there is a saving by a factor of about one-half for large enough M.

The essence of the fast Fourier transform algorithm is to repeat this proceess (4.36), and hence its savings, for $N = 2M = 4J = 8K = \cdots$ and exploit these continued factors of reduction. Indeed if we take the simple case of $N = 2^s$, where s is an integer, we can continue the above reduction process until we reach a one-term sum, that is, a discrete transform of order one,

$$N = 2 \cdot 2^{s-1} = 2^2 \cdot 2^{s-2} = 2^3 \cdot 2^{s-3} = \cdots = 2^{s-1} \cdot 2 = 2^s \qquad (4.39)$$

which needs no operation.

When we reach the smallest system of order one, we must then add all the operations for the total of s decreasing-order systems. We note that the first reduced system has order $M = 2^{s-1}$ since $N = 2M = 2 \cdot 2^{s-1}$, which will require, as we have shown above (4.38),

$$2(M - 1)^2 + M = 2(2^{s-1} - 1)^2 + 2^{s-1} \qquad (4.40)$$

But the number of operations (multiplications) $(2^{s-1} - 1)^2$ can be further reduced when we write the discrete transform of order 2^{s-1} in terms of the next two of the lower order 2^{s-2} as in (4.36), with its new reduced number of multiplications

$$2(2^{s-2} - 1)^2 + 2^{s-2} \qquad (4.41)$$

So if we substitute the result (4.41), instead of $(2^{s-1} - 1)^2$, in (4.40) we obtain

$$2[2(2^{s-2} - 1)^2 + 2^{s-2}] + 2^{s-1} = 2^2(2^{s-2} - 1)^2 + 2 \cdot 2^{s-1} \qquad (4.42)$$

Now we write the discrete transform of order 2^{s-2} in terms of the two of order 2^{s-3} in the same way, replacing the $(2^{s-2} - 1)^2$ multiplications term of (4.42) by its reduced value

$$2(2^{s-3} - 1)^2 + 2^{s-3} \qquad (4.43)$$

and obtain

$$2^2[2(2^{s-3} - 1)^2 + 2^{s-3}] + 2 \cdot 2^{s-1} = 2^3(2^{s-3} - 1)^2 + 2^{s-1} + 2 \cdot 2^{s-1}$$
$$= 2^3(2^{s-3} - 1)^2 + 3 \cdot 2^{s-1} \qquad (4.44)$$

as the total number of operations required for the new system of order 2^{s-3}.

This process of reduction can be continued, and by mathematical induction we can show that the number of computations for the system of order $N = 2^s$ in terms of that for the discrete transform of lower order 2^{s-m} is

$$2^m(2^{s-m} - 1)^2 + m \cdot 2^{s-1} \qquad (4.45)$$

Now if we have $s = m + 1$, this is our last step, which stops at a discrete transform with only two terms. This needs only one multiplication, as shown from $(2 - 1) = 1$ in the term in parentheses in (4.45). In this case the reduced number of multiplications for the discrete transform of order $N = 2^{m+1}$ is obtained from (4.45) as

$$2^m(2^{m+1-m} - 1)^2 + m \cdot 2^{m+1-1} = 2^m + m \cdot 2^m = (m + 1)2^m$$

$$= \frac{(m + 1)2^{m+1}}{2} = \frac{1}{2} 2^{m+1} \log_2 2^{m+1} = \frac{1}{2} N \log_2 N \qquad (4.46)$$

which is the very well-known result of the FFT.

We must stress here that this exciting result is based mainly on the number of multiplications which are $(N - 1)^2$ for the direct evaluation of (4.3). For the actual complete computations, additions will be included; hence we should not be surprised if it takes more operations than indicated by (4.46). Such differences are clear for small N and may become negligible for large enough N, as we shall show in our examples.

We have chosen here $N = 2^s$ for the sake of simplicity. Other versions of the FFT include different factorizations, that is, $N = n_1 n_2 \cdots n_m$, or the same factors with a different base from the above 2.

There are many different methods for proving the FFT result (4.46), including matrix analysis and graphical methods (Brigham, 1974, 1988) One of the most compact methods, which we shall present here, uses mathematical induction on (4.46) with the help of the basic result (4.36) (see Papoulis, 1977).

Let us denote by $\eta(N)$ the number of multiplications necessary for the computation of the discrete transform $\tilde{G}(n/NT)$ of order N. The basic result (4.36), (4.38) states that the number of computations necessary for the evaluation of the discrete transform of order $2M$ is

$$\eta(2M) = 2\eta(M) + M \qquad (4.47)$$

Using this result, we want to prove that $\eta(N) = (N/2)\log_2 N$ of (4.46). For $N = 2$ we know that we need only one multiplication, which is the case in this formula as $\eta(2) = (2/2)\log_2^2 = 1$. Now we suppose that this formula (4.46) is true for M,

$$\eta(M) = \frac{M}{2} \log_2 M \qquad (4.48)$$

and we will use (4.47) to show that (4.48) is also true for $N = 2M$. If we use (4.47) for $\eta(2M)$ and substitute for $\eta(M)$ from (4.48), where we assume this result (4.48) to be true, we have

$$\eta(2M) = 2\eta(M) + M = 2\left[\frac{M}{2} \log_2 M\right] + M$$

$$= M[1 + \log_2 M] = M \log_2 2M = \frac{2M}{2} \log_2 2M$$

which is the same formula (4.48) for $N = 2M$; that is, (4.48) is also true for $N = 2M$. Hence it is valid for all $N = 2^s$ as far as our development is concerned.

The advantage of this important discovery of the FFT algorithm, in reducing the number of computations (multiplications), is illustrated in Table 4.2. To avoid being mislead by this exciting table, we should keep in mind that it gives the comparison for only the number of multiplications involved in the evaluation of the discrete transform and that the actual advantage will still depend on the given problem and the program used for the FFT. That is, we should not be surprised by the statement of large enough N for the clear advantages of the FFT over other methods. For example, the FFT method of computing the Fourier convolution product would show an advantage only after $N = 64$.[†]

This application of the FFT for computing the discrete convolution products, along with the correlation products and the derivatives of functions, will be discussed toward the end of Section 4.1.4.

4.1.4 Construction and Basic Properties of the Discrete Transforms

In the last section we used the form (4.3), (4.4) of the discrete transforms and the important periodicity property of the discrete kernel to establish the basis of the fast Fourier transform. In this section, we will discuss and prove other basic properties of the discrete transforms, which will help in

TABLE 4.2 Number of Multiplications for the FFT

Order of Computations N	FFT Number of Multiplications $\frac{N}{2} \log_2 N$	Direct Method Number of Multiplications $(N - 1)^2$
8	12	49
16	32	225
32	80	961
64	192	3,969
128	448	16,129
256	1,024	66,025

[†]A detailed computer program flow chart and listing of a BASIC and a Pascal program for the FFT are given in Brigham (1988).

understanding the construction of such transforms, and hence the possible errors involved in using them for approximating Fourier integrals or series. The following discrete orthogonality, as a parallel to the orthogonality of the Fourier series basis functions, will prove to be of great importance to our present development.

A. Discrete Orthogonality

For the Fourier series (4.7) of the periodic function $G_s(f)$ on $(-1/2T, 1/2T)$, we know that the continuous kernel constitutes a set of functions $\{e^{-j2\pi kTf}\}_{k=-\infty}^{\infty}$ which are orthogonal on the finite interval $(-1/2T, 1/2T)$. The discrete exponential Fourier transform $\tilde{G}(n/NT)$ has an analogous property in that its discrete kernels $\{e^{-j2\pi nk/N}\}_{k=0}^{N-1}$ are also orthogonal over the finite sequence $n = 0, \ldots, N-1$, that is,

$$\sum_{k=0}^{N-1} e^{j2\pi rk/N} e^{-j2\pi nk/N} = \begin{cases} N, & r = n \\ 0, & r \neq n \end{cases} = N\delta_{r,n} \qquad (4.49)$$

where $\delta_{r,n}$ is the Kronecker delta:

$$\delta_{r,n} = \begin{cases} 1, & r = n \\ 0, & r \neq n \end{cases}$$

This discrete orthogonality property is essential for the discrete inverse transform $\tilde{g}(kT)$ to have the symmetric form (4.4), which we shall verify, but first we will prove the discrete orthogonality (4.49).

To prove the discrete orthogonality (4.49), we recognize that the sum can be written as a finite geometric series,

$$\sum_{k=0}^{N-1} e^{j2\pi k(r-n)/N} = \sum_{k=0}^{N-1} \left[e^{j2\pi(r-n)/N} \right]^k$$

$$= \frac{1 - \left[e^{j2\pi(r-n)/N} \right]^N}{1 - \left[e^{j2\pi(r-n)/N} \right]} = \begin{cases} N, & r = n \\ 0, & r \neq n \end{cases}$$

The zero value for $r \neq n$ is clear since the denominator is not zero while the numerator is zero after using the cyclic property, $\left[e^{j2\pi k/N} \right]^N = e^{j2\pi k} = 1$, of the discrete kernel. The N value for $r = n$ is obtained from adding the N constant terms (of 1) in the above sum.

With the same method of proof, we can show that this discrete orthogonality is valid for the general limits $k = n_0$ to $k = n_0 + N - 1$ and, in particular, the symmetric limits $k = -M$ to M, where $N = 2M + 1$ in

(4.3); that is,

$$\sum_{k=-M}^{M} e^{j2\pi rk/(2M+1)} e^{-j2\pi nk/(2M+1)} = \begin{cases} 2M+1, & r=n \\ 0, & r \neq n \end{cases} \tag{4.49a}$$

(see Exercise 4.9a).

For all the different forms of the discrete orthogonality (4.49)–(4.53), we should observe that the conditions $r = n$ and $r \neq n$ in (4.49) could be generalized to $|r - n| = 0, 2M + 1, 2(2M + 1), \ldots$ and $|r - n| \neq 0, 2M + 1, 2(2M + 1), \ldots$, respectively. This will be helpful when our analysis involves infinite series and hence $(r - n)$ may run through infinitely many multiples of $(2M + 1)$. The following discrete orthogonalities, necessary for the discrete cosine-sine transforms (4.18)–(4.21),

$$\sum_{k=0}^{2N-1} \cos \frac{2\pi rk}{2N} \cos \frac{2\pi nk}{2N} = \begin{cases} 0, & r \neq n \\ N, & r = n \neq 0, N \\ 2N, & r = n = 0, N \end{cases} \tag{4.50}$$

$$\sum_{k=0}^{2N-1} \cos \frac{2\pi rk}{2N} \sin \frac{2\pi nk}{2N} = 0 \tag{4.51}$$

$$\sum_{k=0}^{2N-1} \sin \frac{2\pi rk}{2N} \sin \frac{2\pi nk}{2N} = \begin{cases} 0, & r \neq n \\ N, & r = n \neq 0, N \\ 0, & r = n = 0, N \end{cases} \tag{4.52}$$

can be proved by using the Euler identity (3.14) and the discrete orthogonality of the exponential functions (4.49). A simpler way is to use trigonometric identities; for example, in the case of (4.50) we use

$$\cos A \cos B = \tfrac{1}{2}[\cos(A+B) + \cos(A-B)]$$

then the Euler identity (3.14) and the following single exponential term of (4.49) (with $n = 0$):

$$\sum_{k=0}^{N-1} e^{j2\pi rk/N} = \frac{1 - [e^{j2\pi r/N}]^N}{1 - e^{j2\pi r/N}} = \begin{cases} 0, & r \neq 0, N \\ N, & r = 0, N \end{cases} \tag{4.53}$$

which we have as an exercise (see Exercises 4.9a).

The discrete orthogonality (4.49) will be used for the proofs of most of the following properties of the DFT, including the verification of the discrete transform (4.3) and its inverse (4.4). Besides the discrete orthogonality, we recognize that the discrete transforms (4.3), (4.4) are only finite sums, so we will have the liberty of exchanging double or triple sums (with due care of the involved indices of the particular summation).

To verify that $\tilde{g}(kT)$ in (4.4) is the exact inverse of $\tilde{G}(n/NT)$ in (4.3), we substitute its summation representation from (4.4) in (4.3), exchange the two sums, then use the discrete orthogonality to result in $\tilde{G}(n/NT)$:

$$
\begin{aligned}
\sum_{k=0}^{N-1} \tilde{g}(kT)e^{-j2\pi nk/N} &= \sum_{k=0}^{N-1} \left[\frac{1}{N} \sum_{r=0}^{N-1} \tilde{G}\left(\frac{r}{NT}\right) e^{j2\pi rk/N} \right] e^{-j2\pi nk/N} \\
&= \sum_{r=0}^{N-1} \frac{1}{N} \tilde{G}\left(\frac{r}{NT}\right) \sum_{k=0}^{N-1} e^{j2\pi rk/N} e^{-j2\pi nk/N} \\
&= \sum_{r=0}^{N-1} \frac{1}{N} \tilde{G}\left(\frac{r}{NT}\right) N\delta_{r,n}, \qquad \delta_{r,n} \equiv \begin{cases} 0, & r \neq n \\ 1, & r = n \end{cases} \\
&= \tilde{G}\left(\frac{n}{NT}\right)
\end{aligned}
$$

after using the discrete orthogonality (4.40) for the inner sum in the second line.

The same method can be used to verify the symmetric (4.5), (4.6) and the cosine-sine (4.19)–(4.21) discrete transforms, with the aid of the discrete orthogonalities (4.49) and (4.50)–(4.52), respectively.

As we mentioned at the beginning of this chapter, these discrete transforms (4.3), (4.4) are exact transforms of each other. As such, they may be used on their own to algebraize their compatible equations–the *difference equations*, which we shall present in Section 4.1.5. But in order to give the simplest illustration of such an application, we will need to develop a few pairs of these discrete transforms, which are essential for such an *operational difference calculus* method. Such pairs, as finite sums, will be derived by using the calculus of finite summation in comparison with the integration used for the integral and finite transforms of the last chapters. In that regard we must remember the value of the integration-by-parts method, which has to be replaced by its finite sum analog—*summation by parts*—that we shall develop and validate as (4.93) in Section 4.1.5.

We remark here that, given the discrete orthogonality (4.49), the above verification constitutes a proof of the discrete transform pair (4.3), (4.4). Such a proof may be sufficient if we are to use these transforms for the system representation of a finite set of discrete data. But since our main purpose is to use the discrete transforms for approximating Fourier series and integrals, it is important to have a constructive proof related to these two topics, as we have indicated earlier in relating the discrete transform to the Fourier coefficients (4.22)–(4.24).

If we are satisfied with the above verification of the discrete transforms, (4.3) and (4.4), we must at least inquire into their sample spacing of $\Delta f = 1/NT$ for $\tilde{G}$ and $\Delta T = T$ for $\tilde{g}$. This can be done with the help of the discrete orthogonality (4.49) and the Fourier series (4.7), (4.8). For the latter, we have the set $\{e^{-j2\pi kTf}\}$, which is orthogonal on $[-1/2T, 1/2T]$ with $t_k = kT$, hence a spacing of $\Delta t = T$. If we compare this continuous kernel with that of the discrete transform, we see that $f = n\Delta f = n1/NT$, where $\Delta f = 1/NT$ for $\Delta t = T$.

The full treatment of this required spacing lies with the sampling theorem (see Section 2.4.2, Eq. (2.184)),[†] which we will explore in more detail when we deal with developing new discrete transforms, which will sometimes occur in the absence of a valid or exact discrete orthogonality.

In the following, we will outline a method of constructing the discrete transforms (4.3), (4.4) as they relate to the periodic function of the Fourier series (4.7), (4.8). Then we will relate the samples of the Fourier transforms (4.1), (4.2) to such Fourier series and hence to the discrete transforms. The emphasis will be on the process of the construction, which will reveal clearly the truncation and sampling (aliasing) errors of approximating the Fourier series and Fourier integral by the discrete transforms.

This method resembles the analysis we indicated for (4.19)–(4.24) and it parallels that of Papoulis (1977), where more analysis of discrete systems can be found.

B. Fourier Series and the Discrete Transforms

Consider the Fourier series expansion (4.7) for the continuous function $H(f)$ on $(-1/2T, 1/2T)$,

$$H(f) = \sum_{k=-\infty}^{\infty} c_k e^{-j2\pi kTf}, \qquad -\frac{1}{2T} < f < \frac{1}{2T} \tag{4.7}$$

[†] A detailed method of constructing the discrete transforms (with illustrations) and emphasis on the sampling theorem and Fourier transforms pairs are given in Brigham (1974).

and its samples at $f = n/(2M + 1)T$ as specified by the discrete transform (4.5),

$$H\left(\frac{n}{(2M + 1)T}\right) = \sum_{k=-\infty}^{\infty} c_k e^{-j2\pi kn/(2M+1)} \tag{4.54}$$

Now we can use the discrete orthogonality (4.36) to find a relation between these samples of the periodic function and the discrete transform (4.5). We multiply both sides of (4.54) by $1/(2M + 1)e^{j2\pi rn/(2M+1)}$ and sum from $-M$ to M to have

$$\frac{1}{2M + 1} \sum_{n=-M}^{M} H\left(\frac{n}{(2M + 1)T}\right) e^{j2\pi rn/(2M+1)}$$

$$= \sum_{k=-\infty}^{\infty} c_k \left[\frac{1}{2M + 1} \sum_{n=-M}^{M} e^{j2\pi n(r-k)/(2M+1)}\right] \tag{4.55a}$$

We note that the left side involves the samples of the actual periodic function $H(f)$, which may be different from (4.6), and so we denote it by C_r,

$$C_r = \frac{1}{2M + 1} \sum_{n=-M}^{M} H\left(\frac{n}{(2M + 1)T}\right) e^{j2\pi rn/(2M+1)} \tag{4.55b}$$

For the finite sum of the discrete orthogonality, on the right side of (4.55a), we must observe that $(r - k)$ in the exponent runs through infinitely many multiple of $2M + 1$ as k goes from $-\infty$ to ∞. Hence, according to (4.49a) and the note that followed it, this finite sum vanishes for all $(r - k) \neq l(2M + 1)$ where l is an integer, and it is $(2M + 1)$ when $(r - k) = l(2M + 1)$,

$$C_r \equiv \bar{c}_r = \sum_{l=-\infty}^{\infty} c_{r+l(2M+1)} = c_r + \sum_{|l|=1}^{\infty} c_{r+l(2M+1)} \tag{4.55c}$$

where $\bar{c}_r$ is the superposition of all the translated Fourier coefficients c_k by an integer multiple of $(2M + 1)$. So the infinite sum in (4.55b) represents the overlapping contribution, to the basic finite sequence c_r, $r = -M$ to M, due to all the translations, by $l(2M + 1)$, $|l| = 1$ to ∞, of the infinite new sequence $C_{r+l(2M+1)}$. Such a summation is called the *aliasing error* in approximating c_r by the overlapped (aliased) $C_r \equiv \bar{c}_r$,

$$\epsilon_A c_r \equiv \bar{c}_r - c_r = \sum_{|l|=1}^{\infty} c_{r+l(2M+1)} \tag{4.56}$$

This is illustrated in Fig. 4.8c of Section 4.1.1 for c_k, the Fourier coefficients of the periodic function $G_s(f) = e^{-|f|}$, $-1/2T < f < 1/2T$.

This important relation (4.55c), like (4.22)–(4.24), shows how an infinite number of the Fourier coefficients c_n are "folded" into an infinite sum to give one term C_r, which is called the "aliased" coefficient, and which will prove to be the discrete transform $\bar{h}(rT)$ of (4.6). If we use the discrete orthogonality (4.49a) on (4.55b), we can show how the samples of the periodic function $H(n/(2M+1)T)$ are expressed in terms of only $2M+1$ aliased coefficients C_r,

$$\sum_{r=-M}^{M} C_r e^{-j2\pi rn/(2M+1)} = H\left(\frac{n}{(2M+1)T}\right), \qquad n = -M, \ldots, 0, \ldots, M$$

$$(4.57)$$

This finite sum representation of the $2M+1$ samples of the periodic function is with aliased coefficients C_k and can hardly replace the infinite Fourier series representation (4.54) (with its actual coefficients c_k). This means that (4.57) gives exact sample values of $H(f)$ at the expense of aliased coefficients. However, for these samples to represent the continuous function $H(f)$, a better resolution is needed, which requires large enough M. With this large M we can observe from (4.56) that the aliasing error of c_r would decrease, which makes (4.57) a reasonable representation with large enough M. This brings us again to the importance of the FFT algorithm, which is extremely efficient for large M, as we indicated in Table 4.2.

As to the discrete transform (4.6) equivalence of (4.57), the above use of the discrete orthogonality (4.49), and a simple comparison with (4.5), (4.6), would easily identity C_r with $\bar{h}(rT)$ and $H(n/(2M+1)T)$ with $\bar{H}(n/(2M+1)T)$. Of course, it is clear here that while $\bar{h}(rT)$ is an approximation to the Fourier coefficients c_r, $\bar{H}(n/(2M+1)T)$ are the exact samples of the periodic function $H(f)$.

Next, we will use the above analysis to illustrate the relation between the discrete transforms and the (not necessarily periodic) Fourier transforms $G(f)$ and $g(t)$ of (4.1), (4.2).

C. Fourier Integrals and the Discrete Transforms

The starting point here is to show that the samples $G(n/(2M+1)T)$, of the Fourier transform $G(f)$ of (4.1), can be expressed as the following finite-limit Fourier integral:

$$G\left(\frac{n}{(2M+1)T}\right) = \int_{-(2M+1)T/2}^{(2M+1)T/2} \bar{g}(t) e^{-j2\pi nt/(2M+1)T} \, dt \qquad (4.58)$$

of $\bar{g}(t)$. Here $\bar{g}(t)$ is the sum of all the translations (4.12), by integer multiple of NT, of $g(t)$, the inverse Fourier transform of $G(f)$,

$$\bar{g}(t) = \sum_{n=-\infty}^{\infty} g(t + nNT) \tag{4.12}$$

which is illustrated in Fig. 4.7a,b with a clear indication of the aliasing error $\bar{g}(t) - g(t)$ on $(-NT/2, NT/2)$. This important relation (4.58), which we will prove shortly in (4.59) as the well-known *Poisson summation formula*, puts the computations of the Fourier integral $G(f)$ at the stage of the last section's Fourier series periodic function computation, since its samples $G(n/(2M + 1)T)$ of (4.58) can be recognized as the Fourier coefficients of the aliased function $(2M + 1)T\bar{g}(t)$ on the interval $(-(2M + 1)T/2, (2M + 1)T/2,$

$$\bar{g}(t) = \frac{1}{(2M + 1)T} \sum_{n=-\infty}^{\infty} G\left(\frac{n}{(2M + 1)T}\right) e^{j2\pi nt/(2M+1)T},$$

$$-\frac{(2M + 1)T}{2} < t < \frac{(2M + 1)T}{2} \tag{4.59}$$

If we compare this with the Fourier series (4.7), then (4.54) of the above Fourier series–discrete transforms analysis we note that, to construct a discrete transform for (4.59), we need only sample $\bar{g}(t)$ at $t_k = kT$,

$$\bar{g}(kT) = \frac{1}{(2M + 1)T} \sum_{n=-\infty}^{\infty} G\left(\frac{n}{(2M + 1)T}\right) e^{j2\pi nk/(2M+1)} \tag{4.60}$$

Here $G(n/(2M + 1)T)$ will now play the role of the Fourier coefficients c_n of (4.54), which have to be aliased as $\tilde{G}(n/(2M + 1)T)$ in order to obtain a finite sum representation. To do this for (4.60), we follow the same steps in parallel to that of having (4.57) from (4.54), to multiply both sides of (4.60) by $e^{-j2\pi rk/(2M+1)}$, then sum over from $-M$ to M to have

$$\sum_{k=-M}^{M} \bar{g}(kT)e^{-j2\pi rk/(2M+1)} = \frac{1}{(2M + 1)T} \sum_{k=-M}^{M} \sum_{n=-\infty}^{\infty} G\left(\frac{n}{(2M + 1)T}\right)$$

$$\times e^{j2\pi nk/(2M+1)} e^{-j2\pi rk/(2M+1)} \tag{4.61}$$

If we denote the left-hand side by D_r, in a way similar to that of (4.55b), and follow the same reasoning leading to (4.55c) we have

$$
D_r = \frac{1}{(2M+1)T} \sum_{n=-\infty}^{\infty} G\left(\frac{n}{(2M+1)T}\right) \sum_{k=-M}^{M} e^{[j2\pi k(n-r)]/(2M+1)}
$$

$$
= \frac{1}{T} \left[G\left(\frac{r}{(2M+1)T}\right) + \sum_{|l|=1}^{\infty} G\left(\frac{r+l(2M+1)}{(2M+1)T}\right) \right]
$$

$$
\equiv \frac{1}{T} \overline{G}\left(\frac{r}{(2M+1)T}\right) \tag{4.62}
$$

as the superposition of all the translated $G(r/(2M+1)T)$ by integer multiple of $2M+1$, or what we call the alias of $(1/T)G(r/(2M+1)T)$ (see Fig. 4.6). In deriving (4.62), we used the discrete orthogonality property for the finite inner sum. If we consult the discrete orthogonality for (4.61), (4.62), we recognize that $D_r = (1/T)\overline{G}(r/(2M+1)T)$ is associated with $\tilde{G}(r/(2M+1)T)$, and $\overline{g}(kT)$ is associated with $\tilde{g}(kT)$, giving us a pair of "aliased" (that is, discrete) transforms $\tilde{g}(kT)$ and $\tilde{G}(n/NT)$ as in (4.5), (4.6).

So in the process of establishing the discrete transforms (4.5), (4.6) for the Fourier transforms (4.1), (4.2), we first had to alias $g(t)$ to $\overline{g}(t)$ of (4.12), then sample $G(f)$ to $G(n/(2M+1)T)$ to get (4.58). This was followed by sampling the aliased $\overline{g}(t)$ to $\overline{g}(kT)$ as in (4.60); then $G(n/(2M+1)T)$ has to be aliased to D_r or $\tilde{G}(n/(2M+1)T)$ of (4.62). Figures 4.10–4.13 make clear how both aliasing and sampling are done for the transforms. This is in comparison with only aliasing the Fourier coefficients c_k to $C_k = \tilde{g}(kT)$ and sampling the periodic function $G(f)$ to $G(n/(2M+1)T) = \tilde{G}(n/(2M+1)T)$, for the Fourier series (4.7), (4.8) approximation by the discrete transforms (4.3), (4.4).

The Poisson Summation Formula

We will now prove the important identity (4.58), which we note is in the form of the Fourier series of $\overline{g}(t)$ on the interval $(-(2M+1)T/2, (2M+1)T/2)$,

$$
\overline{g}(t) = \sum_{n=-\infty}^{\infty} g(t+nNT)
$$

$$
= \frac{1}{(2M+1)T} \sum_{n=-\infty}^{\infty} G\left(\frac{n}{(2M+1)T}\right) e^{j2\pi nt/(2M+1)T} \tag{4.59}
$$

This series is recognized as the *Poisson summation formula*, whose proof would be equivalent to proving (4.58). We first establish that the sum on the left side of (4.59) is the result of convolving $g(t)$ with the periodic *impulse train* of (2.156), with period NT, of the *Dirac delta functions*, that is

$$g(t) * \sum_{n=-\infty}^{\infty} \delta(t + nNT) = \int_{-\infty}^{\infty} g(\tau) \sum_{n=-\infty}^{\infty} \delta(t + nNT - \tau) d\tau$$

$$= \sum_{n=-\infty}^{\infty} \int_{-\infty}^{\infty} g(\tau)\delta(t + nNT - \tau) d\tau = \sum_{n=-\infty}^{\infty} g(t + nNT) = \bar{g}(t) \quad (4.63)$$

since

$$\int_{-\infty}^{\infty} g(\tau)\delta(\tau - \tau_0) d\tau = g(\tau_0) \tag{4.64}$$

by the definition of the Dirac delta function in (2.150a) (see also (2.151)). From (4.64) we can easily see that the Fourier transform of $\delta(t)$ is 1, since

$$\int_{-\infty}^{\infty} e^{i\omega t} \delta(t) dt = e^0 = 1$$

So we may now formally write the periodic Fourier series expansion of $\delta(t)$ on $(-NT/2, NT/2)$ with period NT, as the above impulse train in (4.63)

$$\sum_{n=-\infty}^{\infty} \delta(t + nNT) = \frac{1}{(2M + 1)T} \sum_{n=-\infty}^{\infty} e^{j2\pi nt/(2M+1)T} \tag{4.65}$$

where (4.64) is used for obtaining the constant Fourier coefficient $c_n = 1/(2M + 1)T$ of this formal (divergent) Fourier series. If we use the right-hand side of (4.65) for the impulse train of the first line in (4.63), we obtain

$$\bar{g}(t) = \sum_{n=-\infty}^{\infty} g(t + nNT) = g(t) * \sum_{n=-\infty}^{\infty} \delta(t + nNT)$$

$$= \frac{1}{(2M + 1)T} g(t) * \sum_{n=-\infty}^{\infty} e^{j2\pi nt/(2M+1)T}$$

$$= \frac{1}{(2M + 1)T} \int_{-\infty}^{\infty} g(\tau) \left[\sum_{n=-\infty}^{\infty} e^{j2\pi n(t-\tau)/(2M+1)T} \right] d\tau$$

$$= \frac{1}{(2M + 1)T} \sum_{n=-\infty}^{\infty} e^{j2\pi nt/(2M+1)T} \int_{-\infty}^{\infty} g(\tau) e^{-j2\pi n\tau/(2M+1)T} d\tau$$

$$= \frac{1}{(2M + 1)T} \sum_{n=-\infty}^{\infty} e^{j2\pi nt/(2M+1)T} G\left(\frac{n}{(2M + 1)T}\right) \qquad (4.59)$$

after using (4.1) for the last integral.

From this result we can establish the Poisson summation formula (4.58) by simply recognizing $[1/(2M + 1)T]G(n/(2M + 1)T)$ in (4.59) as the Fourier coefficients of $\bar{g}(t)$ on the interval $(-(2M + 1)T/2, (2M + 1)T/2)$, to have

$$\frac{1}{(2M + 1)T} G\left(\frac{n}{(2M + 1)T}\right)$$

$$= \frac{1}{(2M + 1)T} \int_{-(2M+1)T/2}^{(2M+1)T/2} \bar{g}(t)e^{-j2\pi nt/(2M+1)T} \, dt$$

$$G\left(\frac{n}{(2M + 1)T}\right) = \int_{(2M+1)T/2}^{(2M+1)T/2} \bar{g}(t)e^{-j2\pi nt/(2M+1)T} \, dt \qquad (4.58)$$

as the *Poisson summation formula*.

This Poisson summation formula (4.59) is well known in the analysis of Fourier (trigonometric) series and integrals. As we saw in the above derivation, it is very much dependent on the translation property of such transforms kernels, or more precisely it uses the periodicity of the (trigonometric) Fourier series.

An interesting question concerns the possibility of generalizing this very useful formula to other orthogonal expansions, and in the absence of the periodicity of the Fourier trigonometric series. Having such a formula would be instrumental in having a hold on the bound for the aliasing error of general orthogonal expansion series like the Bessel series. It seems that in order to make any progress in this direction, we must overcome the *absence* of the usual translation property of the Fourier exponential kernel. This is done by defining a generalized analog, which we have attempted to present in Section 3.6 and its exercises, as a *generalized translation* in conjunction with the generalized Bessel-sampling series. We shall leave the preliminaries needed for such a *generalized Poisson summation formula* to Section 4.3 and its exercises (see Exercises 4.34–4.40) where a version of a *Bessel-type Poisson summation formula* (4.190) is derived.

D. Basic Properties of the DFT

We now return to develop more of the discrete transform's properties; these and others are summarized in Table 4.3. The additional properties that have not been derived here make good exercises. Also, in Table 4.3 we use the notation $\bar{G}_n = \hat{\mathcal{F}}\{\bar{g}_k\}$ and $\bar{H}_n = \hat{\mathcal{F}}\{\bar{h}_k\}$, and in the detailed

TABLE 4.3 Discrete Fourier Transform Pairs

Sequence	Transform	
$\bar{g}_k = \bar{\mathcal{F}}^{-1}\{\bar{G}_n\}$	$\bar{G}_n = \bar{\mathcal{F}}\{\bar{g}_k\}$; see (4.3)	
(a) *Operation pairs*		
1. $\bar{g}_k, \bar{h}_k$	$\bar{G}_n, \bar{H}_n$	(linearity)
2. $\bar{g}_k + \bar{h}_k$	$\bar{G}_n + \bar{H}_n$	(linearity)
3. $\alpha\bar{g}_k$	$\alpha\bar{G}_n, \; \alpha \in \mathbb{R}$	(linearity)
4. $\bar{g}_k$	$\bar{G}_{n+rN}, \; r \in \mathbb{Z}$	(periodicity)
5. $\dfrac{1}{N}\bar{G}_{-k}$	$\bar{g}_n$	(symmetry)
6. $e^{j2\pi lk/N}\bar{g}_k$	$\bar{G}_{n-l}$	(shift in frequency)
7. $\bar{g}_{k-l}$	$e^{-j2\pi ln/N}\bar{G}_n$	(shift in time)
8. $\overline{\bar{g}}_k$	$\overline{\bar{G}_{-n}}$	(conjugation)
9. $(\bar{g}*\bar{h})_k$	$\bar{G}_n\bar{H}_n$	(convolution)
10. $\bar{g}_k\bar{h}_k$	$(\bar{G}*\bar{H})_n$	(convolution)

(b) *Sequence pairs*

11.	$\delta_{k,\alpha}$	11. $e^{-j2\pi\alpha n/N}$
12.	$\dfrac{1}{2}(\delta_{k,\alpha} + \delta_{k,N-\alpha})$	12. $\cos\dfrac{2\pi\alpha n}{N}$
13.	$\dfrac{j}{2}(\delta_{k,\alpha} - \delta_{k,N-\alpha})$	13. $\sin\dfrac{2\pi\alpha n}{N}$
14.	1	14. $N\delta_{n,0}$
15.	$\dfrac{1}{N}e^{j2\pi\alpha k/N}$	15. $\delta_{n,\alpha}$

16. $\begin{cases} \dfrac{1}{W^k-1}, & k\neq 0 \\[2mm] \dfrac{N-1}{2}, & k=0 \end{cases}$ 16. n

17. $\dfrac{N-1}{W^k-1} - \dfrac{2}{(W^k-1)^2}$ 17. $n^{(2)}$, where

$$n^{(m)} \equiv \begin{cases} n(n-1)\cdots(n-m+1) & \text{for } m \geq 1 \\ n^{(0)} = 1 & \text{for } m = 0 \end{cases}$$

18. $\dfrac{1}{W^k-1}\left(\dfrac{N^{(p)}}{N} - p\bar{\mathcal{F}}\{n^{(p-1)}\}\right)$ 18. $n^{(p)}$ (see entry 17 for $n^{(p)}$)

19. $\begin{cases} \dfrac{1}{2}, & k=0 \\[1mm] 0, & k \text{ even} \\[1mm] \dfrac{2}{N(1-W^k)}, & k \text{ odd} \end{cases}$ 19. $\begin{cases} 0, & 0\leq n < \dfrac{N}{2} \\[2mm] 1, & \dfrac{N}{2} \leq n < N \end{cases}$

derivations we may even refer to $\tilde{G}(n/NT)$ and $\tilde{g}(kT)$ as $\tilde{G}(n)$ and $\tilde{g}(k)$, $\tilde{G}_n$ and $\tilde{g}_k$, and even $\tilde{G}$ and $\tilde{g}$. The development of these basic properties of the DFT will be followed by their application to solving difference equations, such as a traffic flow problem, in Section 4.1.5 and then to computing Fourier series and Fourier integrals in Section 4.1.6.

Periodicity

As mentioned earlier, both the discrete Fourier transform $\tilde{G}(n/NT)$ in (4.3) and its inverse $\tilde{g}(kT)$ in (4.4) are periodic with period N. This is shown for $\tilde{G}$ by substituting for n in $\tilde{G}(n/NT)$ by $n + rN$, where r is an integer, to show that the value of $\tilde{G}$ does not change,

$$
\tilde{G}\left(\frac{n + rN}{NT}\right) = \sum_{k=0}^{N-1} \tilde{g}(kT)e^{-j2\pi k(n+rN)/N}
$$

$$
= \sum_{k=0}^{N-1} \tilde{g}(kT)e^{-j2\pi kn/N}e^{-2j\pi rk}
$$

$$
= \sum_{k=0}^{N-1} \tilde{g}(kT)e^{-j2\pi kn/N}, \qquad e^{-2j\pi rk} = 1
$$

$$
= \tilde{G}\left(\frac{n}{NT}\right)
$$

$$
\tilde{G}\left(\frac{n + rN}{NT}\right) = \tilde{G}\left(\frac{n}{NT}\right), \qquad r = 0, \mp 1, \mp 2, \dots \tag{4.66}
$$

The same can be done for $\tilde{g}(kT)$,

$$
\tilde{g}[(k + rN)T] = \tilde{g}(kT), \qquad r = 0, \mp 1, \mp 2, \dots \tag{4.67}
$$

Besides the required sample spacing $\Delta f = 1/NT$ for $\tilde{G}$ and $\Delta t = T$ for $\tilde{g}$, this periodicity property is a distinguishing characteristic of the discrete Fourier transforms, in that both the transform and its inverse are periodic. In comparison, the Fourier series (4.7) is a periodic function $G_s(f)$, with a sampled infinite sequence (4.8) as the transform (coefficient). The Fourier integral transform, in general, has both the transform (4.1) and its inverse (4.2) as functions, defined on the whole interval $(-\infty, \infty)$ and hence, in general, not periodic.

The other important basic properties of the discrete transforms—such as linearity, symmetry, shifting, Parseval's equality, convolution theorem, and correlation theorem—parallel those of the Fourier integral transforms. For convenience, while discussing these properties, we shall use the operational notation $\tilde{\mathcal{F}}[\tilde{g}]$ and $\tilde{\mathcal{F}}^{-1}[\tilde{G}]$ for the discrete Fourier transform $\tilde{G}(n/NT)$ and its inverse $\tilde{g}(kT)$, respectively.

Linearity

Consider the discrete transforms $\tilde{G}$ and $\tilde{H}$ with their respective inverses $\tilde{g}$ and $\tilde{h}$; the linearity property of the transform

$$\tilde{F}[\tilde{g}(kT) + \tilde{h}(kT)] = \tilde{G}\left(\frac{n}{NT}\right) + \tilde{H}\left(\frac{n}{NT}\right) \tag{4.68}$$

can be proved when we substitute $\tilde{g}(kT) + \tilde{h}(kT)$ for $\tilde{g}(kT)$ in (4.3) and recognize the resulting sum of the two finite series as the two terms $\tilde{G} + \tilde{H}$, after using the definition of the transforms $\tilde{G}$ and $\tilde{H}$ in (4.3).

Symmetry

Just as in the case of the integral transforms, and due to the symmetry of the discrete transform (4.3) and its inverse (4.4), we expect the same type of symmetry, in the sense that the transform of the transform gives, within a minor adjustment, the function back, that is,

$$\tilde{\mathcal{F}}\left\{\frac{1}{N}\tilde{G}\left(\frac{k}{NT}\right)\right\} = \tilde{g}(-nT) \tag{4.69}$$

To prove this we substitute $\tilde{h}(k) = (1/N)\tilde{G}(k/NT)$ for $\tilde{g}(kT)$ in (4.3), then simply use the definition (4.4) for $\tilde{g}(-nT)$,

$$\tilde{\mathcal{F}}\left\{\frac{1}{N}\tilde{G}\left(\frac{k}{NT}\right)\right\} = \sum_{k=0}^{N-1} \frac{1}{N}\tilde{G}\left(\frac{k}{NT}\right)e^{-j2\pi nk/N}$$

$$= \frac{1}{N}\sum_{k=0}^{N-1}\tilde{G}\left(\frac{k}{NT}\right)e^{j2\pi k(-n)} = \tilde{g}(-n)$$

The following two properties of shifting in k (time) and n (frequency) are left for exercises:

Shifting in k (Time)

$$\tilde{F}[\tilde{g}\{(k-i)T\}] = \tilde{G}\left(\frac{n}{NT}\right)e^{-j2\pi ni/N} \tag{4.70}$$

Shifting in n (Frequency)

$$\tilde{F}[\tilde{g}(k)e^{j2\pi ik/N}] = \tilde{G}\left(\frac{n-i}{NT}\right) \tag{4.71}$$

As we will see in Section 4.1.5 when we discuss the application of the discrete transforms in solving difference equations, these two shifting properties play the same role for the mth difference $g(k+m) - g(k)$ as that of the Fourier transforms (2.116)–(2.118) algebraization of the derivative $d^m g/dt^m$ as obtained from (2.118) for $\lambda = 2\pi f$.

$$\mathcal{F}\left\{\frac{d^m g}{dt^m}\right\} = (j2\pi f)^m G(f)$$

This would mean that the pair (4.70), (4.71) is the proper tool for transforming, and hopefully simplifying, the difference equations (see Section 4.1.5). If needed, the samples of the derivative $g'(t)$ can also be approximated by the discrete transform. In this case, we consider $h(t) = g'(t)$ and its Fourier transform $H(f) = j2\pi f G(f)$, then write their corresponding approximate discrete transform representations $\bar{h}(kT) = \bar{g}'(kT)$ and $\tilde{H}(n/NT) = [(j2\pi n/NT)G(n/NT)]^\sim$,

$$\bar{h}(kT) = \bar{g}'(kT) = \frac{1}{N} \sum_{n=0}^{N-1} \left[\frac{j2\pi n}{NT} G\left(\frac{n}{NT}\right)\right]^\sim e^{-j2\pi nk/N} \tag{4.72}$$

where $[\ ]^\sim$ stands for the discrete transform representative of $[\ ]$. This means that in order to approximate the samples of the derivative $g'(t)$, we must first compute $\overline{H(f)} = \overline{j2\pi f G(f)}$ of (4.11) to prepare for the discrete transform approximation $[(j2\pi n/NT)G(n/NT)]^\sim$, necessary for (4.72).

Sometimes we may not have the function $h(t)$ whose derivative is to be computed or approximated, but only some of its sample values, for example, discrete measurement data. In this case, of course, the direct difference or other similar methods may be employed. However, if we are to try the discrete Fourier transform we may use (4.72) with due care. This is in the sense that we may take the data as $\bar{h}(kT)$ to find $\tilde{H}(n/NT)$ from (4.4), but in (4.72) we need $[(j2\pi n/NT)H(n/NT)]^\sim$, where the best we could do now is to approximate it by $(j2\pi n/NT)\tilde{H}(n/NT)$. This point may deserve more analysis as to its error estimate and the possible need of the discrete convolution theorem (4.78), (4.80). To summarize the discrete transform of derivative samples, we have:

Differentiation in k (Time)

$$\tilde{\mathcal{F}}\{\bar{g}'(kT)\} = \left[\frac{j2\pi n}{NT} G\left(\frac{n}{NT}\right)\right]^\sim \tag{4.73}$$

where $[\ ]^\sim$ is the discrete approximation of $[\ ]$.

Differentiation in n (Frequency)

$$\tilde{\mathcal{F}}\left\{\bar{G}'\left(\frac{n}{NT}\right)\right\} = [-j2\pi k T g(kT)]^\sim \tag{4.74}$$

Relation (4.74) is derived in the same way as we derived (4.72) as the pair (4.73).

The same method can be followed to establish (4.73) and (4.74) for higher-derivative samples,

$$\mathcal{F}\{\tilde{g}^{(m)}(kT)\} = \left[\left(\frac{j2\pi n}{NT}\right)^m G\left(\frac{n}{NT}\right)\right]^{\sim} \tag{4.75}$$

$$\mathcal{F}\left\{\tilde{G}^{(m)}\left(\frac{n}{NT}\right)\right\} = [(-j2\pi kT)^m g(kT)]^{\sim} \tag{4.76}$$

These pairs (4.73)–(4.76), of approximately algebraizing the differentiation, are possibly the most used simple applications of the discrete Fourier transforms and their fast algorithm, the FFT. For example, the pair (4.73) is used with the help of the FFT to compute very efficiently the derivative $\tilde{g}'(kT)$. As indicated in (4.73), we first use the discrete Fourier transform on $\tilde{g}(kT)$ to obtain $\tilde{G}(n/NT)$, multiply this result by $j2\pi n/NT$, then use the inverse discrete transform again to transform $(j2\pi n/NT)\tilde{G}(n/NT)$ back to $\tilde{g}'(kT)$, the discrete approximation of the derivative samples $\tilde{g}'(kT)$. It turns out that, because of the FFT's high efficiency, this relatively long indirect process is faster than approximating $g'(kT)$ by direct finite-difference methods.

Next, we define the convolution product of the discrete Fourier transforms and then prove the convolution theorem. The Parseval equality can be obtained as a special case, or we can follow the same method of proof. The correlation theorem also has almost exactly the same proof. This is followed by showing the advantage of using the FFT for the (indirect) evaluation of these *discrete convolution* and *correlation products*.

Discrete Convolution Product
The discrete convolution product $\tilde{G} + \tilde{H}$ of $\tilde{G}(n/NT)$ and $\tilde{H}(n/NT)$ is defined as

$$(\tilde{G} * \tilde{H})\left(\frac{n}{NT}\right) = \frac{1}{N}\sum_{i=0}^{N-1}\tilde{G}\left(\frac{i}{NT}\right)\tilde{H}\left(\frac{n-i}{NT}\right) \equiv \tilde{G} * \tilde{H} \tag{4.77}$$

in the same fashion, the discrete convolution product $(\tilde{g} * \tilde{h})(kT)$ is defined as

$$(\tilde{g} * \tilde{h})(kT) = \sum_{i=0}^{N-1}\tilde{g}(iT)\tilde{h}((k-i)T) \equiv \tilde{g} * \tilde{h} \tag{4.78}$$

Convolution Theorems
Now we state the discrete convolution theorems

$$\sum_{k=0}^{N-1}\tilde{g}(kT)\tilde{h}(kT)e^{-j2\pi nk/N} = \frac{1}{N}\sum_{i=0}^{N-1}\tilde{G}\left(\frac{i}{NT}\right)\tilde{H}\left(\frac{n-i}{NT}\right) \tag{4.79}$$

that is,

$$\bar{F}\{\bar{g}\bar{h}\} = \bar{G} * \bar{H}$$

and

$$\frac{1}{N}\sum_{n=0}^{N-1}\bar{G}\left(\frac{n}{NT}\right)\bar{H}\left(\frac{n}{NT}\right)e^{j2\pi nk/N} = \sum_{i=0}^{N-1}\bar{g}(iT)\bar{h}((k-i)T) \qquad (4.80)$$

that is,

$$\bar{F}^{-1}[\bar{G}\bar{H}] = \bar{g} * \bar{h}$$

We note that the *discrete Parseval's equality*,

$$\sum_{k=0}^{N-1}\bar{g}(kT)\overline{\bar{h}(kT)} = \frac{1}{N}\sum_{i=0}^{N-1}\bar{G}\left(\frac{i}{NT}\right)\overline{\bar{H}\left(\frac{i}{NT}\right)} \qquad (4.81)$$

is a special case of (4.79) with $n = 0$ and after realizing that the complex conjugate $\overline{\bar{H}(-i/NT)}$ is the transform of $\overline{\bar{H}(kT)}$ as seen from (4.3).

The proofs of (4.79), (4.80), or (4.81) follow in the same way and are parallel to those we have for the integral transforms, with the distinct advantage here of not having to worry about convergence, as we exchange only finite sums. This is in comparison with exchanging infinite sums, integrals, or infinite sums with integrals, in the case of Fourier transforms and series, which need some justification to the extent that the final results are still convergent. The proof of (4.79) involves substituting for $\bar{g}(kT)$ from (4.4) in the left side of (4.79) and exchanging the two summations to produce $\bar{H}((n-i)/NT)$,

$$\sum_{k=0}^{N-1}\bar{g}(kT)\bar{h}(kT)e^{-j2\pi nk/N}$$

$$= \sum_{k=0}^{N-1}\left[\frac{1}{N}\sum_{i=0}^{N-1}\bar{G}\left(\frac{i}{NT}\right)e^{j2\pi ik/N}\right]\bar{h}(kT)e^{-j2\pi nk/N}$$

$$= \frac{1}{N}\sum_{i=0}^{N-1}\bar{G}\left(\frac{i}{NT}\right)\left[\sum_{k=0}^{\infty}\bar{h}(kT)e^{-j2\pi(n-i)k/N}\right]$$

$$= \frac{1}{N}\sum_{i=0}^{N-1}\bar{G}\left(\frac{i}{NT}\right)\bar{H}\left(\frac{n-i}{NT}\right) \equiv \bar{G} * \bar{H}$$

after using (4.3) for $\bar{H}((n-i)/NT)$. If we follow the same steps substituting from (4.4) for $\bar{h}$ first, instead of $\bar{g}$, we obtain

$$\frac{1}{N} \sum_{i=0}^{N-1} \bar{G}\left(\frac{n-i}{NT}\right) \bar{H}\left(\frac{i}{NT}\right) \equiv \bar{H} * \bar{G} \tag{4.82}$$

which shows that the convolution product is commutative, i.e., $\bar{G} * \bar{H} = \bar{H} * \bar{G}$. The same is true for $\bar{g} * \bar{h} = \bar{h} * \bar{g}$ in (4.78).

Discrete Correlation

The discrete correlation of $\bar{g}(kT)$ and $\bar{h}(kT)$ is defined as

$$\xi(k) = \sum_{i=0}^{N-1} \bar{g}(iT)\bar{h}(k+i)T \tag{4.83}$$

where we note that there is only a difference of sign in $\bar{h}(k-i)$ from that of the convolution product $\bar{g} * \bar{h}$ of (4.78).

Correlation Theorem

By following the same method of proof for the convolution theorems (4.79), (4.80), we can prove the following correlation theorem:

$$\sum_{i=0}^{N-1} \overline{\bar{g}}(iT)\bar{h}((k+i)T) = \frac{1}{N} \sum_{n=0}^{N-1} \overline{\bar{G}\left(\frac{n}{NT}\right)} \bar{H}\left(\frac{n}{NT}\right) e^{j2\pi nk/N} \tag{4.84}$$

where $\overline{\bar{G}}$ is the complex conjugate of $\bar{G}$.

Computing the Convolution and Correlation Products via the FFT

The fast Fourier transform is used to compute, in an indirect way, these discrete convolution and correlation products, with a clear advantage over the direct methods. For example, the direct computation of the convolution product $\bar{g} * \bar{h}$, as defined on the right side of (4.78), requires a shifting for $\bar{h}(kT)$ to $\bar{h}((k-i)T)$, multiplying this result by $\bar{g}(iT)$, then summing over this product. The very indirect, but surprisingly faster (for large enough numbers of samples N), way of using the FFT will employ the convolution theorem (4.80) to compute its left-hand side for the value of $\bar{g} * \bar{h}$. This means that we use the FFT to transform both $\bar{g}$ and $\bar{h}$ to $\bar{G}$ and $\bar{H}$, respectively; compute the simple product $\bar{Y} = \bar{G}\bar{H}$; then use the FFT again to find $\bar{y}$, the inverse transform of $\bar{Y}$, as the convolution product $\bar{y} = \bar{g} * \bar{h}$ as in (4.80). If we consult Table 4.2 for the speed of the FFT, we can estimate that this relatively long *roundabout* way would have an advantage over the direct method only for large enough N, which happened to be 64 or over. This may be explained roughly in terms of the fact that in this method we have employed the FFT three times for the computations of $\bar{G}$, $\bar{H}$, and then $\bar{y}$ from $\bar{Y}$, with a total of $\frac{3}{2}N \log_2 N$ multiplications. This is to be added to the N^2 multiplications for $\bar{G}\bar{H}$, a total of

$$N^2 + \frac{3}{2}N \log_2 N$$

This is to be compared with the direct method of N^2 shiftings (subtractions) and N^2 multiplications. So the $\frac{3}{2}N \log_2 N$ number of multiplications must compete for speed with the N^2 number of shiftings, which are, of course, easier than multiplications. This balance may explain the need for N over 64 for a clear advantage of the FFT over the direct method of computing the discrete convolution product (4.80).

Such a requirement of $N > 64$ is not a drawback for the FFT if we consider our main application of using the discrete transforms to compute infinite or finite Fourier integrals, where we obviously need large enough N.

A very important point for computing the discrete convolution product, via the FFT, is to recognize that the (finite) domain for the convolution product $\bar{y} = \bar{g} * \bar{h}$ is larger than the (finite) domain of either $\bar{g}$ or $\bar{h}$, and actually is equal to the union of the two domains. This means that we have to guard against the aliasing error in $\bar{y}$ as the result of not carefully observing its resulting larger domain or period. The FFT computations here are carried with N samples on the interval $(0,NT)$ for all the functions concerned, $\bar{g}$, $\bar{h}$, and $\bar{y}$. In Section 4.1.1 we showed and illustrated (Figs. 4.3–4.7) that in order to minimize the aliasing error, of the discrete transform $\bar{g}(kT)$ approximation of $g(kT)$, we should have a large enough interval $(0,NT)$. This is to allow the sequence $g(kT)$ to die out before NT, hence minimizing the overlap with its periodic extensions (with period NT) which is the source of the aliasing in $\bar{g}(kT)$. The same thing applies to $\bar{h}$. To accommodate this, let us assume that the "appreciable" values of $\bar{g}$ and $\bar{h}$ are accommodated, respectively, by $N_1 < N$ and $N_2 < N$ samples on this interval $(0,NT)$. Of course, for each of $\bar{g}$ and $\bar{h}$, we must still use N samples as the total of the appreciable and the close to zero values on $(0,NT)$. Because of its enlarged domain, the convolution product $\bar{y} = \bar{g} * \bar{h}$ would have spread its appreciable values over a large domain which may even exceed our interval $(0,NT)$, hence incurring an aliasing error. To guard against this aliasing error, we should make sure that these appreciable values of $\bar{y}$ are still contained in $(0,NT)$, and preferably it should be allowed to die out toward the ends of this interval. This can be accomplished if we restrict the number of (appreciable) samples N_1 and N_2 for $\bar{g}$ and $\bar{h}$ such that $N_1 + N_2 < N$. This would mean that we have to at least double the number of sample points for $\bar{g}$ and $\bar{h}$ in order to guard against the aliasing of their sought convolution product.

In case one of the functions is constant, we can still cover its (appreciable) constant values by N_1 samples and assign the value zero to the rest of the samples close to the ends of $(0, NT)$.

An FFT subroutine in BASIC for computing the convolution product in the "roundabout" way described above, along with some discussions, may be found in Brigham (1988).

4.1.5 Operational Difference Calculus for the DFT and the z-Transform[†]

In this section, we will have a brief presentation of how the *discrete* (Fourier and Laplace) transforms in this chapter can be used as an operational method to facilitate the solution of *difference equations*. This is analogous to how the Fourier and other transforms algebraize differential equations. This is also related to the fact that the discrete orthogonal kernels, of the present discrete transforms, are solutions of the difference equation approximation to the differential equation that resulted in the orthogonal set of functions for the Fourier series (finite transforms) of the last chapter.

Let us first acquire some familiarity with the notation of *difference calculus*. We will adopt here the forward difference

$$\Delta u_k = u(x_k + h) - u(x_k) \equiv u_{k+1} - u_k \tag{4.85}$$

The second-order difference $\Delta^2 u$ would be

$$\Delta^2 u_k = \Delta(\Delta u_k) = \Delta(u(x_k + h) - u(x_k))$$
$$= [u(x_k + h + h) - u(x_k + h)] - [u(x_k + h) - u(x_k)]$$
$$= u(x_k + 2h) - 2u(x_k + h) + u(x_k) \equiv u_{k+2} - 2u_{k+1} + u_k \tag{4.86}$$

The finite difference approximates the derivative du/dx by $\Delta u/\Delta x = (1/h)\Delta u_k$, and sometimes we may take h to be 1, where then Δu_k becomes the approximation of $u'(x)$.

With this notation, the finite-difference approximation to the first-order differential equation,

$$\frac{du}{dx} + au = f(x) \tag{4.87}$$

[†]A very detailed treatment of this topic with many more applications is the subject of an (unpublished) monograph on "Discrete Transforms for Difference Equations" (Briggs and Jerri, 1985). Few basic elements and illustrations of this subject are presented here with due thanks to Prof. W. L. Briggs.

would be

$$u_{k+1} - u_k + au_k = f_k \tag{4.88}$$

$$u_{k+1} + bu_k = f_k, \qquad b = a - 1$$

and for the general second-order, constant-coefficient, nonhomogeneous differential equation

$$\frac{d^2u}{dx^2} + a\frac{du}{dx} + bu = f(x) \tag{4.89}$$

we have the corresponding difference equation

$$\Delta^2 u_k + a\,\Delta u_k + bu_k = f_k$$

or

$$(u_{k+2} - 2u_{k+1} + u_k) + a(u_{k+1} - u_k) + bu_k = f_k$$

$$u_{k+2} + cu_{k+1} + du_k = f_k$$

$$c = a - 2, \qquad d = b - a + 1 \tag{4.90}$$

As we mentioned at the beginning of this chapter in correspondence with the finite-difference notation in (4.86), we will use $\tilde{G}_n$ for $\tilde{G}(n/NT)$ and $\tilde{g}_k$ for $\tilde{g}(kT)$. Thus we may write the discrete transforms (4.3), (4.4) in the form

$$\tilde{G}_n = \sum_{k=0}^{N-1} \tilde{g}_k W^{-nk} \tag{4.3}$$

$$\tilde{g}_k = \frac{1}{N}\sum_{n=0}^{N-1} \tilde{G}_n W^{nk} \tag{4.4}$$

where $W = e^{j2\pi/N}$ is the Nth root of unity, as it is usually denoted in most books on discrete Fourier transforms. $W = e^{-j2\pi/N}$ is also used, which we may find ourselves using as in the analysis leading to equation (4.98).

We will now present a few of the essential elements of difference and summation calculus, in parallel to that of differential and integral calculus. For example, the formula for the *difference of the product of two sequences* is

$$\Delta(u_k v_k) = u_{k+1}v_{k+1} - u_k v_k$$

$$= u_{k+1}v_{k+1} - u_{k+1}v_k + u_{k+1}v_k - u_k v_k$$

$$= u_{k+1}(v_{k+1} - v_k) + v_k(u_{k+1} - u_k)$$

$$= u_{k+1}\,\Delta v_k + v_k\,\Delta u_k \tag{4.91}$$

(Compare this to the differential of the product of two functions $d(uv) = u\,dv + v\,du$.)

In a similar way, it is possible to find the *difference of the quotient* of two sequences:

$$\Delta\left(\frac{u_k}{v_k}\right) = \frac{v_k\,\Delta u_k - u_k\,\Delta v_k}{v_k v_{k+1}} \tag{4.92}$$

The important operation of *summation by parts* may now be illustrated by returning to the formula for the difference of a product. Summing equation (4.91) between the indices $k = 0$ and $k = N - 1$, we have

$$\sum_{k=0}^{N-1} \Delta(u_k v_k) = \sum_{k=0}^{N-1}(u_{k+1}\,\Delta v_k + v_k\,\Delta u_k)$$

Expanding the left-hand side and summing term by term, we see that all of the interior terms of the sum cancel and only two "boundary terms" survive. This leaves

$$u_N u_N - u_0 v_0 = \sum_{k=0}^{N-1}(u_{k+1}\,\Delta v_k + v_k\,\Delta u_k)$$

Rearranging this expression gives the *summation-by-parts formula*

$$\sum_{k=0}^{N-1} v_k\,\Delta u_k = u_N v_N - u_0 v_0 - \sum_{k=0}^{N-1} u_{k+1}\,\Delta v_k \tag{4.93}$$

The geometric series is indispensable in much of what follows and it serves as a good starting point. If $a \neq 0$ and $a \neq 1$ is a real number and $N > 1$ is a positive integer, then the sum $\sum_{k=0}^{N-1} a^k$ may always be evaluated explicitly. However, it is usually far simpler to use the fact that

$$\sum_{k=0}^{N-1} a^k = \frac{1-a^N}{1-a} = \frac{a^k}{a-1}\bigg|_{k=0}^{N} \tag{4.94}$$

The following example illustrate the "differencing" and summation of the sequence a^{x_k}, as the parallel of differentiating and integrating the function a^x,

$$\Delta a^{x_k} = (a^h - 1)a^{x_k} \tag{4.95}$$

since

$$\Delta a^{x_k} = a^{x_k+h} - a^{x_k} = a^{x_k}(a^h - 1) \tag{4.95}$$

The special case of $h = 1$ in (4.95) gives

$$\Delta a^k = a^k(a-1) \tag{4.95a}$$

Also

$$\sum_{k=0}^{N} a^{x_k} = \sum_{k=0}^{N} a^{kh} = \sum_{k=0}^{N} (a^h)^k = \frac{a^{Nh} - 1}{a^h - 1}, \qquad a^h \neq 1 \qquad (4.96)$$

as a simple geometric series. This is essentially the finite summation formula that we used in Section 4.1.4 to prove the discrete orthogonality (4.49),

$$\sum_{k=0}^{N-1} e^{-j2\pi nk/N} e^{j2\pi mk/N} = \sum_{k=0}^{N-1} [e^{-j2\pi(n-m)/N}]^k$$

To illustrate the formula of summation by parts (4.93), we show that with $h = 1$, $x_0 = 0$, $x_N = N$ and using (4.94), we have

$$\sum_{k=0}^{N-1} x_k a^{x_k} = \sum_{k=0}^{N-1} k a^k = \frac{ka^k}{a-1} \Big|_{k=0}^{N} - \sum_{k=0}^{N-1} \frac{a^{k+1}}{a-1} \cdot 1$$

$$= \frac{Na^N}{a-1} - \frac{1}{(a-1)^2} a^{k+1} \Big|_{k=0}^{N} = \frac{Na^N}{a-1} - \frac{1}{(a-1)^2}[a^{N+1} - a]$$

$$= \frac{a}{(a-1)^2} [(N-1)a^N - Na^{N-1} + 1]$$

$$\sum_{k=0}^{N-1} k a^k = \frac{a}{(a-1)^2} [(N-1)a^N - Na^{N-1} + 1] \qquad (4.97)$$

If we are to develop an operational method for summation and difference calculus, the above sum (4.97) is useful in deriving the discrete transform (4.3) of $x_k = k$, which we will use for solving the difference equation (4.88) with $f_k = k$.

$$\bar{G}_n = \bar{\mathcal{F}}\{x_k\} = \sum_{k=0}^{N-1} k e^{-j2\pi nk/N} = \sum_{k=0}^{N1-} k(W^n)^k,$$

$$W = e^{-j2\pi/N}, \quad W^{kN} = 1$$

$$= \frac{W^n}{(W^n - 1)^2} [(N-1)W^{nN} - NW^{nN-n} + 1], \qquad a = W^n \text{ in } (4.97)$$

$$= \frac{W^n}{(W^n - 1)^2} [N - 1 + 1 - NW^{-n}]$$

$$= \frac{NW^n(1 - W^{-n})}{(W^n - 1)^2} = \frac{N(W^n - 1)}{(W^n - 1)^2} = \frac{N}{W^n - 1}, \qquad 1 \leq n \leq N - 1$$

where we have used (4.97) with $a = W^n$ and the fact that $W^{nN} = e^{-j2\pi nN/N} = 1$. Also, for convenience we used here $W = e^{-j2\pi/N}$ instead of $W = e^{j2\pi/N}$ as in (4.3a), (4.4a). This result holds for $1 \le n \le N - 1$, but clearly does not make sense for $n = 0$. For $n = 0$, we return to the above sum for $\tilde{G}_n$ and set $n = 0$, giving

$$\tilde{G}_0 = \sum_{k=0}^{N-1} k = \frac{N(N-1)}{2}$$

Therefore,

$$\tilde{G}_n = \tilde{\mathcal{F}}\{k\} = \begin{cases} \dfrac{N}{W^n - 1}, & 1 \le n \le N - 1 \\[3mm] \dfrac{N(N-1)}{2}, & n = 0 \end{cases} \tag{4.98}$$

A. Solving Difference Equations with the DFT-Recurrence Relations, and a Traffic Flow Problem

With the help of the discrete transform pair in (4.98), we will attempt to illustrate the use the DFT for solving the difference equation (4.88) with the special case of the nonhomogeneous term $f_k = k$. This is in the sense that the difference equation will reduce to an algebraic equation in the transform (finite sequence!) space, which is easier to solve. Of course, what remains depends on having enough entries in the available DFT tables (see Table 4.3) that would allow us to transform this solution back to the original space as the actual solution of the difference equation.

For the discrete Fourier transforms (4.3), (4.4), and in reference to the differencing (4.95) and the finite summing (4.96),

$$\tilde{G}\left(\frac{n}{NT}\right) = \sum_{k=0}^{N-1} \tilde{g}(kT) e^{-j2\pi nk/N} \tag{4.3}$$

$$\tilde{g}(kT) = \frac{1}{N} \sum_{n=0}^{N-1} \tilde{G}\left(\frac{n}{NT}\right) e^{j2\pi nk/N} \tag{4.4}$$

we recognize that $h = \Delta f = 1/NT$ for the frequency domain of $\tilde{G}(n \cdot 1/NT)$, and $h = \Delta t = T$ for the time domain of $\tilde{g}(kT)$.

In the following few examples we will illustrate the use of the DFT to solving difference equations. Example 4.1 is intended to make very clear the comparison of the present DFT operational difference calculus method with a usual method used for solving difference equations. We will start with solving the simple differential equation (4.87), then use a parallel to the same method to arrive at the solution of its difference equation analog in (4.88).

In Example 4.3 we will use the DFT method on the difference equation (4.88) of Example 4.1 to show how it is transformed to an algebraic equation. As we mentioned before, the particular choice of the nonhomogeneous term $f_k = k$ in (4.88) will enable use to transform (4.88) with the help of the DFT pair (4.98). However, the final solution for the difference equation will depend on the availability of a DFT pair that enables us to transform back. As it turns out for this example, our very modest Table 4.3 does not have such a pair. Still, and even with such a short table, we will present a very interesting illustration of a traffic flow problem (Example 4.4) whose (special case) difference equation representation can be accommodated by the DFT entries of Table 4.3.

EXAMPLE 4.1: Solving a differential equation and its difference equation analog

Consider the differential equation

$$\frac{dy}{dx} + 2y = x \tag{E.1}$$

and its corresponding difference equation approximation,

$$y_{k+1} - y_k + 2y_k = x_k$$

$$y_{k+1} + y_k = x_k \tag{E.2}$$

The solution of the first-order nonhomogeneous differential equation starts with finding the complementary solution $y_c = e^{mx}$; then we search for a particular solution of the form $y_p = ax + b$ for a general solution $y_g = y_c + y_p = Ce^{mx} + ax + b$. If we substitute e^{mx} in the homogeneous equation $y' + 2y = 0$, we have

$$me^{mx} + 2e^{mx} = 0 = e^{mx}(m + 2), \qquad m = -2 \tag{E.3}$$

and so $y_c = e^{-2x}$.

Now if we substitute $y = ax + b$ for the nonhomogeneous equation (E.1), we obtain

$$a + 2(ax + b) = x$$

and if we equate coefficients for x and 1 we have

$$a + 2b = 0, \qquad a = -2b \quad \text{and} \quad 2ax = x, \qquad a = \tfrac{1}{2} \tag{E.4}$$

hence

$$y_p = ax + b = \tfrac{1}{2}x - \tfrac{1}{4} = \tfrac{1}{2}\left(x - \tfrac{1}{2}\right) \tag{E.5}$$

and

$$y_g = Ce^{-2x} + \tfrac{1}{2}(x - \tfrac{1}{2}) \tag{E.6}$$

A similar method can be employed for the difference equation (E.2) with a solution, for the homogeneous equation,

$$y_{k+1} + y_k = 0 \tag{E.7}$$

of the form $y_c = e^{mx_k}$, which when substituted in (E.7) yields

$$e^{mx_{k+1}} + e^{mx_k} = 0 = e^{mx_k}(e^m + 1), \qquad e^m = -1$$

Hence $m = j\pi$ with $j = \sqrt{-1}$, and $y_c = e^{j\pi x_k}$. For the particular solution we also try $y_p = ax_k + b$ for the nonhomogeneous equation (E.2), in the same manner as for the differential equation, to obtain

$$ax_{k+1} + b + ax_k + b = x_k$$

$$= a(x_k + h) + b + ax_k + b = x_k$$

$$(2a - 1)x_k + (2b + ah) = 0$$

If we equate the coefficients of x_k and 1 to zero, we have $a = \tfrac{1}{2}$ and $b = -\tfrac{1}{2}ah = -\tfrac{1}{4}h = -h/4$. So $y_p = \tfrac{1}{2}(x_j - h/2)$, and the general solution to the difference equation (E.2) is

$$y_k = Ce^{j\pi x_k} + \tfrac{1}{2}(x_k - \tfrac{1}{2}h) \tag{E.8}$$

$$= Ce^{j\pi k} + \tfrac{1}{2}(k - \tfrac{1}{2}) = (-1)^k C + \tfrac{1}{2}(k - \tfrac{1}{2}), \qquad h = 1 \tag{E.9}$$

If we compare (E.9), the solution of the difference equation (E.2), with (E.6), the solution of its differential equation analog (E.1), we observe one basic difference in the first part (complementary) of the solution (E.9), where it is oscillatory (and possibly complex) for the difference equation (E.2),

$$e^{\pi j x_k} = \cos \pi x_k + j \sin \pi x_k = (-1)^k \quad \text{when } x_k = k$$

while it is decaying and real for the differential equation. This should not be surprising, since the difference equation, with its fixed difference $\Delta x = h = 1$, is only an approximation to the differential equation (E.1).

In attempting to apply the discrete transforms (4.3), (4.4) for solving difference equations, we must emphasize that here we take the difference equation with its N sample points on its own, without regard to its origin of approximating a differential equation. In this sense, the discrete transform representation is the proper and correct setting, where the N

sample points are transformed via (4.3), and hopefully the resulting system will be easier to work with. Then we transform the sampled solution back exactly via the inverse transform (4.4).

For such a transformation, we note that the difference of order m, y_{k+m} of y_k, represents only a shifting of y_k; hence its discrete transformation would correspond to multiplying by $e^{j2\pi nm/N}$ in the transformed space, i.e.,

$$\tilde{\mathcal{F}}\{\tilde{g}_{k+m}\} = e^{j2\pi nm/N}\tilde{G}_n \equiv W^{nm}\tilde{G}_n \tag{4.100}$$

according to (4.70) with $i = -m$.

As we indicated in the last section for (4.72)–(4.75), there may be another approach for the discrete transform application to differential or difference equations, as indicated by (4.70) and (4.100). This is to approximate the samples of the derivative $g'(kT)$ by the discrete representation $\tilde{g}'(kT)$, as indicated in (4.73), (4.74),

$$\tilde{\mathcal{F}}\{\tilde{g}'(kT)\} = \left[\frac{j2\pi n}{NT}G\left(\frac{n}{NT}\right)\right]^{\sim} \tag{4.73}$$

where [ ]$^{\sim}$ denotes a discrete approximation of [ ]. This reminds us of a parallel to the usual Fourier transform algebraization of the derivatives. In contrast to the first approach for difference equations, this approach with (4.73) carries with it the assumption of a good discrete transform approximation for the continuous transform. Otherwise, we should keep track of the error, or errors, involved in case we have to work with a small number of sample points.

In case we work with the difference approximation of differential equations, we can increase the number of sample points and appeal to the fast Fourier transform algorithm. For such a situation, the two approaches seem to be close, with the difference approach being more natural, when we allow $\Delta x = h$ to get small, and should in principle converge to the solution of the differential equation. However, there may be some situations that need one approach more than the other—for example, when we have a fixed number of measured data points $\{y_k\}$ would like to approximate the derivative. The difference approach would be a clear $y_{k+1} - y_k$, while the second indirect approach of the discrete transform approximation of the derivative samples may offer a smoother result. On the other hand, a suitable example for the difference approach would be in treating *recursion relations* of some special functions where the difference $\Delta x = h$ is fixed. To illustrate this point we will consider a homogeneous difference equation as a recursion relation in the following example.

EXAMPLE 4.2: The DFT Method for Solving Recursion Relations

Consider the following homogeneous recursion relation

$$y_{k+1} - y_k - y_{k-1} = 0 \tag{E.1}$$

with the conditions

$$y_0 = 0 \tag{E.2}$$

$$y_1 = 1 \tag{E.3}$$

We can take y_k as $\tilde{y}_k$ and use the discrete transform (4.3) on (E.1). If we recognize the shifting property (4.70), the transformed result will be the following *algebraic equation* in $\tilde{Y}_n$ (as the DFT of $\tilde{y}_k$)

$$W^n \tilde{Y}_n - \tilde{Y}_n - W^{-n} \tilde{Y}_n = 0$$

whence

$$W^{2n} - W^n - 1 = 0, \qquad W = e^{j2\pi/N} \tag{E.4}$$

as it is a quadratic equation in W^n, with its two roots

$$W^{n_1} = \frac{1 + \sqrt{5}}{2}, \qquad W^{n_2} = \frac{1 - \sqrt{5}}{2} \tag{E.5}$$

This means that we have two values n_1, n_2 of n, which will result in two values $\tilde{G}_{n_1}$ and $\tilde{G}_{n_2}$ to be used in (4.3) to find $\tilde{y}_k$,

$$\tilde{y}_k = \frac{1}{N} \sum_{n_1, n_2} W^{nk} \tilde{G}_n = \frac{1}{N} \left[\tilde{G}_{n_1} W^{n_1 k} + \tilde{G}_{n_2} W^{n_2 k} \right]$$

$$\tilde{y}_k = c_1 \left(\frac{1 + \sqrt{5}}{2} \right)^k + c_2 \left(\frac{1 - \sqrt{5}}{2} \right)^k$$

$$= \frac{1}{\sqrt{5}} \left(\frac{1 + \sqrt{5}}{2} \right)^k - \frac{1}{\sqrt{5}} \left(\frac{1 - \sqrt{5}}{2} \right)^k, \qquad k \geq 0 \tag{E.6}$$

where c_1 and c_2 are two arbitrary constants, which were determined from the boundary conditions (E.2), (E.3) as $1/\sqrt{5}$ and $-1/\sqrt{5}$, respectively. An advantage of this form of solution (E.6) for the recursive relation (E.1) is that we can compute $\tilde{y}_{98}$ directly without having to find all its predecessors as required by the recursion relation (E.1).

Having presented some of the essential preliminaries and identities of difference calculus, we will now illustrate the use of discrete transforms for solving difference equations. This parallels the usual operational method of integral transforms for differential equations.

EXAMPLE 4.3: Solving difference equations by the discrete transforms

Consider the difference equation of our first example (Example 4.1)

$$y_{k+1} + y_k = x_k \qquad \text{(E.1)}$$

Since we are now working in the (natural) discrete setting of difference equations, we will take y_k as $\tilde{y}(kT)$ and Y_n as $\tilde{Y}_n = \tilde{Y}(n/NT)$ of (4.3), (4.4) in the following few examples. Also, for simplicity we let $T = 1$. If we apply the discrete transform to (E.1), we obtain

$$\tilde{\mathcal{F}}\{y_{k+1} + y_k\} = \tilde{\mathcal{F}}\{x_k\} = \tilde{\mathcal{F}}\{k\}$$

$$= W^n Y_n + Y_n = \begin{cases} \dfrac{-N}{1 - W^n}, & 1 \le n \le N - 1 \\[2mm] \dfrac{N(N-1)}{2}, & n = 0 \end{cases} \qquad \text{(E.2)}$$

after using the shifting property (4.70) on y_{k+1} and the discrete transform pair (4.98) for $x_k = k$. If we compare (E.2) with (E.1), we may say, in an analogy to the operational calculus for differential equations, that in (E.2) we have *algebraized* the difference equation (E.1) in the transformed space. The resulting transformed function is

$$Y_n = \frac{\tilde{\mathcal{F}}\{k\}}{(1 + W^n)} = \begin{cases} \dfrac{N}{1 - W^{2n}}, & 1 \le n \le N - 1 \\[2mm] \dfrac{N(N-1)}{4}, & n = 0 \end{cases},$$

$$W \equiv e^{2\pi j/N} \qquad \text{(E.3)}$$

This is to be transformed back via (4.4) to have the solution of the difference equation (E.1),

$$y_k = \frac{N-1}{4} - \sum_{n=1}^{N-1} \frac{1}{1 - W^{2n}} W^{nk} \qquad \text{(E.4)}$$

In comparison with the closed-form solution $y = Ce^{-2x} + \frac{1}{2}(x - \frac{1}{2})$ of the corresponding differential equation [(E.2) in Example 4.1], we realize that the present solution is not in its final simple form, and hence we have to compute the finite sum. However, it is only fair to compare this "difference calculus–operational method" with the usual operational calculus method in light of the fact that the latter cannot give the closed-form solution without necessary pairs like those of the easily accessible integral transform tables. Hence, for this method to be efficient, we need similar tables (see Table 4.3) for the discrete transforms of basic sequences. For example, had we had such pairs for $\bar{g}(kT) = \bar{\mathcal{F}}^{-1}\{1/(1 - W^{2N})\}$, we could have the exact closed-form solution of (E.1) instead of the sum (E.4). Still, the lack of DFT tables is not as bad as it is with the integral transform tables, since in the DFT case our inverse is nothing more than a simple finite sum like (E.4), which is the form ready for its efficient algorithm, the FFT.

The following example illustrates how difference equations arise in the practical modeling of a traffic network to constitute a boundary value problem. With the few tools (and DFT pairs) developed here, we will consider only a very specific network that can be accommodated, with such limited DFT pairs, to a final solution.

EXAMPLE 4.4: Traffic Flow in Equilibrium[†]

The following example may be somewhat idealized, but it does give an indication of the type of problem that can give rise to a difference equation. Consider a traffic network which may be represented as a closed ring, as shown in Fig. 4.14. The nodes (dots) represent intersections. The

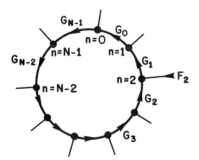

FIG. 4.14   Model for a traffic network.

[†] Examples 4.4–4.6 were supplied mainly by Prof. W. L. Briggs in Briggs and Jerri (1985), with the author's acknowledgment. The main idea of introducing *operational difference calculus*, or the use of discrete transforms for solving difference equations, is that of the author.

arcs (lines) represent streets. There are N nodes numbered from $n = 0$ to $n = N - 1$, where we agree that $n = N$ and $n = 0$ denote the same intersection. We will associate a variable G_n with the arc connecting node n and node $(n + 1)$. We will also choose a direction or positive sense to each arc and mark it by an arrow. This choice is arbitrary, and the figure indicates one particularly simple choice. The variable G_n represents the number of vehicles passing along the nth arc per unit time and is positive when the traffic flow is in the direction of the arrow on that arc. Again, with this natural discrete setting we will use G_n for $\hat{G}_n$ and g_k for $\tilde{g}_k$ as we did in Example 4.3. We will also assume that each intersection may be either an entrance or an exit for traffic to and from the loop. We will denote the traffic flow into or out of intersection n by F_n, where a positive F_n is taken to mean flow into the intersection (entrance) and a negative F_n is taken to mean flow out of the intersection (exit). The goal is to model this traffic network and to find the flow rates, G_n, along the streets when the entire traffic pattern reaches an equilibrium or steady state.

In order to model this system, we must relate the flow along one arc to the flow along neighboring arcs and through neighboring intersections. A few observations make this possible. The key words in the statement of the problem are *equilibrium* and *steady-state*. A steady-state traffic pattern depends on two conditions being met:

1. The total traffic flow into the loop from entrances must balance the total traffic flow out of the loop along exits. A more concise way of stating this is

$$\sum_{n=0}^{N-1} F_n = 0 \tag{E.1}$$

2. At each intersection, traffic flow in must equal traffic flow out.

It should be clear that if either of these conditions is not satisfied, there is the possibility of traffic accumulating somewhere in the loop or perhaps vanishing for the loop. These two conditions amount to a *conservation of traffic law*. The first condition restricts the data in the problem. If the data do not meet the condition, there is no point looking for a steady-state solution. The second condition allows us to find equations that govern the flows, G_n.

Consider intersection n where $0 \leq n \leq N - 1$. Using the sign and arrow conventions we have established, the flow into intersection n is $G_n + F_n$. The flow out of intersection n is G_{n-1}. Condition (2) above can now be expressed as

$$G_n - G_{n-1} + F_n = 0 \tag{E.2}$$

which holds for intersections $0 \leq n \leq N - 1$. To complete the statement of the problem, we include the *periodic boundary condition*, which states that $G_0 = G_N$. The complete problem now appears as

$$G_n - G_{n-1} = -F_n, \qquad 0 \leq n \leq N - 1 \tag{E.3}$$

$$G_0 = G_N \tag{E.4}$$

which is a first-order, constant-coefficient, nonhomogeneous difference equation with periodic boundary condition $G_0 = G_N$.

In regard to this special *periodic boundary condition* $G_0 = G_N$ for the DFT, we may recall other very practical situations where the conditions are specified at each end $n = 0$ and $n = N$. For example, we may be given the sequences $G_0 = 0$ and $G_n = 0$, where we are reminded of the use of the discrete sine transform (DST) in (4.20), (4.21) for this "odd" boundary condition. If we are given vanishing central differences (like a slope) at both ends $G_1 - G_{-1} = 0$ and $G_{N+1} - G_{N-1} = 0$, then we should be reminded of the discrete cosine transform (DCT) in (4.18), (4.19) for such an "even" boundary condition.

Let us now consider a specific traffic network in which the streets which connect to the loop are alternately entrances and exits, each carrying V vehicles per unit time. The sequence $\{F_n\}$ which describes this pattern is $F_n = V(-1)^n = Ve^{j\pi n}$. Note that in order to satisfy condition (1) above we must also assume that N is even.

The job now is to find a sequence of flow rates $\{G_n\}$ which satisfies (E.3), (E.4) with this particular choice of $\{F_n\}$. The periodic boundary condition suggests that we express the solution $\{G_n\}$ in terms of the DFT, since, as we have seen, a sequence so defined has the necessary periodicity. Therefore, we will look for a solution sequence of the form

$$G_n = \sum_{k=0}^{N-1} g_k W^{-nk}, \qquad 0 \leq n \leq N - 1$$

where the g_k's must now be determined. Before proceeding, it is necessary to express the right-hand side sequence $\{F_n\}$ of (E.3) in terms of a DFT. Letting

$$F_n = \sum_{k=0}^{N-1} f_k W^{-nk} = \sum_{k=0}^{N-1} f_k e^{-j2\pi kn/N}$$

and recalling that $F_n = V e^{j \pi n}$, it is evident that

$$
f_k = \begin{cases} 0 & \text{if } k \neq \dfrac{N}{2} \\[2mm] V & \text{if } k = \dfrac{N}{2} \end{cases}
\tag{E.5}
$$

In this case, F_n is a multiple of the $N/2$ harmonic, so all coefficients in the DFT are zero except $f_{N/2}$.

We may now substitute for G_n and F_n in (E.3) and (E.4), giving

$$
\sum_{k=0}^{N-1} g_k W^{-nk} - \sum_{k=0}^{N-1} g_k W^{-(n-1)k} = -\sum_{k=0}^{N-1} f_k W^{-nk}
$$

Canceling common terms and combining sums yields

$$
\sum_{k=0}^{N-1} [g_k(1 - W^k) + f_k] W^{-nk} = 0 \quad \text{for } 0 \leq n \leq N - 1
$$

This sum can vanish for all $0 \leq n \leq N - 1$ only if each term of the sum vanishes independently. That is, we must have

$$
g_k = \frac{-f_k}{1 - W^k} \quad \text{for } 0 \leq k \leq N - 1
$$

Recalling the expression for f_k from (E.5), we have

$$
g_k = \begin{cases} 0 & \text{for } k \neq \dfrac{N}{2} \\[2mm] -\dfrac{V}{1 - W^k} & \text{for } k = \dfrac{N}{2} \end{cases}
\tag{E.6}
$$

Notice that the coefficients $\{g_k\}$ of the solution were found simply by solving an algebraic equation. To recover the solution $\{G_n\}$, it remains only to do the (synthesis or) inverse DFT. First note that

$$
g_{N/2} = \frac{-V}{1 - W^{N/2}} = \frac{-V}{1 - e^{j\pi}} = \frac{-V}{2}
$$

Therefore,

$$
G_n = \sum_{k=0}^{N-1} g_k W^{-nk} = g_{N/2} W^{-nN/2} = \frac{-V}{2} e^{-j\pi n} = \frac{-V}{2} (-1)^n
$$

since the only nonvanishing g_k is $g_{N/2}$, as seen in (E.6).

This says that the flow along each street has a volume of $V/2$ and is along the arrow (counterclockwise) on odd-numbered arcs and opposes

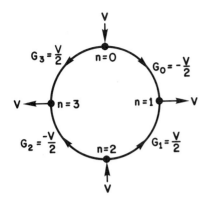

FIG. 4.15 Example of a traffic network with four intersections.

the arrow on even-numbered arcs. The solution for a traffic circle with $N = 4$ intersections is shown in Figure 4.15.

This example illustrates the simplest pattern of inflow and outflow to the loop. More interesting and perhaps more realistic traffic patterns can be handled by the same method. These problems will be investigated in the exercises (see Exercise 4.13).

B. The z-Transform as the Finite (or Discrete) Analog for Laplace Transform—Initial Value Problems and a Problem of Combinatorics

The next example will illustrate the operational difference calculus method for solving difference equations with initial conditions. For such discrete initial value problems, the z-transform (2.274) of Section 2.9.3 will be employed.

EXAMPLE 4.5:   Initial Value problems—the z-Transform

Here we will illustrate the operational difference method for solving difference equations with initial conditions, instead of boundary conditions like those we had in Example 4.4.

Consider the second-order difference equation

$$u_{n+2} - u_{n+1} - 6u_n = 0, \qquad n \geq 0 \tag{E.1}$$

with the initial conditions

$$u_0 = 0, \qquad u_1 = 3 \tag{E.2}$$

The two extra conditions which we have come to expect with second-order difference equations are now given, not at the end points of the interval on which the problem is to be solved, as in the case of boundary conditions, but rather at the "initial end" of the interval on which the problem is to be solved. The initial conditions give the initial configuration of the system which is being modeled, while the difference equation gives the subsequent behavior of the system as it evolves from the initial conditions. We may note in (E.1) that the largest shift in the sequence u_n is that of 2 in u_{n+2}, which determines the *order* of the difference equation (E.1). In that regard it is no coincidence that we are given two independent initial conditions in (E.2). This is seen in parallel to what we expect for initial value problems, associated with second-order differential equations, where we expect the initial value of the function as well as its first derivative. Here in (E.2) we use an initial value of the sequence $u_0 = 0$, and also the forward difference $u_1 - u_0 = 3$, as an approximation of the first derivative at the initial end or point.

One way to solve this initial value problem is to assume that it has solutions of the form $u_n = \lambda^n$, where λ is a constant to be determined. Substituting this trial solution into the difference equation (E.1) gives

$$\lambda^n(\lambda^2 - \lambda - 6) = 0$$

This equation can be satisfied if we choose $\lambda = 0$, which gives very uninteresting (trivial) solutions, or if we require that

$$\lambda^2 - \lambda - 6 = 0 \tag{E.3}$$

This *second*-degree polynomial (arising from a *second*-order difference equation) is called the *characteristic polynomial* of the difference equation. It can be solved to yield the roots $\lambda = 3$ and $\lambda = -2$. Therefore, $u_n = 3^n$ is a solution of the equation and so is $u_n = (-2)^n$. We leave it as an exercise to show that the final solution $c_1 3^n + c_2(-2)^n$ to the initial value problem (E.1), (E.2) is

$$u_n = \tfrac{3}{5}(3)^n - \tfrac{3}{5}(-2)^n \tag{E.4}$$

after using the initial conditions in (E.2) for determining c_1 and c_2.

We will now illustrate the use of the z-transform (2.264) for solving the same initial value problem (E.1), (E.2). As we mentioned in Section 2.9.3, we must first derive a very important operational property of the z-transform before we can apply it to the solution of difference equations. What effect does the z-transform have on the *shifted sequence*

$\{u_{n+1}\}$? The calculation is not difficult

$$Z\{u_{n+1}\} = \sum_{n=0}^{\infty} u_{n+1} z^{-n}$$

$$= z \sum_{n=0}^{\infty} u_{n+1} z^{-n-1}$$

$$= z \sum_{n=1}^{\infty} u_n z^{-n}$$

$$= z \left\{ -u_0 + \sum_{n=0}^{\infty} u_n z^{-n} \right\}$$

$$= -zu_0 = zU(z)$$

$$= zU(z) - zu_0 \tag{E.5}$$

Therefore,

$$Z\{u_{n+1}\} = zU(z) - zu_0 \tag{E.5}$$

A similar calculation shows that

$$Z\{u_{n+2}\} = z^2 U(z) - z^2 u_0 - zu_1 \tag{E.6}$$

and in general it can be shown that

$$Z\{u_{n+k}\} = z^k U(z) - z^k u_0 - z^{k-1} u_1 + \cdots - zu_{k-1} \tag{E.7}$$

(see Exercise 4.14a).

We may call this operation of the z-transform an "algebraization" operation in the sense that it removed the shift from the sequence, where we end up dealing with the z-transform $U(z)$ of the nonshifted sequence u_n. This property is precisely what we need to solve difference equations, since it relates the transform of a shifted sequence to the transform of the original sequence. In addition, the transform of the shifted sequence introduces the initial elements of the sequence $u_0, u_1, u_2, \ldots$. But if we are solving an initial value problem, these initial values are given. This, of course, is exactly how the Laplace transform is used for differential equations with initial conditions.

Now we can start solving the initial value problem (E.1), (E.2) by taking the z-transform of the entire difference equation and using property (E.7)

$$Z\{u_{n+2} - u_{n+1} - 6u_n\} = Z\{0\}, \qquad n \geq 0$$

or

$$Z\{u_{n+2}\} - Z\{u_{n+1}\} - 6Z\{u_n\} = 0$$

or

$$z^2 U(z) - z^2 u_0 - z u_1 - [z U(z) - z u_0] - 6 U(z) = 0 \tag{E.8}$$

Substituting the given initial values $u_0 = 0$, $u_1 = 3$ and solving for $U(z)$ gives

$$U(z) = \frac{3z}{z^2 - z - 6} = \frac{\%}{z - 3} + \frac{\%}{z + 2} \tag{E.9}$$

In anticipation of taking the inverse transform of U, it has been written in terms of partial fractions. The inverse transform comes immediately when we use the z-transform pair in (2.276),

$$u_n = Z^{-1}\{U(z)\} = Z^{-1}\left\{\frac{\%}{z - 3}\right\} + Z^{-1}\left\{\frac{\%}{z + 2}\right\}$$

$$= \%(3)^{n-1} + \%(-2)^{n-1}$$

$$= \%(3^n) - \%(-2)^n \tag{E.10}$$

This solution is valid for $n \geq 2$, but notice that it also satisfies the initial conditions. It is the same result obtained in (E.4).

The use of the z-transform to solve a problem which is this simple hardly offers an improvement over the method that uses the characteristic polynomial. However, the z-transform begins to justify itself for nonhomogeneous problems. Consider the same initial value problem with a right-hand side sequence.

$$u_{n+2} - u_{n+1} - 6u_n = \sin\frac{\pi n}{2}, \qquad n \geq 2 \tag{E.11}$$

$$u_0 = 0, \qquad u_1 = 3 \tag{E.12}$$

We proceed in much the same way as before, and we leave it as an exercise (see Exercise 4.14b).

EXAMPLE 4.6:   The z-transform for an Initial Value Problem in Combinatorics

We close this section by solving a very famous difference equation which appears in many settings. We will introduce the equation by means of a problem in combinatorics. A store manager is putting boxes on a long shelf. He has boxes of two sizes: one type of box occupies one space on the shelf; the other type occupies two spaces on the shelf. Figure 4.16 shows the shelf loaded with 6 boxes occupying 9 spaces. There are many different ways in which boxes could be placed on a shelf 9 spaces long (for example, 9 short boxes, or 4 long boxes and 1 short

1   2
space spaces

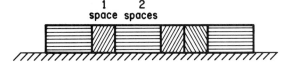

FIG. 4.16 A shelf loaded with six boxes (of two sizes) occupying nine (equal) spaces.

box, or 3 short boxes and 3 long boxes). We ask, how many different combinations of long and short boxes can be arranged on a shelf which is n spaces long? Let C_n denote the number of combinations on a shelf of length n spaces. In particular, we would like to find C_n for any positive integer n and for values of n which may be large. An ingenious argument leads to a difference equation.

Assume we know many ways the boxes may be arranged on shelves of length n and $n + 1$. How can we express C_{n+2}? A shelf of length $n + 2$ can be considered a shelf of length n plus 2 spaces (one large box) or a shelf of length $(n + 1)$ plus one space (one small box). In the first instance, there are C_n combinations of boxes on a shelf of length n with one large box added in each case which makes C_n combinations. In the second instance, there are C_{n+1} combinations of boxes on a shelf of length $n + 1$ with one small box added in each case which makes C_{n+1} combinations. Therefore, the total number of combinations on a shelf of length $n + 2$ spaces is

$$C_{n+2} = C_{n+1} + C_n \quad \text{for } n \geq 1 \tag{E.1}$$

This is a *second-order difference equation*, and if C_1 and C_2 are specified as initial conditions, then we have an initial value problem which can be solved.

The initial values can easily be found. On a shelf which is one space long, there is only one arrangement of boxes, namely one short box. Thus $C_1 = 1$. On a shelf of length two spaces, there are two arrangements, namely two short boxes or one long box. Thus $C_2 = 2$. To put our initial value problem in a standard form, define a new sequence $\{u_n\}$ by letting $u_n = C_{n+1}$ for $n \geq 0$. Then an equivalent initial value problem is

$$u_{n+2} - u_{n+1} - u_n = 0, \qquad n \geq 0 \tag{E.2}$$

$$u_0 = 1, \qquad u_1 = 2 \tag{E.3}$$

We may now proceed by taking a z-transform of the difference equation. Letting $U(z) = Z\{u_n\}$, we have

$$z^2 U(z) - z^2 u_0 - z u_1 - [zU(z) - z u_0] - U(z) = 0 \tag{E.4}$$

If we substitute the initial values $u_0 = 1$ and $u = 2$ from (E.3) in (E.4), we have

$$(z^2 - z - 1)U(z) = z^2 + 1 \tag{E.5}$$

Solving for $U(z)$ gives us

$$U(z) = \frac{z^2 + z}{z^2 - z - 1} = 1 + \frac{1 + 2/\sqrt{5}}{z - \alpha_1} + \frac{1 - 2/\sqrt{5}}{z - \alpha_2} \tag{E.6}$$

where $\alpha_1 = (1 + \sqrt{5})/2$, $\alpha_2 = (1 - \sqrt{5})/2$. Once again, we have expressed $U(z)$ in partial fractions in preparation for taking the inverse transform. Using the table of z-transforms (see Appendix C.17) or the z-transform pairs in (2.275) and (2.276), we have

$$Z^{-1}\{1\} = \delta_{n,0} \quad \text{and} \quad Z^{-1}\left\{\frac{1}{z - \alpha}\right\} = \begin{cases} 0, & n = 0 \\ \alpha^{n-1}, & n \geq 1 \end{cases} \tag{E.7}$$

Thus the solution u_n, as the inverse z-transform of $U(z)$ in (E.6) is

$$u_n = \begin{cases} 1 & \text{if } n = 0 \\ \left(1 + \dfrac{2}{\sqrt{5}}\right)\left(\dfrac{1 + \sqrt{5}}{2}\right)^{n-1} & \\ \quad + \left(1 - \dfrac{2}{\sqrt{5}}\right)\left(\dfrac{1 - \sqrt{5}}{2}\right)^{n-1} & \text{if } n \geq 1 \end{cases} \tag{E.8}$$

This expression cannot be simplified any further. Despite its complexity and the appearance of irrational numbers, it can be checked that this expression generates the sequence of integers $\{u_n\}_{n=0}^{\infty} = \{1, 2, 3, 5, 8, \ldots\}$. The numbers of combinations of boxes on a shelf of length n can be found by readjusting the index such that $C_{n+1} = u_n$ for $n \geq 0$. Clearly, the number of combinations grows rapidly with the length of the shelf, since

$$\{C_n\}_{n=1}^{\infty} = \{C_1, C_2, C_3, \ldots\} = \{1, 2, 3, 5, 8, \ldots\} \tag{E.9}$$

4.1.6 Approximating Fourier Integrals and Series by Discrete Fourier Transforms

The discrete Fourier transform is the normal setting for transforming a finite number of data points in the time (t) space as $\tilde{g}(kT)$ to the same number of points in the frequency (f) space as $\tilde{G}(n/NT)$. This parallels how we use the integral transform for time and frequency representation in signal theory, or coordinate and wave number in quantum mechanics. The advantage is that it is often easier to describe the system in one or the other space. An example is how the transformed differential equations in the past chapters were simplified (algebraized) as we transformed them by using their compatible transforms. In such analysis we often end up with a Fourier integral-type representation for the solution with the simple closed-form solution considered as the exception. The same can be said about Fourier series analysis of periodic functions.

It is in this field of computing infinite and finite (Fourier coefficients) Fourier transforms and Fourier series that the discrete Fourier transform (and hence its efficient algorithm, the FFT) finds its main applications, as we mentioned earlier. In this section we will concentrate on the adequate preparation of these integrals and sums for their approximation by the discrete transform. With such preparation, we are ready to use the FFT algorithm, which has already been written by expert computer scientists, and is now reasonably available, even in hardware (see Section 4.1.3).

Our emphasis all along has been on clarifying the main points of such an approximation and, in particular, the inherited errors of aliasing (sampling) and truncation.

To summarize, let us recall the Fourier integrals (4.1), (4.2); the Fourier series (4.7), (4.8); and the discrete Fourier transforms (4.3), (4.4) and (4.5), (4.6),

$$G(f) = \int_{-\infty}^{\infty} g(t)e^{-j2\pi ft}\, dt \tag{4.1}$$

$$g(t) = \int_{-\infty}^{\infty} G(f)e^{j2\pi ft}\, df \tag{4.2}$$

$$G_s(f) = \sum_{k=-\infty}^{\infty} c_k e^{-j2\pi kTf}, \qquad -\frac{1}{2T} < f < \frac{1}{2T} \tag{4.7}$$

$$c_k = T \int_{-1/2T}^{1/2T} G_s(f)e^{2\pi jkTf}\, df \tag{4.8}$$

$$\tilde{G}\left(\frac{n}{NT}\right) = \sum_{k=0}^{N-1} \tilde{g}(kT)e^{-j2\pi nk/N},$$

$$n = 0, 1, \ldots, N-1, \qquad f_n = n\frac{1}{NT} \tag{4.3}$$

$$\tilde{g}(kT) = \frac{1}{N}\sum_{n=0}^{N-1} \tilde{G}\left(\frac{n}{NT}\right)e^{j2\pi nk/N},$$

$$k = 0, 1, \ldots, N-1, \qquad t_k = kT \tag{4.4}$$

$$\tilde{G}\left(\frac{n}{(2M+1)T}\right) = \sum_{k=-M}^{M} \tilde{g}(kT)\exp\left(-\frac{j2\pi nk}{2M+1}\right),$$

$$n = -M, \ldots, 0, \ldots, M, \qquad f_n = n\frac{1}{(2M+1)T} \tag{4.5}$$

$$\tilde{g}(kT) = \frac{1}{2M+1}\sum_{n=-M}^{M} \tilde{G}\left(\frac{n}{(2M+1)T}\right)\exp\left(\frac{j2\pi nk}{2M+1}\right),$$

$$k = -M, \ldots, 0, \ldots, M, \qquad t_k = kT \tag{4.6}$$

We may recall that the Fourier integral (4.1) converges to $G(f)$ if $g(t)$ is absolutely integrable, that is, $\int_{-\infty}^{\infty} |g(t)|\,dt < \infty$. When $G(f)$ is discontinuous at a point (or points) f_0, then the integral converges to the average of the right- and left-hand limits $\frac{1}{2}[G(f_{0+}) + G(f_{0-})]$. In the case of the Fourier series (4.7), it is sufficient that the periodic function with period $1/T$ is sectionally smooth on the interval $(-1/2T, 1/2T)$ in order that the series converge. For a jump discontinuity at x_0, the series converges to the midvalue of the two right- and left-hand limits $\frac{1}{2}[G_s(x_{0+}) + G_s(x_{0-})]$. Such discontinuities of the Fourier integrals and series are observed in the application of the discrete Fourier transforms, in which we truncate the function $g(t)$ to $(-a, a)$ and then translate the part on the negative interval $(-a, 0)$ to $(a, 2a)$ to prepare it for the form (4.3), (4.4) of the discrete Fourier transforms, as illustrated in Figs. 4.11–4.13 in Section 4.1.2. In the case that the function $g(t)$ is not necessarily even, the above translation will have a jump discontinuity at $a = NT/2$; here the discrete transform is assigned the average of the closest neighboring samples,

$$\tilde{g}\left(\frac{NT}{2}\right) = \frac{1}{2}\left[g\left(\frac{NT}{2} - T\right) + g\left(-\frac{NT}{2} + T\right)\right]$$

$$= \frac{1}{2}\left[\tilde{g}\left(\frac{NT}{2} - T\right) + \tilde{g}\left(-\frac{NT}{2} + T\right)\right]$$

as shown explicitly in the actual computations via the DFT when moving form Fig. 4.34d to Fig 4.34e in Example 4.7. Such an assignment is guided by the convergence of the Fourier series to the middle points of the jump discontinuities in the periodic extension it represents.

We must mention here that this convergence at the midpoint of the jump discontinuity is somewhat theoretical, as it depends on having infinite series or integrals, which is impossible to attain in our actual numerical computation. Indeed, as we use a truncated series or integral for the approximate computations, we face a "chronic" difficulty near the jump discontinuity, where the approximation will have an overshoot that cannot be minimized by a practical enlargement of the truncation window. This effect is termed *Gibbs phenomenon*, which we shall discuss and suggest remedies for in part G of this section (Section 4.1.6G).

The direct practical computation of the Fourier integrals would involves truncating the infinite limits $(-\infty, \infty)$ to $(-A, A)$, discretizing this finite interval, evaluating the sample values, and then computing the sum. A very simplistic way of approaching the value of the integral is to increase the number of samples and increase A until the summation attains a somewhat constant value within the accuracy of the computations. In practice, we often accept this result for the limit. Such summation starts by assuming rectangles to approximate the increments of the integrand or integral. However, the subject of numerical integration, as a branch of numerical analysis, does not leave it at that. The familiar trapezoidal rule, Simpson's rule, and many other more sophisticated methods of numerical integration would necessarily give an estimate of the error involved from the sampling and truncation caused by the numerical approximation. Such methods could be described in simple terms as involving weights (note the 2's and 4's in Simpson's rule, for example) when compared with direct rectangular summation.

For the Fourier infinite series (4.7), we already have sampled values, namely the coefficients c_k, but the series has to be truncated. Modern methods for efficient computation of the series involve particular weights that speed up the convergence with sharp estimates of the error.

Compared with such direct efficient methods of computing infinite integrals or series, the discrete Fourier transform (4.3), (4.4) is a sum with a mere rectangular increment. Then how is it that it can compete with the above methods? The main secret, as we have mentioned earlier, is in its fast computation via the FFT algorithm, where the number of increments N can be taken very large with the "magic" of costing only $(N/2)\log_2 N$ computations, compared to the approximately N^2 computations required for the direct method (see Table 4.2). The efficient direct methods of numerical analysis may, of course, reduce this number, but not nearly to the advantage of $(N/2)\log_2 N$. This is especially important when we have very large N. To give an example, the trapezoidal approximation with 16 points of computation for the integral is equivalent to 32 points for the discrete transform, since it averages the right- and left-side values of the rectangular increments method. So the trapezoidal rule will cost $(16)^2 = 256$ computations, compared to $\frac{32}{2}\log_2 32 = 16 \times 5 = 80$ computations for the discrete transform via its FFT algorithm—a very good saving. Such a saving is clearly observed when we seek a more accurate approximation that needs large N. Table 4.2 illustrates this point for different values of N.

We may mention that if the nature of the problem is to transform a small finite number of samples for a simpler system representation, then the discrete transforms offer an exact direct (4.3) and inverse (4.4) transformation. In case the data are periodic, the period must be estimated carefully before using it for the discrete transform.

In the previous sections, we discussed in some detail the necessary careful preparation for employing the discrete transforms in a variety of approximate computations. We have also attempted to point out clearly the possible errors of *aliasing* and *truncation* that are involved in such computations. In this section we will continue to illustrate such preparations with more examples.

We will present here some general remarks concerning the aliasing and truncation errors, which include a discussion of the important special situations of band-limited or time-limited functions.

A. The Discrete Sum as an Approximation to Integrals or Series

We first note that to approximate the Fourier transform $G(f)$ of the function $g(t)$ in (4.1) by the discrete transform $\tilde{G}(n/NT)$ in (4.3), we must multiply the latter sum by $\Delta t = T$, the increment of the approximate integration,

$$G\left(\frac{n}{NT}\right) \sim T\tilde{G}\left(\frac{n}{NT}\right) \tag{4.101}$$

In the case of approximating $g(t)$ of (4.2) by the sum of $\tilde{g}(kT)$ in (4.3), the increment $\Delta f = 1/NT$ is already there, since if we use $\tilde{G}$ from (4.101), in the sum (4.4) we have

$$\tilde{g}(kT) = \frac{1}{NT}\sum_{n=0}^{N-1}\tilde{G}\left(\frac{n}{NT}\right)e^{j2\pi k/N} \sim g(kT) \tag{4.102}$$

where the sum has $\Delta f = 1/NT$, as seen in Figs. 4.11–4.13.

The approximation of the Fourier series $G_s(f)$ in (4.7) by $\tilde{G}(n/NT)$ of (4.3) obviously does not need an increment of integration. However, the approximation of the integral of the Fourier coefficients c_k of (4.8) by $\tilde{g}(kT)$ of (4.4) needs to be divided by the norm square $1/T$ of the discrete orthogonal kernel,

$$c_k \sim T\tilde{g}(kT) \tag{4.103}$$

Due to the symmetry of the discrete transforms (4.3), (4.4), another alternative is to approximate the Fourier coefficients c_k by $(1/N)\tilde{G}(k/NT)$ of (4.3). In this case, we need to multiply $\tilde{G}$ by $\Delta t = T$ to approximate its respective integral, then to divide by the norm square NT on the interval of integration $(0, NT)$,

$$c_k \sim \frac{T}{NT}\tilde{G}\left(\frac{k}{NT}\right) = \frac{1}{N}\tilde{G}\left(\frac{k}{NT}\right) \tag{4.104}$$

For this case, the Fourier series of $G_s(f)$ can be approximated by $\tilde{g}(kT)$, since we already have $(1/N)\tilde{G}$ in (4.4) approximating c_k as in (4.104),

$$G_s(kT) \sim \tilde{g}(kT) \tag{4.105}$$

For the following remarks covering the aliasing error, we refer to Figs. 4.11–4.13, 4.17–4.19.

B. Aliasing Error of the Fourier Coefficients

As we see in Fig. 4.1a the periodic function of the Fourier series $G_s(f)$ is defined on the interval $(-1/2T, 1/2T)$ where T is fixed for the correct sample spacing kT of c_k. However, $\{c_k\}_{k=1}^{\infty}$ is truncated at a finite value N instead of the infinite number required, and so it is aliased with aliasing error $\bar{c}_k - c_k$, where $\bar{c}_k = C_k$ is given in (4.55c). To improve this situation, we must increase N, the order of computations, to involve more c_k terms and hence decrease the aliasing error. The large number of samples will help the resolution $1/NT$ of the sampled values $G_s(n/NT)$ and, more important, the truncation error for both transforms (4.7) and (4.8) as we take more terms toward the required limits of both the infinite series and the numerical approximation of the integral of the Fourier coefficients. This, depending on the particular function $G_s(f)$, should in general improve the ripples around $\mp(1/2T)$ (see Fig. 4.1c).

In case the periodic function $G_s(f)$ is "essentially" time-limited to KT—that is, c_k almost vanishes for $|k| > K$—then the samples $\tilde{G}(n/NT) = G_s(n/2KT)$ are exact, provided, of course, that the sum in (4.3), (4.4) is taken from $n = 0$ to $N - 1 > 2K - 1$. In this case, there is practically no truncation error and the exact samples $G_s(n/2KT)$,

$$G_s\left(\frac{n}{2KT}\right) = \tilde{G}\left(\frac{n}{2KT}\right) = \sum_{k=0}^{2K-1} \tilde{g}(kT)e^{-j2\pi kn/2K} \qquad (4.106)$$

$$\tilde{g}(kT) = c_k, \quad k = 0, 1, \ldots, 2K - 1 \qquad (4.107)$$

can be used in an N-term sampling series to establish $G_{s,N}(f)$ as a finite Fourier series (trigonometric polynomial),

$$G_{s,N}(f) = \sum_{k=-K}^{K} c_k e^{-j2\pi k Tf}, \qquad -\frac{1}{2T} < f < \frac{1}{2T} \qquad (4.108)$$

with, of course

$$c_k = \tilde{g}(kT) \qquad (4.109)$$

Hence the discrete Fourier transforms (4.3), (4.4) are "almost" exact representation of the samples of the "essentially" time-limited periodic function. Moreover, the discrete transform $\tilde{G}(n/2KT)$ is used to interpolate $G_{s,N}(f)$ as the trigonometric polynomial (4.108) and $\tilde{g}(kT)$ is the exact c_k.

The above discussion should be taken primarily in support of the discrete Fourier transform approximation of both the Fourier series and its coefficients. In case we have the exact coefficients c_k, the above analysis using $2K$ terms of these c_k is a purely truncation error analysis affecting $G_s(n/KT)$.

C. Aliasing Error of the Fourier Integrals

As we have indicated for Figs. 4.3–4.7 and 4.11–4.13, the problem of approximating Fourier integrals by the discrete transforms is more involved than that of the Fourier series. If we look at $\tilde{g}(kT)$ in (4.4), we note that we need NT to be large to minimize the aliasing error in $\tilde{g}(kT)$ near NT. This in turn will help the resolution of $\tilde{G}(n/NT)$. However, increasing N is strictly limited by the necessity of having small T to reduce the aliasing error of $\tilde{G}$ near $1/T$. In addition, N must be large to minimize the truncation error of approximating integrals by a finite series of N terms. This means that we must have small T with large N in such a way that NT is still large, which makes a strong demand for very large N and hence a fast algorithm like the FFT.

Next, we consider the important special case of time-limited functions. Because of the symmetry of the transforms, the case of bandlimited functions will follow in the same way.

D. Time-Limited Functions

In case $G(f)$ is time-limited to b—that is, $g(t)$ vanishes identically beyond $(-b,b)$—then if we choose $T = 2b/N$, there will be no aliasing for $\bar{g}$ of (4.12), i.e.,

$$\bar{g}(kT) = \overline{g}(kT) = g(kT), \qquad k = 0, 1, \ldots, N - 1 \tag{4.110}$$

Hence, we can compute $\overline{G}(n/NT) = \hat{G}(n/NT)$ from (4.3). However, there is still an aliasing $G(n/NT) - \overline{G}(n/NT)$ for $\overline{G}$ which is difficult to improve, unless G is negligible outside $(-1/2T, 1/2T)$. The computation is sometimes limited by the fixed value of N; therefore, if we still want to improve the aliasing of $\overline{G}$, there is a difficult compromise, which basically amounts to a trade-off with an aliasing for $\bar{g}$. This is done by choosing a smaller $T_0 < T = 2b/N$ for a longer period $1/T_0$ of $\overline{G}$, which results in the aliasing of $\bar{g}$

$$\bar{g}(t) = \sum_{k=-\infty}^{\infty} g(t + kT_0) \tag{4.111}$$

In other words, we are now using N samples of $\bar{g}$ that cover only part $(0, NT_0)$ of its domain $(0, NT) = (0, 2b)$, with the remaining part being overlapped. If we note the periodicity of $\bar{g}(t)$ with period T_0, and choose $T_0 = T/K = 2b/NK$, then we have in (4.111) only a finite K-term series which can be evaluated to compute $\overline{G}$ of the larger interval $(-K/2T, K/2T) = (-KN/2b, KN/2b)$, yielding a smaller aliasing error. We must note that in this process of improving the aliasing error of $\overline{G}$, we have damaged the resolution of the samples of $G(f)$ with the present $\Delta f = K(1/NT)$ instead of the smaller one $\Delta f = 1/NT$.

Sometimes it may happen that for fixed N we need better resolution for $\overline{G}$ in order to have a better interpolation of the samples of the discrete transform. If this is more important than improving the aliasing error of $\overline{G}$, then we have to reverse the above process, starting with a larger $T_0 > T = 2b/N$. This means that the N samples of $\bar{g}$ will cover a larger domain $(0, NT_0)$ than that of $(0, 2b)$. For the extra part $(0, NT_0) - (0, 2b)$, where $g(t)$ is identically zero. we assign zeros to the corresponding sample values. As we mentioned before, this better resolution of $\overline{G}$ is done at the expense of worse aliasing, since with $T_0 > T$ we have $\overline{G}$ as the superposition of translations by a smaller $n/NT_0 < n/NT$, i.e., higher aliasing. It follows that such desired improvement of the resolution can be tried only for G that is negligible beyond $(-1/T_0, 1/T_0)$.

This analysis can be applied in the reverse direction for functions $g(t)$ bandlimited to $1/2T$, i.e. when $G(f)$ vanishes identically beyond

$(-1/2T, 1/2T)$. This is due to the symmetry of our exponential Fourier discrete or integral transforms.

This discussion should make clear the need to use a large number of sample points N, which motivated the search for a better computational algorithm to relax the limitation of fixed N and resulted in the discovery of the fast Fourier transform (see Section 4.1.3).

E. Truncation Error, Windowing Effect, Gibbs Effect (Phenomenon), Leakage Error

In all our attempts to approximate the infinite Fourier series and the Fourier integrals, we had to truncate the infinite series of $G_s(f)$ in (4.7) to N terms, and the infinite integrals of $G(f)$ in (4.1) and $g(t)$ in (4.2) to a finite time limits $-b$ to b and a finite frequency limits $-a$ to a, respectively. This means that we had to assume an "almost" time-limited function for $G(f)$ and an "almost" bandlimited function for $g(t)$. In all cases of truncation, we actually employed the gate function $p_a(x)$ with its very abrupt discontinuity at $x = \mp a$. The introduction of this sharp discontinuity in truncating a function has a bad effect on the transform of this truncated function, which we will discuss in this section. The present particular case of truncation is viewed as the result of applying a window of the gate function type $p_a(f)$ to $G(f)$, whose effect on the transform $g(t)$ appears as wiggles around $\mp b$, the ends of its support (or period). This effects is often called the *windowing effect*. The remedy for such an effect is to increase a, as we shall illustrate in Figs. 4.17–4.19. However, in many situations a may be fixed; thus something has to be done about the gate function and its troublesome sharp discontinuities. This means that we have to choose a window that dies out slowly around its truncation edges $\mp a$. We will soon show that the wiggles of the windowing effect are strongly related to the transform of the given window. This suggests choosing an "optimal" window with a transform that has much smaller wiggles than the transform of the gate function window (Fig. 4.17) (see Fig. 4.33a for an optimal window).

In the case of the Fourier series (4.7), its typical truncation to N terms amounts to multiplying the infinite sequence of its Fourier coefficients c_k by the gate function $p_{NT}(kT)$. The resulting discontinuity of c_k has its effect on the transform $G_s(f)$, appearing as wiggles, or ripples, around $\mp 1/2T$, the ends of its period, as illustrated in Figs. 4.1c, 4.11c, and 4.23c. These ripples can be explained in terms of the Fourier transform of the truncated coefficients $c_k P_{NT}(t)$, which is the convolution product $G_s(f) * (\sin 2\pi NTf)/\pi f$, where $(\sin 2\pi NTf)/\pi f$ is the transform of the gate function $p_{NT}(t)$ (see Fig. 4.23).

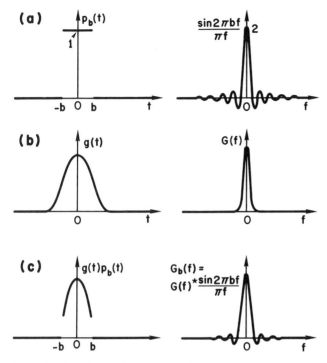

FIG. 4.17 Truncation error (windowing) of Fourier integrals: $G_b(f)$ approximation of $G(f)$ in part (b).

To compare the present truncation error of the Fourier series to our earlier analysis of approximating the Fourier series by the discrete Fourier transform, we note that in the latter we used N aliased coefficients $\bar{c}_k = \tilde{g}(kT)$ to represent the exact N samples $G_s(n/NT) = \tilde{G}(n/NT)$ of the periodic function $G_s(f)$. This means that the discrete transform approximation of $G_s(f)$ contains no error, but note that we have only the discrete N samples of $G_s(n/NT)$ instead of the continuous $G_s(f)$. This apparent lack of error for $G_s(n/NT)$ is not without a price, since in the discrete transform (4.3) we use the (theoretically) *aliased coefficients* $\bar{c}_k = \tilde{g}(kT)$ with aliasing error $\bar{c}_k - c_k$ for the coefficients c_k of (4.8).

To reduce this aliasing error in c_k we must increase the number N of the discrete samples of $\bar{c}_k$ to minimize the overlapping between their extensions as the cause of aliasing. In this sense of having to increase N to reduce the aliasing in c_k, we are also improving the resolution of the samples $G_s(n/NT)$. Still, it is that bit of inaccuracy, like the aliasing error in $\overline{c}_k$, that keeps $G_s(n/NT)$ from being exact. Otherwise, we either have the infinite sequence c_k for an exact $G_s(f)$ or must admit the wiggles as the consequence of the truncation to only a finite number of coefficients of c_k.

The main comparison here is that the truncated Fourier series uses an exact but finite number of coefficients c_k, which results in the wiggles of the truncation error for $G_s(f)$, while the discrete transform approximation of $G_s(f)$ gives its exact sample values by using a finite number of aliased coefficients $\overline{c}_k$. This makes clear the presence of both the truncation and the aliasing error for the general (nonperiodic) Fourier transform approximation by discrete transforms. In this case, the discretization, or sampling, for $g(t)$ would cause aliasing in $G(f)$, as it renders $G(f)$ periodic, hence the aliasing due to the overlapping of this periodic extension (see Fig. 4.6). In turn, the discretization of the aliased G will again cause aliasing to the infinite discrete samples of $g(t)$, as it renders $g(t)$ periodic (see Fig. 4.7).

This is the theoretical picture for the discrete transforms as pairs; in practice, when we approximate a Fourier series $G_s(f)$, we start with the exact N Fourier coefficients c_k and not the theoretical aliased coefficients $\overline{c}_k$. Hence $G_s(f)$ will suffer a truncation error. The same thing happens in approximating the Fourier transform of $g(t)$ as we start with the exact N samples of $g(t)$, which will produce a truncation error in G in addition to its aliasing error.

Since the truncation error, or windowing effect, is common to all Fourier transforms whenever the transformed function is affected by a sharp discontinuity, we will illustrate it here for the truncation of the infinite Fourier integrals. Our main attention will be directed to the convolution product,

$$G_b(f) = G(f) * \frac{\sin 2\pi b f}{\pi f} = \int_{-\infty}^{\infty} \frac{\sin 2\pi b(f-x)}{\pi(f-x)} G(x) \, dx \qquad (4.112)$$

as the transform of $p_b(t)g(t)$, the result of passing $g(t)$ through the abrupt gate function window $p_b(t)$. Figure 4.17a illustrates the gate function and its transform. Figure 4.17b gives an "almost" time-limited function $g(t)$ and its corresponding "almost" bandlimited $G(f)$. Figure 4.17c gives the convolution product $G_b(f) = G(f) * (\sin 2\pi b f)/\pi f$, as the effect on $G(f)$ because of the truncation of $g(t)$ to $p_b(t)g(t)$. Figure 4.17c clearly indicates the windowing effect on $G(f)$ and its usual wiggles, or small peaks (sidelobes), on both sides of the main peak of $G_b(f)$. We must also note the widening of $G_b(f)$ compared to the exact narrower peak of $G(f)$. This is also the result of the above process of convolving $G(f)$ with other than an impulse, which enlarges its domain. It is these small *sidelobes*, or small peaks of Fig. 4.17c, that we would like to investigate further. We wish to find an optimal window to minimize them. Figures 4.18 and 4.19 show the decrease of this windowing effect as we increase b, the width of the gate function window, until we almost reach the time-limited approximation of $G(f)$, where the wiggles practically disappear as shown in Fig. 4.19c. If we compare $G(f)$ with $G_b(f)$ in Fig. 4.19, we note that even after the wiggles disappear, the $G_b(f)$ peak is still *wider* than that of $G(f)$. This, as we mentioned earlier, is an inherited consequence of convolving $G(f)$ with a function other than the impulse function, which enlarges its domain. It is also the type of error that is usually very hard to get rid of. Even when we find optimal windows that practically eliminate the sidelobes, the widening effect will still remain to some extent.

We can see from the above convolution product and its illustrations in Figs. 4.17c and 4.18c that for a given width b of the truncation window, it is very hard to extract the original function $G(f)$ from this convolution

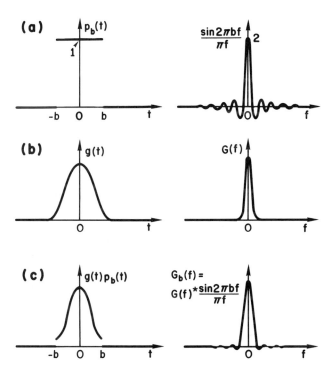

FIG. 4.18   Wider truncation window for lower windowing effect.

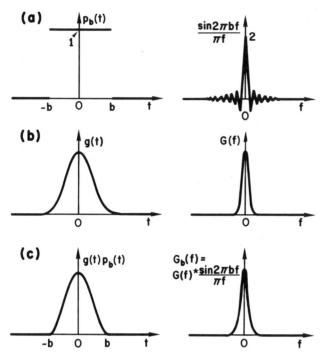

FIG. 4.19 Wide enough truncation window for tolerable windowing effect; note that $G_b(f)$ is still different from $G(f)$.

product $G_b(f)$, especially when we have already decided a fixed time limit b as our best estimate because of some practical considerations. Thus, we reiterate the need for an optimal window to reduce truncation, or windowing error, as signified by the widening of the function and the appearance of sidelobes around its main peak (Fig. 4.17c).

This windowing effect is called by different names for different situations. For example, in the case of a truncated Fourier series, the ripple effect is usually called the *Gibbs effect* or the *windowing effect*, where the window being used is the gate function. The Gibbs phenomenon is usually associated with the difficulty of the Fourier analysis representation near sudden discontinuities. Such an error is common to all types of Fourier transforms (integral, finite, and series) whenever sharp discontinuities are encountered for the transformed function.

As a very simple but illustrative example, we consider the periodic function $g(t) = \cos 2\pi f_0 t$, which can be looked at as a one-term Fourier cosine series on $(-1/2f_0, 1/2f_0)$,

$$g(t) = g_s(t) = \cos 2\pi f_0 t = \tfrac{1}{2}[e^{i2\pi f_0 t} + e^{-i2\pi f_0 t}] \tag{4.112a}$$

The Fourier coefficients, or transform, of this function are impulses $\tfrac{1}{2}[\delta(f+f_0)+\delta(f-f_0)]$ at $\mp f_0$ as illustrated in Fig. 4.20a. The period for this pure cosine function is $1/f_0$. We may add that this period is a natural one in the sense that the above function $\cos 2\pi f_0 t$ is its own periodic extension with the natural period $1/f_0$. The effect of truncating this periodic function

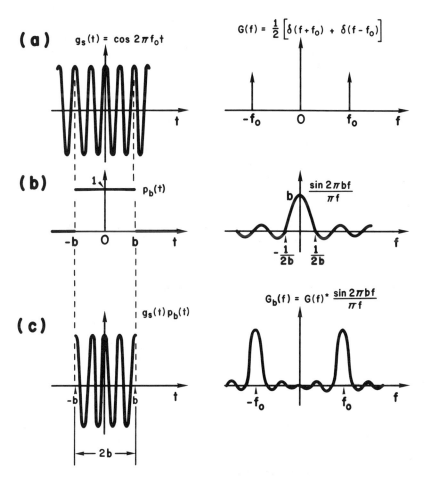

FIG. 4.20   The truncation of a periodic (analytic) function at $b = m(1/f_0)$, an integer multiple of its natural period $1/f_0$.

by the abrupt gate function would result in its Fourier transform as the convolution product,

$$G_b(f) = G(f) * \frac{\sin 2\pi b f}{\pi f} = \frac{1}{2}[\delta(f + f_0) + \delta(f - f_0)] * \frac{\sin 2\pi b f}{\pi f}$$

$$= \frac{1}{2}\left[\frac{\sin 2\pi b(f + f_0)}{\pi(f + f_0)} + \frac{\sin 2\pi b(f - f_0)}{\pi(f - f_0)}\right] \tag{4.113}$$

instead of the two impulses $\frac{1}{2}[\delta(f + f_0) + \delta(f - f_0)]$. This is illustrated in Fig. 4.20c with all the extra frequencies of $(\sin 2\pi b f)/\pi f$ affecting the wide peaks, especially around $\mp f_0$. This should illustrate and isolate very clearly the windowing effect with its wide peaks and sidelobes, which is also called the *leakage error* as discussed further in the next section.

It is important to note that the windowing (or truncation) is done on the function that has the Fourier representation and not on its transform. Such windowing introduces sharp discontinuities in $g_s(t)$ at $t = \mp b$, which will put us face to face with the Gibbs phenomenon *in case this discontinuity remains in the periodic extension of $g_s(t)$.* In the particular case of Fig. 4.20c, we took $b = m(1/f_0)$ an integer multiple of the natural period $1/f_0$ of $g_s(t)$, which makes the periodic extension continuous (see Fig. 4.21a). In case m is not an integer, such as $m = 3.5$ in Fig. 4.21b, we expect jump discontinuities at $b = \mp 3.5n$, n an integer, in the periodic extension which causes a Gibbs phenomenon near these discontinuities, a topic to be discussed after the next section. The periodic extension of the truncated $g_s(t)p_b(t)$ would correspond to discretizing the transform $G_b(f)$ (Figs. 4.20c, 4.21b), where we expect an infinite number of peaks at $\mp k(1/2b)$, k an integer, including the two main peaks at $f = \mp f_0$. In the special case of $b = m(1/f_0)$, m an integer, only the two impulses at $\mp f_0$ will survive as the other samples (coefficients) fall at the zeros of $G_b(f)$ as seen from (4.113), (4.114)–(4.114a) and Fig. 4.20c. Hence we recover the exact representation for the natural (periodic) extension of $\cos 2\pi f_0 t$.

F. Leakage Error

As we mentioned earlier, the appearance of wide peaks instead of the impulses is the consequence of convolving the latter with functions other than impulses. The appearance of the sidelobes on the sides of the main peaks is due to the slowly dying oscillations of $(\sin 2\pi b(f \pm f_0))/\pi(f \pm f_0)$, which are illustrated in Fig. 4.20c. The appearance or leaking of these small peaks to the sides of the desired main peaks earns them the name *sidelobes* or *leakage error*.

In the special case of the discrete transform (4.4), we sample the distorted transform $G_b(f)$ in (4.113) at $f_n = n/NT$, and, as in (4.3), we

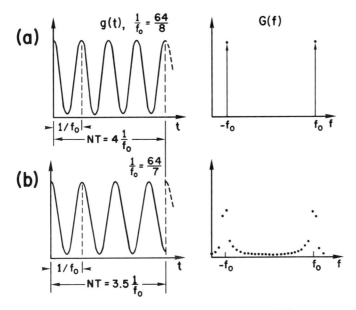

FIG. 4.21a,b The windowing effect where (a) NT, the period of the discrete transform, is equal to an integer multiple of the natural period $NT = 4(1/f_0) = 32$ and (b) NT is not equal to an integer multiple of the natural period, $NT = 3.5(1/f_0) = 32$, $1/f_0 = 64/7$.

use a window $p_b(t) = p_{NT}(t)$ for $\bar{g}(kT)$ to have

$$\frac{1}{2}\left[\frac{\sin 2\pi NT(n/NT + f_0)}{\pi(n/NT + f_0)} + \frac{\sin 2\pi NT(n/NT - f_0)}{\pi(n/NT - f_0)}\right]$$

$$= \frac{1}{2}\left[\frac{\sin 2\pi(n + NTf_0)}{\pi(n + NTf_0)} + \frac{\sin 2\pi(n - NTf_0)}{\pi(n - NTf_0)}\right] \qquad (4.114)$$

If we choose the period of the discrete transform $NT = m/f_0$, that is, an integer multiple of the period $1/f_0$ of $\cos 2\pi f_0 t$, then all values (samples) of (4.114) would vanish except those two at $f_m = m/NT = f_0$ and, $f_{-m} = -m/NT = -f_0$,

$$\frac{1}{2}\left[\frac{\sin 2\pi(n + m)}{\pi(n + m)} + \frac{\sin 2\pi(n - m)}{\pi(n - m)}\right] = \begin{cases} 0, & n \neq \mp m, \\ & \quad n, m \text{ integers} \\ 1, & n = \mp m \end{cases}$$

$$(4.114a)$$

This means that the discrete Fourier transform, as a finite sum, could escape the inherited windowing effect of the finite Fourier series if we choose the discrete transform period to be equal to the approximated function's period. The difficulty we usually face is that, in general, we are not sure of the exact period, hence the windowing effect for the discrete transform approximation of more general periodic functions than $\cos 2\pi f_0 t$. Figure 4.21a shows the discrete Fourier transform of the above function with $NT = 4(1/f_0)$, where we notice no windowing effect for the two impulses. Figure 4.21b shows the windowing effect when NT is not an integer multiple of the natural period $1/f_0$, that is, $NT = 3.5(1/f_0)$. In Fig. 4.21b we notice the sharp discontinuity in stopping this smooth periodic function at $NT = 3.5/f_0$, which is not an integer multiple of the cosine function's natural period $1/f_0$. This sharp discontinuity is due to the windowing of the very smooth (analytic) and periodic $g_s(t)$ by the gate function $p_{NT}(t)$, instead of its natural multiple-period window $p_{m/f_0}(t)$. Figure 4.21c shows the same windowing effect on the two Fourier coefficients of the periodic function $g(t) = \cos t$ of period 2π in Fig. 4.21c(i) when this function is truncated by a window of width 20 which is not an integer multiple of the natural period 2π. We notice here clearly the widening of the two peaks of the Fourier coefficients at $k = \mp 3$ as shown in Fig. 4.21c(ii) by the appearance of pulses around $k = \mp 3$ that represent the sidelobes or the leakage error. The relation between the two names is that the original frequency impulses have *leaked* through the *sidelobes* of $(\sin f)/f$, the transform of a square truncation window compared to that of the impulse $\delta(f)$ before the truncation or windowing. In this particular example, we could have avoided this error by choosing a bit narrower window of width 6π instead of 20.

In the case of the more general smooth periodic function $g_s(t)$ of the Fourier series, with period b, we expect a leakage error for each c_k when we use NT different from the period b. Hence, again, there is a leakage error for c_k because of the sudden discontinuity caused in its transform $g_s(t)$ by not using its natural period b. This error is not so easy to eliminate, since we may have discrete data that we would like to Fourier-transform, with only an estimate of its period. In this case, we have to use a period NT, admit the leakage error for the transform, and be prepared to have some method of minimizing this error for a reasonably accurate transform. As we shall see later, in part I of this section, this will require a more *optimal truncation window* than the very abrupt one of the gate function $p_{NT}(t)$ (see Figs. 4.31 and 4.33 for the optimal *Hanning, Gaussian,* and other truncation windows).

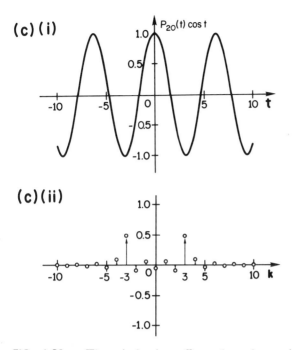

FIG. 4.21c The windowing effect when the period of the discrete transform $NT = b = 20$ is not equal to $n2\pi$, an integer multiple of the period of the function: (i) $g(t) = (\cos t)p_{10}(t)$, $NT = b = 20 \neq n \cdot 2\pi$; (ii) the windowed Fourier coefficients, $NT = b = 20 \neq n \cdot 2\pi$.

G. The Gibbs Phenomenon—Origin, Derivation, and Possible Remedies

All our discussions up to now, regarding the sidelobes, leakage, and Gibbs effect, seem to have ignored the very first familiar Gibbs phenomenon. This is the one noticed in the Fourier series approximation of periodic functions with jump discontinuities inside and/or at the ends of the period like that of Fig. 4.22a for the square wave function with period 2π, and Fig. 4.22b (same as Fig. 3.5b) for $f(x) = 3x$ with period 4.

We shall first attempt to give a heuristic explanation of the noticeable sidelobes on the sides of the jump discontinuities in Fig. 4.22. As compared to the naturally periodic smooth function $\cos 2\pi f_0 t$ of Fig. 4.21a, the function $f(x) = 3x$, for example, though smooth, is not periodic with finite period. Forcing it to be periodic, with period 4, produces the jump

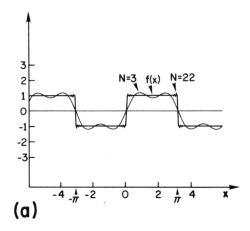

(a)

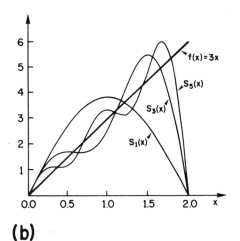

(b)

FIG. 4.22   (a) The Gibbs phenomenon for Fourier series approximation of a square wave (periodic with period 2π, and with jump discontinuities in the interior and at the end points).   (b) The Gibbs phenomena for Fourier series approximation of the periodic functions $f(x) = 3$, $-2 < x < 2$, at the jump discontinuity $x = 2$.   (c) The periodic extension $\bar{f}(x)$ of $f(x)$ in part (b), period 4 and jump discontinuities at $x = \mp 2(2n + 1)$, $n = 0$, 1, 2, ... (the same as Fig. 3.5a).

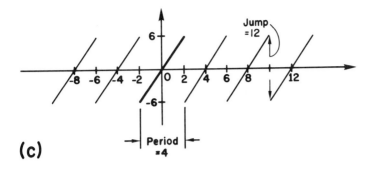

(c)

discontinuities at $t = \mp 2(2n + 1)$, $n = 0, 1, 2, \ldots$, in this particular example as shown in Fig. 4.22c (same as Fig. 3.5a); see also Figs. 3.6 and 3.7 for similar cases. For other examples, the periodic extension may be continuous but not necessarily smooth. In particular, the even functions would have no jump discontinuities in their periodic extension (e.g., $G_s(f) = e^{-|f|}$ in Fig. 4.1a). Figure 4.22a illustrates the Gibbs phenomenon for a square wave near the jump discontinuities located at the ends of the period $\mp\pi$ as well as inside the period at $x = 0$, where we note the overshoots even with $N = 22$ terms in the partial sum of the Fourier series (see also Fig. 4.30a,b and Figs. 3.6 and 3.7).

Intuitively, we may explain this familiar type of Gibbs phenomenon as the result of the difficulty facing the smooth orthogonal functions of the Fourier series to approximate around the discontinuity of the given function. It seems here that we also should attempt to offer some explanation in line with the windowing effect. The main stubborn character of the Gibbs phenomenon is that it represents the difficulty of the Nth partial sum of the Fourier series in converging to the function around the point of discontinuity, even when we take N very large. We may look at this effect as a windowing of the Fourier coefficients $c_k = g(kT)$ by $p_{NT}(kT)$ to give $p_{NT}(kT)g(kT)$, which corresponds in the frequency space to convolving the Fourier series $G_s(f)$ with $(\sin 2\pi NTf)/\pi f$, which as we expect should show wiggles and widening of $G_s(f)$ as shown in Fig. 4.23 and Fig. 4.17. In these terms, we are also to explain why this Gibbs effect is very dependent on the size of the jump discontinuity and is less sensitive to the size of N.

For simplicity, let us take the discontinuity at $f = 0$, for example, the case of the square wave in Fig. 4.24a,

$$G(f) = \begin{cases} -A, & -b < f < 0 \\ A, & 0 < f < b \end{cases}$$

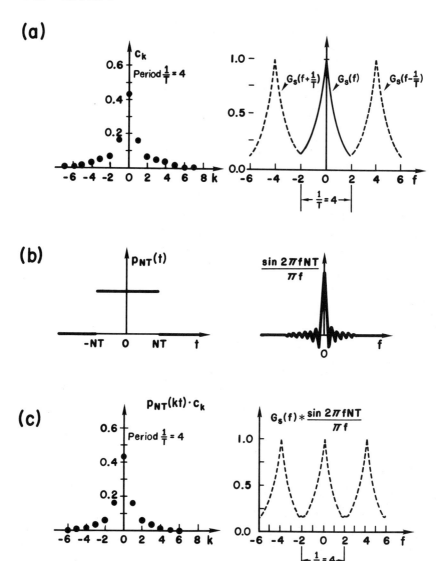

FIG. 4.23 The windowing effect of the Fourier coefficients.

(a)

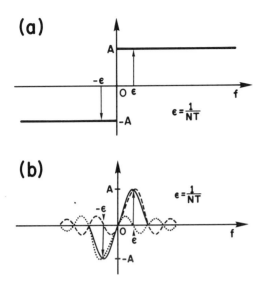

(b)

FIG. 4.24 The Gibbs phenomenon near a jump discontinuity at $f = 0$.

and concentrate only on the two values (samples) at $-\epsilon = -1/NT$ and $\epsilon = 1/NT$ with their respective (opposite) magnitudes $-A$ and A as in Fig. 4.24b. The results of convolving each of these (opposite) samples at $-\epsilon$ and $+\epsilon$ with $(\sin 2\pi NTf)/\pi f$ are shown as dotted lines in Fig. 4.24b. The sum of these convolution products, as the windowing effect of the original two samples of Fig. 4.24a, is shown (solid line) in Fig. 4.24b, where it indicates clearly the preservation of the discontinuity of the original function at $f = 0$ in addition to a distortion in the direction of increasing the jump discontinuity. The two dotted curves are adding in the interval $(-\epsilon, \epsilon)$ for these two samples of opposite sign. We show (Fig. 4.25) in the same way that if the two samples have the same sign— i.e., no jump at $f = 0$—the two dotted curves will subtract, resulting in the two convolved (or windowed) samples being very close to representing no jump trouble. For this continuous case, we consider the gate function (Fig. 4.25a)

$$p_b(f) = \begin{cases} A, & -b < f < b \\ 0, & |f| > b \end{cases}$$

with the same value A for the two samples at $-\epsilon$ and ϵ as shown in Fig. 4.25b. The result of convolving these samples with $(\sin 2\pi NTf)/\pi f$ is shown (dotted lines) in Figs. 4.25b. The sum of these two convolved or

(a)

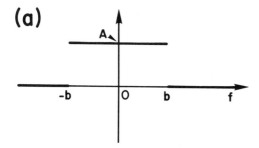

(b)

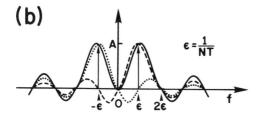

FIG. 4.25 Absence of the Gibbs phenomenon near $f = 0$ when the function is continuous there.

windowed samples is also shown (solid lines) in Fig. 4.25b, where we note that for large enough NT there seems to be no discontinuity-type trouble in the interval $(-1/NT, 1/NT)$ and the two samples would approximate a continuous function in this neighborhood. This is in contrast to the two opposite samples of Fig. 4.24, which resulted in obvious preservation of the discontinuity even if we make N large. Such a preservation and an apparent increase of the jump discontinuity of Fig. 4.24 in this very special case may give us a glimpse at the Gibbs effect. In the following, we will present a very basic treatment of the Gibbs phenomenon associated with integral approximations of discontinuous functions. This is to be followed by a parallel treatment of the Gibbs phenomenon for the Fourier series approximation of discontinuous periodic functions.

The simplest example of a function with jump discontinuity in the interior of its domain is the *signum function* sgn(t), (Fig. 4.26a)

$$\text{sgn}(t) = \frac{t}{|t|} = \begin{cases} 1, & t > 0 \\ -1, & t < 0 \end{cases} \tag{4.115}$$

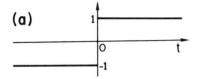

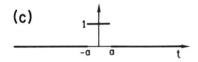

FIG. 4.26 Basic functions with jump discontinuities in the interior of their domains.

This function is very basic to other functions with such discontinuities, since most of them can be expressed in terms of it, or in combination with continuous functions. For example, the *unit step function* $u(t)$ (Fig. 4.26b),

$$u(t) = \begin{cases} 1, & t > 0 \\ 0, & t < 0 \end{cases} \tag{4.116}$$

has a jump discontinuity at the interior point $t = 0$, where it can be expressed in terms of the signum function as

$$u(t) = \tfrac{1}{2} + \tfrac{1}{2}\,\mathrm{sgn}(t) \tag{4.117}$$

The very important *gate function* $p_a(t)$,

$$p_a(t) = \begin{cases} 1, & |t| < a \\ 0, & |t| > a \end{cases} \tag{4.118}$$

with its jump discontinuities at $t \mp a$, can also be expressed in terms of the signum function as (Fig. 4.26c)

$$p_a(t) = \tfrac{1}{2}[\mathrm{sgn}(t + a) - \mathrm{sgn}(t - a)] \tag{4.119}$$

A variety of functions with such jump discontinuities can be expressed with the aid of the above functions (4.115), (4.116), and (4.118). For example a function $f(t)$ with jump discontinuity of size $J = f(t_{0+}) - f(t_{0-})$

at t_0 can be constructed from a continuous function $g_c(t)$ and the unit step function in (4.117) as

$$f(t) = g_c(t) + Ju(t - t_0)$$

$$= g_c(t) + [f(t_{0+}) - g_c(t_{0-})]u(t - t_0) \tag{4.120}$$

as in Fig. 4.27. We note how the part of $f(t)$ for $t > t_0$ is made by raising up the part of $g_c(t)$, $t > t_0$, by the jump $J = f(t_{0+}) - f(t_{0-})$.

In the following discussion, we will concentrate on this basic signum function and the Gibbs phenomenon associated with its truncated Fourier (integral) representation. For the Gibbs phenomenon of the truncated Fourier series we will discuss the signum function analog, i.e., the periodic (period 2π) square wave (Fig. 4.22a)

$$f(t) = \begin{cases} 1, & 0 < t < \pi \\ -1, & -\pi < T < 0 \end{cases}, \ f(t \mp 2n\pi) = f(t) \tag{4.121}$$

most of the detailed illustrations of the Gibbs phenomenon and its possible remedies are done for this square wave function (Fig. 4.30a–g).

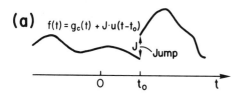

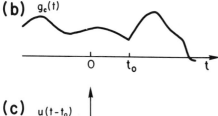

(c) $u(t-t_0)$

FIG. 4.27 Functions with a jump discontinuity expressed with the aid of the signum function. (a) $f(t) = g_c(t) + Ju(t - t_0)$. (b) $g_c(t)$. (c) $u(t - t_0)$.

We note again that after such analysis we should be clear on the windowing effect of the gate function as a consequence of its jump discontinuities at $t = \mp a$.

Let us recall the Fourier integral representation of the gate function $p_a(t)$,

$$p_a(t) = \int_{-\infty}^{\infty} \frac{\sin 2\pi af}{\pi f} e^{j2\pi ft} \, df \tag{4.122}$$

From this we obtain the Fourier transform pair,

$$\int_{-\infty}^{\infty} \delta(f) e^{j2\pi ft} \, df = \mathcal{F}\{\delta(f)\} = 1, \qquad -\infty < t < \infty \tag{4.123}$$

only after defining the *Dirac delta function* $\delta(f)$ as a limit, in the following (general) sense:

$$\lim_{a\to\infty} \int_{-\infty}^{\infty} \frac{\sin 2\pi af}{\pi f} e^{j2\pi ft} \, df = \lim_{a\to\infty} p_a(t) = 1 \tag{4.124}$$

as was done in Section 2.4.1B. This general sense supports what is usually written

$$\lim_{a\to\infty} \frac{\sin 2\pi af}{\pi f} = \delta(f) \tag{4.125}$$

whose justification can come only from comparing (4.123) and (4.122). That is, the constant function $f(t) = 1 = \lim_{a\to\infty} p_a(t)$ for all t, without the jump discontinuities of $p_a(t)$, can have a Fourier integral representation (4.122) only with the aid of the Dirac delta function. $\delta(f)$ is defined in the general sense of distributions which are not the usual functions we work with. On the other hand, the gate function $p_a(t)$ has its Fourier integral representation (4.122) in terms of the usual functions, which can be validated directly by computing its transform $(\sin 2\pi af)/\pi f$ or by direct complex integration. However, in actual numerical computations of this infinite integral representation (4.122), the moment we truncate its infinite limits to $-A$, A, with a gate function $p_A(f)$, we expect a Gibbs phenomenon for the resulting approximation of $p_a(t)$ near its jump discontinuities at $t = \mp a$. This is a windowing effect on $p_a(t)$ with a window $p_A(f)$, which results in convolving $p_a(t)$ with $(\sin 2\pi At)/\pi t$, the transform of $p_A(f)$, to give $p_{a,A}(t)$,

$$\begin{aligned} p_{a,A}(t) &= p_a(t) * \frac{\sin 2\pi At}{\pi t} \\ &= \int_{-\infty}^{\infty} \frac{\sin 2\pi A(t-\tau)}{\pi(t-\tau)} p_a(\tau) \, d\tau \end{aligned} \tag{4.126}$$

This result represents our first exposure to convolving a continuous function $(\sin 2\pi At)/\pi t$ with a function $p_a(t)$ that has jump discontinuities. It is such jump discontinuities that will contribute an extra term $\gamma(t)$ as the source of the Gibbs phenomenon. Such a source of error $\gamma(t)$ can be well understood when isolated in the very basic representation of the *signum function* $\mathrm{sgn}(t)$, after which it is a simple matter to translate our understanding to the case of $p_a(t)$.

We will use the integral

$$\int_0^\infty \frac{\sin x}{x}\, dx = \frac{\pi}{2} \tag{4.127}$$

which can be computed by direct complex integration and which makes the basis for the Fourier representation of $\mathrm{sgn}(t)$. If we let $x = 2\pi ft$ in the above integral, we have

$$\frac{2}{\pi}\int_0^\infty \frac{\sin 2\pi ft}{f}\, df = \begin{cases} 1, & t > 0 \\ -1, & t < 0 \end{cases} \equiv \mathrm{sgn}(t) \tag{4.128}$$

as the Fourier (sine) integral representation of $\mathrm{sgn}(t)$. The two different branches of $\mathrm{sgn}(t)$ in (4.128) are easily obtained from (4.127), where the integrand is positive or negative according to $t > 0$ or $t < 0$.

If we approximate the integral (4.128) by truncating to B in the upper limit of integration, we obtain the following bandlimited function (bandlimited to B), which we denote by $\mathrm{sgn}_B(t)$, as an approximation to $\mathrm{sgn}(t)$:

$$\mathrm{sgn}_B(t) = \frac{2}{\pi}\int_0^B \frac{\sin 2\pi ft}{f}\, df$$

$$= \frac{2}{\pi}\int_0^{2\pi Bt} \frac{\sin x}{x}\, dx = \frac{2}{\pi}\,\mathrm{Si}(2\pi Bt) \tag{4.129}$$

after letting $x = 2\pi ft$, and recognizing the last truncated integral as the known and well tabulated *sine integral* $\mathrm{Si}(2\pi Bt)$ (see Fig. 4.28 for $\mathrm{Si}(t)$).

The study of the behavior of $\mathrm{Si}(2\pi Bt)$, as the truncated approximation of $\mathrm{sgn}(t)$, would give us the first glimpse at the actual reason behind the Gibbs phenomenon of the truncated Fourier approximation of functions with jump discontinuities, especially those in the interior of their domain.

In parallel with (4.112), the first sine integral in (4.129) can be looked at as a representation for the output of passing (our first example) of a discontinuous function $\mathrm{sgn}(t)$ through an ideal low-pass filter with $p_B(t)$ window,

$$\mathrm{sgn}_B(t) = \frac{2}{\pi}\,\mathrm{Si}(2\pi Bt) = \frac{1}{\pi}\int_{-\infty}^\infty \frac{\sin 2\pi B(t-\tau)}{\pi(t-\tau)}\,\mathrm{sgn}(\tau)\, d\tau \tag{4.130}$$

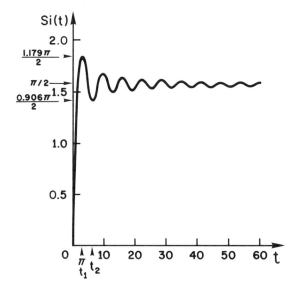

FIG. 4.28   The sine integral

$$\text{Si}(t) = \int_0^t \frac{\sin x}{x}\, dx$$

and the essence of the Gibbs phenomenon near $t = 0$. Note that $\text{sgn}_B(t) = (2/\pi)\,\text{Si}(2\pi Bt)$ in (4.129).

with $G(t) = \text{sgn}(t)$ when compared with (4.112). In (4.112), $G(t)$ is assumed to be continuous, which allows $\lim_{b \to \infty} G_b(t) = G(t)$. In other words, $G_b(t)$ can be brought as close as we desire to the original continuous $G(t)$ by increasing b, the width of the truncating gate window, as seen by comparing Figs. 4.19c and 4.17c. In contrast, as seen in Fig. 4.28, the sine integral representation in $\text{sgn}_B(t)$ of (4.129) cannot be brought closer to $\text{sgn}(t)$ by increasing B. Indeed, as we shall illustrate shortly, it turns out that by increasing B we will only change the time scale in Fig. 4.28, which will bring the peaks together without changing their relative magnitude. In fact, the size of the first maximum above $\pi/2$ of the Gibbs phenomenon in Fig. 4.28 is about 9% of the jump size $J = \pi$ at $t = 0$ regardless of how large we take B to be. To derive and illustrate this phenomenon, we will find the locations and magnitudes of the first maximum and minimum of $\text{Si}(2\pi Bt)$. If we take the first derivative of

Si($2\pi Bt$) in (4.129) and equate it to zero, we have

$$\frac{d}{d2\pi Bt} \operatorname{Si}(2\pi Bt) = \frac{\sin 2\pi Bt}{2\pi Bt} = 0$$

which has zeros at $t_n = n\pi/2\pi B = n/2B$, the locations of the possible maxima and minima of $\operatorname{sgn}_B(t)$. The first maximum is at $t_1 = 1/2B$, and its magnitude can be found by substituting $t = t_1 = 1/2B$ in (4.129) to obtain $(2/\pi)\operatorname{Si}(\pi) = (2/\pi)(\pi/2)(1.17898) = 1.17898$, where $\operatorname{Si}(\pi) = (\pi/2)$ (1.17898) is found from mathematical tables (see Fig. 4.30). Since the size of the jump at $t = 0$ is 2, this maximum would represent 0.17898, which is close to 9% of the jump size as an overshoot from $\operatorname{sgn}(1/2B) = 1$. We must note here that with the location of the extrema $t_n = n/2B$ in (4.129), their magnitudes $(2/\pi)\operatorname{Si}(n\pi)$ for $\operatorname{sgn}_B(t)$ *are independent of B*. This should make clear that the Gibbs phenomenon at a jump discontinuity of any kind cannot be remedied by merely increasing B, the width of the truncating gate function window used for $\operatorname{sgn}_B(t)$ in (4.129).

The next extremum is a minimum at $t_2 = 2/2B$ with a value from (4.129) of

$$\frac{2}{\pi} \int_0^{2\pi} \frac{\sin x}{x}\, dx = \frac{2}{\pi} \operatorname{Si}(2\pi) = \frac{2}{\pi}\frac{\pi}{2}(0.906)$$

which represents $(1 - 0.906)/2 = 0.047$, or about 4.7% of the jump discontinuity (of 2) at $t = 0$, as an undershoot from $\operatorname{sgn}(1/B) = 1$ (see Fig. 4.28).

As we mentioned earlier, the analysis of the Gibbs phenomenon for more general functions, with jump discontinuities in the interior of their domain, will follow in the same way. Such functions can be constructed from the signum function, or its combinations $u(t)$ and $p_a(t)$, by adding them to a continuous function (see Figs. 4.27, 4.26). Thus, in our attempt to find a possible remedy for the Gibbs phenomenon, it is sufficient to stay with the more basic signum function.

The σ-averaging as a Possible Remedy

We have already shown that there is no way for the Gibbs effect to disappear for the (truncated) Fourier representation $\operatorname{sgn}_B(t)$ of the signum function with its jump discontinuity at zero. So the only alternative is to replace $\operatorname{sgn}(t)$ by a *continuous* function which approaches $\operatorname{sgn}(t)$ as $B \to \infty$. A practical justification for such a change of the function is that we may look at two different functions (signals) with the same bandlimit B and may call them equivalent by insisting that their transforms on $(-B, B)$ carry the same energy $\int_{-B}^{B} |G(f)|^2\, df$. The first (an obvious) choice is to replace the part of $\operatorname{sgn}(t)$ on $(-t_1, t_1) = (-1/2B, 1/2B)$ around $t = 0$ in Fig. 4.29 by a straight line, as indicated by $\operatorname{sgn}_{\sigma_1}(t)$ in that figure. The

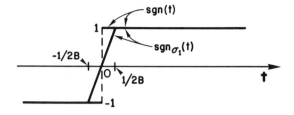

FIG. 4.29 A possible remedy for the Gibbs phenomenon—approximating $\text{sgn}(t)$ by a continuous function $\text{sgn}_{\sigma_1}(t)$.

new function $\text{sgn}_{\sigma_1}(t)$ can be shown as the result of convolving $\text{sgn}(t)$ with the gate function $Bp_{1/2B}(t)$,

$$\text{sgn}_{\sigma_1}(t) = \text{sgn}(t) * Bp_{1/2B}(t) \tag{4.131a}$$

$$= B \int_{-\infty}^{\infty} p_{1/2B}(t - \tau)\,\text{sgn}\,\tau\,d\tau \tag{4.131b}$$

$$= \frac{1}{2(1/2B)} \int_{t-1/2B}^{t+1/2B} \text{sgn}\,\tau\,d\tau$$

$$= \begin{cases} 1, & t > \dfrac{1}{2B} \\[2mm] -1, & t < \dfrac{1}{2B} \\[2mm] 2Bt, & -\dfrac{1}{2B} < t < \dfrac{1}{2B} \end{cases} \tag{4.131c}$$

after noting that $p_{1/2B}(t - \tau) = 0$ for $|t - \tau| > 1/2B$ (or for $\tau < t - 1/2B$ and $\tau > t + 1/2B$), and performing the last integration.

We can see from the last integral in (4.131b) that $\text{sgn}_{\sigma_1}(t)$ is also the result of averaging $\text{sgn}(t)$, with constant weight 1, over intervals of width $2(1/2B) = 1/B$, where $1/2B$ is the location of the first overshoot (maximum) of the Gibbs phenomenon near the jump discontinuity at $t = 0$. In the following we will show how such an averaging process as $\text{sgn}_{\sigma_1}(t)$ results in a reduced Gibbs effect.

The Fourier transform of this averaged function $\text{sgn}_{\sigma_1}(t)$ is the following product of $1/j\pi f$, the transform of $\text{sgn}(t)$, and $B(\sin(\pi/B)f)/\pi f$, the transform of $Bp_{1/2B}(t)$,

$$F_{\sigma_1}(f) = \frac{1}{j\pi f} B \frac{\sin(\pi/B)f}{\pi f} \tag{4.132}$$

$$\text{sgn}_{\sigma_1}(t) = \int_{-\infty}^{\infty} \frac{1}{j\pi f} B \frac{\sin(\pi/B)f}{\pi f} e^{j2\pi ft} \, df \tag{4.133}$$

where we have used $\mathcal{F}\{\text{sgn}(t)\} = 1/j\pi f$ from (4.128), $j = \sqrt{-1}$.

Now if the truncation for this infinite integral representation of (the continuous) $\text{sgn}_{\sigma_1}(t)$ is done with a gate window $p_A(f)$, we have its bandlimited approximation

$$\begin{aligned}\text{sgn}_{\sigma_{1A}}(t) &= \int_{-A}^{A} \frac{1}{j\pi f} B \frac{\sin(\pi/B)f}{\pi f} e^{j2\pi ft} \, df \\ &= \int_{-\infty}^{\infty} \frac{\sin 2\pi A(t-\tau)}{\pi(t-\tau)} \text{sgn}_{\sigma_1}(\tau) \, d\tau \end{aligned} \tag{4.134}$$

In contrast to $\text{sgn}_B(t)$ in (4.129), where the input to the truncation gate window was the discontinuous $\text{sgn}(t)$, we have here its modification (averaged) as a continuous input $\text{sgn}_{\sigma_1}(t)$. This would allow $\text{sgn}_{\sigma_{1A}}$ to get as close as we wish to $\text{sgn}_{\sigma_1}(t)$ by increasing the window width A, hence a reduced Gibbs effect for this replacement $\text{sgn}_{\sigma_1}(t)$ of $\text{sgn}(t)$.

It is important to find the relation between $\text{sgn}_B(t)$ of (4.129), with its Gibbs phenomenon, and $\text{sgn}_{\sigma_{1A}}(t)$, the approximation to $\text{sgn}_{\sigma_1}(t)$, the continuous replacement of $\text{sgn}(t)$.

From the first integral of (4.134) we have

$$\text{sgn}_{\sigma_{1A}}(t) = 2 \int_0^A B \frac{\sin(\pi/B)f}{\pi f} \frac{\sin 2\pi ft}{\pi f} \, df \tag{4.135}$$

since the Fourier-transformed function in the first integral in (4.134) is an odd function. If we choose $A = B$ we can compare the above result with that of $\text{sgn}_B(t)$ in (4.129),

$$\text{sgn}_B(t) = \frac{2}{\pi} \int_0^{2\pi Bt} \frac{\sin x}{x} \, dx \tag{4.129}$$

With $A = B$ and $2\pi ft = x$ in (4.135) we have

$$\text{sgn}_{\sigma_{1B}}(t) = \frac{2}{\pi} \int_0^B \frac{\sin(\pi/B)f}{(1/B)f} \frac{\sin 2\pi ft}{\pi f} \, df \tag{4.136a}$$

$$= \frac{2}{\pi} \int_0^{2\pi Bt} \frac{\sin(1/2Bt)x}{(1/2Bt)x} \frac{\sin x}{x} \, dx \tag{4.136b}$$

If we compare (4.129) with (4.136b) we note that the above averaging or smoothing process for managing the Gibbs phenomenon of $\text{sgn}(t)$, as in (4.136b), introduces a decaying factor $(\sin(1/2Bt)x)/(1/2Bt)x$ inside the sine integral. This factor seems to have brought the sine integral to a balance close to $\text{sgn}(t)$. For example, at $t_1 = 1/2B$, where (4.130) has a maximum $\text{sgn}_B(1/2B) = (2/\pi)\,\text{Si}(\pi)$, we now have from (4.136b)

$$\text{sgn}_{\sigma_{1,B}}\left(\frac{1}{2B}\right) = \frac{2}{\pi}\int_0^\pi \left(\frac{\sin x}{x}\right)^2 dx \qquad (4.136b)$$

where the integrand decays much faster than the $(\sin x)/x$ of $\text{sgn}_B(t)$ in (4.129), hence yielding a lower value than that of $(2/\pi)\,\text{Si}(\pi)$.

The first integral (4.134) suggests that the averaging process seems to have made the Fourier transform of the original function tend more toward a bandlimited function (see the extra $1/f$ factor in the integrand). This is in general due to the convolution by $p_{1/2B}(t)$ in the time space, which enlarges the domain of the resulting function (see (4.131)) and, according to the uncertainty principle, causes a narrowing in the frequency space domain. This is in the direction of helping the truncation process, which is good as long as it does not change the main character of the original function, in particular the fast rise from 0 to 1 near $t = 0$. In the present case the averaging is done with the flat top $p_{1/2B}(t)$, which affected the original function only to close the discontinuity at $t = 0$, where it replaced the jump discontinuity by a line of slope $2B$.

It remains to see whether, in the light of the above analysis for the Gibbs phenomenon, we can suggest a further improvement to this averaging remedy. We precede such improvement by stating that the above averaging process moves the location of the extrema farther away from the discontinuity, as can be shown by differentiating $\text{sgn}_{\sigma_{1,B}}(t)$ in (4.136). (See (4.129).)

Higher Order σ-averaging

The most direct approach to further improvement is to repeat the averaging process, which should decrease the magnitude of the extrema. The n-times repeated averaging of $\text{sgn}(t)$ would amount to convolving it n times with $Bp_{1/2B}(t)$. This is equivalent to convolving $\text{sgn}(t)$ with the *hill function* $\phi_n(t)$ (B spline) of order n, since the latter is defined as the $n-1$ self-convolution of the gate function $p_{1/2B}(t)$. We could also look at this process as averaging with a weight which is a spline of order n instead of (the flat top) $B\phi_1(t) = Bp_{1/2B}(t)$ used in the above analysis. The convolution of $\text{sgn}(t)$ with $B\phi_n(t)$ would amount to multiplying inside its integral in (4.136a) by $\sigma_n = [\sin(\pi/B)f/(\pi/B)f]^n$ as the transform of $B^n\phi_n(t) = B^n(p_{1/2B} * \overset{n-1}{\cdots} * p_{1/2B})(t)$.

There is no doubt that such a self-truncating factor inside the integral would improve its convergence to give smoother results. However, the wider domain $(-n/2B, n/2B)$ of the higher-order hill function $\phi_n(t)$ would cause a slower rise for the approximate function near the jump discontinuity. This is not an extremely difficult problem, since it can be remedied by increasing B, which of course will increase the number of computations. The latter is not so big a price to pay when we are using the FFT. Figure 4.30a–g illustrates such a remedy for the most familiar case of the Gibbs phenomenon in the (truncated) Fourier series approximation of discontinuous functions such as the square wave in (4.137), where similar detailed analysis follows.

In the next section, we will treat the Gibbs phenomenon for the Fourier series approximation of (periodic) functions with jump discontinuities. The analysis will largely parallel the one above of the integrals, where the σ_n factor improvements are illustrated for the (periodic) square wave in Fig. 4.30a–g. In essence, the numerical computations show a larger rise time near $t = 0$, but much smoother approximation with σ_2 than that of σ_1 (Fig. 4.30d). With σ_6 the rise time is double that of σ_3 but the accuracy is orders of magnitude better when an 80-term series was used for both cases (Fig. 4.30g).

The Gibbs Phenomenon in Fourier Series Approximation

Now we turn to the usual Gibbs phenomenon of the truncated Fourier series for a periodic function with jump discontinuities. The example we take is that of the square wave of unit amplitude on $(-\pi, \pi)$, as illustrated in Fig. 4.30a,

$$f(t) = \begin{cases} 1, & 0 < t < \pi \\ -1, & -\pi < t < 0 \end{cases} \tag{4.137}$$

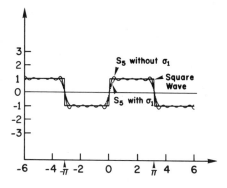

FIG. 4.30a The square wave, its Fourier series approximation and Gibbs phenomenon, and the σ_1 (averaging) remedy, $N = 5$.

whose Fourier sine series is

$$f(t) \sim \frac{4}{\pi} \sum_{n=1}^{\infty} \frac{1}{2n-1} \sin(2n-1)t \tag{4.138}$$

and whose Nth partial sum

$$S_N(t) = \frac{4}{\pi} \sum_{n=1}^{N} \frac{1}{2n-1} \sin(2n-1)t \tag{4.139}$$

is illustrated for $N = 5$ in Fig. 4.30a.

To find the locations of the extrema of the Gibbs phenomenon of $S_N(t)$ in (4.139), we will try to express the $S_N(t)$ as an integral and then follow the last section's analysis of the Fourier integral approximation of the signum function in (4.129).

Inside the partial sum $S_N(t)$ in (4.139) we can write

$$\frac{\sin(2n-1)t}{(2n-1)} = \int_0^t \cos(2n-1)x \, dx \tag{4.140}$$

to have

$$S_N(t) = \frac{4}{\pi} \sum_{n=0}^{N} \int_0^t \cos(2n-1)x \, dx \tag{4.141}$$

and after interchanging the integral and sum we have

$$S_N(t) = \frac{4}{\pi} \int_0^t \left[\sum_{n=0}^{N} \cos(2n-1)x \right] dx \tag{4.142}$$

But we can also show that the sum inside (4.142) is $(\sin 2Nx)/(2\sin x)$ by simply writing

$$2\sin x \sum_{n=1}^{N} \cos(2n-1)x = \sum_{n=1}^{N} 2\sin x \cos(2n-1)x$$

$$= \sum_{n=1}^{N} [\sin(2n-1+1)x + \sin(1-2n+1)x]$$

$$= \sum_{n=1}^{N} [\sin 2nx - \sin(2n-2)x]$$

$$= \sin 2x + \sin 4x - \sin 2x + \cdots$$

$$\quad + \sin 2Nx - \sin(2N-2)x$$

$$= \sin 2Nx \tag{4.143}$$

after using simple trigonometric identities that helped in the cancellation of all terms except $\sin 2Nx$. From (4.142) and (4.143), we have

$$S_N(t) = \frac{2}{\pi} \int_0^t \frac{\sin 2Nx}{\sin x}\, dx \qquad (4.144)$$

To search for the extrema of the Gibbs phenomenon, as we did for the truncated Fourier integrals in (4.129), we differentiate $S_N(t)$ with respect to t to have

$$\frac{dS_N(t)}{dt} = \frac{2}{\pi} \frac{\sin 2Nt}{\sin t}$$

which has its extrema at $t_k = k\pi/2N$ with the first maximum at $t_1 = \pi/2N$, as illustrated in Fig. 4.30b with $N = 10$.

The magnitude of the first maximum can be evaluated from (4.144) with $t_1 = \pi/2N$ as

$$S_N\left(\frac{\pi}{2N}\right) = \frac{2}{\pi} \int_0^{\pi/2N} \frac{\sin 2Nx}{\sin x}\, dx \qquad (4.145)$$

which can be computed numerically. If N is large enough, then we are integrating in (4.145) over a small interval $(0, \pi/2N)$ where x is small enough to allow the approximation of $\sin x \sim x$; thus the integral (4.145)

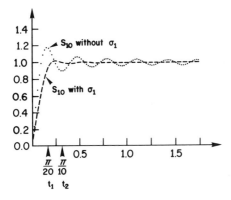

FIG. 4.30b The Gibbs phenomenon (of the square wave) with its (continuous) σ_1 (averaging) remedy ($N = 10$).

can be approximated by the following sine integral:

$$S_N\left(\frac{\pi}{2N}\right) \sim \frac{2}{\pi} \int_0^{\pi/2N} \frac{\sin 2Nx}{x}\, dx, \qquad N \text{ large}$$

$$= \frac{2}{\pi} \int_0^{\pi} \frac{\sin y}{y}\, dy = \frac{2}{\pi}\, \mathrm{Si}(\pi) = \frac{2}{\pi}\frac{\pi}{2}(1.17898) = 1.17898$$

as we found before using $\mathrm{Si}(\pi)$. It is interesting to note that for the integrals as in (4.129) the maximum truncation error was 9% and was independent of the size $2\pi B$ of the truncation window, while here it is a lower bound of 9% attained as $N \to \infty$. For smaller N, we see that $\sin x < x$, which makes $S_N(\pi/2N)$ larger than 1.18, hence an even larger overshoot of the Gibbs effect. The same type of analysis can be done for the Gibbs phenomenon near the discontinuity at $t = \pi$, as seen in Fig. 4.30a. Figure 4.30b illustrates this Gibbs phenomenon with $N = 10$ as well as the following σ_1 averaging for reducing it.

As to the remedy for removing this Gibbs phenomenon, it should be of the same nature as the one for the Fourier integral representation, namely to remove the jump discontinuity. This means that we approximate the discontinuous function by a function which is continuous. We replace the part of the square wave near the discontinuity at $x = 0$ by a straight line on the interval $(-\pi/2N, \pi/2N)$ as we did for the signum function $\mathrm{sgn}(t)$ around $t = 0$ in Fig. 4.29. This again would mean averaging the function over intervals of width π/N or convolving the function with $p_{\pi/2N}(t)$. The latter modification is also equivalent to multiplying the Fourier coefficients $(4/\pi)(1/(2n - 1))$ of the original function in (4.139) by

$$\frac{\sin(2n - 1)\pi/2N}{(2n - 1)\pi/2N}$$

(as Fourier coefficients of the Fourier expansion of $Np_{\pi/2N}(t)$ on $(-\pi, \pi)$). Therefore, the average $S_{\sigma_{1,N}}(t)$ of the partial sum

$$S_{\sigma_{1,N}}(t) = \frac{N}{\pi} \int_{t-\pi/2N}^{t+\pi/2N} S_N(x)\, dx$$

$$= \frac{N}{\pi} \int_{t-\pi/2N}^{t+\pi/2N} \frac{4}{\pi} \sum_{n=1}^{N} \frac{1}{2n - 1} \sin(2n - 1)x$$

$$= \frac{4}{\pi} \cdot \frac{N}{\pi} \sum_{n=1}^{N} \frac{1}{2n - 1} \int_{t-\pi/2N}^{t+\pi/2N} \sin(2n - 1)x\, dx$$

$$= \frac{4}{\pi} \cdot \frac{N}{\pi} \sum_{n=1}^{N} \frac{1}{2n-1} \left. \frac{-\cos(2N-1)x}{(2n-1)} \right|_{x=t-\pi/2N}^{x=t+\pi/2N}$$

$$= \frac{4}{\pi} \sum_{n=1}^{N} \frac{1}{2n-1} \left[\frac{\sin(2n-1)\pi/2N}{(2n-1)\pi/2N} \right] \sin(2n-1)t \tag{4.146}$$

Observe that the averaging process introduces a "decaying" factor

$$\frac{\sin(2n-1)\pi/2N}{(2n-1)\pi/2N}$$

to the original coefficients of the series (4.139), which is exactly in parallel to the development of the Fourier integrals in (4.136). This σ_1 averaging of the Gibbs phenomenon is illustrated and magnified near the jump discontinuity $t = 0$ in Fig. 4.30b, where a 10-term $(N = 10)$ series was used. Figure 4.30c shows two periods of the square wave along with the above σ_1 averaged series, where a 22-term series was used.

As to further improvements in getting rid of the Gibbs phenomenon, we suggest the same repeated averaging that we employed in the previous case of the Fourier integral representation. The m repeated averagings are equivalent to convolving the function m times with $Np_{\pi/2N}(t)$, which in turn will introduce the self-truncating factor

$$\sigma_m = \left[\frac{\sin(2n-1)\pi/2N}{(2n-1)\pi/2N} \right]^m \tag{4.147}$$

to the Fourier coefficients of $S_N(t)$ in (4.139). Figure 4.30d illustrates the

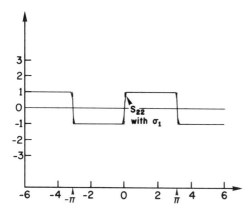

FIG. 4.30c The square wave with the σ_1 averaging of the Gibbs phenomenon, $N = 22$.

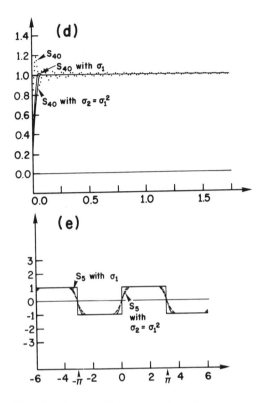

FIG. 4.30d,e (d) The Gibbs phenomenon and its remedies of one (σ_1) and two (σ_2) averagings, $N = 40$. (e) The square wave and the σ_1, σ_2 remedies of its Gibbs phenomenon, $N = 5$.

Gibbs phenomenon and the improvements attained by applying σ_1 and $\sigma_2 = \sigma_1^2$ averagings with $N = 40$, while Fig. 4.30e illustrates the same two averagings with $N = 5$. Before the present suggestion of σ_m repeated averagings, the well-known case was that of σ_1, which is due to Lancos (see Hamming, 1973). Figure 4.30f clarifies how smooth the higher averagings (σ_2 and σ_3) are compared to the σ_1 averaging.

We note that with the σ_1 factor, the oscillations of the Gibbs phenomenon decrease; moreover, with σ_3, in Fig. 4.30f, they have almost disappeared. The only minor drawback of σ_3 and its very good accuracy is that it has about double the rise time of σ_1. However, this can be made up for by doubling N to have a rise time the same as or better than that of σ_1, as illustrated in Fig. 4.30g for σ_1 with $N = 20$ and σ_3 with $N = 40$.

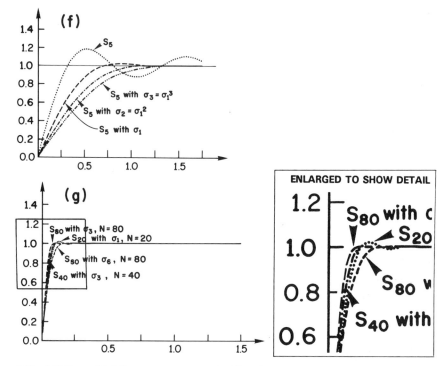

FIG. 4.30f,g (f) The square wave, its Fourier series approximation $S_5(x)$, and σ_1, σ_2, and σ_3 averagings as remedies for the Gibbs phenomenon. (g) Higher-order averaging to reduce the Gibbs phenomenon. The rise time for σ_3 is reduced by increasing N.

In Fig. 4.30g we also illustrate the σ_6 case with $N = 80$, which still has double the rise time of σ_3 with $N = 40$, but it also has very high accuracy for approximating the original square wave function.

We remark again that the high-order σ_m averaging is equivalent to convolving the function with a high-order *hill (B-spline) function* $\phi_{m+1}(t)$ compared to the gate function $p_a(t) = \phi_1(t)$ of σ_1. This point should be worth investigating when looking for optimal truncation windows, since the Fourier transform of high-order $\phi_{m+1}(t)$ will have very rapidly dying sidelobes compared with that of $\phi_1(t)$, as seen in (4.147).

With this detailed analysis, illustrations, and remedies for the Gibbs phenomena of the truncated Fourier trigonometric series and integral representations, we may inquire about possible such phenomena for the truncated general orthogonal expansion, such as the Fourier–Bessel series

(3.100). The answer is known to be in the affirmative (Gottleib and Orszag (1977)), but detailed investigation seems to be lacking. For the interested reader we refer to some illustrations of these phenomena with reference to a possible similar remedy in Jerri (1992), where the generalized hill functions ψ_{R+1} in (3.188) are used instead of ϕ_{R+1} in (2.190).

H. Deciding the Period for the Discrete Transforms

We discussed earlier the discrete transform approximation of periodic functions and stressed the importance of using the exact period of the smooth periodic functions to avoid the leakage error to its transform (Fig. 4.21a–c). Here we will discuss briefly the common practice of choosing or estimating this important period. In the next section we will discuss the possible remedies to correct for the error in such estimates.

In general, it is a common practice to look at most functions as "almost" time-limited to $(-b,b)$ and *also* bandlimited to $(-a,a)$. This is done, even though it is a violation of the uncertainty principle. To make up for the latter difficulty, a conservative practice is to use 10 times the bandlimit $A = 10a$ for sampling purposes. This means that we first find a very good estimate of the time limit b for the given function $g(t)$; then we guess at its band limit a, but use $10a$ for the purpose of the sample spacing $\Delta t = \pi/10a$. To be even more careful, the number of samples N, for the discrete transform, would have to be chosen in such a way that these N samples cover a time domain 10 times that of $(-b,b)$, i.e.,

$$N\Delta t = N \frac{\pi}{10a} > 10(2b)$$

In practice, it is more difficult to know the band limit a, so the above common intuitive method is followed where the sample spacing Δt is chosen such that the smooth curve passing through the samples $g(k\Delta t)$ would very closely resemble the given function $g(t)$, and the N must be large enough for $N\Delta t$ to cover $20b$,

$$N\Delta t > 10(2b)$$

This conservative choice of large N for the sake of lessening the uncertainty principle violation seems to be somewhat wasteful! However, it may not be such a big price for the accuracy we seek, if we recall the efficiency of the FFT algorithm that is used for these computations via the discrete Fourier transforms.

Reducing the Leakage Error

In all the situations illustrated in Fig. 4.1, we used the correct sample spacing required of the sampling theorem by sampling at T and $1/NT$

for $\bar{g}(kT)$ and $\bar{G}(n/NT)$, respectively. In the case of the periodic function $G_s(f)$, we have the truncation at one period $(0, N/NT) = (0, 1/T)$. This corresponds to exact Fourier coefficients, whose later truncation to N values causes the ripples (oscillations) of $\bar{G}(n/NT)$ around $\mp(1/2T)$ as seen in Fig. 4.1c. This is what we discussed in part E as the *windowing effect*, whose special case is the *leakage error* which arises when we violate the sampling theorem requirement—for example, if the periodic function $G_s(f)$ with period $1/T$ is approximated by the discrete transform (4.4) with different period $2a < 1/T$. Usually the leakage error analysis is done, as in the last section, for the periodic $g_s(t)$ instead of $G_s(f)$, hence the sidelobes or (frequency) leakage appearance on the sides of the main peaks in the frequency domain as in Fig. 4.20c. To emphasize this subject again, and help in the comparison with our analysis in the very early sections, we shall (for most of the illustrations here) stay with Fig. 4.1, where $G_s(f)$ is the periodic function, and the leakage error, if any, is analyzed in the time domain. As we discussed earlier, these sidelobes can be explained since the truncation of $G_s(f)$ to $(-a, a)$ amounts to its multiplication by the gate function $p_a(f)$, which corresponds to convolving its (finite) transform c_k with the (finite) transform of $p_a(f)$. In general, this will result in a distortion around every c_k, as we illustrated in Figs. 4.20c, 4.21b, and 4.21c in the frequency domain. However, this distorted picture for representing c_k will be sampled later at kT as in (4.3), so the sidelobe as shown from (4.114) will contribute especially to the neighboring samples and distort them when $a \neq 1/2T$ or its multiples. *In case $a = 1/2T$, the zeros of the transform of $p_a(f)$ will coincide with the neighboring samples, causing no distortion or sidelobes for the final discrete transform* (see Fig. 4.21a; note that the truncation is done on $g(t)$).

Assume that we are not sure of the actual period $(-a, a)$ of $G_s(f)$. However, we guessed it as $(-1/2T, 1/2T)$ and hence sampled $\bar{g}$ at kT in (4.3), (4.4). In this case, we have a leakage or sidelobe error for $\bar{g}(kT)$ due to the truncation to $(-1/2T, 1/2T)$ instead of the actual value $(-a, a)$. However, a modification can be made to minimize this error in the sense that we choose an "optimal" truncation function instead of the abrupt gate function $p_{1/2T}(f)$, with a transform that has smaller sidelobes than $(\sin(\pi t/T)/\pi t)$. There are many optimal truncation functions, and we will consider one of them, the *Hanning function*.

$$H(f) = \begin{cases} \dfrac{1}{2} + \dfrac{1}{2} \cos \dfrac{2\pi f}{a}, & |f| \leq a/2 \\ 0, & \text{otherwise} \end{cases} \tag{4.148a}$$

and its transform

$$h(t) = \frac{1}{2} \frac{\sin \pi at}{\pi t} + \frac{1}{4} \left[\frac{\sin \pi a(t + 1/a)}{\pi(t + 1/a)} + \frac{\sin \pi a(t - 1/a)}{\pi(t - 1/a)} \right] \qquad (4.149)$$

which are illustrated in Fig. 4.31a for the two cases $a = 1$ and $a = 1.5$.

For the important comparison we have added the gate function and its transform in Fig. 4.31b.

From Fig. 4.31a, we see that the transform of the Hanning function has much smaller sidelobes than that of the gate function in Fig. 4.31b. If the Hanning window was used instead of the gate function window in our examples of Fig. 4.17, we could easily have predicted much smaller sidelobes, as the result of convolving the given function with the above transform of the Hanning function and its very small sidelobes (see Exercise 4.21 f(i)).

If we compare the appearance of the leakage sidelobes to the Gibbs phenomenon of the last section, we note that both are the result of the abrupt truncation of the transform of the function (or sequence c_k). The difference is that the leakage error sidelobes appear to the sides

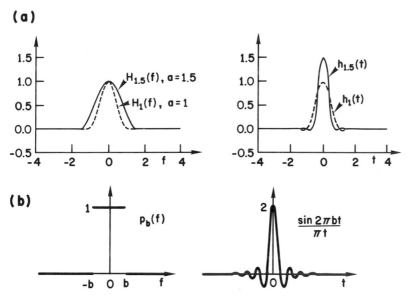

FIG. 4.31 (a) The Hanning function and its transform. (b) The gate function and its transform.

of the main peaks expected for the function, whereas the Gibbs phenomenon appears near the jump discontinuity of the function. The other important difference is that while the leakage error can be reduced by using a smoother truncation window in the transform space, the Gibbs phenomenon requires *making the function itself smoother*, that is, approximating it by a function without a jump discontinuity (as in Fig. 4.29). The latter smoothing corresponds to using a smoother window via the σ_m-averaging factor of (4.147).

Of course, this correction, with the Hanning function for reducing the leakage error, means that we have to modify our discrete transforms (4.3), (4.4) by including the weight of the Hanning function for $\hat{G}(n/NT)$ and convolve its transform with $\tilde{g}(kT)$.

In Fig. 4.21 we illustrated the leakage error due to the use of the wrong period $NT = 3.5/f_0$ for the discrete transform, instead of the natural integer multiple $(NT = m(1/f_0))$ of the period $1/f_0$, and where the gate function window was used. In Fig. 4.32 we consider the periodic function

$$g_s(t) = \cos \frac{\pi}{2}t - 2\cos 3\pi t$$

of period 4 and its four Fourier coefficients as illustrated in Fig. 4.32a and b, respectively. Figure 4.32c illustrates the sidelobes of the leakage error to those Fourier coefficients, when a gate function window of width $^{31}\!/_8$ (slightly different from the natural period of 4) was used. We shall leave it as an exercise to show the diminishing of the leakage error of Fig. 4.32c when the Hanning window is used. Also that when the actual period of 4 is greatly violated by using a period of 6.75, even with the help of the Hanning function, the leakage is still apparent (see Exercise 4.21 f(i), (ii)).

Another well-known optimal window is the *Gaussian window* that is characterized by its oscillation-free Gaussian transform as illustrated in Fig. 4.33a, which is to be compared with the highly oscillating transform of the gate function window in Fig. 4.31b. These two windows represent the extremes among a new class of windows called the *super-Gaussian windows* defined by

$$SG(x, A, N) = e^{-|x/A|N} \tag{4.148b}$$

where the *Gaussian function window* corresponds to the case $N = 2$. It can be shown that the gate function is a limiting case as $N \to \infty$.

Although the Gaussian window eliminates the sidelobes, it will not correct for the widening of the impulses which is caused by their convolution with the transform of this window. This is a drawback to almost all finite-

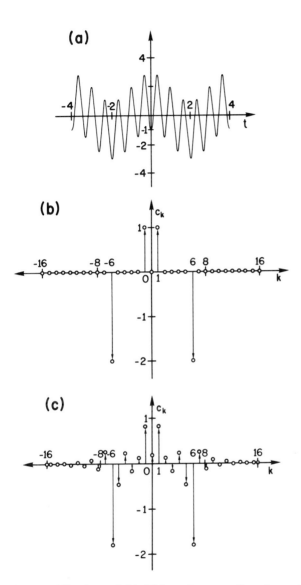

FIG. 4.32 (a and b) Using the gate function window with correct and slightly incorrect period. (a) $g_s(t) = \cos(\pi/2)t - 2\cos 2\pi t$, with the natural period of 4. (b) c_k computations with correct period 4—no aliasing error. (c) c_k computations with the slightly incorrect period $3\frac{1}{8}$ instead of 4 and with a gate function window.

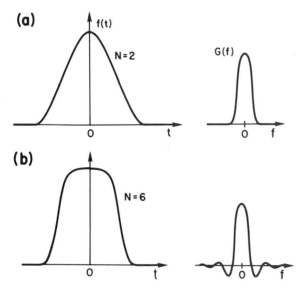

FIG. 4.33 (a) Gaussian ($N = 2$) and (b) super-Gaussian ($N = 6$) windows with their respective transforms.

width truncation windows. We will leave it as an exercise to show that the super-Gaussian window of order close to $N = 6$ as shown in Fig. 4.33b may reduce the convolution widening of the peaks, as well as diminish the sidelobes (see Exercise 4.21 for use of optimal windows).

In general, such optimal windows will not be of much help if we use a period that is far off. We shall leave it as an exercise to illustrate this point for the periodic function

$$g_s(t) = \cos \frac{\pi}{2} t - 2 \cos 3\pi t$$

with period 4 (with its two exact Fourier cosine coefficients $a_1 = 1$ and $a_6 = -2$). The Hanning window and the Gaussian window are of great help in reducing the sidelobes when we use slightly incorrect period such as $3\frac{1}{8}$ instead of 4. On the other hand, both windows are of no help when we use an incorrect period like 6.75 instead of 4 (see Exercise 4.21 f(i)).

4.1.7 Examples of Computing Fourier Integrals and Series

A. Computations of the Fourier Integrals

EXAMPLE 4.7: The discrete transform approximation of Fourier integrals

Our first example is to illustrate the use of the discrete Fourier transform of the *causal* function

$$g(t) = \begin{cases} 0, & -\infty < t < 0 \\ e^{-t}, & 0 < t < \infty \end{cases} \tag{E.1}$$

of Fig. 4.34a, then compare the discrete approximation with the exact (complex) transform

$$G(f) = \frac{1}{1 + (2\pi f)^2} - j\frac{2\pi f}{1 + (2\pi f)^2} = G_r(f) + jG_i(f) \tag{E.2}$$

Figure 4.34b and c show the real and imaginary parts of $G(f)$ in (E.2), respectively.

We will first use the N samples of $g(t)$ on the interval $(-b,b) = (-4,4)$. We realize here that a time limit of $b = 4$ is large enough for the "almost" time-limited approximation. In this illustration, we will not consider the special feature of causal functions $(f(t) \equiv 0$ for $t < 0)$, but rather use it as an example of neither odd nor even functions. Also, in the transform space we may graph $|G(f)|$ as in Fig. 4.10a. To follow our development in this chapter, as illustrated in Figs. 4.11 and 4.10, we sample the function with $\Delta t = T = \frac{8}{32} = \frac{1}{4}$ with 32 samples on the interval $(-4,4)$ as in Fig. 4.34d, where we note the assigned value of $\frac{1}{2}[0 + 1] = \frac{1}{2}$ for the jump discontinuity at $t = 0$. Next we translate the part on $(-4,0)$ to $(4,8)$ as in Fig. 4.34e, where we use the 32 samples of $g(kT)$ with $k = 0$ (at $t = 0$) to $k = 31$ in (4.3). Here we also note the jump discontinuity at $t = 4$ in Fig. 4.34e, where we assigned it the average value of the two neighboring samples $\frac{1}{2}[g(15T) + g(-15T)] = \frac{1}{2}e^{-3.75}$. The real and imaginary samples of the resulting transform $\tilde{G}(n/NT)$ from (4.3) are illustrated in Figs. 4.34f and g, where we have also graphed the real and imaginary parts of the exact transform in (E.2) as a solid line. For comparison, we note that the discrete transform approximation of the real part gives very good agreement, while there is a clear error for the imaginary part. Of course, in what we see in Fig. 4.34f and g, we should translate the part on $(2,4) = (1/2T, 1/T)$ to $(-1/2T, 0) = (-2,0)$ to have the actual picture of the transform as illustrated in Fig. 4.34b,c. The main error in (the discrete approximation of) $G_i(f)$ is close to the ends of the

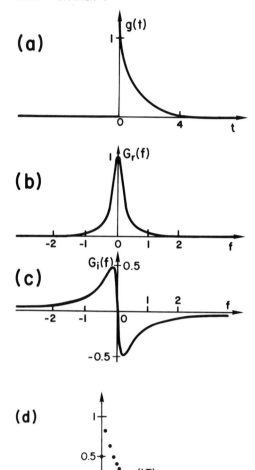

FIG. 4.34 (a–d) The causal function $g(t)$ and its transform $G_r(f) + jG_i(f)$. (a) The causal function

$$g(t) = \begin{cases} 0, & t < 0 \\ e^{-t}, & t > 0 \end{cases}$$

(b) The real part $G_r(f)$ of $G(f)$, $G_r(f) = 1/[1+(2\pi f)^2]$. (c) The imaginary part $G_i(f)$ of $G(f)$, $G_i(f) = -2\pi f/[1 + (2\pi f)^2]$. (d and e) Sampling and then translating the samples on $(-4,0)$ to $(4,8)$: (d) The samples $g(kT)$ of $g(t)$.

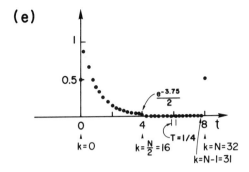

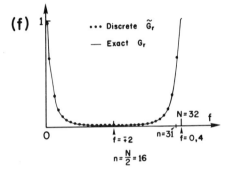

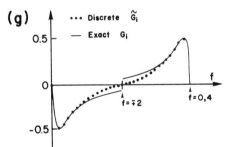

FIG. 4.34 (e–g) Sampling and translation of the function and its transform. (e) $g(kT)$ on $(-4,0)$ translated to $(4,8)$. (f and g) The real and imaginary parts of the discrete $(\cdots)$ and exact (———) transform of $g(t)$.

period $\mp 1/T = \mp 2$, which can be attributed mainly to the aliasing error of $\tilde{G}_i(f)$.

To further explain the larger error in the discrete transform approximation of the imaginary part $G_i(f)$ compared to the real part, we note that in Fig. 4.34d the truncation domain $(-4,4)$ for $g(t)$ was large enough to make $g(kT) \sim \tilde{g}(kT)$; that is, $g(kT)$ has small aliasing error. To calculate the real part $G_r(f)$, using this $g(kT)$, the resulting aliased $\tilde{G}_r(n/NT)$ would also be very close to $G_r(n/NT)$, since the interval $(-2,2)$ is large enough for G_r to die out considerably as seen in Fig. 4.34b. In comparison, this interval $(-2,2)$ for G_i (as seen in Fig. 4.34c) was not large enough, as witnessed by the clear jump discontinuity at ∓ 2, yielding an aliasing error.

We will leave it as an exercise to increase the number of samples to 128 on the same time interval $(-4,4)$ for better resolution of $g(t)$. This would also result in a larger frequency domain $(-8,8)$, which should obviously improve the aliasing of $G_i(f)$ in addition to further improving it for $G_r(f)$ (see Exercise 4.7c). Another exercise is to use the same 32 samples but increase the time interval to $(-8,8)$, thus diminishing the aliasing of $g(t)$ but increasing the aliasing for both G_r and G_i, as they are now taken on a smaller frequency interval $(-1,1)$ (see Exercise 4.7c).

To further illustrate the discrete transform approximation of Fourier integrals, we used the samples of the transform $G(f)$ of (E.2) in (4.4) to approximate the inverse transform $g(t)$ in (E.1). The samples of the real and imaginary parts used (after translating the parts on $(-2,0)$ to $(2,4)$) are illustrated in Fig. 4.35a,b with due attention to the jump discontinuity at $f = 2$ of the imaginary (odd) part (where the average of a zero value is assigned to the sample at $f = 2$) and the continuity at the (even) real part. The resulting approximate discrete transform $\tilde{g}(kT)$ along with its exact values $g(kT)$ are illustrated in Fig. 4.35c, and with these samples on $(4,8)$ translated to their natural position on $(-4,0)$ as in Fig. 4.35d. We notice an error around the end points in Fig. 4.35d, where the periodic extension has a small jump discontinuity and even larger error around $t = 0$, where the function itself has (an interior) jump discontinuity. If we follow our analysis, we see first that such an error at the ends of the periods is due more to the windowing of the imaginary part of $G(f)$, with its sizable jump discontinuity at $f = \mp 2$ in Fig. 3.35a, than to that of the continuous real part in Fig. 4.35b. The second reason for this error, especially at $t = 0$, is the usual *Gibbs phenomenon* associated with discontinuities of the function being approximated, which cannot easily be improved simply by increasing N (see part G of this section, particularly Fig. 4.30). We should also note that the error in the samples around $k = 30$ in Fig. 4.35c are of the samples to the left of $t = 0$, where the main jump discontinuity occurs. This is seen clearly in Fig. 4.35d when compared to the exact (causal) function in Fig. 4.34a,d.

We must emphasize that the major error that we noted around $t = 0$ in Fig. 4.35d is clearly a Gibbs phenomenon due to jump discontinuity at $t = 0$ of the causal function $g(t)$, and it should not be confused with a windowing error at the end point as it may seem (incorrectly) if we are

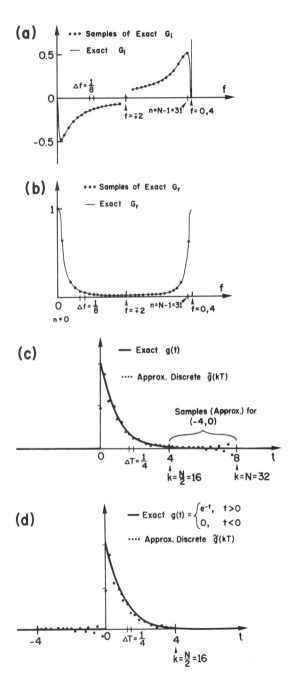

FIG. 4.35 (a and b) Samples of the exact real and imaginary parts of $G(f)$. (c and d) The inverse discrete transform $\tilde{g}$.

to interpret the output from the format of Fig. 4.35c. The latter misinterpretation often happened, as the major error appears to be around end points such as $t = 0$ and $t = 8$ in Fig. 4.35c. As indicated earlier, the proper and safer picture of the output is that of Fig. 4.35d, where we are again looking at the (causal) function in its natural setting with its troublesome discontinuity (at $t = 0$) to the Fourier analysis attempt of approximating it, and which causes the latter (stubborn) Gibbs phenomenon. This problem becomes more acute when the Fourier analysis in terms of the DFT is used for the numerical inversion of the Laplace transform, as we alluded to briefly at the end of Section 2.8. As seen from (2.266), if the Laplace transform $F(s)$ is valid for $s > \gamma > 0$, then a DFT approximation for its inverse $f(x)$ has an additional factor $e^{\gamma x}$ besides the Fourier sum, resulting in a serious magnification of the Gibbs error of the DFT sum of Fig. 4.35d.

As discussed in part G, the remedy for reducing the Gibbs phenomenon of the jump at $t = 0$ is to average out its oscillations or, equivalently, convolve the present function with $Np_{a/N}(t)$. To minimize the windowing effect, we must resort to a truncation window with smoother edges than that of the abrupt gate function window, i.e., a truncation window that has a Fourier transform with small sidelobes like the Hanning window in (4.148a) or the Gaussian window in (4.148b) (see also Figs. 4.31 and 4.33).

For more such DFT computations of Fourier transforms, see the detailed Exercises 4.1, 4.5, 4.7, and 4.8.

Computations of the Fourier Series

EXAMPLE 4.8: The Discrete Fourier Transform Approximation of Fourier Series and its Coefficients

In this example, we consider the periodic function

$$G_s(f) = \begin{cases} -1, & -4 < f < -2 \\ 1, & -2 < f < 2 \\ -1, & 2 < f < 4 \end{cases}$$

with period 8, as illustrated in Fig. 4.36a. Since this function is real and even, its Fourier coefficients are also real and even and are illustrated in Fig. 4.36b as the discrete samples $c_k = g(kT)$, where in this particular example $c_{2k} = 0$.

For a clear illustration, we chose $\Delta f = 1/NT = \frac{1}{4}$ with $N = 32$ for the interval $(-4, 4)$ to have $T = \frac{1}{8}$. We first sample $G_s(f)$ at $f_n = n/NT$, $n = 0, 1, 2, \ldots, 31$, and at the same time translate the samples on $(-4, 0)$

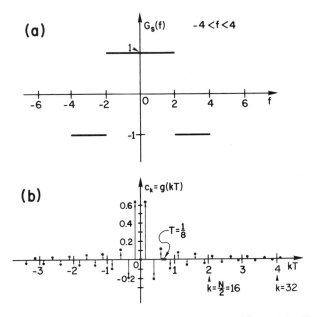

FIG. 4.36a,b The periodic function $G_s(f)$ and its Fourier coefficients c_k: (a) $G_s(f)$, $-4 < f < 4$; (b) $c_k = g(kT)$, $T = \frac{1}{8}$.

to $(4, 8)$ as seen in Fig. 4.36c. We should note the jump discontinuities at $f = \mp 2$, where a sample value of zero, as the average of the two neighboring samples on both sides of each jump, is assigned at the positions $f = \mp 2$ of these two jumps. With such (large) jump discontinuities we should be prepared for the Gibbs effect of Fourier (series) analysis around such discontinuities. In contrast, the Fourier coefficients c_k have no such bad discontinuity even at $\mp NT/2 = 16 \times \frac{1}{8} = \mp 2$, where we are truncating it, as seen in Fig. 4.36b, and for emphasis we repeat that in Fig. 4.36d, where the samples on $(2, 4)$ are those translated from $(-2, 0)$ in Fig. 4.36b.

If we use these samples of c_k in Fig. 4.36d as $\bar{g}(kT)$ in (4.4), we obtain $G(n/NT)$ in Fig. 4.36e as the discrete transform approximation of the 32 samples of the periodic function presented $G_s(f)$, and these are also compared with the exact values of Fig. 4.36c. Note that the exact values of the function $G_s(f)$ in Fig. 4.36a, c, and e are equivalent, since they all represent the same (periodic) function on a complete period of length 8. This is even though the picture on one period may not look the same as that of Fig. 4.36a, but in the final analysis the periodic extension $\bar{G}_s(f)$, which is what the Fourier series actually represents, is the same!

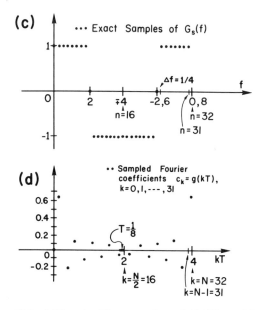

FIG. 4.36c,d The samples of $G_s(f)$ and its truncated Fourier coefficients $N = 32$: (c) samples of $G_s(f)$ with the part on $(-4,0)$ translated $(4,8)$; (d) the truncated Fourier coefficients with the part on $(-2,0)$ translated to $(2,4)$.

Here we can see the Gibbs phenomenon near $f = \mp 2$ where we have the jump discontinuities. The windowing effect is minimal due to the almost time-limited situation for $G_s(f)$ where c_k dies out toward the limits ∓ 2 (Fig. 4.36b). This can be further reduced by increasing NT. In contrast, the Gibbs effect error of G_s in Fig. 4.36e would be insensitive to N (see also Fig. 4.30). We leave it as an exercise to illustrate these two points by using $N = 128$ with the same T.

As discussed before, we expect the discrete transform approximation to be much better for the Fourier coefficients c_k; that is, in this example we don't expect the stubborn Gibbs phenomenon.

If we use the exact sample values of $G(f)$ at n/NT (as shown in Fig. 4.36c) in (4.3), we have the discrete approximation of the Fourier coefficients, which we illustrate in Fig. 4.36f, where we also compare it with the exact values. As anticipated, the error is small and can easily be improved by, in turn, improving the resolution of $\tilde{G}(n/NT)$ i.e., by increasing N, which improves the computation of the Fourier integral representation of c_k.

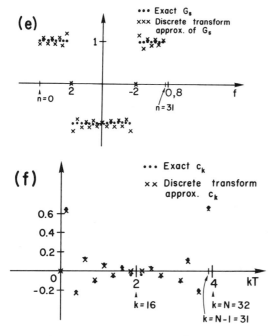

FIG. 4.36e,f The exact and discrete transform approximation of $G_s(f)$ and its Fourier coefficients c_k.

The Fourier coefficients on the interval $(2,4)$ in Fig. 4.36f can, of course, be translated back by a period of 4 to $(-2,0)$, to look like those of the shape of the exact (even) ones in the original Fig. 4.36b.

For more on such DFT computations of Fourier series and their coefficients see Exercises 4.2, 4.3, 4.4, and 4.6.

4.2 Discrete Orthogonal Polynomial Transforms

In this section, we will present a brief discussion of the discrete version of the classical orthogonal polynomial transform (3.143), (3.142) of Section 3.5. After such an introduction we will state their basic properties and leave their, relatively simple, proofs for exercises with clear guiding hints. This will be followed by discussing the applications of discrete Legendre and Laguerre transforms. The computations will be left for the interest of the reader.

These discrete polynomial transforms are considered as advanced topics compared to the discrete Fourier transforms of the last section. Our purpose here is not to make a thorough study, but rather to present the analysis and advantages of such general discrete transforms, apart from the very familiar Fourier trigonometric ones. We shall attempt to make this presentation, and the exercises that support it, clear and self-contained.

Such discrete transforms are defined as finite sums (4.159) over the zeros $\{x_{n,m}\}$ of the classical polynomial $p_n(x)$. The inverse of this transform will be derived as another finite sum (4.160). These general discrete transforms are illustrated here for their *discrete Legendre* and *Laguerre transforms* with emphasis on their basic properties and applications (see Sections 3.5.1 and 3.5.2 for the *finite* transform versions). The illustration and use of other discrete orthogonal transforms depend primarily on the availability of the zeros of their orthogonal polynomial kernels (see Abramowitz and Stegun, 1964). These applications parallel the use of the discrete Fourier transforms of Section 4.1.

The present discrete *classical* orthogonal polynomial transform should be distinguished from transforms based on the "discrete orthogonal polynomials," which are defined at equidistant values of the argument of the polynomial $p_n(x)$ instead of its (more natural) zeros $\{x_{n,m}\}$ (see Elliot and Rao, 1982).

Of utmost importance to any discrete transform is its associated discrete orthogonality, which we will present first.

4.2.1 Basic Properties and Illustrations

A. Discrete Orthogonality

Let $\{p_n(x)\}$ be the set of classical orthogonal polynomials on the interval (a,b) with the weight function $\rho(x)$ as presented in Section 3.5,

$$p_n(x) = k_n x^n + k_n' x^{n-1} + \cdots \tag{3.140), (4.150}$$

$$\int_a^b \rho(x) p_n^2(x)\, dx = h_n \tag{3.141), (4.151}$$

and consider the n zeros $\{x_{n,m}\}_{m=1}^n$ of $p_n(x)$,

$$p_n(x_{n,m}) = 0, \qquad m = 1, 2, \ldots, n \tag{4.152}$$

which are real, distinct, and lie in the interior of the interval $[a,b]$. These zeros $\{x_{n,m}\}$ are found in Abramowitz and Stegun (1964). The *first discrete orthogonality* property of $p_n(x)$ with respect to its zeros $\{x_{n,m}\}$ can be derived as (see Exercise 4.28)

$$\sum_{j=0}^{n-1} \frac{p_j(x_{n,m})p_j(x_{n,m'})}{h_j} = \begin{cases} 0, & m \neq m' \\ -\dfrac{k_n}{h_n k_{n+1}} p_n'(x_{n,m})p_{n+1}(x_{n,m}), & m = m' \end{cases}$$

(4.153)

This discrete orthogonality is easily illustrated for the case of *Legendre polynomials* $P_n(x)$, with $(a,b) = (-1,1)$,

$$\rho(x) = 1$$

$$h_n = \frac{2}{2n+1}, \qquad k_n = \frac{2^n}{n!\sqrt{\pi}}\left(n + \frac{1}{2}\right)$$

(4.154)

$$P_n'(x_{n,m}) = \frac{(n+1)P_{n+1}(x_{n,m})}{x_{n,m}^2 - 1}$$

(4.155)

to give (see Exercise 4.26)

$$\sum_{j=0}^{n-1} \frac{(2j+1)}{2} P_j(x_{n,m})P_j(x_{n,m'}) = \begin{cases} 0, & m \neq m' \\ \dfrac{(n+1)^2}{2(1-x_{n,m}^2)} P_{n+1}^2(x_{n,m}), & m = m' \end{cases}$$

(4.156)

We should stress here that this discrete orthogonality is with respect to the zeros of the polynomial, which is different from the usual orthogonality (3.144), for the finite Legendre transforms, which is with respect to the degree n of $P_n(x)$. The discrete analog of the latter will be presented as a *second* type of discrete classical polynomial orthogonality in (4.167).

B. Discrete Classical Polynomial Transforms

In Section 3.5 we presented the orthogonal polynomial expansion of $f(x)$ in (3.142), (3.143) on the interval (a,b) as a finite transform,

$$f(x) = \sum_{j=0}^{\infty} \frac{1}{h_j} F(j)p_j(x)$$

(4.157)

$$F(j) = \int_a^b \rho(x)f(x)p_j(x)\,dx$$

(4.158)

Now we define the *discrete classical orthogonal polynomial* transform as

$$\bar{F}(j) = -\frac{h_n k_{n+1}}{k_n} \sum_{m=1}^{n} \frac{1}{p_n'(x_{n,m})p_{n+1}(x_{n,m})} \tilde{f}(x_{n,m})p_j(x_{n,m}) \qquad (4.159)$$

with its inverse

$$\tilde{f}(x_{n,m}) = \sum_{j=0}^{n-1} \frac{1}{h_j} \bar{F}(j)p_j(x_{n,m}) \qquad (4.160)$$

This can be verified by substituting $\bar{F}(j)$ from (4.159) into (4.160), interchanging the two summations, and using the discrete orthogonality property (4.153). It is clear that the discrete inverse transform $\tilde{f}$ in (4.160) is only an approximation to the samples of the function $f(x)$ as represented by the (infinite) Fourier series (4.157). This is in the same line with the discrete Fourier transform $\bar{G}(n/NT)$ in (4.3) versus the Fourier series $G_s(f)$ in (4.7), which we discussed in detail in Section 4.1.1A. However, it can be shown that the transform (4.159) is exact when $f(x)$ is a polynomial of degree $\leq n - 1$ (see Exercise 4.27a).

In the following, we present the cases of the *discrete Laguerre* and *Legendre* polynomial transforms and follow this with an outline of the basic properties of the discrete Legendre transform with reference to its application in solving boundary value problems and computing (Legendre) Fourier series. The proofs are outlined in the form of exercises.

C. Discrete Laguerre Transforms

From (4.159), (4.160) we can easily obtain the following discrete Laguerre transform:

$$\bar{F}(j) = -\frac{1}{n(n+1)} \sum_{m=1}^{n} \frac{x_{n,m}}{L_{n-1}(x_{n,m})L_{n+1}(x_{n,m})} \tilde{f}(x_{n,m})L_j(x_{n,m}) \qquad (4.161)$$

and its inverse,

$$\tilde{f}(x_{n,m}) = \sum_{j=0}^{n-1} \bar{F}(j)L_j(x_{n,m}), \qquad L_n(x_{n,m}) = 0, \; m = 1,2,\ldots,n \qquad (4.162)$$

after consulting the properties of the Laguerre polynomials in Appendix A.5 (see Exercise 4.29b).

D. Discrete Legendre Transforms

From (4.159), (4.160) we have the discrete Legendre transform

$$\tilde{F}(j) = \frac{2}{(n+1)^2} \sum_{m=1}^{n} \frac{1-x_{n,m}^2}{[P_{n+1}(x_{n,m})]^2} \tilde{f}(x_{n,m}) P_j(x_{n,m}) \tag{4.163}$$

and its inverse,

$$\tilde{f}(x_{n,m}) = \sum_{j=0}^{n-1} \frac{2j+1}{2} \tilde{F}(j) P_j(x_{n,m}), \qquad P_n(x_{n,m}) = 0, \ m = 1, 2, \ldots, n \tag{4.164}$$

as follows from properties of Legendre polynomials in Appendix A.5 (see Exercise 4.29c).

4.2.2 Properties of the Discrete Legendre Transforms

Most of the properties of the discrete Legendre transforms parallel those of the finite Legendre transforms of Section 3.5.1:

$$F(j) = \int_{-1}^{1} f(x) P_j(x) \, dx \tag{3.146}, (4.165)$$

$$f(x) = \sum_{j=0}^{\infty} \left(\frac{2j+1}{2} \right) F(j) P_j(x) \tag{3.147}, (4.166)$$

We shall use the same operational notation as in Section 3.5.1: $l[\tilde{f}]$ for the discrete Legendre transform $\tilde{F}(j)$ and $l^{-1}[\tilde{F}]$ for its inverse transform $\tilde{f}(x_{n,m})$. The proofs of most of these transform properties parallel those of their finite transform analog of Section 3.5.1, and we shall leave them for a simple exercise (see Exercises 4.30 and 4.31). This includes linearity, symmetry, Parseval's equality, two convolution products (one for the transform and another for its inverse) and of course a convolution theorem.

As we mentioned, when commenting on the (first) discrete orthogonality (4.156) of the Legendre transforms, with respect to the zeros $\{x_{n,m}\}$ of $P_n(x)$, there is a *second discrete orthogonality* in j, the degree of $P_j(x)$, which can be derived as

$$\frac{2}{(n+1)^2} \sum_{m=1}^{n} \frac{1-x_{n,m}^2}{P_{n+1}^2(x_{n,m})} P_{j'}(x_{n,m}) P_j(x_{n,m}) = \begin{cases} 0, & j \neq j' \\ \dfrac{2}{2j+1}, & j = j' \end{cases} \tag{4.167}$$

This version of the discrete orthogonality parallels the usual orthogonality of the Legendre polynomials as we had it in (3.144) for the finite Legendre transforms (3.146), (3.147).

The Parseval Equality
The discrete version of Parseval's equality, of the finite Legendre transform in (3.146), (3.147),

$$\int_{-1}^{1} f(x)g(x)\,dx = \sum_{j=0}^{\infty} \frac{2j+1}{2} F(j)G(j) \tag{4.168}$$

is

$$\frac{2}{(n+1)^2} \sum_{m=1}^{n} \frac{1-x_{n,m}^2}{P_{n+1}^2(x_{n,m})} \tilde{f}(x_{n,m})\tilde{g}(x_{n,m}) = \sum_{j=0}^{n-1-} \frac{2j+1}{2} \bar{F}(j)\bar{G}(j) \tag{4.169}$$

where $\bar{F}(j) = l[\tilde{f}(x_{n,m})]$ and $\bar{G}(j) = l[\tilde{g}(x_{n,m})]$. The proof is simple and is left for an exercise (see Exercise 4.31a).

Convolution Products
As discussed in Section 3.6.1, we are dealing here with *nonexponential*-type kernel transforms like the present Legendre polynomial transform. Hence, we don't expect the simple translation property of the convolution product of the Fourier and Laplace transform (in (2.127) and (2.9), respectively), which is a consequence of their special exponential kernels. So instead of the simple translation $\bar{F}(j-k)$, in the case of the discrete Fourier transforms, we define, parallel to what we did for the generalized transforms as in (3.185) of Section 3.6.1, *general translations* $\bar{F}(j\theta k)$,

$$\bar{F}(j\theta k) \equiv \frac{2}{(n+1)^2} \sum_{m=1}^{n} \frac{1-x_{n,m}^2}{P_{n+1}^2(x_{n,m})} \tilde{f}(x_{n,m})P_j(x_{n,m})P_j(x_{n,m})P_k(x_{n,m}) \tag{4.170}$$

and $\tilde{f}(x_{n,m\theta m'})$,

$$\tilde{f}(x_{n,m\theta m'}) \equiv \sum_{j=0}^{n-1} \frac{2j+1}{2} \bar{F}(j)P_j(x_{n,m'})P_j(x_{n,m}) \tag{4.171}$$

These definitions are not surprising, since the translation for the Fourier transform was also the result of multiplying the transformed function by an exponential functions as in (2.111) which is of the same type as the transform kernel.

The *convolution product* $\tilde{f} * \bar{g}$ of $\tilde{f}$ and $\bar{g}$ is defined as

$$(\tilde{f} * \bar{g})(x_{n,k}) \equiv \frac{2}{(n+1)^2} \sum_{m=1}^{n} \frac{1-x_{n,m}^2}{P_{n+1}^2(x_{n,m})} \tilde{f}(x_{n,m}) \bar{g}(x_{n,m\theta k}) \tag{4.172}$$

This convolution product is commutative, i.e., $\tilde{f} * \bar{g} = \bar{g} * \tilde{f}$ (see Exercise 4.31b(i)). We define another type of convolution product $\bar{F} \times \bar{G}$ for the transforms as

$$(\bar{F} \times \bar{G})(k) \equiv \sum_{j=0}^{n-1} \frac{2j+1}{2} \bar{F}(j) \bar{G}(j\theta k) \tag{4.173}$$

which is also commutative, i.e., $\bar{F} \times \bar{G} = \bar{G} \times \bar{F}$ (see the hint Exercise 4.31b(i)):

With these two kinds of convolution products, we are now in a position to state two *convolution theorems* (see Exercise 4.31b(ii)):

$$\mathfrak{l}[\tilde{f} * \bar{g}] = \bar{F}(j)\bar{G}(j) \tag{4.174}$$

and

$$\mathfrak{l}[\tilde{f}(x_{n,m})\bar{g}(x_{n,m})] = \bar{F} \times \bar{G} \tag{4.175}$$

The discrete Legendre transform and its inverse for the special cases of even and odd functions are left for exercises.

4.2.3 The Use of the Discrete Orthogonal Polynomial Transforms

The first illustration of the use of such discrete transforms is in computing the orthogonal polynomials–Fourier series for a number of functions. This is followed by computing for a simple classical boundary value problem, namely the potential distribution inside a unit sphere, whose solution by the finite Legendre transform was discussed in Example 3.15 of Section 3.5.1.

A. Orthogonal Polynomial Series Computations

The advantage of using the discrete Legendre transform for representing a given function on n points lies, in general, in its better accuracy than the nth partial sum of the above Legendre–Fourier series representation of the same function in (4.166). By this comparison we mean that we first consider functions $f(x)$ with known Fourier coefficients $F(j)$, in a closed form, to use in (4.166). Then we compute the truncated Fourier series

$f_n(x)$,

$$f_n(x) = \sum_{j=0}^{n} (2j + 1)F(j)P_j(x) \tag{4.176}$$

at the zeros of the classical polynomials, where we have a truncation error $\epsilon_n \equiv |f(x_{n,m}) - f_n(x_{n,m})|$. This error is to be compared with the discrete transform error $\tilde{\epsilon}_n \equiv |f(x_{n,m}) - \tilde{f}(x_{n,m})|$ in approximating $f(x)$ at $\{x_{n,m}\}$. In Exercise 4.32 we present three functions with known Legendre series coefficients $F(j)$ (see Appendix C.13). Exercise 4.33 deals with three functions with known Laguerre series coefficients $F(j)$ (see Appendix C.15). Most of these functions are relatively far from being special cases of the Fourier series or the discrete transform. To indicate the accuracy, we give a representative case in which $\tilde{\epsilon}_n$ is around 10^{-13} while ϵ_n is around 10^{-5} (see more computational results in the answers to Exercises 4.32 and 4.33). Such better accuracy for the discrete transform over the nth partial sum of the corresponding Fourier series stems from the theory of *Gauss mechanical quadrature*, which was used indirectly in deriving the discrete orthogonal polynomial transforms (4.159), (4.160) and especially their "exact" discrete orthogonality (4.153). We have the verification of these results for Exercises 4.32 and 4.33.

Of course, when the Fourier coefficients $F(j)$ are not available in closed form, they have to be evaluated numerically for the truncated Fourier series, which increases the time of computation and will increase the error. This is not the case for the discrete transform representation, since $\bar{F}(j)$ is evaluated as the nth sum of (4.163) which is different from $F(j)$ in (4.165). The discrete transform representation, of course, becomes exact when the function is a polynomial of degree $\leq n - 1$ (see Exericse 4.29a). This may be noticed in the examples of $f(x) = x^3$ and $f(x) = x^5$ in Exercises 4.32a and 4.33a for the discrete Legendre and Laguerre transforms, respectively. We find that the accuracy of both the Laguerre Fourier series (3.159), (3.158) and the discrete Laguerre transform (4.162) is very good when $n = 5$ to $n = 10$ terms are used for both (finite sum) representations of this $f(x) = x^5$. The same can be said about the example of $f(x) = x^3$ for the case of the discrete Legendre transform (see Exercises 4.33d and 4.32d).

B. Boundary Value Problems

EXAMPLE 4.9: Potential Distribution Inside a Sphere—the Discrete Legendre Transform Approximation

As a simple example of using the discrete Legendre transform for solving boundary value problems, we will consider the same problem as in Example 3.15 in Section 3.5.1. Consider the potential distribution $v(\rho, \cos\theta)$ inside a unit sphere where the potential on the surface $v(1, \cos\theta) = f(\cos\theta)$ is a function of θ only (see Fig. 3.12 in Chapter 3). We let $x = \cos\theta$ to have the boundary value problem

$$\frac{1}{\rho^2}\left\{\frac{\partial}{\partial\rho}\left(\rho^2\frac{\partial v}{\partial\rho}\right) + \frac{\partial}{\partial x}\left[(1-x^2)\frac{\partial v}{\partial x}\right]\right\} = 0,$$

$$0 \le \rho < 1, \quad -1 < x < 1 \tag{E.1}$$

$$v(1,x) = f(x), \qquad -1 < x < 1, \ x = \cos\theta \tag{E.2}$$

which was presented as (E.1) and (E.2) in Example 3.15. Then we use the finite Legendre transform (3.146), (3.147) to find the solution as the following infinite Legendre series:

$$v(\rho, \cos\theta) = \frac{1}{2}\sum_{j=0}^{\infty}(2j+1)F(j)\rho^j P_j(\cos\theta) \tag{E.3}$$

where $F(j)$ is the finite Legendre transform (3.146) of $f(x)$,

$$F(j) = \int_{-1}^{1} f(x)P_j(x)\,dx \tag{E.4}$$

This method, of course, depends on the availability of $F(j)$ in closed form from existing tables of finite Legendre transforms (see Appendix C.13, for example). Otherwise, we must either evaluate $F(j)$ numerically or possibly resort to an "involved" convolution product of the finite Legendre transforms as a double sum (see (E.7) below).

To illustrate the use of the discrete Legendre transform, we consider $\tilde{v}(\rho, x_{n,m})$ as an approximation to the solution of the *Dirichlet* problem (E.1), (E.2) at $\{x_{n,m}\}$ (the zeros of the Legendre polynomial $P_n(x)$) and apply the discrete Legendre transform (4.163) on the Laplace equation (E.1) to obtain a differential equation in $\tilde{V}(\rho,j) = l[\tilde{v}(\rho, x_{n,m})]$,

$$\frac{1}{\rho^2}\left[\frac{d}{d\rho}\left(\rho^2\frac{d\tilde{V}}{d\rho}(\rho,j)\right) + j(j+1)\tilde{V}(\rho,j)\right] = 0 \tag{E.5}$$

following the same steps as in Example 3.15, where we find a similar solution

$$\tilde{V}(\rho,j) = \rho^j \tilde{F}(j) \tag{E.6}$$

where

$$\tilde{F}(j) = l[\widetilde{f}(x_{n,m})]$$

The approximate solution $\tilde{v}(\rho,x_{n,m})$ to the original boundary value problem is the inverse discrete Legendre transform (4.164) of $\tilde{V}(\rho,j)$, which can be obtained formally via (4.172) as the discrete convolution product $\tilde{f}(x_{n,m}) * \tilde{g}(x_{n,m})$ with $\tilde{g}(x_{n,m}) = l^{-1}[\rho^j]$,

$$\tilde{v}(\rho,x_{n,k}) = \tilde{f}(x_{n,k}) * \tilde{g}(x_{n,k})$$

$$= \frac{2}{(n+1)^2} \sum_{m=1}^{n} \left[\frac{1-x_{n,m}^2}{P_{n+1}^2(x_{n,m})} \right] \tilde{f}(x_{n,m})\tilde{g}(x_{n,m\theta k}) \tag{E.7}$$

Here $\tilde{g}(x_{n,m\theta k})$ is the *k-generalized translation* of $\tilde{g}(x_{n,m})$ as we defined it in (4.171),

$$\tilde{g}(x_{n,m\theta k}) = \sum_{j=0}^{n-1} \left(\frac{2j+1}{2} \right) \rho^j P_j(x_{n,k})P_j(x_{n,m}),$$

$$m = 1,2,\ldots,n, \; k = 1,2,\ldots,n \tag{E.8}$$

What we are showing here is an illustration of the operational method of discrete orthogonal polynomial transforms, and there are a number of subtleties that must be noticed. For example, in (E.7) we seem to have the solution as a finite sum instead of the infinite Legendre–Fourier series (E.3). The point here is that with the discrete transform method, we assumed that the x-variation of $v(\rho,x)$ in equation (E.1) can be approximated by $\tilde{v}(\rho,x_{n,m})$, the finite sum (inverse) discrete Legendre transform (4.164) at the points $\{x_{n,m}\}$,

$$v(\rho,x_{n,m}) \sim \tilde{v}(\rho,x_{n,m}) = \sum_{j=0}^{n-1} \frac{2j+1}{2} \tilde{V}(\rho,j)P_j(x_{n,m}) \tag{E.9}$$

Such an approximation is the subject of the following numerical computation. We may also add that with such a discrete approximation at $\{x_{n,m}\}$, the differential term

$$\frac{\partial}{\partial x} \left[(1-x^2) \frac{\partial v}{\partial x} \right]$$

of the Laplace equation (E.1) should be correctly approximated as a *difference operator*. This concept was discussed in detail for the discrete Fourier transform in Section 4.1.5. In the present case, it would be more difficult as the sample spacings, or differences, $x_{n,m+1} - x_{n,m}$ are not constants.

To compare the discrete Legendre transform method with the classical finite transform method, we choose for the boundary condition (E.2) the function

$$v(1,x) = f(x) = -\tfrac{1}{2}\ln(1-x) \qquad (E.10)$$

whose finite Legendre transform $F(j)$ is known in closed form,

$$F(j) = \frac{1}{j(j+1)} \qquad (E.11)$$

which gives the infinite Legendre series solution (E.3) as

$$v(\rho, \cos\theta) = \frac{1}{2}\sum_{j=0}^{\infty}\frac{(2j+1)}{j(j+1)}\rho^{j}P_{j}(\cos\theta) \qquad (E.12)$$

When the truncation Legendre series of (E.12) with 16 terms was used, the truncation error ranged between 1.25×10^{-15} and 1.53×10^{-1} for $\rho = 0.1$ and $\rho = 0.999$, respectively. This good accuracy at $\rho = 0.1$ was destroyed when, instead of using the exact $F(j)$ of (E.11), it was evaluated numerically via its integral representation (E.4), using Simpson's rule with 200 terms. The error became of order 10^{-2}, and hence the truncation error for the 16-term Fourier series (E.3) at $\rho = 0.1$. When the discrete Legendre transform (E.7) was used with the same 16 terms, the difference between the exact solution $v(\rho, x_{n,m})$ and the approximate discrete transform solution ranged between 1.57×10^{-3} and 6.64×10^{-3} as ρ ranged between 0.1 and 0.999. This indicates a more stable solution, as a function of ρ, than that of the truncated classical Fourier series (E.12).

More illustrations of the analysis and computations for the present discrete orthogonal transforms may be found in Jerri (1979). For general references, see Luke (1975), Luke et al. (1975), and Szego (1959).

We should mention that the advantage illustrated here for the use of the discrete Legendre transform is a direct consequence of the power of the classical polynomial Gaussian quadratures in approximating functions. Such properties produced important formulas for the derivation of the discrete polynomial transforms, such as the Christoffel–Darboux formula (see Exercise 4.28), which was used for deriving the important discrete orthogonality (4.153) of these transforms. These formulas will be missed when we deal with general transforms with nonpolynomial kernel. An example will be an "approximate" discrete Hankel (Bessel) transform, which we present briefly in the following section.

4.3 Bessel-type Poisson Summation Formula (for the Bessel-Fourier Series and Hankel Transforms)

Like the discrete Fourier transforms (4.3), (4.4) of Section 4.1 and the discrete classical orthogonal polynomial transforms (4.159), (4.160) of Section 4.2, the discrete Hankel transform (when established!) would be defined as a similar finite sum. However, due to the lack of an exact discrete orthogonality, we cannot have an (exact) discrete Hankel transform pair like the Fourier one in (4.3), (4.4). The best that can be done at the present is to define the discrete Hankel transform $\bar{H}_0$ as a finite sum, as we will present it at the end of this section in (4.192). However, its discrete inverse $\bar{h}_0$ can be written only as an "approximate" sum, whose discussion we shall leave for the exercises (see Exercise 4.41 and 4.42). This is in the sense that we cannot transform N points of $\bar{h}_0$ to N points of $\bar{H}_0$ and be able to transform back to the exact N points of $\bar{h}_0$.

In the following we will derive the (new) *Bessel-type Poisson sum formula* (4.190), where we will depend on the *impulse train* (3.204) of the Bessel series that we introduced in Section 3.6.2. These two important tools—the impulse train and the Poisson summation formula—are the basic steps toward any attempt at a discrete Hankel transform pair. In fact, the following Poisson summation formula has already paved the way for the first estimate of the aliasing error of the Bessel sampling expansion (3.181). Thus it helped us avoid the major assumption of periodicity, which the aliasing error bounds of the (Fourier) sampling series had to rely heavily on.

It is instructive first to introduce the familiar Poisson sum formula of Section 4.1.4D with its simple derivation and show its importance in reducing the infinite Fourier integral to a finite-limit one. This is followed by deriving a (new) general Poisson-type sum formula, which is illustrated for the Hankel (Bessel) transform.

Let $y(t)$ be the Fourier transform of $Y(\omega)$,

$$y(t) = \int_{-\infty}^{\infty} Y(\omega)e^{i\omega t}\, d\omega \tag{4.177}$$

$$Y(\omega) = \frac{1}{2\pi} \int_{-\infty}^{\infty} y(t)e^{-i\omega t}\, dt \tag{4.178}$$

Consider $\bar{y}(t)$, the superposition of all the translations of $y(t)$ by nT (see (4.11) and Fig. 4.6),

$$\bar{y}(t) = \sum_{n=-\infty}^{\infty} y(t + nT) \tag{4.179}$$

which is, of course, periodic with period T. Now we write the Fourier series expansion of $\bar{y}(t)$,

$$\bar{y}(t) \equiv \sum_{n=-\infty}^{\infty} y(t + nT) = \frac{1}{T} \sum_{k=-\infty}^{\infty} c_k e^{i2\pi kt/T} \tag{4.180}$$

$$c_k = \int_{-T/2}^{T/2} \bar{y}(t) e^{-i2\pi kt/T} \, dt = \int_{-T/2}^{T/2} \sum_{-\infty}^{\infty} y(t + nT) e^{-i2\pi kt/T} \, dt \tag{4.181}$$

If we exchange the integration and summation, make the simple change of variables $x = t + nT$, and use the periodicity of $e^{-i2\pi kt/T}$, the integral in (4.181) becomes an infinite integral,

$$c_k = \int_{-\infty}^{\infty} y(t) e^{-i2\pi kt/T} \, dt \equiv Y\left(\frac{2\pi k}{T}\right) \tag{4.182}$$

Hence (4.180) and (4.182) give the usual *Poisson summation formula for the Fourier transform*,

$$\sum_{n=-\infty}^{\infty} y(t + nT) = \frac{1}{T} \sum_{n=-\infty}^{\infty} Y\left(\frac{2\pi n}{T}\right) e^{i2\pi nt/T} \tag{4.183}$$

With this method we can see clearly how the samples $Y(2\pi k/T)$, as an infinite integral of $y(t)$ in (4.182), can be expressed as the finite integral of $\bar{y}(t)$ in (4.181),

$$Y\left(\frac{2\pi k}{T}\right) = \int_{-\infty}^{\infty} y(t) e^{-i2\pi kt/T} \, dt = \int_{-T/2}^{T/2} \bar{y}(t) e^{-i2\pi kt/T} \, dt \tag{4.184}$$

This is very important when we attempt to approximate the infinite integral, which is now a finite integral, by the discrete Fourier transform and its fast algorithm, the fast Fourier transform. We may remark here that, according to the above method, obtaining (4.182) from (4.181) depends entirely on the direct application of the periodicity, and hence the translated replicas from all the equal intervals (with length T) of the real line $(-\infty, \infty)$ to the basic finite interval $(-T/2, T/2)$. Thus if we are to move to other orthogonal expansion, which in general are not periodic, like the Bessel series expansion, we have to dispense with the periodicity and should be contented with some *generalized* form of repetition such as what we introduced in (3.185) and illustrated in (3.185a). Anticipating such a difficulty, we present another well-known approach for obtaining (4.183), which is more in the direction of our general development. Here we use the *impulse train* as defined by the divergent Fourier series,

$$\sum_{n=-\infty}^{\infty} \delta(t + nT) \equiv \frac{1}{T} \sum_{n=-\infty}^{\infty} e^{i2\pi nt/T} \tag{4.185}$$

and convolve it with $y(t)$ to obtain $\bar{y}(t)$ and the *Poisson summation formula* (4.183)

$$\bar{y}(t) \equiv \sum_{n=-\infty}^{\infty} y(t + nT) = y(t) * \sum_{n=-\infty}^{\infty} \delta(t + nT)$$

$$= y(t) * \frac{1}{T} \sum_{n=-\infty}^{\infty} e^{i2\pi nt/T}$$

$$= \frac{1}{T} \sum_{n=-\infty}^{\infty} \int_{-\infty}^{\infty} y(x) e^{i2\pi n(t-x)/T} \, dx$$

$$= \frac{1}{T} \sum_{n=-\infty}^{\infty} e^{i2\pi nt/T} \int_{-\infty}^{\infty} y(x) e^{-i2\pi nx/T} \, dx$$

$$= \frac{1}{T} \sum_{n=-\infty}^{\infty} Y\left(\frac{2\pi n}{T}\right) e^{i2\pi nt/T} \tag{4.183}$$

Such an approach will be very useful in establishing the generalization of (4.183) to other integral transforms like the Hankel (Bessel) transform. As we shall see next, this will depend on the generalized translation concept in (3.185) and even more so on the *generalized impulse train* that we introduced in (3.196) and (3.197) and illustrated for Bessel kernels in (3.204) and Fig. 3.13.

Our main example here would be a Hankel (Bessel) transform in conjunction with Fourier–Bessel series. This will exemplify the contrast of the *nonperiodic* nature of the Fourier–Bessel series expansion with the periodic Fourier (trigonometric) series.

Consider the J_0-Hankel (Bessel) transform $H_0(\omega)$,

$$H_0(\omega) = \int_0^{\infty} t h_0(t) J_0(t\omega) \, dt \tag{4.186}$$

and its inverse $h_0(t)$,

$$h_0(t) = \int_0^{\infty} \omega H_0(\omega) J_0(\omega t) \, d\omega \tag{4.187}$$

where J_0 is the Bessel function of the first kind of order zero. The same can be done for $J_m(x)$, but we stay with $J_0(x)$ to simplify the first illustration. The zero subscript in $H_0(\omega)$ and $h_0(t)$ was used to specify the J_0-Hankel transforms, but in the sequel it will be dropped for simplicity. In parallel to the development of using (4.185) to derive the (usual) Poisson summation formula (4.183), we will introduce $\overline{H}(\omega)$, the superposition of all the "generalized translations" (3.185) of $H(\omega)$ in (4.186). This can be obtained by convolving $H(\omega)$ (in the sense of the convolution product (3.186)) with the impulse train $X(\omega)$ in (3.204) to get $\overline{H}(\omega) = (H * X)(\omega)$,

$$\overline{H}(\omega) = H(\omega) * \left[\frac{\delta(\omega)}{\omega} + \sum_{m=1}^{\infty} d_m \frac{\delta(\omega - 2mb)}{\omega} \right], \qquad |d_1| = 1$$

$$\overline{H}(\omega) = \int_0^{\infty} x H(\omega \theta x) \frac{\delta(x)}{x} \, dx + \sum_{m=1}^{\infty} d_m \int_0^{\infty} x H(\omega \theta x) \frac{\delta(x - 2mb)}{x} \, dx$$

$$\overline{H}(\omega) = H(\omega) + \sum_{m=1}^{\infty} d_m H(\omega \theta 2mb), \qquad |d_1| = 1 \qquad (4.188)$$

Before we convolve $H(\omega)$ with the Bessel-type impulse train, as the Bessel series $X_s(\omega)$ of the right-hand side of (3.204), we should mention that the sought $(H * X_s)(\omega)$ is again a J_0-Bessel series,

$$(H * X_s)(\omega) = H(\omega) * \frac{2}{b^2} \sum_{n=1}^{\infty} \frac{J_0(\omega j_{0,n}/b)}{J_1^2(j_{0,n})}$$

$$= \frac{2}{b^2} \sum_{n=1}^{\infty} \frac{\int_0^{\infty} x H(\omega \theta x) J_0(x j_{0,n}/b) \, dx}{J_1^2(j_{0,n})}$$

$$= \frac{2}{b^2} \sum_{n=1}^{\infty} \frac{h(j_{0,n}/b) J_0(\omega j_{0,n}/b)}{J_1^2(j_{0,n})}$$

which would obviously vanish at the end point $\omega = b$, while in general there is no good reason for the above $\overline{H}(\omega)$ to vanish at $\omega = b$. To be certain about this point, we have verified it numerically for a number of Hankel transform pairs, as we shall discuss in the exercises. For this reason, we will seek the same type of Bessel series on $(0,b)$ for the function $\hat{H}(\omega) \equiv \overline{H}(\omega) - \overline{H}(b) = (H * X)(\omega) - (H * X)(b)$, which vanishes at $\omega = b$. This requires convolving $H(\omega)$ with both sides of (3.204), then evaluating the resulting expression at ω and b to get

$$\hat{H}(\omega) = \overline{H}(\omega) - \overline{H}(b) = (H * X)(\omega) - (H * X)(b)$$

$$= (H * X_s)(\omega) - (H * X_s)(b) = H(\omega) * \frac{2}{b^2} \sum_{n=1}^{\infty} \frac{J_0(\omega j_{0,n}/b)}{J_1^2(j_{0,n})} \Bigg|_{\omega=b}^{\omega}$$

$$= \frac{2}{b^2} \sum_{n=1}^{\infty} \frac{\int_0^{\infty} x H(\omega\theta x) J_0(x j_{0,n}/b)\, dx}{J_1^2(j_{0,n})} \Bigg|_{\omega=b}^{\omega}$$

$$= \frac{2}{b^2} \sum_{n=1}^{\infty} \frac{h(j_{0,n}/b) J_0(\omega j_{0,n}/b)}{J_1^2(j_{0,n})} \tag{4.189}$$

If we combine this result with the expression for $\overline{H}(\omega)$ in (4.188), we obtain the desired *Bessel-type Poisson summation formula*,

$$\tilde{H}(\omega) = \overline{H}(\omega) - \overline{H}(b) = \sum_{m=0}^{\infty} d_m [H(\omega\theta 2mb) - H(b\theta 2mb)]$$

$$\tilde{H}(\omega) = H(\omega) - H(b) + \sum_{m=1}^{\infty} d_m [H(\omega\theta 2mb) - H(b\theta 2mb)]$$

$$= \frac{2}{b^2} \sum_{n=1}^{\infty} \frac{h(j_{0,n}/b) J_0(\omega j_{0,n}/b)}{J_1^2(j_{0,n})} \tag{4.190}$$

As a relatively new result, (4.190) was verified numerically with a number of J_0-Hankel transform pairs directly in (4.190), whose discussion and computations we leave for the exercises. The exercises will also include the basic elements of establishing the first bound on the aliasing error for the Bessel-sampling series (3.181), which is based primarily on the above Bessel-type Poisson summation formula (4.190) (see Exercises 4.36–4.40, where you may need the detailed analysis and illustrations given in Jerri (1988, 1992).

We should remark here that even though we wrote $\delta(\omega\theta x) = \delta(\omega - x)/x$ for the purpose of locating the pulses (or sample values), it is essential in the above derivation of (4.189), (4.190) that we use $H(\omega\theta x)$, with its *generalized translation* for the above process of convolving. This is necessary for bringing about the main feature of a Poisson-type summation formula, which is to involve the samples $h(j_{0,n}/b)$ of the (infinite limit) integral transform of $H(\omega)$ in (4.190). This $h(j_{0,n}/b)$ will be in the position of the required finite integral of the Fourier–Bessel coefficients of $\tilde{H}(\omega) = \overline{H}(\omega) - \overline{H}(b)$, the modified (or aliased) $H(\omega)$ on $(0,b)$. This $\tilde{H}(\omega)$ in (4.190) can be considered as an "aliased" version of $H(\omega)$ on $(0,b)$ because of the use of the (infinite) integral of $h(j_{0,n}/b)$ for the Fourier coefficients of the series for $\tilde{H}(\omega)$ on $(0,b)$. This means that if the series on the right-hand side of (4.190) is to be used to approximate the original (infinite) transform $H(\omega)$ on $(0,b)$, there will be an *aliasing error* ϵ_A in the general Fourier series expansion

$$\epsilon_A = |H(\omega) - \tilde{H}(\omega)| = \left| H(b) - \sum_{m=1}^{\infty} d_m [H(\omega\theta 2mb) - H(b\theta 2mb)] \right|$$

(4.191)

(See Exercises 4.37–4.39 for $\epsilon_{A,S}$, the aliasing error of the Bessel—sampling series.)

An Approximate Discrete Hankel Transform

In the absence of an exact discrete orthogonality for a discrete Hankel transform pair—or, in other words, if and until the above development leads to such discrete transform pairs—we will resort to approximate discrete Hankel transforms. This is what we shall present next.

For the J_0-Hankel transform pair in (4.186), (4.187), the *discrete J_0-Hankel transform* $\tilde{H}_0$ (DHT) is defined as

$$\tilde{H}_0 \left(\frac{j_{0,n}}{c} \right) = \frac{2}{b^2} \sum_{k=1}^{K \ni j_{0,K} = bc} \frac{h(j_{0,k}/b) J_0(j_{0,k} j_{0,n}/bc)}{J_1^2(j_{0,k})}$$

(4.192)

$$n = 1, 2, \ldots, N, \ j_{0,N} = bc, \ J_0(j_{0,n}) = 0$$

The discrete inverse J_0-Hankel transform (DIHT) $\tilde{h}_0$ can be written only as "an approximate," which we leave along with its "approximate" discrete orthogonality for the exercises (see Exercise 4.41).

Generalized Translation

As in the case of the discrete orthogonal polynomial transforms of Section 4.2, since we are dealing with Bessel functions as kernels of the transforms, we cannot expect the usual translation as in the convolution product of the Fourier or Laplace transform, which is the consequence of their special exponential kernels. So instead of the simple translation $f(j - k)$, as in the case of the discrete Fourier transform, we have a general translation $f(j\theta k)$ in parallel to that in (3.185). In the last section, we applied the same concept of generalized translation to the discrete classical polynomial transforms. In the case of the discrete J_0-Hankel transforms we define the *general translation* of $\tilde{h}_0(j_{0,k}/b)$ by k' as

$$\tilde{h}_0 \left(\frac{j_{0,k}\theta k'}{b} \right) = \frac{2}{b^2} \sum_{n=1}^{N \ni j_{0,N} = bc} \frac{\tilde{H}_0(j_{0,n}/c) J_0(j_{0,n} j_{0,k}/bc) J_0(j_{0,n} j_{0,k'}/bc)}{J_1^2(j_{0,n})},$$

$$k, k' = 1, 2, \ldots, K \ni j_{0,K} = bc$$

(4.193)

This definition is not surprising, since the translation for the Fourier transform is also the result of multiplying the transformed function by an exponential function of the same type as the transform kernel.

Computations of Hankel Integral Transforms
A total of 20 different functions with known continuous J_0-Hankel transforms were used as h_0 in (4.192) to evaluate the discrete transform $\bar{H}_0$. The results were compared with those of evaluating such integrals, using Simpson's rule, as truncated ones $H_{0,S}$ on $(0,c)$ and with the same number of terms N. The results of these computations and their comparisons are the subject of Exercise 4.42. Such functions included e^{-t}, e^{-2t^2}, $1/\sqrt{1+t^2}$, $1/t$, and

$$h_0(t) = \begin{cases} 1 - t^2, & 0 \le t \le 1 \\ 0, & t > 1 \end{cases}$$

Discrete J_m-Hankel transforms and other Extensions
The discrete J_m-Hankel transform associated with the Bessel function $J_m(x)$ of the first kind of order m (DH_mT) can easily be written, in parallel to that of the discrete J_0-Hankel transform in (4.192), as

$$\bar{H}_m\left(\frac{j_{m,n}}{c}\right) = \frac{2}{b^2} \sum_{k=1}^{K \ni j_{m,K}=bc} \frac{\bar{h}_m(j_{m,k}/b)J_m(j_{m,k}j_{m,n}/bc)}{J_{m+1}^2(j_{m,k})},$$

$$n = 1,2,\ldots,N, \quad j_{m,N} = bc, \quad J_m(j_{m,n}) = 0^\dagger \qquad (4.194)$$

The approximate inverse $\bar{h}_m(j_{m,k})$ of this discrete transform has the same form (see Exercise 4.40 for detailed analysis.)

Discrete Hankel transform, like that of (4.194), could also be derived in the same way with samples at the zeros of the equation

$$aJ_m(x) + hJ_m'(x) = 0 \qquad (4.195)$$

instead of the special case $J_m(x) = 0$, which makes the transforms suitable for solving other more general boundary value problems.

Relevant References for Chapter 4

The basic references, among others, for this chapter are Papoulis (1977), Brigham (1974, 1988), Hamming (1973), Cooley et al. (1969), Weaver (1983, 1988), Briggs and Jerri (1985), Ziemer et al. (1983), Atkinson (1964), Luke (1975), Luke et al. (1975), Elliot and Rao (1982), and Jerri (1988, 1977a, 1992).

---

†Twenty zeros $\{j_{m,n}\}_{n=1}^{20}$ of $J_m(x)$ for $m = 0,1,2,3,4,5$ are listed in Table A.1 of Appendix A.5.

Exercises

Section 4.1   Discrete Fourier Transforms (DFT)

4.1. *Approximating Fourier integrals by the discrete Fourier transforms.* (See also Exercises 4.5, 4.7, and 4.8.) Consider the function

$$g(t) = \frac{2}{1 + (2\pi t)^2}, \qquad -\infty < t < \infty \qquad \text{(E.1) (see (4.1))}$$

and its Fourier transform as in Figs. 4.4, 4.3.

$$G(f) = e^{-|f|}, \qquad -\infty < f < \infty \qquad \text{(E.2) (see (4.2))}$$

For the numerical computations take $N = 32$, $T = \frac{1}{4} = \Delta t$, so $\Delta f = 1/NT = \frac{1}{8}$. Note: For (4.5), (4.6) we have $N = 2M + 1$. This pair was used in the text, but we still want it to serve as a lead for doing the following analysis for computing *Fourier transform pairs*, in Exercises 4.5, 4.7, and 4.8.

(a) Graph $g(t)$, then $G(f)$. For the scale of the graph see Figs. 4.4, 4.3, and for the rest of the problem see Figs. 4.5–4.7.

(b) With the same scale as in (a), graph $\bar{g}(t)$ of (4.12) showing *three* translations in the graph. Note that the nontranslated part (for $n = 0$ in (4.12)) corresponds exactly to that of part (a). However it will suffer here from overlappings. For $K \to \infty$, you may take $K \to 40$ (see Fig. 4.7).

(c) Do the same for $\bar{G}(f)$ of (4.11) as you did for $\bar{g}(t)$ in part (c).

(d) Evaluate and then graph $\bar{g}(kT) = \tilde{g}(kT)$ of (4.12).

(e) Use this $\bar{g}(kT)$ of part (d) in (4.3) to evaluate and then graph $\bar{G}(n/NT)$. We expect it to be real function, but graph the imaginary part (as error) if any. Compare the discrete transforms $\bar{g}$ and $\bar{G}$ that you found with their corresponding integral transforms g and G of (E.1), (E.2). See Figs. 4.5, and 4.6.

(f) Compare this discrete result $\bar{G}(n/NT)$ with the samples $G(n/NT)$. Graph both in one figure and compare. See Fig. 4.6.

(g) Use (4.11), (4.11a), and (4.11b) to check the error in part (f) as a function of K. See if you can estimate some behavior of the aliasing error as a function of K.

(h)   Study the aliasing error $\tilde{g}(kT) - g(kT)$, as you did in parts (f) and (g) for $\tilde{G}(n/NT) - G(n/NT)$, See (4.12), (4.12a), and (4.13) and Figs. 4.4 and 4.7a,b.

(i)   Research parts (a) to (h) with different N and T, repeating all your steps. Draw your conclusions about the approximation and its error.

(j)   Repeat Example 4.7 with $N = 128$ and compare your results with the conclusions from Example 4.7.

4.2. *Approximating Fourier series by the discrete Fourier transform.* (See also Exercises 4.3, 4.4, and 4.6.)   Consider the Fourier series expansion (4.7) of the periodic function

$$G_s(f) = e^{-|f|}, \qquad -\frac{1}{2T} < f < \frac{1}{2T} \tag{E.1}$$

with its Fourier coefficients (4.8) as in Fig. 4.1a,b

$$c_k = 2T \frac{1 + (-1)^{k+1} e^{-1/2T}}{1 + (2\pi kT)^2} \tag{E.2}$$

This pair was used in the text, and we want it to serve as a lead for doing the following analysis, for computing *the Fourier series and its coefficients*, in Exercises 4.3, 4.4 and 4.6.

(a)   Graph $G_s(f)$ and c_k (with the same scale as in Exercise 4.1a). See Figs. 4.1a,b and 4.2.

(b)   With the same scale as in (a), graph the periodic extension $\tilde{G}_s(f)$ of (4.9) with period $1/T$, showing three translations (see Fig. 4.2a, $1/T = 4$).

(c)   Show the three translations of c_k in (4.55c) corresponding to $l = 0$, 1 and -1, that is, c_k, c_{k+N}, and c_{k-N}, on the same graph. Graph $\bar{c}_k = C_k$ in (4.55c) with $|l| = 1$ to 40; i.e., take $K \to 40$ instead of ∞.

(d)   Use $\tilde{G}(n/NT) = G_s(n/NT)$ and graph it. This is an advantage of Fourier series.

(e)   Let $C_k = \tilde{g}(kT)$. Use $\tilde{G}(n/NT)$ of part (d) in (4.3) to evaluate and graph $\tilde{g}(kT)$, which should be exactly equal to $C_k = \tilde{g}(kT)$. Compare these $\tilde{G}$ and $\tilde{g}$ with G_s and c_k in part (a).

(f)   Note the advantage of the Fourier series, over the Fourier integral approximation by a discrete Fourier transform, in that we have $\tilde{G}(n/NT) = G_s(n/NT)$ when $\tilde{g}(kT) = C_k \equiv \bar{c}_k$.

(g) See part (c) to check the error in part (a), $\epsilon_A = C_k - c_k = \bar{c}_k - c_k$ as a function of K, and compare with that given by (4.55c). Attempt to estimate the error behavior as a function of K.

(h) Research parts (a) to (g) with different N and T. Draw your conclusions about the approximation and its error.

4.3. Consider the periodic functions and their corresponding Fourier coefficients,

(i) $f(x) = \begin{cases} 1, & -a < x < 0 \\ 0, & 0 < x < a \end{cases}$

$c_k = \begin{cases} 0, & k \text{ even} \\ \dfrac{-j}{\pi k}, & k \text{ odd}, \ j = \sqrt{-1} \end{cases}$

(ii) $f(x) = e^x, \quad -l < x < l$

$c_k = \dfrac{\sinh al \cos k\pi}{al - n\pi j}, \ j = \sqrt{-1}$

(iii) $f(x) = \cos x, \quad -l < x < l$

$c_k = \dfrac{l \sin l \cos k\pi}{l^2 - k^2\pi^2}$

(iv) $f(x) = \begin{cases} -1, & -l < x < 0 \\ 1, & 0 < x < l \end{cases}$

$c_k = \begin{cases} 0, & k = 0 \\ \dfrac{1 - \cos k\pi}{k\pi j}, & k \neq 0 \end{cases}$

Repeat parts (a) to (g) of Exercise 4.2 for each of the above Fourier series pairs.

4.4. Repeat steps (a) to (g) of Exercise 4.2 using the symmetric form of the discrete transform (4.5), (4.6) with $M = 32$, $2M + 1 = 65$, $T = \frac{1}{4}$. Observe the change in (4.55c) and (4.62) for $N \to 2M + 1$.

4.5. Repeat parts (a) to (i) of Exercise 4.1 using the symmetric form (4.5), (4.6) with $M = 32$, $2M + 1 = 65$, $T = \frac{1}{4}$. Observe the change in (4.55c) and (4.62) for $N \to 2M + 1$.

4.6. Try the discrete Fourier sine-cosine transform (4.15)–(4.17) for Exercise 4.2a–g, as much as it pertains to the important question of approximating Fourier series by discrete Fourier transforms.

4.7. Consider the function

$$g(t) = \begin{cases} e^{-t}, & t > 0 \\ 0, & t < 0 \end{cases} \tag{E.1}$$

of Example 4.7 and its complex-valued Fourier transform (see Fig. 4.9a–d and Fig. 4.34a–c)

$$G(f) = \frac{1}{(2\pi f)^2 + 1} - j\frac{2\pi f}{(2\pi f)^2 + 1} \equiv G_r(f) + jG_i(f) \tag{E.2}$$

Note that the absolute value of $G(f)$ is

$$|G(f)| = \frac{1}{\sqrt{1 + (2\pi f)^2}} \tag{E.3}$$

(a) Graph $g(t)$, $|G(f)|$, the real part $G_r(f) = 1/(1 + (2\pi f)^2)$, and the imaginary part $G_i(f) = -2\pi f/(1 + (2\pi f)^2)$ of $G(f)$ (see Fig. 4.34).

(b) Use the symmetric form (4.5), (4.6) for part (a) with $N = 2M + 1 = 33$, $T = \frac{1}{4}$, repeating the steps in Exercise 4.1a–i.

(c) For $g(t)$ in (E.1) use: (i) the interval $(-4, 4)$ and 128 samples and (ii) the interval $(-8, 8)$ and 32 samples to compute the DFT approximation of $G_r(f)$ and $G_i(f)$ in (E.2). Compare your results with the conclusions in Example 4.7.

4.8. (a) Repeat Exercise 4.7 for the Fourier transform pair

$$g(t) = (1 - t)e^{-|t|}, \qquad -\infty < t < \infty,$$

$$G(f) = \frac{2}{1 + (2\pi f)^2} + j\frac{8\pi f}{1 + (2\pi f)^2}, \qquad -\infty < f < \infty$$

(b) Repeat Exercise 4.7 for the functions in Fig. 4.26.

4.9. (a) Prove the following *discrete orthogonality properties* for the trigonometric sine and cosine functions:

$$\sum_{k=0}^{N-1} \cos\left(\frac{2\pi\alpha k}{N}\right) \cos\left(\frac{2\pi\beta k}{N}\right) =$$

$$= \begin{cases} 0, & \alpha \neq \beta \\ \dfrac{N}{2}, & \alpha = \beta \neq 0, N \\ N, & \alpha = \beta = 0, N \end{cases} \tag{E.1}$$

$$\sum_{k=0}^{N-1} \sin\left(\frac{2\pi\alpha k}{N}\right) \sin\left(\frac{2\pi\beta}{N}\right)$$

$$= \begin{cases} 0, & \alpha \neq \beta \\ \dfrac{N}{2}, & \alpha = \beta \neq 0, N \\ 0, & \alpha = \beta = 0, N \end{cases} \tag{E.2}$$

$$\sum_{k=0}^{N-1} \sin\left(\frac{2\pi\alpha k}{N}\right) \cos\left(\frac{2\pi\beta k}{N}\right) = 0 \tag{E.3}$$

(Use the identity: $\cos A \cos B = \frac{1}{2}[\cos(A+B) + \cos(A-B)]$ and its relatives.)

(b) Use the above discrete orthogonality of the trigonometric functions in (E.1)–(E.3) to derive the sine-cosine form of the DFT as in (4.15)–(4.17) without resorting to the complex form (4.3), (4.4).

Hint: Write (4.15) and then multiply by one of the trigonometric functions, sum over, from $k = 0$ to $k = N - 1$, interchange the double sum, and use the suitable discrete orthogonality from (E.1)–(E.3).

4.10. Consider the discrete sine-cosine form of the DFT in (4.15)–(4.17).

(a) Show that if the sequence $\tilde{G}_n$ is real, then we have the following useful properties for its discrete transform:

(i) A_n and B_n are real.

   Hint: See Section 4.1.2 and Table 4.1.

(ii) $\tilde{g}_k$ has the symmetry $\tilde{g}_{-k} = \tilde{g}_{2M-k} = \overline{\tilde{g}}_k$, where the bar stands for complex conjugation; thus we can write

$$A_k = 2\operatorname{Re}\{\tilde{g}_k\} \quad \text{and} \quad B_k = 2\operatorname{Im}\{\tilde{g}_k\}$$

   Hint: $A_k = \tilde{g}_k + \tilde{g}_{2M-k}$

(iii) $A_{-k} = A_k$

$$A_{N-k} = A_k, \qquad 0 \le k \le \frac{N}{2}$$

$$B_{-k} = -B_k$$

$$B_{N-k} = -B_k, \qquad 0 \le k \le \frac{N}{2}$$

$$B_0 = B_{N/2} = 0$$

$$|\tilde{g}_k| = \tfrac{1}{2}\sqrt{A_k^2 + B_k^2}$$

(b) (i) Show that the period of the kth harmonic $e^{j2\pi nk/N}$ (or $\cos(2\pi nk/N)$ and $\sin(2\pi nk/N)$ on $0 \le n \le N$ is N/k.

  (ii) From the information in part (a), justify that "any sequence of N points can be exactly represented by using roughly the first $N/2$ harmonics."

4.11. (a) Show that $\tilde{G}_n = \{n\}$ is the DFT of

$$\tilde{g}_k = \begin{cases} \dfrac{N-1}{2}, & k = 0 \\[2mm] \dfrac{1}{W^k - 1}, & 1 \le k \le N-1, \ W \equiv e^{j2\pi/N} \end{cases} \qquad \text{(E.1)}$$

Hint: See (4.4) and the symmetric derivation of the DFT pair in (4.98).

(b) Find the DFT of the sequences:

  (i) $\tilde{G}_n = \left\{ \cos \dfrac{2\pi ln}{N} \right\}_{n=0}^{N-1}$, l an integer, $0 \le l \le N-1$

  (ii) $\tilde{G}_n = \{ e^{-i2\pi ln/N} \}_{n=0}^{N-1}$, l an integer, $0 \le l \le N-1$

  (iii) $\tilde{G}_n = \{1\}_{n=0}^{N-1}$

(c) For the constant sequence $\tilde{G}_n = \{1\}$, show that it needs only one harmonic $k = 0$.

Hint: Use the geometric series. (See parts (b) iii and (b) i and their answers.)

(d) For the DFT pair in (E.1), take $N = 16$, and illustrate $\text{Re}\{\tilde{g}_k\}$, $\text{Im}\{\tilde{g}_k\}$, and the spectrum $|\tilde{g}_k| = \tfrac{1}{2}\sqrt{A_k^2 + B_k^2}$ for the DFT complex representation in (4.3) and its sine-cosine representation in (4.15). Relate the results to those in part (a) and (bii) of Exercise 4.10.

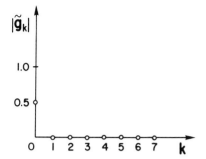

FIG. 4.37 Spectrum $|\tilde{g}_k|$ for answer of Exercise 4.11c.

ANS. (b) (i) $\tilde{g}_k = \frac{1}{2}[\delta_{-l,k} + \delta_{l,k}] = \frac{1}{2}[\delta_{l,k} + \delta_{l,N-k}]$

(ii) $\tilde{g}_k = \delta_{k,l}$

(iii) $\tilde{g}_k = \delta_{k,0} = \begin{cases} 1, & k = 0 \\ 0, & k \neq 0 \end{cases}$

(c) See part (b iii). Spectrum $|\tilde{g}_k| = \frac{1}{2}\sqrt{A_k^2 + B_k^2}$ for $\{G_n\} = \{1\}$ (Fig. 4.37).

(d) See Figs. 4.38–4.40.

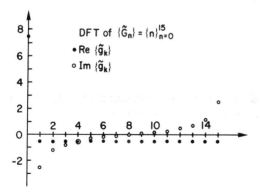

FIG. 4.38 DFT of $\{\bar{G}_n\} = \{n\}_{n=0}^{15}$, $\mathrm{Re}\,\{\tilde{g}_k\}$, $\mathrm{Im}\,\{\tilde{g}_k\}$ (answer to Exercise 4.11d).

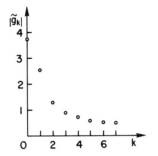

FIG. 4.39 Spectrum $|\tilde{g}_k| = \frac{1}{2}\sqrt{A_k^2 + B_k^2}$ for $\{\bar{G}_n\} = \{n\}_{n=0}^{15}$. (answer to Exercise 4.11d).

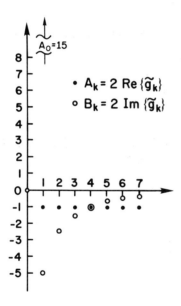

FIG. 4.40 Sine and cosine coefficients A_k, B_k for $\{\bar{G}_n\} = \{n\}_{n=0}^{15}$ (answer to Exercise 4.11d).

4.12. (a) Derive the geometric series in (4.94).

Hint: Note $S_N - S_{N-1} = a^N$.

(b) Use the integrating factor to solve the differential equation in (E.1) of Example 4.1. Can you think of a parallel to this method for the finite-difference equation analog (E.2) of (E.1)?

4.13. In Example 4.4 of the traffic flow, what assumptions are required to ensure that an intersection loop has a steady-state traffic flow when the entrance/exit traffic at intersection n is given by $\sin(n\pi/3)$? Find the flow of traffic through the streets of this network.

ANS. Closed, entrance with $V/2$ (vehicles), entrance with $V/2$, exit with $V/2$, exit with $V/2$, closed ....

4.14. (a) Prove the general algebraization property of the z-transform in (E.7) of Example 4.5.

Hint: See the two special cases in (E.5) and (E.6) and continue the same process.

(b) Use the z-transform to solve the initial value problem with nonhomogeneous difference equation in (E.11), (E.12) of Example 4.5.

Hint: Use the result of $Z\{\sin(\pi n/2)\} = z/(z^2+1)$, then consult the z-transform table in Appendix B.17 to find the inverse z-transform after doing some partial fractions.

(c) Verify the answer to part (b) by showing that it produces the same sequence by evaluating the difference equation in (E.11) of Example 4.5 recursively.

(d) Attempt to generate the final result in (E.8) of Example 4.6 by relating it to the general solution in (E.6) of Example 4.2 subject to the present initial values $u_0 = 1$ and $u_1 = 2$ in (E.3) of Example 4.6.

ANS. (b) $u_n = \frac{31}{50}(3)^n - \frac{16}{15}(-2)^n$

$$-\frac{1}{50}\begin{cases} (-1)^{(n/2)-1} & \text{if } n \text{ is even} \\ 7(-1)^{(n-1)/2} & \text{if } n \text{ is odd} \end{cases}$$

4.15. Consider Figs. 4.17–4.19 illustrating the effect of widening the truncation window to minimize the windowing effect.

(a) Use the function

$$G(f) = \frac{4 \sin^2 \omega b}{2b\omega^2} \tag{E.1}$$

and its transform

$$g(t) = q_{2b}(t) = \begin{cases} 1 - \dfrac{|t|}{2b}, & |t| < 2b \\ 0, & t > 2b \end{cases} \tag{E.2}$$

with $b = 4$ to illustrate the windowing effect using the (same gate function type) truncation window of (half) width $a = 4$, 6, and 7. Graph these three results and compare them to Figs. 4.17, 4.18, and 4.19, respectively.

Hint: Note that $q_{2b}(t)$ in (E.1) is the self-convolution of the gate function $p_b(t)$, that is, $q_{2b}(t) = (p_b * p_b)(t)$, which makes $G(f)$ bandlimited with band limit of $2b$, where a window of (half) width $a = 8 = 2b$ should give an exact result.

(b) Repeat the illustration of reducing the windowing effect in part (a) for the following Fourier transforms:

$$G(f) = \frac{2\pi^2 \sin 2\pi f b}{2\pi f (\pi^2 - 4\pi^2 b^2 f^2)} \tag{E.3}$$

$$g(t) = \begin{cases} 1 + \cos \dfrac{\pi}{b}, & |t| < b \\ 0, & |t| > b \end{cases} \tag{E.4}$$

with $b = 4$ and window widths of 2, 4, 6, and 8.

Hint: Note that $G(f)$ in (E.3) is a bandlimited function with a band limit of b as indicated in (E.4). Graph and compare your results for the four window widths to those in Figs. 4.17, 4.18, and 4.19.

(c) The $g(t)$ in (E.4) of part (b) is itself a useful truncating window with width $2b$ (see Fig. 4.31a). It is characterized by a much less abrupt drop at $\mp b$ than that of the gate function window used in Figs. 4.17–4.19 and parts (a) and (b). Use this window instead of the gate function $p_a(t)$ for parts (a) and (b) and compare your results.

Hint: You may see Example 4.7 for the important use of this type of smoother truncation window in (E.4).

4.16. (a) Consider Figure 4.20c, where $\cos 2\pi f_0 t$ is truncated by a window of (half) width b as integer multiple of its period, $b = m(1/f_0)$. Use (4.113) to show that two main contributions to the Fourier coefficients of $(\cos 2\pi f_0 t) p_b(t)$ are the two peaks at $f = \mp f_0$. For the other coefficients use $f = k f_0$, $k = \mp 2, \mp 3, \ldots$, in (4.113) to show that all the rest of the Fourier coefficients vanish, or in other words they fall at the zeros, of the (wider) convolved $G(f)$ with the Fourier transform of the gate function window.

(b) Repeat part (a) for the periodic function $f(t) = \sin 2\pi f_0 t$ starting with producing the analog to equation (4.113) and Fig. 4.20.

(c) Repeat part (a) with a window of (half) width $(m + \frac{1}{2})(1/f_0)$, i.e., not an integer multiple of the period $1/f_0$ of $f(t) = \cos 2\pi f_0 t$. In Fig. 4.20 we used $m = 4$; use the same m here for $b = 4.5(1/f_0)$ and compare the result with Fig. 4.20. Explain the difference (see Fig. 4.21b).

(d) Repeat what you did in part (c) for the sine function in part (b).

(e) In Fig. 4.20c we had according to (4.113), for $b = m(1/f_0)$, all coefficients vanish except those at $\mp f_0$. What is the reason behind the small "sidelobes" around the main peaks at $\mp f_0$? In Fig. 4.21a we have a truncation window of (half) width $4(1/f_0)$ for the same function $\cos 2\pi f_0 t$, but we don't see sidelobes for the corresponding transform. Why?

Hint: Compare (4.114) with (4.113) for Fig. 4.20c, since we are dealing here with only sample points for the discrete Fourier transform, where according to (4.114) only the two samples $f = \mp m/NT$ will appear. This is in contrast to Fig. 4.20c, which is the finite Fourier transform (or Fourier series) of the continuous function $\cos 2\pi f_0 t$.

(f) Repeat the computations and illustrations for Fig. 4.21c to illustrate the windowing effect on truncating the periodic function $\cos t$ with window of (half) width $NT = b = 20$, not an integer multiple of the period 2π of $\cos t$. Compare this to what we have in Fig. 4.20c regarding the sidelobes. Increase

b to 32, which is closer to an integer multiple of 2π, and compare your results with those for $b = 20$.

4.17. (a) Repeat the computations and illustrations of Fig. 4.21a, where the discrete Fourier transform has N samples, $NT = m(1/f_0)$, that is, an integer multiple of the period $1/f_0$ of $f(t) = \cos 2\pi f_0 t$. Verify that in this problem there are no sidelobes around $\mp f_0$ and give your reason (see Exercise 4.16(e).)

(b) In part (a), the DFT representation escaped the windowing effect when we took $NT = m(1/f_0)$. Here take $NT = 3.5(1/f_0)$ and repeat the computations to verify Fig. 4.21b. Attempt to explain the widening of the two peaks around $\mp f_0$ in light of (4.114).

(c) Repeat parts (a) and (b) for the periodic function $f(t) = \sin 2\pi f_0 t$.

(d) Repeat part (c) for $1/f_0 = 16$, and find how this wider window with $NT = 3.5(1/f_0)$ affects the windowing and the widening of the peaks around $\mp f_0$.

4.18. (a) Repeat the computations for the Fourier series representation of the square wave in Fig. 4.22b with N values $N = 1, 5, 22, 76$ to illustrate the Gibbs phenomenon in this problem.

(b) For the function of part (a) in Fig. 4.22b, repeat the computations for the σ_1 and σ_2 averaging for reducing the Gibbs phenomenon. See Fig. 4.30. Try $N = 5, 10,$ and 22. See Exercise 4.19a and Fig. 4.22b.

(c) Do the same as in part (a) for the function in Fig. 4.22a.

(d) Repeat the computations of part (b) for averaging the Gibbs phenomenon of the function in Fig. 4.22a (see Fig. 4.30.)

4.19. (a) Perform the computations for Fig. 4.23c to verify the effect of windowing (truncating) the Fourier coefficients on the (exact) Fourier series representation of the function $G_s(f)$ in Fig. 4.23a.

(b) What is the basic difference for the maximum overshoot of the Gibbs phenomenon of the truncated (windowed) Fourier integrals (Figs. 4.17–4.19) and the truncated Fourier series (Fig. 4.23c)?

ANS. (b) See the comments following the computation of (4.145).

4.20. Consider the square wave function of Fig. 4.22b on the interval (0,2).

(a) Write its J_0-Bessel–Fourier series on (0,2).

(b) Compute the truncated Bessel series of part (a) with $N = 3$, 5, and 20 and show the possible Gibbs type phenomenon at $x = 0$ and $x = 2$.

(c) In Fig. 4.30a we illustrated the σ_1 averaging process for reducing the Gibbs phenomenon of Fig. 4.22b of the Fourier series expansion of the square wave near the function's jump discontinuities. Attempt an averaging process for reducing the Bessel series Gibbs phenomenon that you have already observed in part (b).

Hint: Try multiplying the coefficients of the series by the Fourier coefficients of a *narrow* gate function window around the jump discontinuity.

(d)† Attempt to develop the analysis for the locations and magnitudes for the maximum overshoot near $x = 0$ and $x = 2$ for the Bessel–Fourier series expansion. Note that this problem should sound new and challenging.

Hint: You may follow an analysis parallel to what we did for the Fourier (trigonometric) integrals starting at (4.126) and then for the Fourier (trigonometric) series starting at (4.137) and ending after (4.145). See Jerri (1992).

4.21. (a) What is the relation and main difference between the windowing effect in general and the leakage error? Refer to the specific figures that illustrate the leakage error.

(b) How did we avoid the leakage error in the finite sum form of the discrete Fourier transform representation (4.3)? Give the explicit condition.

Hint: See (4.114), (4.114a), and the discussion around them.

---

†Optional.

(c) In case the condition in part (b) is violated, what can be done about the leakage error? What is the essence of such attempts?

(d) In comparing the optimal Hanning window in Fig. 4.31, for reducing the leakage error, with the gate function window in Fig. 4.31b, there must be a trade-off. What is it? How involved is the effort?

(e) The leakage error and the Gibbs phenomenon both result from the abrupt truncation of the transform (or coefficients) of the represented function. What are the main differences between them or in the treatment for reducing them?

(f) (i) Illustrate the use of the *Hanning window* in (4.148a) and then the *Gaussian window* in (4.148b) with $N = 2$ in reducing the leakage error for the periodic function

$$f(t) = \cos \frac{\pi}{2}t - 2\cos 3\pi t$$

of Fig. 4.32a when:

(a) A slightly incorrect period of $^{31}\!/_8$ is used instead of its natural period of 4. Compare your results to that of using the gate function window as illustrated in Fig. 4.32c.

(b) Repeat the computations and illustrations with a very incorrect period of 6.75 to show that in this case both the Hanning window and the Gaussian window are not of much help.

(ii) Repeat the illustrations of the leakage error and its remedies in Fig. 4.32a–f for the periodic function

$$g_s(t) = \cos \frac{\pi}{3}t - \cos 5\pi t$$

Try the correct period of 6, a wrong but a close one of $^{35}\!/_6$, and another one of period 5.

(g) How do these smooth windows fare in the widening of the expected narrow impulses?

(h) Through your own attempts and illustrations, discuss how the super-Gaussian windows in (4.148b) of Fig. 4.33 could be optimal for $N = 6$.

ANS. (a) Violating the exact period of a periodic function.

   (b) We choose $NT = m/f_0$. See (4.114), (4.114a).

   (c) Find a less abruptly truncated window than the gate function one.

   (d) While the gate function window has the simple value of 1, other (optimal!) windows like the Hanning window have a variable value, which has to be involved in a modification of the discrete Fourier transforms (4.3), (4.4).

   (e) The leakage error's sidelobes appear to the sides of the main peaks expected for the function, while the Gibbs phenomenon appears near the jump discontinuities of the function or its periodic extension. The leakage error can be reduced by modifying the window to a smoother one; the apparent remedy for the Gibbs phenomenon requires changing (averaging) the function itself to a smoother one.

   (f) (i) For period 4: exact c_k (Fig. 4.32a). Period $^{31}/_8$ with gate function window: clear sidelobes especially around $k = \mp 6$ (Fig. 4.32c). Period $^{31}/_8$ with Hanning window; clear pulses at $k = \mp 6$; the Gaussian window gives a similar answer. With the incorrect period of 6.75, the Hanning window is of no help: dominant sidelobes.

4.22. Consider the function $g(t)$ and its transform $G(f)$ in Example 4.7 for the illustrations of computing these Fourier integrals via the DFT as illustrated in Figs. 4.34a–g and 4.35a–d for computing $G(f)$ and $g(t)$, respectively. In these illustrations we used a truncated $g(t)$ to the interval $(-4, 4)$, which was large enough for $g(t)$ to (relatively) diminish at $t = 4$ for the sake of a low aliasing error. With the use of 32 samples for the DFT on this interval, the corresponding interval for $G(f)$ was $(-2, 2)$, which was about enough for the real part $G_r(f)$ of $G(f)$ to die out for a small aliasing error, but not enough for the slowly varying imaginary part $G_i(f)$, where there are clear jumps $G_i(\mp 2)$ at $f = \mp 2$ for a serious aliasing error. The following is an attempt to improve the situation, which is along the line of what we suggested at the end of Example 4.7.

(a) Use 128 samples for Example 4.7 with $g(t)$ defined on the same interval $(-4, 4)$ but $G(f)$ now defined on the (larger) interval $(-8, 8)$. Follow the steps in Fig. 4.34 to show an improvement in the accuracy of the computations as a result of this larger number of samples.

(b) Use 32 sample points on a larger interval $(-8, 8)$ for $g(t)$, which of course means a smaller interval for $G(f)$. Compare your results with Fig. 4.34 and the results in part (a).

(c) From the experience of Example 4.7 and parts (a) and (b) of this exercise, try the best possible (economical) choices for the number of samples and the truncated domains for $g(t)$ that would give a satisfactory resolution as well as a very low aliasing error.

(d) Parts (a), (b), and (c) were for computing $G(f)$, the transform of $g(t)$, via the DFT as in Fig. 4.34a–f. Figure 4.35a–c are for computing $g(t)$, the inverse Fourier transform of $G(f)$, via the DFT. Do the same type of analysis and computations as in parts (a), (b), and (c) [for $G(f)$ in Fig. 4.34] to compute $g(t)$ and compare your results to those in Figs. 4.35.

4.23. Repeat the computations in parts (a) to (d) of Exercise 4.22 for the computation of the Fourier transform and its inverse for the functions in Exercises 4.1 and 4.8.

4.24. Consider Example 4.8, where the DFT with 32 sample points is used to compute the Fourier series representation of the periodic function $G_s(f)$ on $(-4, 4)$ with period 8 and its Fourier coefficients c_k. The particular $G_s(f)$ here has two jump discontinuities inside its interval, located at $f = \mp 2$, where we anticipated and found a Gibbs phenomenon as in Fig. 4.36e for the DFT approximation of this $G_s(f)$. At that point we suggested increasing N from 32 to 128 for possible improvement of this Gibbs phenomenon, which is the subject of part (a) of this exercise.

On the other hand, the 32-sample DFT approximation of the Fourier coefficients in Example 4.8 was quite good, as shown in Fig. 4.35f. This is so because we truncate at $\mp(N/2)T = \mp 16(\frac{1}{8}) = \mp 2$, where c_k of this example practically dies out near $kT = \mp 2$, which results in very small jump discontinuities there and a minimal windowing effect. Nevertheless, by increasing N, we can improve the resolution of $G(n/NT)$ for a more accurate computation of the

Fourier integral representing the Fourier coefficients, which is the subject of part (b) of this exercise.

(a) In Example 4.8, use $N = 64$, then 128 samples with the same $T = \frac{1}{8}$ for the DFT to approximate $G_s(f)$ of Fig. 4.36a. Compare your result with that of using $N = 32$ in Fig. 4.36e. In particular, comment on any improvements in the Gibbs phenomenon near the jump discontinuities at $f = \mp 2$. *Note*: For both parts of this exercise, it is extremely helpful first to redo the computations of Example 4.8 with $N = 32$, and it is better yet if you have already redone Example 4.7, for the DFT approximation of Fourier integrals, and its related Exercise 4.22 with its very detailed computational instructions. See also Exercise 4.23.

(b) For the DFT approximation of the Fourier coefficients of Fig. 4.36b in Example 4.8, use $N = 64$ and then 128, and compare your results with those of Fig. 4.36f for $N = 32$. In particular, note how, with larger N, a better resolution of $G(n/NT)$ will affect a more accurate c_k, by comparing these results with the exact values and the approximate ones (with $N = 32$) as depicted in Fig. 4.36f.

(c) For parts (a) and (b) vary your N and T for a better or best results.

Hint: Consult parallel instructions in the different parts of Exercise 4.22.

4.25. Repeat parts (a) to (c) of Exercise 4.24 for the DFT approximation of the Fourier series $G_s(f)$ and its coefficients c_k in Exercises 4.2, 4.3, and 4.6.

4.26. *Two-dimensional DFT.* Consider the two-dimensional function $g(x,y)$ when it is sampled in the x and y directions with sample intervals T_1 and T_2, respectively, and with N samples in the x direction and M samples in the y direction; $g(pT_1,qT_2)$, $p = 0$, 1, 2, ..., $N - 1$; $q = 0$, 1, ..., $M - 1$. In a way similar to the one-dimensional DFT in (4.3), (4.4), establish the double discrete Fourier transform of $g(pT_1,qT_2)$,

$$G\left(\frac{n}{NT_1}, \frac{m}{MT_2}\right) = \sum_{q=0}^{M-1}\left[\sum_{p=0}^{N-1} g(PT_1,qT_2)e^{-j2\pi np/N}\right]e^{-j2\pi mq/M}$$

$$n = 0, 1, 2, \ldots, N - 1; \quad m = 0, 1, 2, \ldots, M - 1$$

Hint: Follow steps parallel to the development of the double Fourier transform (2.112) in Section 2.6, where the function of two variables $f(x_1, x_2)$ is Fourier-transformed in (2.111) with respect to the first variable x_1 only, leaving x_2 as a constant parameter, and then repeating the same operation with respect to x_2 to arrive at the double Fourier transform $F_{(2)}(\lambda_1, \lambda_2)$ of $f(x_1, x_2)$ in (2.212). For the FFT computation of the double DFT, see Brigham (1988).

4.27. *Interpolation (or sampling) for the DFT.*

(a) Assume that you have eight equidistant samples of a wave on the time interval $(0, 8)$, that is, $g_0, g_1, \ldots, g_7$, and you find that its DFT of (4.4) vanishes for $n = 3$, 4, and 5; that is, you have $G_0, G_1, G_2, 0, 0, 0, G_6, G_7$. What can you conclude about the wave?

(b) Can you interpolate for 32 equidistant samples of g_k on the same time interval $(0, 8)$? How?

ANS. (a) The wave may be considered as bandlimited to the frequency $f_2 = \frac{2}{8} \cdot 1/T = \frac{1}{4}$, since $T = 1$.

(b) Yes, since the spectrum $G(f_n)$ vanishes beyond f_2, we can assume it vanishes for f_3 to f_{29} by adding zero samples there for (4.4) with $N = 32$. This will enlarge $1/T$ to $1/T' = 4$ for (4.4), which gives $T' = 0.25$. Then we sample g_k at $kT' = 0.25k$, $k = 0$, 1, 2, $\ldots$, 31 in (4.4) instead of the original $kT = k$, $k = 0$, 1, 2, $\ldots$, 7. So use (4.4) with $T' = 0.25$ and $N = 32$.

Section 4.2   Discrete Orthogonal Polynomials Transforms

For the problems of this section you may consult Appendix A.5 for the orthogonal polynomials and their properties.

4.28. Use the following *Christoffel-Darboux formula*:

$$K_n(x, y) \equiv \sum_{j=0}^{n-1} \frac{p_j(x) p_j(y)}{h_j}$$

$$= \frac{k_n}{h_n k_{n+1}} \frac{p_{n+1}(x) p_n(y) - p_n(x) p_{n+1}(y)}{x - y} \tag{E.1}$$

and its special case

$$K_n(x,x) = \sum_{j=0}^{n-1} \frac{p_j^2(x)}{h_j} = \frac{k_n}{h_n k_{n+1}} [p_n(x)p'_{n+1}(x) - p'_n(x)p_{n+1}(x)]$$

(E.2)

to derive the *first discrete orthogonality* (4.153) of the classical orthogonal polynomials $\{p_n(x_{n,m})\}$. (See (4.167) for the second discrete orthogonality.)

4.29. (a) Show that the discrete orthogonal polynomial $\tilde{F}(j)$ in (4.159) is exact when $f(x)$ of (4.157) is a polynomial of degree $\leq n-1$.

Hint: Consult the theory of mechanical quadratures (see, for example, Szego (1959, pp. 43–47).

(b) Use (4.159), (4.160) to derive its special case (4.161), (4.162) for the discrete Laguerre transforms. (See Appendix A.5 and its exercises for the necessary properties of the Laguerre polynomials.)

(c) Derive the discrete Legendre transform pair (4.163), (4.164). (See part a of this problem and Appendix A.5 for the necessary properties of Legendre polynomials.)

4.30. (a) *Linearity property*: Show that the discrete Legendre transformation in (4.163) is linear; that is, if

$$\tilde{F}(j) = l[\widetilde{f}(x_{n,m})] \quad \text{and} \quad \tilde{G} = l[\tilde{g}(x_{n,m})]$$

(E.1)

then

$$l[c_1\widetilde{f} + c_2\tilde{g}] = c_1 l[\widetilde{f}] + c_2 l[\tilde{g}] = c_1\tilde{F} + c_2\tilde{G}$$

(E.2)

In the same way, show that the inverse transform (4.164) is also linear.

(b) *Symmetry*: Prove the following symmetry property of the discrete Legendre transforms:

$$l[\widetilde{f}(-x_{n,m})] = (-1)^j \tilde{F}(j)$$

(E.3)

Hint: Replace $x_{n,m}$ by $-x_{n,m}$ in (4.164) and use $P_j(-x) = (-1)^j P_j(x)$.

(c) Prove that

$$\tilde{F}(-j) = \tilde{F}(j-1)$$

(E.4)

Hint: See (4.163) and $P_{-j}(x) = P_{j-1}(x)$.

(d) Derive the following discrete Legendre transform pair:

$$l\left[f(x_{n,m'}) = \begin{cases} 0, & m \neq m' \\ \dfrac{(n+1)^2 P_{n+1}^2(x_{n,m})}{2(1-x_{n,m}^2)}, & m = m' \end{cases} \right.$$

$$= P_j(x_{n,m'}) \tag{E.5}$$

Hint: See the discrete orthogonality (4.122) with respect to the zeros $\{x_{n,m}\}$ of the Legendre polynomial $P_n(x)$.

(e) Derive the following discrete Legendre pairs:

$$l[P_{j'}(x_{n,m})] = \tilde{F}(j) = \begin{cases} 0, & j \neq j' \\ \dfrac{2}{2j+1}, & j = j' \end{cases} \tag{E.6}$$

with its particular cases for $j' = 0$ in (E.6)

$$l[1] = \tilde{F}(j) = \begin{cases} 0, & j \neq 0 \\ 2, & j = 0 \end{cases} \tag{E.7}$$

and for $j' = 1$ in (E.6)

$$[x_{n,m}] = \tilde{F}(j) = \begin{cases} 0, & j \neq 1 \\ \tfrac{2}{3}, & j = 1 \end{cases} \tag{E.8}$$

Hint: For (E.6) see the second discrete Legendre transform orthogonality in (4.167).

(f) Use the following recursion relation of the Legendre polynomials:

$$xP_j(x) = \frac{1}{2j+1}[(j+1)P_{j+1}(x) + jP_{j-1}(x)] \tag{E.9}$$

to derive the following discrete Legendre transform pair:

$$l[x_{n,m}\tilde{f}(x_{n,m})] = \frac{1}{2j+1}\{(j+1)\tilde{F}(j+1) + j\tilde{F}(j-1)\} \tag{E.10}$$

(g) Obtain the result (E.8) from (E.10).

4.31. (a) Prove (4.169), the discrete version of *Parseval's equality* of the discrete Legendre transform.

Hint: Substitute for $\tilde{f}(x_{n,m})$ and $\tilde{g}(x_{n,m})$ from (4.164) in (4.169), interchange the triple sum, and use the discrete orthogonality property (4.167) for the inner sum.

(b) (i) Prove that the (first type) generalized convolution product $\tilde{f} * \tilde{g}$, of the discrete Legendre transforms in (4.173), is commutative: $\tilde{f} * \tilde{g} = \tilde{g} * \tilde{f}$.

Hint: Note that both $\tilde{f} * \tilde{g}$ and $\tilde{g} * \tilde{f}$ are the transforms of the same product $\tilde{F}\tilde{G} = \tilde{G}\tilde{F}$.

(ii) Prove the discrete Legendre transform's first and second convolution theorems in (4.174) and (4.175), respectively.

Hint: To prove (4.174), we substitute $\tilde{F}(j)\tilde{G}(j)$ for $\tilde{F}(j)$ in (4.164), use (4.163) for $\tilde{F}(j)$, interchange the double series, and use the definition of the "generalized" translation in (4.171). To prove (4.175), we follow the same types of steps, starting with $\tilde{f}(x_{n,m})\tilde{g}(x_{n,m})$ for $\tilde{f}(x_{n,m})$ in (4.163).

4.32. (a) Find the Legendre–Fourier series coefficients $F(j)$ as in (3.146) for the following functions (see also Exercise 3.45 in Section 3.5):

(i) $f(x) = \sqrt{\dfrac{1-x}{2}}, \qquad -1 < x < 0$ (E.1)

(ii) $f(x) = \begin{cases} 0, & -1 \le x < 1 \\ \frac{1}{2}, & x = 0 \\ 1, & 0 < x \le 1 \end{cases}$

(iii) $f(x) = x^3, \qquad 0 < x \le 1$

(b) For each function write the nth partial sum $f_n(x)$ of its Legendre–Fourier series, then compute its error $\epsilon_n = |f(x_{n,m}) - f_n(x_{n,m})|$ in representing $f(x)$ at $x_{n,m}$, $m = 0, 1, 2, \ldots, n$, for $n = 10$.

(c) Compute the discrete Legendre transform representation $\tilde{f}(x_{n,m})$ with $n = 10$, and find its error $\tilde{\epsilon}_n = |f(x_{n,m}) - \tilde{f}(x_{n,m})|$ in representing $f(x)$ at $\{x_{n,m}\}$.

(d) Compare the error $\tilde{\epsilon}_n$ in part (c), of the discrete Legendre transform, with ϵ_n of the nth partial sum of the Legendre–Fourier series in part (b). Give the basic reason for the low error $\tilde{\epsilon}_n$.

(e) Repeat parts (b), (c), and (d) with $n = 16$.

ANS. (b and c) See answer to part (d).

    (d)   (i) $\epsilon_n \sim 10^{-3}\text{--}10^{-4}$, $\tilde{\epsilon}_n \sim 10^{-13}\text{--}10^{-14}$

        (ii) $\epsilon_n \sim 10^{-2}\text{--}10^{-4}$, $\tilde{\epsilon}_n \sim 10^{-13}\text{--}10^{-14}$

The basic reason for the better accuracy of the discrete Legendre transform approximation is that its derivation is based on the theory of Gauss mechanical quadrature. This is one of the best means of approximating an integral by a finite sum.

4.33. (a) Find the Laguerre–Fourier series coefficients $F(j)$ as in (3.156) of the following functions:

    (i) $f(x) = e^{-x}$, $0 < x < \infty$

    (ii) $f(x) = x \sin x$, $0 < x < \infty$

    (iii) $f(x) = x^5$, $0 < x < \infty$

   (b) Repeat parts (b) to (d) of Exercise 4.32.

ANS. (b and c) See the answer to part (d).

    (d)   (i) $\epsilon_n \sim 10^{1}\text{--}10^{-15}$ i) $\tilde{\epsilon}_n \sim 10^{-8}\text{--}10^{-12}$

        (ii) $\epsilon_n \sim 10^{4}\text{--}10^{-2}$ ii) $\tilde{\epsilon}_n \sim 10^{-7}\text{--}10^{-12}$

        (iii) $\epsilon_n \sim 10^{-8}\text{--}10^{-15}$ iii) $\tilde{\epsilon}_n \sim 10^{-5}\text{--}10^{-11}$

4.34. (a) Verify the numerical results of Example 4.4, i.e., the discussion of the numerical results following (E.11).

   (b) Repeat the computations with $n > 16$, and compare your results to those in part (a).

Section 4.3 Bessel-type Poisson Summation Formula (for the Bessel–Fourier Series and Hankel Transforms)

Note: For these problems, the Bessel functions' series representation, their needed properties, and their zeros are found in Table A.1 may be found in Appendix A.5 and the references to it.

4.35. Verify numerically that $\bar{H}(\omega)$, the "generalized repetition type" extension in (4.189), vanishes at $x = b$.

4.36. (a) Verify the Bessel-type Poisson sum formula (4.190) for the following pairs of J_0-Hankel transforms:

(i) $h(t) = \dfrac{1}{t} e^{-at}$, $\quad H(\omega) = \dfrac{1}{\sqrt{\omega^2 + a^2}}$

(ii) $h(t) = \dfrac{1}{(t^2 + a^2)^{1/2}}$, $\quad H(\omega) = \dfrac{e^{-a\omega}}{\omega}$

(iii) $h(t) = \dfrac{1}{(t^2 + 1)^{3/2}}$, $\quad H(\omega) = e^{-\omega}$

(iv) $h(t) = \tfrac{1}{2} e^{-t^2/4}$, $\quad H(\omega) = e^{-\omega^2}$

Hint: See Jerri (1988) or (1992).

(b) Compute for the divergent Bessel series in (3.204) to verify its representation of the (new) Bessel-type pulse train in Fig. 3.13. Use 20 to 40 terms in the series.

4.37. Write an expression for the (sampling series) aliasing error $\epsilon_{A,S}$ incurred in applying the J_0-sampling series to $f(t)$ which is not necessarily a bandlimited J_0-Hankel transform $f_b(t)$.

Hint: See the reference to Exercise 3.36.

ANS. $\epsilon_{A,S} = \left| f(t) - \displaystyle\sum_{n=1}^{\infty} f\left(\dfrac{j_{0,n}}{b}\right) \dfrac{2 j_{0,n} J_0(bt)}{b^2 (j_{0,n}^2/b^2 - t^2) J_1(j_{0,n})} \right|$

$= \left| \displaystyle\int_b^{\infty} \omega J_0(\omega t) F(\omega)\, d\omega + \dfrac{b J_1(bt)}{t} F(b) \right.$

$- \displaystyle\sum_{m=1}^{\infty} d_m \int_0^b \omega J_0(\omega t)[F(\omega \theta 2mb)$

$\left. - F(b\theta 2mb)]\, d\omega \right|$ $\qquad\qquad$ (E.2)

4.38. Clearly, the form for the aliasing error bound $\epsilon_{A,S}$ of the J_0-Bessel sampling series in (E.2) of Exercise 4.37 is not very practical. At this stage we limit ourselves to the J_0-Hankel transform of the class of *nonnegative monotonically decreasing functions* $H(\omega)$. For such functions, show that the error bound can be obtained in a form similar to that of the Fourier transforms (see Exercise 3.28

in Chapter 3), i.e., in terms of the integral of the ignored part of the spectrum $H(\omega)$ on (a, ∞),

$$|\epsilon_A| \leq K \int_a^\infty H(\omega) \, d\omega$$

Find the constant K.

Hint: For the full derivation of this new result, see Jerri (1988), which is also helpful for Exercises 4.39.

4.39. (a) For the J_0-Hankel transform pairs in Exercise 4.36ii–iv, calculate the analytic expression for the upper bound on the aliasing error $\epsilon_{A,S}$ of the J_0-sampling series, as discussed in Exercise 4.37, and for the class of nonnegative monotonically decreasing $H_0(\omega) \geq 0$.

Hint: See the reference to Exercise 4.38.

(b) Verify your results in part (a) by comparing these aliasing error bounds to the actual aliasing error as expressed in the answer to Exercise 4.37.

ANS. (a) (ii) $\epsilon_{A,S} \leq \dfrac{3}{2a} e^{-ab} + \dfrac{b}{2} e^{-ab}$

(iii) $\epsilon_{A,S} \leq \dfrac{3}{2} e^{-b^2}(b + 1) + \dfrac{b^2}{2} e^{-b}$

(iv) $\epsilon_{A,S} \leq \dfrac{3}{4} e^{-b^2} + \dfrac{b^2}{2} e^{-b^2}$

4.40. (a) The Bessel-type Poisson sum formula in (4.190) is for the J_0-Hankel transforms (4.186), (4.187). Develop the same type of Poisson formula for the J_1-Hankel transform.

Hint: Follow the steps from (4.188) to (4.190). You may also want to verify the present generalization, of the impulse train $X(\omega)$ of (3.204) and its d_m values before using it for a "generalized repetition" type extension such as $\tilde{H}(\omega)$ in (4.188).

(b) Develop an expression for the aliasing error of the J_1-Bessel sampling series.

Hint: See Exercise 4.37.

4.41. Consider a discrete Hankel transform like that of $\tilde{H}_0(j_{0,n}/c)$ in (4.192). Assume an approximate inverse of the same form as $\bar{h}_0(j_{0,k}/b)$.

(a) Show how the exactness of $\tilde{H}_0$ depends on an exact discrete orthogonality, which we still don't have.

Hint: Substitute $\tilde{H}_0(j_{0,n}/c)$ in the summation for $\bar{h}_0$, interchange the summation, and examine the inner sum of all the factors of $\bar{h}_0$.

(b) Examine how the above inner sum depends on N, the number of terms in the sum, to approach a Kronecker delta and thus the sought discrete orthogonality.

ANS. (a) $\bar{h}\left(\dfrac{j_{0,k}}{b}\right) \simeq \dfrac{2}{c^2} \displaystyle\sum_{n=1}^{N \ni j_{0,N} = bc} \dfrac{\tilde{H}_0(j_{0,n}/c)J_0(j_{0,n}j_{0,k}/bc)}{J_1^2(j_{0,n})}$

$k = 1, 2, \ldots, k, \quad j_{0,k} = bc, \quad J_0(j_{0,n}) = 0$

4.42. (a) Use the discrete Hankel transform with $N = 20$ terms to approximate the Hankel transform $H_0(x) = e^{-x}$ of the function $h_0(t) = (1 + t^2)^{-3/2}$.

(b) Use Simpson's rule with $N = 20$ terms to approximate the Hankel transform $(H_{0,S})$ of $h_0(t)$, which is truncated to $t \in (0, c)$ as in part (a), where $bc = j_{0,N}$, $N = 20$. Compare your results $H_{0,S}$ with those of $\tilde{H}_0$ in part (a).

(c) Test the approximate discrete orthogonality of the discrete Hankel transforms for the pair in part (a), using $N = 20$; then try $N = 5, 10$.

Hint: Use the 20 samples of H_0 to evaluate their discrete Hankel transform $\bar{h}_0$ in the sum of the answer to Exercise 4.41a; then use these $\bar{h}_0$ values to evaluate $\tilde{H}_0$ in (4.192). The test is whether these $\tilde{H}_0$ values will give $\bar{h}_0$ as in the answer to Exercise 4.41a with good accuracy.

ANS. (a and b) See the accompanying Table 4.4.

(c) The accuracy is within 10^{-7}.

TABLE 4.4 Comparison of the Continuous and Discrete J_0-Hankel Transforms for $h_0(t) = (1 + t^2)^{3/2}$ with $H_0(x) = e^{-x}$, $x \in (0,c)$, $bc = j_{0,20}$; $b = 3$. ($H_{0,s}$ is the approximation using Simpson's rule.)

$\dfrac{j_{0,n}}{c}$	$H_0\left(\dfrac{j_{0,n}}{c}\right)$	$\tilde{H}_0\left(\dfrac{j_{0,n}}{c}\right)$	$H_0 - \tilde{H}_0$	$H_{0,s}$	$H_0 - H_{0,s}$
0.8018084	0.4486068	0.4845392	−0.0359323	0.4885446	−0.0389378
1.8400259	0.1588133	0.1439307	0.0148826	0.1455301	0.0132832
2.8845758	0.0558785	0.0641985	−0.0083200	0.0631620	−0.0072836
3.9305115	0.0196336	0.0141891	0.0054445	0.0124673	0.0071664
4.9769723	0.0068949	0.0108107	−0.0039158	0.0100016	−0.0031067
6.0236867	0.0024207	−0.0005778	0.0029985	0.0004765	0.0019442
7.0705414	0.0008498	0.0032504	−0.0024007	0.0048881	−0.0040383
8.1174876	0.0002983	−0.0016891	0.0019874	−0.0004937	0.0007920
9.1644897	0.0001047	0.0017939	−0.0016892	0.0016707	−0.0015660
10.2115328	0.0000367	−0.0014297	0.0014865	−0.0017470	0.0017837
11.2586060	0.0000129	0.0013087	−0.0012958	0.0017301	−0.0017172
12.3056946	0.0000045	−0.0011578	0.0011623	0.0008892	−0.0008846
13.3528035	0.0000016	0.0010578	−0.0010562	0.0041093	−0.0041077
14.3999278	0.0000006	−0.0009705	0.0009711	0.0024197	−0.0024192
15.4470622	0.0000002	0.0009024	−0.0009022	0.0047147	−0.0047145
16.4942017	0.0000001	−0.0008460	0.0008460	0.0041843	−0.0041843
17.5413462	0.0000000	0.0008004	−0.0008003	0.0094328	−0.0094327
18.5885010	0.0000000	−0.0007635	0.0007635	0.0100578	−0.0100578
19.6356608	0.0000000	0.0007340	−0.0007340	0.0107984	−0.0107984
20.6828206	0.0000000	−0.0007109	0.0007109	−0.0221143	0.0221143

Appendix A

BASIC SECOND-ORDER DIFFERENTIAL EQUATIONS AND THEIR (SERIES) SOLUTIONS

Special Functions

A.1 Introduction

The purpose of this section is to review and discuss the methods of solving the basic linear second-order differential equations that are of most concern to us in this book. In particular, we shall discuss in detail the *power series method* of solution, which results in the special functions of interest and which may have been glossed over in the elementary differential equations preparation. Also of importance is the *method of variation of parameters*. It is the main tool for solving nonhomogeneous differential equations with general nonhomogeneous terms. Also, it is this method that is used in deriving the Green's function, which is used for solving nonhomogeneous differential equations associated with initial or boundary value problems.

Other very basic differential equations such as homogeneous ones with constant coefficients and nonhomogeneous equations with a very particular form of the nonhomogeneous term—a sum of products of polynomials, exponential and trigonometric functions (UC functions)—are left to be reviewed in the exercises (see Exercises A.1 to A.6 for a review of these elementary problems).

Many of the partial differential equations frequently encountered in physics and engineering are linear and of second order. The application of the integral transform method on all the variables except one, or the method of products, results in linear second-order differential equations as indicated throughout this book. Some of these ordinary differential equations can be solved in terms of elementary functions whose properties are well known to us. For example, the constant-coefficient equation

$$y'' + \lambda^2 y = 0 \tag{A.1.1}$$

has the general solution

$$y(x) = A \sin \lambda x + B \cos \lambda x \tag{A.1.2}$$

For such simple functions we know their analytic properties and, in particular, their power series representations about $x = 0$, that is,

$$\sin x = \sum_{m=0}^{\infty} \frac{(-1)^m x^{2m+1}}{(2m+1)!} \tag{A.1.3}$$

$$\cos x = \sum_{m=0}^{\infty} \frac{(-1)^m x^{2m}}{(2m)!} \tag{A.1.4}$$

The solutions of other ordinary differential equations are not such familiar functions, but equations which are most frequently encountered have been systematically studied, and their solutions are called *special functions*. An example is *Bessel's differential equation*

$$R'' + \frac{1}{r} R' + \left(\lambda^2 - \frac{\nu^2}{r^2} \right) R = 0 \tag{A.1.5}$$

which often results, as the radial part, when we apply separation of variables to some partial differential equations written in cylindrical coordinates. Also, it appears when the integral transform method is applied to such a partial differential equation to algebraize the differential operations with respect to all variables except the radial variable r.

Of course, our reason here for studying and emphasizing the Bessel functions is that it makes the kernel of the *Hankel transform* of Section 2.7 that algebraizes the differential part $R'' + (1/r)R'$ of the Bessel differential equation in (A.1.5) (see (1.24) and (2.249)–(2.251)), $\nu = n = 0$.

We will consider the solutions of ordinary differential equations with emphasis on the power series method for solving homogeneous equations. The two linearly independent solutions, when obtained via the power series method, can then be used for the method of *variation of parameters* to obtain the general solution for the associated nonhomogeneous equation with general nonhomogeneous term. Also, in case we find only one solution for the homogeneous problem, the variation-of-parameters method helps in constructing the second linearly independent one. This particular application is termed the *reduction-of-order method*. For these reasons we shall first review the method of variation of parameters for nonhomogeneous equations, before we consider the detailed power series method for solving most of the well-known (associated) homogeneous equations in the remaining sections.

There is, of course, a very basic reason for studying the method of solving nonhomogeneous equations, which stems from the nature of the development of this book as it relies on the transform methods of solving boundary and initial value problems that are often associated with nonhomogeneous equations. This is the fact that the transform method will reduce the nonhomogeneous partial differential equation to a nonhomogeneous (ordinary) differential equation, when we use transforms that are compatible with the differential operations with respect to all variables except the one remaining in the resulting (transformed) differential equation.

A.2 Method of Variation of Parameters

This method is commonly used to generate a linearly independent solution from a known solution of a homogeneous differential equation (*the method of reduction of order*) and to construct the solution of a nonhomogeneous ordinary differential equation from the general solution of its homogeneous form. As the procedure is similar in both cases, we shall illustrate the latter application and leave the illustration of the former (reduction of order) for the exercises (see Exercise A.7a).

Consider the second-order nonhomogeneous equation

$$y'' + p(x)y' + q(x)y = f(x) \tag{A.2.1}$$

The general solution of the homogeneous form,

$$y'' + p(x)y' + q(x)y = 0 \tag{A.2.2}$$

or, as it is termed, the *complementary solution* y_c of (A.2.1) is

$$y_c = c_1 y_1(x) + c_2 y_2(x) \tag{A.2.3}$$

where c_1 and c_2 are arbitrary constants and $y_1(x)$ and $y_2(x)$ are the two linearly independent solutions of (A.2.2). To construct a solution of (A.2.1), or what is termed the *particular solution* y_p of (A.2.1), we treat c_1 and c_2 in (A.2.3) as functions of x. Thus, let us look for a solution of a form like (A.2.3), except that the constant amplitudes c_1 and c_2 are replaced now by variable ones $\alpha(x)$ and $\beta(x)$,

$$y = \alpha(x)y_1(x) + \beta(x)y_2(x) \tag{A.2.4}$$

Differentiating (A.2.4), dropping the argument (x), we obtain

$$y' = \alpha y_1' + \alpha' y_1 + \beta y_2' + \beta' y_2 \tag{A.2.5}$$

Since $\alpha(x)$ and $\beta(x)$ are yet to be determined, *we may simplify (A.2.5) by choosing*

$$\alpha' y_1 + \beta' y_2 = 0 \tag{A.2.6}$$

which makes y' independent of derivatives of the unknowns α and β

$$y' = \alpha y_1' + \beta y_2' \tag{A.2.5a}$$

This choice is made in order to have the new unknowns $\alpha(x)$ and $\beta(x)$ involved in a system of two first-order differential equations (A.2.6), (A.2.8) instead of second-order equations (A.2.1), (A.2.2) in $y(x)$. So, from the $y' = \alpha y_1' + \beta y_2'$ of (A.2.5a), we find the simpler y'',

$$y'' = \alpha y_1'' + \alpha' y_1' + \beta' y_2' + \beta y_2'' \tag{A.2.7}$$

Substituting y' from (A.2.5a) and this y'' from (A.2.7) into (A.2.1) and simplifying the result, we obtain the *second first-order equation* in $\alpha(x)$, $\beta(x)$,

$$\alpha' y_1' + \beta' y_2' = f \tag{A.2.8}$$

Equations (A.2.6) and (A.2.8) are solved simultaneously to give

$$\alpha' = -\frac{fy_2}{(y_1 y_2' - y_2 y_1')} \tag{A.2.9}$$

$$\beta' = \frac{fy_1}{(y_1 y_2' - y_2 y_1')} \tag{A.2.10}$$

Here we recognize the denominator in both (A.2.9) and (A.2.10) as the *Wronskian*

$$W(y_1, y_2) \equiv \begin{vmatrix} y_1(\xi) & y_2(\xi) \\ y_1'(\xi) & y_2'(\xi) \end{vmatrix} = y_1(\xi)y_2'(\xi) - y_1'(\xi)y_2(\xi) \tag{A.2.11}$$

which is not zero, since we assumed $y_1(\xi)$ and $y_2(\xi)$ are two linearly independent solutions of (A.2.2).

Finally, $\alpha(x)$ and $\beta(x)$ are obtained by integration of (A.2.9) and (A.2.10), respectively. Thus with x_0 as a particular initial value, we can integrate to have the two unknown amplitudes $\alpha(x)$ and $\beta(x)$,

$$\alpha(x) = -\int_{x_0}^{x} \frac{f(\xi)y_2(\xi)}{y_1(\xi)y_2'(\xi) - y_2(\xi)y_1'(\xi)} \, d\xi \qquad (\text{A.2.12})$$

$$\beta(x) = \int_{x_0}^{x} \frac{f(\xi)y_1(\xi)}{y_1(\xi)y_2'(\xi) - y_2(\xi)y_1'(\xi)} \, d\xi \qquad (\text{A.2.13})$$

Hence the general solution y_g to (A.2.1) is

$$y_g = y_c + y_p = c_1 y_1(x) + c_2 y_2(x) + \alpha(x)y_1(x) + \beta(x)y_2(x) \qquad (\text{A.2.14})$$

where c_1 and c_2 are constants and $\alpha(x)$ and $\beta(x)$ are functions to be determined from (A.2.12) and (A.2.13), respectively.

We must note here that the particular solution to the nonhomogeneous problem (A.2.1) can be written as an integral that involves the nonhomogeneous term $f(x)$ weighed by the *Green's function* $G(x,\xi)$,

$$y_p(x) = \alpha(x)y_1(x) + \beta(x)y_2(x)$$

$$= -\int_{x_0}^{x} \frac{f(\xi)y_2(\xi)y_1(x)}{W(y_1(\xi),y_2(\xi))} \, d\xi + \int_{x_0}^{x} \frac{f(\xi)y_1(\xi)y_2(x)}{W(y_1(\xi),y_2(\xi))} \, d\xi \qquad (\text{A.2.15})$$

where we have entered $y_1(x)$ and $y_2(x)$ inside the integrals since they are viewed as constants relative to the integration operation done with respect to ξ

$$y_p(x) = \int_{x_0}^{x} \frac{y_1(\xi)y_2(x) - y_2(\xi)y_1(x)}{W(y_1(\xi),y_2(\xi))} f(\xi) \, d\xi$$

$$\equiv \int_{x_0}^{x} G(x,\xi)f(\xi) \, d\xi \qquad (\text{A.2.16})$$

where $G(x,\xi)$ is the Green's function

$$G(x,\xi) \equiv \frac{y_1(\xi)y_2(x) - y_2(\xi)y_1(x)}{W(y_1(\xi),y_2(\xi))} \qquad (\text{A.2.17})$$

which is constructed in terms of $y_1(x)$ and $y_2(x)$, the solutions of the associated homogeneous equation (A.2.2). We will illustrate this method of variation of parameters with the following example. We leave other illustrations along with constructing the Green's function, for initial and boundary value problems, as exercises (see Exercise A.11).

EXAMPLE A.1:

Find the general solution to the nonhomogeneous equation,

$$y'' + y = \sec x \qquad (\text{E.1})$$

The general solution of the homogeneous equation

$$y'' + y = 0 \tag{E.2}$$

which is the y_c of (E.1) is

$$y_c = c_1 \sin x + c_2 \cos x \tag{E.3}$$

since $\sin x$ and $\cos x$ are the two linearly independent solutions of (E.2). To construct the particular solution y_p of (E.1) we follow the method of variation of parameters and let

$$
\begin{aligned}
y_p &= \alpha(x) y_1(x) + \beta(x) y_2(x) \\
&= \alpha(x) \sin x + \beta(x) \cos x
\end{aligned}
\tag{E.4}
$$

To determine $\alpha(x)$ and $\beta(x)$ we use (A.2.12) and (A.2.13), respectively, with $f(x) = \sec x$, $y_1(x) = \sin x$, $y_2(x) = \cos x$, and so

$$
\begin{aligned}
y_1 y_2' - y_2 y_1' &= \sin x (-\sin x) - \cos x (\cos x) \\
&= -\sin^2 x - \cos^2 x = -1,
\end{aligned}
$$

$$\alpha(x) = \int \sec x \cos x \, dx = \int dx = x \tag{E.5}$$

$$\beta(x) = -\int \sec x \sin x = -\int \frac{\sin x}{\cos x} \, dx = \ln|\cos x| \tag{E.6}$$

Hence from (E.4), (E.5), and (E.6) we have

$$y_p = x \sin x + \cos x \ln|\cos x| \tag{E.7}$$

and so the general solution to (E.1) is

$$y_g = y_c + y_p = c_1 \sin x + c_2 \cos x + x \sin x + \cos x \ln|\cos x| \tag{E.8}$$

In Exercise A.11, we will use the Green's function in (A.2.17), (A.2.16) to arrive at $y_p(x)$.

The analogous *method of reduction of order* will be outlined and illustrated in Exercise A.7a.

A.3   Power Series Method of Solution

Properties of Power Series

Before we illustrate the power series and the *Frobenius* form of solutions, we shall list without proof a few of the most important algebraic and analytic properties of the convergent power series about a point $x = x_0$ with radius of convergence R

$$\sum_{n=0}^{\infty} a_n(x - x_0)^n, \qquad |x - x_0| < R \qquad \text{(A.3.1)}$$

Differentiation of Series

The series obtained by differentiating (A.3.1) term by term converges with the same radius of convergence as the original series. For example,

$$\sin x = \sum_{n=0}^{\infty} \frac{(-1)^n x^{2n+1}}{(2n + 1)!}, \qquad -\infty < x < \infty \qquad \text{(A.3.2)}$$

$$= x - \frac{x^3}{3!} + \frac{x^5}{5!} + \cdots + (-1)^n \frac{x^{2n+1}}{(2n + 1)!} + \cdots$$

and its derivative

$$\frac{d \sin x}{dx} = 1 - \frac{3x^2}{3!} + \frac{5x^4}{5!} + \cdots$$

$$= 1 - \frac{x^2}{2!} + \frac{x^4}{4!} + \cdots$$

$$= \sum_{n=0}^{\infty} \frac{(-1)^n x^{2n}}{(2n)!} = \cos x, \qquad -\infty < x < \infty \qquad \text{(A.3.3)}$$

both converge for all x.

Addition of Series

Two convergent power series

$$\sum_{n=0}^{\infty} a_n(x - x_0)^n = f(x), \qquad |x - x_0| < R_1, \qquad \text{(A.3.4)}$$

and

$$\sum_{n=0}^{\infty} b_n(x - x_0)^n = g(x), \qquad |x - x_0| < R_2, \qquad \text{(A.3.5)}$$

can be added term by term to result in a convergent power series

$$f(x) + g(x) = \sum_{n=0}^{\infty} (a_n + b_n)(x - x_0)^n, \qquad |x - x_0| < \min(R_1, R_2)$$

$$\text{(A.3.6)}$$

in the interior of the smallest convergence interval of both of them ($R = $ minimum of R_1 and R_2).

Another property which follows from the property of addition of power series is that a power series with positive radius of convergence whose sum is identically zero,

$$\sum_{n=0}^{\infty} c_n(x - x_0)^n \equiv 0, \qquad |x - x_0| < R \tag{A.3.7}$$

must have each coefficient zero, that is $c_n = 0$, $n = 0, 1, 2, 3, \ldots$. We shall depend great deal on this property. It is in the sense of finding relations between the different coefficients, and possibly a general form of c_n for the particular differential equation, which is solved via the power series method.

Multiplication of Series

The product of two convergent power series

$$\begin{aligned}
f(x)g(x) &= a_0b_0 + (a_0b_1 + a_1b_0)(x - x_0) \\
&\quad + (a_0b_2 + a_1b_1 + a_2b_0)(x - x_0)^2 + \cdots \tag{A.3.8} \\
&= \sum_{n=0}^{\infty} (a_0b_n + a_1b_{n-1} + \cdots + a_nb_0)(x - x_0)^n,
\end{aligned}$$

$$|x - x_0| < \min(R_1, R_2) \tag{A.3.9}$$

which in this form is called the *Cauchy product*, converges within the smallest circle of convergence of the two power series (R = minimum of R_1, R_2).

We will concentrate here on the power series representation of the two (linearly independent) solutions $y_1(x)$ and $y_2(x)$ of the homogeneous equation (A.2.2),

$$y'' + p(x)y' + q(x)y = 0 \tag{A.2.2}$$

from which we can compute numerical values for each of $y_1(x)$ and $y_2(x)$ and derive their general properties such as integrals and derivatives.

There are two types of power series expansions for $y(x)$ about a point $x = x_0$, depending on how well-behaved the coefficients $p(x)$ and $q(x)$ are at $x = x_0$. The point $x = x_0$ is called an *ordinary point* if $p(x_0)$ and $q(x_0)$ are finite, and in this case the solution $y(x)$ assumes the power series expansion

$$y(x) = \sum_{m=0}^{\infty} c_m(x - x_0)^m \tag{A.3.10}$$

When the coefficients $p(x)$ and $q(x)$ are not finite at $x = x_0$ we call x_0 a *singular point*. There are two types of singularities that result in two different forms for the solution. The singularity at x_0 is called a *removable singularity* (or *regular-singular* point) if

$$\lim_{x \to x_0} (x - x_0)p(x) \quad \text{and} \quad \lim_{x \to x_0} (x - x_0)^2 q(x) \quad \text{exist} \tag{A.3.11}$$

In this case, we may apply the *Frobenius method*, which gives the solution as

$$u(x) = (x - x_0)^\beta \sum_{m=0}^{\infty} c_m (x - x_0)^m \tag{A.3.12}$$

where β is not necessarily an integer.

When one of the above limits in (A.3.11) does not limit, then the point x_0 is termed an *irregular singular point*, and we cannot conclude that a general solution has the form of either (A.3.10) or (A.3.12) (see Coddington and Levinson, 1955).

Power Series Expansion About a Regular Point

To illustrate this method we consider the simple differential equation with constant coefficients (A.1.1) with $\lambda = 1$,

$$y'' + y = 0 \tag{A.3.13}$$

We wish to show that the power series expansion

$$y(x) = \sum_{n=0}^{\infty} c_n x^n = c_0 + c_1 x + c_2 x^2 + \cdots + c_n x^n + \cdots \tag{A.3.14}$$

for $y(x)$ about the obviously ordinary point $x = 0$ results in a linear combination of $\sin x$ and $\cos x$ as represented in (A.1.3) and (A.1.4), respectively.

The procedure then is to take $y(x)$ from (A.3.14), insert it in (A.3.13) and then use (A.3.7) to equate to zero the coefficients of similar powers of x and attempt to find the explicit form of c_n for (A.3.14). Differentiating once, we obtain

$$y'(x) = c_1 + 2c_2 x + 3c_3 x^2 + 4c_4 x^3 + 5c_5 x^4 + \cdots + n c_n x^{n-1} + \cdots$$

$$= \sum_{n=1}^{\infty} n c_n x^{n-1} \tag{A.3.15}$$

Note that the limit of summation is $n = 1$ in (A.3.15) instead of $n = 0$ as in (A.3.14). It is useful to differentiate the infinite series term by term as we did here and then condense the result using the summation notation.

$$y''(x) = 2c_2 + 6c_3 x + 12c_4 x^2 + 20c_5 x^3 + \cdots + n(n-1)c_n x^{n-2} + \cdots$$

$$= \sum_{n=2}^{\infty} n(n-1)c_n x^{n-2} \tag{A.3.16}$$

We insert the $y''(x)$ of (A.3.16) and y of (A.3.14) in (A.3.13) to obtain

$$2c_2 + 6c_3 x + 12c_4 x^2 + 20c_5 x^3 + \cdots + n(n-1)c_n x^{n-2} + \cdots$$
$$+ c_0 + c_1 x + c_2 x^2 + c_3 x^3 + \cdots + \cdots + c_{n-2} x^{n-2} + \cdots = 0$$

If we collect terms of equal powers in x as arranged above,

$$(2c_2 + c_0) + (6c_3 + c_1)x + (12c_4 + c_2)x^2 + (20c_5 + c_3)x^3 + \cdots + \cdots$$
$$[n(n-1)c_n + c_{n-2}]x^{n-2} + \cdots = 0 \tag{A.3.17}$$

and equate each coefficient to zero we obtain

$$2c_2 + c_0 = 0 \qquad\qquad 6c_3 + c_1 = 0$$
$$12c_4 + c_2 = 0, \qquad (20c_5 + c_3) = 0, \ldots, n(n-1)c_n + c_{n-2} = 0$$

From these relations we have

$$c_2 = -\tfrac{1}{2}c_0, \qquad c_4 = -\tfrac{1}{12}c_2 = \tfrac{1}{24}c_0 = \frac{c_0}{4!}$$

$$c_3 = -\tfrac{1}{6}c_1, \qquad c_5 = -\tfrac{1}{20}c_3 = \tfrac{1}{120}c_1 = \frac{c_1}{5!}$$

$$c_n = -\frac{c_{n-2}}{n(n-1)}, \qquad n = 2, 3, 4, \ldots$$

which express the even-indexed coefficients c_{2n} in terms of c_0 and the odd-indexed coefficients c_{2n+1} in terms of c_1.

If we examine more terms in (A.3.17) and note the above relations, we find the general explicit forms or *recurrence relations* for c_{2n} and c_{2n+1}

$$c_{2n} = \frac{(-1)^n}{(2n)!}c_0, \qquad c_{2n+1} = \frac{(-1)^n}{(2n+1)!}c_1$$

We must point out here that for more general differential equations, the general form of c_n is not as easily obtained as the above ones.

If we insert the above forms for c_{2n} and c_{2n+1} in (A.3.14) we obtain

$$y(x) = c_0\left(1 - \frac{x^2}{2!} + \frac{x^4}{4!} + \cdots + \frac{(-1)^n x^{2n}}{(2n)!} + \cdots\right)$$
$$+ c_1\left(x - \frac{x^3}{3!} + \frac{x^5}{5!} + \cdots + \frac{(-1)^n x^{2n+1}}{(2n+1)!} + \cdots\right)$$
$$= c_0 \sum_{m=0}^{\infty} \frac{(-1)^m x^{2m}}{(2m)!} + c_1 \sum_{m=0}^{\infty} \frac{(-1)^m x^{2m+1}}{(2m+1)!} \tag{A.3.18}$$
$$= c_0 \cos x + c_1 \sin x$$

after consulting the power series expansions of $\sin x$ and $\cos x$ in (A.1.3) and (A.1.4), respectively. So the general solution to (A.3.13) is a linear combination of the two linearly independent solutions $\sin x$ and $\cos x$, where c_0 and c_1 serve as the arbitrary constants.

Let us note here that the series expansions in (A.1.3) and (A.1.4) are well-known representations for $\sin x$ and $\cos x$, respectively, and hence we immediately recognized the series of (A.3.18) as $\sin x$ and $\cos x$. When the power series expansion for the solution of a differential equation results in such familiar and well-studied functions, we term it a *closed-form solution*.

A.4 Frobenius Method of Solution—Power Series Expansion About a Regular Singular Point

We shall illustrate this method to solve *Bessel's differential equation*,

$$r^2 R'' + r R' + (\lambda^2 r^2 - \nu^2) R = 0 \qquad (A.4.1)$$

which has a *regular singularity* at $r = 0$. The Frobenius form of solution about $r = 0$ is

$$R(r) = r^\beta \sum_{k=0}^{\infty} c_k r^k = r^\beta (c_0 + c_1 r + c_2 r^2 + \cdots + c_k r^k + \cdots)$$

$$= c_0 r^\beta + c_1 r^{\beta+1} + c_2 r^{\beta+2} + \cdots + c_k r^{\beta+k} + \cdots \qquad (A.4.2)$$

where the parameter β will be determined by equating the coefficients of r^β to zero after we use (A.4.2) in (A.4.1).

Now we use (A.4.2) to find $R'(r)$ and $R''(r)$, substitute in (A.4.1), and then equate to zero the coefficients of the terms of equal powers of r as we did in the previous problem,

$$R'(r) = \beta c_0 r^{\beta-1} + (\beta + 1)c_1 r^\beta + (\beta + 2)c_2 r^{\beta+1} + \cdots$$

$$+ (\beta + k)c_k r^{\beta+k-1} + \cdots \qquad (A.4.3)$$

$$R''(r) = \beta(\beta - 1)c_0 r^{\beta-2} + (\beta + 1)\beta c_1 r^{\beta-1} + (\beta + 2)(\beta + 1)c_2 r^\beta + \cdots$$

$$+ (\beta + k)(\beta + k - 1)c_k r^{\beta+k-2} + \cdots = 0 \qquad (A.4.4)$$

Let us write the series expansion for all of the terms in (A.4.1), arranging them according to powers of r

$$r^2 R'' = \beta(\beta - 1)c_0 r^\beta + (\beta + 1)\beta c_1 r^{\beta+1} + (\beta + 2)(\beta + 1)c_2 r^{\beta+2} + \cdots$$

$$+ c_k(\beta + k)(\beta + k - 1)r^{\beta+k} + \cdots$$

$$rR' = c_0\beta r^\beta + (\beta + 1)c_1 r^{\beta+1} + (\beta + 2)c_2 r^{\beta+2} + \cdots$$
$$+ c_k(\beta + k)r^{\beta+k} + \cdots$$

$$-\nu^2 R = -\nu^2 c_0 r^\beta - \nu^2 c_1 r^{\beta+1} - \nu^2 c_2 r^{\beta+2} + \cdots$$
$$- \nu^2 c_k r^{\beta+k} + \cdots$$

$$\lambda^2 r^2 R = \lambda^2 c_0 r^{\beta+2} + \lambda^2 c_1 r^{\beta+3} + \cdots + \lambda^2 c_{k-2} r^{\beta+k} + \cdots$$

If we substitute these expansions in (A.4.1), recognizing the terms with equal powers of r as indicated above, we have

$$r^2 R'' + rR' - \nu^2 R + \lambda^2 r^2 R = [\beta(\beta - 1)c_0 + c_0\beta - \nu^2 c_0]r^\beta$$
$$+ [(\beta + 1)\beta c_1 + (\beta + 1)c_1 - \nu^2 c_1]r^{\beta+1}$$
$$+ [(\beta + 2)(\beta + 1)c_2 + (\beta + 2)c_2 - \nu^2 c_2 + \lambda^2 c_0]r^{\beta+2} + \cdots + \cdots$$
$$+ [(\beta + k)(\beta + k - 1)c_k + (\beta + k)c_k - \nu^2 c_k + \lambda^2 c_{k-2}]r^{\beta+k} + \cdots$$
$$= 0 \tag{A.4.5}$$

and after simplifying the coefficients we obtain

$$c_0[\beta^2 - \nu^2]r^\beta + c_1[(\beta + 1)^2 - \nu^2]r^{\beta+1} + \cdots$$
$$+ [c_k\{(\beta + k)^2 - \nu^2\} + \lambda^2 c_{k-2}]r^{\beta+k} + \cdots + \cdots = 0$$

Now we equate the coefficients of $r^{\beta+n}$, $n = 0, 1, \ldots, k$, to zero to obtain

$$c_0[\beta^2 - \nu^2] = 0 \tag{A.4.6}$$
$$c_1[(\beta + 1)^2 - \nu^2] = 0 \tag{A.4.7}$$

$$\vdots$$

$$c_k\{(\beta + k)^2 - \nu^2\} + \lambda^2 c_{k-2} = 0, \qquad k = 2, 3, 4, \ldots \tag{A.4.8}$$

from which we may obtain the form of c_k. The first equation (A.4.6), which is the coefficient of r^β, the lowest power of r, requires either $c_0 = 0$ or $\beta^2 - \nu^2 = 0$. To avoid having a trivial solution, $R = 0$, as will become clear later, we choose $c_0 \neq 0$ and hence the *indicial equation*

$$\beta^2 - \nu^2 = 0, \qquad \beta_1 = \nu, \qquad \beta_2 = -\nu$$

must be satisfied, which determines the two values of β in the Frobenius method. From the theory of differential equations we have that if $\beta_1 - \beta_2 \neq 0$ or integer then

$$R_1 = r^{\beta_1} \sum_{k=0}^{\infty} c_k r^k \quad \text{and} \quad R_2 = r^{\beta_2} \sum_{k=0}^{\infty} c_k r^k$$

are the two linearly independent solutions of the Bessel equation (A.4.1). If $\beta_1 - \beta_2 = 0$ or integer, then the above solutions are not linearly independent but we may now use the method of variation of parameters (reduction of order) to construct a new function in terms of R_2 which will be linearly independent of R_1. In our case here $\beta_1 - \beta_2 = 2\nu$ will guarantee two linearly independent solutions provided that $2\nu \neq 0$ or integer.

(i) $\beta = \nu$, $\quad 2\nu \neq 0$ or integer

Let us consider the first case for $\beta = \nu$ of the indicial equation where equation (A.4.7) reduces to $c_1[(\nu + 1)^2 - \nu^2] = (2\nu + 1)c_1 = 0$ and since 2ν is not an integer, $2\nu \neq -1$, then $c_1 = 0$. If we use this result $c_1 = 0$ in (A.4.8), we find that all the odd-indexed coefficients c_{2n+1} must vanish; that is,

$$c_3 = -\frac{\lambda^2}{(\nu + 3)^2 - \nu^2}c_1 = \frac{-\lambda^2 c_1}{6\nu + 9} = 0$$

$$c_5 = -\frac{\lambda^2}{(\nu + 5)^2 - \nu^2}c_3 = 0, \ldots, c_{2n+1} = 0, \ldots$$

Now it is clear why we choose $c_0 \neq 0$, since if $c_0 = 0$ then from (A.4.8)

$$c_2 = -\frac{\lambda^2}{(\nu + 2)^2 - \nu^2}c_0 = \frac{-\lambda^2}{4\nu + 4}c_0 = 0, \ c_4 = 0, \ldots, c_{2n} = 0, \ldots$$

which will result in a *trivial* solution, since in this case all the odd- and even-indexed coefficients will vanish.

So, with $c_0 \neq 0$ we will attempt to relate all the even-indexed coefficients c_{2n} to c_0 by using the *recursion relation* (A.4.8),

$$c_k = -\frac{\lambda^2}{(\nu + k)^2 - \nu^2}c_{k-2} \tag{A.4.9}$$

as follows:

$$c_2 = -\frac{2\lambda^2}{2(2\nu - 2)}c_0$$

$$c_4 = \frac{\lambda^2}{4(2\nu + 4)}c_2 = \frac{\lambda^4}{2 \cdot 4(2\nu + 2)(2\nu + 4)}c_0, \ldots$$

If we continue this, we will arrive at the general form

$$c_{2k} = \frac{(-1)^k (\lambda/2)^{2k}}{k!(\nu + 1)(\nu + 2) \cdots (\nu + k)}c_0 \tag{A.4.10}$$

So the first solution to (A.4.1) corresponding to $\beta = \nu$ where $2\nu \neq 0$ or integer is

$$c_0 r^\nu \sum_{k=0}^{\infty} \frac{(-1)^k (\lambda r/2)^{2k}}{k!(\nu + 1)(\nu + 2)\ldots(\nu + k)} = AJ_\nu(\lambda r) \tag{A.4.11}$$

and with the proper choice of c_0 (or A) (see (A.4.14)) this series represents $J_\nu(\lambda r)$, the *Bessel function of the first kind of order* ν.

(ii) $\beta = -\nu, \qquad 2\nu \neq 0$ or integer

This case of $\beta = -\nu$ and other cases will be left for Section A.5, where we discuss other types of solutions for the Bessel differential equation. (See (A.5.5) where $\beta = -\nu$ an integer.)

The form (A.4.11) for $J_\nu(\lambda r)$ can be written in a more compact way when we introduce the *gamma function*

$$\Gamma(\nu + 1) + \int_0^\infty x^\nu e^{-x}\, dx, \qquad \nu > 0 \tag{A.4.12}$$

which has the very important property

$$\Gamma(\nu + 1) = \nu\Gamma(\nu) \tag{A.4.13}$$

which is proved in Exercise 2.6a in Chapter 2. This shows that the gamma function is a generalization of the factorial; i.e., when n is a nonnegative integer, $\Gamma(n + 1) = n! = n(n - 1)!$. The proper choice of c_0 in (A.4.11) for the series to represent $J_\nu(\lambda r)$ where $J_0(0) = 1$ is

$$c_0 = \left(\frac{\lambda}{2}\right)^\nu \frac{1}{\Gamma(\nu + 1)} \tag{A.4.14}$$

where then (A.4.11) becomes

$$J_\nu(\lambda r) = \sum_{k=0}^{\infty} \frac{(-1)^k (\lambda r/2)^{2k+\nu}}{k!\Gamma(\nu + k + 1)} \tag{A.4.15}$$

after using $(\nu + 1)(\nu + 2)\cdots(\nu + k)\Gamma(\nu + 1) = \Gamma(\nu + k + 1)$. (See Fig. A.1 for $J_0(x)$, $J_1(x)$, and $J_2(x)$ in the answer to Exercise A.22e.)

A.5 Special Differential Equations and Their Solutions

In the following we examine some of the most familiar differential equations of physics and engineering, using the power series method to develop their solutions.

A.5.1 Bessel's Equation

As in the last section, the Bessel differential equation

$$r^2R'' + rR' + (r^2 - \nu^2)R = 0 \tag{A.5.1}$$

is singular at $r = 0$, and hence we used the Frobenius method to obtain solutions

$$J_\nu(r) = \sum_{k=0}^{\infty} \frac{(-1)^k (r/2)^{2k+\nu}}{k!\Gamma(\nu + k + 1)} \tag{A.5.2}$$

and

$$J_{-\nu}(r) = \sum_{k=0}^{\infty} \frac{(-1)^k (r/2)^{2k-\nu}}{k!\Gamma(-\nu + k + 1)} \tag{A.5.3}$$

When 2ν is nonzero and not an integer, the general solution to (A.5.1) is

$$AJ_\nu(r) + BJ_{-\nu}(r) \tag{A.5.4}$$

If $\nu = n$, an integer, $J_n(r)$ and $J_{-n}(r)$ are linearly related, as in (A.5.5a), since

$$J_n(r) = \sum_{k=0}^{\infty} \frac{(-1)^k}{k!(n+k)!} \left(\frac{r}{2}\right)^{2k+n}$$

and

$$J_{-n}(r) = \sum_{k=0}^{\infty} \frac{(-1)^k}{k!(k-n)!} \left(\frac{r}{2}\right)^{2k-n}$$

$$= \sum_{k=0}^{\infty} \frac{(-1)^k k \cdots (k-n+2)(k-n+1)}{(k!)^2} \left(\frac{r}{2}\right)^{2k-n} \tag{A.5.5}$$

But for $k = 0, 1, \ldots, n-1$ the coefficients in (A.5.5) vanish. The first nonzero coefficient is for $k = n$, so we may change the summation variable from k to $p = k - n$ to give

$$J_{-n}(r) = \sum_{p=0}^{\infty} \frac{(-1)^{p+n}}{(p+n)!p!} \left(\frac{r}{2}\right)^{2p+n}$$

$$= (-1)^n \sum_{p=0}^{\infty} \frac{(-1)^p}{p!(p+n)!} \left(\frac{r}{2}\right)^{2p+n} = (-1)^n J_n(r)$$

$$J_{-n}(r) = (-1)^n J_n(r), \qquad n = 0, 1, 2, \ldots \tag{A.5.5a}$$

We can find a linearly independent solution by applying the method of variation of parameters (reduction of order) to give

$$Y_n(r) = \lim_{\nu \to n} \frac{\cos \nu \pi J_\nu(r) - J_{-\nu}(r)}{\sin \nu \pi} \tag{A.5.6}$$

which is called the *Bessel function of order n* of the *second kind*. Thus the general solution to (A.5.1) when n is an integer is

$$R(r) = AJ_n(r) + BY_n(r) \tag{A.5.7}$$

When $\nu = n/2$, where $n = 1, 2, \ldots$, the two solutions will also be linearly independent; for example, when $\nu = \frac{1}{2}$ we can easily show from (A.5.2), (A.1.3), (A.5.3), and (A.1.4) that

$$J_{1/2}(x) = \left(\frac{2}{\pi x}\right)^{1/2} \sin x \tag{A.5.8}$$

$$J_{-1/2}(x) = \left(\frac{2}{\pi x}\right)^{1/2} \cos x \tag{A.5.9}$$

In the following we present a summary of the properties of the Bessel functions $J_n(x)$ along with other important kinds of Bessel functions, where we leave most of their proofs for the exercises. For n as nonnegative integer,

$$J_n(x) = \sum_{k=0}^{\infty} \frac{(-1)^k (x/2)^{n+2k}}{k!(n+k)!} \tag{A.5.10}$$

$$J_{-n}(x) = (-1)^n J_n(x) \tag{A.5.11}$$

$$J_n(x) = \frac{1}{\pi} \int_0^\pi \cos(x \cos u - nu) \, du \tag{A.5.12}$$

$$J_{n-1}(x) + J_{n+1}(x) = \frac{2n}{x} J_n(x) \tag{A.5.13}$$

$$\frac{dJ_n(x)}{dx} = -\frac{n}{x} J_n(x) + J_{n-1}(x) = \frac{n}{x} J_n(x) - J_{n+1}(x)$$
$$= \frac{1}{2}[J_{n-1}(x) - J_{n+1}(x)] \tag{A.5.14}$$

$$\frac{d}{dx}[x^n J_n(cx)] = cx^n J_{n-1}(cx) \tag{A.5.15}$$

$$\frac{d}{dx}[x^{-n} J_n(cx)] = -cx^{-n} J_{n+1}(cx) \tag{A.5.16}$$

$$\int x J_n(ax) J_n(bx) \, dx = \frac{bx J_n(ax) J_{n-1}(bx) - ax J_{n-1}(ax) J_n(bx)}{a^2 - b^2} \tag{A.5.17}$$

$$\int x[J_n(ax)]^2 \, dx = \frac{x^2}{2} \left\{ [J_n(ax)]^2 - J_{n-1}(ax)J_{n+1}(ax) \right\} \tag{A.5.18}$$

The *generating function* of Bessel functions $J_n(x)$:

$$e^{(x/2)(t-1/t)} = \sum_{n=-\infty}^{\infty} J_n(x)t^n \tag{A.5.19}$$

The solution to the Bessel equation (A.5.1) can also be written as a linear combination of the Hankel functions $H_\nu^{(1)}(x)$ and $H_\nu^{(2)}(x)$ (*Bessel functions of the third kind*).

$$H_\nu^{(1)}(z) = J_\nu(z) + iY_\nu(z) \tag{A.5.20}$$

$$H_\nu^{(2)}(z) = J_\nu(z) - iY_\nu(z) \tag{A.5.21}$$

Modified Bessel Functions I, K

$$I_\nu(z) = e^{-\nu\pi i/2}J_\nu(ze^{\pi i/2}) = i^{-\nu}J_\nu(iz), \qquad \left(-\pi < \arg z < \frac{\pi}{2}\right) \tag{A.5.22}$$

is the *modified Bessel function of the first kind of order* ν, which is a solution to (the modified) Bessel differential equation

$$z^2 R''(z) + zR'(z) + (-z^2 - \nu^2)R(z) = 0 \tag{A.5.23}$$

Macdonald's Functions $K_\nu(z)$

$$K_\nu(z) = \frac{1}{2}\pi \frac{I_{-\nu}(z) - I_\nu(z)}{\sin\nu\pi} \tag{A.5.24}$$

Integral Representation of $J_\nu(z)$

$$J_\nu(z) = \frac{2(\frac{1}{2}\pi z)}{\pi^{\frac{1}{2}}\Gamma(\nu + \frac{1}{2})} \int_0^1 (1-t^2)^{\nu-1/2} \cos zt \, dt, \qquad \text{Re}\,\nu > -\frac{1}{2} \tag{A.5.25}$$

The Zeros of the Bessel functions $J_m(x)$
In Table A.1 we list 20 of the zeros $j_{m,n}$ of the Bessel function $J_m(x)$, that is, $J_m(j_{m,n}) = 0$, $n = 1, 2, \ldots, 20$, for $m = 1, 2, 3, 4$, and 5. They are definitely needed for our analysis and illustration of the Bessel-type Poisson summation formula and a discrete Hankel transform in Section 4.3.

A.5.2 Legendre's Equation

We have seen in (3.155) how the following (special case of) the Legendre equation:

$$(1-x^2)X'' - 2xX' + \nu(\nu + 1)X = 0 \tag{A.5.26}$$

TABLE A.1 Twenty Zeros $j_{m,n}$ for the Bessel Functions $J_m(x)$, $m = 1, 2, 3, 4, 5$; $J_m(j_{m,n}) = 0$, $n = 1, 2, \ldots, 20$, $m = 1, 2, \ldots, 5$

n	$j_{0,n}$	$j_{1,n}$	$j_{2,n}$	n	$j_{3,n}$	$j_{4,n}$	$j_{5,n}$
1	2.40482 55577	3.83171	5.13562	1	6.38016	7.58834	8.77148
2	5.52007 81103	7.01559	8.41724	2	9.76102	11.06471	12.33860
3	8.65372 79129	10.17347	11.61984	3	13.01520	14.37254	15.70017
4	11.79153 44391	13.32369	14.79595	4	16.22347	17.61597	18.98013
5	14.93091 77086	16.47063	17.95982	5	19.40942	20.82693	22.21780
6	18.07106 39679	19.61586	21.11700	6	22.58273	24.01902	25.43034
7	21.21163 66299	22.76008	24.27011	7	25.74817	27.19909	28.62662
8	24.35247 15308	25.90367	27.42057	8	28.90835	30.37101	31.81172
9	27.49347 91320	29.04683	30.56920	9	32.06485	33.53714	34.98878
10	30.63460 64684	32.18968	33.71652	10	35.21867	36.69900	38.15987
11	33.77582 02136	35.33231	36.86286	11	38.37047	39.85763	41.32638
12	36.91709 83537	38.47477	40.00845	12	41.52072	43.01374	44.48932
13	40.05842 57646	41.61709	43.15345	13	44.66974	46.16785	47.64940
14	43.19979 17132	44.75932	46.29800	14	47.81779	49.32036	50.80717
15	46.34118 83717	47.90146	49.44216	15	50.96503	52.47155	53.96303
16	49.48260 98974	51.04354	52.58602	16	54.11162	55.62165	57.11730
17	52.62405 18411	54.18555	55.72963	17	57.25765	58.77184	60.27025
18	55.76551 07550	57.32753	58.87302	18	60.40322	61.91925	63.42205
19	58.90698 39261	60.46946	62.01622	19	63.54840	65.06700	66.57289
20	62.04846 91902	63.61136	65.15927	20	66.69324	68.21417	69.72289

resulted from the separation of variables of the solution to Laplace's equation in spherical coordinates (ρ, θ, ϕ) for the variable $X(x)$, $x = \cos\theta$. Since $x = \cos\theta$, $0 \le \theta \le \pi$ we are interested in a solution for $-1 \le x \le 1$, and so we look for power series expansions about either the ordinary point $x = 0$ or the regular singular points $x = \mp 1$. We shall develop a power series expansion about $x = 0$. Let

$$X(x) = \sum_{k=0}^{\infty} c_k x^k \tag{A.5.27}$$

and substitute in (A.5.26) to obtain

$$(1 - x^2) \sum_{k=2}^{\infty} k(k-1)c_k x^{k-2} - 2x \sum_{k=1}^{\infty} k c_k x^{k-1} + \nu(\nu+1) \sum_{k=0}^{\infty} c_k x^k = 0$$

$$= \sum_{k=2}^{\infty} k(k-1)c_k x^{k-2} - \sum_{k=2}^{\infty} k(k-1)c_k x^k$$

$$- 2\sum_{k=1}^{\infty} kc_k x^k + \nu(\nu+1)\sum_{k=0}^{\infty} c_k x^k = 0$$

Now we write the series at length and collect the terms of the same powers in x in order to equate their coefficients to zero to obtain relations for c_k in terms of the coefficients of lower-order terms.

$$[2c_2 + \nu(\nu+1)c_0] + [6c_3 - 2c_1 + \nu(\nu+1)c_1]x$$

$$+ [12c_4 - 2c_2 - 4c_2 + \nu(\nu+1)c_2]x^2 + \cdots$$

$$+ [(m+1)(m+2)c_{m+2} - m(m-1)c_m - 2mc_m + \nu(\nu+1)c_m]x^m$$

$$+ \cdots + \cdots = 0 \qquad (A.5.28)$$

By equating the coefficients of each power of x to zero, we obtain after simplification

$$2c_2 + \nu(\nu+1)c_0 = 0, \qquad c_2 = \frac{-\nu(\nu+1)}{2}c_0$$

$$6c_3 + [-2 + \nu(\nu+1)]c_1 = 0, \qquad c_3 = \frac{(\nu-1)(\nu+2)}{6}c_1$$

$$12c_4 + [-6 + \nu(\nu+1)]c_2 = 0, \qquad c_4 = -\frac{(\nu-2)(\nu+3)}{12}c_2$$

or, by eliminating c_2

$$c_4 = \frac{(\nu-2)\nu(\nu+1)(\nu+3)}{4!}c_0$$

Also,

$$20c_5 + [-12 + \nu(\nu+1)]c_3 = 0, \qquad c_5 = -\frac{(\nu-3)(\nu+4)}{20}c_3$$

or, by eliminating c_3

$$c_5 = \frac{(\nu-3)(\nu-1)(\nu+2)(\nu+4)}{5!}c_1$$

If we use the mth term in (A.5.28) we may write the *recursion relation*

$$c_{m+2} = -\frac{(\nu-m)(\nu+m+1)}{(m+2)(m+1)}c_m \qquad (A.5.29)$$

Note that the power series separates into series involving even powers and odd powers; that is, we may write the solution in the form

$$X(x) = c_0 \left[1 - \frac{\nu(\nu+1)}{2!}x^2 + \frac{\nu(\nu-2)(\nu+1)(\nu+3)}{4!}x^4 + \cdots \right]$$

$$+ c_1 \left[x - \frac{(\nu-1)(\nu+2)}{3!}x^3 \right.$$

$$+ \frac{(\nu-1)(\nu-3)(\nu+2)(\nu+4)}{5!}x^5 + \cdots \right]$$

$$= c_0 u_\nu(x) + c_1 v_\nu(x) \tag{A.5.30}$$

It is clear that $u_\nu(x)$ and $v_\nu(x)$ are linearly independent, and hence (A.5.30) represents the general solution. The functions $u_\nu(x)$ and $v_\nu(x)$ are called *Legendre functions*. We also note that both series converge for $-1 < x < 1$ and that they diverge at $x \mp 1$, which corresponds to $\theta = 0$ and π since $x = \cos\theta$.

An important special case, $\nu = n =$ integer, arises in physical applications. When $\nu = n$, a positive integer, we note that the first series of (A.5.30) terminates at $m = n$ when n is even and the second series terminates at $m = n$ when n is odd. Thus for $\nu = n =$ integer either u_ν or v_ν reduces to a polynomial of degree n, depending on whether n is even or odd, respectively. This polynomial is then finite for $-1 \le x \le 1$, and when normalized to 1 at $x = 1$ is called $P_n(x)$ the *Legendre polynomial* of degree n. The remaining nonterminating series together with a conveniently defined coefficient is called $Q_n(x)$, the *Legendre function of the second kind*. For example, when $\nu = 1$ the polynomial solution is given by $v_1(x) = x = P_1(x)$, where $P_1(1) = 1$, and the infinite series solution is

$$u_1(x) = 1 - x^2 - \frac{8}{4!}x^4 \cdots$$

$$= - \left[-1 + x^2 + \frac{8}{4!}x^4 + \cdots \right] = -Q_1(x) \tag{A.5.31}$$

where the conventional choice for the coefficient of $Q_1(x)$ is -1 and thus the general solution for (A.5.26) with $\nu = 1$ is

$$X(x) = -c_0 Q_1(x) + c_1 P_1(x) \tag{A.5.32}$$

From (A.5.29) we obtain the general form of the coefficients c_n for the power series $P_n(x)$. For $\nu = n$ we have $c_{n+2} = 0$, $c_{n+4} = 0$, ..., $c_{n+2j} = 0$. Now we write $c_{n-2}, c_{n-4}, \ldots, c_{n-2j}$ in terms of c_n, where we choose

$$c_n = \frac{(2n-1(2n-3)\cdots 3 \cdot 1}{n!}$$

in order that $P_n(1) = 1$. From (A.5.29) we have

$$c_{n-2} = -\frac{n(n-1)}{2(2n-1)}c_n$$

$$c_{n-4} = -\frac{(n-2)(n-3)}{4(2n-3)}\cdot c_{n-2} = \frac{n(n-1)(n-2)(n-3)}{2\cdot 4(2n_1)(2n-3)}c_n$$

If we continue in this manner, we obtain the general case

$$c_{n-2j} = (-1)^j\frac{(2n-2j)!}{2^n j!(n-j)!(n-2j)!} \tag{A.5.33}$$

and $P_n(x)$ becomes

$$P_n(x) = \sum_{j=0}^{J}(-1)^j\frac{(2n-2j)!}{2^n j!(n-j)(n-2j)!}x^{n-2j} \tag{A.5.34}$$

where $J = n/2$ when n is even and $(n-1)/2$ when n is odd. For example, when $n = 0, 1, 2, 3, 4$, and 5 then $J = 0, 0, 1, 1, 2$, and 2, respectively, and from (A.5.34) we have

$$P_0(x) = 1, \qquad P_1(x) = x, \qquad P_2(x) = \tfrac{1}{2}(3x^2 - 1)$$
$$P_3(x) = \tfrac{1}{2}(5x^3 - 3x), \qquad P_4(x) = \tfrac{1}{8}(35x^4 - 30x^2 + 3)$$
$$P_5(x) = \tfrac{1}{8}(63x^5 - 700x^3 + 15x)$$

In the case of the nonterminating series $Q_n(x)$ the c_n coefficient is chosen, by convention again, according to whether n is even or odd; thus

$$Q_n(x) = \frac{(-1)^{n/2}2^n[(n/2)!]^2}{n!}\left[x - \frac{(n-1)(n+2)}{3!}x^3\right.$$
$$\left. + \frac{(n-1)(n-3)(n+2)(n+4)}{5!}x^5 + \cdots\right], n \text{ even} \tag{A.5.35}$$

$$Q_n(x) = \frac{(-1)^{(n+1)/2}\{[(n-1)/2]!\}^2 2^{n-1}}{n!}\left[1 - \frac{n(n+1)}{2!}x^2\right.$$
$$\left. + \frac{n(n-2)(n+1)(n+3)}{4!}x^4\cdots\right], n \text{ odd} \tag{A.5.36}$$

The symbol $Q_n(x)$ is usually used for this (infinite series) Legendre function, and it is not to be confused with the polynomial $Q_n(x)$ of degree n in (3.70) that we used for Theorem 3.8 concerning the Weierstrass polynomial approximation of continuous functions.

As an example,

$$Q_0(x) = x + \frac{2}{3!}x^3 + \frac{1 \cdot 2 \cdot 3 \cdot 4}{5!}x^5 + \frac{1 \cdot 3 \cdot 5 \cdot 2 \cdot 4 \cdot 6}{7!}x^7 + \cdots$$

$$= x + \frac{x^3}{3} + \frac{x^5}{5} + \frac{x^7}{7} + \cdots = \frac{1}{2}\ln\left(\frac{1+x}{1-x}\right)$$

and

$$Q_1(x) = -\left[1 - \frac{2}{2!}x^2 - \frac{8}{4!}x^4 + \frac{144}{6!}x^6 + \cdots\right]$$

$$= -1 + x\left[x + \frac{x^3}{3} + \frac{x^5}{5} + \cdots\right]$$

$$= \frac{x}{2}\ln\left(\frac{1+x}{1-x}\right) - 1 = xQ_0(x) - 1 = P_1(x)Q_0(x) - 1 \qquad \text{(A.5.37)}$$

The *Rodrigues formula* for generating the Legendre polynomials $P_n(x)$

$$P_n(x) = \frac{1}{2^n n!}\frac{d^n(x^2-1)^n}{dx^n}, \qquad P_n(1) = 1 \qquad \text{(A.5.38)}$$

can easily be derived (see Exercise A.26) and used to generate the Legendre polynomials; for example,

$$P_0(x) = 1, \qquad P_1(x) = x, \qquad P_2(x) = \frac{1}{2}(3x^2 - 1)$$

$$P_3(x) = \frac{1}{2}(5x^3 - 3x), \qquad P_4(x) = \frac{1}{8}(35x^4 - 30x^2 + 3) \qquad \text{(A.5.39)}$$

For the orthogonality of $\{P_n(x)\}$ on $[-1, 1]$ we can show that $\|P_n\|^2 \equiv h_n = 2/(2n+1)$ (see Exercise A.28).

Zeros of the Legendre Polynomials $P_n(x)$

In Table A.2 we list a number of zeros $\{x_{n,m}\}$ of $P_n(x)$, that is, $P_n(x_{n,m}) = 0$, $m = 1, 2, 3, \ldots, n$. These are needed for the discussion of the *discrete Legendre polynomial transforms* in Section 4.2.

The rest of these tabulated zeros for $P_n(x)$ of degree n up to 48 may be found in Abramowitz and Stegun (1964). Such zeros are needed for the analysis and illustration of the discrete Legendre transform in Sections 4.2.1D and 4.2.3.

A.5.3 Other Special Equations

Laguerre Equation

The following equation, attributed to Laguerre,

$$xy'' + (1 - x)y' + \nu y = 0 \qquad \text{(A.5.40)}$$

TABLE A.2 Zeros $x_{n,m}$ of the Legendre Polynomials $P_n(x)$; $P_n(x_{n,m}) = 0$, $m = 0, 1, 2, 3, \ldots, n$

Degree n	Zeros $\mp x_{n,m}$ of $P_n(x)$	Degree n	Zeros $\mp x_{n,m}$ of $P_n(x)$
1	0	7	0.74153 11855 99394
2	0.57735 02691 89626		0.94910 79123 42759
3	0.00000 00000 00000	8	0.18343 46424 95650
	0.77459 66692 41483		0.52553 24099 16329
			0.79666 64774 13627
4	0.33998 10435 84856		0.96028 98564 97536
	0.86113 63115 94053		
		9	0.00000 00000 00000
5	0.00000 00000 00000		0.32425 34234 03809
	0.53846 93101 05683		0.61337 14327 00590
	0.90617 98459 38664		0.83603 11073 26636
			0.96816 02395 07626
6	0.23861 91860 83197		
	0.66120 93864 66265	10	0.14887 43389 81631
	0.93246 95142 03152		0.43339 53941 29247
			0.67940 95682 99024
7	0.00000 00000 00000		0.86506 33666 88985
	0.40584 51513 77397		0.97390 65285 17172

has a regular singular point at $x = 0$, so we use the Frobenius method for the solution. If we follow the same steps as in the case of Bessel's equation (A.4.1), we find a solution $L_\nu(x)$, the Laguerre function, and for $\nu = n$ a positive integer, the solution is a polynomial of degree n, $L_n(x)$, the *Laguerre polynomial*.

$$L_n(x) = (-1)^n \left[x^n - \frac{n^2}{1!} x^{n-1} + \frac{n^2(n-1)^2}{2!} x^{n-2} + \cdots + (-1)^n n! \right]$$
(A.5.41)

For example,

$$L_0(x) = 1, \qquad L_1(x) = -x + 1, \qquad L_2(x) = x^2 - 4x + 2$$

$$L_3(x) = -x^3 + 9x^2 - 18x + 6$$
(A.5.42)

Next we present a few of the basic properties of the Laguerre polynomials $L_n(x)$. They are orthogonal on the *semi-infinite* interval $(0, \infty)$ with respect to $\rho(x) = e^{-x}$, and are normalized with $h_n = 1 \equiv \|L_n\|^2$. Also, its

Rodrigues formula

$$L_n(x) = \frac{e^x}{n!} \frac{d^n}{dx^n} (x^n e^{-x}) = \sum_{k=0}^{n} \frac{(-1)^k n! x^k}{(k!)^2 (n-k)!} \tag{A.5.43}$$

$$L_0(x) = 1, \qquad L_1(x) = 1 - x, \qquad L_2(x) = 1 - 2x + \tfrac{1}{2}x^2 \tag{A.5.44}$$

$$L_3(x) = \tfrac{1}{6}[-x^3 + 9x^2 - 18x + 16]$$

$$\frac{d}{dx} [xe^{-x} L_n'(x)] = -ne^{-x} L_n(x) \tag{A.5.45}$$

Hermite Equation

The following *Hermite* equation:

$$y'' - 2xy' + 2\nu y = 0 \tag{A.5.46}$$

has an ordinary point at $x = 0$, and if we use the power series expansion about $x = 0$ we find a solution $H_\nu(x)$, the Hermite function. For the case of $\nu = n$, a positive integer, one solution is a polynomial of degree n, $H_n(x)$, the *Hermite polynomial* of degree n, which has the following representation according to whether n is even or odd:

$$H_{2n}(x) = \sum_{k=0}^{n} (-1)^k 2n(2n-1) \cdots (2n - 2k + 1)(2x)^{2n-2k} \tag{A.5.47}$$

or

$$H_{2n+1}(x) = \sum_{k=0}^{n} (-1)^k (2n+1)(2n) \cdots (2n - 2k + 1)(2x)^{2n+1-2k}$$

$$\tag{A.5.48}$$

Examples are $H_0(x) = 1$, $H_1(x) = 2x$, $H_2(x) = 4x^2 - 2$, $H_3(x) = 8x^3 - 12x$, and $H_4(x) = 16x^4 - 48x^2 + 12$.

The following are the Rodrigues formula and a recursion relation for the Hermite polynomials $H_n(x)$ which are orthogonal on the *infinite interval* $(-\infty, \infty)$ with respect to $\rho(x) = e^{-x^2}$. Also, they are normalized by $h_n = \sqrt{\pi} 2^n n! \equiv \|H_n\|^2$.

$$H_n(x) = (-1)^n e^{x^2} \frac{d^n}{dx^n} e^{-x^2} \tag{A.5.49}$$

$$H_{n+1}(x) - 2xH_n(x) + 2nH_{n-1}(x) = 0 \tag{A.5.50}$$

See the other Hermite polynomial $He_n(x) \equiv 2^{-n/2} H_n(x/2)$ before the exercises of Section A.5b.

Other special functions and some of the properties of the special functions of this appendix will be subjects of the exercises.

Exercises

Section A.1 Review of Simple Differential Equations

A.1. For the *nth-order linear homogeneous* differential equation with *constant coefficients* in $y(x)$,

$$a_0 \frac{d^n y}{dx^n} + a_1 \frac{d^{n-1} y}{dx^{n-1}} + \cdots + a_{n-1} \frac{dy}{dx} + a_n y = 0 \qquad (E.1)$$

we know that it has a solution of the form $y(x) = e^{mx}$.

(a) Show that for the differential equation (E.1) to have a non-trivial solution, the parameter m must satisfy the following *characteristic equation* of (E.1):

$$a_0 m^n + a_1 m^{n-1} + \cdots + a_{n-1} m + a_n = 0 \qquad (E.2)$$

(b) In general, for (E.2) we expect n roots (or zeros). Use this method to solve the second-order differential equation for the two possible solutions.

$$\frac{d^2 y}{dx^2} - 3 \frac{dy}{dx} + 2y = 0 \qquad (E.3)$$

(c) The two solutions $y_1(x)$ and $y_2(x)$ of the equation

$$a_0 \frac{d^2 y}{dx^2} + a_1 \frac{dy}{dx} + a_2 y = 0 \qquad (E.4)$$

are *linearly independent* if and only if their *Wronskian* $W(y_1, y_2) \equiv y_1 y_2' - y_2 y_1'$ is not identically zero on the domain of the equation (E.4). Use this for equation (E.3) to show whether or not its two solutions are linearly independent.

(d) For (E.4) show that if the two roots of its characteristic equation are distinct, then the two corresponding solutions are linearly independent.

(e) The equation

$$\frac{d^2 y}{dx^2} - 2 \frac{dy}{dx} + y = 0 \qquad (E.5)$$

has repeated roots $m_1 = m_2 = 1$ from $m^2 - 2m + 1 = (m - 1)^2 = 0$. Verify this result of $y(x) = e^x$ by direct substitution in (E.5). Then verify that $u(x) = xe^x$ is the second linearly independent solution (see the method of reduction of order in Exercise A.7a for obtaining $u(x)$ by multiplying $y(x) = e^x$ by x).

(f) For a homogeneous second-order differential equation with two linearly independent solutions $y_1(x)$ and $y_2(x)$, the *general solution* is a linear combination $y(x) = c_1y_1(x) + c_2y_2(x)$ of $y_1(x)$ and $y_2(x)$, where c_1 and c_2 are arbitrary constants. Write the general solutions for the equations (E.3) in part (b) and (E.5) in part (e).

(g) For the general solution in part (f), we must supply (reasonable) conditions to determine the arbitrary constants c_1 and c_2 to have a unique solution. Consider the equation (E.3) in part (b) with the two initial conditions $y(0) = -1$ and $y'(0) = 3$ to find its unique solution.

(h) Solve the following initial value problem:

$$\frac{d^2y}{dt^2} + 2\frac{dy}{dt} - 3y = 0, \qquad 0 < t < \infty \tag{E.6}$$

$$y(0) = 3 \tag{E.7}$$

$$y'(0) = 1 \tag{E.8}$$

(i) Use the above method of letting $y(x) = e^{mx}$, and another method, to solve the first-order differential equation

$$\frac{dy}{dx} + by = 0 \tag{E.9}$$

Hint: Note that it is separable.

(j) For the same first-order differential equation on the domain $0 < t < \infty$

$$\frac{dy}{dt} + by = 0 \tag{E.9}$$

find the (bounded) solution by imposing the necessary condition.

(k) Solve the following atomic decay initial value problem in $N(t)$,

$$\frac{dN}{dt} = -\lambda N, \qquad \lambda > 0, \ t > 0 \tag{E.10}$$

$$N(0) = N_0 \tag{E.11}$$

ANS. (b) $m^2 - 3m + 2 = 0, \ m_1 = 1, \ m_2 = 2$

$y_1(x) = e^x, \ y_2(x) = e^{2x}$

(c) $W(e^x, e^{2x}) = 2e^{3x} - e^{3x} = e^{3x} \neq 0$, hence e^x and e^{2x} are linearly independent.

(d) For $m_1 \neq m_2$, $W(e^{m_1x}, e^{m_2x}) = (m_2 - m_1)e^{(m_1+m_2)x} \neq 0$; thus e^{m_1x} and e^{m_2x} are linearly independent.

(e) $m_1 = m_2 = 1$, $y_1(x) = y_2(x) = e^x$

(f) (i) $y(x) = c_1e^x + c_2e^{2x}$ for (E.3) in part (b)

    (ii) $y(x) = c_1e^x + c_2xe^x$ for (E.5) in part (e)

(g) $y(x) = -5e^x + 4e^{2x}$

(h) $y(t) = \frac{5}{2}e^t + \frac{1}{2}e^{-3t}$

(i) (i) $m = -b$, $y = e^{-bx}$

    (ii) Also it is a simple separable equation that can be integrated easily as $\int dy/y = -b \int dx$, $\ln y = -bx + c$, $y = e^{-bx}e^c = y = Ae^{-bx}$

(j) $y(t) = Ae^{-bt}$, $b > 0$, $0 < t < \infty$

    or $y(t) = Ae^{-\alpha^2 t}$, α real

(k) $N(t) = N_0e^{-\lambda t}$, $\lambda > 0$, $t > 0$

A.2. Consider the *Euler–Cauchy* equation with its very particular variable coefficients,

$$a_0x^2 \frac{d^2y}{dx^2} + a_1x \frac{dy}{dx} + a_2y = 0, \qquad x > 0 \tag{E.1}$$

where a_0, a_1, and a_2 are constants. It is known that the change of variable $x = e^t$ will reduce (E.1) to a constant-coefficient equation in $y(x) = y(e^t) \equiv u(t)$. Then for $u(t)$ we can use $u(t) = e^{mt} = (e^t)^m = x^m = y(x)$; i.e., it is equivalent to letting $y(x) = x^m$ in (E.1).

(a) Use $y(x) = x^m$ in (E.1) to obtain the characteristic equation for finding m.

(b) Use this method to solve the equation

$$x^2y'' + 3xy' - 3y = 0, \qquad x > 0 \tag{E.2}$$

by finding the characteristic equation. Show that for the two distinct roots you have two linearly independent solutions; then write the general solution of (E.2).

(c) Consider the differential equation (E.2) on the line $1 < x < \infty$, and find the (bounded) solution.

Hint: Note that the solution $y_1(x) = x$ in (E.1) is not bounded, so we should choose the arbitrary constant $c_1 = 0$ to avoid this problem.

(d) What happens if (E.2) is defined in a disc of radius 3, where x denotes the radial distance from the origin?

Hint: Watch for the center of the disc at $x = 0$.

(e) Solve the equation

$$x^2 y'' + 5xy' + 4y = 0, \qquad x > 0 \tag{E.3}$$

Hint: Watch for double roots, and then consult part (e) of Exercise A.1 for generating the second linearly independent solution (of E.5). We must also remember that multiplying by x (in the constant-coefficient equation (E.5)) is equivalent to multiplying in the present problem by $t = \ln x$. Here the Euler–Cauchy equation (E.3) becomes one with constant coefficients in $u(t) = y(e^t) \equiv y(x)$. Thus the second linearly independent solution is $u_2(t) = tu_1(t) = (\ln x)y_1(x)$.

(f) (i) Solve the differential equation in $R(r)$,

$$r^2 R'' + rR' - n^2 R = 0,$$

$$0 < r_1 < r < r_2, \; n = 0, 1, 2, \ldots \tag{E.4}$$

on the ring $0 < r_1 < r < r_2$, where n is an integer.

(ii) What is the solution outside the unit disc $r \le 1$?

(g) (i) Solve the differential equation in $R(\rho)$

$$\rho^2 \frac{d^2 R}{d\rho^2} + 2\rho \frac{dR}{d\rho} - n(n + 1)R = 0,$$

$$0 < a < \rho < b, \; n = 0, 1, 2, \ldots$$

(ii) What would the solution be on the domain $0 < b < \rho < \infty$?

(iii) What is the solution on the domain $0 \le \rho < a$?

ANS. (a) $a_0 m(m - 1) + a_1 m + a_2 = 0$.

(b) $m^2 + 2m - 3 = 0$, $m_1 = 1$, $m_2 = -3$, $y(x) = c_1 x + c_2/x^3$

(c) $y(x) = c_2/x^3$

(d) $y(x) = c_1 x$

(e) $m^2 + 4m + 4 = (m + 2)^2$, $m_1 = m_2 = -2$, $y_1(x) = x^{-2}$, and by the method of reduction of order (see Exercises A.7a, A.1e) we can have $y_2(x) = x^{-2} \ln x$ as the second linearly independent solution of (E.3), $y(x) = c_1 x^{-2} + c_2 x^{-2} \ln x$.

(f) (i) $R(r) = c_1 r^n + c_2 r^{-n}$

    (ii) $R(r) = c_2 r^{-n}$, $r > 1$, $n = 0, 1, 2, 3, \ldots$

(g) (i) $R(\rho) = c_1 \rho^n + c_2 \rho^{-n-1}$

    (ii) $R(\rho) = c_2 \rho^{-n-1}$

    (iii) $R(\rho) = c_1 \rho^n$, $n = 0, 1, 2, 3, \ldots$

A.3. (a) Solve the following second-order differential equation:

$$\frac{d^2 y}{dx^2} + \lambda^2 y = 0, \qquad \lambda^2 > 0 \quad \text{(or } \lambda \text{ is real)} \tag{E.1}$$

and write its general solution in two equivalent forms, one in terms of trigonometric functions and the other in terms of (complex) exponential functions. For a review of basic elements of complex numbers see Exercise 1.6 in Chapter 1.

Hint: Use the Euler identities

$$e^{\mp ix} = \cos x \mp i \sin x \tag{E.2}$$

(b) Solve the following differential equation:

$$\frac{d^2 y}{dx^2} - \lambda^2 y = 0, \qquad \lambda^2 > 0 \tag{E.3}$$

and write its general solution in two equivalent forms, one in terms of hyperbolic functions ($\cosh x, \sinh x$) and the other in terms of exponential functions.

Hint: Use the Euler identities

$$e^{\mp x} = \cosh x \mp \sinh x \tag{E.4}$$

(c) Considering the differential equation in part (b) on the domain $0 < x < \infty$, which of the two forms of the solution would you prefer? Why? Write the solution with the restriction to guarantee that it is bounded as $x \to \infty$.

(d) As in part (c), what form would you choose for the domain $0 < x < l$ and with the one boundary condition $y(0) = 0$?

(e) In part (c) or (d), would your choice have made a big difference?

(f) Consider the differential equation in part (a). Choose the form of solution that most suits the domain $0 < x < \infty$ and the boundary condition $y'(0) = 0$.

(g) Use the same method with $y(x) = e^{\gamma x}$ to solve the following fourth-order differential equation

$$\frac{d^4y}{dx^4} - \lambda^4 y = 0$$

Hint: The characteristic equation is $\gamma^4 - \lambda^4 = (\gamma^2 - \lambda^2)(\gamma^2 + \lambda^2) = 0$, $\gamma_1 = \lambda$, $\gamma_2 = -\lambda$, $\gamma_3 = i\lambda$, and $\gamma_4 = -i\lambda$.

ANS. (a) (i) $y(x) = c_1 \cos \lambda x + c_2 \sin \lambda x$

(ii) $y(x) = c_3 e^{i\lambda x} + c_4 e^{-i\lambda x}$

(b) (i) $y(x) = c_1 \cosh \lambda x + c_2 \sinh \lambda x$

(ii) $y(x) = c_3 e^{\lambda x} + c_4 e^{-\lambda x}$

(c) The second one, since it is transparent to the important condition of having a bounded solution on this semi-infinite domain. With $\lambda > 0$, $c_3 = 0$, we have a bounded $y(x) = c_4 e^{-\lambda x}$.

(d) The first form, since its compatibility ($\sinh 0 = 0$) with this boundary condition is clear. With $c_1 = 0$, $y(x) = c_2 \sinh \lambda x$.

(e) No, but we avoided unnecessary algebra.

(f) For the same reason as in parts (c) and (d), the first one is more convenient, since with $c_2 = 0$, $y(x) = c_1 \cos \lambda x$ satisfies the boundary condition $y'(0) = c_1 \lambda \sin 0 = 0$, and, of course, it is bounded as $x \to \infty$, $y(x) = c_1 \cos \lambda x$.

(g) $y(x) = c_1 e^{\lambda x} + c_2 e^{-\lambda x} + c_3 e^{i\lambda x} + c_4 e^{-i\lambda x}$

$$= A \cosh \lambda x + B \sinh \lambda x + C \cos \lambda x + D \sin \lambda x$$

A.4. Solve the following *boundary value problems*, associated with the second-order differential equation (E.1) in part (a) or (b) of Exercise A.3 on the indicated domain and with the indicated boundary conditions. Note that for infinite domain we must consider the boundedness of the solution as a condition

(a) $y'' + \lambda^2 y = 0,$ $\quad 0 < x < l$

$\quad y(0) = y(l) = 0$

(b) $y'' + \lambda^2 y = 0,$ $\quad 0 < x < l$

$\quad y'(0) = y'(l) = 0$

(c) $y'' + \lambda^2 y = 0,$ $\quad 0 < x < l$

$\quad y'(0) = 0, \quad y(l) = 0$

(d) $y'' + \lambda^2 y = 0,$ $\quad 0 < x < l$

$\quad y(0) = 0, \quad y'(l) = 0$

(e) $y'' + \lambda^2 y = 0,$ $\quad 0 < x < \infty$

$\quad y(0) = 0, \quad |y(x)| < \infty$

(f) $y'' + \lambda^2 y = 0,$ $\quad 0 < x < \infty$

$\quad y'(0) = 0, \quad |y(x)| < \infty$

(g) $y'' + \lambda^2 y = 0,$ $\quad -\infty < x < \infty$

$\quad |y(x)| < \infty$

(h) $y'' + \lambda^2 y = 0,$ $\quad 0 < x < l$

$\quad y(0) = 0, \quad y(l) + hy'(l) = 0$

(i) $y'' - \lambda^2 y = 0,$ $\quad 0 < x < \infty$

$\quad y(0) = 1, \quad |y(x)| < \infty$

(j) $y'' - \lambda^2 y = 0,$ $\quad 0 < x < l$

$\quad y(0) = 1, \quad y(l) = 0$

ANS. (a) A set of solutions $y_n(x) = A \sin(n\pi x/l),$ $n = \mp 1, \mp 2,$ $\mp 3, \ldots,$ where A is an arbitrary constant.

(b) $y_n(x) = A \cos(n\pi x/l),$ $n = 0, \mp 1, \mp 2, \mp 3, \ldots$

(c) $y_n(x) = A \cos{(n + \frac{1}{2})} \pi x/l$, $n = 0, \mp 1, \mp 2, \mp 3, \ldots$

(d) $y_n(x) = A \sin{(n + \frac{1}{2})} \pi x/l$, $n = 0, \mp 1, \mp 2, \mp 3, \ldots$

(e) $y(x) = A \sin \lambda x$, λ real

(f) $y(x) = A \cos \lambda x$, λ real

(g) $y(x) = A \cos \lambda x + B \sin \lambda x$, λ real

(h) $y_n(x) = A \sin \lambda_n x$, where $\{\lambda_n\}$ are the solutions of $\tan \lambda_n l = -h \lambda_n$

(i) $y(x) = e^{-\alpha^2 x}$, $\alpha^2 = \lambda$, α real

(j) $y(x) = \dfrac{\sinh \lambda(l - x)}{\sinh \lambda l}$

Nonhomogeneous Equations

First-Order (Linear) Equations

A.5. For the first-order, linear and nonhomogeneous equation

$$\frac{dy}{dx} + p(x)y = f(x) \tag{E.1}$$

we use the *integrating factor* $\mu(x) = \exp(\int^x p(x) dx)$ to multiply (E.1) by, and reduce its left side to an *exact differential*

$$\frac{d}{dx}[\mu(x)y(x)] = \mu(x)f(x) \tag{E.2}$$

whose direct integration gives the solution to (E.1) in terms of an integral of the nonhomogeneous term $f(x)$,

$$y(x) = \frac{1}{\mu(x)} \int^x \mu(\xi)f(\xi) d\xi + C \tag{E.3}$$

(a) Verify the result in (E.2) by showing that it is equivalent to (E.1).

Hint: Do the differentiation in (E.2) and remember that

$$\frac{d}{dx} \int^x \rho(\xi) d\xi = \rho(x)$$

which is needed for

$$\frac{d}{dx} \mu(x) = \frac{d}{dx} \exp\left(\int^x \rho(\xi) d\xi\right)$$

$$= \exp\left(\int^x \rho(\xi) d\xi\right) \rho(x) = \mu(x)\rho(x) \tag{E.4}$$

(b) Solve the following problems by first finding the integrating factor for the differential equation:

(i) $\dfrac{dy}{dx} - 2xy = x$

(ii) $\dfrac{dy}{dx} - \dfrac{1}{x}y = x^2$

(iii) $\dfrac{dy}{dx} + y = \sin x$,

$y(\pi) = 1$

ANS. (b) (i) $\mu(x) = \exp\left(\displaystyle\int -2x\,dx\right) = e^{-x^2}$, $y(x) = \tfrac{1}{2} + Ce^{x^2}$

(ii) $\mu(x) = \exp\left(\displaystyle\int -\dfrac{1}{x}\,dx\right) = \dfrac{1}{x}$, $y(x) = \tfrac{1}{2}x^3 - \tfrac{5}{2}x$

(iii) $\mu(x) = e^x$, $y(x) = \tfrac{1}{2}(\sin x - \cos x + e^{\pi - x})$

Second-Order Linear Constant-Coefficient Equations, with a Special UC (Undetermined Coefficients) Nonhomogeneous Term

A.6. Consider the differential equation

$$a_0\frac{d^2y}{dx^2} + a_1\frac{dy}{dx} + a_2 y = f(x) \tag{E.1}$$

where a_0, a_1 and a_2 are constants and $f(x)$ has a very special form, which, simply put, is a combination of products of polynomials and trigonometric and exponential functions—in other words, those functions on which all differentiation should result in a finite number of terms. This would allow us to guess a (particular) solution, for the nonhomogeneous equation (E.1), as a linear combination of these *carefully* prepared finite terms. The coefficients of this linear combination are found by equating similar terms on both sides of equation (E.1). This is the particular solution $y_p(x)$ of (E.1). The general solution of (E.1) is $y = y_p(x) + y_c(x)$, where $y_c(x)$ is the (general) solution

$$y_c(x) = c_1 y_1(x) + c_2 y_2(x) \tag{E.2}$$

of the (associated) homogeneous equation,

$$a_0\frac{d^2y}{dx^2} + a_1\frac{dy}{dx} + a_2 y = 0 \tag{E.3}$$

and where $y_1(x)$ and $y_2(x)$ are the two linearly independent solutions of the homogeneous equation (E.3).

(a) Solve the nonhomogeneous equation

$$\frac{d^2y}{dx^2} - \frac{dy}{dx} - 2y = \sin 2x \qquad (E.4)$$

Hint: The carefully prepared (review this process!) UC set of $f(x) = \sin 2x$ is $S = \{\cos 2x, \sin 2x\}$, and so we can write $y_p(x) = A \cos 2x + B \sin 2x$.

(b) Find the solution to the above differential equation with the initial conditions $y(0) = 0$ and $y'(0) = 1$.

Hint: Apply these conditions to the general solution of (E.4).

(c) Solve the nonhomogeneous equation

$$\frac{d^2y}{dx^2} + y = \cos x \qquad (E.5)$$

Hint: Note that $y_c(x) = c_1 \cos x + c_2 \sin x$, which intersects with the UC set $S = \{\cos x, \sin x\}$ of $\cos x$, so the (modified) UC set for $f(x) = \cos x$ is $S' = \{x \cos x, x \sin x\}$, which presents the nominee for a particular solution of (E.5) as $u_p(x) = Ax \cos + Bx \sin x$.

(d) Solve the initial value problem

$$\frac{d^2y}{dx^2} + 4\frac{dy}{dx} + 8y = 4\cos x + 7\sin x,$$

$$y(0) = 1, \ y'(0) = -1$$

Hint: $u_p(x) = A \cos x + B \sin x$

ANS. (a) $y_p(x) = \frac{1}{20} \cos 2x - \frac{3}{20} \sin 2x, \ y_c(x) = c_1 e^{-x} + c_2 e^{2x}$

$y(x) = c_1 e^{-x} + c_2 e^{2x} + \frac{1}{20} \cos 2x - \frac{3}{20} \sin 2x$

(b) $y(x) = -\frac{7}{15} e^{-x} + \frac{5}{12} e^{2x} + \frac{1}{20} \cos 2x - \frac{3}{20} \sin 2x$

(c) $y(x) = c_1 \cos x + c_2 \sin x + \frac{1}{2} x \sin x$

(d) $y(x) = e^{-2x} \cos 2x + \sin x$

Section A.2 Variation of Parameters

A.7. *The Method of Reduction of Order.* Consider the second-order nonhomogeneous differential equation with variable coefficients

$$a_0(x)\frac{d^2y}{dx^2} + a_1(x)\frac{dy}{dx} + a_2(x)y = f(x) \tag{E.1}$$

and its associated homogeneous equation (with $f(x) \equiv 0$)

$$a_0(x)\frac{d^2y}{dx^2} + a_1(x)\frac{dy}{dx} + a_2(x)y = 0 \tag{E.2}$$

The method of *variation of parameters* starts with the two linearly independent solutions $y_1(x)$ and $y_2(x)$ of the homogeneous equation (E.2) to consider $y_p(x) = \alpha(x)y_1(x) + \beta(x)y_2(x)$ for (E.2) as in (A.2.4), where the method involves the determination of the variable amplitudes $\alpha(x)$ and $\beta(x)$ from *two first-order differential equations* (lower order than (E.1)) in $\alpha(x)$ and $\beta(x)$. The method of *reduction of order* is very analogous to variation of parameters in that it uses a variable amplitude and results in a lower-order differential equation. However, it relies only on having one nontrivial solution $y_1(x)$ of the homogeneous equation (E.2). Then it assumes a solution $y_p(x) = v(x)y_1(x)$ for the nonhomogeneous (E.1) that results in a *first-order nonhomogeneous* equation in $w(x) \equiv dv/dx$,

$$a_0(x)y_1(x)\frac{dw}{dx} + \left[2a_0(x)\frac{dy_1}{dx} + a_1(x)y_1(x)\right]w = f(x) \tag{E.3}$$

After (E.3) is solved for $w(x)$, a simple integration gives $v(x)$ and thus $y_p(x) = v(x)y_1(x)$. Also, the method determines the second linearly independent solution $y_2(x)$ to the homogeneous equation (E.2) by letting $y_2(x) = v(x)y_1(x)$, where it again results in a first-order (homogeneous) equation in $w(x) \equiv dv/dx$,

$$a_0(x)y_1(x)\frac{dw}{dx} + [2a_0(x)\frac{dy_1}{dx} + a_1(x)y_1(x)]w(x) = 0 \tag{E.4}$$

This, of course, is the convenient way of finding the second linearly independent solution $y_2(x)$, when we have repeated roots for the characteristic equation of the (constant-coefficient) homogeneous equation (E.2).

(a) In Exercise A.1e with repeated roots, where $y_1(x) = y_2(x) = e^x$, use the method of reduction of order to justify that the second linearly independent solution is indeed $y_2(x) = xe^x$.

(b) Verify that $y_1(x) = x$ is a solution of the (Euler–Cauchy) equation

$$2x^2 \frac{d^2y}{dx^2} - x\frac{dy}{dx} + y = 0, \qquad x > 0 \tag{E.5}$$

Use the method of reduction of order to find the second linearly independent solution $y_2(t) = x^{1/2}$, $y(x) = c_1x + c_2x^{1/2}$.

(c) Try the method for the Euler–Cauchy equation in Exercise A.2 by letting $y(x) = x^m$ in (E.5) to verify the answer $y(x) = c_1x + c_2x^{1/2}$ in part (b).

(d) Derive (E.4) and (E.3)

(e) Verify that $y_1(x) = x^{-1/2}\cos x$ is a solution to the differential equation

$$\frac{d}{dx}\left(x\frac{dy}{dx}\right) + \left(x - \frac{1}{4x}\right)y = 0$$

Then find the second solution $y_2(x)$.

ANS. (e) $y_2(x) = x^{-1/2}\sin x$.

A.8. Solve the following equation:

$$\frac{d^2y}{dx^2} + y = \tan x, \qquad -\frac{\pi}{2} < x < \frac{\pi}{2} \tag{E.1}$$

ANS. $y_c(x) = c_1\cos x + c_2\sin x$

$y_p(x) = (-\cos x)\ln|\sec x + \tan x|$

Hint: Use the method of variation of parameters with $y_1(x) = \cos x$, $y_2(x) = \sin x$.

A.9. Use the method of variation of parameters to solve Exercise A.6d.

A.10. Solve the equation

$$\frac{d^2y}{dx^2} - 2\frac{dy}{dx} + y = \frac{e^x}{x}, \qquad x > 0$$

ANS. $y_c(x) = c_1e^x + c_2xe^x$

$y_p(x) = -xe^x + xe^x\ln|x|$

A.11. In Example A.1, solve the problem by constructing the Green's function in (A.2.17) and then using the integral representation of the solution in (A.2.16). Note that you may need to consult the tables of integrals to get to the final result for $y_p(x)$ in (E.7) of Example A.1.

A.12. (a) Solve the following problem:

$$\frac{d^2y}{dx^2} + \lambda^2 y = f(x)$$

(b) Do part (a) for $f(x) = \sin \lambda x$.

ANS. (a) $y_p(x) = \dfrac{1}{\lambda} \displaystyle\int_0^x \sin \lambda(x - \xi) f(\xi) d\xi$

(b) $y(x) = y_c(x) + y_p(x) = c_1 \cos \lambda x + c_2 \sin \lambda x - \dfrac{x}{2\lambda} \cos \lambda x$

Section A.3 Power Series Method of Solution

A.12. Verify the statement around (A.3.4) to (A.3.6) to show that the power series for $f(x) + g(x) = 1/(1 - x) + \cos x$ is convergent for $|x| < 1 = \min(R_1, R_2)$, where $R_1 = 1$ for the power series of $1/(1-x)$ and $R_2 = \infty$ for $\cos x$.

Hint: Use the infinite geometric series for $1/(1 - x)$, and help yourself with the ratio test for the convergence.

A.13. Use the ratio test to verify the theory that the sine series in (A.3.2) and the result of its differentiation as the cosine series (A.3.3) are both convergent for all x.

A.14. (a) Verify by direct differentiation that the power series representation of $\sin x$ and $\cos x$ in (A.1.3) and (A.1.4), respectively, do satisfy differential equation (A.1.1) with $\lambda = 1$.

Hint: Differentiate the power series term by term twice, since this operation is permitted on these particular power series by the statement following (A.3.1). Then combine the two series to see that the coefficients of the combined series vanish for all n.

(b) Use your experience in part (a) to verify the result used for (A.3.7), that for an identically vanishing power series we have each coefficient vanish for all n.

A.15. Use power series expansion about $x_0 = 0$ to solve the equation

$$\frac{d^2y}{dx^2} - y = 0$$

Hint: Note that $e^x = \sum_{n=0}^{\infty} \frac{x^n}{n!} = \cosh x + \sinh x$.

ANS. $y(x) = c_0 \cosh x + c_1 \sinh x$

$$= Ae^x + Be^{-x}$$

A.16. Use power series expansion about $x_0 = 0$ to solve the equation

$$\frac{d^2y}{dx^2} - 3\frac{dy}{dx} + 2y = 0$$

ANS. $y(x) = c_0 + c_1 x + (\tfrac{3}{2}c_1 - c_0)x^2 + (\tfrac{7}{6}c_1 - c_0)x^3 + \cdots$

Let $c_0 = A + B$ and $c_1 = A + 2B$ to obtain $y(x) = Ae^x + Be^{2x}$.

A.17. Use power series expansion about $x_0 = 0$ to solve the following equation. See if you can recognize any of its series solutions as a familiar (elementary) function.

$$\frac{d^2y}{dx^2} + x\frac{dy}{dx} + y = 0$$

ANS. $y(x) = c_0 \left[1 - \dfrac{x^2}{2} + \dfrac{x^4}{8} - \dfrac{x^6}{24} + \cdots \right]$

$$+ c_1 \left[x - \dfrac{x^3}{3} + \dfrac{x^5}{15} - \dfrac{x^7}{105} + \cdots \right]$$

The first solution is recognized as $y_1(x) = e^{-x^2/2}$, since it can be rewritten as

$$y_1(x) = 1 + \left(-\frac{x^2}{2} \right) + \frac{1}{2!}\left(-\frac{x^2}{2} \right)^2 + \frac{1}{3!}\left(-\frac{x^2}{2} \right)^3 + \cdots$$

which is the Taylor series of $e^{-x^2/2}$ expansion about $x_0 = 0$.

Section A.4   Frobenius Method of Solution

A.18. (a) Use the ratio test to show that the series (A.4.15) for the Bessel function $J_n(r)$, $n = 0, 1, 2, \ldots$, converges for all r.

(b)   Use the series representation of the Bessel function $J_\nu(\lambda r)$ in (A.4.15) to verify by direct differentiation that it does satisfy the Bessel differential equation (A.4.1).

A.19.   (a)   Show that the following differential equation has a regular singular point at $x_0 = 0$:

$$x^2 \frac{d^2y}{dx^2} + 2x \frac{dy}{dx} - 2y = 0 \tag{E.1}$$

(b)   Use the Frobenius method to find the solution.

ANS.   (a)   For $y'' + \dfrac{2}{x}y' - \dfrac{2}{x^2}y = 0$,

$$\lim_{x \to 0} x \frac{2}{x} = 2 \quad \text{and} \quad \lim_{x \to 0} x^2 \left(-\frac{2}{x^2}\right) = -2$$

That is, both limits exists as required in (A.3.11), thus $x_0 = 0$ is a regular singular point of (E.1).

(b) $y(x) = c_0 x + c_1 \dfrac{1}{x^2}$

A.20.   (a)   Show that the following differential equation has a regular singular point at $x_0 = 0$:

$$2x \frac{d^2y}{dx^2} + (x + 1)\frac{dy}{dx} + 3y = 0$$

(b)   Use the Frobenius method to find the solution

ANS.   (a)   For $y'' + \dfrac{x+1}{2x}y' + \dfrac{3}{2x}y = 0$,

$$\lim_{x \to 0} x \frac{x+1}{2x} = \frac{1}{2} \quad \text{and} \quad \lim_{x \to 0} x^2 \left(\frac{3}{2x}\right) = 0$$

That is, both limits exist.

(b) $y(x) = c_0 x^{1/2}\left[1 - \dfrac{7}{6}x + \dfrac{21}{40}x^2 + \cdots\right]$

$$+ c_1[1 - 3x + 2x^2 + \cdots]$$

A.21.   Use the Frobenius method to solve for the following differential equation near $x = 0$:

$$2x^2 \frac{d^2y}{dx^2} + x\frac{dy}{dx} + (x^2 - 1)y = 0$$

ANS. $y(x) = c_1 \left(1 - \frac{x^2}{14} + \frac{x^4}{616} - \cdots\right) + c_2 x^{-1/2} \left(1 - \frac{x^2}{2} + \frac{x^4}{40} - \cdots\right)$

Section A.5a Special Differential Equations and Their Solutions

Bessel Functions

A.22. (a) Show that $J_n(-r) = (-1)^n J_n(r)$, $n = 0, 1, 2, \ldots$.

Hint: Substitute $-r$ for r in (A.5.2), the series of $J_n(r)$ with $n = 0, 1, 2, \ldots$.

(b) Show that $J_0(0) = 1$ and $J_\nu(0) = 0$ for $\nu > 0$.

Hint: Use (A.5.2).

(c) Prove the identities in (A.5.8) and (A.5.9).

Hint: Use the series representations of the functions concerned in (A.5.2), (A.1.3) for (A.5.8) and in (A.5.3), (A.1.4) for (A.5.9).

(d) Write the series for $J_0(x)$ and $J_1(x)$ and show by direct differentiation that $dJ_0/dx = -J_1(x)$.

(e) Use the tabulated values and properties of $J_0(x)$, $J_1(x)$ and $J_2(x)$ to sketch their graphs.

ANS. (e) See Fig. A.1.

A.23. Prove identities (E.1)–(E.3) of (A.5.14) and (E.4) of (A.5.13)

(a) $x J_n'(x) = n J_n(x) - x J_{n+1}(x)$ (E.1)

(b) $x J_n'(x) = -n J_n(x) + x J_{n-1}(x)$ (E.2)

(c) $2 J_n'(x) = J_{n-1}(x) - J_{n+1}(x)$ (E.3)

(d) $\dfrac{2n}{x} J_n(x) = J_{n-1}(x) + J_{n+1}(x)$ (E.4)

Hint: For (E.1) use the series representation (A.5.10) for $J_n(x)$ and $J_{n+1}(x)$, and make sure to change the variable of summation k, for the resulting series of $x J_{n+1}(x)$, to $k = j - 1$; then combine the two series to have one for $x J_n'(x)$. For (E.2) combine the two series and work on simplifying the coefficients. (E.3) and (E.4) are obtained from adding (E.2) to (E.1) and subtracting (E.1) from (E.2) respectively.

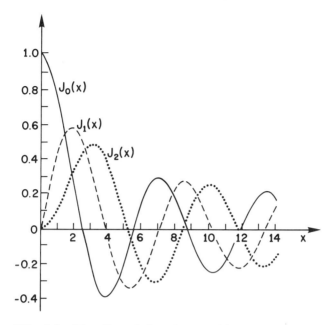

FIG. A.1 The Bessel functions $J_n(x)$ of the first kind of order $n = 0, 1, 2$.

A.24. (a) Prove (A.5.15).

(b) Prove (A.5.16).

Hint: For (A.5.15) multiply inside the series (A.5.10) of $J_n(cx)$ by x; then differentiate term by term, and simplify the coefficients in the series. For (A.5.16), see the hint for (E.1) in Exercise A.23, where you will need to make a change of summation variable of the resulting series from k to $k = j + 1$.

A.25. Show that

(a) $\dfrac{d}{dx}[xJ_1(x)] = xJ_0(x)$ (E.1)

(b) $\dfrac{dJ_0(x)}{dx} = -J_1(x)$ (E.2)

(c) $\dfrac{d}{dx}\left[\dfrac{1}{2}x^2\left\{J_0^2(x) + J_1^2(x)\right\}\right] = xJ_0^2(x)$ (E.3)

(d) $\displaystyle\int_0^a x J_0^2(x)\,dx = \frac{a^2}{2}[J_0^2(a) + J_1^2(a)]$ (E.4)

Hint: For (E.1) and (E.2) use (A.5.15) with $n = 1$ and (A.5.16) with $n = 0$, respectively. For (E.3) differentiate and use (E.1), (E.2) above and (E.2) of Exercise A.23.

(e) Show that $J_n(x)$ is bounded as $x \to 0$ and as $x \to \infty$.

Hint: For $x \to 0$, see the series (A.5.2) for the Bessel function $J_n(x)$. For $x \to \infty$, use the asymptotic relation

$$J_n(x) \sim \sqrt{\frac{2}{\pi x}}\,\cos\left(x - \frac{2n+1}{4}\pi\right)$$

for large x. Also for $x \to 0$ you may use the asymptotic relation

$$J_n(x) \sim \frac{1}{\Gamma(n+1)}(x/2)^n$$

for small x.

Legendre Polynomials

A.26. Use the Legendre polynomial representation in (A.5.34) to derive its Rodrigues formula in (A.5.38).

Hint: Note the *binomial expansion*

$$(x^2 - 1)^n = \sum_{j=0}^n \frac{n!}{j!(n-j)!}(x^2)^{n-j}$$ (E.1)

and under the differentiation $d^n(x^2 - 1)^n/dx^n$, the upper limit on the sum can be written as J of (A.5.34) instead of n. (Why?) Also

$$\frac{d^n}{dx^n}(x^{2n-2j}) = (2n - 2j)\frac{d^{n-1}}{dx^{n-1}}(x^{2n-j}) = \cdots$$

$$= \cdots = \frac{(2n - 2j)!}{(n - 2j)!}x^{n-2j}$$ (E.2)

A.27. (a) Use the Rodrigues formula (A.5.38) to derive the recurrence relation

$$\frac{dP_{n+1}}{dx} = (2n + 1)P_n + \frac{dP_{n-1}}{dx}$$ (E.1)

(b) Derive the recurrence relation

$$\frac{dP_{n+1}}{dx} = (n+1)P_n + x\frac{dP_n}{dx} \tag{E.2}$$

Hint: For

$$\frac{d^{n+2}}{dx^{n+2}}[(x^2-1)^{n+1}] = 2(n+1)\frac{d^{n+1}}{dx^{n+1}}[x(x^2-1)^n]$$

use the Leibniz formula for repeated differentiation of a product (see (E.1) of Exercise A.29 with $u = x$, $v = f(x)$).

$$\frac{d^n}{dx^n}[xf(x)] = x\frac{d^nf}{dx^n} + n\frac{d^{n-1}}{dx^{n-1}}f(x) \tag{E.3}$$

(c) Use the above two recurrence relations (E.1) and (E.2) to derive the recurrence relation

$$nP_n(x) = (2n+1)xP_{n-1}(x) - (n-1)P_{n-2}(x) \tag{E.4}$$

which is very useful for generating Legendre polynomials (see part d).

Hint: Use (E.1), (E.2) to find a recurrence relation (E.5) that relates P_n, P_n', and P_{n-1}'; then use (E.1), (E.2), (E.5) to eliminate the derivatives to obtain the desired (E.4).

(d) (i) With $P_0(x) = 1$ and $P_1(x) = x$, use the recurrence relation (E.4) in part (c) to find $P_2(x)$, $P_3(x)$, and $P_4(x)$.

    (ii) Use the Rodrigues formula (A.5.38) to find $P_2(x)$, $P_3(x)$, and $P_4(x)$.

ANS. (d) (i and ii): See (A.5.39).

A.28. Use the recursion relation of the Legendre polynomials (E.4) in part (c) of Exercise A.27,

$$lP_l(x) = (2l+1)xP_{l-1}(x) - (l-1)P_{l-2}(x) \tag{E.1}$$

to show that the norm square $\|P_n\|^2$ on $(-1,1)$ is

$$\int_{-1}^{1}[P_l(x)]^2 dx = \frac{2}{2l+1} \tag{E.2}$$

Hint: Multiply (E.1) by $P_l(x)$ and integrate from -1 to 1 to generate $\|P_l\|^2$; then replace l by $l+1$ in (E.1), multiply by $P_{l-1}(x)$, and integrate to generate $\|P_{l-1}\|^2$ and relate the two results to have

$$\|P_l\|^2 = \frac{2l-1}{2l+1} \|P_{l-1}\|^2$$

Then proceed from P_l to P_0, where we know that $\|P_0\|^2 = \int_{-1}^{1} dx = 1$.

A.29. *Leibniz rule for finding the nth derivative of a product* $d^n(u(x)v(x))/dx^n$, *with the notation* $u^{(n)} \equiv d^n u/dx^n$, *is*

$$\frac{d^n}{dx^n}(uv) = uv^{(n)} + nu'v^{(n-1)} + \frac{n(n-1)}{2!}u''v^{(n-2)}$$

$$+ \cdots + \frac{n(n-1)\cdots(n-j+1)}{j!}u^{(j)}v^{(n-j)}$$

$$= \sum_{j=0}^{n} \frac{n!}{(n-j)!j!}u^{(j)}v^{(n-j)} + u^{(n)}v, \quad u^{(n)} \equiv \frac{d^n u}{dx^n} \quad \text{(E.1)}$$

(a) Verify this result for $n = 2, 3$.

(b) Use the Leibniz rule to find $d^{n+1}[x(x^2-1)^n]/dx^{n+1}$ in terms of derivatives of $(x^2-1)^n$.

ANS. (b) $\quad x \dfrac{d^{n+1}}{dx^{n+1}}(x^2-1)^n + (n+1)\dfrac{d^n}{dx^n}(x^2-1)^n \qquad$ (E.2)

A.30. Consider $\phi(x,h)$, the *generating function for Legendre polynomials*,

$$\phi(x,h) = (1 - 2xh + h^2)^{-1/2}, \qquad |h| < 1 \qquad \text{(E.1)}$$

in the sense that

$$\phi(x,h) = \sum_{l=0}^{\infty} h^l P_l(x) \qquad \text{(E.2)}$$

Verify (E.2) by showing the first three terms in the expansion of the series in (E.2), i.e.,

$$(1 - 2xh + h^2)^{-1/2} = P_0(x) + hP_1(x) + h^2 P_2(x) + \cdots \qquad \text{(E.3)}$$

Hint: Let $2xh - h^2 = \gamma$, expand $(1-\gamma)^{-1/2}$, and then collect powers of h to get the right side of (E.3).

A.31. The following is a more general Legendre equation, called the *associated* Legendre equation:

$$(1-x^2)y'' - 2xy' + \left[l(l+1) - \frac{m^2}{1-x^2}\right]y = 0 \qquad (E.1)$$

with the Legendre equation (A.5.26) as its special case corresponding to $m = 0$. Show that the solution to (E.1), the *associated Legendre* polynomial $P_l^m(x)$, can be expressed in terms of the derivative of the Legendre polynomial $P_l(x)$,

$$P_l^m(x) = (1-x^2)^{m/2} \frac{d^m}{dx^m} P_l(x) \qquad (E.2)$$

Hint: Let $y(x) = (1-x^2)^{m/2}u(x)$ and substitute in (E.1) to get a result (E.3) that has $u(x)$ in place of $y(x)$. If you differentiate this new result you get another one (E.4), where u' is now in place of u and m is replaced by $m + 1$. *Conclusion:* The first says that $u(x) = P_l(x)$ is a solution of (E.3) when $m = 0$, while the second says that $u' = dP_l/dx$ is a solution of (E.3) when $m = 1$, and we proceed to differentiate until we reach $u(x) = d^m P_l(x)/dx^m$ as the solution of (E.3),

$$y(x) = (1-x^2)^{m/2} \frac{d^m}{dx^m} P_l(x)$$

for (E.1)

Section A.5b   Other More General Special Functions

The exercises on this topic will mainly involve showing that the special functions which we have already presented, and many more, are no more than special cases of a more general class of special functions. We will first present a summary of such general functions as the *hypergeometric* functions and the *confluent hypergeometric* functions, list their familiar special cases, and leave it for the exercises to verify such a recognition. See Exercises A.32 to A.34 that will follow this summary.

Here we will use the binomial notation

$$(a)_\nu \equiv \frac{\Gamma(a+\nu)}{\Gamma(a)}, \qquad \binom{\alpha}{\beta} \equiv \frac{\Gamma(\alpha+1)}{\Gamma(\beta+1)\Gamma(\alpha-\beta+1)} \qquad (E.1)$$

Gamma Function $\Gamma(z)$

$$\Gamma(z) = \int_0^\infty t^{z-1}e^{-t}\,dt, \qquad \text{Re}\, z > 0 \qquad (E.2)$$

$$\Gamma(z+1) = z\Gamma(z) = z!, \qquad \Gamma(\tfrac{1}{2}) = \sqrt{\pi} \qquad (E.3)$$

$$\Gamma(z)\Gamma(1-z) = \pi \cosec(\pi z) \tag{E.4}$$

$$\Gamma(nz) = (2\pi)^{(1-n)/2} n^{nz-1/2} \prod_{k=0}^{n-1} \left(z + \frac{k}{n} \right) \tag{E.5}$$

Beta Function B(z, w)

$$B(z,w) = \int_0^\infty \frac{t^{z-1}}{(1+t)^{z+w}} \, dt \qquad (\mathrm{Re}\, z > 0, \ \mathrm{Re}\, w > 0) \tag{E.6}$$

$$B(z,w) = \frac{\Gamma(z)\Gamma(w)}{\Gamma(z+w)} = B(w,z) \tag{E.7}$$

Hypergeometric Functions $_2F_1(a, b; c; x)$

The differential equation

$$x(1-x)\frac{d^2y}{dx^2} + [c - (a+b+1)x]\frac{dy}{dx} - aby = 0 \tag{E.8}$$

has one solution regular at the origin. When $y(0)$ is normalized to one, this solution $y(x)$ is called the *hypergeometric function* and is denoted by $_2F_1(a,b;c;x)$,

$$\begin{aligned}
_2F_1(a,b;c;x) = 1 &+ \frac{ab}{1\cdot c}x + \frac{a(a+1)b(b+1)}{1\cdot 2\cdot c(c+1)}x^2 \\
&+ \frac{a(a+1)(a+2)b(b+1)(b+2)}{1\cdot 2\cdot 3\cdot c(c+1)(c+2)}x^2 + \cdots,
\end{aligned}$$
$$|x| < 1 \tag{E.9}$$

Many special functions of interest are special cases of $_2F_1(a,b;c;x)$ or can be related to it. Examples are the Jacobi polynomial (E.12) and its special cases, the Legendre polynomials (E.22), Tchebychev polynomials (E.18) (E.20), Gegenbauer polynomials (E.15), and associated Legendre polynomials and functions (E.29).

The Confluent Hypergeometric Functions $_1F_1(a, b; x)$

The solution of the differential (Kummer) equation

$$x\frac{d^2y}{dx^2} + (b-x)\frac{dy}{dx} - ay = 0 \tag{E.10}$$

is called the *confluent hypergeometric function* and is denoted by $_1F_1(a;b;x)$, $M(a,b,x)$, and $\Phi(a;b;x)$.

$$_1F_1(a,b;x) = M(a,b,x) = 1 + \frac{a}{b}\cdot\frac{x}{1!} + \frac{a(a+1)}{b(b+1)}\cdot\frac{x^2}{2!} +$$

$$+ \frac{a(a+1)(a+2)}{b(b+1)(b+2)} \cdot \frac{x^3}{3!} + \cdots \tag{E.11}$$

Examples are the Hermite polynomials in (E.26) and the generalized Laguerre polynomials in (E.23).

Orthogonal Polynomials

The following are polynomials $f_n(x)$ of degree n which satisfy a differential equation of the form

$$g_2(x)f_n'' + g_1(x)f_n' + a_n f_n = 0$$

They are orthogonal on the interval $[a,b]$ with respect to weight function $\rho(x)$ and with the normalization factor h_n as indicated.

Jacobi Polynomial $P_n^{(\alpha,\beta)}(x)$

$$P_n^{(\alpha,\beta)}(x) = \binom{n+\alpha}{n} {}_2F_1\left((-n, n+\alpha+\beta+1; \alpha+1); \frac{1-x}{2}\right) \tag{E.12}$$

$[a,b]; [-1,1], \ \rho(x) = (1-x)^\alpha (1+x)^\beta$

$$(1-x^2)y'' + [\beta - \alpha - (\alpha+\beta+2)x]y' + n(n+\alpha+\beta+1) = 0 \tag{E.13}$$

$$h_n = \frac{2^{\alpha+\beta+1}}{2n+\alpha+\beta+1} \cdot \frac{\Gamma(n+\alpha+1)\Gamma(n+\beta+1)}{n!\Gamma(n+\alpha+\beta+1)},$$

$$\alpha > -1, \ \beta > -1 \tag{E.14}$$

Gegenbauer Polynomial $C_n^{(\alpha)}(x)$

$$C_n^{(\alpha)}(x) = \frac{\Gamma(\alpha+\frac{1}{2})\Gamma(2\alpha+n)}{\Gamma(2\alpha)\Gamma(\alpha+n+\frac{1}{2})} \cdot P_n^{(\alpha-1/2,\alpha-1/2)}(x), \qquad \alpha \neq 0 \tag{E.15}$$

$[a,b] = [-1,1], \ \rho(x) = (1-x^2)^{\alpha-1/2},$

$$h_n = \frac{\pi 2^{1-2\alpha}\Gamma(n+2\alpha)}{n!(n+\alpha)[\Gamma(\alpha)]^2}, \qquad \alpha \neq 0, \ \alpha > -\frac{1}{2} \tag{E.16}$$

$$(1-x^2)y'' - (2\alpha+1)xy' + n(n+2\alpha)y = 0 \tag{E.17}$$

Tchebychev Polynomial of the First Kind $T_n(x)$

$$T_n(x) = \frac{n!\sqrt{\pi}}{\Gamma(n+\frac{1}{2})} P_n^{(-1/2,-1/2)}(x) = \frac{n}{2} C_n^{(0)}(x) \tag{E.18}$$

$$[a,b] = [-1,1], \ \rho(x) = (1-x^2)^{-1/2}, \ h_n = \begin{cases} \dfrac{\pi}{2}, & n \neq 0 \\ \pi, & n = 0 \end{cases}$$

$$(1-x^2)y'' - xy' + n^2 y = 0 \tag{E.19}$$

Tchebychev Polynomial of the Second Kind $U_n(x)$

$$U_n(x) = \frac{(n+1)!\sqrt{\pi}}{2\Gamma(n+\frac{3}{2})} P_n^{(1/2,1/2)}(x) = C_n^{(1)}(x) \tag{E.20}$$

$[a,b] = [-1,1]$, $\rho(x) = (1-x^2)^{1/2}$, $h_n = \pi/2$,

$$(1-x^2)y'' - 3xy' + n(n+2)y = 0 \tag{E.21}$$

Legendre Polynomials $P_n(x)$

$$P_n(x) = P_n^{(0,0)}(x) = C_n^{(1/2)}(x) \tag{E.22}$$

Generalized Laguerre Polynomials $L_n^{(\alpha)}(x)$

$$L_n^{(\alpha)}(x) = \binom{n+\alpha}{n} {}_1F_1(-n;\alpha+1;x) \tag{E.23}$$

on $(0,\infty)$, $\rho(x) = e^{-x}x^{\alpha}$, $h_n = \Gamma(\alpha+n+1)/n!$, $\alpha > -1$

$$xy'' + (\alpha+1-x)y' + ny = 0 \tag{E.24}$$

Laguerre Polynomial $L_n(x)$

$$L_n(x) = L_n^{(0)}(x) \tag{E.25}$$

Hermite Polynomial $He_n(x)$ (See Hermite polynomial $H_n(x)$ in A.5.47–A.5.50

$$He_n(x) = 2^{-n/2}H_n\left(\frac{x}{\sqrt{2}}\right) = \frac{(2n)!(-\frac{1}{2})^n}{n!} {}_1F_1\left(-n;\frac{1}{2};\frac{1}{2}x^2\right) \tag{E.26}$$

on $(-\infty,\infty)$, $\rho(x) = e^{-x^2/2}$, $h_n = \sqrt{2\pi}n!$

$$y'' - xy' + ny = 0 \tag{E.27}$$

Associate Legendre Functions

We consider the Legendre equation

$$(1-z^2)\frac{d^2u}{dz^2} - 2z\frac{du}{dz} + \left[\nu(\nu+1) - \frac{\mu^2}{1-z^2}\right]u = 0 \tag{E.28}$$

where z, ν, μ are unrestricted. The solution is defined in terms of the associated Legendre function $P_\nu^\mu(z)$ and in terms of the indicated hypergeometric function:

$$u(z) = P_\nu^\mu(z)$$

$$= \frac{1}{\Gamma(1-\mu)}\left(\frac{z+1}{z-1}\right)^{\mu/2} {}_2F_1(-\nu,\nu+1;1-\mu;\frac{1}{2}-z/2) \tag{E.29}$$

$$|1-z| < 2$$

For n, m integers,

$$\int_{-1}^{1} |P_n^m(x)|^2 \, dx = \frac{(n+m)!}{(n-m)!\,(n+\frac{1}{2})}, \qquad n, m \text{ integers} \tag{E.30}$$

A.32. (a) Evaluate:

   (i) $\dfrac{\Gamma(6)}{\Gamma(3)}$

   (ii) $\Gamma(-\frac{1}{2})$

   Hint: Use (E.2) for $z = -\frac{1}{2}$, and (E.3)

   (iii) $\int_0^\infty x^6 e^{-2x} \, dx$

   Hint: Let $y = 2x$ and use (E.1).

   (b) Use the definition of the beta function to evaluate the definite interval

   $$\int_0^b x^4 \sqrt{b^2 - x^2} \, dx$$

   Hint: Let $x^2 = b^2 y$.

ANS. (a) (i) 60

   (ii) $-2\sqrt{\pi}$

   (iii) $\frac{45}{8}$

   (b) $\pi b^6 / 32$.

A.33. (a) Verify that the Jacobi polynomial, as written in (E.12), is represented in terms of the hypergeometric function $_2F_1$ in (E.9).

   Hint: Compare their corresponding differential equations in (E.13) and (E.8), respectively.

   (b) Verify the representations of the following polynomials (or functions) in terms of the Jacobi polynomial (E.12): the Legendre polynomials in (E.22); the Tchebychev polynomials in (E.18), (E.20); the Gegenbauer polynomials in (E.15); and the associated Legendre functions and polynomials in (E.29).

   (c) Show that (E.9) gives the geometric series as a special case.

Hint: Let $a = 1$ and $b = 0$ to get $1 - x + x^2 + \cdots = 1/(1-x) \equiv {}_2F_1(1,b,b;x)$, $|x| < 1$.

(d) Verify the way the (other) Hermite polynomial $He_n(x)$ in (E.26) and the generalized Laguerre polynomials in (E.23) are expressed in terms of the confluent hypergeometric function ${}_1F_1$ in (E.11).

A.34. (a) Show that the hypergeometric equation (E.8) has regular singular points at $x = 0$ and $x = 1$.

(b) Use the Frobenius method as in (A.4.2) to solve for (E.8) by first finding the *indicial equation*, the general recursion relation of the coefficients, and then the two linearly independent solutions with $y_1(x)$ as in (E.9).

(c) Show that the confluent hypergeometric or "Kummer" equation (E.10) has a regular singularity at $x = 0$. Then use the Frobenius method as in (A.4.2) to find the general solution with $y_1(x) = M(a,b,x) = {}_1F_1(a,b;x)$ as in (E.11).

Hint: Follow the same steps as for the above hypergeometric equation in parts (a) and (b) where $\beta_1 = 0$ and $\beta_2 = 1 - b$,

(d) In the above part (c) show that the special case $a = -n$, integer, makes $M(-n,b,x)$ a polynomial of degree n. Also, the Laguerre polynomial $L_n(x)$ is recovered for $a = -n$ and $b = 1$. Show that, for example, $M(-2,1,x) = \frac{1}{2}L_2(x)$.

ANS. (b) $\beta_1 = 0$, $\beta_2 = 1 - c$

$$c_k(\beta + k)(\beta + k + c - 1)$$
$$= c_{k-1}(\beta + k - 1 + a)(\beta + k - 1 + b) \qquad \text{(E.1)}$$

(i) For $\beta_1 = 0$, with the choice of $c_0 = 1$,

$$y_1 = \left[1 + \frac{ab}{1!c}x + \frac{a(a+1)b(b+1)}{2!c(c+1)}x^2 + \cdots \right.$$
$$\left. + \frac{a(a+1)\cdots(a+k-1)b(b+1)\cdots(b-k+1)}{k!c(c+1)\cdots(c+k-1)}x^k + \cdots \right]$$
$$\equiv {}_2F_1(a,b;c;x)$$

or $= {}_2F_1(a,b;c;x)$, $|x| < 1$

$$y = c_1 {}_2F_1(a,b;c;x)$$
$$+ c_2 x^{1-c} {}_2F_1(a-c+1,b-c+1;2-c;x),$$
$$-1 < x < 1$$

Note that 2 and 1 in ${}_2F_1(a,b;c;x)$ are used to indicate the two parameters a and b which appear in the numerator of the above coefficients and the one parameter c which appears in the denominator. Here we shall use either notation ${}_2F_1(a,b,c;x)$ of $F(a,b,c;x)$

(ii) For $\beta_2 = 1 - c$

$$y_2 = c_0 x^{1-c} \left[1 + \sum_{k=1}^{\infty} \right.$$

$$\left. \frac{(a-c+1)(a-c+2)\cdots(a-c+k)(b-c+1)(b-c+2)\cdots(b-c+k)}{k!(2-c)(3-c)\cdots(k+1-c)} x^k \right]$$

$$y_2 = c_0 x^{1-c} F(a-c+1,b-c+1;2-c;x),$$
$$|x| < 1$$

(c) $y = A \left[1 + \dfrac{a}{b} x + \dfrac{a(a+1)}{b(b+1)} \dfrac{x^2}{2!} + \dfrac{a(a+1)(a+2)}{b(b+1)(b+2)} \right.$

$\left. \cdot \dfrac{x^3}{3!} + \cdots \right] + B x^{1-b} \left[1 + \dfrac{a-b+1}{(2-b)} x \right.$

$\left. + \dfrac{(a-b+1)(a-b+2)}{2!(2-b)(3-b)} x^2 + \cdots \right]$

$$y = AM(a,b,x) + B x^{1-b} M(a-b+1;2-b;x)$$

Appendix B

MATHEMATICAL MODELING OF PARTIAL DIFFERENTIAL EQUATIONS

Boundary and Initial Value Problems

The rigor with which a mathematical analysis of a physical process or system can be said to apply depends on the exactness with which a model describes the physical situation, as well as on the mathematical treatment of the model. Mathematical descriptions of processes may be developed in some cases by application of first principles, such as conservation of mass or energy, but in other situations phenomenological laws must be introduced. Examples of phenomenological laws encountered are Fourier's law of heat conduction, which states that the flux of heat by molecular conduction is proportional to the temperature gradient; Fick's law of diffusion, which states that the flux associated with molecular diffusion is proportional to the concentration gradient of the diffusing species; and Hooke's law, which states that the stress on a surface of a body is proportional to the strain.

Even though the partial differential equation describing an event is an accurate description, the boundary conditions may be only an approximation to the physical situation. A commonly used, though infrequently attained, boundary condition associated with heat transfer systems is that of constant uniform surface temperature. Constant heat flux boundary conditions can be attained by electrical or nuclear heating, but the boundary conditions appropriate to a specific system are established only by analysis of the physics of the system.

In this limited space we will formulate some representative cases of the most important classical partial differential equations by using first principles and by applying phenomenological laws. The choice is determined by the equations used in this book, which are primarily the diffusion (heat) equation, the potential (Laplace) equation, and the vibrating systems (wave) equation. These three partial differential equations are the simplest representatives of the three main classes of second-order partial differential equations, namely the parabolic, elliptic, and hyperbolic equations. We leave a brief discussion of this important classification for the exercises (see Exercises B.18).

B.1 Partial Differential Equations for Vibrating Systems

The Vibrating String—Wave Equation

The classical problem of the vibration of a stretched elastic string was one of the first problems to be studied through the use of partial differential equations. Let us consider the deviation $y(x,t)$ of a vertical motion (in one plane only) of a flexible string (which transmits force only in the direction of its length) that is attached under tension T to two points a distance l apart as shown in Fig. B.1. Newton's law, written for vertical motion in the xy plane, is

$$T \sin \alpha \, |_{x+\Delta x} - T \sin \alpha \, |_x - \Delta s \rho g = \Delta s \rho \frac{\partial^2 y}{\partial t^2} \tag{B.1.1}$$

where Δs is the element of the string undergoing the motion, ρ is the mass of the string per unit length, and g is the acceleration due to gravity. Let us note that $\Delta s \simeq [1 + (\Delta y/\Delta x)^2]^{1/2}$; thus it is only for very small deviations, i.e., $(dy/dx)^2 \ll 1$, that we accept the approximation $\Delta s \sim \Delta x$. Also, to be very accurate, the acceleration $\partial^2 y/\partial t^2$ in (B.1.1) must be looked at as a mean value at some ξ where $x < \xi < x + \Delta x$, which approaches x as $\Delta x \to 0$. Since the motion is assumed to be only

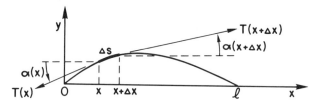

FIG. B.1 Forces on a segment of a vibrating string.

vertical, the horizontal force on the segment Δs is zero

$$T(x + \Delta x)\cos \alpha \mid_{x+\Delta x} - T(x)\cos \alpha \mid_x = 0$$

$$T(x + \Delta x)\cos \alpha \mid_{x+\Delta x} = T(x)\cos \alpha \mid_x = T_0 \tag{B.1.2}$$

which implies a constant horizontal tension T_0, since the above equality (B.1.2) is valid for the arbitrary x and Δx, i.e., for any segment Δs along the length of the string.

Let us note that we are after a partial differential equation in $y(x,t)$, and we already see in (B.1.1) a second-order partial derivative with respect to time $\partial^2 y/\partial t^2$ as the acceleration. The possible partial derivative with respect to the spatial variable x lies in the geometric relation $\tan \alpha = \partial y/\partial x$. This can be generated in (B.1.1) if we divide the corresponding terms in (B.1.1) by the suitable terms, out of the three identical ones in (B.1.2), with the appropriate position of $\partial y/\partial x$ in mind,

$$\frac{T(x + \Delta x)\sin \alpha \mid_{x+\Delta x}}{T(x + \Delta x)\cos \alpha \mid_{x+\Delta x}} - \frac{T(x)\sin \alpha \mid_x}{T(x)\cos \alpha \mid_x} - \frac{\Delta s \rho g}{T_0} = \frac{\Delta s \rho}{T_0}\frac{\partial^2 y}{\partial t^2}$$

$$= \tan \alpha \mid_{x+\Delta x} - \tan \alpha \mid_x = \frac{\Delta s \rho g}{T_0} + \frac{\Delta s \rho}{T_0}\frac{\partial^2 y}{\partial t^2} \tag{B.1.3}$$

What remains now is to divide by Δx and take the limit as $\Delta x \to 0$ to get the derivative of $\tan \alpha$, i.e.,

$$\frac{\partial}{\partial x}\tan \alpha = \frac{\partial^2 y}{\partial x^2}$$

on the left side. However, on the right side we will have

$$\lim_{\Delta x \to 0}\frac{\Delta s}{\Delta x} = \frac{\partial s}{\partial x} = \sqrt{1 + \left(\frac{\partial y}{\partial x}\right)^2}$$

which results in a *nonlinear wave equation*,

$$\frac{\partial^2 y}{\partial x^2} = \frac{\rho}{T_0}\left[1 + \left(\frac{\partial y}{\partial x}\right)^2\right]^{1/2}\left[g + \frac{\partial^2 y}{\partial t^2}\right] \tag{B.1.4}$$

As we have indicated above, if we assume very small deviations $y(x,t)$, which implies $(\partial y/\partial x)^2 \ll 1$, then the wave equation (B.1.4) becomes the usual linear wave equation for the (ideal) small vibrations of a perfectly flexible string,

$$\frac{\partial^2 y}{\partial x} = \frac{\rho}{T_0}g + \frac{\rho}{T_0}\frac{\partial^2 y}{\partial t^2} = \frac{1}{c^2}\frac{\partial^2 y}{\partial t^2} + \frac{1}{c^2}g \tag{B.1.5}$$

where $c = \sqrt{T_0/\rho}$ is the velocity of the wave. This linear wave equation (B.1.5) is nonhomogeneous because of the gravity term g/c^2, and of course this could have been a general external force $F(x,t)$. In the present case of (B.1.5), we often make the approximation that the gravitational acceleration is negligible compared to the acceleration of the string, i.e., $g \ll \partial^2 y/\partial t^2$, so we can neglect the term g/c^2 in (B.1.5) to arrive at the (idealized) one-dimensional linear homogeneous wave equation of the vibrating string,

$$\frac{\partial^2 y}{\partial x^2} = \frac{1}{c^2}\frac{\partial^2 y}{\partial t^2} \tag{B.1.6}$$

We must also note that the above derivation of all forms of the wave equation was done for an arbitrary segment Δs along the length of the string, which did not involve what is happening at the boundaries $x = 0$ and $x = l$ of the string in Fig. B.1. Thus we can have these wave equations valid for a semi-infinite or an infinite string, which is also perfectly flexible, and for small deviations $y(x,t)$.

The Vibrating Membrane

The wave equation in two dimensions, i.e., for $u(x,y,t)$ as a small deviation from the xy-plane for a flexible membrane, can be derived in the same way, to get

$$\frac{\partial^2 u}{\partial x^2} + \frac{\partial^2 u}{\partial y^2} = \frac{1}{c^2}\frac{\partial^2 u}{\partial t^2} \tag{B.1.7}$$

$$\nabla^2 u = \frac{1}{c^2}\frac{\partial^2 u}{\partial t^2}$$

The derivation also starts by adding the vertical forces, which here are due to the changes in the tension in the x direction as well as the y direction (see Exercise B.16 and Fig. B.5).

For a membrane vibration, we usually have a circular membrane. If we remember the boundary conditions for such a problem and the importance of the simple description of the boundary of the domain of the problem, we choose the compatible polar coordinates for the displacement $u(r,\theta,t)$. If we use the chain rule on (B.1.7) we can arrive at the wave equation in $u(r,\theta,t)$,

$$\frac{1}{r}\frac{\partial}{\partial r}\left(r\frac{\partial u}{\partial r}\right) + \frac{1}{r^2}\frac{\partial^2 u}{\partial \theta^2} = \frac{1}{c^2}\frac{\partial^2 u}{\partial t^2}$$

$$= \nabla^2 u(r,\theta,t) = \frac{1}{c^2}\frac{\partial^2 u(r,\theta,t)}{\partial t^2} \tag{B.1.8}$$

For the direct modeling of the wave equation (B.1.8) see Exercise B.17 and Fig. B.6.

The problem of the vibrating string, of finite length l and with fixed ends as in Fig. B.1, is fully described once we provide information on the initial position and velocity of the string at all points along the string, recognizing that the end points of the string remain stationary. The information concerning the initial position $y(x,0) = f(x)$ and velocity $\partial y(x,0)/\partial t = g(x)$ of the string is referred to as the *initial conditions* of the problem, and the (particular) information for Fig. B.1 that the string is stationary at the points $x = 0$ and $x = l$ is referred to as the *boundary conditions*: $y(0,t) = 0$ and $y(l,t) = 0$. For example, the partial differential equation (B.1.6), the above initial conditions, and the boundary conditions for Fig. B.1 constitute the *initial and boundary value problem* to be solved:

1. *Partial differential equation (PDE)—the wave equation*:

$$\frac{\partial^2 y}{\partial x^2} = \frac{1}{c^2} \frac{\partial^2 y}{\partial t^2}, \qquad 0 < x < l, \ t > 0 \tag{B.1.9}$$

2. *Initial conditions (IC)*:

   IC1: $y(x,0) = f(x), \qquad 0 < x < l$ (B.1.10)

   IC2: $\dfrac{\partial y}{\partial t}(x,0) = g(x), \qquad 0 < x < l$ (B.1.11)

3. *Boundary conditions (BC)*:

   BC1: $y(0,t) = 0, \qquad t > 0$ (B.1.12)

   BC2: $y(l,t) = 0, \qquad t > 0$ (B.1.13)

Other boundary conditions besides those of fixed ends in Fig. B.1 can be described. For example, we may speak of a free end at $x = l$, and we describe this by the absence of a vertical force at $x = l$ or $\partial y(l,t)/\partial x = 0$. Also, as we mentioned earlier, the wave equation we derived in (B.1.6) is based on a segment of a very flexible string with small deviations $y(x,t)$ and is independent of the string being of finite length and of the boundary conditions at the ends (boundary) $x = 0$, $x = l$ of the finite string depicted in Fig. B.1. So we may have an initial and boundary value problem for a semi-infinite $(0 < x < \infty)$ or an infinite $(-\infty < x < \infty)$ flexible string with small vibrations, which are governed by the wave equation (B.1.6) on the respective domains. Examples of initial and boundary value problems of the vibrating string on the semi-infinite domain $0 < x < \infty$ or the infinite domain $-\infty < x < \infty$ are worked out in Chapter 2, where Fourier transforms are used (see Example 2.19). Examples of the vibrations of a finite string or circular membrane are suitable for Chapter 3, where they are treated with finite transforms (see Exercise 3.4.2).

The Telephone and Telegraph Equations

It is known that the Maxwell equations are the basis of electromagnetic field theory, but electrical circuit theory proceeds from analysis of the resistive, capacitive, and inductive characteristics of the circuit.

In many applications of electric circuit theory the resistance, capacitance and inductance of the system can be considered as *lumped parameters*, and ordinary differential equations can be derived to describe the system; but when these parameters are *continuously distributed*, partial differential equations result. In this section we derive two important equations related to the transmission of electricity. If a transmission line involves both capacitance and current leakage to ground because of imperfect insulation, the system is a *distributed parameter* system. Consider the transmission line of Fig. B.2. Let $e(x,t)$ and $i(x,t)$ be the electrical potential and current, respectively. The potential difference across the element of length Δx can be written as the sum of the potential difference due to the resistance of the element and the inductance as follows:

$$e(x + \Delta x, t) - e(x, t) = -(R\,\Delta x)i - (L\,\Delta x)\frac{\partial i}{\partial t} \tag{B.1.14}$$

where R is the resistance of the line per unit length and L is the inductance per unit length. Dividing the equation by Δx and taking the limit as Δx approaches zero, we obtain

$$\frac{\partial e}{\partial x} = -Ri - L\frac{\partial i}{\partial t} \tag{B.1.15}$$

Due to leakage of current to ground and capacitance related loss associated with the varying change on the element, there is a current change across the element Δx, which is given by

$$i(x + \Delta x, t) - i(x, t) = -(G\,\Delta x)e - (C\,\Delta x)\frac{\partial e}{\partial t} \tag{B.1.16}$$

where G is the conductance to ground per unit length of wire and C is the capacitance to ground per unit length. Dividing the equation by Δx

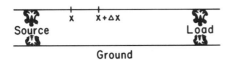

FIG. B.2 An electrical transmission line.

and taking the limit as Δx approaches zero, the equation becomes

$$\frac{\partial i}{\partial x} = -Ge - C\frac{\partial e}{\partial t} \tag{B.1.17}$$

Equations (B.1.15) and (B.1.17) can be combined in such a way as to give equations involving only e or only i as the dependent variable. To accomplish this, differentiate (C.1.15) with respect to x and (C.1.17) with respect to t to give

$$\frac{\partial^2 e}{\partial x^2} = -R\frac{\partial i}{\partial x} - L\frac{\partial^2 i}{\partial x\, \partial t} \tag{B.1.18}$$

$$\frac{\partial^2 i}{\partial t\, \partial x} = -G\frac{\partial e}{\partial t} - C\frac{\partial^2 e}{\partial t^2} \tag{B.1.19}$$

Since the current and its partial derivatives here are assumed to be continuous, the order of differentiation can be interchanged $\partial^2 i/\partial t\, \partial x = \partial^2 i/\partial x\, \partial t$, and $\partial^2 i/\partial x\, \partial t$ can be eliminated from the two equations. Then if $\partial i/\partial x$ is eliminated using (B.1.17) we obtain one *telephone equation*

$$\frac{\partial^2 e}{\partial x^2} = LC\frac{\partial^2 e}{\partial t^2} + (RC + GL)\frac{\partial e}{\partial t} + RGe \tag{B.1.20}$$

Another of the *telephone equations* can be obtained by differentiating (B.1.15) with respect to t and (B.1.17) with respect to x to give

$$\frac{\partial^2 e}{\partial x\, \partial t} = -R\frac{\partial i}{\partial t} - L\frac{\partial^2 i}{\partial t^2} \tag{B.1.21}$$

$$\frac{\partial^2 i}{\partial x^2} = -G\frac{\partial e}{\partial x} - C\frac{\partial^2 e}{\partial t\, \partial x} \tag{B.1.22}$$

Eliminating $\partial^2 e/\partial x\, \partial t$ in the manner in which $\partial^2 i/\partial x\, \partial t$ was eliminated above, and then eliminating the derivatives of e using (B.1.15), we obtain the second telephone equation

$$\frac{\partial^2 i}{\partial x^2} = LC\frac{\partial^2 i}{\partial t^2} + (RC + GL)\frac{\partial i}{\partial t} + RGi \tag{B.1.23}$$

If leakage and inductance are negligible, i.e., if $G = L = 0$, the telephone equations reduce to the *telegraph equations*

$$\frac{\partial^2 e}{\partial x^2} = RC\frac{\partial e}{\partial t} \tag{B.1.24}$$

$$\frac{\partial^2 i}{\partial x^2} = RC\frac{\partial i}{\partial t} \tag{B.1.25}$$

In relation to the diffusion equations that we shall derive in Section B.2, we note that these equations have the same form as the one-dimensional diffusion or unsteady-state heat conduction equation (B.2.5).

Another important special case of the telephone equations applies at high frequencies when the terms involving i and $\partial i/\partial t$ (or e and $\partial e/\partial t$) are negligible compared with $\partial^2 i/\partial t^2$ (or $\partial^2 e/\partial t^2$). In this case the telephone equations reduce to the wave equations in the current $i(x,t)$ and the potential $e(x,t)$, respectively,

$$\frac{\partial^2 i}{\partial x^2} = LC \frac{\partial^2 i}{\partial t^2} \tag{B.1.26}$$

$$\frac{\partial^2 e}{\partial x^2} = LC \frac{\partial^2 e}{\partial t^2} \tag{B.1.27}$$

B.2 Diffusion (or Heat Conduction) Equation

The diffusion, or heat conduction, equation will be derived here for a rod with uniform cross section, which is insulated laterally. Thus it is assumed that the temperature varies only in the x direction, as $T(x,t)$, where the time dependence is for the general case of an unsteady-state temperature distribution. Consider a segment of the uniform rod between x and $x + \Delta x$ as shown in Fig. B.3, which, of course, does not have to be of circular cross section.

In our formulation of this diffusion of energy (or heat) process, we will rely on the law of conservation of energy, as opposed to the derivation of the wave equation (B.1.6) of the vibrations, which relied, naturally, on Newton's law of force. Let $q(x,t)$ be the heat flux (in cal/cm^2-sec) at any point x in the rod, and let A, ρ, c, κ be the cross-sectional area, density, specific heat, and conductivity of the rod, respectively. Let us recall Fourier's law, which relates the heat flux q to the gradient of temperature $\partial T/\partial x$ as

$$q = -\kappa \frac{\partial T}{\partial x} \tag{B.2.1}$$

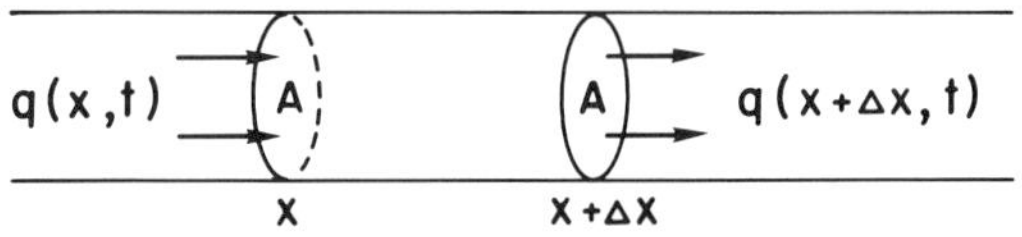

FIG. B.3 Heat flow in a rod.

To make an energy balance in this element of the rod we have the heat entering at x with a rate of $Aq(x,t)$ (in cal/sec) and the heat leaving at $x + \Delta x$ with a rate $Aq(x + \Delta x, t)$. If we assume that there is a source (or sink) giving heat at a rate of $s(x,t)$ (in cal/cm^3-sec) by means other than the above heat diffusion in the x direction, we have a net heat rate (in cal/sec) of $Aq(x,t) + A \Delta x \, s(x,t) - Aq(x + \Delta x, t)$ in the indicated element of the rod. This heat rate will result in a temperature rate of change $\partial T / \partial t$ according to the energy balance in this element,

$$Aq(x,t) - Aq(x + \Delta x, t) + A \Delta x \, s(x,t) = A \Delta x \, \rho c \, \frac{\partial T}{\partial t} \tag{B.2.2}$$

As we did for the derivation of the wave equation in (B.1.3), we can see, in the first two terms on the left side of this equation, the main ingredient of making a partial derivative $\partial q / \partial x$ with respect to the spatial variable x. Thus, if we divide both sides of (B.2.2) by $A \, \Delta x$ and then take the limit as $\Delta x \to 0$, we have

$$-\frac{\partial q}{\partial x} + s(x,t) = \rho c \, \frac{\partial T}{\partial t} \tag{B.2.3}$$

Now if we use Fourier's law (C.2.1) for $q(x,t)$ we have

$$\frac{\partial}{\partial x} \left[\kappa \, \frac{\partial T}{\partial t} \right] = -s(x,t) + \rho c \, \frac{\partial T}{\partial t} \tag{B.2.4}$$

In this derivation we may say that κ, ρ, and c can still be dependent on x and t. However, we should also remember that in (B.2.2) the term $\rho c \, \partial T / \partial t$ is taken as a mean value at some ξ: $x < \xi < x + \Delta x$, where $\xi \to x$ as $\Delta x \to 0$. There is another method that would avoid this difficulty of $\partial T / \partial t$, which is to write the energy balance on a whole domain D bounded by boundary B. In this case D is the whole rod and B is the two faces at the ends of the rod, since it is assumed that there is no heat flow through the (insulated) lateral surface of the rod (see Exercise B.14b).

If we assume no heat source and constant values for κ, ρ, and c, the heat equation is simplified to

$$\frac{\partial^2 T}{\partial x^2} = \frac{\rho c}{\kappa} \, \frac{\partial T}{\partial t}$$

$$\frac{\partial^2 T}{\partial x^2} = \frac{1}{k} \, \frac{\partial T}{\partial t} \tag{B.2.5}$$

where $k = \kappa / \rho c$ is called the *diffusivity* of the material of the rod.

Temperature Distribution in a Plate

To derive the heat equation for a rectangular plate, we can apply the above analysis for the net heat rate in both the x and y directions to arrive at the heat equation for the temperature distribution in two dimensions $T(x,y,t)$,

$$\frac{\partial^2 T}{\partial x^2} + \frac{\partial^2 T}{\partial y^2} = \frac{1}{k} \frac{\partial T}{\partial t} \tag{B.2.6}$$

$$\nabla^2 T = \frac{1}{k} \frac{\partial T}{\partial t}$$

For the detailed derivation see Exercise B.15 and its Fig. B.4.

The same can be done for the temperature distribution $T(x,y,z,t)$ in a solid,

$$\frac{\partial^2 T}{\partial x^2} + \frac{\partial^2 T}{\partial y^2} + \frac{\partial^2 T}{\partial z^2} = \frac{1}{k} \frac{\partial T}{\partial z}$$

$$\nabla^2 T = \frac{1}{k} \frac{\partial T}{\partial t} \tag{B.2.7}$$

Laplace Equations

The steady-state temperature is the temperature that is no longer dependent on time, that is $T(x,y,z,t) = u(x,y,z)$, where the above diffusion equations (B.2.6) and (B.2.7) reduce to their corresponding Laplace equations

$$\frac{\partial^2 u}{\partial x^2} + \frac{\partial^2 u}{\partial y^2} = 0 \tag{B.2.8}$$

$$\frac{\partial^2 u}{\partial x^2} + \frac{\partial^2 u}{\partial y^2} + \frac{\partial^2 u}{\partial z^2} = 0 \tag{B.2.9}$$

The Laplace equation is usually known for its governing of the electrostatic or gravitational potential $u(x,y,z)$ in a space that is free of charge. In case there is a charge distribution with density ρ, the governing equation of the potential $u(x,y,z)$ is the *Poisson equation*

$$\nabla^2 u = -\rho \tag{B.2.10}$$

The Laplace equation in $u(r,\theta,z)$ for cylindrical coordinates becomes

$$\nabla^2 u = \frac{1}{r} \frac{\partial}{\partial r} \left(r \frac{\partial u}{\partial r} \right) + \frac{1}{r^2} \frac{\partial^2 u}{\partial \theta^2} + \frac{\partial^2 u}{\partial z^2} = 0, 0 < \theta < 2\pi \tag{B.2.11}$$

and in spherical coordinates for $u(\rho,\phi,\theta)$ (see Fig. 3.12) is

$$\nabla^2 u = \frac{1}{\rho^2} \frac{\partial}{\partial \rho} \left(\rho^2 \frac{\partial u}{\partial r} \right) + \frac{1}{\rho^2 \sin\theta} \frac{\partial}{\partial \theta} \left(\sin\theta \frac{\partial u}{\partial \theta} \right) + \frac{1}{\rho^2 \sin^2\theta} \frac{\partial^2 u}{\partial \phi^2} = 0,$$

$$0 < \theta < \pi, \ 0 < \phi < 2\pi \tag{B.2.12}$$

The derivation of (B.2.11) from (B.2.10) is done by "somewhat lengthy" computations involving the chain rule, and the computations are even lengthier for (B.2.12).

Boundary Conditions for Heat Transfer

We emphasize again that it is not enough to know the governing equation, for we must also specify auxiliary or boundary conditions of the physical system. There is, of course, an important question of how many and what type of boundary conditions are required for the partial differential equation at hand, but we shall limit ourselves to examining the physical basis for boundary conditions associated with heat transfer.

Three types of boundary conditions are frequently encountered in the analysis of heat transfer, and there are many other special cases. The three most common boundary conditions that are applied at a solid-solid or fluid-solid interface are

1. Constant surface temperature
2. Constant heat flux at the surface or interface
3. A linear relation between the surface heat flux and the surface temperature

As an example of the boundary condition of the first kind, as it is called, suppose that a red-hot cannonball is plunged into a pool of water. Vigorous boiling of the water all over the surface of the sphere could cause the surface temperature of the sphere to equal that of the boiling water as long as the vigorous boiling occurs. Constant surface temperature could be a good approximation in this case.

The boundary condition of the second kind, constant heat flux, could be achieved by electrical heating. For example, a fluid could be heated by passing it through a thin-walled tube heated either by using the tube wall as a resistance heater or by wrapping the tube with high-resistance electrical wire through which a current is passed. If the energy production per unit area of tube surface is constant, the heat flux at the fluid-solid interface would be constant. Since the heat transfer to the fluid adjacent to the wall is by conduction, we can write Fourier's law for the radial heat conduction as $-\kappa(\partial T/\partial r)_w = q_w$, where κ is the thermal conductivity of the fluid and q_w is the heat flux at the wall (solid-fluid interface). Thus the boundary condition of the second kind is equivalent to specifying the temperature gradient at the boundary.

The boundary condition of the third kind arises by applying the phenomenological law referred to as *Newton's law of cooling*, which states that the heat flux or heat loss from a hot surface is proportional to the difference in temperature between the surface and the surrounding medium. For example, the heat loss from the human body (assuming the skin temperature to be approximately constant) is observed to depend on the temperature of the surroundings. Mathematically we can state $-\kappa(\partial T/\partial x)_s = h(T_s - T_a)$, where $-\kappa(\partial T/\partial x)_s$ is the heat flux at the surface of the skin, T_s is the surface temperature, T_a is the ambient temperature, and h is a proportionality constant usually called the *heat transfer coefficient*. The coefficient h is not a physical property of the system but depends on additional factors such as the wind velocity. If we write $\theta = T - T_a$, then $\partial T/\partial x = \partial \theta/\partial x$, and the boundary condition can be written as

$$h\theta_s + \kappa \frac{\partial \theta}{\partial x}\bigg|_s = 0 \qquad (B.2.13)$$

Equation (B.2.13) is a linear relation between the surface temperature and the temperature gradient at the surface. The previous two boundary conditions can be considered as special cases of the third boundary condition, for if we divide each term of equation (B.2.13) by h and take the limit as h approaches infinity, we recover the first boundary condition, i.e., $\theta_s = T_s - T_a = 0$, and if we divide by κ and take the limit as κ approaches infinity, we recover a special case of the second boundary condition, i.e., $(\partial \theta/\partial x)_s = 0$.

In some applications, either the surface temperature or the surface heat flux is a known function of time and position on the surface, so it is convenient to summarize the most common heat transfer boundary condition as

1. $T_B = f(x,y,z,t)\,|_B \quad$ or $T_B =$ constant $\qquad (B.2.14)$

2. $(\partial T/\partial n)_B = g(x,y,z,t)\,|_B \quad$ or $(\partial T/\partial n)_B =$ constant $\qquad (B.2.15)$

3. $k_1 T_B + k_2(\partial T/\partial n)_B = 0 \qquad (B.2.16)$

where $\partial T/\partial n$ refers to the derivative normal to the boundary, and k_1 and k_2 are constants.

Exercises

Section B.1 Vibration Problems

B.1. *The wave equation for a semi-infinite string.* State the initial and boundary value problem for the displacement $u(x,t)$ of a *semi-*

infinite string with zero initial displacement and velocity, where its end at $x = 0$ is given a time-dependent displacement $h(t)$.

ANS. $\dfrac{\partial^2 u}{\partial x^2} = \dfrac{1}{c^2} \dfrac{\partial^2 u}{\partial t^2}$, $0 < x < \infty, \ t > 0$

$u(x,0) = \dfrac{\partial u}{\partial t}(x,0) = 0$, $0 < x < \infty$

$u(0,t) = h(t)$, $t > 0$

B.2. (a) Verify that $u(x,t) = \phi(x + ct) + \psi(x - ct)$ is a solution to the wave equation $\partial^2 u/\partial x^2 = (1/c^2)\partial^2 u/\partial t^2$ where ϕ and ψ are arbitrary twice-differentiable functions.

Hint: $\partial\phi/\partial x = \phi' \cdot 1$, $\partial\phi/\partial t = \phi' \cdot c$, ...

(b) Verify that $u(x,t) = \frac{1}{2}[f(x+ct)+f(x-ct)]$ is a solution of the wave equation in (a) with initial displacement $u(x,0) = f(x)$ and zero initial velocity $u_t(x,0) = 0$.

B.3. (a) Verify that

$$y(x,t) = [C_1 \cos \lambda x + C_2 \sin \lambda x + C_3 \cosh \lambda x + C_4 \sinh \lambda x]$$
$$\times [C_5 \cos a\lambda^2 t + C_6 \sin a\lambda^2 t]$$

as the *deflection in the transverse direction of a beam*, is a solution of the following equation of vibration of a beam:

$$\dfrac{\partial^4 y}{\partial x^4} = -\dfrac{1}{a^2}\dfrac{\partial^2 y}{\partial t^2}, \qquad a^2 = \dfrac{EI}{AP}$$

where A is the *cross-sectional area*, ρ the *density*, I the *moment of inertia* of the cross-sectional area of the beam about the z-axis, and E is the *Young's modulus* or modulus of elasticity.

(b) State the boundary conditions when both ends of the beam (of length l) are fixed, and with zero moment (zero curvature: let $u_{xx} = 0$ at both ends).

ANS. (b) $u(0,t) = u(l,t) = 0$

$u_{xx}(0,t) = u_{xx}(l,t) = 0$

B.4. (a) Derive the telephone equation (B.120) in $e(x,t)$.

Hint: See (B.1.18), (B.1.19), and (B.1.17).

(b)   Derive the second telephone equation (B.1.23) in $i(x,t)$.

Hint: See (B.1.21), (B.1.22), and (B.1.15).

B.5.   *Poisson, Laplace, and wave equations via Maxwell's equations.* Now let **E** be the *electric field*, **H** the *magnetic field*, **D** the *electric displacement* (where $\mathbf{D} = \epsilon\mathbf{E}$, and ϵ is the dielectric constant), **B** the *magnetic induction* (where $\mathbf{B} = \mu\mathbf{H}$, and μ is the permeability), **J** the current density, and ρ the charge density. The *Maxwell equations* that relate these quantities, where we choose here the *MKS* system of units, are

$$\nabla \cdot \mathbf{D} = \rho \tag{E.1}$$

$$\nabla \cdot \mathbf{B} = 0 \tag{E.2}$$

$$\nabla \times \mathbf{E} = -\frac{\partial \mathbf{B}}{\partial t} \tag{E.3}$$

$$\nabla \times \mathbf{H} = \mathbf{J} + \frac{\partial \mathbf{D}}{\partial t} \tag{E.4}$$

Consider the electrostatic potential in a homogeneous medium (constant ϵ) with charge density ρ. The electric field is defined to be the negative of the potential gradient, i.e.,

$$\mathbf{E} = -\nabla\phi \tag{E.5}$$

(a)   Show that the electrostatic potential ϕ satisfies the *Poisson equation*:

$$\nabla^2\phi = \frac{-\rho}{\epsilon} \tag{E.6}$$

and in the absence of charge ($\rho = 0$) the potential satisfies the *Laplace equation*

$$\nabla^2\phi = 0 \tag{E.7}$$

Hint: Use $\mathbf{D} = \epsilon\mathbf{E}$, and consult (E.1), (E.5).

(b)   Consider Maxwell's equations (E.1)–(E.4) in free space (vacuum) and in the absence of charge, i.e., $\epsilon = \epsilon_0$, $\mu = \mu_0$, $\rho = 0$, and $J = 0$. Also, let us recall the vector identity:

$$\nabla \times (\nabla \times \mathbf{v}) = \nabla(\nabla \cdot \mathbf{v}) - \nabla^2\mathbf{v} \tag{E.8}$$

where $\nabla^2\mathbf{v}$ is defined for $\mathbf{v} = \mathbf{i}v_1 + \mathbf{j}v_2 + \mathbf{k}v_3$, as

$$\nabla^2\mathbf{v} \equiv \mathbf{i}\frac{\partial^2 v_1}{\partial x^2} + \mathbf{j}\frac{\partial^2 v_2}{\partial y^2} + \mathbf{k}\frac{\partial^2 v_3}{\partial z^2} \tag{E.9}$$

Show that the electric field **E** and the magnetic induction **B** satisfy the wave equations

$$\nabla^2 \mathbf{E} = \frac{1}{c^2} \frac{\partial^2 \mathbf{E}}{\partial t^2} \tag{E.10}$$

and

$$\nabla^2 \mathbf{B} = \frac{1}{c^2} \frac{\partial^2 \mathbf{B}}{\partial t^2} \tag{E.11}$$

respectively, where $c = 1/\sqrt{\mu_0 \epsilon_0}$ is the speed of light in vacuo.

Hint: To establish (E.10), we take the curl of the two sides of (E.3) i.e.,

$$\nabla \times (\nabla \times \mathbf{E}) = -\nabla \times \left(\frac{\partial \mathbf{B}}{\partial t} \right) = -\frac{\partial}{\partial t} \nabla \times \mathbf{B}$$

and then appeal to (E.8) for the left side of this result and to (E.4) for the right side. For (E.11) we do similar steps starting with the curl of (E.4).

B.6. For the vibrations of a perfectly flexible string, derive a more complete (nonlinear) wave equation without the assumption of small vibrations, i.e., without the use of $\partial y/\partial x \ll 1$, and without using (B.1.2).

Hint: $\sin \alpha = \tan \alpha \cos \alpha = \dfrac{\partial y}{\partial x} \left[1 + \left(\dfrac{\partial y}{\partial x} \right)^2 \right]^{-1/2}$.

ANS. $T \dfrac{\partial}{\partial x} \left(\dfrac{\partial y/\partial x}{[1 + (\partial y/\partial x)^2]^{1/2}} \right) = \rho \left(g + \dfrac{\partial^2 y}{\partial t^2} \right) \left[1 + \left(\dfrac{\partial y}{\partial x} \right)^2 \right]^{1/2}$

$$\tag{E.1}$$

Carrying out the differentiation on the left-hand side and rearranging, the equation becomes

$$\frac{\partial^2 y}{\partial x^2} = \frac{\rho}{T} \left(g + \frac{\partial^2 y}{\partial t^2} \right) \left[1 + \left(\frac{\partial y}{\partial x} \right)^2 \right]^2 \tag{E.2}$$

Section B.2 Diffusion Problems

B.7. *Heat equation in a uniform rod.* State the initial and/or boundary value problem that describes the temperature distribution in

(a) A uniform rod of length l and with constant conductivity where both ends are kept at zero temperature, when its initial temperature distribution is given by $u(x,0) = 3x$.

(b) As in part (a) except that both ends are insulated,

(c) For the steady-state temperature $(\partial u/\partial t = 0)$ when the left end is kept at constant temperature T_0, the right end is insulated, and the initial temperature is $u(x,0) = f(x)$. Attempt to find a solution.

(d) Solve for the steady-state temperature in part (a).

ANS. (a) $\dfrac{\partial u}{\partial t} = k\dfrac{\partial^2 u}{\partial x^2}$, $\qquad 0 < x < l,\ t > 0$ $\qquad$ (E.1)

$u(0,t) = u(l,t) = 0, \qquad t > 0$ $\qquad$ (E.2)

$u(x,0) = 3x, \qquad 0 < x < l$ $\qquad$ (E.3)

(b) Replace (E.2) in part (a) by

$$\frac{\partial u}{\partial x}(0,t) = \frac{\partial u(l,t)}{\partial x} = 0 \qquad (E.4)$$

(c) As steady-state temperature u is independent of time $v(x)$, it is independent of the initial condition $u(x,0) = f(x)$.

$\dfrac{d^2 v}{dx^2} = 0, \qquad 0 < x < l$

$v(0) = T_0$

$v'(l) = 0$

$v(x) = T_0, \qquad 0 < x < l$

(d) $\dfrac{d^2 v}{dx^2} = 0$

$v(0) = v(l) = 0$

$v(x) = 0, \qquad 0 < x < l$

B.8. (a) Verify that $u(x,t) = (A\cos\lambda x + B\sin\lambda x)e^{-k\lambda^2 t}$ is a solution of the heat equation in one dimension

$$\frac{\partial u}{\partial t} = \frac{1}{k}\frac{\partial^2 u}{\partial x^2} \qquad (E.1)$$

(b) Verify that

$$u_n(x,t) = B \sin \frac{n\pi x}{l} e^{-k(n\pi/l)^2 t}, \qquad n = 0, \mp 1, \mp 2, \ldots$$

is a set of solutions of the heat equation in (E.1) along with the boundary conditions of the zero temperature of both ends $(u(0,t) = u(l,t) = 0)$ of a rod of length l.

(c) Verify that

$$u_n(x,t) = A \cos \frac{n\pi x}{l} e^{-k(n\pi/l)^2 t}$$

is a solution of the heat equation in part (a) along with the boundary conditions that both ends of the bar are perfectly insulated, i.e.,

$$\frac{\partial u(0,t)}{\partial x} = \frac{\partial u(l,t)}{\partial x} = 0$$

B.9. State the initial and boundary value problem that describes the temperature distribution in

(a) A uniform plate of length l, width b, and with constant diffusivity k. The lower and left sides are kept at zero temperature, and the upper and right sides are insulated. The rod is initially at temperature described by $u(x,y,0) = f(x,y)$ and an external source of heat $F(x,y)$ is being applied.

(b) A semi-infinite slab (or rod) with the end $x = 0$ kept at zero temperature and the initial temperature given by $f(x)$.

(c) An infinite slab with initial temperature given by $f(x)$.

ANS. (a) $\quad \dfrac{\partial u}{\partial t} = k \left(\dfrac{\partial^2 u}{\partial x^2} + \dfrac{\partial^2 u}{\partial y^2} \right) - F(x,y),$

$$0 < x < l, \ 0 < y < b, \ t > 0$$

$$u(x,0) = u(0,y) = 0$$

$$\frac{\partial u}{\partial y}(x,b) = \frac{\partial u}{\partial x}(l,y) = 0$$

$$u(x,y,0) = f(x,y), \qquad 0 < x < l, \ 0 < y < b$$

(b) $\quad \dfrac{\partial u}{\partial t} = k \dfrac{\partial^2 u}{\partial x^2}, \qquad 0 < x < \infty, \ t > 0$

$$u(0,t) = 0, \qquad t > 0$$

$$|u(x,t| < \infty$$

$$u(x,0) = f(x), \qquad 0 < x < \infty$$

(c) $\quad \dfrac{\partial u}{\partial t} = k \dfrac{\partial^2 u}{\partial x^2}, \qquad -\infty < x < \infty, \; t > 0$

$$|u(x,t)| < \infty$$

$$u(x,0) = f(x), \qquad -\infty < x < \infty$$

B.10. *Potential distribution in a semi-infinite slot.* State the boundary value problem describing the potential distribution inside a semi-infinite slot of width a, with the two sides held at zero potential and the base at potential described by $f(x) = x^2(a - x)$.

ANS. $\quad \dfrac{\partial^2 u}{\partial x^2} + \dfrac{\partial^2 u}{\partial y^2} = 0, \qquad 0 < x < a, \; y > 0$

$$u(0,y) = u(a,y) = 0, \qquad y > 0$$

$$|u(x,y)| < \infty$$

$$u(x,0) = x^2(a - x), \qquad 0 < x < a$$

B.11. *Laplace equation in a disc.* State the boundary value problem for the potential distribution u in a unit disc when its boundary is kept at a potential described by $f(\theta)$.

ANS. $\quad \dfrac{\partial^2 u}{\partial r^2} + \dfrac{1}{r} \dfrac{\partial u}{\partial r} + \dfrac{1}{r^2} \dfrac{\partial^2 u}{\partial \theta^2} = 0, \quad 0 < r < 1, \quad 0 < \theta < 2\pi$

$$u(1,\theta) = f(\theta), \qquad 0 \le \theta \le 2\pi$$

B.12. *Laplace equation in a sphere.* State the boundary value problem for the potential distribution $u(\rho, \phi, \theta)$ in a sphere of radius R (see Fig. 3.12 and Eq. 3.154) when the surface is kept at a potential given by: (i) $g(\phi, \theta)$, (ii) $f(\phi)$, (iii) $h(\theta)$

ANS. (i) $\quad \dfrac{\partial^2 u}{\partial \rho^2} + \dfrac{2}{\rho} \dfrac{\partial u}{\partial \rho} + \dfrac{1}{\rho^2 \sin\theta} \dfrac{\partial}{\partial \theta} \left(\sin\theta \dfrac{\partial u}{\partial \theta} \right) +$

$$+ \frac{1}{\rho^2 \sin^2 \theta} \frac{\partial^2 u}{\partial \phi^2} = 0,$$

$$0 < \rho < R, \ 0 < \theta < \pi, \ 0 < \phi < 2\pi$$

$$u(R, \phi, \theta) = g(\phi, \theta), \qquad 0 \le \theta \le \pi, \ 0 \le \phi \le 2\pi$$

(ii) $\quad \dfrac{\partial^2 u}{\partial \rho^2} + \dfrac{2}{\rho} \dfrac{\partial u}{\partial \rho} + \dfrac{1}{\rho^2} \dfrac{\partial^2 u}{\partial \phi^2} = 0,$

$$0 < \rho < R, \ 0 < \phi < 2\pi$$

$$u(R, \phi, \theta) = f(\phi), \qquad 0 \le \phi \le 2\pi$$

(iii) $\quad \dfrac{\partial^2 u}{\partial \rho^2} + \dfrac{2}{\rho} \dfrac{\partial u}{\partial \rho} + \dfrac{1}{\rho^2 \sin \theta} \dfrac{\partial}{\partial \theta} \left(\sin \theta \dfrac{\partial u}{\partial \theta} \right) = 0,$

$$0 < \rho < R, \ 0 < \theta < \pi$$

$$u(R, \phi, \theta) = h(\theta), \qquad 0 < \theta < \pi$$

B.13. *Laplace equation in a cube.* State the mathematical model for the potential distribution in a cube of side π with the top, front, and back sides grounded; the left and right sides insulated. Also, the bottom side is kept at potential given by $u(x,y,0) = f(x,y) = xy$.

ANS. $\quad \dfrac{\partial^2 u}{\partial x^2} + \dfrac{\partial^2 u}{\partial y^2} + \dfrac{\partial^2 u}{\partial z^2} = 0,$

$$0 < x < \pi, \ 0 < y < \pi, \ 0 < z < \pi$$

$$u(x,y,\pi) = u(0,y,z) = u(\pi,y,z) = 0$$

$$\frac{\partial u}{\partial y}(x,0,z) = \frac{\partial u}{\partial y}(x,\pi,z) = 0, \qquad 0 < z < \pi$$

$$u(x,y,0) = xy, \qquad 0 < x < \pi, \ 0 < y < \pi$$

B.14. *Diffusion (or heat) equation via the divergence theorem.* The divergence theorem states that for a solid V bounded by a smooth or piecewise smooth surface S,

$$\iint_S \mathbf{v} \cdot \mathbf{n} \, dS = \iiint_V \nabla \cdot \mathbf{v} \, dV \qquad \text{(E.1)}$$

for the smooth vector field **v**, and where **n** is the outward unit normal to the surface S. *Also note that for continuous* $f(x,y,z)$,

$$\iiint_V f(x,y,z)\,dV = 0 \quad \text{implies } f(x,y,z) \equiv 0, \qquad (x,y,z) \in V$$

$$\text{(E.2)}$$

(a)   Prove the result in (E.2).

   Hint: Use a contradiction argument, since if the continuous $f(x,y,z)$ is not identically zero on the arbitrary V, let it be not zero (positive) at a point; thus by its continuity it must be nonzero (positive) in a neighborhood of a ball V' centered at that point. This would render the integral in (E.2) not equal to zero.

(b)   Derive the heat equation (B.2.5) for the temperature distribution $T(x,t)$ in the rod.

   Hint: Use the divergence theorem to express the rate of heat leaving the surface as a volume integral, remembering that the surface that counts (for heat flow) is the two cross sections of the rod, since the lateral surface is assumed insulated for (B.2.5). This heat rate is equated to that stored (or lost) to the body as $\partial H/\partial t$, where $H = \iiint_V c\rho T\,dV$. Then after combining the two volume integrals, use (E.2), assuming the continuity of all variables involved to conclude the vanishing of the integrand, which gives the heat equation for the rod.

B.15.   *Heat equation in a plate.* To derive the heat equation for the temperature distribution $T(x,y,t)$ in a plate, consider the rectangular increment in the plate as depicted in Fig. B.4. The plate has unit thickness δ, conductivity κ, density ρ, and specific heat c.

(a)   Apply the energy balance analysis that we used for the one-dimensional heat flow in (B.2.2), in both x and y directions, to write the two-dimensional heat flow analog of (B.2.2).

(b)   After dividing by $\Delta x \Delta y$ in the result of part (a), take the limits $\Delta x \to 0$ and $\Delta y \to 0$ to derive the heat equation in two dimensions

$$\frac{\partial^2 T}{\partial x^2} + \frac{\partial^2 T}{\partial y^2} = \frac{1}{k}\frac{\partial^2 T}{\partial t^2}, \qquad k = \frac{\kappa}{c\rho} \qquad \text{(E.1)}$$

ANS.   (a) $\delta\,\Delta y[q(x,y,t) - q(x+\Delta x,y,t)] + \delta\,\Delta x[q(x,y+\Delta y,t)+$

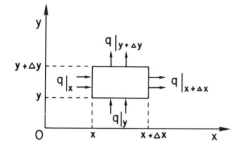

FIG. B.4 Heat flow in a plate—using Cartesian coordinates.

$$-q(x,y,t)] = c\rho\delta\,\Delta x\,\Delta y\,\frac{\partial T}{\partial t} \tag{E.2}$$

B.16. *The vibrating membrane.* To derive the wave equation for the displacement $u(x,y,t)$ of a membrane, consider the displaced increment ΔS of the membrane as shown in Fig. B.5.

(a) Apply the analysis that we used for the one-dimensional wave equation in (B.1.1)–(B.1.3) to derive the basic balance of forces in the vertical direction of the displacement $u(x,y,t)$. The membrane has the same properties, T is tension per unit

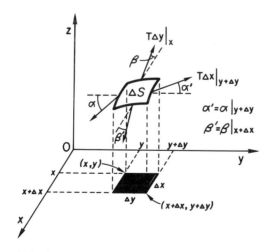

FIG. B.5 Forces on a segment of a vibrating membrane— using Cartesian coordinates.

length; and α and β are the angles that the tensions in the x and y directions make with the xy plane. ΔS is the area of the element, which is approximated by $\Delta A = \Delta x \, \Delta y$.

(b) Use the limit process as Δx, $\Delta y \to 0$ on the result in part (a) to derive the wave equation in two dimensions:

$$\frac{\partial^2 u}{\partial x^2} + \frac{\partial^2 u}{\partial y^2} = \frac{1}{c^2}\frac{\partial^2 u}{\partial t^2}, \qquad c^2 = \frac{T}{\rho} \tag{E.1}$$

ANS. $T \, \Delta x [\tan \alpha \, |_{y+\Delta y} - \tan \alpha \, |_y] + T \, \Delta y [\tan \beta \, |_{x+\Delta x} - \tan \beta \, |_x]$

$$= \rho \, \Delta s \, \frac{\partial^2 u}{\partial t^2} \tag{E.2}$$

where $\tan \alpha = \partial u / \partial y$ and $\tan \beta = \partial u / \partial x$.

B.17. *Vibrations of circular membrane.*

(a) Consider the increment ΔS in polar coordinates of a vibrating membrane. Follow the analysis in (B.1.1)–(B.1.3) and Exercise B.16 for the rectangular membrane to find the balance of the vertical forces on the increment ΔS of the membrane as shown in the Fig. B.6. Let the tension per unit length in the radial direction be T_r, which makes an angle α with

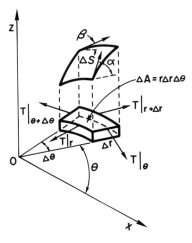

FIG. B.6 Forces on a segment of a vibrating membrane—using polar coordinates.

the xy plane, while T_θ is the tension in the θ direction and makes an angle β with the xy plane. Note also that for a small oscillations $\sin\alpha \approx \tan\alpha \approx \Delta u/\Delta r$, $\sin\beta \approx \tan\beta \approx (1/r)\Delta u/r\Delta\theta$, $\Delta S \sim \Delta A = r\,\Delta r\,\Delta\theta$.

(b) Take the limits as Δr, $\Delta\theta \to 0$, after dividing (E.2) below by $Tr\,\Delta r\,\Delta\theta$, to find the two-dimensional wave equations in polar coordinates,

$$\frac{1}{r}\frac{\partial}{\partial r}\left(r\frac{\partial u}{\partial r}\right) + \frac{1}{r^2}\frac{\partial^2 u}{\partial\theta^2} = \frac{1}{c^2}\frac{\partial^2 u}{\partial t^2}, \qquad c^2 = \frac{T}{\rho} \qquad \text{(E.3)}$$

ANS. (a) $[Tr\,\Delta\theta\tan\alpha\,|_{r+\Delta r} - Tr\,\Delta\theta\tan\alpha\,|_r] + [T\,\Delta r\tan\beta\,|_{\theta+\Delta\theta}$

$\qquad - T\,\Delta r\tan\beta\,|_\theta]$

$$= \rho\Delta S\frac{\partial^2 u}{\partial t^2} = \rho r\,\Delta r\,\Delta\theta\frac{\partial^2 u}{\partial t^2} \qquad \text{(E.1)}$$

where $\tan\alpha = \partial u/\partial r$ and $\tan\beta = (1/r)\partial u/\partial\theta$

$$= T\,\Delta\theta\left[\left(r\frac{\partial u}{\partial r}\right)\bigg|_{r+\Delta r} - \left(r\frac{\partial u}{\partial r}\right)\bigg|_r\right]$$

$$+ T\,\Delta r\left[\frac{1}{r}\frac{\partial u}{\partial\theta}\bigg|_{\theta+\Delta\theta} - \frac{1}{r}\frac{\partial u}{\partial\theta}\bigg|_\theta\right]$$

$$= \rho r\,\Delta r\,\Delta\theta\frac{\partial^2 u}{\partial t^2} \qquad \text{(E.2)}$$

B.18. *Classification of partial differential equations.* Consider the most general *linear second-order partial differential equation* in $u(x,y)$

$$a(x,y)u_{xx} + 2b(x,y)u_{xy} + c(x,y)u_{yy}$$

$$+ d(x,y)u_x + e(x,y)uy + f(x,y)u = g(x,y) \qquad \text{(E.1)}$$

The theory of classifying such linear second-order partial differential equations is analogous to what we have in analytic geometry. The partial differential equation (E.1) is called *hyperbolic* if the discriminant $b^2 - ac > 0$, *elliptic* if $b^2 - ac < 0$, and *parabolic* if $b^2 = ac$. Clearly the wave equation $u_{xx} = (1/c^2)u_{tt} = 0$ in (B.1.6) is hyperbolic, since $b^2 - ac = 0 - (-1/c^2) = 1/c^2 > 0$. The heat equation $u_{xx} - (1/k)u_t = 0$ in (B.2.5) is parabolic for $b^2 - ac = 0 - (1)(0) = 0$, while the Laplace equation $u_{xx} + u_{yy} = 0$ in (C.2.8) is elliptic since

$b^2 - ac = 0 - 1 = -1 < 0$. Classify the following partial differential equations in two variables

(i) $u_{rr} + \dfrac{1}{r}u_r + \dfrac{1}{r^2}u_{\theta\theta} = 0$ (in $u(r,\theta)$)

(ii) $\dfrac{\partial^2 u}{\partial x\,\partial y} = 0$ (in $u(x,y)$)

(iii) $u_{xx} + u_{yy} + u = 0$ (in $u(x,y)$)

(iv) $u_{xx} - xu_{yy} = 0$ (in $u(x,y)$)

ANS.   (i)   elliptic, $r > 0$

        (ii)   hyperbolic

        (iii)   elliptic

        (iv)   hyperbolic for $x > 0$, elliptic for $x < 0$

Appendix C

TABLES OF TRANSFORMS

For the convenience of the reader we have included in this appendix a reasonable collection of tables for the various integral and finite transforms discussed in this book with a few entries for the z-transform (in C.17).

Basic references: As we indicated in the text, the most complete tables of integral transforms are found in Erdelyi et al. (1954), Ditkin and Prudnikov (1965), and few other references that can be found in the bibliography. For the symbols used in these tables, consult their listing at the end of this book, and for most of their definitions see Appendix A, Section A.5 on special functions, and the Index of Symbols and Notations preceding the Subject Index.

TABLE C.1 Laplace Transforms

$f(t)$	$\mathcal{L}\{f\} \equiv F(s) = \int_0^\infty e^{-st}f(t)dt, \quad \mathrm{Re}\,s > \gamma$
$\dfrac{1}{2\pi i}\lim\limits_{L\to\infty}\displaystyle\int_{\gamma-iL}^{\gamma+iL} e^{iz}F(z)dz$	$F(s), \quad \mathrm{Re}\,s > \gamma$
$\dfrac{df}{dt}$	$sF(s) - f(0)$
$\dfrac{d^n f}{dt^n}$	$s^n F(s) - s^{n-1}f(0) - s^{n-2}f'(0) - \cdots - f^{(n-1)}(0)$
$\displaystyle\int_0^t f(\tau)\,d\tau$	$\dfrac{1}{s}F(s)$
$\begin{cases} f(t-b) & \text{if } t > b \\ 0 & \text{if } t < b \end{cases}$	$e^{-bs}F(s), \quad b > 0$
$\displaystyle\int_0^t f(\tau)g(t-\tau)\,d\tau$	$F(s)G(s)$
$-tf(t)$	$\dfrac{dF}{ds}$
$(-1)^n t^n f(t)$	$\dfrac{d^n F}{ds^n}$
$\dfrac{1}{t}f(t)$	$\displaystyle\int_s^\infty F(x)\,dx$

$f(t)$	$F(s)$
$e^{at}f(t)$	$F(s-a)$
$\dfrac{1}{c}f\left(\dfrac{t}{c}\right)$	$F(cs),\quad c>0$
$f(t)\quad\text{if } f(t+a)=f(t)$	$\dfrac{\int_0^a e^{-st}f(t)\,dt}{1-e^{-as}},\quad a>0$
1	$s^{-1},\quad \text{Res}>0$
$t^\nu,\ \nu>-1$	$s^{-\nu-1}\Gamma(\nu+1)$
e^{at}	$(s-a)^{-1},\quad \text{Res}>a$
$\cosh at$	$s(s^2-a^2)^{-1},\quad \text{Res}>a$
$\sinh at$	$a(s^2-a^2)^{-1},\quad \text{Res}>a$
$t\cosh at$	$(s^2+a^2)(s^2-a^2)^{-2},\quad \text{Res}>a$
$t\sinh at$	$2as(s^2-a^2)^{-2},\quad \text{Res}>a$
$\cosh at + \tfrac{1}{2}at\sinh at$	$s^3(s^2-a^2)^{-2},\quad \text{Res}>a$
$\cos at$	$s(s^2+a^2)^{-1}$
$\sin at$	$a(s^2+a^2)^{-1}$
$t\cos at$	$(s^2-a^2)(s^2+a^2)^{-2}$
$t\sin at$	$2as(s^2+a^2)^{-2}$
$t^{-1}\sin(at)$	$\cot^{-1}\left(\dfrac{s}{a}\right)$

$$\text{Si}(t) = \int_0^t \frac{\sin x}{x}\, dx \qquad s^{-1}\cot^{-1} s$$

$$\text{Ci}(t) = -\int_t^\infty \frac{\cos x}{x}\, dx \qquad -\left(\frac{1}{2s}\right)\log(1+s^2)$$

$$e^{-b^2 t^2/4} \qquad \pi^{1/2} b^{-1} e^{s^2/b^2}\,\text{Erfc}\left(\frac{s}{b}\right)$$

$$\text{erf}\left(\frac{t}{2a}\right) \qquad s^{-1} e^{a^2 s^2}\,\text{Erfc}(as)$$

$$\text{erf}(a\sqrt{t}) \qquad as^{-1}(s+a^2)^{-1/2}$$

$$\text{erfc}(a\sqrt{t}) \qquad s^{-1}[1 - a(s+a^2)^{-1/2}]$$

$$e^{a^2 t}\,\text{erfc}(a\sqrt{t}) \qquad s^{-1/2}(s^{1/2}+a)^{-1}$$

$$\text{erfc}\left(\frac{a}{\sqrt{t}}\right) \qquad s^{-1} e^{-2a\sqrt{s}}, \quad \text{Res} > 0$$

$$t^{-1/2} e^{-c/t} \qquad (\pi/s)^{1/2} e^{-2\sqrt{cs}}, \quad \text{Res} > 0$$

$$t^\nu J_\nu(at), \quad \nu > -\tfrac{1}{2} \qquad (2a)^\nu \pi^{-1/2}\Gamma(\nu + \tfrac{1}{2})(s^2 + a^2)^{-\nu-1/2}, \quad \text{Res} > 0$$

$$t^n J_n(at) \qquad \frac{(2n)!}{2^n n!} a^n (s^2 + a^2)^{-n-1/2}, \quad \text{Res} > 0$$

$$t^{n+1} J_n(at) \qquad \frac{(2n+1)!}{2^n n!} a^n (s^2 + a^2)^{-n-1/2}, \quad \text{Res} > 0$$

$$J_\nu(t) \quad \nu > -\tfrac{1}{2} \qquad (s^2 + 1)^{-1/2}\left\{s + \sqrt{(s^2+1)}\right\}^{-\nu}, \quad \text{Res} > 0$$

$t^\nu I_\nu(at)$, $\nu > -1/2$	$(2a)^\nu \pi^{-1/2}\Gamma(\nu + 1/2)(s^2 - a^2)^{-\nu - 1/2}$, $\text{Res} > a > 0$
$t^{-1}\sin(a\sqrt{t})$	$rf\left(1/2 as^{-1/2}\right)$, $\text{Res} > 0$
$\sin(a\sqrt{t})$	$1/2 a\pi^{1/2}s^{-3/2}e^{-a^2/4s}$, $\text{Res} > 0$
$L_n(t)$	$s^{-n-1}(s - 1)^n$, $\text{Res} > 0$

TABLE C.2 Fourier Exponential Transforms

$f(x)$, $-\infty < x < \infty$	$\mathcal{F}_e\{f\} \equiv F_e(\lambda)^* = \displaystyle\int_{-\infty}^{\infty} e^{-i\lambda x}f(x)\,dx$, $\quad -\infty < x < \infty$
$\displaystyle\lim_{L\to\infty}\frac{1}{2\pi}\int_{-L}^{L}F_e(\lambda)e^{i\lambda x}\,d\lambda$	$F(\lambda)$
$\dfrac{d^m f}{dx^m}$	$(i\lambda)^m F_e(\lambda)$, $\quad m = 1,2,\ldots$
$xf(x)$	$i\dfrac{dF_e}{d\lambda}$
$f(x + c)$	$e^{ic\lambda}F_e(\lambda)$, $\quad c$ real
$e^{icx}f(x)$	$F_e(\lambda - c)$, $\quad c$ real
$\dfrac{1}{|c|}f\left(\dfrac{x}{c}\right)$	$F_e(c\lambda)$, $\quad c$ real, $c \neq 0$

$\begin{cases} f(-x) \\ f(x) \end{cases}$	$\begin{aligned} &F_e(-\lambda) \\ &F_e(-\lambda) \end{aligned}$
$\displaystyle\int_{-\infty}^{\infty} f(t)g(x-t)\,dt$	$F_e(\lambda)G_e(\lambda)$
$e^{-cx}H(x)^{**}$	$\dfrac{1}{c+i\lambda}, \quad c > 0$
$e^{cx}H(-x)^{**}$	$\dfrac{1}{c-i\lambda}, \quad c > 0$
$xe^{-cx}H(x)^{**}$	$\dfrac{1}{(c+i\lambda)^2}, \quad c > 0$
1	$2\pi(\delta(\lambda))$
$p_a(x) = \begin{cases} 1, & |x| < a \\ \tfrac{1}{2}, & x = a \\ 0, & |x| > a \end{cases}$	$\dfrac{2\sin\lambda a}{\lambda}$

$$\left(\frac{\sin ax}{x}\right)^{2m}, \quad m = 1, 2, 3, \ldots$$

$$2(-1)^m 2^{-2m} m\pi \left\{ (m!)^{-2}\lambda^{2m-1} + \right.$$
$$\left. + \sum_{n=1}^{m} \frac{(-1)^n[(2an+\lambda)^{2m-1}+(|2an-\lambda|)^{2m-1}]}{(m+n)!(m-n)!} \right\}$$

for $\lambda \le 2am$,

0 for $\lambda \ge 2am$

$$\left(\frac{\sin ax}{x}\right)^{2m+1}, \quad m = 0, 1, 2, \ldots$$

$$(-1)^m \pi 2^{-2m-1}(2m+1)F(a),$$

where

$$F(a) = \sum_{n=0}^{m}(-1)^n \frac{[(2n+1)a+\lambda]^{2m}+[(2n+1)a-\lambda]^{2m}}{(m+1+n)!(m-n)!}$$

for $0 \le \cdot \le a,$

$$F(a) = \sum_{n=0}^{k-1}(-1)^n \frac{[\lambda+(2n+1)a]^{2m}-[\lambda-(2n+1)a]^{2m}}{(m+1+n)!(m-n)!} +$$
$$+ \sum_{n=k}^{m}(-1)^n \frac{[(2n+1)a+\lambda]^{2m}+[(2n+1)a-\lambda]^{2m}}{(m+1+n)!(m-n)!}$$

for $(2k-1)a \le \lambda \le (2k+1)a$, $k = 1, 2, 3, \ldots, m$

$F(a) = 0$ for $\ge (2m+1)a$

$\begin{cases} -1, & -a < x < 0 \\ 1, & 0 < x < a \\ 0, & \|x\| > a \end{cases}$	$2\dfrac{1-\cos\lambda a}{i\lambda}, \quad a > 0$
x^{-1}	$-\pi i \, \mathrm{sgn}\,\lambda$
$(a^2 + x^2)^{-1}, \quad \mathrm{Re}(a) > 0$	$\pi a^{-1}e^{-a\|\lambda\|}$
$x(a^2 + x^2)^{-1}, \quad \mathrm{Re}(a) > 0$	$\dfrac{-\pi i}{2}a^{-1}\lambda e^{-a\|\lambda\|}$
$P_n(x)H(1 - \|x\|)$**	$(-i)^n\lambda^{-1/2}J_{n+1/2}(\lambda)$
$e^{i\alpha x} \quad \alpha \text{ real}$	$2\pi\delta(-\lambda + \alpha)$
$e^{-a\|x\|}$	$\sqrt{(2/\pi)a(a^2 + \lambda^2)^{-1}}$
$xe^{-a\|x\|}$	$-4ai\lambda(a^2 + \lambda^2)^{-2}$
$\|x\|e^{-a\|x\|}$	$2(a^2 - \lambda^2)(a^2 + \lambda^2)^{-2}$
$e^{-a^2x^2}$	$\sqrt{\pi}a^{-1}e^{-\lambda^2/4a^2}$
$xe^{-a^2x^2}$	$-i\sqrt{\pi}\lambda e^{-\lambda^2/4a^2}, \quad a > 0$
$x^{-1/2}J_{n+1/2}(x)$	$(-i)^nP_n(\lambda)H(1 - \|\lambda\|)$
$\delta(x)$	1
$\delta(bx + \mu), \quad b \neq 0$	$\|\lambda\|^{-1}e^{-i\mu\lambda/b}$
$\delta^{(k)}(x)$	$(i\lambda)^k$

x^k	$2\pi i^{-k}\delta^{(k)}(\lambda)$
$|x|^\alpha$, $\quad \alpha < -1$ but not an integer	$2\Gamma(\alpha+1)^{-\alpha-1}\times\cos\{\tfrac{1}{2}\pi(\alpha+1)\}$
$|x|^\alpha\operatorname{sgn}x$, $\quad \alpha < -1$ but not an integer	$-2i\operatorname{sgn}\lambda\sqrt{(2/\pi)}\Gamma(\alpha+1)|\lambda|^{-\alpha-1}\times\sin\{\tfrac{1}{2}\pi(\alpha+1)\}$
$x^k\operatorname{sgn}x$	$2(i\lambda)^{-k-1}k!$
$x^kH(x)^{**}$	$\sqrt{(2\pi)}i^k\left[\,\tfrac{1}{2}\delta^{(k)}(\lambda)+\dfrac{(-1)^kk!}{2\pi i\lambda^{k+1}}\,\right]$
$\begin{cases}\cos^2ax, & \|x\| < \dfrac{\pi}{2a}\\[2mm] 0, & \|x\| > \dfrac{\pi}{2a}\end{cases}$	$\dfrac{4a^2}{\lambda(4a^2-\lambda^2)}\sin\dfrac{\pi\lambda}{2a},\quad a>0$

*Note that in the text and Sec. 2.2 we often used $F(\lambda)$ for $F_e(\lambda)$.

**$H(x) = u_0(x) \equiv \begin{cases} 1, & x>0 \\ 0, & x<0 \end{cases}$

TABLE C.3 Fourier Sine Transforms

$f(x),\quad x>0$	$\mathcal{F}_s\{f\} \equiv F_s(\lambda) = \displaystyle\int_0^\infty f(x)\sin\lambda x\,dx,\quad \lambda>0$
$\dfrac{2}{\pi}\displaystyle\int_0^\infty F_s(\lambda)\sin\lambda x\,d\lambda = \dfrac{2}{\pi}\mathcal{F}_s\{F_s(\lambda)\}$	$F_s(\lambda)$
$f''(x)$	$-\lambda^2F_s(\lambda) + \lambda f(0)$

$f(x)$	$\dfrac{i}{2}\mathcal{F}_e\{f_1(x)\}^*$
$\dfrac{1}{k}f\left(\dfrac{x}{k}\right)$	$F_s(\lambda k),\quad k>0$
$-x^2 f(x)$	$F_s''(\lambda)$
$\displaystyle\int_0^\infty f(r)\int_{|x-r|}^{x+r} g(t)\,dt\,dr$	$\dfrac{2}{\lambda}F_s(\lambda)G_s(\lambda)$
$f_1(x+k)+f_1(x-k)^*$	$2F_s(\lambda)\cos\lambda k$
$2f(x)\cos kx$	$F_s(\lambda+k)+F_s(\lambda-k)$
$-xf(x)$	$F_c'(\lambda)\quad (F_c\text{ in Table C.4})$
$-f'(x)$	$\lambda F_c(\lambda)$
$f_2(x-k)-f_2(x+k)^{**}$	$2F_c(\lambda)\sin\lambda k$
$2f(x)\sin kx$	$F_c(\lambda-k)-F_c(\lambda+k)$
$\displaystyle\int_0^\infty f(r)[g(|x-R|)-g(x+r)]\,dr$ $=\displaystyle\int_0^\infty g(t)[f(x+t)+f_1(x-t)]\,dt$	$2F_s(\lambda)G_c(\lambda)$
x^{-1}	$\dfrac{1}{2}\pi\,\mathrm{sgn}\,\lambda$
$x^{p-1},\quad 0<p<1$	$\lambda^{-p}\Gamma(p)\sin(\tfrac{1}{2}\pi p)$

$\begin{cases} 1, & 0 < x < a \\ 0, & x > a \end{cases}$ $\qquad \dfrac{1 - \cos a\lambda}{\lambda}, \quad a > 0$

$x^{-1/2}$ $\qquad \sqrt{\dfrac{\pi}{2}}\,\lambda^{-1/2}$

$x(a^2 + x^2)^{-1}$ $\qquad \tfrac{1}{2}\pi e^{-a\lambda}$

$\dfrac{x}{4 + x4}$ $\qquad \dfrac{\pi}{4} e^{-\lambda}\sin\lambda$

$x(a^2 + x^2)^{-2}$ $\qquad \tfrac{1}{2} a^{-1}\lambda e^{-a\lambda}, \quad \lambda > 0$

$x^{-1}(a^2 + x^2)^{-1}$ $\qquad \tfrac{1}{2}\pi a^{-2}(1 - e^{-a\lambda}), \quad \lambda > 0$

$e^{-ax}, \quad a > 0$ $\qquad \lambda(a^2 + \lambda^2)^{-1}$

xe^{-ax} $\qquad 2a\lambda(a^2 + \lambda^2)^{-2}$

$x^{n-1}e^{-ax}$ $\qquad (n-1)!\, r^{-n}\sin(n\theta),$
$\qquad r = \sqrt{(a^2 + \lambda^2)}, \quad 0 = \tan^{-1}(\lambda/a)$

$x(1 + ax)e^{-xa}, \quad a > 0$ $\qquad \dfrac{8a^3\lambda}{(a^2 + \lambda^2)^3}$

$e^{-x/\sqrt{2}}\sin\dfrac{x}{\sqrt{2}}$ $\qquad \dfrac{\lambda}{1 + \lambda4}$

$\log\dfrac{x + a}{|x - a|}$ $\qquad \pi\dfrac{\sin\lambda a}{\lambda}, \quad a > 0$

$xe^{-a^2x^2}$	$\dfrac{\sqrt{\pi}}{4}a^{-3}\lambda e^{-(1/4)\lambda^2/a^2}$
$x^{-1}e^{-ax}$	$\tan^{-1}(\lambda/a)$
$x(a^2-x^2)^{\nu-3/2}H(a-x),{}^{***}\quad \nu>\tfrac12$	$2^{\nu-3/2}a^\nu\lambda^{1-\nu}\Gamma(\nu-\tfrac12)J_\nu(\lambda a)$
$\dfrac{\cosh(\alpha x)}{\sinh(\beta x)},\quad 0<\alpha<\beta$	$\tfrac12\pi\beta^{-1}\dfrac{\sinh(\pi\lambda/\beta)}{\cosh(\pi\lambda/\beta)+\cos(\pi\alpha/\beta)}$
$\operatorname{erf}\dfrac{x}{2\sqrt{a}}$	$\dfrac{1-e^{-a\lambda^2}}{\lambda},\quad a>0$
$\arctan\dfrac{a}{x}$	$\dfrac{\pi}{2}\dfrac{1-e^{-a\lambda}}{\lambda},\quad a>0$
$x^{1-\nu}J_\nu(ax),\quad \nu>\tfrac12$	$\sqrt{\dfrac{\pi}{2}}\dfrac{\lambda(a^2-\lambda^2)^{\nu-3/2}H(\nu-\lambda)}{2^{\nu-3/2}a^\nu\Gamma(\nu-\tfrac12)}$
$x^{-1}J_0(ax)$	$\begin{cases}\tfrac12\pi, & 0<a<\lambda\\[4pt] \sin^{-1}(\lambda/a), & 0<\lambda<a\end{cases}$
$x(b^2+x^2)^{-1}J_0(ax)$	$\tfrac12\pi e^{-b\lambda}I_0(ab),\quad \lambda>a$

---

*$f_1(x)$ is the odd extension of $f(x)$.
**$f_2(x)$ is the even extension of $f(x)$.
***$H(x)=u_0(x)$, see footnote on p. 773.

TABLE C.4 Fourier Cosine Transforms

$f(x), \quad x > 0$	$\mathcal{F}_c\{f\} \equiv F_c(\lambda) = \displaystyle\int_0^\infty f(x)\cos\lambda x\,dx, \quad \lambda > 0$
$\dfrac{2}{\pi}\displaystyle\int_0^\infty F_c(\lambda)\cos\lambda x\,d\lambda = \dfrac{2}{\pi}F_c\{F_c(\lambda)\}$	$F_c(\lambda)$
$f''(x)$	$-\lambda^2 F_c(\lambda) - f'(0)$
$f(x)$	$\tfrac{1}{2}\mathcal{F}_e\{f(|x|)\}$
$\dfrac{1}{k}f\left(\dfrac{x}{k}\right)$	$F_c(\lambda k), \quad k > 0$
$-x^2 f(x)$	$F_c''(\lambda)$
$\displaystyle\int_0^\infty f(r)[g(x+r) + g(|x-r|)]\,dr$	$2F_c(\lambda)G_c(\lambda)$
$f(|x-k|) + f(x+k)$	$2F_c(\lambda)\cos\lambda k, \quad k > 0$
$2f(x)\cos kx$	$F_c(\lambda+k) + F_c(\lambda-k)$
$xf(x)$	$F_s'(\lambda) \quad (F_s \text{ in Table C.3})$
$\displaystyle\int_x^\infty f(r)\,dr$	$\dfrac{1}{\lambda}F_s(\lambda)$
$f(x+k) - f_1(x-k)$	$2F_s(\lambda)\sin\lambda k, \quad k > 0$
$2f(x)\sin kx$	$F_s(\lambda+k) - F_s(\lambda-k)$
$\displaystyle\int_0^\infty f(r)[g(x+r) - g_1(x-r)]\,dr$	$2F_s(\lambda)G_s(\lambda)$

$$x^{p-1}, \quad 0 < p < 1 \qquad\qquad \lambda^{-p}\Gamma(p)\cos\left(\tfrac{1}{2}p\pi\right)$$

$$\begin{cases} 1, & o < x < a \\ 0, & x > a \end{cases} \qquad\qquad \frac{\sin \lambda a}{\lambda}, \quad a > 0$$

$$(a^2 + x^2)^{-1} \quad \text{Re}(a) > 0 \qquad\qquad \tfrac{1}{2}\pi a^{-1} e^{-a\lambda}$$

$$(a^2 + x^2)^{-2}, \quad \text{Re}(a) > 0 \qquad\qquad \tfrac{1}{4}\pi a^{-3} e^{-a\lambda}(1 + a\lambda)$$

$$\frac{1 - x^2}{(1 + x^2)^2} \qquad\qquad \frac{\pi}{2}\lambda e^{-\lambda}$$

$$e^{-ax}, \quad a > 0 \qquad\qquad a(a^2 + \lambda^2)^{-1}$$

$$xe^{-ax} \qquad\qquad (a^2 - \lambda^2)(a^2 + \lambda^2)^{-2}$$

$$x^{n-1}e^{-ax} \qquad\qquad (n-1)! \, r^{-n} \cos(n\theta), \quad r = \sqrt{(a^2 + \lambda^2)}, \quad \theta = \tan^{-1}(\lambda/a)$$

$$\frac{1}{\sqrt{x}} \qquad\qquad \sqrt{\frac{\pi}{2}} \frac{1}{\sqrt{\lambda}}$$

$$(1 + x)e^{-x} \qquad\qquad \frac{2}{(1 + \lambda^2)^2}$$

$$\frac{2}{x} e^{-x} \sin x \qquad\qquad \arctan \frac{2}{\lambda^2}$$

$$e^{-x/\sqrt{2}} \sin\left(\frac{\pi}{4} + \frac{x}{\sqrt{2}}\right) \qquad\qquad \frac{1}{1 + \lambda 4}$$

$$e^{-x/\sqrt{2}} \cos\left(\frac{\pi}{4} + \frac{x}{\sqrt{2}}\right) \qquad\qquad \frac{\lambda^2}{1 + \lambda 4}$$

$e^{-a^2x^2}$, $a > 0$ $\qquad$ $\dfrac{\sqrt{\pi}}{2}a^{-1}e^{-(1/4)\lambda^2/a^2}$

$x^{-1}(e^{-bx} - e^{-ax})$, $0 < b < a$ $\qquad$ $\dfrac{1}{2}\log\dfrac{a^2 + \lambda^2}{b^2 + \lambda^2}$

$\cos(\tfrac{1}{2}x^2)$ $\qquad$ $\dfrac{\sqrt{\pi}}{2}\{\cos(\tfrac{1}{2}\lambda^2) + \sin(\tfrac{1}{2}\lambda^2)\}$

$\sin(\tfrac{1}{2}x^2)$ $\qquad$ $\dfrac{\sqrt{\pi}}{2}\{\cos(\tfrac{1}{2}\lambda^2) - \sin(\tfrac{1}{2}\lambda^2)\}$

$(a^2 - x^2)^{\nu - 1/2}H(a-x)$,* $\nu > -\tfrac{1}{2}$ $\qquad$ $\sqrt{\pi}\,2^{\nu-1}\Gamma(\nu + \tfrac{1}{2})a^{\nu}\lambda^{-\nu}J_\nu(a\lambda)$

$\dfrac{\sinh(\alpha x)}{\sinh(\beta x)}$, $0 < \alpha < \beta$ $\qquad$ $\tfrac{1}{2}\pi\beta^{-1}\dfrac{\sin(\pi\alpha/\beta)}{\cosh(\pi\lambda/\beta) + \cos(\pi\alpha/\beta)}$

$(b^2 + x^2)^{-1}J_0(ax)$ $\qquad$ $\tfrac{1}{2}\pi b^{-1}e^{-b\lambda}I_0(ab)$, $\lambda > a$

$x^{-\nu}J_\nu(ax)$, $\nu > -\tfrac{1}{2}$ $\qquad$ $\sqrt{\dfrac{\pi}{\sqrt{2}}}\,\dfrac{(a^2 - \lambda^2)^{\nu-1/2}H(a - \lambda)}{2^{\nu-1/2}a^\nu\Gamma(\nu + \tfrac{1}{2})}$

$K_0(ax)$ $\qquad$ $\tfrac{1}{2}\pi(a^2 + \lambda^2)^{1/2}$

---

*$H(x) = u_0(x)$, see footnote on p. 773.

TABLE C.5 Hankel Transforms

$f(r)$, $\quad 0 < r < \infty$	$\mathcal{H}_n\{f\} \equiv F_n(\lambda) = \displaystyle\int_0^\infty r J_n(\lambda r) f(r)\, dr$
$F_n(r)$	$f(\lambda)$
$\dfrac{d^2 f}{dr^2} + \dfrac{1}{r}\dfrac{df}{dr} - \dfrac{n^2}{r^2}f$	$-\lambda^2 F_n(\lambda)$
$f(ar)$	$\dfrac{1}{a^2} F_n\left(\dfrac{1}{a}\lambda\right)$
$\dfrac{1}{r}f(r)$	$\dfrac{\lambda}{2n}\left[F_{n-1}(\lambda) + F_{n+1}(\lambda)\right], \quad n \neq 0$
$r^{n-1}\dfrac{d}{dr}\left(r^{1-n}f(r)\right)$	$-\lambda F_{n-1}(\lambda)$
$r^{-n-1}\dfrac{d}{dr}\left(r^{n+1}f(r)\right)$	$\lambda F_{n+1}(\lambda)$
$r^\nu H(a-r)$	$\lambda^{-1}a^{\nu+1}J_{\nu+1}(a\lambda)$
r^{s-1}, $\quad -\nu-1 < \text{Re}\, s < \nu+1$	$\dfrac{2^s\Gamma\left(\frac{1}{2}s + \frac{1}{2}\nu + \frac{1}{2}\right)}{\lambda^{s+1}\Gamma\left(\frac{1}{2}\nu - \frac{1}{2}s + \frac{1}{2}\right)}$
$r^\nu(a^2-r^2)^{\mu-\nu-1}H(a-r)$,* $\quad \mu > \nu > 0$	$2^{\mu-\nu-1}\Gamma(\mu-\nu)a^\mu\lambda^{\nu-\mu}J_\mu(\lambda a)$
$r^{\nu-1}e^{-ar}$	$(2\lambda)^\nu\pi^{-1/2}\Gamma\left(\nu+\frac{1}{2}\right)(a^2+\lambda^2)^{-\nu-1/2}$
$r^\nu e^{-ar}$	$2^{\nu+1}\lambda^\nu a\pi^{-1/2}\Gamma\left(\nu+\frac{3}{2}\right)(a^2+\lambda^2)^{-\nu-3/2}$

$r^{-1}e^{-ar}$

$$\lambda^\nu(a^2+\lambda^2)^{-1/2}\left\{a+\sqrt{(a^2+\lambda^2)}\right\}^{-\nu}$$

e^{-ar}

$$\lambda^\nu(a^2+\lambda^2)^{-3/2}\left\{a+\sqrt{(a^2+\lambda^2)}\right\}^{-\nu}$$
$$\times\left\{a+\nu\sqrt{(a^2+\lambda^2)}\right\}$$

$r^\nu e^{-r^2/a^2}$

$$\left(\tfrac{1}{2}a^2\right)^{\nu+1}\lambda^\nu e^{-(1/4)\lambda^2 a^2}$$

$r^\mu e^{-r^2/a^2}$

$$\frac{\lambda^\nu a^{\mu+\nu+2}}{2^{\nu+1}\Gamma(1+\nu)}\times\Gamma\left(1+\tfrac{1}{2}\mu+\tfrac{1}{2}\nu\right)$$
$$\times {}_1F_1\left(1+\tfrac{1}{2}\mu+\tfrac{1}{2}\nu;\ \nu+1;\ -\tfrac{1}{4}a^2\lambda^2\right)^{**}$$

$r^{-1}e^{-(1/4)a^2r^2}$

$$\sqrt{\frac{\pi}{a}}\,e^{-\lambda^2/2a}I_{n/2}\left(\frac{\lambda^2}{2a}\right)$$

$$f(r)=\begin{cases}J_n(ar),&0<r<b\\[4pt]0,&r>b\end{cases}$$

$$\frac{b\lambda J_{n+1}(b\lambda)J_n(ba)-baJ_n(b\lambda)J_{n+1}(ba)}{\lambda^2-a^2}$$

$^*H(x)=u_0(x)$, see footnote on p. 773.

$^{**}{}_1F_1(a;b;x)$ is the confluent hypergeometric function; see Appendix A.5.

TABLE C.6   Mellin Transforms

$f(x),\quad x > 0$	$\mathcal{M}\{f\} = F(s) = \int_0^\infty x^{s-1}f(x)\,dx$
$\dfrac{1}{2\pi i}\displaystyle\int_{\gamma-i\infty}^{\gamma+i\infty} F(s)x^{-s}\,ds$	$F(s)$
$f(ax),\quad a > 0$	$\dfrac{1}{a^s}F(s)$
$x^a f(x)$	$F(s+a)$
$f\left(\dfrac{1}{x}\right)$	$F(-s)$
$\dfrac{df}{dx}$	$-(s-1)F(s-1)$
$\dfrac{d^n f}{dx^n}$	$(-1)^n(s-n)F(s-n)$
$x^n\dfrac{d^n f}{dx^n}$	$(-1)^n s(s+1)\cdots(s+n-1)F(s)$
$\displaystyle\int_0^\infty f_1(\xi)f_2\left(\dfrac{x}{\xi}\right)\dfrac{d\xi}{\xi}$	$F_1(s)F_2(s)$
$x^a\displaystyle\int_0^\infty \xi^\beta f_1(x\xi)f_2(\xi)\,d\xi$	$F_1(s+a)F_2(1-s-a+\beta)$
$(1-x)^{-1}$	$\pi\cot(\pi s),\quad 0 < \operatorname{Re}s < 1$
$(1+x)^{-1}$	$\pi\operatorname{cosec}(\pi s),\quad 0 < \operatorname{Re}s < 1$
$(1+x^a)^{-b}$	$\dfrac{\Gamma(s/a)\Gamma(b-s/a)}{a\Gamma(b)},\quad 0 < \operatorname{Re}s < ab$
e^{-x}	$\Gamma(s),\quad \operatorname{Re}s > 0$
e^{-x^2}	$\tfrac{1}{2}\Gamma(\tfrac{1}{2}s),\quad \operatorname{Re}s > 0$
$\cos x$	$\Gamma(s)\cos(\tfrac{1}{2}\pi s),\quad 0 < \operatorname{Re}s < 1$
$\sin x$	$\Gamma(s)\sin(\tfrac{1}{2}\pi s),\quad 0 < \operatorname{Re}s < 1$
$x^{-\nu}J_\nu(x),\quad \nu > -\tfrac{1}{2}$	$\dfrac{2^{s-\nu-1}\Gamma(\tfrac{1}{2}s)}{\Gamma(\nu-\tfrac{1}{2}s+1)},\quad 0 < \operatorname{Re}s < 1$

TABLE C.7   Hilbert Transforms

$f(x)$	$\mathcal{H}\{f\} \equiv F(\lambda) = \dfrac{1}{\pi} P \displaystyle\int_{-\infty}^{\infty} \dfrac{f(x)}{x - \lambda}\, dx^{*}$
$\mathcal{F}\{\mathcal{H}\{f\}\} = -i \operatorname{sgn} \mathcal{F}\{f\}$	
$F(x)$	$-f(\lambda)$
$f(a + x), \quad a$ is a real number	$F(a + \lambda)$
$f(at), \quad a > 0$	$F(a\lambda)$
$f(-at), \quad a > 0$	$-F(-a\lambda)$
$xf(x)$	$\lambda F(\lambda) + \dfrac{1}{\pi} \displaystyle\int_{-a}^{\infty} f(x)\, dx$
$\dfrac{df}{dx}$	$\dfrac{dF}{d\lambda}$
1	0
$f(x) = \begin{cases} 0, & -\infty < x < a \\ 1, & a < x < b \\ 0, & b < x < \infty \end{cases}$	$\dfrac{1}{\pi} \log\left| \dfrac{b - \lambda}{a - \lambda} \right|$
$\dfrac{1}{x^2 + a^2}, \quad \operatorname{Re} a > 0$	$-\dfrac{\lambda}{a(\lambda^2 + a^2)}$
$\dfrac{x}{x^2 + a^2}, \quad \operatorname{Re} a > 0$	$\dfrac{a}{\lambda^2 + a^2}$
$\sin ax, \quad a > 0$	$\cos a\lambda$
$\dfrac{\sin ax}{x}, \quad a > 0$	$\dfrac{\cos a\lambda - 1}{\lambda}$
$\sin ax J_1(ax), \quad a > 0$	$\cos a\lambda J_1(a\lambda)$
$\cos ax J_1(ax), \quad a > 0$	$-\sin a\lambda J_1(a\lambda)$

*P stands for Cauchy principal value, see (1.4) on p. 8.

TABLE C.8 Finite Exponential Transforms

$f(x), \quad -\pi < x < \pi$	$\mathfrak{f}\{f\} = F(n) = \displaystyle\int_{-\pi}^{\pi} e^{-inx} f(x)\,dx$
$\dfrac{1}{2\pi} \displaystyle\sum_{n=-\infty}^{\infty} F(n)e^{inx}, \quad -\pi < x < \pi$	$F(n)$
$\dfrac{df}{dx}$	$(-1)^n \{f(\pi) - f(-\pi)\} + inF(n)$

TABLE C.9 Finite Sine Transforms

$f(x), \quad 0 < x < \pi$	$f_s\{f\} = F_s(n) = \int_0^\pi \sin nx\, f(x)\,dx, \quad n = 1,2,\ldots$
$\dfrac{2}{\pi}\displaystyle\sum_{n=1}^\infty F_s(n)\sin nx$	$F_s(n)$
$\dfrac{d^2 f}{dx^2}$	$-n^2 F_s(n) + n[f(0) - (-1)^n f(\pi)]$
$\dfrac{x}{\pi}\displaystyle\int_0^\pi (\pi - r)f(r)\,dr - \int_0^x (x - r)f(r)\,dr$	$\dfrac{1}{n^2}F_s(n)$
$f(\pi - x)$	$(-1)^{n+1}F_s(n)$
$f_1(x + c) + f_1(x - c)^*$	$2F_s(n)\cos nc$
$\displaystyle\int_{-\pi}^\pi f_1(x - r)h_2(r)\,dr^{**}$	$2F_s(n)H_c(n)$
$\pi - x$	$\dfrac{\pi}{n}$
$\dfrac{x}{\pi}$	$\dfrac{1}{n}(-1)^{n+1}$
1	$\dfrac{1}{n}[1 - (-1)^n]$
$-x \quad$ if $x < c$ $\pi - x \quad$ if $x > c$	$\dfrac{\pi}{2}\cos nc, \quad 0 < c < \pi$

$(\pi - c)x \quad$ if $x \le c$

$(\pi - x)c \quad$ if $x \ge c \qquad\qquad \dfrac{\pi}{n^2}\sin nc, \quad 0 < c < \pi$

$x(\pi - x)(2\pi - x) \qquad\qquad \dfrac{6\pi}{n^3}$

$x(\pi - x) \qquad\qquad \dfrac{2}{n^3}[1-(-1)^n]$

$x^2 \qquad\qquad \dfrac{\pi^2}{n}(-1)^{n+1} - \dfrac{2}{n^3}[1-(-1)^n]$

$x^3 \qquad\qquad \pi(-1)^n\left(\dfrac{6}{n^3} - \dfrac{\pi^2}{n}\right)$

$e^{\alpha} \qquad\qquad \dfrac{n}{n^2+c^2}[1-(-1)^n e^{c\pi}]$

$\dfrac{\sinh c(\pi - x)}{\sinh c\pi} \qquad\qquad \dfrac{n}{n^2+c^2}, \quad c \ne 0$

$\cosh cx \qquad\qquad \dfrac{n}{n^2+c^2}[1-(-1)^n \cosh c\pi]$

$\dfrac{\sin k(\pi - x)}{\sin k\pi} \qquad\qquad \dfrac{n}{n^2-k^2}, \quad k \ne 0, +1, +2,\dots$

$\sin mx, \quad m = 1,2,\dots \qquad\qquad 0$ if $n \ne m$; $F_s(m) = \dfrac{\pi}{2}$

$\cos kx, \quad k \ne \pm1, \pm2,\dots \qquad\qquad \dfrac{n}{n^2-k^2}[1-(-1)^n \cos k\pi]$

$\cos mx, \quad m = 1,2,\dots \qquad\qquad n\dfrac{1-(-1)^{m+n}}{n^2-m^2}$ if $n \ne m$; $F_s(m) = 0$

$$\frac{\partial}{\partial k}\left[\frac{\sin k(\pi - x)}{\sin k\pi}\right] \qquad \frac{2kn}{(n^2-k^2)^2}, \quad k \neq 0, \pm 1, \pm 2, \ldots$$

$$\frac{\partial}{\partial c}\left[\frac{\sinh c(x-\pi)}{\sinh c\pi}\right] \qquad \frac{2cn}{(n^2+c^2)^2}, \quad c \neq 0$$

$$\sin cx \sinh c(2\pi - x) - \sin c(2\pi - x)\sinh cx \qquad \frac{4c^2(\sin^2 c\pi + \sinh^2 c\pi)n}{n^4 + 4c^4}, \quad c \neq 0$$

$$\frac{\sinh kx}{\sinh k\pi} - \frac{\sin kx}{\sin k\pi} \qquad \frac{2k^2 n(-1)^n}{n^4 - k^4}, \quad k \neq 0, \pm 1, \pm 2, \ldots$$

$$\frac{\sinh cx}{\sinh c\pi} - \frac{x}{\pi} \qquad \frac{c^2(-1)^n}{n(n^2+c^2)}, \quad c \neq 0$$

$$\frac{x}{\pi} - \frac{\sin kx}{\sin k\pi} \qquad \frac{k^2(-1)^n}{n(n^2-k^2)}, \quad k \neq 0, \pm 1, \pm 2, \ldots$$

$$\frac{2}{\pi}\frac{b\sin x}{1+b^2 - 2b\cos x} \qquad b^n, \quad -1 < b < 1$$

$$\frac{1}{\pi}\frac{\sin x}{\cosh y - \cos x} \qquad e^{-ny}, \quad y > 0$$

$$\frac{2}{\pi}\arctan\frac{b\sin x}{1 - b\cos x} \qquad \frac{b^n}{n}, \quad -1 < b < 1$$

$$\frac{2}{\pi}\arctan\frac{\sin x}{e^y - \cos x} \qquad \frac{1}{n}e^{-ny}, \quad y > 0$$

$$\frac{2}{\pi}\arctan\frac{2b\sin x}{1 - b^2} \qquad \frac{1-(-1)^n}{n}b^n, \quad -1 < b < 1$$

$$\frac{2}{\pi}\arctan\frac{\sin x}{\sinh y} \qquad \frac{1-(-1)^n}{n}e^{-ny}, \quad y>0$$

*$f_1(x)$ is the odd extension of $f(x)$.
**$f_2(x)$ is the even extenstion of $f(x)$.

TABLE C.10 Finite Cosine Transforms

$f(x)$, $\quad 0<x<\pi$	$f_c\{f\}\equiv F_c(n)=\displaystyle\int_0^\pi \cos nx\, f(x)\,dx, \quad n=0,1,2,\ldots$
$\dfrac{F_c(0)}{\pi}+\dfrac{2}{\pi}\displaystyle\sum_{n=1}^{\infty}F_c(n)\cos nx$	$F_c(n)$
$\dfrac{d^2f}{dx^2}$	$-n^2F_c(n)-f'(0)+(-1)^nf'(\pi)$
$\displaystyle\int_0^\pi(x-r)f(r)dr+\dfrac{F_c(0)}{2\pi}(x-\pi)^2+A$	$\dfrac{1}{n^2}F_c(n)$ if $n=1,2,\ldots$
$f(\pi-x)$	$(-1)^nF_c(n)$
$f_2(x+c)+f_2(x-c)^*$	$2F_c(n)\cos nc$
$f_1(x+c)-f_1(x-c)^{**}$	$2F_s(n)\sin nc$
$-\displaystyle\int_0^x f(r)dr+A$	$\dfrac{1}{n}F_s(n)$ if $n=1,2,\ldots$

$\displaystyle\int_{-\pi}^{\pi} f_2(x-r)h_2(r)\,dr$	$2F_c(n)H_c(n)$
$f(x) + A$	$\begin{cases} F_c(n) & \text{if } n \neq 0 \\ F_c(0) + A\pi & \text{if } n = 0 \end{cases}$
1	$\begin{cases} 0 & \text{if } n = 1,2,\ldots \\ F_c(0) = \pi \end{cases}$
$-\dfrac{2}{\pi}\log\left(2\sin\dfrac{x}{2}\right)$	$\begin{cases} \dfrac{1}{n} & \text{if } n = 1,2,\ldots \\ F_c(0) = 0 \end{cases}$
$\begin{cases} 1 & \text{if } 0 < x < c \\ -1 & \text{if } c < x < \pi \end{cases}$	$\begin{cases} \dfrac{2}{n}\sin nc & \text{if } n \neq 0 \\ F_c(0) = 2c - \pi \end{cases}$
x	$\begin{cases} \dfrac{(-1)^n - 1}{n^2} & \text{if } n \neq 0 \\ F_c(0) = \dfrac{\pi^2}{2} \end{cases}$
x^2	$\begin{cases} \dfrac{2\pi}{n^2}(-1)^n & \text{if } n \neq 0 \\ F_c(0) = \dfrac{\pi^3}{3} \end{cases}$

$$\frac{(\pi-x)^2}{2\pi} - \frac{\pi}{6} \qquad\qquad \begin{cases} \dfrac{1}{n^2} & \text{if } n \neq 0 \\[2mm] F_c(0) = 0 \end{cases}$$

$$x^4 - 2\pi^2 x^2 + \frac{7}{15}\pi^4 \qquad\qquad \begin{cases} -\dfrac{24\pi}{n^4}(-1)^n & \text{if } n \neq 0 \\[2mm] F_c(0) = 0 \end{cases}$$

$$\frac{\cosh c(\pi-x)}{c\sinh c\pi}, \quad c \neq 0 \qquad\qquad \frac{1}{n^2+c^2}, \quad c \neq 0$$

$$\frac{1}{c}e^{cx}, \quad c \neq 0 \qquad\qquad \frac{(-1)^n e^{c\pi}-1}{n^2+c^2}, \quad c \neq 0$$

$$-\frac{\cos k(\pi-x)}{k\sin k\pi} \qquad\qquad \frac{1}{n^2-k^2}, \quad k \neq 0, \pm1, \pm2, \ldots$$

$$\frac{\sin kx}{k} \qquad\qquad \frac{(-1)^n \cos k\pi - 1}{n^2 - k^2}, \quad k \neq 0, \pm1, \pm2, \ldots$$

$$\frac{1}{\pi}x\sin x \qquad\qquad \begin{cases} \dfrac{(-1)^{n+1}}{n^2-1} & \text{if } n \neq 1 \\[2mm] F_c(1) = -\frac{1}{4} & \text{if } n = 1 \end{cases}$$

$$\frac{\sin mx}{m}, \quad m = 1,2,\ldots \qquad\qquad \begin{cases} \dfrac{(-1)^{m+n}-1}{n^2-m^2} & \text{if } n \neq m \\[2mm] F_c(m) & \text{if } n = m \end{cases}$$

$$\cos mx, \quad m = 1, 2, \ldots \qquad \begin{cases} 0 & \text{if } n \neq m \\[4pt] F_c(m) = \dfrac{\pi}{2} \end{cases}$$

$$\frac{\partial}{\partial c}\left(\frac{\cosh cx}{c \sinh c\pi}\right) \qquad \frac{2c(-1)^{n+1}}{(n^2+c^2)^2}, \quad c \neq 0$$

$$\frac{1}{\pi}\,\frac{1-b^2}{1+b^2-2b\cos x} \qquad b^n, \quad -1 < b < 1,\ b \neq 0$$

$$\frac{1}{\pi}\,\frac{\sinh y}{\cosh y - \cos x} \qquad e^{-ny}, \quad y > 0$$

$$-\frac{1}{\pi}\log(1+b^2-2b\cos x) \qquad \begin{cases} \dfrac{b^n}{n} & \text{if } n \neq 0 \\[4pt] F_c(0) = 0, & -1 < b < 1 \end{cases}$$

$$y - \log(2\cosh y - 2\cos x) \qquad \begin{cases} \dfrac{\pi}{n}e^{-ny} & \text{if } n \neq 0,\ y > 0 \\[4pt] F_c(0) = 0 \end{cases}$$

$$\frac{1}{\pi}\,\frac{\partial}{\partial y}\left(\frac{\sinh y}{\cos x - \cosh y}\right) \qquad ne^{-ny}, \quad y > 0$$

$$\exp(b\cos x)\cos(b\sin x) \qquad \begin{cases} \dfrac{\pi}{2}\dfrac{b^n}{n!} & \text{if } n \neq 0 \\[4pt] F_c(0) = \pi \end{cases}$$

*$f_2(x)$ is the even extension of $f(x)$.

**$f_1(x)$ is the odd extension of $f(x)$.

TABLE C.11 Finite (First) Hankel Transforms, $J_n(\lambda_k a) = 0$

$f(r)$, $0 < r < a$	$h_n\{f\} \equiv F_n(\lambda_k) = \int_0^a r J_n(\lambda_k r) f(r)\, dr$, $\quad J_n(\lambda_k a) = 0$, $k = 1, 2, \ldots$
$\dfrac{2}{a^2} \displaystyle\sum_{k=1}^{\infty} \dfrac{F(\lambda_k) J_n(\lambda_k r)}{J_{n+1}^2(a\lambda_k)}$	$F_n(\lambda_k)$
$\dfrac{1}{r}\dfrac{d}{dr}(rf(r)) - \dfrac{n^2}{r^2} f(r)$	$-\lambda_k^2 F_n(\lambda_k) - a\lambda_k J_{n+1}(a\lambda_k) f(a\lambda)$
$\dfrac{df}{dr}$	$\dfrac{n+1}{2n}\lambda_k F_{n-1}(\lambda_k) - \dfrac{n-1}{2n}\lambda_k F_{n+1}(\lambda_k)$, $\quad n \neq 0$
$\dfrac{f(r)}{r}$	$\dfrac{\lambda_k}{2n}\{F_{n-1}(\lambda_k) + F_{n+1}(\lambda_k)\}$, $\quad n \neq 0$
r^n	$a^{n+1}\dfrac{1}{\lambda_k}J_{n+1}(\lambda_k a)$
$\dfrac{J_n(mr)}{J_n(ma)}$	$a\lambda_k[\lambda_k^2 - m^2]J_{n+1}(\lambda_k a)$

TABLE C.12 Finite (Second) Hankel Transforms, $\lambda_k J_n'(\lambda_k a) + h J_n(\lambda_k a) = 0$

$$h_n\{f\} = F_n(\lambda_k) = \int_0^a r J_n(\lambda_k r) f(r)\, dr, \quad \lambda_k J_n'(\lambda_k a) + h J_n(\lambda_k a) = 0,$$

$$k = 1, 2, \ldots$$

$f(r)$	$F_n(\lambda_k)$
$\dfrac{2}{a^2} \displaystyle\sum_{k=1}^{\infty} F_n(\lambda_k) \dfrac{\lambda_k^2 J_n(\lambda_k r)}{[\lambda_k^2 + h^2 - n^2/a^2][J_n(\lambda_k a)]^2}$	
$\dfrac{1}{r}\dfrac{d}{dr}\left(r\dfrac{df}{dr}\right) - \dfrac{n^2}{r^2} f(r)$	$-\lambda_k^2 F_n(\lambda_k) - \dfrac{a\lambda_k J_n'(\lambda_k a)}{h}[f'(a) + hf(a)]$
$\dfrac{df}{dr}$	$af(a) J_n(\lambda_k a) + \dfrac{n+1}{2n}\lambda_k F_{n-1}(\lambda_k) - \dfrac{n-1}{2n}\lambda_k F_{n+1}(\lambda_k)$
$\dfrac{f(r)}{r}$	$\dfrac{\lambda_k}{2n}[F_{n-1}(\lambda_k) + F_{n+1}(\lambda_k)], \quad n > 0$
$\dfrac{1}{n+ha}\left(\dfrac{r}{a}\right)^n$	$\dfrac{1}{\lambda_k^2} J_n(\lambda_k a)$
$\dfrac{J_n(mx)}{a[hJ_n(ma) + mJ_n'(ma)]}$	$\dfrac{1}{\lambda_k^2 - m^2} J_n(\lambda_k a)$

TABLE C.13 Finite Legendre Transforms

$f(x)$	$l\{f\} \equiv F(j) = \int_{-1}^{1} f(x)P_j(x)\,dx$
$\dfrac{1}{2}\displaystyle\sum_{j=0}^{\infty}(2j+1)F(j)P_j(x)$	$F(j)$
$f(-x)$	$(-1)^j F(j)$
$\dfrac{d}{dx}\left[(1-x^2)\dfrac{df}{dx}\right]$	$-j(j+1)F(j)$
$\displaystyle\int_{-1}^{x} f(t)\,dt$	$\dfrac{1}{2j+1}[F(j-1)-F(j+1)]$
$\dfrac{df}{dx},\quad j=2k$	$f(1)-f(-1)-\displaystyle\sum_{r=0}^{k-1}(4k-4r-1)F(2k-2r-1)$
$\dfrac{df}{dx},\quad j=2k+1$	$f(1)-f(-1)-\displaystyle\sum_{r=0}^{k}(4k-4r+1)F(2k-2r)$
$xf(x)$	$\dfrac{1}{2j+1}[(j+1)F(j+1)+jF(j-1)]$
$x^2 f(x)$	$\dfrac{(j+1)(j+2)}{(2j+1)(2j+3)}F(j+2)+\dfrac{2j^2+2j-1}{(2j-1)(2j+3)}F(j)+\dfrac{j(j-1)}{(2j-1)(2j+1)}F(j-2)$

$H(x)^*$	$\dfrac{1}{2j+1}[P_{j-1}(0) - P_{j+1}(0)]$, $\quad j \geq 1$
$\dfrac{1}{1-2ax+a^2}$	$\dfrac{2a^j}{2j+1}$
$\log(1-x)$	$-\dfrac{2}{j(j+1)}$
e^{ax}	$\sqrt{\dfrac{2\pi}{a}}\, I_{j+1/2}(a)$
$\cos ax$	$\begin{cases} 0 & \text{if } j \text{ is odd} \\ (-1)^{-j}\sqrt{\dfrac{2\pi}{a}}\, J_{j+1/2}(a) & \text{if } j \text{ is even} \end{cases}$
$\sin ax$	$\begin{cases} (-1)^{-(j-1)}\sqrt{\dfrac{2\pi}{a}}\, J_{j+1/2}(a) & \text{if } j \text{ is odd} \\ 0 & \text{if } j \text{ is even} \end{cases}$

$^*H(x) = u_0(x)$, see footnote on p. 773.

TABLE C.14 Finite Tchebychev Transforms

$f(x)$	$t\{f\} \equiv F(j) = \int_{-1}^{1} (1-x^2)^{-1/2} T_j(x) f(x)\, dx$
$\dfrac{1}{\pi} F(0) + \dfrac{2}{\pi} \displaystyle\sum_{j=1}^{\infty} F(j) T_j(x)$	$F(j)$
$(1-x^2)\dfrac{d^2 f}{dx^2} - x\dfrac{df}{dx}$	$-j^2 F(j)$

TABLE C.15 Finite Laguerre Transforms

$f(x)$	$L_G\{f\} \equiv F(j) = \int_{0}^{\infty} e^{-x} L_j(x)^* f(x)\, dx$
$\displaystyle\sum_{j=0}^{\infty} F(j) L_j(x)$	$F(j)$
$x\dfrac{d^2 f}{dx^2} + (1-x)\dfrac{df}{dx}$	$-j F(j)$

*Orthonormal $L_j(x)$.

TABLE C.16 Finite Hermite Transforms

$f(x)$	$h\{f\} \equiv F(j) = \int_{-\infty}^{\infty} e^{-x^2} H_j(x)^{**} f(x)\, dx$
$\displaystyle\sum_{j=0}^{\infty} F(j) H_j(x)$	$F(j)$
$\dfrac{d^2 f}{dx^2} - 2x\dfrac{df}{dx} + 2jf$	$-2j F(j)$

**Orthonormal $H_j(x)$.

TABLE C.17 z-Transforms

$\{u_n\}_{n=0}^{\infty}$	$Z\{u_n\} = U(z) = \displaystyle\sum_{n=0}^{\infty} u_n z^{-n}$
$\{u_n\}$	$U(z)$
$\{\alpha u_n\}, \quad \alpha \in \mathbb{C}$	$\alpha U(z)$
$\{u_{1,n} + u_{2,n}\}$	$U_1(z) + U_2(z)$
$\{1\}$	$\dfrac{z}{z-1}, \quad |z| > 1$
$\{a^n\}, \quad a \in \mathbb{C}$	$\dfrac{z}{z-a}, \quad |z| > a$
$\{\cos n\theta\}$	$\dfrac{z - \cos\theta}{z - 2\cos\theta + z^{-1}}, \quad |z| > 1$
$\{\sin n\theta\}$	$\dfrac{\sin\theta}{z - 2\cos\theta + z^{-1}}, \quad |z| > 1$
$\left. \begin{cases} 0, & n = 0 \\ a^{n-1}, & n \geq 1 \end{cases} \right\}, \quad a \in \mathbb{C}$	$\dfrac{1}{z-a}, \quad |z| > |a|$
$\left. \begin{cases} 0, & n = 0,\ n \text{ odd} \\ (-1)^{(n/2)+1} a^{n-2}, & n \text{ even} \end{cases} \right\}$	$\dfrac{1}{z^2 + a^2}, \quad |z| > |a|$
$\left. \begin{cases} 0, & n \text{ even} \\ (-1)^{(n-1)/2} a^{n-1}, & n \text{ odd} \end{cases} \right\}$	$\dfrac{z}{z^2 + a^2}, \quad |z| > |a|$
$\delta n, 0 = \begin{cases} 1, & n = 0 \\ 0, & n > 0 \end{cases}$	1

BIBLIOGRAPHY

Abramowitz, M. and I. E. Stegun (1964) *Handbook of Mathematical Functions*, Applied Mathematics Series No. 55. National Bureau of Standards, Gaithersburg, MD.

Akhiezer, N. I. (1988). *Lectures on Integral Transforms*, translated by H. H. McFaden. American Mathematical Society, Providence, R.I.

Alexits, G. (1961) *Convergence Problems of Orthogonal Series*. Pergamon Press, New York.

Ames, W. F. (1964) *Nonlinear Problems in Engineering*. Academic Press, New York.

Ames, W. F. (1968) *Nonlinear Ordinary Differential Equations in Transparent Processes*. Academic Press, New York.

Andrews, L. C. and B. K. Shivamoggi (1988) *Integral Transforms for Engineers and Applied Mathematicians*. Collier Macmillan, London.

Arsac, J. (1966) *Fourier Transforms and the Theory of Distributions*. Prentice-Hall, Englewood Cliffs, N.J.

Askey, R. (1975) *Orthogonal Polynomials and Special Functions*, Reg. Conf. Lec. Appl. Math., vol. 21, SIAM, Philadelphia.

Atkinson, F. V. (1964) *Discrete and Continuous Boundary Problems*. Academic Press, New York.

Berg, L. (1967) *Introduction to Operational Calculus*. North Holland, Amsterdam.

Bergland, G. D. (1969) A guided tour of the fast Fourier transform. IEEE Spectrum, vol. 6, no. 7, pp. 41–52.

Boas, R. P. and R. C. Buck (1964) *Polynomial Expansions of Analytic Functions*. Academic Press, New York.

Bracewell, R. N. (1965) *The Fourier Transform and Its Applications*. McGraw-Hill, New York.

Bremerman, A. V. (1965) *Distributions, Complex Variables, and Fourier Transforms.* Addison-Wesley, Reading, Mass.

Briggs, W. L. and A. J. Jerri (1985) Operational Difference Calculus (monograph).

Brigham, E. O. (1974) *The Fast Fourier Transform.* Prentice-Hall, Englewood Cliffs, N.J.

Brigham, E. O. (1988) *The Fast Fourier Transform and Its Applications.* Prentice-Hall, Englewood Cliffs, N.J.

Brown, J. L. Jr. (1968a) A least upper bound for aliasing error. IEEE Trans. Autom. Control, vol. AC-13, no. 6, pp. 754–755.

Brown, J. L. Jr. (1968b) Sampling theorem for finite energy signals. IEEE Trans. Inform. Theory, vol. IT-14, pp. 818–819.

Brychkov, Yu. A., and A. P. Prudnikov (1989) *Integral Transforms of Generalized Functions.* Gordon & Breach, New York.

Budak, B. B., and S. V. Fomin (1973) *Multiple Integrals, Field Theory and Series,* translated by V. M. Vosolow. Mir, Moscow.

Burrus, C. S., and T. W. Parks (1981) *Discrete Fourier Transforms, Fast Fourier Transforms and Convolution Algorithms; Theory and Implementation.* Wiley, New York.

Butzer, P. S., and R. J. Nessel (1971) *Fourier Analysis and Approximation.* Academic Press, New York.

Campbell, L. L. (1964) A Comparison of the sampling theorems of Kramer and Whittaker. J. Soc. Ind. Appl. Math. vol. 12, pp. 117–130.

Carslaw, H. S. (1952) *Introduction of the Theory of Fourier's Series and Integrals,* 3rd ed. Dover, New York.

Churchill, R. V. (1954) The operational calculus of Legendre transforms. J. Math. Phys., vol. 33, p. 165.

Churchill, R. V. (1972) *Operational Mathematics.* McGraw-Hill, New York.

Churchill, R. V., and J. W. Brown (1963) *Fourier Series and Boundary Value Problems,* 3rd ed. McGraw-Hill, New York.

Coddington, E. A., and N. Levinson (1955) *Theory of Ordinary Differential Equations,* McGraw-Hill, New York, 1955.

Colombo, S. (1959) *The Transforms of Mellin and Hankel, Their Applications to Mathematical Physics* (in French). Centre National de la Recherche Scientifique, Paris.

Conte, S. D. (1955) Gegenbauer transforms. Quart. J. Math. Oxford (2), vol. 6, p. 48.

Cooley, J. W., and J. W. Tukey (1965) An algorithm for the machine calculations of complex Fourier series. Math. Comput., vol. 19, no. 90, pp. 297–301.

Cooley, J. W., P. A. W. Lewis, and P. D. Welch (1969). The finite Fourier transform. IEEE Trans. Audio Electroacoust., vol. AV-17, no. 2, pp. 77–85.

Craig, E. J. (1964) *Laplace and Fourier transforms for Electrical Engineers.* Holt, Rinehart & Winston, New York.

Davies, B. (1985) *Integral Transforms and Their Applications,* 2nd ed. Springer-Verlag, New York.

Debnath, L. (1960) On Laguerre transform. Bull. Calcutta Math. Soc., vol. 52, p. 69.

Debnath, L. (1964) On Hermite transform. Mat. Vesnik, vol. 1, p. 285.

Ditkin, V. A., and A. P. Prudnikow (1965) *Integral Transforms and Operational Calculus.* Pergamon, New York.

Dodd, R. K., et al. (1982) *Solitons and Nonlinear Waves.* Academic Press, New York.

Doetsch, G. (1950, 1951) *Handbook of the Laplace Transform* (in German), vols. 1 and 2. Birkhauser, Basel.

Doetsch, G. (1961) *Guide to the Applications of the Laplace and z-transforms.* Van Nostrand Reinhold, New York.

Edwards, R. E. (1967a) *Fourier Series, a Modern Introduction*, vol. 1. Holt, Rinehart & Winston, New York.

Edwards, R. E. (1967b) *Fourier Series, a Modern Introduction*, vol. 2. Holt, Rinehart & Winston, New York.

Elliot, D. F., and K. R. Rao (1982) *Fast Transforms, Analysis, Applications.* Academic Press, New York.

Erdelyi, A., et al. (1954a) *Tables of Integral Transforms*, vol. 1. McGraw-Hill, New York.

Erdelyi, A., et al. (1954b) *Tables of Integral Transforms*, vol. 2. McGraw-Hill, New York.

Erdelyi, A., et al. (1953a) *Higher Transcendental Functions*, vol. 1. McGraw-Hill, New York.

Erdelyi, A., et al. (1953b) *Higher Transcendental Functions*, vol. 2. McGraw-Hill, New York.

Erdelyi, A., et al. (1955) *Higher Transcendental Functions*, vol. 3. McGraw-Hill, New York.

Eringen, A. C. (1954) The Finite Sturm–Liouville Transform. Quart. J. Math. Oxford (2), vol. 5, p. 120.

Eves, H. (1983) *An Introduction to the History of Mathematics*, 5th ed. Saunders, New York.

Fadeeva, V. N., and M. K. Gauvrin (1950) Tables of Bessel functions, $J_n(x)$ of integer numbers from 0 to 120 (in Russian). Izv. Akad. Nauk SSSR, Moscow.

Goldberg, R. R. (1961) *Fourier Transforms.* Cambridge University Press, Cambridge, England.

Gottlieb, D., and S. A. Orszag (1977) *Numerical Analysis of Spectral Methods; Theory and Applications*, SIAM Regional Conf. Series in Appl. Math. SIAM, Philadelphia.

Greenberg, M. D. (1988) *Advanced Engineering Mathematics.* Prentice-Hall, Englewood Cliffs, N.J.

Grossman, S. I., and W. R. Derrick (1988) *Advanced Engineering Mathematics.* Harper & Row, New York.

Haimo, D. T. (1958) Integral equations associated with Hankel convolutions. Trans. Amer. Math. Soc., vol. 116, pp. 330–375.

Hamming, R. W. (1973) *Numerical Methods for Scientists and Engineers.* McGraw-Hill, New York.

Hardy, G. H., and W. Rogosinski (1968) *Fourier Series.* Cambridge University Press, New York.

Harmuth, H. W. (1972) *Transmission of Information by Orthogonal Functions*, 2nd ed. Springer-Verlag, New York.

Harmuth, H. W. (1977) *Sequencing Theory: Foundation and Applications*. Academic Press, New York.

Helms, H. D., and J. B. Thomas (1962) Truncation error of sampling theorem expansion. Proc. IRE, vol. 50, pp. 179–184, February.

Higgins, J. R. (1976) *Completeness and Basis Properties of Sets of Special Functions*. Cambridge University Press, London.

Higgins, J. R. (1985) Five short stories about the cardinal series. Bull. (New Ser.) Amer. Math. Soc., vol. 12, p. 45.

Hirschmann, I. I. (1963) Laguerre transforms. Duke Math. J., vol. 30, p. 495.

Hirschmann, I. I., and D. V. Widder (1955) *The Convolution Transform*. Princeton University Press, Princeton, N.J.

Hobson, E. W. (1957) *The Theory of Functions of a Real Variable and the Theory of Fourier Series*, vol. 1. Dover, New York.

Hobson, E. W. (1931) *The Theory of Spherical and Ellipsoidal Harmonics*. Cambridge University Press, Cambridge, England.

Hochstadt, H. (1966) *Special Functions of Mathematical Physics*. Holt, Rinehart & Winston, New York.

Hong, J. P. (1970) Fast two-dimensional Fourier transform. 3rd Hawaii International Conference on System Science, pp. 990–993.

Jagadeesan, M. (1970) n-Dimensional fast Fourier transform. Proc. IEEE, 13th Midwest Symposium on Circuit Theory, Minneapolis, vol. 8, pp. 7–8.

Jerri, A. J. (1969a) On the application of some interpolating functions in physics. J. Res. Natl. Bur. Stand. B Math. Sci., vol. 73B, no. 3, July–September.

Jerri, A. J. (1969b) On the equivalence of Kramer's and Shannon's sampling theorems. IEEE Trans. Inform. Theory, vol. IT-15, no. 4, July.

Jerri, A. J. (1972) Application of the sampling theorem to time varying systems. J. Franklin Inst., vol. 293, no. 1, pp. 53–58.

Jerri, A. J. (1973) Sampling for not necessarily finite energy signals, Int. J. System Sci., vol. 4, no. 2, pp. 255–260.

Jerri, A. J. (1976) Sampling for the L_ν^α-Laguerre transforms. J. Res. Natl. Bur. Stand. B Math. Sci., vol. 80B, pp. 415–418, September.

Jerri, A. J. (1977a) The Shannon sampling theorem—its various extensions and applications: A tutorial review. Proc. IEEE, vol. 65, no. 11, pp. 1565–1596.

Jerri, A. J. (1977b) A property and computations of the hill functions (B-splines). Math. of Computations, vol. 31, no. 138, pp. 481–484.

Jerri, A. J. (1979) The application of general discrete transforms to computing orthogonal series and solving boundary value problems. Bull. Cal. Math. Soc., vol. 71, pp. 177–187.

Jerri, A. J. (1982) A note on sampling expansion for a transform with parabolic cylinder kernel. Inform. Sci., vol. 26, pp. 155–158.

Jerri, A. J. (1983a) Application of the transform-iterative method to nonlinear concentration boundary value problems. J. Chem. Eng. Commun., vol. 23, pp. 101–113.

Jerri, A. J. (1983b) A definite integral, SIAM Rev., vol. 24, p. 77.

Jerri, A. J. (1984). A transform-iterative method for nonlinear or variable velocity waves. In *Advances in Computer Methods for Partial Differential Equations* (R. Vichnevetsky and R. S. Stepleman, eds.), pp. 313–317, IMACS.

Jerri, A. J. (1985) *Introduction to Integral Equations with Applications*. Marcel Dekker, New York.

Jerri, A. J. (1986) Elements and applications of integral equations. UMAP J., vol. 7, pp. 45–80. (Also UMAP Module 609, 1982.)

Jerri, A. J. (1988) An extended Poisson sum formula for the generalized integral transforms and aliasing error bound for the generalized sampling theorem. J. Appl. Anal., vol. 26, pp. 199–221.

Jerri, A. J., and M. L. Glasser (1980) An infinite series. SIAM Rev., vol. 22, no. 2, p. 230.

Jerri, A. J., and I. Joslin (1982) Truncation error for the generalized Bessel type sampling series. J. Franklin Inst., vol. 314, pp. 323–328.

Jerri, A. J., R. Weiland, and R. Herman (1987a) The modified iterative method for nonlinear concentration in cylindrical and spherical pellets. J. Chem. Eng. Commun., pp. 379–404.

Jerri, A. J., and K. W. Tse (1987b) A modified iterative method—applications to nonlinear waves and nonlinear chemical problems. Proc. Appl. Math. Conf. Cairo Univ., Cairo, January 3–6, pp. 329–344.

Jerri, A. J. (1991) A recent modification of iterative methods for solving nonlinear problems. In *The Mathematical Heritage of C. F. Gauss* (G. M. Rassias, ed.), an invited paper, pp. 379–404. World Scientific Publishing, Singapore.

Jerri, A. J. (1992) Error analysis for application of generalization to the sampling theorem. In *Shannon Sampling and interpolation Theory* (R. J. Marks II, ed.), vol. II, an invited Chapter VII, pp. 255–338. Springer-Verlag, New York.

Jury, E. I. (1964) *Theory and Applications of the z-Transform*. Wiley, New York.

Korner, T. W. (1988) *Fourier Analysis*. Cambridge University Press, New York.

Kramer, H. P. (1959) Generalized sampling theorem. J. Math. Phys., vol. 38, pp. 68–72.

Krylov, V. I., and N. S. Skolidya (1977) *A Handbook of Methods of Approximate Fourier Transformation and Inversion of the Laplace Transformation*. Mir, Moscow.

Kuhtiffig, P. K. F. (1978) *Introduction to the Laplace Transform*. Plenum, New York.

Kulathinal, J. (1989) *Transform Analysis and Electronic Networks with Applications*. Merril, New York.

Laskshmanarao, S. K. (1954) Gegenbauer transforms. Math. Sludut, vol. 22, p. 161.

Lebedev, N. N. (1949) Analogue of the Parseval theorem for a particular integral transform (in Russian). Dokl. Akad. Nauk SSSR, vol. 68, pp. 653–656.

Lebedev, N. V. (1965) *Special Functions and Their Applications*. Prentice-Hall, Englewood Cliffs, N.J.

Ledermann, W., and S. Vajda (1982) *Handbook for Applicable Mathematics*. Wiley, New York.

Lemon, D. (1962) Application of the Mellin transform to boundary-value problem. Proc. Iowa Acad. Sci., vol. 69, pp. 436–442.

Lighthill, J. M. (1964) *Fourier Analysis and Generalized Functions*. Cambridge University Press, Cambridge, England.

Lowndes, J. S. (1964) Note on generalized Mehler transform. Proc. Cambridge Philos. Soc., vol. 60, p. 57.

Luke, Y. L. (1969) *The Special Functions and Their Approximations*, 2 vols. Academic Press, New York.

Luke, Y. L. (1975) On the error in certain interpolation formulas and in the Gaussian integration formula. J. Austr. Math. Soc., Ser. A, vol. 19, pp. 196–209.

Luke, Y. L., B. Y. Ting, and M. I. Kemp (1975) On generalized Gaussian quadrature. Math. Comput., vol. 29, no. 132, pp. 1083–1093.

Marks, R. J. II (1991) *Shannon Sampling and Interpolation Theory*, vols. I and II. Springer-Verlag, New York.

McCollum, P. A., and B. F. Brown (1965) Laplace Transforms Tables. Holt, Rinehart & Winston, New York.

McCully, J. (1960) The Laguerre transforms. SIAM Rev., vol. 2, p. 185.

Mikusiniski, J. (1983) *Operational Calculus*, 2nd ed. Pergamon, New York.

Miles, J. W. (1971) *Integral Transforms in Applied Mathematics*. Cambridge University Press, Cambridge, England.

Naylor, D. (1963). On a Mellin type integral transform. J. Math. Mech., vol. 12, p. 265.

Neuman, C. P., and D. I. Schonbach (1974) Discrete orthogonal (Legendre) polynomials—a survey. Int. J. Numer. Methods Eng., vol. 8, pp. 743–770.

Newton, R. G. (1966) *Scattering Theory of Waves and Particles*. McGraw-Hill, New York.

Nixon, F. E. (1965) *Handbook of Laplace Transformation*. Prentice-Hall, Englewood Cliffs, N.J.

Oberhettinger, F. (1971) *Tables of Bessel Transforms*. Springer-Verlag, New York.

Oberhettinger, F. (1973a) *Fourier Expansions: A Collection of Formulas*. Academic Press, New York.

Oberhettinger, F. (1973b) *Fourier Transforms of Distributions and Their Inverses: Collection of Tables*. Academic Press, New York.

Oberhettinger, F. and L. Baddi (1973) *Tables of Laplace Transforms*. Springer-Verlag, New York.

Oberhettinger, F., and L. Baddi (1974) *Tables of Mellin Transforms*. Springer-Verlag, New York.

Oberhettinger, F., and T. P. Higgins (1961) Tables of Lebedev, Mehler and generalized Mehler transforms. Boeing Sci. Res. Lab. Math. Note 246, Seattle.

Oberhettinger, F., W. Magnus, and P. Soni (1964) *Formulas of Mathematical Physics*. Springer-Verlag, New York.

Olevskii, A. M. (1975) *Fourier Series with Respect to General Orthogonal Functions*. Springer-Verlag, Berlin.

Olver, F. W. J. (1974) Asymptotics and Special Functions. Academic Press, New York.

Openheim, A. V., and R. W. Schafer (1974) *Digital Signal Processing*. Wiley, New York.

Osler, T. J. (1973) A further extension of the Leibnitz rule for fractional derivatives and its relation to Parseval's formula. SIAM J. Math. Anal., vol. 4, pp. 456–459.

Paley, R. E. A. C., and N. Wiener (1934) *Fourier Transforms in the Complex Domain*, vol. 19, Colloq. Publ. American Mathematical Society, Providence, R.I.

Papoulis, A. (1962) *The Fourier Integral and Its Applications*. McGraw-Hill, New York.

Papoulis, A. (1968) *Systems and Transforms with Applications in Optics*. McGraw-Hill, New York.

Papoulis, A. (1977) *Signal Analysis*. McGraw-Hill, New York.

Papoulis, A. (1980) *Circuits and Systems, a Modern Approach*. Holt, Rinehart & Winston, New York.

Pearson, C. E. (1983) *Handbook of Applied Mathematics, Selected Results and Methods*. Van Nostrand, New York.

Pipes, L. A. (1965) *Operational Methods in Nonlinear Mechanics*. Dover, New York.

Pipes, L. A., and L. R. Harvill (1970) *Applied Mathematics for Engineers and Physicists*. McGraw-Hill, New York.

Powers, D. L. (1987) *Boundary Value Problems*, 3rd ed. Academic Press, New York.

Rader, C. M. (1968) Discrete Fourier transforms when the number of data samples is prime. Proc. IEEE, vol. 56, pp. 1107–1108.

Rainville, E. D. (1960) *Special Functions*. Macmillan, New York.

Rao, K. R., L. C. Mrig, and N. Ahmed (1973) A modified generalized discrete transform. Proc. IEEE Lett., vol. 9, no. 61, pp. 668–669.

Roberts, G. E., and H. Kaufman (1966). *Tables of Laplace Transforms*. Saunders, Philadelphia.

Rooney, P. G. (1963) A generalization of an inversion formula for the Gauss transform. Canad. Math. Bull., vol. 6, pp. 45–53.

Royden, H. L. (1963) *Real Analysis*. Macmillan, New York.

Scott, E. J. (1953) Jacobie transforms. Quart. J. Math. Oxford, vol. 4, p. 36.

Shannon, C. E. (1949) Communications in the presence of noise. Proc. IRE, vol. 37, pp. 10–21.

Sharma, K. C. (1963) Theorems relating Hankel and Meijer's Bessel transform. Proc. Glasgow Math. Assoc., vol. 6, pp. 107–112.

Sharma, K. C. (1965) On an integral transform. Math. Z., vol. 89, pp. 94–97.

Shum, F. Y. Y., A. R. Elliot, and W. O. Brown (1973) Speech processing with Walsh–Hadamard transforms. IEEE Trans. Audio Electroacoust., vol. AV-21, no. 3, pp. 174–179.

Smith, M. G. (1966) *Laplace Transform Theory*. Van Nostrand Reinhold, New York.

Sneddon, I. N. (1946) *Finite Hankel transforms*. Philos. Mag., vol. 37, p. 17.

Sneddon, I. N. (1951) *Fourier Transforms*. McGraw-Hill, New York.

Sneddon, I. N. (1960) The elementary solution of dual integral equations. Proc. Glasgow Math. Assoc., vol. 4, p. 108.

Sneddon, I. N. (1969) The inversion of Hankel transforms of order zero and unity. Glasgow Math. J., vol. 10, p. 156.

Sneddon, I. N. (1972) *The Use of the Integral Transforms*. McGraw-Hill, New York.

Sneddon, I. N. (1980). *Special Functions*. 3rd ed. Longman, New York.

Splettstosser, W. (1978) On generalized sampling sums based on convolution integrals. Arch. Elek. Ubert., vol. 32, pp. 267–275.

Splettstosser, W. (1983) 75 years aliasing error in the sampling theorem, Signal Processing II: Theories and Applications. Proc. Eusipco-83, September. North-Holland, Amsterdam.

Srivastava, K. N. (1965) On Gegenbauer transforms. Math. Student, vol. 33, p. 129.

Stakgold, I. (1979) *Green's Functions and Boundary Value Problems*. Wiley, New York.

Stark, H. (1982) *Applications of Optical Fourier Transforms*. Academic Press, New York.

Stieglitz, K. (1974) *An Introduction to Discrete Systems*. Wiley, New York.

Stockham, T. G. (1966) High-Speed convolution and correlation. In *Proceedings, 1966 Spring Joint Computer Conference*, Washington, D. C. pp. 229-233. Spartan, New York.

Szego, G. (1959) *Orthogonal polynomials*. Amer. Math. Soc. Colloq. Publ., vol. 23.

Sz-Nagy, B. (1965) *Introduction of Real Functions and Orthogonal Expansions*. Oxford University Press, New York.

Tichmarsh, E. C. (1937) *Introduction to the Theory of Fourier Integrals*. Oxford University Press, New York.

Tichmarsh, E. C. (1958) *Eigenfunction Expansions Associated with Second Order Differential Equations*, part II. Oxford University Press, New York.

Tolstov, G. P. (1962) *Fourier Series*. Prentice-Hall, Englewood Cliffs, N.J.

Tranter, C. J. (1950) Legendre transforms. Quart. J. Math. Oxford, vol. 2, p. 1.

Tranter, C. J. (1951) *Integral Transforms in Mathematical Physics*. Wiley, New York.

Tricomi, F. G. (1951) On the finite Hilbert transformation. Quart. J. Math. Oxford, vol. 2, p. 199.

Tricomi, F. G. (1955). *Lectures on Orthogonal Series* (in German). Springer, Berlin.

Verma, C. B. L. (1959) On some properties of K-transforms involving Meijer function. Proc. Natl. Acad. Sci. India, vol. A28, pp. 200–207.

Vichnevetsky, R., and J. B. Bowles (1982) *Fourier Analysis of Numerical Approximations of Hyperbolic Equations*. SIAM, Philadelphia.

Walker, J. S. (1988) *Fourier Analysis*. Oxford University Press, New York.

Watson, G. N. (1966) *A Treatise on the Theory of Bessel Functions*. Cambridge University Press, Cambridge, England.

Weaver, H. J. (1983) *Applications of Discrete and Continuous Fourier Analysis*. Wiley, New York.

Weaver, H. J. (1988) *Theory of Discrete and Continuous Fourier Analysis*. Wiley, New York.

Weinberger, H. F. (1965) *Partial Differential Equations with Complex Variables and Transform Methods*. Blaisdell, Waltham, Mass.

Weiss, P. (1957) Sampling theorems associated with Sturm–Liouville systems, Bull. Am. Math. Soc., vol. 63, p. 242.

Weiss, P. (1963) An estimate of error arising from misapplication of the sampling theorem. Amer. Math. Soc. Notices 10.351 (Abstract No. 601–54), 1.

Widder, D. V. (1941) *The Laplace Transform.* Princeton University Press, Princeton, N.J.

Widder, D. V. (1956) Integral transforms related to heat conduction. Ann. Mat., vol. 42, pp. 279–305.

Widder, D. V. (1975) *The Heat Equation.* Academic Press, New York.

Wiener, N. (1933) *The Fourier Integral and Certain of Its Applications.* Cambridge University Press, Cambridge, England.

Whittaker, E. T. and G. N. Watson (1963) *A Course in Modern Analysis.* Cambridge University Press, Cambridge, England.

Wolf, K. B. (1978) *Integral Transforms in Science and Engineering.* Plenum, New York.

Zemanian, A. H. (1965) *Distribution Theory and Transform Analysis.* McGraw-Hill, New York.

Zemanian, A. H. (1968) *Generalized Integral Transformations.* Interscience, New York.

Zhdanov, M. S. (1988) *Integral Transforms in Geophysics.* Springer-Verlag, New York.

Ziemer, R. E., W. H. Tranter and D. R. Famin (1983) *Signals and Systems, Continuous and Discrete.* Macmillan, New York.

Zygmund, A. (1959) *Trigonometrical Series*, vol. 1, Cambridge University Press, New York.

Index of Notations

Subject Index